Lecture Notes in Computer Science 13833

The series Lecture Notes in Computer Science (LNCS), including its subseries Lecture Notes in Artificial Intelligence (LNAI) and Lecture Notes in Bioinformatics (LNBI), has established itself as a medium for the publication of new developments in computer science and information technology research, teaching, and education.

LNCS enjoys close cooperation with the computer science R & D community, the series counts many renowned academics among its volume editors and paper authors, and collaborates with prestigious societies. Its mission is to serve this international community by providing an invaluable service, mainly focused on the publication of conference and workshop proceedings and postproceedings. LNCS commenced publication in 1973.

Duc-Tien Dang-Nguyen · Cathal Gurrin ·
Martha Larson · Alan F. Smeaton ·
Stevan Rudinac · Minh-Son Dao ·
Christoph Trattner · Phoebe Chen
Editors

MultiMedia Modeling

29th International Conference, MMM 2023
Bergen, Norway, January 9–12, 2023
Proceedings, Part I

Editors
Duc-Tien Dang-Nguyen
University of Bergen
Bergen, Norway

Martha Larson
Radboud University Nijmegen
Nijmegen, The Netherlands

Stevan Rudinac
University of Amsterdam
Amsterdam, The Netherlands

Christoph Trattner
Department of Information Science
and Media Studies
University of Bergen
Bergen, Norway

Cathal Gurrin
Dublin City University
Dublin, Ireland

Alan F. Smeaton
Dublin City University
Dublin, Ireland

Minh-Son Dao
National Institute of Information
and Communications Technology
Tokyo, Japan

Phoebe Chen
La Trobe University
Melbourne, VIC, Australia

ISSN 0302-9743 ISSN 1611-3349 (electronic)
Lecture Notes in Computer Science
ISBN 978-3-031-27076-5 ISBN 978-3-031-27077-2 (eBook)
https://doi.org/10.1007/978-3-031-27077-2

This Springer imprint is published by the registered company Springer Nature Switzerland AG
The registered company address is: Gewerbestrasse 11, 6330 Cham, Switzerland

Preface

This two-volume proceedings contains the papers accepted at MMM 2023, the 29th International Conference on MultiMedia Modeling.

Organized for more than 25 years, MMM is a well-established international conference bringing together excellent researchers from both academia and industry. MMM is officially ranked as a core-B conference. During the conference, novel research work from MMM-related areas (especially multimedia content analysis; multimedia signal processing and communications; and multimedia applications and services) are shared along with practical experiences, results, and exciting demonstrations. The 29th instance of the conference was organized in Norway on January 9–12, 2023. MMM 2023 received a large number of submissions organized in different tracks.

Specifically, 267 papers were submitted to MMM 2023. Papers were reviewed by three reviewers from the Program Committee, while the TPC chairs and special event organizers acted as meta-reviewers and assisted in the decision-making process. Out of 218 regular papers, 86 were accepted for the proceedings. In particular, 28 papers were accepted for oral presentation (acceptance rate of 12.8%) and 58 papers for poster presentation (acceptance rate of 26.6%). Regarding the remaining tracks, 27 special session papers were submitted, with 19 accepted for oral presentation. Additionally, 2 papers were accepted (from 4 submitted) for a new Brave New Ideas track, one paper was accepted for the Research2Biz track, and 4 demonstration papers were accepted. Finally, there were 13 papers accepted for participation at the Video Browser Showdown 2023.

The special sessions are traditionally organized to extend the program with novel challenging problems and directions. The MMM 2023 program included three special sessions: MDRE, Multimedia Datasets for Repeatable Experimentation; ICDAR, Intelligent Cross-Data Analysis and Retrieval; and SNL, Sport and Nutrition Lifelogging.

Besides the three special sessions, the Video Browser Showdown represented an important highlight of MMM 2023 with 13 participating systems in this exciting (and challenging) competition. In addition, three highly respected speakers were invited to MMM 2023 to present their impressive talks and results in multimedia-related topics: "Real-Time Visual Exploration of Very Large Image Collections" by Kai Uwe Barthel, HTW Berlin - University of Applied Sciences; "Multimodal Augmented Homeostasis" by Ramesh Jain, University of California, Irvine; and "Multi-Perspective Modeling of Complex Concepts - The Case of Olfactory History & Heritage" by Marieke van Erp, Knaw Humanities Cluster.

We acknowledge the significant effort from all authors of submitted papers, all reviewers, and all members of the MMM 2023 organization team for their great work and support. Finally, we would like to thank all members of the MMM community who contributed to the MMM 2023 event. They all helped MMM 2023 to be an exciting and inspiring international event for all participants!

January 2023

Duc-Tien Dang-Nguyen
Cathal Gurrin
Martha Larson
Alan F. Smeaton
Stevan Rudinac
Minh-Son Dao
Christoph Trattner
Phoebe Chen

Organizing Committee

General Chairs

Duc-Tien Dang-Nguyen	University of Bergen, Norway
Cathal Gurrin	Dublin City University, Ireland
Martha Larson	MediaEval, The Netherlands
Alan Smeaton	Dublin City University, Ireland

Community Direction Chairs

Björn Þór Jónsson	IT University of Copenhagen, Denmark
Mehdi Elahi	University of Bergen, Norway
Minh-Triet Tran	University of Science, VNU-HCM, Vietnam
Liting Zhou	Dublin City University, Ireland

Technical Program Chairs

Stevan Rudinac	University of Amsterdam, The Netherlands
Minh-Son Dao	National Institute of Information and Communications Technology (NICT), Japan
Phoebe Chen	La Trobe University, Australia
Christoph Trattner	University of Bergen, Norway

Demo Chairs

Wallapak Tavanapong	Iowa State University, USA
Michael Riegler	University of Oslo, Norway

Technical Program Coordinator

Marc Gallofré Ocaña	University of Bergen, Norway
Khanh-Duy Le	University of Science, VNU-HCM, Vietnam

Local Arrangement Chairs

Knut Risnes	University of Bergen, Norway
Fredrik Håland Jensen	University of Bergen, Norway
Audun Håkon Klyve Gulbrandsen	University of Bergen, Norway
Ragnhild Breisnes Utkilen	University of Bergen, Norway

Video Browser Showdown Chairs

Klaus Schoeffmann	Klagenfurt University, Austria
Werner Bailer	Joanneum Research, Austria
Jakub Lokoč	Charles University, Prague, Czech Republic
Cathal Gurrin	Dublin City University, Ireland

Publication Chairs

Khanh-Duy Le	University of Science, VNU-HCM, Vietnam
Minh-Triet Tran	University of Science, VNU-HCM, Vietnam

Steering Committee

Phoebe Chen	La Trobe University, Australia
Tat-Seng Chua	National University of Singapore, Singapore
Kiyoharu Aizawa	University of Tokyo, Japan
Cathal Gurrin	Dublin City University, Ireland
Benoit Huet	Eurecom, France
Klaus Schoeffmann	Klagenfurt University, Austria
Richang Hong	Hefei University of Technology, China
Björn Þór Jónsson	IT University of Copenhagen, Denmark
Guo-Jun Qi	University of Central Florida, USA
Wen-Huang Cheng	National Chiao Tung University, Taiwan
Peng Cui	Tsinghua University, China

Web Chairs

Anh-Duy Tran	University of Science, VNU-HCM, Vietnam
Ruben Caldeira	University of Bergen, Norway

Organizing Agency

Minh-Hung Nguyen-Tong	CITE, Vietnam

Special Session Organizers

Multimedia Datasets for Repeatable Experimentation (MDRE)

Cathal Gurrin	Dublin City University, Ireland
Duc-Tien Dang-Nguyen	UiB and Kristiania, Norway
Adam Jatowt	University of Innsbruck, Austria
Liting Zhou	Dublin City University, Ireland
Graham Healy	Dublin City University, Ireland

Intelligent Cross-Data Analysis and Retrieval (ICDAR)

Minh-Son Dao	NICT, Japan
Michael A. Riegler	SimulaMet and UiT, Norway
Duc-Tien Dang-Nguyen	UiB and Kristiania, Norway
Uraz Yavanoglu	Gazi University, Ankara, Turkey

Sport and Nutrition Lifelogging (SNL)

Pål Halvorsen	SimulaMet, Norway
Michael A. Riegler	SimulaMet and UiT, Norway
Cise Midoglu	SimulaMet, Norway
Vajira Thambawita	SimulaMet, Norway

Program Committees and Reviewers (Regular and Special Session Papers)

Aakash Sharma	UiT, Norway
Aaron Duane	ITU Copenhagen, Denmark
Adam Jatowt	UIBK, Austria
Alan F. Smeaton	Dublin City University, Ireland
Alexandros Oikonomidis	The Centre for Research & Technology, Hellas, Greece
Anastasios Karakostas	DRAXIS Environmental, Greece
Andrea Marheim Storås	SimulaMet, Norway
Andreas Leibetseder	Alpen-Adria University, Klagenfurt, Austria
Anh-Duy Tran	University of Science, VNU-HCM, Vietnam

Anh-Khoa Tran	NICT, Japan
Annalina Caputo	Dublin City University, Ireland
Ari Karppinen	Finnish Meteorological Institute, Finland
Athanasios Efthymiou	University of Amsterdam, The Netherlands
Athina Tsanousa	ITI-CERTH, Greece
Benjamin Bustos	University of Chile, Chile
Bin Wang	Ningxia University, China
Bingchun Luo	Harbin Institute of Technology, China
Bingxin Zhao	Chongqing University, China
Binh Nguyen	University of Science, Ho Chi Minh City, Vietnam
Bo Zhang	Dalian Maritime University, China
Bogdan Ionescu	University Politehnica of Bucharest, Romania
Bowen Wan	Shenzen University, China
Boyang Zhang	Ningxia University, China
Cathal Gurrin	Dublin City University, Ireland
Cem Direkoglu	Middle East Technical University - Northern Cyprus Campus, Cyprus
Chaoqun Niu	Sichuan University, China
Chengfeng Ruan	Nanjing University, China
Chenglong Fu	Huzhou University, China
Chenglong Zhang	Zhengzhou University, China
Chih-Wei Lin	Fujian Agriculture and Forestry University, China
Chongjian Zhang	University of Science and Technology Beijing, China
Christian Timmerer	Klagenfurt University, ITEC – MMC, Austria
Chutisant Kerdvibulvech	National Institute of Development Administration (NIDA), Thailand
Claudiu Cobarzan	Babes-Bolyai University, Romania
Cong-Thang Truong	Aizu University, Japan
David Bernhauer	Charles University, Czech Republic
De-ning Di	Zhejiang Dahua Technology, China
Dehong He	Southwest University, China
Despoina Touska	Centre for Research and Technology Hellas, Greece
Duc Tien Dang Nguyen	University of Bergen, Norway
Edgar Chavez	CICESE, Mexico
Elissavet Batziou	CERTH-ITI, Greece
Eric Arazo Sanchez	Dublin City University, Ireland
Esra Açar Çelik	Fortiss Research Institute, Germany
Evlampios Apostolidis	CERTH-ITI, Greece
Fabrizio Falchi	Consiglio Nazionale delle Ricerche, Italy

Licheng Zhang	University of Science and Technology of China, China
Lifeng Sun	Tsinghua University, China
Linlin Shen	Shenzhen University, China
Liting Zhou	Dublin City University, Ireland
Longlong Zhou Zhou	Jiangnan University
Louliangshan Lou	University of Chinese Academy of Sciences, China
Lu Lin	Xiamen University, China
Luca Rossetto	University of Zurich, Switzerland
Lucia Vadicamo	ISTI-CNR, Italy
Lux Mathias	University of Klagenfurt, Austria
Ly-Duyen Tran	Dublin City University, Ireland
Maarten Michiel Sukel	University of Amsterdam, The Netherlands
Marc Gallofré Ocaña	University of Bergen, Norway
Marcel Worring	University of Amsterdam, The Netherlands
Mario Taschwer	Klagenfurt University, Austria
Markus Koskela	CSC - IT Center for Science Ltd., Espoo, Finland
Martin Winter	Joanneum Research, Austria
Mehdi Elahi	University of Bergen, Norway
Meining Jia	Ningxia University, China
Mengping Yang	East China University of Science and Technology, China
Mengqin Bai	Guangxi Normal University, China
Michael A. Riegler	SimulaMet, Norway
Mihai Datcu	DLR/UPB, Romania
Ming Gao	Anhui University, China
Minh-Khoi Nguyen-Nhat	University of Science, VNU-HCM, Vietnam
Minh-Son Dao	National Institute of Information and Communications Technology, Japan
Minh-Triet Tran	University of Science, VNU-HCM, Vietnam
Minyan Zheng	Shenzhen University, China
Muhamad Hilmil Pradana	National Institute of Information and Communications Technology, Japan
Mukesh Saini	Indian Institute of Technology Ropar, India
Naoko Nitta	Mukogawa Women's University, Japan
Naushad Alam	Dublin City University, Ireland
Naye Ji	Communication University of Zhejiang, China
Nengwu Liu	Qilu University of Technology, China
Nese Cakici Alp	Kocaeli University, Turkey
Nhat-Tan Bui	University of Science, VNU-HCM, Vietnam
Nikolaos Gkalelis	ITI/CERTH, Greece

Nitish Nagesh	University of California, Irvine, USA
Omar Shahbaz Khan	IT University of Copenhagen, Denmark
Pål Halvorsen	SimulaMet, Norway
Pan Li	Ningxia University, China
Peiguang Jing	Tianjin University, China
Peijie Dong	National University of Defense Technology, China
PengCheng Shu	Wuhan Institute of Technology, China
Pengcheng Yue	Zhejiang Lab, China
PengJu Wang	Inner Mongolia University, China
Pengwei Tang	Renmin University of China, Beijing, China
Phivos Mylonas	Ionian University, Greece
Pierre-Etienne Martin	Max Planck Institute for Evolutionary Anthropology, Germany
Ping Feng	Guangxi Normal University, China
Qi Sun	Shanghai Institute of Microsystem and Information Technology, Chinese Academy of Sciences, China
Qiao Wang	Southeast University, China
Qiaowei Ma	South China University of Technology, China
Quang-Trung Truong	Hong Kong University of Science and Technology, China
Ralph Ewerth	TIB - Leibniz Information Centre for Science and Technology, Germany
Robert Mertens	BHH University of Applied Sciences, Germany
Runtao Xi	CCTEG Changzhou Research Institute, Changzhou, China
Sai-Kit Yeung	HKUST, China
Sanbi Luo	University of Chinese Academy of Sciences, Beijing, China
Shahram Ghandeharizadeh	USC, USA
Shanshan Xiang	Wuhan Textile University, China
Shanshan Zhong	School of Computer Science and Engineering, Sun Yat-Sen University, China
Shaodong Li	Guangxi University, China
Shengwei Zhao	Xi'an Jiaotong University, China
Shih-Wei Sun	Taipei National University of the Arts, Taiwan
Shilian Wu	USTC, China
Shin'Ichi Satoh	National Institute of Informatics, Japan
Shingo Uchihashi	Fujifilm Business Innovation Corp., Japan
Shinohara Takayuki	Pasco, Japan
Shintami Chusnul Hidayati	Institut Teknologi Sepuluh Nopember, Indonesia
Shiqi Shen	Guilin University of Electronic Technology, China

Shu Xinyao	Nanjing University of Information Science and Technology, China
Shufan Dai	Zhengzhou University, China
Shuo Chen	Fudan University, China
SiDi Liu	Qilu University of Technology, China
Silvan Heller	University of Basel, Switzerland
Song Chen	Ningbo University, China
Stefanie Onsori-Wechtitsch	Joanneum Research, Austria
Stefanos Vrochidis	CERTH ITI, Greece
Stephane Marchand-Maillet	Viper Group - University of Geneva, Switzerland
Stevan Rudinac	University of Amsterdam, The Netherlands
Suping Wu	Ningxia University, China
Takayuki Nakatsuka	National Institute of Advanced Industrial Science and Technology (AIST), Japan
Tan-Sang Ha	HKUST, China
Tao Peng	UT Southwestern Medical Center, USA
Tao Wen	Wuhan University, China
Theodora Pistola	ITI-CERTH, Greece
Thi-Oanh Nguyen	Hanoi University of Science and Technology, Vietnam
Thi-Phuoc-Van Nguyen	USQ, Australia
Thitirat Siriborvornratanakul	National Institute of Development Administration (NIDA), Thailand
Thittaporn Ganokratana	King Mongkut's University of Technology Thonburi, Thailand
Tianrun Chen	Zhejiang University, China
Tianxing Feng	Peking University, China
Tiberio Uricchio	University of Florence, Italy
Tien-Tsin Wong	Chinese University of Hong Kong, China
Ting Zou	Soochow University, China
Tom van Sonsbeek	University of Amsterdam, The Netherlands
Tomas Skopal	Charles University, Czech Republic
Tor-Arne Schmidt Nordmo	UiT The Arctic University of Norway, Norway
Toshihiko Yamasaki	University of Tokyo, Japan
Tse-Yu Pan	National Tsing Hua University, China
Tu Van Ninh	Dublin City University, Ireland
Tuan Linh Dang	HUST, Vietnam
Ujjwal Sharma	University of Amsterdam, The Netherlands
Vajira Thambawita	SimulaMet, Norway
Vasileios Mezaris	CERTH, Greece
Vassilis Sitokonstantinou	National Observatory of Athens, Greece
Vijay Cornelius Kirubakaran John	RIKEN, Japan

Xunyu Zhu	Institute of Information Engineering, Chinese Academy of Sciences, China
Yan Wang	Beijing Institute of Technology, China
Yang Yang	University of Electronic Science and Technology of China, China
Yannick Prié	LS2N - University of Nantes, Italy
Yanrui Niu	Wuhan University, China
Yanxun Jiang	Southeast University, China
Yasutomo Kawanishi	NAIST, Japan
Yaxin Ma	Wuhan University, China
Yibo Hu	Beijing University of Posts and Telecommunications, China
Yidan Fan	Tianjin University, China
Yihua Chen	Guangxi Normal University, China
Yijia Zhang	Dalian Maritime University, China
Yimao Xiong	Hunan University of Science and Technology, China
Ying Zang	Huzhou University, China
Yingnan Fu	East China Normal University, China
Yingqing Xu	Future Lab at Tsinghua University, China
Yiyang Cai	Tongji University, China
Yizhe Zhu	Shanghai Jiao Tong University, China
Yong Ju Jung	Gachon University, South Korea
Yu-Kun Lai	Cardiff University, UK
Yuan Gao	Donghua University, China
Yuan Lin	Kristiania University College, Norway
Yuan Zhang	South China University of Technology, China
Yuanhang Yin	Shanghai Jiao Tong University, China
Yuchen Cao	Zhejiang University, China
Yuchen Xie	Aisainfo Technologies, China
Yue Yang	Beijing Institute of Technology, China
Yue Zhang	Shanghai University, China
Yuewang Xu	Wenzhou University, China
Yufei Shi	National University of Singapore, Singapore
Yuhang Li	Shanghai University, China
Yuki Hirose	Osaka University, Japan
Yukun Zhao	Beijing Jiaotong University, China
Yun Lan	Wuhan University, China
Yunhong Li	Ningxia University, China
Yunshan Ma	National University of Singapore, Singapore
Yuqing Zhang	Shenzhen University, China
Yuxin Peng	Ningxia University, China

Yuzhe Hao	Tokyo Institute of Technology, Japan
Zarina Rakhimberdina	Tokyo Institute of Technology, Japan
Ze Xu	Huaqiao University, China
ZeXu Zhang	Ningxia University, China
Zeyu Cui	USTC, China
Zhaoquan Yuan	Southwest Jiaotong University, China
Zhaoyong Yan	Shanghai University, China
Zheng Guang Wang	Shangwang Network Co. Ltd., Shanghai, China
Zhenhua Yu	Ningxia University, China
Zhenzhen Hu	Hefei University of Technology, China
Zhi-Yong Huang	University of Science and Technology Beijing (USTB), China
Zhichen Liu	Shenzhen University, China
Zhihang Ren	Dalian University, China
Zhihuan Liu	Zhengzhou University, China
Zhiqi Yan	Tongji University, China
Zhixiang Yuan	Ningxia University, China
Zhiying Lu	University of Science and Technology of China, China
Zhiyong Cheng	Shandong Artificial Intelligence Institute, China
Zhiyong Zhou	Jilin University, China
Zhiyuan Chen	Guangxi Normal University, China
Zi Chai	Peking University, China
Ziqiao Shang	Huazhong University of Science and Technology, China
Ziyang Zhang	South China University of Technology, China
Ziyu Guan	Xidian University, China

Contents – Part I

Image Quality Assessment and Enhancement

Multimedia Analytics Application

Multimedia Content Generation

ICDAR: Intelligent Cross-Data Analysis and Retrieval

MDRE: Multimedia Datasets for Repeatable Experimentation

SNL: Sport and Nutrition Lifelogging

VBS: Video Browser Showdown

Contents – Part II

Multimedia Processing and Applications

BNI: Brave New Ideas

Research2Biz

Demo

Detection, Recognition and Identification

MMM-GCN: Multi-Level Multi-Modal Graph Convolution Network for Video-Based Person Identification

Ziyan Liao, Dening Di, Jingsong Hao(✉), Jiang Zhang, Shulei Zhu, and Jun Yin

Dahua Technology Co., Ltd., Hangzhou, China
hao_jingsong2022@163.com

Abstract. Video-based multi-modal person identification has attracted rising research interest recently to address the inadequacies of single-modal identification in unconstrained scenes. Most existing methods model video-level and multi-modal-level information of target video respectively, which suffer from separation of different levels and insufficient information contained in a specific video. In this paper, we introduce extra neighbor-level information for the first time to enhance the informativeness of target video. Then a Multi-Level(neighbor-level, multi-modal-level, and video-level) and Multi-Modal GCN model is proposed, to capture correlation among different levels and achieve adaptive fusion in a unified model. Experiments on iQIYI-VID-2019 dataset show that MMM-GCN significantly outperforms current state-of-the-art methods, proving its superiority and effectiveness. Besides, we point out feature fusion is heavily polluted by noisy nodes that result in a suboptimal result. Further improvement could be explored on this basis to approach the performance upper bound of our paradigm.

Keywords: Person identification · Multi-modal · Multi biometrics · GCN · Feature fusion

1 Introduction

Person identification refers to confirming the identity of target person via biometric features, such as face recognition, person re-identification (Re-ID), speaker recognition, etc. With the development of deep learning, all these methods have achieved great success. For face recognition, ArcFace [4] reached 99.83% precision on LFW benchmark [9]. For Re-ID, Rank-1 accuracy raised to 97.1% [23]. For speaker recognition, the classification error rate is merely 0.85% [6].

Everything seems alright until trying to apply these methods to real unconstrained scenes. Face recognition is sensitive to pose, blur, occlusion, etc. Re-ID can't handle clothes changing yet. Person are not always speaking, resulting in a lack of audio features for speaker recognition. Therefore, it is necessary

D.-T. Dang-Nguyen et al. (Eds.): MMM 2023, LNCS 13833, pp. 3–15, 2023.
https://doi.org/10.1007/978-3-031-27077-2_1

to combine all these single-modal methods together to complement each other. Recently, the largest dataset for video-based multi-modal person identification task was proposed by [17]. Compared with single-modal person identification, this dataset has two additional levels of information available: (1) it contains multi-modal information to resolve the limitation of single-modal identification (multi-modal-level); (2)video with frames and richer content is used to replace still image which is commonly used in person identification (video-level). Consequently, approaches utilizing above two levels of information have been explored and achieved more meaningful results.

Current general paradigm [15,17] for video-based multi-modal person identification task models video-level and multi-modal-level information independently, then cascades them for joint training, as shown in Fig. 1. However, separate modeling for different levels leads to a restricted view, which is not conducive to capture correlations between video-level and multi-modal-level. Intuitively, it will be better to fuse information of different levels adaptively in a unified model.

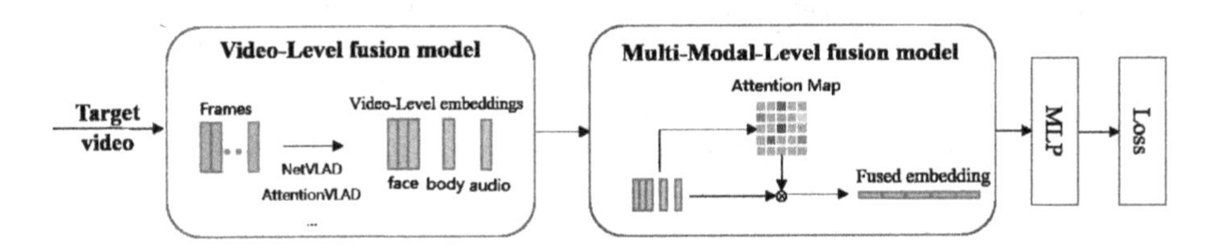

Fig. 1. Current paradigm for video-based multi-modal person identification task.

Although precision is improved by the fusion of video-level and multi-modal-level, information contained in a specific video is insufficient to determine the identity of target person, especially when visual features are broken or invisible throughout the video. Inspired by using neighbor information in re-ranking task [25], we introduce extra k-nearest neighbors for each video as its neighbor-level information. By fusing auxiliary information from neighbors, more discriminative and integral embeddings would be reconstructed.

In summary, we consider that there is some certain correlation among neighbor-level, video-level and multi-modal-level information, and hope to design an effective modeling method, which can sufficiently capture the correlation and adaptively fuse them into a unified model. Therefore, the Multi-Level Multi-Modal Graph Convolution Network (MMM-GCN) is proposed, which introduces a new paradigm for video-based multi-modal person identification task. Figure 2 shows the overall framework of MMM-GCN.

The main contributions of our work can be summarized as follows.

- For video-based multi-modal person identification task, we introduce extra neighbor-level information for the first time, to enhance the informativeness of target video.

- The proposed MMM-GCN achieves an adaptive fusion of neighbor-level, multi-modal-level and video-level in a unified model, which explores a new paradigm for video-based multi-modal person identification task.
- State-of-the-art performances are achieved by MMM-GCN on iQIYI-VID-2019 dataset [18], surpassing previous ones by a large margin. While verifying the superiority of MMM-GCN, the embedding pollution problem by noisy nodes is also pointed out, offering a clue for further research.

2 Related Work

2.1 Video-Based Multi-Modal Person Identification

Multi-modal person identification task was first proposed by [13], which also released PIPA dataset. Later, CSM dataset [10] was proposed for multi-modal person retrieval task. Due to the lack of video information or rich modalities, related approaches often use special contextual cues such as social relationships, geographical and temporal information. Therefore, the above two datasets are not suitable for video-based multi-modal person identification task. Recently, iQIYI-VID-2019 dataset [18] was released in the 2019 iQIYI Celebrity Video Identification Challenge[1]. To the best of our knowledge, it is the largest and most challenging video dataset for multi-modal person identification task. Liu et al. [17] extracted video features through NetVLAD [1] and proposed a multi-modal attention module (MMA) to fuse different modal of features adaptively. Li et al. [15] improved video-level model and multi-modal-level model respectively based on the framework of Liu et al. [17], and proposed the frame aggregation and multi-modal fusion (FAFM) method, which achieved state-of-the-art (SOTA). There are also some methods in the 2019 iQIYI Celebrity Video Identification Challenge [2,5,12]. However, most of them re-extracted features, or adopted model ensemble methods, which have less exploration on multi-modal fusion technology. In addition, all above methods modeled video-level and multi-modal-level information separately (Fig. 1), which is not conducive to exploiting correlations between different levels.

2.2 Graph Convolution Networks

Graph Convolutional Networks (GCNs) [7,11,14] extend the convolutional idea of CNNs to deal with graph data in non-Euclidean space. A graph consists of a set of objects (nodes) and relationships between these objects (edges). The basic idea of GCN is to generate a more discriminative embedding for target node by aggregating features from its neighbor nodes along edges. Due to its effectiveness and simplicity, GCNs have shown impressive capabilities on a variety of tasks [8,21,22]. Recently, some works applied GCNs to multi-modal fusion [8,20,22]. Binh et al. [20] modeled visual features and attribute labels of the body, utilized GCN to learn the topological structure of visual signature of a person. Hu et al.

[1] http://challenge.ai.iqiyi.com/detail?raceId=5c767dc41a6fa0ccf53922e7.

[8] modeled contextual information, multi-modal information and inter-speaker information in a dialogue, and used GCN to complete the task of sentiment classification of the dialogue. All these methods modeled context interaction and achieved feature aggregation through GCNs for different tasks, which verified the applicability of GCN for information fusion task.

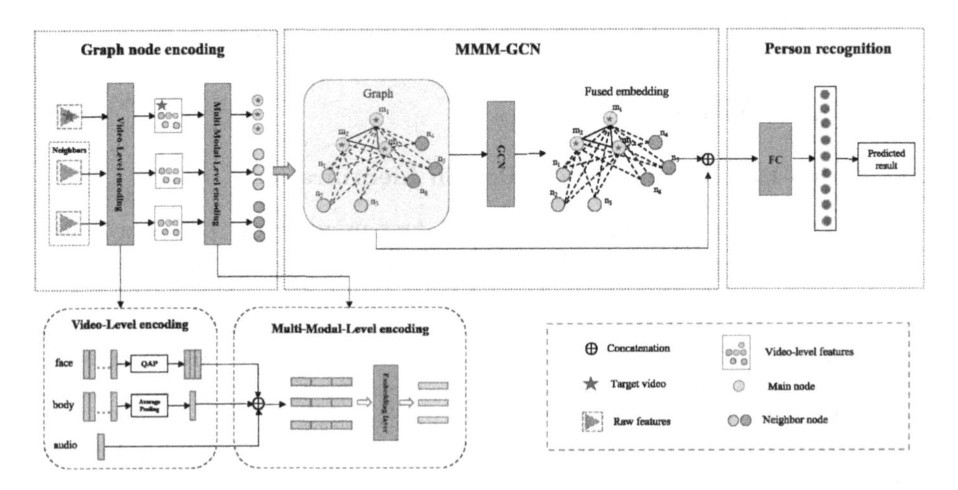

Fig. 2. Illustration of our proposed framework including three parts: (1) graph node encoding; (2) multi-level multi-modal GCN(MMM-GCN); (3) person identification.

3 Methodology

3.1 Overview

Figure 2 shows the overall framework of our method, which is sequentially divided into three parts: graph node encoding, multi-level fusion model (MMM-GCN) and person identification. Firstly, graph node encoding module(Section 3.3) encodes neighbor-level, video-level, and multi-modal-level information and generates graph node features for target video. Then, a local graph is constructed (Sect. 3.2.1) based on encoded features and adjacency containing three-level correlations, to achieve adaptive fusion of these levels through feature aggregation and update mechanism of GCN (Sect. 3.2.2). Finally, the person identification task is implemented based on the fused embedding output by GCN(Section 3.4).

3.2 Graph Construction and Learning

Different graph modeling methods have a direct effect on the information aggregation of GCNs, which further affects feature fusion. In this paper, we propose to model neighbor-level, video-level and multi-modal-level information in a unified graph, enabling target person take full advantage of the correlation among them and achieve adaptive fusion.

Graph Construction. Given a certain video, we model it with a local graph $\mathcal{G} = (\mathcal{V}, \mathcal{E})$, where $\mathcal{V}$ denotes nodes composed of target video and its neighbor videos, $\mathcal{E} \subset \mathcal{V} \times \mathcal{V}$ is a set of edges containing correlations among video-level, neighbor-level and multi-modal-level.

Nodes. The graph is constructed as shown in Fig. 3. For each video, video-level encoding module are used to generate k_2 diverse features, which are represented by nodes with same color. Nodes with red star represent target video and the rest denote k_1 nearest neighbor videos of target video selected by neighbor-level encoding module. For each node, it obtains multi-modal information through multi-modal-level encoding module. For more details on encoding module, please refer to session 3.3. We refer to nodes of target video as main nodes, and nodes of neighbor videos as neighbor nodes, denoted by $\mathcal{M} = \{m_1, m_2, ..., m_{k_2}\}$, $\mathcal{N} = \{n_1, n_2, ..., n_{k_1 \times k_2}\}$, respectively. Thus, for a local graph, the number of graph nodes $|\mathcal{V}| = k_2 \times (k_1 + 1)$, where 1 represents target video itself.

Adjacency matrix. The adjacency matrix $\mathbf{A} \in \mathbb{R}^{|\mathcal{V}| \times |\mathcal{V}|}$ contains edge weights between each two nodes. Due to the introduction of neighbor-level information, the adjacency matrix $\mathbf{A}$ can be divided into four correlation matrices, as shown in Eq. 1

$$\mathbf{A} = \begin{bmatrix} \mathbf{MM} \; \mathbf{MN} \\ \mathbf{NM} \; \mathbf{NN} \end{bmatrix} \tag{1}$$

where $\mathbf{MM} \in \mathbb{R}^{k_2 \times k_2}$, $\mathbf{MN} \in \mathbb{R}^{k_2 \times (k_2 \times k_1)}$, $\mathbf{NM} \in \mathbb{R}^{(k_2 \times k_1) \times k_2}$ and $\mathbf{NN} \in \mathbb{R}^{(k_2 \times k_1) \times (k_2 \times k_1)}$ denotes the correlation matrix of $\mathcal{M} - \mathcal{M}$, $\mathcal{M} - \mathcal{N}$, $\mathcal{N} - \mathcal{M}$ and $\mathcal{N} - \mathcal{N}$, separately.

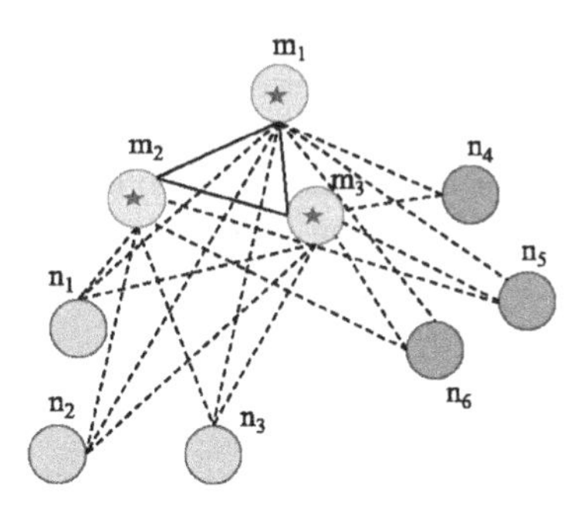

Fig. 3. Illustration of graph construction

According to the construction of graph nodes, **MM** contains both multi-modal-level and video-level correlations (solid line between nodes in Fig. 3), and **MN** adds additional neighbor-level correlations (dotted line between nodes in Fig. 3) on the basis of **MM**. When computing the elements of **MM** and **MN**, we use exponential cosine similarity and set a balance factor ω to control the fusion degree of neighbor-level information, as shown in Eq. 2, where γ denotes temperature parameter.

$$\mathbf{MM_{ij}} = exp(\frac{sim(m_i, m_j)}{\gamma}) \qquad \mathbf{MN_{ij}} = \omega \; exp(\frac{sim(m_i, n_j)}{\gamma}) \tag{2}$$

Moreover, to keep model focus on the fusion of target video, we suppress the message pass to neighbor nodes. That means all elements in **NM** and **NN** are set to 0 (no edge), except for self-connection in **NN**. In fact, we have done comparison experiments and the result shows a slight decrease in precision without suppressing the propagation of the neighbor nodes. We guess preventing neighbor nodes from aggregation could act as stabilizer and anchor to improve model stability, especially if neighbors are noisy nodes.

The adjacency matrix **A** calculated based on the above description is a non-symmetric matrix, whose degree matrix is calculated as:

$$\mathbf{D}_{row}(i,i) = \sum_j \mathbf{A}_{ij} \qquad \mathbf{D}_{col}(j,j) = \sum_i \mathbf{A}_{ij} \tag{3}$$

The normalized graph Laplacian matrix [14] is calculated as:

$$\hat{\mathbf{A}} = \mathbf{D}_{row}^{-\frac{1}{2}} \mathbf{A} \mathbf{D}_{col}^{-\frac{1}{2}} \tag{4}$$

Graph Learning. Following chen et al. [3], we use Eq. 5 to update node embedding of the network.

$$\mathbf{F}^{l+1} = \sigma(((1-\alpha)\hat{\mathbf{A}}\mathbf{F}^l + \alpha\mathbf{F}^0)((1-\beta^l)\mathbf{I} + \beta^l\mathbf{W}^l)) \tag{5}$$

where α and β^l are hyperparameters, σ denotes the activation function, $\mathbf{W}^l$ is a learnable weight matrix. A residual connection to the first layer $\mathbf{F}^0$ is added to $\hat{\mathbf{A}}\mathbf{F}^l$, and an identity mapping $\mathbf{I}$ is added to the weight matrix $\mathbf{W}^l$. We set β^l as shown in Eq. 6, where λ is also a hyperparameter to ensure the decay of weight matrix adaptively increases when stacking more layers.

$$\beta^l = \log(\frac{\lambda}{l} + 1) \tag{6}$$

3.3 Graph Node Encoding

Neighbor-level Encoding. In this paper, we introduce extra neighbor-level information to enhance the informativeness of target video. Specifically, we construct KNN graph for target video, and encode neighbor-level information into target embeddings through message aggregating mechanism of GCN. Since face modal has the best identification effect and robustness compared to other modalities, we use face feature to calculate the k-nearest neighbors of target video based on the cosine similarity.

Video-level Encoding. Video has a large number of similar frames compared to still image. To ensure the completeness of context information while reducing redundant information and computational burden, we propose a quality-based average pooling (QAP) method to extract k_2 video-level features for each video. Specifically, due to the strong correlation between face quality and similarity, average pooling operation is performed respectively on feature groups with different quality to generate diverse video-level features. Please refer to session 4.1 for parameter settings of QAP.

Multi-Modal-Level Encoding. In order to make full use of the complementarity between multi-modal features and reduce redundancy, we designed a multi-modal-level encoding module. For a given video, we concatenate multi modal features from each video-level, then map the concatenated multi-modal feature into a unified and low-dimensional space through a fully connected layer, to get a more informative and compact feature.

3.4 Person Identification

We concatenate origin node feature n_t with the output of GCN f_t to generate the final representation for each node refer to [8]. Predicted probability is obtained through a FC layer and $softmax$ layer, as shown in Eq. 7, where $\|$ is the concatenation operation. We use categorical cross-entropy along with L2-regularization as the loss function.

$$P_t = softmax(FC(f_t \parallel n_t)) \tag{7}$$

4 Experiment

4.1 Experiment Setups

Dataset. We evaluate our proposed MMM-GCN on iQIYI-VID-2019 dataset [18]. To our best knowledge, iQIYI-VID-2019 dataset is the largest benchmark for video based multi-modal person identification. It contains 200k videos of 10k celebrities under unconstrained scenes, which makes person identification task more challenging. To ensure a fair comparison of different fusion methods, we use multi-modal features (face, head, body and audio) and face quality scores provided by official.

Implementation Details. In our best model, face, body and audio feature are used. When calculating edge weights, ω is set as 2 and γ as 1 through cross-validation. Three layers of GCN are used. The dimension of all hidden layers is set to 512. The hyperparameters α and λ are 0.1 and 0.5, respectively. We trained for a total of 40 epochs. Initial learning rate is set to 0.001, decreases by a factor of ten every 20 epochs. The Adam optimizer is used, weight decay is 0.00001.

When constructing a local graph, k_1 is set to 2. For QAP, we use a certain quality threshold 40 to classify features into high and low quality, to obtain AP_{high}, AP_{low}. The same operation is performed on all features to obtain AP_{mid}. Thus $k_2 = 3$. During testing, prediction result of AP_{high} from target video was taken as the final result. Due to small discrimination of body modal, average pooling operation is directly performed to obtain video-level features of body modal.

Evaluation Metrics. To evaluate the retrieval results, we use the Mean Average Precision (MAP) [19], which is the official metric of iQIYI-VID-2019 dataset:

$$MAP(Q) = \frac{1}{|Q|} \sum_{i=1}^{|Q|} \frac{1}{m_i} \sum_{j=1}^{n_i} Precision(R_{i,j}) \tag{8}$$

where Q is the set of person IDs to retrieve, m_i is the number of positive examples for the i-th ID, n_i is the number of positive examples within the top k retrieval results for the i-th ID, and $R_{i,j}$ is the set of ranked retrieval results from the top until you get j positive examples. Following the official evaluation, only top 100 retrievals are kept for each person ID.

4.2 Comparison to the State of the Art

Table 1 compares MMM-GCN to the current methods on iQIYI-VID-2019 dataset. The experimental results show that MMM-GCN significantly outperforms other methods. Compared with FAMF [15] (row 4), our method improves MAP by 2.47%, reaching 85.41% corresponds to row 6 in Table 1.

Table 1. Comparison with the state-of-the-art method on iQIYI-VID-2019 dataset.

Method	Level	MAP
GhostVLAD [24]	v+m	0.8109
Liu et,al [17]	v+m	0.8246
NeXtVLAD [16]	v+m	0.8283
FAMF [15]	v+m	0.8294
MMM-GCN	v+m	**0.8406** (+ 1.12%)
MMM-GCN	v+m+n	**0.8541** (+ 2.47%)

Table 2. MAP result of MMM-GCN under different number of neighbors.

k_1	MAP
0	0.8406
1	0.8508
2	**0.8541**
4	0.8502
9	0.8381
19	0.8281

Table 3. MAP result of MMM-GCN under different video-level features.

Video-Level Features			MAP
AP^m_{high}	AP^m_{mid}	AP^m_{low}	
✓			0.8426
	✓		0.8418
		✓	0.6859
✓	✓		0.8522
✓		✓	0.8462
✓	✓	✓	**0.8541**

Table 4. MAP result of MMM-GCN under different combinations of multi-modal.

Modal				MAP
Face	Head	Body	Audio	
✓				0.8476
	✓			0.6190
		✓		0.3949
			✓	0.2613
✓	✓			0.8448
✓		✓		0.8497
✓			✓	0.8482
✓		✓	✓	**0.8541**
✓	✓	✓	✓	0.8499

4.3 Ablation Study

Effect of MMM-GCN. All current methods (rows 1–4 in Table 1) model video-level and multi-modal-level information independently for person identification. To further verify the superiority of MMM-GCN, we experimented MMM-GCN with only video-level and multi-modal-level fusion, as shown in row 5 in Table 1. Compared with the current SOTA [15], our method improves MAP by 1.12%, which verifies the effectiveness of our proposed framework for adaptive fusion on different levels of information.

Effect of Neighbor-Level Fusion. When fusing neighbor-level information, k_1 is a hyperparameter. We hope that the selected set of neighbors contains more samples with same label, making the information richer and more comprehensive, while not introducing too much noise from neighbors with different labels that may pollute target features during fusion. The effect of neighbor selection on feature fusion will be described in detail in Sect. 4.4.

Table 2 shows effect of neighbor-level fusion, where $k_2 = 0$ means neighbor-level information is not used. It shows that $1 \leq k_2 \leq 4$ achieves a higher MAP than $k_2 = 0$, and the best result is obtained at $k_2 = 2$. When k_2 becomes larger, for example, k_2 is 9 and 19 respectively, noisy information gradually increases, which has a negative impact on MMM-GCN and reduces the fusion performance.

Effect of Video-Level Fusion. Table 3 shows the effect of video-level fusion. When using a single video-level feature, AP_{high} performs best, followed by AP_{mid} and AP_{low}, respectively. Although AP_{low} does not work well alone, it still has some unique information to improve feature diversity and raise MAP, when cooperating with other video-level features. Above results indicate that MMM-GCN could extract complementary information from video-level features adaptively, which verifies its effectiveness. In addition, MMM-GCN can also learn some valuable information between local feature, represented by AP_{high}, AP_{low}, and global feature, represented by AP_{mid}.

Effect of Multi-Modal Level Fusion. Table 4 shows the effect of multi-modal-level fusion. It can be seen that face modal performs best compared with other single modals. We take single face modal as baseline and combine it with head, body and audio respectively, results are shown in rows 5–7. Except for head modal, other modals combined with face modal could improve identification precision compared to the single face modal. We guess that the high redundancy between head and face makes the fusion invalid, while the head feature is less robust than face due to some unstable factors (hairstyle, etc.), leading to a lower MAP when fusion. Finally, we fused face, body and audio modal together, and obtained the best result with a MAP of 85.41%.

Table 5. Distribution of true and false samples under different neighbor purity (NP). Numeric in bracket represents the number of samples.

Neighbor purity	True samples	False samples
1	0.9974 (36556)	0.0026 (94)
0.5	0.7688 (2583)	0.2312 (777)
0	0.1880 (1019)	0.8120 (4402)

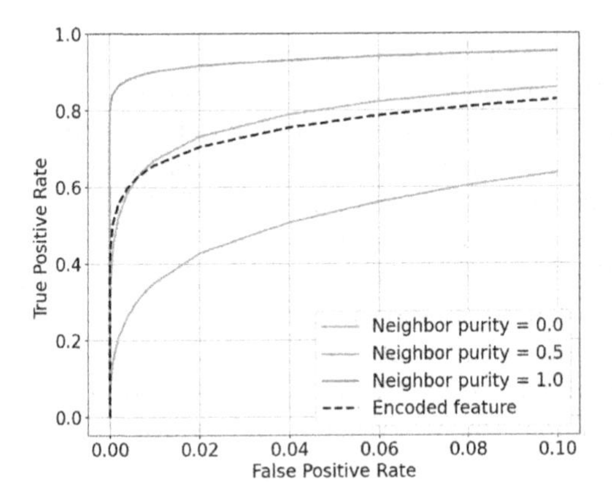

Fig. 4. ROC curves of 1:1 feature verification from randomly selected pairs, where encoded feature means the feature before fusion, the others represent fused embeddings with different neighbor purity.

4.4 Discussion

Our ablation experiments verified the effectiveness of MMM-GCN on neighbor-level, video-level and multi-modal-level fusion. However, for neighbor-level, the number of neighbors will influence performance, as shown in Table 2.

We calculated neighbor purity(NP) of true and false samples from identification results, as shown in Table 5, where NP is defined as the proportion of neighbors with same label as target video. A smaller NP means a larger number of noisy nodes (neighbors with different labels). It can be seen that with the decrease of NP, the ratio of false samples increases quickly. When $NP = 0$, false samples accounted for a majority of ratio, reaching 81.2%. The opposite performance occurs when NP increases. Above results illustrate that noisy nodes directly increase the risk of misidentification. When neighbors are selected incorrectly, the noisy information propagated by noisy nodes will confuse target features during the aggregation of GCN, thus degrading performance.

Then, we plot the ROC curves for different values of NP, as shown in Fig. 4. We use encoded feature as baseline. It is observed that performance decreases with the decrease of NP. Compared with encoded feature, $NP = 1$ shows a significant improvement, while $NP = 0$ degrades model performance.

Finally, we use official label information to manually set NP of all target videos to 1. The MAP improved by 6.86% compared with our best result, reaching 92.27%, which reconfirms that noisy nodes heavily pollute fusion, while correct neighbors can achieve an extra significant improvement.

Based on results above, we conclude that GCN is not robust enough to noisy nodes. The target feature would be heavily polluted by noisy nodes, and decreased distance between them, which makes the performance degrade. This conclusion points out a promising direction for subsequent research.

5 Conclusion

In this paper, we propose a novel framework named MMM-GCN for video-based multi-modal person identification task. Compared with existing methods, we introduce extra neighbor-level information, and achieve adaptive fusion among three levels in a unified modal. We validate the effectiveness of our proposed method on the iQIYI-VID-2019 dataset. The results show that MMM-GCN outperforms current state-of-the-art method, improving MAP by 1.12% when fusing same level information (v+m) as SOTA, and by 2.47% after introducing neighbor-level information. Finally, we point out feature fusion is heavily polluted by noisy nodes in neighbors and explore the upper bound on this direction.

References

1. Arandjelovic, R., Gronat, P., Torii, A., Pajdla, T., Sivic, J.: Netvlad: CNN architecture for weakly supervised place recognition. In: Proceedings of the IEEE Conference on Computer Vision and Pattern Recognition, pp. 5297–5307 (2016)

2. Chen, J., Yang, L., Xu, Y., Huo, J., Shi, Y., Gao, Y.: A novel deep multi-modal feature fusion method for celebrity video identification. In: Proceedings of the 27th ACM International Conference on Multimedia, pp. 2535–2538 (2019)
3. Chen, M., Wei, Z., Huang, Z., Ding, B., Li, Y.: Simple and deep graph convolutional networks. In: International Conference on Machine Learning, pp. 1725–1735. PMLR (2020)
4. Deng, J., Guo, J., Xue, N., Zafeiriou, S.: ArcFace: additive angular margin loss for deep face recognition. In: Proceedings of the IEEE/CVF Conference on Computer Vision and Pattern Recognition, pp. 4690–4699 (2019)
5. Dong, C., Gu, Z., Huang, Z., Ji, W., Huo, J., Gao, Y.: DeepMEF: a deep model ensemble framework for video based multi-modal person identification. In: Proceedings of the 27th ACM International Conference on Multimedia, pp. 2531–2534 (2019)
6. Garofolo, J.S., Lamel, L.F., Fisher, W.M., Fiscus, J.G., Pallett, D.S.: DARPA TIMIT acoustic-phonetic continous speech corpus CD-ROM. NIST speech disc 1-1.1. NASA STI/Recon technical report n 93, 27403 (1993)
7. Hamilton, W., Ying, Z., Leskovec, J.: Inductive representation learning on large graphs. In: Advances in Neural Information Processing Systems, vol. 30 (2017)
8. Hu, J., Liu, Y., Zhao, J., Jin, Q.: MMGCN: multimodal fusion via deep graph convolution network for emotion recognition in conversation. arXiv preprint arXiv:2107.06779 (2021)
9. Huang, G.B., Mattar, M., Berg, T., Learned-Miller, E.: Labeled faces in the wild: a database forstudying face recognition in unconstrained environments. In: Workshop on Faces in'Real-Life'Images: Detection, Alignment, and Recognition (2008)
10. Huang, Q., Liu, W., Lin, D.: Person search in videos with one portrait through visual and temporal links. In: Proceedings of the European Conference on Computer Vision (ECCV), pp. 425–441 (2018)
11. Huang, W., Zhang, T., Rong, Y., Huang, J.: Adaptive sampling towards fast graph representation learning. In: Advances in Neural Information Processing Systems, vol. 31 (2018)
12. Huang, Z., Chang, Y., Chen, W., Shen, Q., Liao, J.: Residual dense network: a simple approach for video person identification. In: Proceedings of the 27th ACM International Conference on Multimedia, pp. 2521–2525 (2019)
13. Joon Oh, S., Benenson, R., Fritz, M., Schiele, B.: Person recognition in personal photo collections. In: Proceedings of the IEEE International Conference on Computer Vision, pp. 3862–3870 (2015)
14. Kipf, T.N., Welling, M.: Semi-supervised classification with graph convolutional networks. arXiv preprint arXiv:1609.02907 (2016)
15. Li, F., Wang, W., Liu, Z., Wang, H., Yan, C., Wu, B.: Frame aggregation and multi-modal fusion framework for video-based person recognition. In: Lokoč, J., et al. (eds.) MMM 2021. LNCS, vol. 12572, pp. 75–86. Springer, Cham (2021). https://doi.org/10.1007/978-3-030-67832-6_7
16. Lin, R., Xiao, J., Fan, J.: NeXtVLAD: an efficient neural network to aggregate frame-level features for large-scale video classification. In: Proceedings of the European Conference on Computer Vision (ECCV) Workshops (2018)
17. Liu, Y., et al.: iQIYI-VID: a large dataset for multi-modal person identification. arXiv preprint arXiv:1811.07548 (2018)
18. Liu, Y., et al.: iQIYI celebrity video identification challenge. In: Proceedings of the 27th ACM International Conference on Multimedia, pp. 2516–2520 (2019)
19. Manning, C., Raghavan, P., Schütze, H.: Introduction to information retrieval. Nat. Lang. Eng. **16**(1), 100–103 (2010)

20. Nguyen, B.X., Nguyen, B.D., Do, T., Tjiputra, E., Tran, Q.D., Nguyen, A.: Graph-based person signature for person re-identifications. In: Proceedings of the IEEE/CVF Conference on Computer Vision and Pattern Recognition, pp. 3492–3501 (2021)
21. Shen, S., et al.: Structure-aware face clustering on a large-scale graph with 107 nodes. In: Proceedings of the IEEE/CVF Conference on Computer Vision and Pattern Recognition, pp. 9085–9094 (2021)
22. Tao, Z., Wei, Y., Wang, X., He, X., Huang, X., Chua, T.S.: MGAT: multimodal graph attention network for recommendation. Inf. Process. Manag. **57**(5), 102277 (2020)
23. Wang, G., Yuan, Y., Chen, X., Li, J., Zhou, X.: Learning discriminative features with multiple granularities for person re-identification. In: Proceedings of the 26th ACM International Conference on Multimedia, pp. 274–282 (2018)
24. Zhong, Y., Arandjelović, R., Zisserman, A.: GhostVLAD for set-based face recognition. In: Jawahar, C.V., Li, H., Mori, G., Schindler, K. (eds.) ACCV 2018. LNCS, vol. 11362, pp. 35–50. Springer, Cham (2019). https://doi.org/10.1007/978-3-030-20890-5_3
25. Zhong, Z., Zheng, L., Cao, D., Li, S.: Re-ranking person re-identification with k-reciprocal encoding. In: Proceedings of the IEEE Conference on Computer Vision and Pattern Recognition, pp. 1318–1327 (2017)

Feature Enhancement and Reconstruction for Small Object Detection

Chong-Jian Zhang[1,2], Song-Lu Chen[1,2], Qi Liu[1,2], Zhi-Yong Huang[1,2], Feng Chen[2,3], and Xu-Cheng Yin[1,2](✉)

[1] University of Science and Technology Beijing, Beijing 100083, China
{chongjianzhang,qiliu7,huang.zhiyong}@xs.ustb.edu.cn,
{songluchen,xuchengyin}@ustb.edu.cn
[2] USTB-EEasyTech Joint Lab of Artificial Intelligence, Beijing 100083, China
[3] EEasy Technology Company Ltd., Zhuhai 519000, China
cfeng@eeasytech.com

Abstract. Due to the small size and noise interference, small object detection is still a challenging task. The previous work can not effectively reduce noise interference and extract representative features of the small object. Although the upsampling network can alleviate the loss of features by enlarging feature maps, it can not enhance semantics and will introduce more noises. To solve the above problems, we propose CAU (Content-Aware Upsampling) to enhance feature representation and semantics of the small object. Moreover, we propose CSA (Content-Shuffle Attention) to reconstruct robust features and reduce noise interference using feature shuffling and attention. Extensive experiments verify that our proposed method can improve small object detection by 2.2% on the traffic sign dataset TT-100K and 0.8% on the object detection dataset MS COCO compared with the baseline model.

Keywords: Small object detection · Content-aware upsampling · Content-shuffle attention

1 Introduction

Small object detection has a wide range of applications, such as traffic sign detection, face recognition, and remote sensing image analysis. However, due to the small size and noise interference, generic object detectors are not effectively applicable to small object detection. A common practice to detect small objects is to enlarge the feature map using upsampling. Traditional upsampling methods include nearest-neighbor interpolation and bilinear interpolation, which can enlarge the image resolution but introduce more noises. Deep-learning-based upsampling method DUpsampling [23] can enlarge the feature map based on the relationship between pixels. However, the downsampling operation in the network can cause the loss of object information, especially for the small object,

D.-T. Dang-Nguyen et al. (Eds.): MMM 2023, LNCS 13833, pp. 16–27, 2023.
https://doi.org/10.1007/978-3-031-27077-2_2

thus leading to missing detection of the small object. CARAFE [25] is an upsampling method via content sensing and feature recombination, which can reduce the information loss of small objects via context modeling. However, it does not consider multi-scale features during feature recombination and is not conducive to detecting small objects. To solve the above problems, we propose an upsampling module CAU to reduce the loss of object information by aggregating the global context information and extracting multi-scale features, thus improving small object detection. Moreover, small objects are easily affected by background noises. The attention mechanism can suppress noise interference by focusing on essential areas and ignoring irrelevant information. SENet [10] and ECANet [26] can capture the relationship between channels with channel attention, suppressing background noises using global context modeling. However, the channel attention can not capture local information around the object, affecting small object detection. Bilinear attention mechanism GSoP-Net [8] and Fang et al. [6] propose to capture local feature interactions within each channel via spatial attention. Although the spatial attention can effectively utilize local relationships, the feature interaction brings heavy computational complexity. To solve the above problems, we propose a CSA module to reconstruct features via feature shuffling and attention, which combines channel attention and feature shuffling for robust representation. CSA can improve small object detection with little computation added. In this paper, CAU is added to the detection neck, and CSA is added to the detection head. By adding a few parameters and calculations, small object detection can be improved significantly. Our main contributions can be summarized as follows.

1. To reduce the loss of object information, we propose an upsampling module CAU, which can enhance feature representation via global context aggregation and multi-scale feature extraction.
2. To reduce background noise interference, we propose an attention module CSA, which significantly improves small object detection via robust feature reconstruction.
3. Our proposed method achieves 66.1% and 21.9% on the small object of TT-100K and MS COCO, respectively, 2.2% and 0.8% higher than the baseline model by adding a few parameters and calculations.

2 Related Work

2.1 Upsampling Method

Upsampling is used to restore the resolution of the image or feature map to the original resolution. Traditional upsampling methods include nearest-neighbor upsampling, bilinear upsampling, and bicubic upsampling, which only improve the image resolution according to its image signal. However, these methods bring many side effects, such as increased noise and computational complexity. To solve the above problems, deep-learning-based upsampling methods are proposed. Sub-pixel layer [20] is an end-to-end learnable layer that generates and recombines multiple channels to perform upsampling. DUpsampling [23] performs downsampling operations for low-level feature representations and then concatenate them with high-level features to complete feature fusion and upsampling. CARAFE [25] performs feature upsampling by recombining features in the region centered at each location by weighted combination, which can aggregate contextual information of large receptive fields. However, small objects will be missed due to the lack of multi-scale features in feature recombination. We propose an upsampling module that can aggregate global context and extract multi-scale features to improve small object detection.

2.2 Multi-scale Feature Extraction

SPPNet [9] and GoogLeNet [21] propose a parallel branch to extract features at different spatial scales based on its receptive field, i.e., spatial pyramid. Liu et al. propose RFBNet [16] with dilated convolution and fuse three branches to improve the poor representation of small objects. Zhao et al. [30] use global average pooling to extract multi-scale information and achieve competitive results in semantic segmentation. Chen et al. [3] design multiple parallel dilated convolution modules with different sampling rate to extract multi-scale features. We propose to extract global information by adding a multi-scale feature extraction module.

2.3 Attention Mechanism

The attention mechanism can focus on the essential area of the object and suppress irrelevant information. Channel attention includes SENet [10] and ECANet [26], which focuses on the relationship between channels and automatically learns the importance of different channel features. Bilinear attention mechanism GSoP-Net [8] and Fang et al. [6] propose to enhance the local pairwise feature interaction in each channel while retaining spatial information, improving feature representation via local relationship. SANet [28] introduces the random channel mixing operation, which uses spatial and channel attention mechanisms in parallel. Ying et al. [27] propose a multi-scale global attention mechanism to alleviate the loss of small object information caused by downsampling operations. We propose to combine feature reconstruction and attention mechanism to enhance object feature representation.

3 Method

3.1 Network Architecture

The overall network architecture is shown in Fig. 1. We choose the lightweight YOLOv5 as our baseline model. YOLOv5 uses CSPDarknet53 [18] as the backbone network for feature extraction. The detection neck adopts a Path Aggregation Network (PANet) [15] for feature fusion, where the low-resolution features are upsampled by nearest-neighbor upsampling to be concatenated with the high-resolution features. In this paper, we introduce an upsampling module CAU to replace the nearest-neighbor upsampling module in the detection neck. CAU can reduce information loss by global context aggregation and multi-scale feature extraction. Moreover, we propose an attention module CSA in the detection head to reconstruct features for robust representation. CAU and CSA can improve small object detection with a few parameters and calculations added.

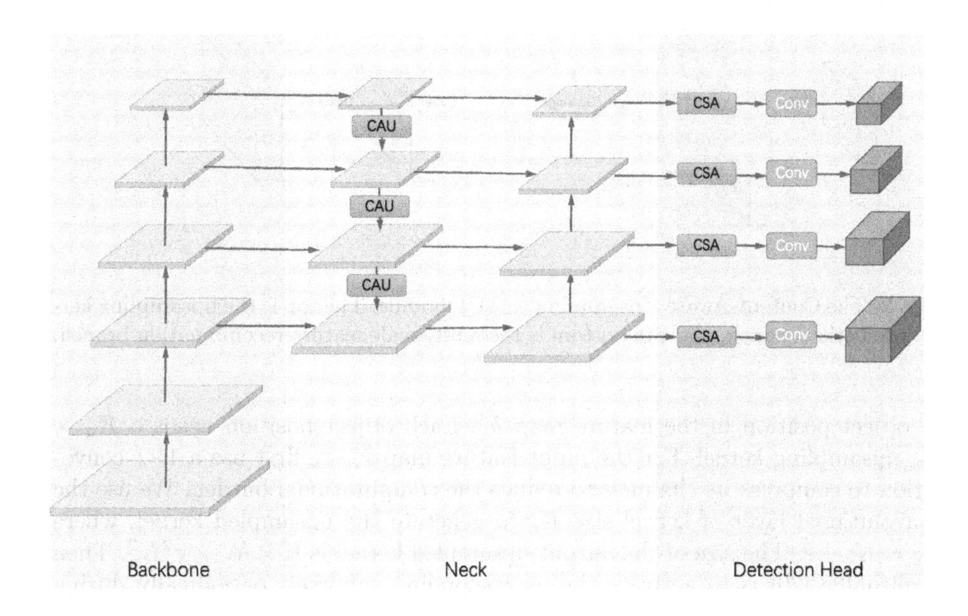

Fig. 1. Overall architecture.

3.2 Content-Aware Upsampling (CAU)

As shown in Fig. 2, CAU is divided into two branches: the upsampling kernel prediction branch and the multi-scale feature recombination branch. The former branch can automatically predict the upsampled kernel corresponding to the object location. The latter branch can extract multi-scale features for robust representation. We predict the upsampling kernel and then use the multi-scale feature recombination module to complete the upsampling operation. Given an input feature map F of H $\times$ W $\times$ C and an upsampling rate r, the size of the output feature map F' will be rH $\times$ rW $\times$ C. For any object location $l' = (i',$ $j')$ of the output F', there is a corresponding source location $l = (i, j)$ at the input F, where $i = \lfloor i'/r \rfloor$, $j = \lfloor j'/r \rfloor$.

Upsampling Kernel Prediction. We use a content-aware method to predict the upsampled kernel. Each position on the feature map F corresponds to the

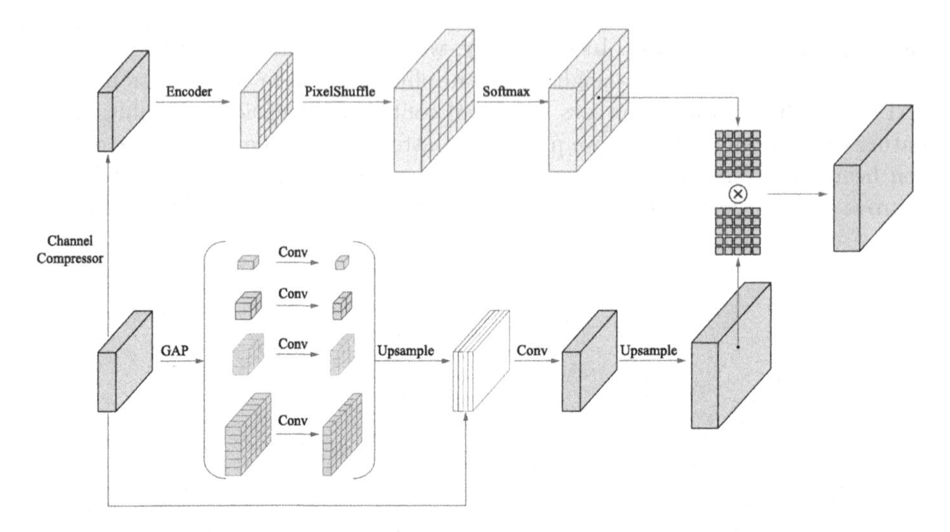

Fig. 2. The Content-Aware Upsampling (CAU) module. The top is the upsampling kernel prediction branch, and the bottom is the multi-scale feature recombination branch.

r^2 object position in the feature map F'. Each object position needs a $K_u \times K_u$ upsampling kernel. For the input feature map F, we first use a 1×1 convolution to compress its channels to reduce the computational burden. We use the convolutional layer of kernel size K_e to generate the upsampled kernel, where $K_e = K_u - 2$. The size of the output upsampled kernel is H × W × $r^2 K_u^2$. Then pixelshuffle [20] is to activate the corresponding subpixels periodically during convolution according to different subpixel positions of the low-resolution feature map to complete the construction of the high-resolution feature map. Finally, the size of the final output upsampled kernel is rH × rW × K_u^2, and the softmax function is applied to each $K_u \times K_u$ for normalization. As shown in Eq. (1), the upsampling kernels at different positions are predicted adaptively.

$$K_{l'} = \psi(N(F_l, K_e)) \tag{1}$$

where the kernel prediction module ψ predicts a location-wise kernel $K_{l'}$ for each location l' based on the neighbor of F_l. Here we denote N(F_l, K_e) as the $K_e \times K_e$ sub-region of F centered at the location l, i.e., the neighbor of F_l.

Multi-scale Feature Recombination. For the input feature map F, we divide the feature map into different sub-regions through adaptive pooling operations, obtaining 1×1, 2×2, 3×3, and 6×6 feature maps. Then we perform 1×1 convolution on each feature map to reduce the number of channels to $\frac{1}{4}$ of F. The sub-region features are then upsampled and concatenated with the input features, which aggregate global context information of different scales to improve small object detection,the output feature map f. As shown in Eq. (2), for each position l' in the output feature map, we map it to the feature map f_l and dot product the region centered by $K_u \times K_u$ with $K_{l'}$ to obtain the output value.

$$F'_{l'} = \phi(N(f_l, K_u), K_{l'}) \tag{2}$$

where ϕ is the multi-scale content-aware reassembly module that reassembles the neighbor of f_l with the kernel $K_{l'}$.

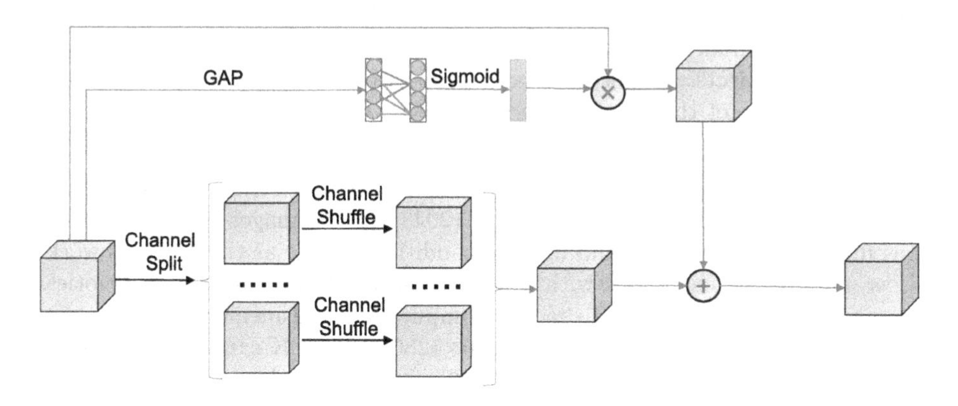

Fig. 3. The Channel Shuffle Attention (CSA) module.

3.3 Channel Shuffle Attention (CSA)

As shown in Fig. 3, we introduce the CSA module, which combines feature reconstruction and attention mechanism. Through feature reconstruction, the network can effectively suppress background noise interference; By adding the attention mechanism, the network can focus on critical objects and filter out useless information. CSA is mainly divided into two branches. In the bottom feature reconstruction branch, CSA divides the channels of the input feature X into multiple groups and performs corresponding interception operations. Then channel shuffle is used for each group to enhance the feature robustness. This branch can disrupt the internal relationship between features and force the network to learn more subtle features. In the top attention branch, global average pooling is carried out on the input feature X to aggregate global spatial information. As shown in Eq. (3), the 1-dimensional convolution operation is carried out to capture the local cross-channel information interaction, where we use the sigmoid activation function to calculate the weight of each channel. This way, we can extract the dependencies between channels, improving small object detection. As shown in Eq. (4), the output feature M' is obtained by multiplying the weights with the original input features. As shown in Eq. (5), the final feature M'' is obtained via adding the top and bottom output.

$$W = \sigma(Conv1D(GAP(X))) \tag{3}$$

$$M' = W \cdot X \tag{4}$$

$$M'' = \Phi(M', X') \tag{5}$$

4 Experiments

4.1 Datasets

MS COCO. [14] Microsoft COCO (MS COCO) is a widely used dataset for object detection, including 118k training images, 5k validation images, and 40k test images. MS COCO has 80 categories, approximately containing 41% small objects, 34% medium objects, and 24% large objects. MS COCO is challenging because small objects account for a large proportion with complex background. In this paper, we set the input image size as 640×640.

Tsinghua-Tencent 100K (TT-100K). [32] TT-100K is a traffic sign detection dataset, containing 100k high-resolution (2048×2048) images and 30k traffic signs under various weather and lighting conditions. Same as the previous work [5], we remove the categories with a few samples and only keep 45 categories of more than 200 categories. To reduce computation and memory overflow, we crop the original image to the size of 1280×1280. TT-100K can be divided into three scales, i.e., the area lower than 32×32 as the small-sized object, the area between 32×32 and 96×96 as the middle-sized object, and the area greater than 96×96 as the large-sized object.

4.2 Evaluation Metrics

Same as the evaluation metrics in the MS COCO competition [14], we use Average Precision (AP) to evaluate the detection performance, which is a comprehensive metric of precision and recall. AP is calculated as Eq. (6), where p represents precision, and r represents recall. As shown in Eq. (7), mean Average Precision (mAP) denotes average AP of multiple categories, where n represents the number of categories. Moreover, we use Floating Point Operations (FLOPs) to measure the computation complexity, which represents the total number of calculation operations of a detection model.

$$AP = \int_0^1 p(r)dr \tag{6}$$

$$mAP = \frac{1}{n}\sum_{i=1}^{n} AP_i \tag{7}$$

4.3 Implementation Details

We conduct all the experiments with Pytorch 1.11.0 and CUDA 11.3. All the networks are trained with 4 NVIDIA GeForce 2080Ti GPUs. We use the SGD optimizer to train the models for 300 epochs. The initial learning rate is set to 0.01, and the cosine annealing strategy is used to reduce the learning rate. The weight decay, batch size, and momentum are set to 0.0005, 8, and 0.937, respectively.

4.4 Ablation Study

As shown in Table 1, we perform ablation experiments on the TT-100K dataset to verify the proposed CAU and CSA modules. We use YOLOv5l6 as the baseline detection model. Experiments prove that CAU and CSA can improve the detection performance of all sizes with little computation added, achieving 1.7% mAP improvement compared with the baseline model. Specially, the proposed method can sinificantly improve small object detection, 2.2% higher than YOLOv5l6. CAU can improve the AP of small objects by 1.9%, which proves its ability to reduce information loss.

Table 1. Ablaton study on TT-100K

Method	AP_s	AP_m	AP_l	AP_{50}	mAP	params	GFLOPs
YOLOv5l6	63.9	81.3	87.1	95.5	76.5	76.5M	**110.8**
YOLOv5l6+CAU	65.8	`81.9	88.2	95.9	77.5	80.5M	116.6
YOLOv5l6+CAU+CSA	**66.1**	**82.8**	**88.5**	**96.4**	**78.2**	84.9M	121.9

4.5 Comparative Results

As shown in Table 2, we compare our method with other state-of-the-art detectors on the TT-100K dataset, including prestigious one-stage SSD [17], RetinaNet [13], YOLO [1,18] and two-stage Faster R-CNN [19], FPN [12]. Our proposed method can achieve the best detection performance for all evaluation metrics. Especially for the small object, our method can acquire a large performance gain. Moreover, Table 3 shows the detection results on MS COCO. Due to the

Table 2. Comparison with state-of-the-art methods on TT-100K.

Method	AP_s	AP_m	AP_l	AP_{50}	mAP
Faster R-CNN [19]	50.0	82.0	88.0	90.3	72.1
Cascade R-CNN [2]	55.7	85.4	90.4	92.5	74.9
FPN [12]	40.6	63.7	63.0	43.5	64.1
SSD [17]	25.3	67.8	81.5	83.5	59.8
RetinaNet [13]	60.9	79.5	81.2	92.4	70.9
Efficientdet-D0 [22]	-	74.4	83.6	85.4	66.7
YOLOv3 [18]	60.4	76.7	81.1	91.6	70.2
YOLOv4 [1]	62.8	79.8	85.4	94.7	74.9
YOLOv5l6	63.9	81.3	87.1	95.5	76.5
Ours	**66.1**	**82.8**	**88.5**	**96.4**	**78.2**

limited computing resources, we use YOLOv5s as the baseline model[1], replacing the upsampling module with CAU in the detection neck and adding CSA to the detection head. Our method achieves the best detection performance, proving its effectiveness and generality.

Table 3. Comparison with state-of-the-art methods on MS COCO.

Method	AP_s	AP_m	AP_l	AP_{50}	mAP
Faster R-CNN [19]	15.7	35.8	44.3	55.2	34.1
Cascade R-CNN [2]	19.6	38.9	48.0	55.8	36.8
Deformable R-FCN [4]	19.4	40.1	52.5	58.0	37.5
CoupleNet [31]	11.6	36.3	50.1	53.5	33.1
CornerNet [11]	17.0	39.0	50.5	53.7	37.8
RefineDet [29]	16.5	38.8	51.5	57.0	36.1
DSSD513 [7]	13.0	35.4	51.1	53.3	33.2
RetainNet [13]	18.5	37.2	45.4	53.4	35.1
DeNet [24]	12.3	36.1	50.8	53.4	33.8
YOLOv3 [18]	20.5	39.3	45.9	54.8	35.1
YOLOv5s	21.1	42.3	47.5	56.7	37.1
Ours	**21.9**	**42.9**	**47.8**	**57.1**	**37.5**

4.6 Qualitative Results

In Fig. 4, we visualize the feature map after using CAU and CSA. Compared with the original detection neck, CAU can highlight the features of small traffic signs, which proves CAU can enhance feature representation. Compared with the original detection head, CSA can effectively suppress background noise interference and extract robust features. This way, our method can improve small object detection.

Figure 5 shows some detection examples of TT-100K and MS COCO. Our method can accurately detect objects, especially for small objects.

[1] Please refer to https://github.com/ultralytics/yolov5. For TT-100K, we use the large YOLOv5l6 as the baseline model. For MS COCO, we use the small YOLOv5s as the baseline model. Except for CSA and CAU, our model is the same as the official model. The parameters of YOLOv5l6 are about 10× of YOLOv5s.

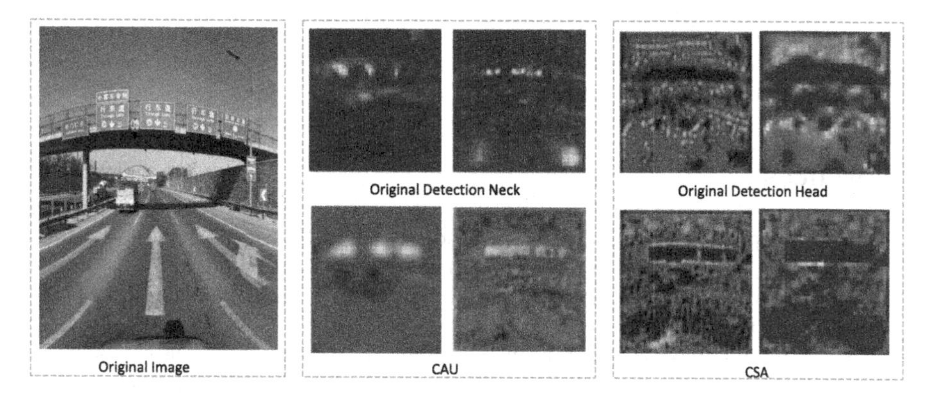

Fig. 4. Feature visualization. CAU can enhance object features. CSA can suppress background noises for robust feature extraction.

Fig. 5. Qualitative results on TT-100K (1st row) and MS COCO (2nd & 3rd rows)

5 Conclusion

This paper proposes a feature enhancement module CAU and a feature reconstruction module CSA for small object detection. CAU can enhance feature representation with less information loss, and CSA can suppress background noise interference to extract robust features. Comprehensive experiments prove that our method can improve object detection, especially for small objects. In the future, we will validate the generality of the proposed modules with more networks.

Acknowledgment. The research is supported by National Key Research and Development Program of China (2020AAA0109701), National Natural Science Foundation of China (62076024, 62006018).

References

1. Bochkovskiy, A., Wang, C.-Y., Mark Liao, H.-Y.: YOLOv4: optimal speed and accuracy of object detection. arXiv preprint arXiv:2004.10934 (2020)
2. Cai, Z., Vasconcelos, N.: Cascade R-CNN: delving into high quality object detection. In: CVPR (2018)
3. Chen, L.-C., Papandreou, G., Schroff, F., Adam, H.: Rethinking atrous convolution for semantic image segmentation. arXiv preprint arXiv:1706.05587 (2017)
4. Dai, J., et al.: Deformable convolutional networks. In: ICCV, pp. 764–773 (2017)
5. Deng, C., Wang, M., Liu, L., Liu, Y., Jiang, Y.: Extended feature pyramid network for small object detection. IEEE Trans. Multimedia **24**, 1968–1979 (2021)
6. Fang, P., Zhou, J., Kumar Roy, S., Petersson, L., Harandi, M.: Bilinear attention networks for person retrieval. In: ICCV, pp. 8029–8038 (2019)
7. Fu, C.-Y., Liu, W., Ranga, A., Tyagi, A., Berg, A.C.: DSSD: deconvolutional single shot detector. arXiv preprint arXiv:1701.06659 (2017)
8. Gao, Z., Xie, J., Wang, Q., Li, P.: Global second-order pooling convolutional networks. In: CVPR, pp. 3024–3033 (2019)
9. He, K., Zhang, X., Ren, S., Sun, J.: Spatial pyramid pooling in deep convolutional networks for visual recognition. IEEE Trans. Pattern Anal. Mach. Intell. **37**(9), 1904–1916 (2015)
10. Hu, J., Shen, L., Sun, G.: Squeeze-and-excitation networks. In: CVPR, pp. 7132–7141 (2018)
11. Law, H., Deng, J.: CornerNet: detecting objects as paired keypoints. In: Ferrari, V., Hebert, M., Sminchisescu, C., Weiss, Y. (eds.) Computer Vision – ECCV 2018. LNCS, vol. 11218, pp. 765–781. Springer, Cham (2018). https://doi.org/10.1007/978-3-030-01264-9_45
12. Lin, T.-Y., Dollár, P., Girshick, R.B., He, K., Hariharan, B., Belongie, S.J.: Feature pyramid networks for object detection. In: CVPR, pp. 936–944 (2017)
13. Lin, T.-Y., Goyal, P., Girshick, R.B., He, K., Dollár, P.: Focal loss for dense object detection. In: ICCV, pp. 2999–3007 (2017)
14. Lin, T.-Y., et al.: Microsoft COCO: common objects in context. In: Fleet, D., Pajdla, T., Schiele, B., Tuytelaars, T. (eds.) ECCV 2014. LNCS, vol. 8693, pp. 740–755. Springer, Cham (2014). https://doi.org/10.1007/978-3-319-10602-1_48
15. Liu, S., Qi, L., Qin, H., Shi, J., Jia, J.: Path aggregation network for instance segmentation. In: CVPR, pp. 8759–8768 (2018)
16. Liu, S., Huang, D., Wang, Y.: Receptive field block net for accurate and fast object detection. In: Ferrari, V., Hebert, M., Sminchisescu, C., Weiss, Y. (eds.) ECCV 2018. LNCS, vol. 11215, pp. 404–419. Springer, Cham (2018). https://doi.org/10.1007/978-3-030-01252-6_24
17. Liu, W., et al.: SSD: single shot multibox detector. In: Leibe, B., Matas, J., Sebe, N., Welling, M. (eds.) ECCV 2016. LNCS, vol. 9905, pp. 21–37. Springer, Cham (2016). https://doi.org/10.1007/978-3-319-46448-0_2
18. Redmon, J., Farhadi, A.: YOLOv3: an incremental improvement. arXiv preprint arXiv:1804.02767, 2018

19. Ren, S., He, K., Girshick, R.B., Sun, J.: Faster R-CNN: towards real-time object detection with region proposal networks. IEEE Trans. Pattern Anal. Mach. Intell. **39**(6), 1137–1149 (2017)
20. Shi, W., et al.: Real-time single image and video super-resolution using an efficient sub-pixel convolutional neural network. In: CVPR, pp. 1874–1883 (2016)
21. Szegedy, C., et al.: Going deeper with convolutions. In: CVPR, pp. 1–9 (2015)
22. Tan, M., Pang, R., Le, Q.V.: EfficientDet: scalable and efficient object detection. In: CVPR, pp. 10778–10787 (2020)
23. Tian, Z., He, T., Shen, C., Yan, Y.: Decoders matter for semantic segmentation: data-dependent decoding enables flexible feature aggregation. In: CVPR, pp. 3126–3135 (2019)
24. Tychsen-Smith, L., Petersson, L.: Denet: scalable real-time object detection with directed sparse sampling. In: ICCV, pp. 428–436 (2017)
25. Wang, J., et al.: CARAFE: content-aware reassembly of features. In: ICCV, pp. 3007–3016 (2019)
26. Wang, Q., Wu, B., Zhu, P., Li, P., Zuo, W., Hu, Q.: ECA-Net: efficient channel attention for deep convolutional neural networks. In: CVPR, pp. 11531–11539 (2020)
27. Ying, X., et al.: Multi-attention object detection model in remote sensing images based on multi-scale. IEEE Access **7**, 94508–94519 (2019)
28. Zhang, Q.-L., Yang, Y.-B.: SA-Net: shuffle attention for deep convolutional neural networks. In: ICASSP, pp. 2235–2239 (2021)
29. Zhang, S., Wen, L., Bian, X., Lei, Z., Li, S.Z.: Single-shot refinement neural network for object detection. In: CVPR, pp. 4203–4212 (2018)
30. Zhao, H., Shi, J., Qi, X., Wang, X., Jia, J.: Pyramid scene parsing network. In: CVPR, pp. 6230–6239 (2017)
31. Zhu, Y., Zhao, C., Wang, J., Zhao, X., Wu, Y., Lu, H.: CoupleNet: coupling global structure with local parts for object detection. In: ICCV, pp. 4146–4154 (2017)
32. Zhu, Z., Liang, D., Zhang, S.-H., Huang, X., Li, B., Hu, S.-M.: Traffic-sign detection and classification in the wild. In: CVPR, pp. 2110–2118 (2016)

Toward More Accurate Heterogeneous Iris Recognition with Transformers and Capsules

Zhiyong Zhou[1,2], Yuanning Liu[1,2], Xiaodong Zhu[1,2(✉)], Shuai Liu[1,2], Shaoqiang Zhang[1,2], and Zhen Liu[3]

[1] College of Computer Science and Technology, Jilin University, Changchun, China
zhuxd@jlu.edu.cn

[2] Key Laboratory of Symbolic Computation and Knowledge Engineering of Ministry of Education, Changchun, China

[3] Graduate School of Engineering, Nagasaki Institute of Applied Science, Nagasaki, Japan

Abstract. As diverse iris capture devices have been deployed, the performance of iris recognition under heterogeneous conditions, e.g., cross-spectral matching and cross-sensor matching, has drastically degraded. Nevertheless, the performance of existing manual descriptor-based methods and CNN-based methods is limited due to the enormous domain gap under heterogeneous acquisition. To tackle this problem, we propose a model with transformers and capsules to extract and match the domain-invariant feature effectively and efficiently. First, we represent the features from shallow convolution as vision tokens by spatial attention. Then we model the high-level semantic information and fine-grained discriminative features in the token-based space by a transformer encoder. Next, a Siamese transformer decoder exploits the relationship between pixel-space and token-based space to enhance the domain-invariant and discriminative properties of convolution features. Finally, a 3D capsule network not only efficiently considers part-whole relationships but also increases intra-class compactness and inter-class separability. Therefore, it improves the robustness of heterogeneous recognition. Experimental validation results on two common datasets show that our method significantly outperforms the state-of-the-art methods.

Keywords: Heterogeneous iris recognition · Transformers · Capsules

1 Introduction

The irises from billions of citizens worldwide have been collected under near-infrared(NIR) illumination to establish national ID database [19]. Visible(VIS) light sensors can acquire identifiable long-range images and are cheaper than NIR sensors [7]. However, drastic performance degradation is anticipated when VIS

D.-T. Dang-Nguyen et al. (Eds.): MMM 2023, LNCS 13833, pp. 28–40, 2023.
https://doi.org/10.1007/978-3-031-27077-2_3

iris imaged by popular high-resolution smartphones and surveillance devices are matched with the NIR iris stored in the national ID database [8]. In addition, to reduce the re-enrollment costs and time in large-scale applications, enrollment and authentication are performed through different sensors, which leads to reduced performance [13]. In cross-spectral and cross-sensor heterogeneous iris recognition, the disparities in imaging cause the large gap in domain distribution between iris images. Thus, intra-class compactness is weakened, significantly affecting recognition performance [20]. Cross-spectral and cross-sensor recognition is a well-known challenge task, and related competitions have been held at BIOSIG 2016 [16] and IJCB 2017 [17].

Some traditional manual descriptor-based methods for reducing the domain gap in heterogeneous irises can be roughly classified into integrated and non-integrated methods. Oktiana et al. [11] proposed a multiscale weberface integrated with Gabor local binary pattern. [10] integrated Gradientface-based normalization and the texture descriptors. Integrated methods are more capable of overcoming domain variances relative to the particular one among them. However, redundant prior knowledge and computational complexity are increased. Gradually non-integrated methods have been developed to significantly reduce the domain gaps. [13] alined the domain distribution by a kernel learning framework. [12] learned a common sparse dictionary representation to mitigate the distribution discrepancy. Nevertheless, such traditional methods require extensive prior knowledge with low upper performance. CNNs have been successfully applied to iris recognition tasks and surpassed traditional methods [14]. [9] extracted domain-invariant features using pre-trained off-the-shelf CNNs. [4] learned paired CNNs to adapt heterogeneous iris features. [19] applied Siamese network and Triplet to extract features and proposed a supervised discrete hash to improve the matching rate. However, intra-class variations in iris images are minuter than in natural images [21]. For heterogeneous iris recognition, like fine-grained image recognition tasks [18], pure CNNs are inherently limited to the size of the reception field. Thus pure CNNs model limited contextual global information and struggle to prevent performance degradation. Generating more fine-grained domain-invariant discriminative features is still a challenging task.

To address the above challenge, following [19], we propose a new feature extraction and matching architecture with transformers and capsules. We introduce a transformer that models rich contextual global information in token-based space, using a series of vision tokens to represent high levels of intra-class subtle variances. Further, the Siamese transformer decoder establishes links between the token-based space represented by the transformer encoder and the pixel-space represented by the CNN to refine the CNN features. Therefore, the size of the receptive field is greatly enlarged to obtain finer-grained domain-invariant discriminative features. In the feature matching learning phase, [19] successfully utilized multilayer perceptrons. However, these studies lose spatial information and do not consider the part-whole relationship of the features. Capsule networks [15] can be used as an advanced alternative to CNNs. Different capsules can represent various properties of entities, and the routing mechanism can learn the part-whole relationship [3]. Motivated by such, we propose an improved 3D capsule network as feature matching to support different instantiation parameters such as specific spectrums, specific sensors, rotations, etc., which improves the robustness of unconstrained and different domains. Furthermore, the intra-class features are pulled closer to each other and inter-class features are kept away from each other, thereby improving the performance.

In summary, our main contributions are as follows:

(1) A transformer is proposed to model rich contextual spatial relationships in token-based space. It enhances the domain-invariant properties of features and obtains finer-grained discriminative features for heterogeneous iris recognition.
(2) A modified 3D capsule network is proposed as a feature matcher to represent the part-whole relationship between features while ignoring inter-domain differences, which improves the robustness of heterogeneous iris recognition.
(3) Our approach yields state-of-the-art results on two commonly used datasets in this community.
(4) To our knowledge, this work pioneers the use of transformers and capsules to improve the performance of heterogeneous iris recognition.

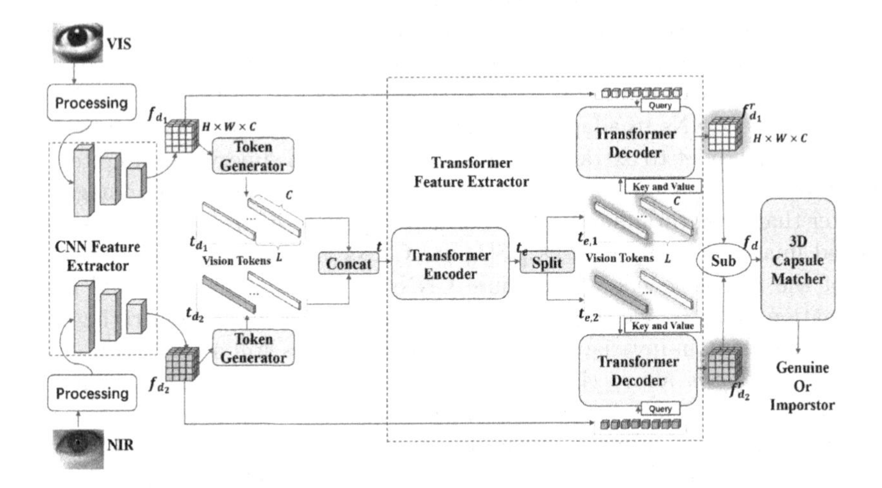

Fig. 1. Illustration of overall framework: preprocessing, twin CNN feature extractor, vision token generator, transformer feature extractor, and 3D capsule matcher.

2 Technical Details

The overall framework of our proposed method is summarized in Fig. 1. Heterogeneous iris pairs from two different domains (i.e., NIR against VIS or one sensor against another different sensor) are preprocessed to obtain normalized iris images. The CNN feature extractor outputs feature maps for each normalized iris. Thereafter, a vision token generator is used to obtain vision tokens from corresponding feature maps. Subsequently, the transformer feature extractor comprising a transformer encoder and a Siamese transformer decoder takes vision tokens to generate refined feature maps. Finally, an efficient 3D capsule matcher makes the decision for iris image pairs (i.e., genuine or imposter).

2.1 Vision Token Generator

The illustration of the vision token generator is shown in Fig. 2, where a spatial attention mechanism generates feature maps that interact with CNN feature maps to generate vision tokens. Consider d_1 and d_2 be two different domains(i.e., spectrum or sensor). Let $f_{d_1}, f_{d_2} \in \mathbb{R}^{H\times W\times C}$ be feature maps from the CNN feature extractor corresponding to d_1 and d_2 domain, where H, W, C denotes height, width, channels of the feature maps. Suppose $t_{d_1}, t_{d_2} \in \mathbb{R}^{H\times W\times C}$ represents tokens from vision token generator, where L denotes length of tokens, i.e., numbers of the element in tokens. The 1×1 convolution with a softmax function on the $H \times W$ dimension takes $f_{d_i} (i = 1, 2)$ as input to obtain the spatial attention maps $s_{d_i} \in \mathbb{R}^{H\times W\times C}$. Subquently, a weighted sum operation between f_{d_i} and s_{d_i} is used to obtain vision tokens $t_{d_i} \in \mathbb{R}^{L\times C}$.

The vision token generator is formulated as follows:

$$t_{d_i} = (s_{d_i})^T f_{d_i} = (\sigma(\psi(f_{d_i}; W)))^T f_{d_i}, \tag{1}$$

where $\psi(\cdot)$ denotes 1×1 convolution, $W \in \mathbb{R}^{1\times 1\times C\times L}$ indicates weight parameter of 1×1 convolution, $\sigma(\cdot)$ denotes softmax function operating on $H \times W$ dimension, and T indicates transpose operation.

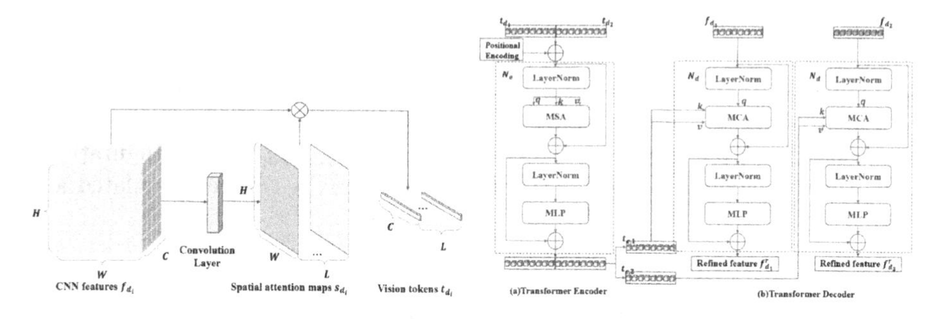

Fig. 2. Illustration of the token generator.

Fig. 3. Illustration of the transformer feature extractor.

2.2 Transformer Encoder

As shown in Fig. 3 (a), the transformer encoder has N_e layers consisting of residual connection, Layernorm, multi-head self-attention(MSA), and multilayer perceptron(MLP). t_{d_1} and t_{d_2} are concatenated as $t \in \mathbb{R}^{2L \times C}$. Furthermore, t plus a learnable positional coding $W_p \in \mathbb{R}^{2L \times C}$ is fed to the transformer encoder to output $t_e \in \mathbb{R}^{2L \times C}$. An attention head in MSA can be formulated as follows:

$$\begin{aligned} Q^n &= t^n W_q^n, K^n = t^n W_k^n, V^n = t^n W_v^n, \\ A\left(Q^n, K^n, V^n\right) &= \sigma\left(\frac{Q^n (K^n)^T}{\sqrt{d}}\right) V^n, \end{aligned} \tag{2}$$

where n denotes the nth layer of transformer encoder. Q^n, K^n, and V^n are query matrix, key matrix, and value matrix of the self-attention. $W_q^n, W_k^n, W_v^n \in \mathbb{R}^{C \times d}$ are three learnable transform matrices, t^n denotes forwarding tokens, and d denotes the column dimension of Q^n, K^n, V^n. MSA maps the input to different representation subspaces using multiple sets of (Q^n, K^n, V^n) in parallel. Then the results in these subspaces are concatenated and passed through an additional transform matrix to obtain the final output. MSA is formulated as follows:

$$\begin{aligned} \mathrm{MSA}\left(t^n\right) &= \mathrm{Cat}\left(a_1, a_2, \ldots, a_{h_s}\right) W_O, \\ a_j &= A\left(t^n W_{q_j}^n, t^n W_{k_j}^n, t^n W_{v_j}^n\right), \end{aligned} \tag{3}$$

where $\mathrm{Cat}(\cdot)$ denotes the concatenation operation in the vector dimension, $a_j \in \mathbb{R}^{2L \times d}$ is the output of the jth self-attention head, h_s denotes the number of self-attention heads, and $W_O \in \mathbb{R}^{(H \times d) \times C}$ denotes the transform matrix of obtaining the MSA's output. MLP has three layers comprising C neurons in the input and output layers and $2C$ neurons in the hidden layer with a GELU function.

2.3 Transformer Decoder

As shown in Fig. 3 (b), we separate t_e into $t_{e,1}, t_{e,2} \in \mathbb{R}^{L \times C}$, and a Siamese transformer decoder converts tokens into pixel features. The final refined feature maps $f_{d_1}^r, f_{d_2}^r \in \mathbb{R}^{H \times W \times C}$ with global context are obtained. The transformer decoder has N_d layers consisting of residual connection, Layernorm, multi-head cross-attention(MCA), and MLP. Different from transformer encoder, we remove positional coding and substitute MSA with MCA. The key and value of MCA are obtained by $t_{e,1}, t_{e,2}$, and the query is obtained by f_{d_1}, f_{d_2}. The configuration of MLP is the same as in the transformer encoder. MCA can be formulated as follows:

$$\begin{aligned} \mathrm{MCA}\left(f_{d_i}^n, t_{e,i}\right) &= \mathrm{Cat}\left(a_1, a_2, \ldots, a_{h_c}\right) W_O, \\ a_j &= A\left(f_{d_i}^n W_{q_j}^n, t_{e,i} W_{k_j}^n, t_{e,i} W_{v_j}^n\right), \end{aligned} \tag{4}$$

where $a_j \in \mathbb{R}^{2L \times d}$ denotes the output of the jth cross-attention head, h_c denotes the number of cross-attention heads, and $W_O \in \mathbb{R}^{(H \times d) \times C}$ denotes the transform matrix of obtaining the MCA's output.

2.4 3D Capsule Matcher

The structure of our 3D capsule matcher is shown in Fig. 4, containing capsule convolution layer, 3D capsule layer, and digital capsule layer. The difference

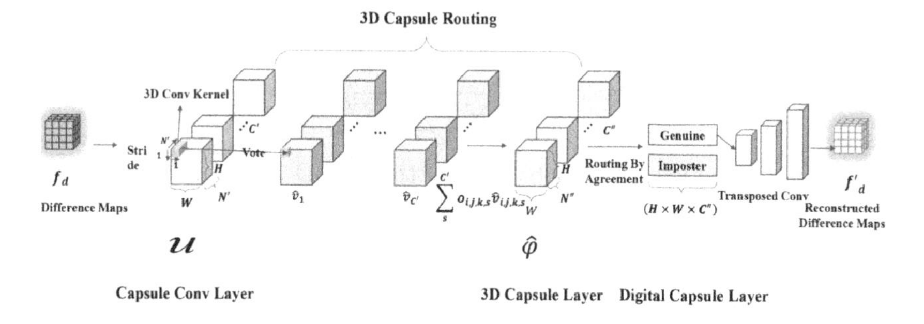

Fig. 4. Illustration of the 3D capsule matcher and reconstruction decoder.

maps $f_d \in \mathbb{R}^{H \times W \times C}$ are obtained by performing L2 Norm operation on the sub of $f_{d_1}^r, f_{d_2}^r$ from transformer decoder. Capsules are then routed to the 3D capsule layer using 3D capsule routing and finally routed to the digital capsule layer using routing-by-agreement mechanisms. The length of the final output capsule indicates the probability of whether the image pairs belong to the same class.

A Capsule Convolution Layer. The capsule convolution layer takes $f_d \in \mathbb{R}^{H \times W \times C}$ as input to obtain $u \in \mathbb{R}^{H \times W \times C' \times N'}$. The capsule convolution layer contains convolution, reshaping operation, and nonlinear activation. First, f_d is convolved with $C' \times N'$ filters, 3×3 kernels, stride of 1, and padding of 1 to obtain the feature maps with the shape $(H, W, C' \times N')$. Secondly, they are reshaped into 3D capsules $S \in \mathbb{R}^{H \times W \times C' \times N'}$, where N' denotes the dimension of the capsule and C' denotes the number of capsules. Finally, these capsules are applied with a nonlinear activation function *squash* to obtain $u \in \mathbb{R}^{H \times W \times C' \times N'}$. The *squash* is formulated as follows:

$$u_j = squash\,(s_j) = \frac{\|s_j\|^2}{1 + \|s_j\|^2} \frac{s_j}{\|s_j\|}, \tag{5}$$

where s_j denotes the jth capsule in 3D capsules, u_j denotes the nonlinear activation output of the jth capsule.

B 3D Capsule Layer. Our proposed new 3D capsule routing algorithm takes u as input to predict the new output capsules. The entire routing algorithm is shown in Fig. 4 . Firstly, We reshape u into $\hat{u} \in \mathbb{R}^{H \times W \times (C' \times N') \times 1}$. Secondly, $\hat{u}$ is convoluted using 3D convolution with $C'' \times N''$ filters, $3 \times 3 \times 3$ kernels, stride of $(1, 1, N')$, padding of $(1, 1, 0)$ to obtain intermediate votes $v \in \mathbb{R}^{H \times W \times C' \times (C'' \times N'')}$. Next, v is reshaped into $\hat{v} \in \mathbb{R}^{H \times W \times C'' \times N'' \times C'}$, which means a set of capsules can represent entities and routing. The 3D convolution obtains localized features allowing neighboring capsules to share similar information. Hence, such significantly reduces the number of parameters compared to dynamic routing in [15]. Subsequently, log probability $b \in \mathbb{R}^{H \times W \times 1 \times C'' \times C'}$ is initialized to 0 and passed through $3d\_softmax$ function, which is formulated as follows:

$$o_s = 3d\_softmax\left(b_s\right) = \frac{\exp\left(b_s\right)}{\sum_i^H \sum_j^W \sum_k^{C''} \exp\left(b_{i,j,k,s}\right)}, \tag{6}$$

where b_s denotes the log prior probability of the sth 3D capsule, and o_s denotes the coupling coefficients of the sth 3D capsule. The $(i, j, k)^{th}$ capsule in the digital capsule layer has C' possible outcomes from C' 3D capsules in the previous layer. Therefore, the intermediate votes v and the coupling coefficients o_s perform weighted summation to obtain inactivated output capsules $\varphi \in \mathbb{R}^{H\times W\times N''\times C''\times 1}$, which is formulated as follows:

$$\varphi_{i,j,k} = \sum_s^{C'} o_{i,j,k,s}\hat{v}_{i,j,k,s}, \tag{7}$$

where $\varphi_{i,j,k}$ denotes the $(i, j, k)^{th}$ inactivated output capsule in the 3D capsule layer. A modified 3D nonlinear activation function compresses the values of the capsule between 0 and 1 to obtain the capsules $\hat{\varphi} \in \mathbb{R}^{H\times W\times N''\times C''\times 1}$, which is formulated as follow:

$$\hat{\varphi}_{i,j,k} = 3d\_squash\left(\varphi_{i,j,k}\right) = \frac{\left\|\varphi_{i,j,k}\right\|^2}{0.5 + \left\|\varphi_{i,j,k}\right\|^2} \frac{\varphi_{i,j,k}}{\left\|\varphi_{i,j,k}\right\|}, \tag{8}$$

where $\hat{\varphi}_{i,j,k}$ denotes the activated output capsules in the 3D capsule layer. The agreement between $\hat{v}$ and $\hat{\varphi}$ is determined by their inner product. If the inner product of two capsules is positive, the coupling coefficient between these two capsules will increase. The update of b is formulated as follows:

$$b_{i,j,k,s} = b_{i,j,k,s} + \hat{\varphi}_{i,j,k} \cdot \hat{v}_{i,j,k,s}, \tag{9}$$

where $b_{i,j,k,s}$ denotes the log probability between the sth capsule in the previous layer and the $(i, j, k)^{th}$ capsule in the next layer. Following [15], we use 3 routing iterations and achieve better convergence.

C Digital capsule layer. The output capsules $\hat{\varphi}$ in the 3D capsule layer are reshaped into $\hat{\varphi}_r \in \mathbb{R}^{\left(H\times W\times C''\right)\times N''}$. Then using a routing-by-agreement mechanism in [15], $\hat{\varphi}_r$ are mapped into digital capsules $g \in \mathbb{R}^{2\times N''}$, which denotes imposter and genuine capsules in the case of binary classification.

2.5 Loss Function

The coupling coefficients between capsules are learned by a routing algorithm. But the back-propagation algorithm learns other weight parameters through a separable function called margin loss, which increases intra-class compactness and inter-class separability. The margin loss of the jth digital capsule is formulated as follows:

$$L_j = T_j \max\left(0, m^+ - \left\|g_j\right\|\right)^2 + \lambda_m \left(1 - T_j\right) \max\left(0, \left\|g_j\right\| - m^-\right)^2, \tag{10}$$

where T_j is 1 when class j is activated, otherwise T_j is 0, m^+ denotes the lower boundary of $\|g_j\|$, m^- denotes the upper boundary of $\|g_j\|$, λ_m is a down-weighting parameter, and $\|g_j\|$ denotes the probability of existence of class . To prevent overfitting of the 3D capsule matcher, we use a regularization term loss called reconstruction loss to guide the capsule decoder and 3D capsule matcher for finer-grained feature judgments. As shown in Fig. 4, we use transposed convolutional layers as capsule decoder to reconstruct difference maps. The first 4 layers employ BatchNorm and ReLU. The tanh is employed in the last layer. In the training phase, we reshape the digital capsule $g_r \in \mathbb{R}^{1\times N''}$ corresponding to GroudTruth into $\hat{g}_r \in \mathbb{R}^{1\times 1\times N''}$. Then it is fed to the capsule decoder to obtain reconstructed difference maps $f'_d \in \mathbb{R}^{H\times W\times C}$. The reconstruction loss is formulated as follows:

$$L_r = \left\|f_d - f'_d\right\|^2. \tag{11}$$

The total loss of our proposed method is the summation of margin loss and reconstruction loss scaled by a hyperparameter λ_r. It is formulated as follows:

$$L_t = \sum_{j}^{2} L_j + \lambda_r L_r. \tag{12}$$

3 Experiments and Results

To fairly and comprehensively evaluate the effectiveness of our proposed method, we use two general datasets including a cross-spectral PolyU iris dataset [19] and a cross-sensor IIIT-D Contact Lens Iris(CLI) dataset [2]. Genuine acceptance rate(GAR), false acceptance rate(FAR), and equal error rate(EER) are used to quantitatively evaluate the performance.

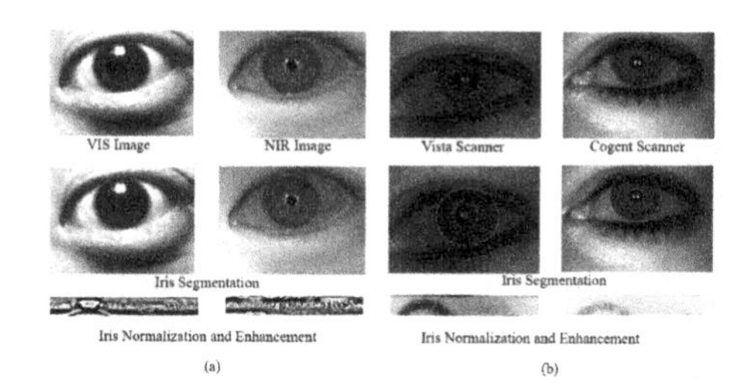

Fig. 5. Some samples from two datasets: (a) PolyU iris dataset and (b) IIIT-D CLI dataset.

3.1 Datasets

We use a public algorithm [22] for each dataset to accurately segment, normalize, and enhance the iris images. The normalized iris images resized into 256×256 are taken as input by the subsequent network. Two datasets are summarized as follows: (1) PolyU iris database, containing 12540 iris images from 209 subjects, is acquired in the NIR and VIS wavelengths from the left and right eyes. Each subject consists of 15 different instances of right and left eye for both VIS and NIR spectrum. Samples are shown in Fig. 5 (a). (2) IIIT-D CLI database, containing 6570 iris images from 101 subjects, is obtained using two different NIR sensors. It is developed to evaluate the performance of contact lens presentation attacks. Because it contains 2163 images without contact lenses, it is also used to evaluate the cross-sensor performance. Samples are shown in Fig. 5 (b).

3.2 Network Structure

CNN Feature Extractor. We modify the pretrained ResNet18, set stride of last two stages to 1, and add a 1×1 convolution with 32 filters and the upsample with bilinear interpolation.

Transformer Feature Extractor. The layer numbers N_e of the transformer encoder are set to 1. The layer numbers N_d of the transformer decoder are set to 4. The number of heads h_s in MSA and h_c in MCA are set to 8. The dimension d of each head in MSA and MCA is set to 8.

3D Capsule Matcher. The convolution in the capsule convolution layer has $C' \times N' = 8 \times 16$ filters. The 3D convolution in the 3D capsule layer has $C' \times N' = 8 \times 16$ filters.

Capsule Decoder. It consists of 5 transposed convolution layers with filter numbers $\{32, 64, 128, 256, 32\}$, 4×4 kernels, stride of $\{1, 2, 2, 2, 2\}$, padding of $\{0, 1, 1, 1, 1\}$.

3.3 Training Setup

Our method is implemented by using torch 1.6.0. The training process lasts 100 epochs with a batch-size of 32. We use the SGD optimizer for training on two NVIDIA GeForce GTX 1080 TI. The momentum is set to 0.99, the weight decay is set to 0.0005 , and the learning rate is set to 0.001. In the loss function, m^- is set to 0.1, m^+ is set to 0.9, λ_m is set to 0.5, and λ_r is set to 0.1.

3.4 Evaluation Protocol

To fairly compare with other state-of-the-art methods, we follow the data partitioning protocol in [8]. In the PolyU iris database, the first ten images per eye from the first 140 subjects are selected as the training set and the remaining five images as the test set. Such generates $2800(10 \times 140 \times 2)$ genuine scores and $1953000(6975 \times 140 \times 2)$ imposter scores from the PolyU iris database to evaluate the cross-spectral performance. In the IIIT-D CLI database, the first 3 images per eye from 110 subjects without contact lenses are selected as the training set, the fourth image as the valuation set, and the remaining one as the test set. Such generates $202(1\times1\times101\times2)$ genuine scores and $40602(1\times1\times202\times201)$ imposter scores from the IIIT-D CLI database to evaluate the cross-sensor performance.

3.5 Comparison with Other State-of-the-art Heterogeneous Iris Recognition Algorithms

Experiments on Cross-spectral Scenario. In the cross-spectral scenario, VIS iris images are matched with NIR iris images. We compare two traditional methods: MRF [8] and IrisCode [6], and two deep learning methods: CNN with SDH [19] and DenseSENet [1], because these methods report competitive results on PolyU iris database. As is shown in Table 1, our proposed method obtains the best performance compared with other state-of-the-art methods in GAR and EER. Besides, our method significantly outperforms the state-of-the-art traditional method MRF and the deep learning method DenseSENet. Our method reduces the EER metric by 43.33% compared to DenseSENet and by 74.77% compared to CNN-SDH. The results show that our method models more contextual relationships in the feature extraction space and obtains more fine-grained discriminative features. In addition, our 3D capsule routing decreases the loss of spatial information and is robust to domain variation and non-ideal conditions. However, other deep learning methods don't consider the part-whole correlation and loss of spatial information.

Table 1. Comparison of other methods on cross-spectal scenario.

Methods	0.1* (%)	0.01* (%)	EER(%)
IrisCode [6]	51.89	33.12	30.81
MRF [8]	64.91	47.74	24.50
CNN-SDH [19]	96.41	90.71	5.39
DenseSENet [1]	99.05	96.81	2.40
Ours	**99.72**	**98.45**	**1.36**

*: GAR at FAR=0.1

Table 2. Comparison of other methods on cross-sensor scenario.

Methods	0.1(%)	0.01(%)	EER(%)
ESG [5]	49.20	21.53	29.59
DA-DBNN [8]	89.92	85.04	10.02
CNN-SDH	96.41	90.71	5.39
DenseSENet	98.94	98.07	1.73
Ours	**99.56**	**98.91**	**0.68**

*: GAR at FAR=0.01

Experiments on Cross-sensor Scenario. In the cross-sensor scenario, iris images from one sensor are matched with iris images from the other different sensor. Our method is compared with two traditional methods: Even symmetric Gabor(ESG) [5] and DA-DBNN [8], and a deep learning method: DenseSENet on IIIT-D CLI database. As is shown in Table 2, our proposed method obtains the best results: GAR = 99.56% at FAR=0.1 and GAR=98.91% at FAR=0.01. Compared with DenSENet and DA-DBNN, our method decreases by 60.69%

and 93.21%, respectively, in EER. It is worth noting that we reimplement DenseSENet due to possible errors in the introduction and partition of the dataset they claim. The results show our method improves the cross-sensor performance. Because the transformer models the context in the token-based space to refine the features, which are more domain-invariant than those of pure convolution and traditional descriptors. Moreover, compared with fully connected networks and distance metrics, the 3D capsule matcher considers the most informative partial representation while overlooking the domain gap.

Table 3. Ablation on different modules.

TG	TFE	CM	0.1* (%)	0.01* (%)	EER(%)
✗	✗	✗	72.52	61.43	19.26
✗	✓	✗	96.14	94.87	2.76
✓	✓	✗	97.90	96.34	2.46
✗	✗	✓	80.89	75.76	12.84
✓	✓	✓	**99.56**	**98.91**	**0.68**

*: GAR at FAR=0.1

Table 4. Ablation on λ_m and λ_r.

λ_m	λ_r	0.1(%)	0.01(%)	EER(%)
0.1	0.1	97.89	95.63	2.21
0.25	0.1	98.46	97.04	1.94
0.5	0.1	**99.56**	**98.91**	**0.68**
0.1	0.25	97.72	95.54	2.30
0.1	0.5	97.32	95.12	2.56

*: GAR at FAR=0.01

3.6 Ablation Experiments

Ablation on Token Generator(TG), Transformer Feature Extractor(TFE) and 3D Capsule Matcher(CM). The ablation experiments are performed on IIIT-D CLI database. When CM is removed, we add a fully connected network with softmax as the feature matcher. When TG is removed, the input to TFE comes from the chunked sequence of features. As seen in Table 3, the proposed token generator, transformer feature extractor, and 3D capsule matcher all contribute to improving the performance.

Ablation on Hyperparameters. We performed ablation experiments on the IIIT-D CLI database to select the hyperparameters in the loss function. From Table 4, we can note that the proposed approach achieves optimal performance when it is trained with $\lambda_m = 0.5$ and $\lambda_r = 0.1$.

4 Conclusions

In this paper, we use transformers and capsules to ease the difficulty of heterogeneous iris recognition. Leveraging vision tokens, transformer encoder and transformer decoder models the relationship between token-based space and pixel-space to refine the low semantic features. And 3D capsule routing efficiently models the part-whole correlation. Extensive experiments on cross-spectral and cross-sensor scenarios show our proposed method significantly improves the performance of heterogeneous iris recognition.

Acknowledgements. This work was supported by the National Natural Science Foundation of China (grant number 61471181) and the Natural Science Foundation of Jilin Province (grant number YDZJ202101ZYTS144).

References

1. Chen, Y., Zeng, Z., Zeng, Y., Gan, H., Chen, H.: DenseSENet: more accurate and robust cross-domain iris recognition. J. Electron. Imaging **30**(6), 063024 (2021)
2. Kohli, N., Yadav, D., Vatsa, M., Singh, R.: Revisiting iris recognition with color cosmetic contact lenses. In: 2013 International Conference on Biometrics (ICB), pp. 1–7. IEEE (2013)
3. Kosiorek, A., Sabour, S., Teh, Y.W., Hinton, G.E.: Stacked capsule autoencoders. In: Advances in Neural Information Processing Systems, vol. 32 (2019)
4. Liu, N., Zhang, M., Li, H., Sun, Z., Tan, T.: Deepiris: learning pairwise filter bank for heterogeneous iris verification. Pattern Recogn. Lett. **82**, 154–161 (2016)
5. Ma, L., Tan, T., Wang, Y., Zhang, D.: Personal identification based on iris texture analysis. IEEE Trans. Pattern Anal. Mach. Intell. **25**(12), 1519–1533 (2003)
6. Masek, L.: Matlab source code for a biometric identification system based on iris patterns. http://peoplecsse.uwa.edu.au/pk/studentprojects/libor/ (2003)
7. Mostofa, M., Mohamadi, S., Dawson, J., Nasrabadi, N.M.: Deep GAN-based cross-spectral cross-resolution iris recognition. IEEE Trans. Biometrics Behav. Identity Sci. **3**(4), 443–463 (2021)
8. Nalla, P.R., Kumar, A.: Toward more accurate iris recognition using cross-spectral matching. IEEE Trans. Image Process. **26**(1), 208–221 (2016)
9. Nguyen, K., Fookes, C., Ross, A., Sridharan, S.: Iris recognition with off-the-shelf CNN features: a deep learning perspective. IEEE Access **6**, 18848–18855 (2017)
10. Oktiana, M., et al.: Advances in cross-spectral iris recognition using integrated gradientface-based normalization. IEEE Access **7**, 130484–130494 (2019)
11. Oktiana, M., Saddami, K., Arnia, F., Away, Y., Munadi, K.: Improved cross-spectral iris matching using multi-scale weberface and gabor local binary pattern. In: 2018 10th International Conference on Electronics, Computers and Artificial Intelligence (ECAI), pp. 1–6. IEEE (2018)
12. Pillai, J.K., Patel, V.M., Chellappa, R., Ratha, N.K.: Secure and robust iris recognition using random projections and sparse representations. IEEE Trans. Pattern Anal. Mach. Intell. **33**(9), 1877–1893 (2011)
13. Pillai, J.K., Puertas, M., Chellappa, R.: Cross-sensor iris recognition through kernel learning. IEEE Trans. Pattern Anal. Mach. Intell. **36**(1), 73–85 (2013)
14. Proença, H., Neves, J.C.: Irina: Iris recognition (even) in inaccurately segmented data. In: CVPR, pp. 538–547 (2017)
15. Sabour, S., Frosst, N., Hinton, G.E.: Dynamic routing between capsules. In: Advances in Neural Information Processing Systems, vol. 30 (2017)
16. Sequeira, A., et al.: Cross-eyed-cross-spectral iris/periocular recognition database and competition. In: 2016 International Conference of the Biometrics Special Interest Group (BIOSIG), pp. 1–5. IEEE (2016)
17. Sequeira, A.F., et al.: Cross-eyed 2017: cross-spectral iris/periocular recognition competition. In: IJCB,, pp. 725–732. IEEE (2017)
18. Wang, J., et al.: Learning fine-grained image similarity with deep ranking. In: CVPR, pp. 1386–1393 (2014)
19. Wang, K., Kumar, A.: Cross-spectral iris recognition using CNN and supervised discrete hashing. Pattern Recogn. **86**, 85–98 (2019)

20. Wei, J., Wang, Y., Li, Y., He, R., Sun, Z.: Cross-spectral iris recognition by learning device-specific band. IEEE Trans. Circ. Syst. Video Technol. (2021)
21. Zhao, B., Tang, S., Chen, D., Bilen, H., Zhao, R.: Continual representation learning for biometric identification. In: WACV, pp. 1198–1208 (2021)
22. Zhao, Z., Ajay, K.: An accurate iris segmentation framework under relaxed imaging constraints using total variation model. In: ICCV, pp. 3828–3836 (2015)

MCANet: Multiscale Cross-Modality Attention Network for Multispectral Pedestrian Detection

Xiaotian Wang[1,2](✉), Letian Zhao[1,2], Wei Wu[1,2], and Xi Jin[1,2]

[1] State Key Laboratory of Particle Detection and Electronics, University of Science and Technology of China, Hefei, China
wxtdsg@mail.ustc.edu.cn
[2] Institute of Microelectronics, Department of Physics, University of Science and Technology of China, Hefei, China

Abstract. Multispectral pedestrian detection is an important and challenging task, that can provide complementary information of visible images and thermal images for high-precision and robust object detection results. To fully exploit the different modalities, we propose a Multiscale Cross-Modality Attention (MCA) module to efficiently extract and fuse features. In this module, the transformer architecture is used to extract features of two modalities. Based on these features, we design a novel spatial attention mechanism that can adaptively enhance object details and suppress background. Finally, the features of each branch are fused using the channel attention mechanism and sent to the detector. To verify the effect of the MCA module, we propose the MCANet. The MCA modules are embedded at different depths of the two-stream network and interconnected to share multiscale information. Extensive experimental results demonstrate that MCANet achieves state-of-the-art detection accuracy on the challenging KAIST multispectral pedestrian dataset.

Keywords: Multispectral detection · Cross-modality feature fusion · Attention mechanism

1 Introduction

In recent years, with the rapid development of computer technology, major breakthroughs have been made in many fields of computer vision. As one of the important object detection tasks, pedestrian detection has a wide spectrum of application prospects, including autonomous driving, surveillance, and search and rescue [1,2]. In real-world applications, the environment is often complex and changeable, such as rain, fog, and low light. In these cases, pedestrian detectors that only focus on visible images have difficulty achieving sufficient accuracy. Thermal images can overcome these difficulties because they can be imaged

D.-T. Dang-Nguyen et al. (Eds.): MMM 2023, LNCS 13833, pp. 41–53, 2023.
https://doi.org/10.1007/978-3-031-27077-2_4

clearly without relying on illumination conditions. However, thermal images lose the color and texture information, cannot be imaged through transparent objects such as glass, and do not work well when the ambient temperature is close to the target temperature. Visible and thermal images have their own advantages and disadvantages in different scenarios. Therefore, researchers have raised interest in multispectral pedestrian detection technology.

Fig. 1. It can be observed that the illumination conditions are poor and thermal images are more suitable for pedestrian detection than visible images. However, as shown in the red box, the reflection of thermal radiation on the marble surface creates a ghostly image of pedestrians. Adjusting the detector's propensity for visible or thermal images only based on illumination conditions can easily lead to false detection results. (Color figure online)

From the difference in imaging principles, it is natural to realize that different illumination conditions have a great impact on the imaging quality of visible images. Therefore, some works [3,4] design an illumination-aware network to evaluate the illumination conditions of an entire image and generate weights to fuse the features extracted from different modalities. However, there are difficulties in some scenarios, mainly for two reasons. First, the light source in the real world is complex. In addition to sunlight, there are street lights, car lighting and various reflections. In different spatial locations of an image, the illumination conditions are often inconsistent, so the contribution of visible images and thermal images to features cannot be simply represented by a single value. Second, illumination conditions are not the only standard to judge whether visible or thermal features are more conducive to pedestrian detection. For example, the specular reflection of visible light by glass, and the reflection of thermal radiation by the smooth marble surface will cause interference with pedestrian detection. A specific example is shown in Fig. 1.

Based on the above considerations, we propose a novel Multiscale Cross-Modality Attention (MCA) module to efficiently extract and fuse features. We embed the module into the two-stream backbone, and introduce MCANet for multispectral pedestrian detection. The main contributions of our work are as follows:

- We introduce an end-to-end Multiscale Cross-Modality Attention Network (MCANet) for multispectral pedestrian detection and validate the effectiveness of fusion for learning cross-modality features.
- We propose the Multiscale Cross-Modality Attention (MCA) module. We develop a novel attention fusion mechanism and combine it with transformer to enhance the saliency of objects and suppress the background. A new loss item is introduced to the loss function for training the attention weights.
- MCANet conducts extensive experiments on the KAIST dataset, obtaining state-of-the-art performance.

2 Related Work

2.1 Multispectral Pedestrian Detection

In order to fuse visible and thermal images and greatly improve the accuracy and robustness of pedestrian detection algorithms in different scenarios, researchers have made many efforts. Some challenging multispectral pedestrian detection datasets have been proposed, such as KAIST [5], LLVIP [6], etc., which have become important references to verify the performance of the algorithm. With the development of deep learning, convolutional neural networks have gradually been used in multispectral pedestrian detection tasks. Liu et al. [7] design four ConvNet fusion architectures, which fuse channel features at different ConvNet stages and prove that the halfway fusion model can achieve better performance. CIAN [8] proposes the cross-modality interactive attention to explicitly model the importance of feature channels and introduces the context enhancement blocks (CEBs) to further augment contextual information. Illumination-aware Faster R-CNN [3] introduces an illumination-aware weighting mechanism to adaptively weight the detection confidence of two modalities according to the illumination measure and adaptively merge the two sub-networks to obtain final detections. MBNet [4] designs an illumination-aware feature alignment module to align two modality features and induce the network to be optimized adaptively according to illumination conditions. Fang et al. [9] propose a transfomer-based fusion approach, named Cross-Modality Fusion Transformer (CFT), to enhance the representation capability of two-stream CNNs. Li et al. [10] propose the dense fusion strategy to fuse information at the feature level and use Dempster's combination rule to fuse the results of different branches according to the uncertainty.

2.2 Attention Mechanism

Multispectral fusion can be further divided into two questions: 1. How can complementary features be extracted between different modalities? 2. How can the extracted features be fused into the previous branch efficiently? For the first question, the previous works [7] directly use the feature maps of their respective branches for addition or concatenation operations. [4,8] adopt global pooling and linear layers to squeeze the spatial dimension, and then channel-wise attention based weighting is applied to the cross-modality features. [11] use convolutional layers to extract features, and apply the spatial-wise attention mechanism by element-wise multiplication. Considering that it is difficult to determine whether the pixel value of the pedestrian is larger or smaller than the background, we believe that the direct use of element-wise multiplication may not be the best fusion method. In this paper, combined with the confidence information of pedestrians, a novel spatial attention mechanism using supervised learning is proposed, which enhances the saliency of objects and suppresses the background. This method can make full use of the complementary features of different modalities.

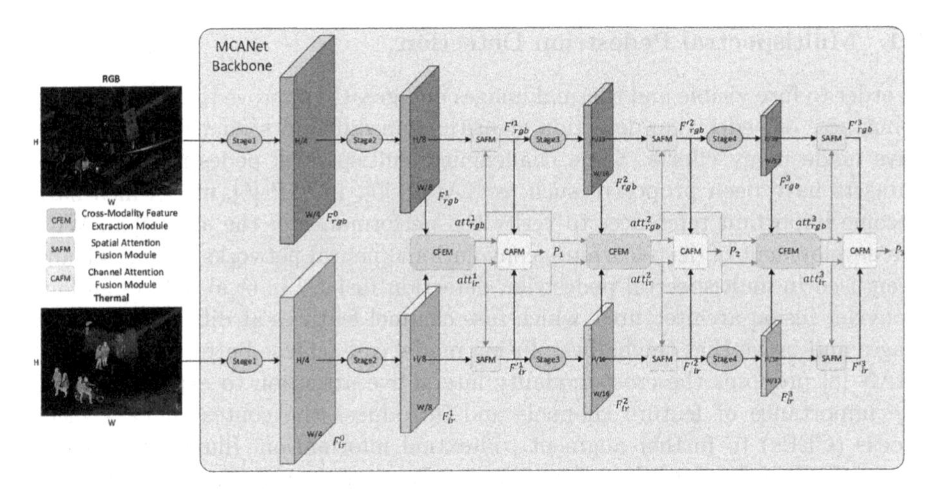

Fig. 2. Overview of Multiscale Cross-Modality Attention Network's backbone. Stage l represents the convolution module of each branch. F^l_{rgb} and F^l_{ir} are the feature maps of RGB and Thermal modalities. att^l_{rgb} and att^l_{ir} are the extracted weights after CFEM. F'^l_{rgb} and F'^l_{ir} are the fused feature maps after SAFM and P_l are the fused feature maps after CAFM. P_l will be sent to YOLOv5 detectors for prediction.

3 Proposed Method

The MCANet extends the framework of YOLOv5 [12], to enable multispectral object detection. An illustration of Multiscale Cross-Modality Attention Backbone is shown in Fig. 2. The MCA module consists of three basic components: Cross-Modality Feature Extraction Module(CFEM), Spatial Attention Fusion Module (SAFM) and Channel Attention Fusion Module (CAFM), as shown in detail in Fig. 3. They are embedded at different depths of the network and share information at different scales through interconnections.

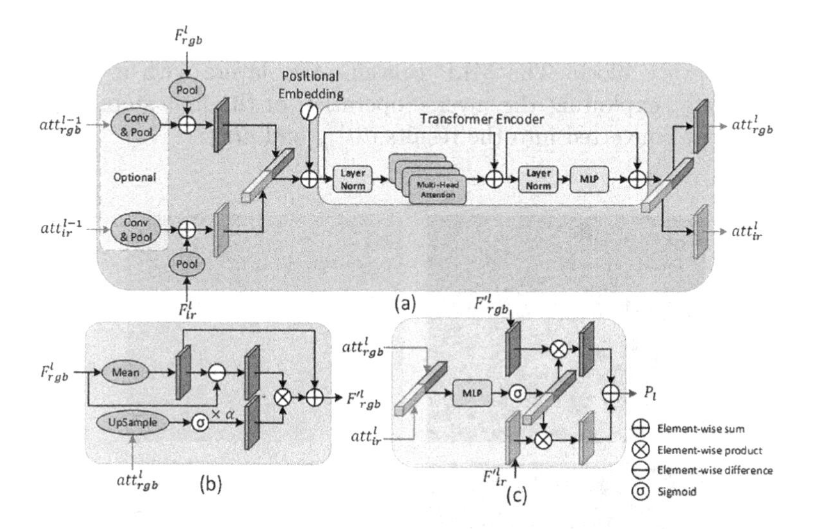

Fig. 3. The architecture of the proposed MCA module: (a) Cross-Modality Feature Extraction Module. (b) Spatial Attention Fusion Module. (c) Channel Attention Fusion Module.

3.1 Cross-Modality Feature Extraction Module

In CFEM, we want to obtain complementary features between different modalities to provide spatial weights for later attention mechanisms. Unlike convolution, which only has a local receptive field, the transformer can take into account global spatial information. Inspired by [8], we use Transformer for cross-modality feature extraction. The details are shown in Fig. 3(a).

In order to reduce the parameters and computation, taking the lth layer as an example, the RGB feature map F^l_{rgb} and thermal feature map F^l_{ir} are first downsampled to $f^l_{rgb} \in R^{C \times H \times W}$ and $f^l_{ir} \in R^{C \times H \times W}$ by max pooling and average pooling. If the previous CFEM exists, a convolution and a pooling operation will be performed on the previous outputs att^{l-1}_{rgb} and att^{l-1}_{ir}. Then the results will be added to f^l_{rgb} and f^l_{ir} respectively. The sequence of embedded patches $x^l_{p-rgb} \in R^{HW \times C}$ and $x^l_{p-ir} \in R^{HW \times C}$ is achieved by flattening the spatial dimensions of the feature map and projecting to the transformer dimension. We concatenate the patch embeddings of each modality and position embeddings are added to the patch embeddings to retain positional information. The fusion patch embeddings x^l_p are obtained and passed to the Transformer Encoder which consists of alternating layers of multiheaded self-attention (MSA) (1) and MLP (2) blocks:

$$Z'_l = x^l_p + MSA(LayerNorm(x^l_p)) \tag{1}$$

$$Z_l = Z'_l + MLP(LayerNorm(Z'_l)) \tag{2}$$

Layernorm (LN) is applied before every block, and residual connections are applied after every block. The MLP contains two layers with a GELU non-linearity. Finally, exploiting the inverse operation of the first step, the output sentences Z_l are converted into the results att_{rgb}^l and att_{ir}^l.

Fig. 4. (a) The grayscale image example. (b) The mean of image (a) is shifted to 127. (c) The partitioned grids and bboxes. (d) The fused image after SAFM and the mean of it is also shifted to 127.

Specifically, we want att_{rgb}^l and att_{ir}^l to give the approximate position where the object is located in the feature map. For the sake of illustration and visualization, we take Fig. 4 as an example. We convert the image from RGB to gray as the extracted feature map F_{rgb}^l in Fig. 4(a). Suppose we resize it to f_{rgb}^l with shape $h \times w$ and obtain output att_{rgb}^l. Then each element of att_{rgb}^l corresponds to the region with shape $H/h \times W/w$ in feature map F_{rgb}^l, i.e. the green grid in Fig. 4(c). The red box represents the bounding box of the object. The value of the green grids that intersect with the bounding box is set to 1, and the rest are set to 0. We take it as the ground truth, denoted as gt_att^l. During training, we apply a piecewise activation function $\sigma'(x)$ on att_{rgb}^l to restrict its range to (0,1). The $\sigma'(x)$ is formulated as:

$$\sigma'(x) = \begin{cases} 1 & \sigma(x) > 0.7 \\ \sigma(x) & 0.3 < \sigma(x) < 0.7 \\ 0 & \sigma(x) < 0.3 \end{cases} \tag{3}$$

where $\sigma(x)$ denotes the Sigmoid function. When $\sigma(x)$ is greater than 0.7 or less than 0.3, the value is directly set to 1 or 0, respectively. In this way, during backward propagation, while having the ability to roughly describe the object location, att_{rgb}^l can concentrate on receiving the gradients propagated by the backbone network and learn more representations.

Then we compute a Binary Cross Entropy Loss with gt_att^l. The same operation is performed for att_{ir}^l and the final loss is the average of them.

$$BCELoss(x,y) = -\frac{1}{whc}\sum_{i=1}^{w}\sum_{j=1}^{h}\sum_{k=1}^{c} -[y_{i,j}\cdot \log x_{i,j,k} + (1-y_{i,j})\cdot \log(1-x_{i,j,k})] \quad (4)$$

$$L_{att} = \sum_{l=1,2,3}\frac{1}{2}[BECLoss(\sigma'(att_{rgb}^l), gt\_att^l) + BECLoss(\sigma'(att_{ir}^l), gt\_att^l)] \quad (5)$$

The total loss function uses 4 components: box, class, objectness and attention as follows:

$$L = \lambda_{obj}L_{obj} + \lambda_{box}L_{box} + \lambda_{cls}L_{cls} + \lambda_{att}L_{att} \quad (6)$$

The first three items are the same as in YOLOv5. Four parameters are used to balance different losses. In the task of multispectral pedestrian detection, the classification loss is equal to 0. In our experiment, λ_{obj}, λ_{box}, and λ_{att} are set to 0.05, 1, and 0.015, respectively.

3.2 Spatial Attention Fusion Module

In the spatial attention fusion module (SAFM), we do not use the traditional spatial attention mechanism in which the feature maps are multiplied directly with the weights. We note that pedestrians in visible images may have lower pixel values than the background due to wearing dark clothes, or higher pixel values than the background at night due to insufficient illumination conditions. In the thermal image, the pixel values of pedestrians are generally higher than that of the background. However, it is also possible that the pixel values of pedestrians are lower due to wearing thick clothes, or the ground and vehicle shell being exposed to high temperature for a long time, especially in summer. The pixel values of different parts of pedestrians may also vary greatly. Therefore, it is not always possible to extract better features if the spatial weights are multiplied with feature maps directly. We think it may be better to zoom in on the difference between the object and the background.

Therefore we design the SAFM as shown in Fig. 3(b). Similar to Sect. 3.1, we divide the input feature F_{rgb}^l into $w \times h$ green grids, as shown in Fig. 4(c). Within each grid, we average all pixel values and reassign all pixels with the mean. The offsets are obtained by subtracting the mean from the original feature maps as follows:

$$Offset_{rgb}^l = F_{rgb}^l - Mean(F_{rgb}^l) \quad (7)$$

Table 1. Comparisons with the state-of-the-art methods on the KAIST dataset

Methods	Reasonable			All		
	Rea.	Rea.Day	Rea.Night	All	Day	Night
ACF [5]	47.32	42.57	56.17	67.74	64.31	75.06
Halfway Fusion [7]	25.75	24.88	26.59	49.18	47.58	52.35
FusionRPN+BF [13]	18.29	19.57	16.27	51.70	52.33	51.09
IAF R-CNN [3]	15.73	14.55	18.26	44.23	42.46	47.70
IATDNN+IASS [14]	14.95	14.67	15.72	48.96	49.02	49.37
CIAN [8]	14.12	14.77	11.13	35.53	36.02	32.38
MSDS-RCNN [15]	11.34	10.53	12.94	34.15	32.06	38.83
AR-CNN [16]	9.34	9.94	8.38	34.95	34.36	36.12
MBNet [4]	**8.13**	**8.28**	7.86	31.87	32.37	30.95
CMPD [10]	8.16	8.77	7.31	28.98	28.3	30.56
MCANet	8.24	8.97	**7.00**	**26.07**	**27.07**	**24.3**

In Sect. 3.1, we already have the information about the possible locations of the objects. Then we resize it to $W \times H$ by nearest-neighbor sampling and apply a Sigmoid activation to it, denoting the result as Att^l_{rgb}.

$$Att^l_{rgb} = \sigma(UpSampling(att^l_{rgb})) \tag{8}$$

The offset is scaled by multiplying by Att^l_{rgb} and coefficient α, and then added to the mean of the input feature to obtain the final output F'^l_{rgb}.

$$F'^l_{rgb} = \alpha \cdot Att^l_{rgb} \cdot Offset^l_{rgb} + Mean(F^l_{rgb}) \tag{9}$$

Take Fig. 4(b) and Fig. 4(d) as an intuitive comparison of the fusion effect. Figure 4(b) is the grayscale image and Fig. 4(d) is the ideal fusion result F'^l_{rgb}. Due to the poor illumination conditions, we move the mean values of both Fig. 4(b) and Fig. 4(d) to 127 for easier comparison. Ideally, the value of pixels which represent the background in gt^l_{att} is 0. As a result, the background values after fusion are all equal to the mean of each grid. In contrast, the variance of pixel values becomes larger for regions containing objects. The pedestrians in Fig. 4(d) are more salient than those in Fig. 4(b), and the background regions are greatly suppressed. The two pedestrians on the left are much brighter than the background, and the clothes of the pedestrian in the middle of the picture are much darker and easier to distinguish. It should be noted that since the pixel values of Fig. 4(d) may be out of range (0–255) after being scaled up, in order to display normally, the excess parts have to be truncated, resulting in some details in Fig. 4(d) becoming blurred. In fact, there is no truncation operation during the training and inference, so all texture details are preserved without concerns about the loss of information.

3.3 Channel Attention Fusion Module

In the Channel Attention Fusion Module (CAFM), we make a small modification to the traditional channel attention mechanism for multi-modality fusion, as shown in Fig. 3(c). We apply the Global Average Pooling (GAP) to att^l_{rgb} and att^l_{ir} and concatenate them. After that, they are sent to the MLP block and a Sigmoid function to generate the weights of channels as follows:

$$ch\_att^l = \sigma(MLP(Concat(GAP(att^l_{rgb}), GAP(att^l_{ir})))) \tag{10}$$

The ch_att^l is separated into two parts $ch\_att^l_{rgb}$ and $ch\_att^l_{ir}$. The fusion result P_l for later prediction can be formulated as:

$$P_l = ch\_att^l_{rgb} \cdot F'^l_{rgb} + ch\_att^l_{ir} \cdot F'^l_{ir} \tag{11}$$

Table 2. Evaluations on the KAIST dataset under six subsets.

Methods	Near	Medium	Far	None	Partial	Heavy	All
ACF [5]	28.74	53.67	88.2	62.94	81.40	88.08	67.74
Halfway Fusion [7]	8.13	30.34	75.70	43.13	65.21	74.36	49.18
FusionRPN+BF [13]	0.04	30.87	88.86	47.45	56.10	72.20	51.70
IAF R-CNN [3]	0.96	25.54	77.84	40.17	48.40	69.76	44.23
IATDNN+IASS [14]	0.04	28.55	83.42	45.43	46.25	64.57	48.96
CIAN [8]	3.71	19.04	55.82	30.31	41.57	62.48	35.53
MSDS-RCNN [15]	1.29	16.19	63.73	29.86	38.71	63.37	34.15
AR-CNN [16]	0.00	16.08	69.00	31.40	38.63	55.73	34.95
ASPFFNet [11]	0.01	16.27	45.42	25.60	34.90	57.53	-
MBNet [4]	0.00	16.07	55.99	27.74	35.43	59.14	31.87
CMPD [10]	0.00	12.99	51.22	24.04	33.88	59.37	28.98
MCANet	**0.00**	**12.22**	**42.40**	**21.37**	**29.78**	**56.52**	**26.07**

4 Experiments

4.1 Dataset and Metric

The KAIST multispectral pedestrian detection dataset [5] contains 95,328 pairs of aligned visible and thermal images. It contains a variety of scenes acquired during the day and night to cover changes in diverse lighting conditions. The test set consists of 2, 252 frames sampled every 20th frame from video, among which 1,455 images are captured during daytime and the remaining 797 images are captured during nighttime. Due to the problematic annotations in the original training data, we adopt the annotations improved by Zhang et al. [16] for training. All the detection performances are evaluated on the KAIST test set with annotations improved by Liu et al. [7]. The evaluation metric follows the standard KAIST evaluation [5]: log-average Miss Rate over False Positive Per Image (FPPI) range of $[10^{-2}, 10^{0}]$ (denoted as MR^{-2}). A lower score indicates better performance.

4.2 Implementation Details

Throughout this paper, we extend the framework of YOLOv5l, to enable multispectral object detection. The anchors are set to [10,13, 16,30, 33,23], [30,61, 62,45, 59,119], and [116,90, 156,198, 373,326] on three detectors with different scales. We use the Stochastic Gradient Descent (SGD) optimizer with an initial learning rate of 1e-2, a momentum of 0.937, and a weight decay of 0.0005. To avoid optimization instabilities, we use the first three epochs for warmup. The warmup initial momentum is set to 0.8 and the warmup initial bias learning rate is set to 0.1. For data augmentation, we use the mosaic method which mixes four training images into one image. The MCANet is developed on an Ubuntu 18.04 platform with PyTorch 1.12.0 and two NVIDIA 3090 GPUs. The network is trained for 20 epochs and the batch size is 32.

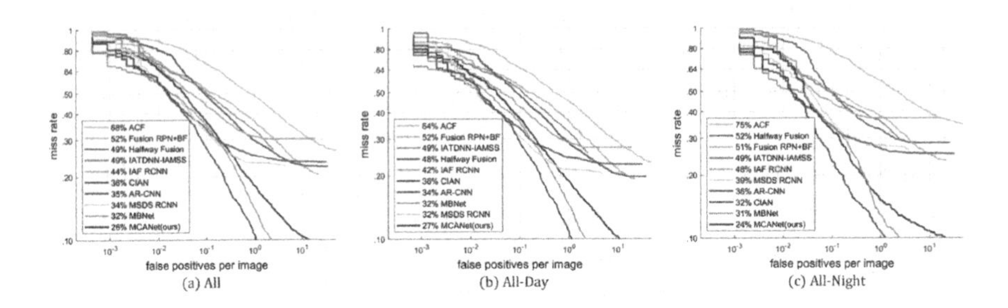

(a) All (b) All-Day (c) All-Night

Fig. 5. Comparisons of detection results on KAIST.

4.3 Quantitative Evaluation

As shown in Table 1, we present the experimental results in terms of MR^{-2} under reasonable and all-dataset settings, respectively, as in existing works. In reasonable setting, only pedestrians taller than 55 pixels under no or partial occlusions are considered in the evaluation. Instead, all the labels, including small pedestrians and heavy occlusions, are used for evaluation in all-dataset setting. Therefore, it is obvious that all-dataset setting is more challenging than the reasonable setting. In the all-dataset setting, Table 1 shows that our proposed method achieves approximately 2.91% lower on MR^{-2} which implies that the MCANet has a substantially better localization accuracy compared with CMPD. The results show that our model can better extract and fuse complementary features of multiple modalities to improve the detection accuracy. The FPPI-MR curve on the all-dataset setting is shown in Fig. 5, which also demonstrates the superiority of our method.

In order to have a comprehensive understanding of detector performance, we also make an evaluation under other six subsets including the pedestrian distances and occlusion levels. As shown in Table 2, the MCANet ranks first in all six subsets. Especially in the Far subset, MCANet achieves 8.82% lower on MR^{-2}, which indicates that MCANet performs satisfactorily in detecting small targets.

Fig. 6. Comparisons of detection results on KAIST with MBNet.

Table 3. The ablation experiments of the MCANet on the KAIST dataset

Method	All	Day	Night
CFEM+element-wise product+element-wise sum	9.39	10.29	7.45
CFEM+SAFM+element-wise sum	8.55	9.11	7.24
CFEM+SAFM+CAFM	**8.24**	**8.97**	**7.00**

4.4 Qualitative Evaluation

In order to further demonstrate the effectiveness of the proposed MCANet, the detection results are compared with MBNet, which is one of the state-of-the-art algorithms that have been open sourced, as shown in Fig. 6. The first row is the original images and the green rectangles are manually labeled ground truth. The second and third rows are the detection results of MBNet and our method, respectively. It can be clearly seen from Fig. 6 that MCANet can effectively solve the problem of missed detection. Due to the efficient feature extraction and saliency enhancement functions of MCANet, the detection accuracy of small targets can be effectively improved.

4.5 Ablation Study

Ablation experiments are performed on the KAIST dataset to demonstrate the effectiveness of the components of our MCA module. We test three different fusion strategies as shown in Table 3. All of them use CFEM to extract the features. The first method replaces the SAFM and CAFM with element-wise product and element-wise sum respectively. And the second method replaces the CAFM with element-wise sum. The third method uses all of the proposed components. The experiments show that the final version with all three designed components outperforms the other versions. The results of the ablation study demonstrate the effectiveness of the proposed components.

5 Conclusion

In this work, we propose a novel MCA module to efficiently extract and fuse features. The MCA modules are embedded into two-stream backbone and the MCANet is introduced. We explore how to effectively enhance the saliency of objects and suppress the background. Specifically, the transformer architecture is used to extract Cross-Modality complementary features. Then a novel attention fusion mechanism is developed. The multiscale information is shared between different depths of the network to ensure the robustness of the detector. The experiments demonstrate that the proposed MCANet outperforms the state-of-the-art on the challenging KAIST dataset in terms of the accuracy.

References

1. Torabi, A., Massé, G., Bilodeau, G.-A.: An iterative integrated framework for thermal-visible image registration, sensor fusion, and people tracking for video surveillance applications. Comput. Vis. Image Underst. **116**(2), 210–221 (2012)
2. Wu, B., Iandola, F., Jin, P.H., Keutzer, K.: SqueezeDet: unified, small, low power fully convolutional neural networks for real-time object detection for autonomous driving. In: Proceedings of the IEEE conference on computer vision and pattern recognition workshops, pp. 129–137 (2017)
3. Li, C., Song, D., Tong, R., Tang, M.: Illumination-aware faster R-CNN for robust multispectral pedestrian detection. Pattern Recogn. **85**, 161–171 (2019)
4. Zhou, K., Chen, L., Cao, X.: Improving multispectral pedestrian detection by addressing modality imbalance problems. In: Vedaldi, A., Bischof, H., Brox, T., Frahm, J.-M. (eds.) ECCV 2020. LNCS, vol. 12363, pp. 787–803. Springer, Cham (2020). https://doi.org/10.1007/978-3-030-58523-5_46
5. Hwang, S., Park, J., Kim, N., Choi, Y., So Kweon, I.: Multispectral pedestrian detection: benchmark dataset and baseline. In: Proceedings of the IEEE Conference on Computer Vision and Pattern Recognition, pp. 1037–1045 (2015)
6. Jia, X., Zhu, C., Li, M., Tang, W., Zhou, W.: LLVIP: a visible-infrared paired dataset for low-light vision. In: Proceedings of the IEEE/CVF International Conference on Computer Vision, pp. 3496–3504 (2021)
7. Liu, J., Zhang, S., Wang, S., Metaxas, D.N.: Multispectral deep neural networks for pedestrian detection. arXiv preprint arXiv:1611.02644 (2016)

8. Zhang, L., et al.: Cross-modality interactive attention network for multispectral pedestrian detection. Inf. Fusion **50**, 20–29 (2019)
9. Qingyun, F., Dapeng, H., Zhaokui, W.: Cross-modality fusion transformer for multispectral object detection. arXiv preprint arXiv:2111.00273 (2021)
10. Li, Q., Zhang, C., Hu, Q., Fu, H., Zhu, P.: Confidence-aware fusion using dempster-shafer theory for multispectral pedestrian detection. In: IEEE Trans. Multimedia (2022)
11. Fu, L., Gu, W.-B., Ai, Y.-B., Li, W., Wang, D.: Adaptive spatial pixel-level feature fusion network for multispectral pedestrian detection. Infrared Phys. Technol. **116**, 103770 (2021)
12. Jocher, G.: YOLOv5 release v5.0 (2022). https://github.com/ultralytics/yolov5/releases/tag/v5.0
13. Konig, D., Adam, M., Jarvers, C., Layher, G., Neumann, H., Teutsch, M.: Fully convolutional region proposal networks for multispectral person detection. In: Proceedings of the IEEE Conference on Computer Vision and Pattern Recognition Workshops, pp. 49–56 (2017)
14. Guan, D., Cao, Y., Yang, J., Cao, Y., Yang, M.Y.: Fusion of multispectral data through illumination-aware deep neural networks for pedestrian detection. Inf. Fusion **50**, 148–157 (2019)
15. Li, C., Song, D., Tong, R., Tang, M.: Multispectral pedestrian detection via simultaneous detection and segmentation. arXiv preprint arXiv:1808.04818 (2018)
16. Zhang, L., Zhu, X., Chen, X., Yang, X., Lei, Z., Liu, Z.: Weakly aligned cross-modal learning for multispectral pedestrian detection. In: Proceedings of the IEEE/CVF International Conference on Computer Vision, pp. 5127–5137 (2019)

Human Action Understanding

Overall-Distinctive GCN for Social Relation Recognition on Videos

Yibo Hu, Chenyu Cao, Fangtao Li, Chenghao Yan, Jinsheng Qi, and Bin Wu(✉)

Beijing University of Posts and Telecommunications, Beijing, China
{huyibo,ccyu,lift,yanch,qijs,wubin}@bupt.edu.cn

Abstract. Recognizing social relationships between multiple characters from videos can enable intelligent systems to serve human society better. Previous studies mainly focus on the still image to classify the relationships while ignoring the important data source of the video. With the prosperity of multimedia, the methods of video-based social relationship recognition gradually emerge. However, those methods either only focus on the logical reasoning between multiple characters or only on the direct interaction in each character pair. To that end, inspired by the rules of interpersonal social communication, we propose Overall-Distinctive GCN (OD-GCN) to recognize the relationships of multiple characters in the videos. Specifically, we first construct an overall-level character heterogeneous graph with two types of edges and rich node representation features to capture the implicit relationship of all characters. Then, we design an attention module to find mentioned nodes for each character pair from feature sequences fused with temporal information. Further, we build distinctive-level graphs to focus on the interaction between two characters. Finally, we integrate multimodal global features to classify relationships. We conduct the experiments on the MovieGraphs dataset and validate the effectiveness of our proposed model.

Keywords: Social relationship recognition · Graph convolutional network · Video understanding

1 Introduction

Social relation is an essential part of human society. Most human behaviors are motivated by their social networks. Recognition of human relationships from videos can enable the intelligent system to serve the human community better. Besides, social relation recognition can enlighten various applications, such as human behavior prediction, event analysis, event warning, etc. In recent computer vision research, the understanding of human relationships of video is still in its nascent stage. For social relationship recognition in videos, previous studies

D.-T. Dang-Nguyen et al. (Eds.): MMM 2023, LNCS 13833, pp. 57–68, 2023.
https://doi.org/10.1007/978-3-031-27077-2_5

have explored joint inference-based methods, knowledge graph-based methods, etc. For example, Kukleva [8] *et al.* proposed a method to utilize the interactions of social character pairs to recognize relationships. However, this method ignores the connection of multiple characters. The characteristics of characters in a family atmosphere are very different from those in a workplace atmosphere. So it is necessary to get information from a holistic perspective for character relationship identification first. Yan [18] *et al.* proposed a multi-entity reasoning framework to extract relationships of multiple character pairs, which considers the situation features of the video and constructs a relation knowledge graph with all character entities. Still, this method can not embody the special features of different character pairs. It is equally important to focus on the key information, such as the action and dialogue of each character pair in the video. Liu [12] *et al.* focused on the spatial and temporal inspection to reason the logic, but this method only obtains the character relationships at a video level which means each video obtains one relationship. Our method is p2p granularity. Social relation recognition of video faces huge challenges. Firstly, videos often contain multiple people. Characters can appear at any time, and interacting character pairs may not appear in the same frame, which makes it difficult to capture the interaction between multiple characters. In addition, some character pairs have very similar relationships, such as the interactions between siblings and lovers may be relatively close. It is tough to distinguish such relationships from just one video clip.

To this end, we propose an Overall-Distinctive Graph Convolutional Network (OD-GCN) framework for social relation recognition from videos. We build two-level graphs to represent the universal connection in all characters for logic reasoning and the unique connection in each character pair for partial attention. Specifically, we use two different types of edges $(inter, intra)$ in the overall-level graph to associate all the characters and update the character representation by heterogeneous GCN. Furthermore, we construct the distinctive-level graphs by designing an attention mechanism that captures the most relevant information for a specific character pair to obtain the unique relation for the character pairs. Specifically, we first integrate the temporal cues by LSTMs before the attention mechanism. Then, we find the top-k most valuable clues for each character pair. Finally, we combine the multimodal global features for character pairs and put them into the classifier for relation recognition. In this way, we can recognize the relationship of character pairs from a global view to a slight view.

We summarize our contribution as follows:

- We propose a novel heterogeneous graph construction method to represent connections between characters in videos.
- We propose an Overall-Distinctive Graph Convolutional Network (OD-GCN) framework to recognize multiple character relationships in videos with p2p granularity.
- We conduct validation experiments of our framework on the MovieGraphs dataset and demonstrate its effectiveness.

2 Related Work

In recent years, social relation recognition in visual content has attracted wide attention. Existing works mainly focus on still images [5,10,19]. For this task, previous works have collected two big datasets called People in Photo Album (PIPA) [20] and People in Social Context (PISC) [9]. However, the still image always contains all the character pairs simultaneously, but the character may appear in any video frame. So this task has some differences from our problem.

Some research attempted to recognize social relationships from videos. Most of them construct and utilize the graph of characters to infer the relationships. Wu *et al.* proposed HC-GCN framework to generate a frame-level graph, and further aggregate them along the temporal trajectory [17]. Teng *et al.* constructed a character and relationship reasoning graph to combine video contextual information for social relation reasoning [14]. Yan *et al.* designed a representation of three kinds of triples to build a relationship knowledge graph and associated situation information for relation prediction and relation recognition. Liu *et al.* proposed a VTF framework to integrate multimodal information into a representation model for the multiple relation extraction [13]. However, none of them grasp the connection of the overall characters while paying attention to the particularity of each specific character pair, which is different from our method.

3 Methodology

3.1 Framework Overview

Our Overall-Distinctive GCN framework mainly contains two parts, which are illustrated in Fig. 1. The first part is the overall-level character GCN. We believe that the other characters in the video also have an impact on the relationship recognition of specific character pairs. To capture the overall connection of characters, we construct a heterogeneous graph with two different types of edges. The nodes of overall-level graph are the appearance entity of all characters. Further, we utilize GCN to propagate the representation of each character. The second part is distinctive-level character GCN, which consists of two modules. For a certain character pair, we attempt to find the most relevant mentioned nodes for recognizing relationships. Considering that video is a long sequence process, we use LSTM to encode the input features. Then, each character pair obtained by overall-level character GCN is fed to the attention mechanism to find the top-k mentioned nodes. Along this line, we construct distinctive heterogeneous graph for each character pair and obtain a deeper representation of characters. Finally, we aggregate the multimodal global features like text information and visual information to recognize the relationship via the relation classification module.

3.2 Overall-level Character GCN

In this section, we will introduce the details of overall-level character GCN. Specifically, we have a video clip v, containing sequence frames of length m

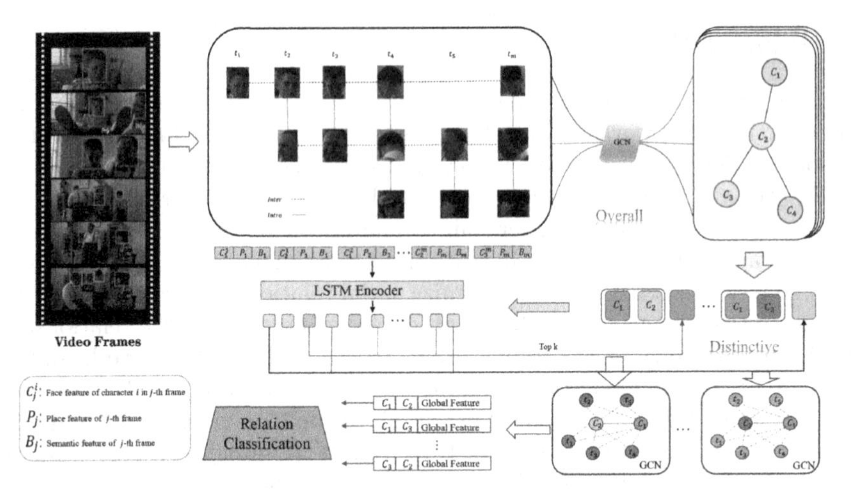

Fig. 1. The overall architecture of the OD-GCN framework

and M characters. The overall-level character GCN is aimed to construct a heterogeneous graph along with the whole clip, which can be defined by $\mathcal{G} = (\mathcal{V}, \mathcal{E})$. The $\mathcal{V}$ denotes the node-set, and $\mathcal{E}$ denotes the edge set. In our framework, we consider each node as an entity of character appearance in each frame, which means the number of the graph nodes N equals the number of appearances of all characters. We initialize the representation of each node as:

$$F_n = (C_j^i, P_j, B_j)$$

where C_j^i is the face feature of i-th character appear in j-th frame extracted by Arcface [2], P_j is the background feature of j-th frame extracted by ResNeXt-101 [6], and B_j is the semantic feature of j-th frame extracted by Deeplabv3+ [1].

To describe the connection of characters, we adopt two kinds of edges in the overall-level graph, i.e., *intra* and *inter*. We define the *intra* edge that connects all the characters in the same frame so that we can capture the subtle relationship atmosphere of all characters. And the *inter* edge connects the same person appearing in different frames, which involves the continuous change of a character. Specifically, the adjacency matrix of the overall-level character graph is defined as $\widetilde{A} \in \mathbb{R}^{N \times N}$. The graph construction rule is as below:

$$A(N_1, N_2) = \begin{cases} intra & (N_1, N_2) \in \{(C_j^i, C_j^k) | j \leq m \& i \neq k\}, \\ inter & (N_1, N_2) \in \{(C_t^i, C_k^i) | i \leq M \& t \neq k\}, \\ 0 & otherwise \end{cases}$$

To obtain the information of the overall-level character graph comprehensively, we use GCN to propagate the representation of nodes. Considering that the characters always complete an action in several continuous frames, and characters in the same frame always interact with each other. GCN can capture the features of nodes neighbors in the graph and performs message propagation well. Specifically, the overall-level character graph contains N nodes, which feature representations are F_n. Given node u at the l-th layer, we denote the graph convolution layer in our model as:

$$H_u^{(l+1)} = \text{ReLU}(\sum_{k \in \mathcal{K}} \sum_{v \in \mathcal{N}_k(u)} W_k^{(l)} H_v^{(l)} + b_k^{(l)})$$

where $\mathcal{K}$ are different types of edges *intra* and *inter*, $\mathcal{N}_k(u)$ are neighbors for node u connected in k-th type edge, $W_k^{(l)} \in \mathbb{R}^{N \times N}$ and $b_k^{(l)} \in \mathbb{R}^N$ are trainable parameters, and ReLU is an activation function performed as $\text{ReLU}(x) = \max(0, x)$. Along the overall-level character GCN, we obtain the updated representation of each character which aggregate the connections between all characters as:

$$Z = \text{GCN}(F_n, A)$$
$$Z_i = \frac{1}{|N_i|} \sum_{n \in N_i} Z_i^n$$

where Z is the updated representation obtained by GCN, Z_i is i-th character overall-level representation, N_i denotes the node set of character i, and Z_i^n is the n-th entity feature of i-th character.

3.3 Distinctive-level Character GCN

Pairwise Attention Mechanism. After obtaining the overall connection of all characters, we believe that different character pairs should focus on different keyframes to recognize the relationships. Different from the still image, the temporal cue in the video is significant for understanding the video clips. We use LSTM [7] to capture the continuous scene information in the video. The updated sequence $L \in \mathbb{R}^{N \times D}$ representation denotes as:

$$L = \text{LSTM}(F_1, F_2, ..., F_n)$$

Further, for each character pair obtained by the overall-level character GCN, we attempt to find the top-k mentioned nodes by pairwise attention mechanism. The attention mechanism is designed to automatically learn and calculate the contribution of input data to output data, which is widely used in both natural language processing (NLP) and CV. Generally, the key-query-value is always adopted to describe the attention operation. Specifically, We concat the representation of characters Z_i in pairs and send them to query the sequence respectively. Then, we utilize the learnable parameter matrix to obtain the scores of sequence for the character pair. The score denotes as:

$$S_L = \sigma(LW[Z_i; Z_j])$$

where $L \in \mathbb{R}^{N \times D}$ is the sequence obtained by LSTM, $Z_i, Z_j \in \mathbb{R}^D$ are character representation from overall-level character GCN, $[;]$ is the concat operation, σ

is an activation function like ReLU, and $W \in \mathbb{R}^{D\times 2D}$ is a learnable parameter matrix. Along this line, we obtain the scores $S_L \in \mathbb{R}^N$ and use $Softmax$ operation to normalize the score. Character pairs always have more than one key mentioned node for relation recognition. We select the top-k nodes with the highest scores as the mentioned nodes for each character pair, respectively.

Distinctive Graph. In this part, we will introduce the graph construction for each character pair. We have obtained the character representation and the mentioned nodes for each character pair. We consider that the selected mentioned nodes are seriously related to each character pair. Therefore, we also define two kinds of edges for distinctive graphs, i.e., *solid* and *mentioned*. The *solid* edge connects the character nodes so that we can build the interaction between character pairs directly. The *mentioned* edge connects the mentioned nodes selected by paiwise attention mechanism and each character node so that the key information can be involved. Specifically, the adjacency matrix of a distinctive graph $\widetilde{A} \in \mathbb{R}^{(k+2)\times(k+2)}$ denote as:

$$A(N_1, N_2) = \begin{cases} solid & (N_1, N_2) \in \{(Z_i, Z_j)\}, \\ mentioned & (N_1, N_2) \in \{(Z_i, M_t), (Z_j, M_t) | t \le k\}, \\ 0 & otherwise \end{cases}$$

where M_t is mention node, and Z_i, Z_j is character pair node, respectively. In the distinctive graph module, we also use GCN to propagate the representation of each node. Specifically, we obtain the character representation as:

$$C_i = \mathrm{GCN}(Z_i, M_t, A)$$

Considering the multimodal information contained in the video, we finally concat the global feature for each character pair. Since the conversation between characters always contains rich information for relation recognition, such as "dad" for "parents", "darling" for "couple", "work" for "colleague", etc. Therefore, we obtain the subtitle features extracted by Bert [3] as one of the global features. Also, we integrate all the frame features to represent the global visual feature of the video. Then, the concat features are fed to the relation classification module.

3.4 Relation Classification

We define our problem as a classification task, although recent research solves the problem from the reasoning perspective. We suppose that our OD-GCN framework has included the logic reasoning cues between characters. In our relation classification module, we organize the basic architecture Multi-Layer Perceptron (MLP) as the classifier. Unfortunately, the relation in the movie always has an unbalanced distribution. For example, friends and colleagues appear in the movies frequently, while service and opponent only have several sample character pairs. In this case, we consider that focal loss [11], which is an improved method of cross entropy loss (CE) but applies a modulating term to the CE, focuses on handling the imbalanced data by reducing the weight for the numerous easy negatives. Therefore, we utilize the focal loss function to compute the loss of our model. More formally, the loss denotes as:

$$\mathcal{L}_{\mathrm{FL}} = -\frac{1}{N}\sum_{i}\sum_{r\in R} y_{ir}(1-p_{ir})^{\gamma}\log(p_{ir})$$

where N is the sum of training samples, i is the i-th sample, R is the set of all relationships, p_{ir} is the predicted probability of i-th sample in r type relation, γ is a hyperparameter, and y_{ir} is a indicative function which denotes as:

$$y_{ir} = \begin{cases} 1 & \text{if the true relation of sample } i \text{ is } r, \\ 0 & otherwise \end{cases}$$

4 Experiments

4.1 Dataset

We conduct experiments of our method on the MovieGraphs [15] dataset. The dataset contains over 7600 clips of 51 movies. We split all movies into train set (35 movies), val set (7 movies), and test set (9 movies) according to the ratio of 10:2:3. Each clip has comprehensive annotations, such as subtitles between characters, the bounding box of characters and objects, relationships of character pairs, etc. Different from the ViSR [12] dataset, which only has one character pair relationship of a clip, MovieGraphs has multiple character pairs in one clip. However, the annotation of relationship types is too fine-grained, which includes 106 categories. In the video relation recognition task, we follow the previous settings [18] to reduce the number of relationship types from 106 to 8 or 15.

4.2 Implementation Details

Features Extraction. The representation of graph nodes consists of face feature, frame place feature, and frame semantic feature. We extract 1024-dim character face feature according to the annotations of the character bounding box by ArcFace. For each frame in clip, we utilize ResNeXt-101 pre-trained on kinetics-400 dataset [6] to extract 512-dim place features. And we use Deeplabv3+ pre-trained on VOC2012 dataset [4] with Xception back-bone to extract 512-dim semantic features. As to the subtitles of the clips, we use Bert to extract the 1024-dim features as the global textual feature.

Model Settings. After the construction of the overall-level character graph, we send the graph into 2 hidden layers GCN without dropout operation. Meanwhile, the sequence node features are fed into 3 layers Bi-LSTM with 0.01 dropout operation. Then, we construct distinctive graphs for each character pair and sent them into 2 hidden layers GCN separately. Finally, we concat the global features and classify the relationship by the classification module, which consists of 3 fully connected layers with the normalization layer before each ReLU activation function. The hyperparameter γ of focal loss is set to 2.5. The learning rate is set to 1e-3 with the Adam optimizer.

4.3 Baseline Methods

To evaluate the effectiveness of our proposed framework OD-GCN, we compare it with the following state-of-the-art methods on the MovieGraphs dataset. The methods are all based on the videos, and the details are as follows:

- LiRec [8]. Learning Interactions and Relationships between Movie Characters focus on using the interaction between two characters to recognize the relationship, which utilizes visual and textual information to jointly predict the interactions and relationships.
- PGCN [12]. The Pyramid Graph Convolution Networks model is designed to learn multi-scale dynamics receptive fields to perform the temporal reasoning from Triple Graphs. For each graph convolutional layer in GCNs, it inserts temporal pyramid branches.
- MSTR [12]. The Multi-scale Spatial-Temporal Reasoning model combines PGCN and TSN [16], which aims to learn both temporal cues and global spatial features. Further, classifying the character pairs relation after the weighted consensus model.
- MRR [18]. The Multi-entity Relation Reasoning model utilizes Graph Attention Networks to propagate the character entity knowledge graph and use the situation to aid in inferring the relationships between character pairs.

4.4 Experiment Results

Table 1. Comparison to the state-of-the-art methods on MovieGraphs dataset. The best score is in bold

	Top-1 Accuracy								
	Leader-sub	Colleague	Service	Parent-offs	Sibling	Couple	Friend	Opponent	Average
LiRec [8]	0.115	0.294	0.103	**0.492**	0.055	0.269	0.289	0.063	0.210
PGCN [12]	0.313	0.374	0.29	0.137	0.32	0.25	0.407	0.375	0.308
MSTR [12]	0.409	0.392	0.342	0.407	0.434	0.326	0.373	0.357	0.380
MRR [18]	0.454	**0.423**	0.428	0.385	0.392	**0.446**	**0.439**	0.365	0.417
OD-GCN (Ours)	**0.463**	0.394	**0.442**	0.415	**0.457**	0.436	0.428	**0.423**	**0.432**

We follow the previous research to show the performance of top-1 accuracy and average on the relation recognition task. The results on MovieGraphs are summarized in Table 1. From the results, we can see that only focusing on the interaction between character pair works poorly in multiple relation videos with 0.21 average accuracy. Conversely, other methods that consider the global character relationship have significantly improved the effect. It teaches us that all the characters in the videos are interconnected, and the extra characters have a positive effect on the relationship recognition of character pairs.

Also, MSTR has an additional TSN module than PGCN, and the effect has been greatly improved. This phenomenon teaches us that we need to combine the temporal cues in videos and fine-grained features at the video frame level to recognize the relationships. In comparison, the MRR model further improves the previous method, which utilizes a two-stage structure with a situation recognition part to solve the task. It inspires us that global scene information like situation features in the video is also beneficial.

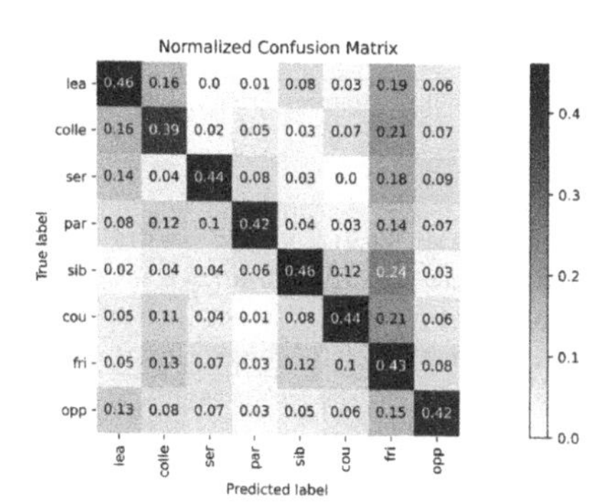

Fig. 2. The Normalized Confusion Matrix of OD-GCN framework on MovieGraphs Dataset

Table 2. Comparison to the state-of-the-art methods on 15 relationship recognition tasks.

	Top-1 Accuracy
LiRec [8]	0.281
MRR [18]	0.305
OD-GCN (Ours)	**0.312**

Correspondingly, our OD-GCN model utilizes two-level graphs to understand the character logic in the video and combine global features to recognize multiple character pairs' relationships, which achieves the highest score of 0.432 on the MovieGraphs dataset. Specifically, we notice that our approach performs well in those less intimate relationship categories of opponent, leader-sub, sibling, and service. However, the classification results in friend and colleague relationships are not ideal. This may be because the overall video is too smooth to figure out the mentioned nodes for each character pair. On the contrary, characters in an opponent relationship always have obvious conflict fragments, and characters in a leader-sub relationship or a service relationship always have typical actions or gestures, such as bending over, ending plate, etc. Besides, we draw the normalized confusion matrix of 8 categories of relationships on the MovieGraphs dataset. The Fig. 2 shows that our model is prone to mistakes on similar relations such as *(colleague, friend)* and *(couple, friend)*. But these mistakes are acceptable to a certain extent as it is difficult even for humans to judge strongly similar relationships from just a 30 s video clip. As we know, friends and couples sometimes have the same interactions, such as hugging or leaning on each other.

We also conduct 15 categories of relationships experiments on MovieGraphs and make a comparison of state-of-art methods. The results are shown as Table 2. Although we have made improvements on the 15 categories of relationships, the effect is not significant. We realize that fine-grained character relationship categories remain a challenging task that contains many similar pairwise emotional relationships.

5 Ablation Study

Table 3. The Ablation Studies for OD-GCN Framework

Module			*Top-1 Accuracy*								
Overall Graph	Discri-minitive Graph	LSTM	Leader-sub	Colle-ague	Service	Parent-offs	Sibling	Couple	Friend	Oppo-nent	Ave-rage
✓			0.034	0.163	0	0.087	0.231	0.217	0.562	0.012	0.163
✓	✓		0.435	0.365	0.316	0.387	0.269	0.419	0.407	0.394	0.374
✓		✓	0.16	0.297	0.083	0.164	0.316	0.279	0.486	0.29	0.259
✓	✓	✓	0.463	0.394	0.442	0.415	0.457	0.436	0.428	0.423	0.432

To explore the different contributions of each module in the OD-GCN framework to the relation recognition task, we design several ablation experiments between modules. The experimental results of the combination of different modules are shown in the Table 3. From the results, we can find that each module has different degrees of contribution to the classification effect. One of the interesting phenomena is when we only use the overall graph module, we will find that except for the friend relation category, which has a very high accuracy rate, the accuracy rate of the other relationship categories is very low or even 0. We think it may be because the state of friends can also represent the atmosphere of the social relationship of other overall characters. Moreover, when we add the mentioned nodes in the video to each character pair, the effect of the experiment across different categories becomes more even and effective. The temporal cue in videos is equally important, according to the results.

Besides, we conduct the global features ablation experiments with different modalities. The results are shown in the Table 4. From the results, we can find that the situation information and textual cues both benefit the recognition performance. It may be because the dialogues of characters and scenes have implicit relationship information. For example, the characters in a hospital scene are most likely related to the service relationship. Therefore, It is necessary to pay attention to the global features.

Table 4. The Global Features Ablation Studies for OD-GCN

Combinations	Average
No Global Features	0.359
Place Global Features	0.407
Text Global Features	0.396
(Text + Place) Global Features	0.432

6 Conculusion

In this paper, we proposed a novel Overall-Distinctive GCN framework to analyze the social relationship between multiple character pairs from the videos. Specifically, we first take the entities where the characters appear as the nodes of the graph and integrate the frame-level visual cues as the features of the nodes to construct an overall-level graph that includes all characters. Among them, we define two different edges to better represent the connection between characters. Then, we designed an attention mechanism to find the most valuable mention nodes for each character pair and further build distinctive-level graphs. Finally, combing with the global multimodal information of the video, we perform multiple character relationship recognition on the MovieGraphs dataset. Our experimental results validate the effectiveness of the proposed OD-GCN framework, and we uncover many phenomena worth exploring about video understanding on this basis. Moreover, dynamic relationships of character pairs in the long-term video and precise discrimination of similar relationships are worth exploring in future work.

Acknowledgments. This work is supported by the NSFC-General Technology Basic Research Joint Funds under Grant (U1936220), the National Natural Science Foundation of China under Grant (61972047), the National Key Research and Development Program of China (2018YFC0831500).

References

1. Chen, L.C., Zhu, Y., Papandreou, G., Schroff, F., Adam, H.: Encoder-decoder with atrous separable convolution for semantic image segmentation. In: Proceedings of the European Conference on Computer Vision (ECCV), pp. 801–818 (2018)
2. Deng, J., Guo, J., Xue, N., Zafeiriou, S.: ArcFace: additive angular margin loss for deep face recognition. In: Proceedings of the IEEE/CVF Conference on Computer Vision and Pattern Recognition, pp. 4690–4699 (2019)
3. Devlin, J., Chang, M.W., Lee, K., Toutanova, K.: BERT: pre-training of deep bidirectional transformers for language understanding. arXiv preprint arXiv:1810.04805 (2018)
4. Everingham, M., Eslami, S., Van Gool, L., Williams, C.K., Winn, J., Zisserman, A.: The pascal visual object classes challenge: a retrospective. Int. J. Comput. Vis. **111**(1), 98–136 (2015)

5. Goel, A., Ma, K.T., Tan, C.: An end-to-end network for generating social relationship graphs. In: Proceedings of the IEEE/CVF Conference on Computer Vision and Pattern Recognition, pp. 11186–11195 (2019)
6. Hara, K., Kataoka, H., Satoh, Y.: Can spatiotemporal 3D CNNs retrace the history of 2D CNNs and imagenet? In: Proceedings of the IEEE Conference on Computer Vision and Pattern Recognition, pp. 6546–6555 (2018)
7. Hochreiter, S., Schmidhuber, J.: Long short-term memory. Neural Comput. **9**(8), 1735–1780 (1997)
8. Kukleva, A., Tapaswi, M., Laptev, I.: Learning interactions and relationships between movie characters. In: Proceedings of the IEEE/CVF Conference on Computer Vision and Pattern Recognition, pp. 9849–9858 (2020)
9. Li, J., Wong, Y., Zhao, Q., Kankanhalli, M.S.: Dual-glance model for deciphering social relationships. In: Proceedings of the IEEE International Conference on Computer Vision, pp. 2650–2659 (2017)
10. Li, J., Wong, Y., Zhao, Q., Kankanhalli, M.S.: Visual social relationship recognition. Int. J. Comput. Vis. **128**(6), 1750–1764 (2020)
11. Lin, T.Y., Goyal, P., Girshick, R., He, K., Dollár, P.: Focal loss for dense object detection. In: Proceedings of the IEEE International Conference on Computer Vision, pp. 2980–2988 (2017)
12. Liu, X., et al.: Social relation recognition from videos via multi-scale spatial-temporal reasoning. In: Proceedings of the IEEE/CVF Conference on Computer Vision and Pattern Recognition, pp. 3566–3574 (2019)
13. Liu, Z., Hou, W., Zhang, J., Cao, C., Wu, B.: A multimodal approach for multiple-relation extraction in videos. Multimedia Tools Appl. **81**(4), 4909–4934 (2022)
14. Teng, Y., Song, C., Wu, B.: Toward jointly understanding social relationships and characters from videos. Appl. Intell. **52**(5), 5633–5645 (2022)
15. Vicol, P., Tapaswi, M., Castrejon, L., Fidler, S.: MovieGraphs: towards understanding human-centric situations from videos. In: Proceedings of the IEEE Conference on Computer Vision and Pattern Recognition, pp. 8581–8590 (2018)
16. Wang, L., et al.: Temporal segment networks: towards good practices for deep action recognition. In: Leibe, B., Matas, J., Sebe, N., Welling, M. (eds.) ECCV 2016. LNCS, vol. 9912, pp. 20–36. Springer, Cham (2016). https://doi.org/10.1007/978-3-319-46484-8_2
17. Wu, S., Chen, J., Xu, T., Chen, L., Wu, L., Hu, Y., Chen, E.: Linking the characters: video-oriented social graph generation via hierarchical-cumulative GCN. In: Proceedings of the 29th ACM International Conference on Multimedia, pp. 4716–4724 (2021)
18. Yan, C., Liu, Z., Li, F., Cao, C., Wang, Z., Wu, B.: Social relation analysis from videos via multi-entity reasoning. In: Proceedings of the 2021 International Conference on Multimedia Retrieval, pp. 358–366 (2021)
19. Zhang, M., Liu, X., Liu, W., Zhou, A., Ma, H., Mei, T.: Multi-granularity reasoning for social relation recognition from images. In: 2019 IEEE International Conference on Multimedia and Expo (ICME), pp. 1618–1623. IEEE (2019)
20. Zhang, N., Paluri, M., Taigman, Y., Fergus, R., Bourdev, L.: Beyond frontal faces: improving person recognition using multiple cues. In: Proceedings of the IEEE Conference on Computer Vision and Pattern Recognition, pp. 4804–4813 (2015)

Weakly-Supervised Temporal Action Localization with Regional Similarity Consistency

Haoran Ren, Hao Ren, Hong Lu(✉), and Cheng Jin

Shanghai Key Lab of Intelligent Information Processing, School of Computer Science, Fudan University, Shanghai, China
{hrren20,hren17,honglu,jc}@fudan.edu.cn

Abstract. The weakly-supervised temporal action localization task aims to train a model that can accurately locate each action instance in the video using only video-level class labels. The existing methods take into account the information of different modalities (primarily RGB and Flow), and present numerous multi-modal complementary methods. RGB features are obtained by calculating appearance information, which are easy to be disrupted by the background. On the contrary, Flow features are obtained by calculating motion information, which are usually less disrupted by the background. Based on this phenomenon, we propose a Regional Similarity Consistency (RSC) constraint between these two modalities to suppress the disturbance of background in RGB features. Specifically, we calculate the regional similarity matrices of RGB and Flow features, and impose the consistency constraint through L_2 loss. To verify the effectiveness of our method, we integrate the proposed RSC constraint into three recent methods. The comprehensive experimental results show that the proposed RSC constraint can boost the performance of these methods, and achieve the state-of-the-art results on the widely-used THUMOS14 and ActivityNet1.2 datasets.

Keywords: Temporal action localization · Weakly-supervised learning · Multi-modal complementary learning

1 Introduction

Temporal Action Localization (TAL) is the task of predicting the boundary (start and end time) of potential action events and categorizing them for untrimmed videos. It is now widely applied in video security [26] and retrieval [5]. Many fully-supervised methods [4,18,25,34] have been presented in the past for this task, which require the training video to contain the labels of the boundary of each action instance. It is very labor-intensive to construct huge datasets. To address this issue, some weakly-supervised research [6,19,22,27] have begun to use datasets that only contain video-level labels.

H. Ren, H. Ren—Contributed equally to this work.

D.-T. Dang-Nguyen et al. (Eds.): MMM 2023, LNCS 13833, pp. 69–81, 2023.
https://doi.org/10.1007/978-3-031-27077-2_6

Because Weakly-supervised TAL (WTAL) only relies on video-level annotation, current mainstream methods typically employ Multiple Instance Learning (MIL) to handle this task. In the training stage, the video is broken into numerous non-overlapped segments firstly, and features are extracted by a pre-trained network. Then the action category of each segment is predicted and aggregated to build video-level prediction. Lastly, the model is trained by the supervision of video-level labels. In the testing stage, these segments are firstly filtered by multiple pre-defined thresholds, and then post-processed by temporal Non-Maximum Suppression (NMS) to obtain the final boundaries for each action instance [16].

How to accurately locate action boundaries has always been a problem. The attention mechanism [13] has been popular to solve this problem in recent years, *e.g.*, BaS-Net [14] refines the boundary by suppressing the background's response and enhancing the foreground's response through different attention maps. On the other hand, some methods [7,29] extract both RGB and Flow features, and then fuse them to contain additional perspective characteristics, *e.g.*, CO_2-Net [7] proposes a module that takes both features as global and local features to enhance each other. Inspired by these methods, we discover that RGB features are easy to be disrupted by the background, whereas Flow features are usually less disrupted by the background. Based on this observation, we propose a Regional Similarity Consistency (RSC) constraint mechanism. By calculating the regional similarity matrices of RGB and Flow features and imposing the consistency constraint through L_2 loss to suppress the disturbance of background in RGB features. In addition, the mechanism can be easily embedded into the existing WTAL methods.

Our main contributions can be summarized as follows:

- A RSC constraint mechanism is proposed to suppress the disturbance of background in RGB features.
- We conduct detailed ablation studies to verify the effectiveness of our RSC mechanism.
- The comprehensive WTAL experiments are conducted on the widely-used THUMOS14 [9] and ActivityNet1.2 [1] datasets with three recent methods. The experimental results demonstrate the superiority of the proposed method.

2 Related Work

2.1 Weakly-supervised Temporal Action Localization

WTAL only uses video-level annotation, which eliminates the issue of onerous frame-level annotation in Fully-supervised TAL (FTAL). In order to overcome the problem of no fine-grained annotation, the usual practice is to segment the video first, then classify each segment by multiple instance learning, and finally aggregate all segments to obtain the prediction results of the whole video. UntrimNet [27] is the first method to formally propose WTAL task and solve it using attention mechanism and multiple instance learning. STAR [28] uses recurrent neural network to model the temporal relationship between action instances, but its modeling effect on long-term dependence is poor. Liu *et al.* [16] propose a multi-branch neural network to model the completeness of action instances, prompting it to discover missing parts of actions.

2.2 Multi-modal Complementary Learning

Recently, multi-modal complementary methods [2,7,29,31] have gained popularity in WTAL task, which mainly leverage the features of different modalities to enhance each other or eliminate redundant information. For example, TSCN [31] uses a two stream consensus network to iteratively refine the boundaries of action instances. UGCT [29] improves their respective features through RGB and Flow collaborative training. On the basis of Flow information, DMP-Net [2] introduces motion map to model motion, in order to fully utilize motion information. Inspired by the fact that the Flow feature is not easily disrupted by background, we propose a RSC mechanism, which can effectively filter the task-irrelevant information (primarily background information) in RGB features by modeling the regional consistency relationship between RGB and Flow similarity matrices, thereby improving the final action localization accuracy.

3 Proposed Method

In this section, we elaborate on the proposed Regional Similarity Consistency (RSC) mechanism, which devotes to suppress the disturbance of background information in RGB features by imposing the consistency constraint with the Flow features. Specifically, the problem formulation and notations for WTAL task are described in Sect. 3.1. Then we recap the pipeline of recent WTAL methods in Sect. 3.2 for better demonstrate our methods. Finally, in Sect. 3.3, we show the proposed RSC mechanism in detail.

3.1 Problem Formulation

We define a set of N untrimmed videos noted as $\mathcal{V} = \{V^{(n)}\}_{n=1}^{N}$, where $V^{(n)}$ indicates the n-th video. For each video $V^{(n)}$, we have the video-level label $\mathbf{y}^{(n)} \in \mathbb{R}^C$, where $\mathbf{y}_i^{(n)} \in \{0, 1\}$ represents whether the instance of i-th action category exists in $V^{(n)}$, and C is the total number of action labels. The WTAL task aims to learn a model that can predict a set of action instances $\{t_s, t_e, c, q\}$ for each video, where t_s and t_e represent the start time and end time of each action instance, c denotes the predicted action category, and q shows the confidence score of the action instance.

3.2 WTAL Methods Pipeline Recap

Segment-level Feature Extraction. According to the recent WTAL methods [6–8,10,22,32], the untrimmed video V is firstly divided into L non-overlapped segments (16-frames as one segment). Then a pre-trained network (*i.e.*, I3D [3]) is used to extract RGB and Flow features. We denote the RGB and Flow features as $\mathbf{X}^R = [\mathbf{x}_1^R, \mathbf{x}_2^R, \cdots, \mathbf{x}_L^R]^T \in \mathbb{R}^{L\times D}$ and $\mathbf{X}^F = [\mathbf{x}_1^F, \mathbf{x}_2^F, \cdots, \mathbf{x}_L^F]^T \in \mathbb{R}^{L\times D}$ respectively, where $\mathbf{x}_i^R \in \mathbb{R}^D$ or $\mathbf{x}_i^F \in \mathbb{R}^D$ denotes the D-dimensional feature vector of the i-th segment, D is normally 1024. Current methods usually use both RGB features $\mathbf{X}^R$ and Flow features $\mathbf{X}^F$ for WTAL task.

Segment-level Action Classification. As current WTAL methods exploit features from both modalities for prediction, the quality of the modality features directly affects the prediction accuracy. Therefore a key issue to be addressed is how to enhance the features $\mathbf{X}^R$ and $\mathbf{X}^F$ to obtain the better performance. We use $\tilde{\mathbf{X}}$ to denote the features after enhancement. For example, CO_2-Net [7] introduces cross-modal consensus networks for feature calibration, which improves RGB features by using global information from Flow features to remove task-irrelevant information in RGB features, and vice versa. Next, a classifier head f_{cls} is used to obtain the category logits $Q = f_{cls}(\tilde{\mathbf{X}}) \in \mathbb{R}^{L\times C}$ for the given video $V^{(n)}$. The logits Q are called the temporal Class Activation Sequences (CAS), which give the probability of each segment belonging to an action category.

Video-level Action Classification. Since only the video-level label $\mathbf{y} \in \{0,1\}^C$ is available, the WTAL methods aggregate the segment-level category logits Q to obtain the video-level predictions $\tilde{\mathbf{y}}$. In practice, to obtain the video-level logits $\mathbf{q} \in \mathbb{R}^C$ from Q, most WTAL methods use a top-k selection strategy:

$$\mathbf{q}_i = \max_{l\subset\{1,\cdots,L\}} \frac{1}{k}\sum_{k\in l} Q_{k,i}, \quad s.t.\ |l| = k. \tag{1}$$

Then the video-level prediction is calculated as $\tilde{\mathbf{y}} = Softmax(\mathbf{q})$. Finally the whole model is supervised by the classification loss:

$$\mathcal{L}_{cls} = -\sum_{i=1}^{C} \mathbf{y}_i \log(\tilde{\mathbf{y}}_i). \tag{2}$$

To improve the accuracy of action localization, some additional losses are introduced, please refer [7,8,10,22] for more details.

Action Instances Localization. The recent WTAL methods [7,8,10,29] utilize a outer-inner contrastive strategy to localize the action instances in a video following Liu *et al.* [16]. Specifically, the video-level prediction $\tilde{\mathbf{y}}$ is first used to detect existing actions with a given threshold. Then, for each existing action c, CAS is used to localize the action instance by computing its start time t_s, end time t_e and confidence score q with multiple pre-defined thresholds and temporal NMS to form the final prediction.

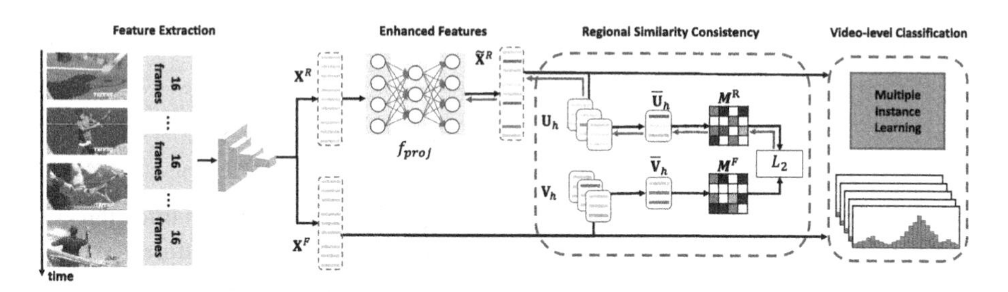

Fig. 1. The mechanism of Regional Similarity Consistency (RSC). Firstly, the RGB features are projected by a simple MLP to obtain $\tilde{\mathbf{X}}^R$. Then the projected RGB features $\tilde{\mathbf{X}}^R$ are enhanced by the constraint between the regional similarity matrices of M^R and M^F through L_2 loss. The red line shows the back-propagation direction of the gradient. Note that the gradient of L_2 loss is only back-propagated to $\mathbf{X}^R$. (Color figure online)

3.3 Regional Similarity Consistency Mechanism

In this section, we present the details of the proposed RSC mechanism, which effectively filters out redundant information from RGB features by modeling the regional relationship between RGB and Flow modalities. This region similarity consistency constraint results in the enhanced RGB features $\tilde{\mathbf{X}}^R$, and can be easily integrated with the existing methods.

As shown in Fig. 1, we receive the RGB and Flow features $\mathbf{X}^R$ and $\mathbf{X}^F$ from the pre-trained model firstly. Then we feed $\mathbf{X}^R$ into a projection module f_{proj} to get the projected features $\tilde{\mathbf{X}}^R = f_{proj}(\mathbf{X}^R)$. This projection module consists of two fully-connected layers and *ReLU* activation function. It can be regarded as a simple MLP structure, which first reduces the feature dimension and then restores it.

To ensure the projection module could reduce the redundant information and obtain the enhanced RGB features, we introduce the constraint of Regional Similarity Consistency (RSC). Specifically, the projected RGB features $\tilde{\mathbf{X}}^R$ and the original Flow features $\mathbf{X}^F$ are first divided into H regions $\{\mathbf{U}_h\}_{h=1}^H$ and $\{\mathbf{V}_h\}_{h=1}^H$ along the temporal dimension evenly. Each region $\mathbf{U}_h = [\mathbf{u}_1^h, \mathbf{u}_2^h, \cdots, \mathbf{u}_m^h]^T$ is called region-level RGB features, where $\mathbf{u}_i^h \in \mathbb{R}^D$, $i = 1, \cdots, m$, $m = \lfloor L/H \rfloor$, and each region $\mathbf{V}_h = [\mathbf{v}_1^h, \mathbf{v}_2^h, \cdots, \mathbf{v}_m^h]^T$ is region-level Flow features. Next, we average the features within the same region to obtain the representative regional RGB and Flow features $\bar{\mathbf{U}}_h$ and $\bar{\mathbf{V}}_h$, which are then been l_2-normalized. Afterwards the regional similarity matrices of these regional RGB and Flow features are calculated respectively:

$$M^R = \bar{\mathbf{U}}_h^T * \bar{\mathbf{U}}_h, \quad M^F = \bar{\mathbf{V}}_h^T * \bar{\mathbf{V}}_h. \tag{3}$$

Based on the regional similarity matrices M^R and M^F, we use L_2 loss to constrain the two modalities to be as similar as possible. Specifically, we minimize the Euclidean distance of the regional similarity matrices in the element-wise manner:

$$L_2(M^R, M^F) = \frac{1}{n}\sqrt{\sum_{i=1}^{n}(M_i^R - M_i^F)^2}, \quad n = H * H. \tag{4}$$

This allows the projection module f_{proj} to take advantage of the information in $\mathbf{X}^F$ to filter the redundant information in $\mathbf{X}^R$ by the regional similarity consistency constraint. Our proposed RSC mechanism is based on the fact that the Flow feature is not easy to be disrupted by the background information, hence we believe it can be used to lessen the background disturbance in the RGB feature. Simultaneously, to prevent the influence of outliers, the loss is calculated in region-level features. And this constraint is a result of the assumption that the same video should show regional consistency across all modalities.

4 Experiments

In this section, we first introduce the datasets, then describe our method's implementation details and provide the evaluation metrics. Then we compare our method with the most recent methods. Lastly, extensive ablation studies are carried out to investigate the impact about the number of regions and the projection module.

4.1 Datasets

The proposed method is evaluated on two well-known datasets (THUMOS14 [9] and ActivityNet1.2 [1]) containing untrimmed videos with varied degrees of action duration.

For each of the 20 classes in the THUMOS14 dataset, there are temporal annotations for a subset of videos in the validation and testing sets. We employ 200 videos in the validation set for training and 213 videos in the testing set for evaluation, as in earlier work [7,10,32]. This dataset is difficult due to the finely annotated action instances. On average, each video comprises 15.5 action clips. The duration of the action ranges from a few seconds to minutes.

The ActivityNet1.2 dataset contains 100 activity classes and 4819 training videos, 2383 validation videos, and 2480 testing videos (whose labels are withheld). On average, each video has 1.5 occasions of action. We utilize the training set to train our model and the validation set to evaluate it, following the previous work [7,10,32].

4.2 Implementation Details

In this paper, we use I3D network [3] pre-trained on the Kinetics-400 [3] dataset to extract features. To extract optical flow features, the TV-L1 algorithm [30] is employed. To validate the success of our proposed method, we re-train the three most recent models as baselines on our own machine using the official

HAM-Net, CoLA, and CO2-Net implementations. In the same experimental environment, the proposed RSC is merged with these models and then trained with the Adam [12] optimizer. We use grid search [15] to determine other introduced hyper-parameters. We chose these three methods because they give open source code and a solid baseline for our later experimentation. Our code is available on https://github.com/leftthomas/RSC, implemented with PyTorch [23] library, and the experiments are run on an NVIDIA Geforce GTX 1080 Ti GPU.

Table 1. Performance comparisons with state-of-the-art WTAL methods on THUMOS14 dataset. AVG is the average mAP under the thresholds 0.1:0.1:0.7, while † means the use of additional information, such as action frequency or human pose, * indicates using I3D features, - means that corresponding results are not reported in the original papers, and (Rerun) represents the results are obtained by retraining the official code on our own machine. The best results for each baseline method are bold.

Supervision	Method	mAP@IoU (%)							
		0.1	0.2	0.3	0.4	0.5	0.6	0.7	AVG
Weak†	3C-Net* [21], ICCV'19	59.1	53.5	44.2	34.1	26.6	–	8.1	-
	SF-Net* [20], ECCV'20	71.0	63.4	53.2	40.7	29.3	18.4	9.6	40.8
Weak	UntrimNet [27], CVPR'17	44.4	37.7	28.2	21.1	13.7	–	–	–
	W-TALC* [24], ECCV'18	55.2	49.6	40.1	31.1	22.8	–	7.6	–
	ASSG* [33], MM'19	65.6	59.4	50.4	38.7	25.4	15.0	6.6	37.3
	BaS-Net* [14], AAAI'20	58.2	52.3	44.6	36.0	27.0	18.6	10.4	35.3
	EM-MIL* [19], ECCV'20	59.1	52.7	45.5	36.8	30.5	22.7	16.4	37.7
	HAM-Net* [10], AAAI'21	65.9	59.6	52.2	43.1	32.6	21.9	12.5	41.1
	CoLA* [32], CVPR'21	66.2	59.5	51.5	41.9	32.2	22.0	13.1	40.9
	CO_2-Net* [7], MM'21	70.1	63.6	54.5	45.7	38.3	26.4	13.4	44.6
	RSKP* [8], CVPR'22	71.3	65.3	55.8	47.5	38.2	25.4	12.5	45.1
	ASM-Loc* [6], CVPR'22	71.2	65.5	57.1	46.8	36.6	25.2	13.4	45.1
	HAM-Net* [10] (Rerun)	65.8	59.3	**51.1**	41.4	31.3	20.4	10.9	40.0
	HAM-Net* [10] with RSC	**66.9**	**60.2**	51.0	**42.0**	**31.7**	**22.1**	**12.0**	**40.9**
	CoLA* [32] (Rerun)	66.4	60.3	51.8	43.0	34.0	23.5	12.8	41.7
	CoLA* [32] with RSC	**67.2**	**61.5**	**52.9**	**43.9**	**34.8**	**24.9**	**13.0**	**42.6**
	CO_2-Net* [7] (Rerun)	70.1	63.6	54.5	45.7	38.3	**26.4**	13.4	44.6
	CO_2-Net* [7] with RSC	**70.6**	**64.2**	**55.9**	**47.7**	**38.9**	26.0	**13.6**	**45.3**

4.3 Evaluation Metrics

We analyze the temporal action localization performance using mean Average Precision (mAP) values at various levels of IoU thresholds, as per the conventional evaluation protocol [7,10,32]. For the THUMOS14 and ActivityNet1.2 datasets, the IoU threshold sets are [0.1:0.1:0.7] and [0.5:0.05:0.95], respectively. Greater mAP suggests improved performance. The results are calculated using the benchmark code supplied by ActivityNet for fair comparison.

4.4 Comparison with State-of-the-art Methods

We compare our method against state-of-the-art weakly-supervised methods on both THUMOS14 [9] and ActivityNet1.2 [1] datasets. The three recent models (HAM-Net [10], CoLA [32] and CO_2-Net [7]) are re-trained as our baseline, and the quantitative results are listed in Tables 1 and 2. The results show that when the proposed RSC is applied to these three methods, the performances of these methods are improved (0.7%∼0.9% on THUMOS14 dataset in terms of average mAP). CO_2-Net with RSC achieves a new state-of-the-art performance, outperforming the previous best methods (45.3% *vs* 45.1% and 27.1% *vs* 26.9% on THUMOS14 and ActivityNet1.2 datasets in terms of average mAP, respectively). This demonstrates that the proposed RSC is a generic method that can be easily incorporated with existing WTAL methods to boost performance even more. It also implies that our method is capable of enhancing RGB features by imposing the constraint through Flow features.

Table 2. Performance comparisons with state-of-the-art WTAL methods on ActivityNet1.2 dataset. AVG is the average mAP under the thresholds 0.5:0.05:0.95, while † means the use of action counts, * indicates using I3D features, - means that corresponding results are not reported in the original papers, and (Rerun) represents the results are obtained by retraining the official code on our own machine. The best results for each baseline method are bold.

Supervision	Method	mAP@IoU (%)			
		0.5	0.75	0.95	AVG
Weak†	3C-Net* [21], ICCV'19	37.2	–	–	21.7
Weak	UntrimNet [27], CVPR'17	7.4	3.9	1.2	3.6
	W-TALC* [24], ECCV'18	37.0	14.6	–	18.0
	BaS-Net* [14], AAAI'20	38.5	24.2	5.6	24.3
	HAM-Net* [10], AAAI'21	41.0	24.8	5.3	25.1
	ACSNet* [17], AAAI'21	40.1	26.1	6.8	26.0
	CoLA* [32], CVPR'21	42.7	25.7	5.8	26.1
	CO_2-Net* [7], MM'21	43.3	26.3	5.2	26.4
	CSCL* [11], MM'21	43.8	26.9	5.6	26.9
	HAM-Net* [10] (Rerun)	**41.1**	24.6	**5.0**	25.0
	HAM-Net* [10] with RSC	40.8	**24.8**	**5.0**	**25.2**
	CoLA* [32] (Rerun)	40.6	27.6	3.1	26.3
	CoLA* [32] with RSC	**41.0**	**27.9**	**2.6**	**26.7**
	CO_2-Net* [7] (Rerun)	43.4	25.9	5.3	26.1
	CO_2-Net* [7] with RSC	**44.4**	**26.8**	**5.7**	**27.1**

We also offer a visualization comparison in Fig. 2 between CO_2-Net [7] and CO_2-Net with RSC on THUMOS14 dataset in order to more intuitively explain the benefits of our method. As we can see, CO_2-Net mis-localizes two parts (marked by yellow boxes), but both are corrected with the help of RSC. Specifically, the correct segments should be the cliff diving action of two different

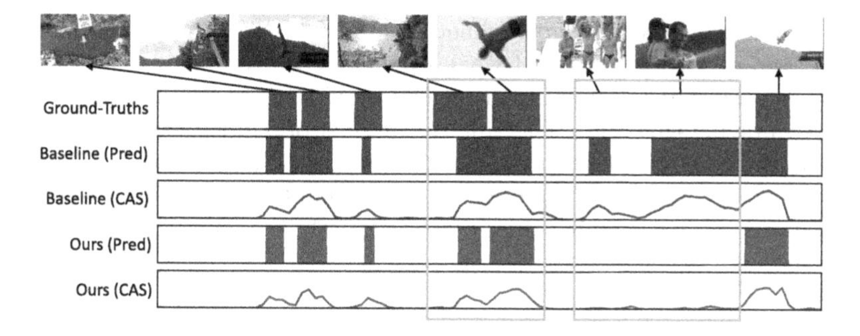

Fig. 2. Qualitative visualization results on an example of the *CliffDiving* action in the THUMOS14 dataset. The horizontal axis denotes the timestamps. The results of Ground-Truths, CO_2-Net (Baseline) and CO_2-Net with RSC (Ours) are shown in green, blue and red, respectively. "Pred" means predicted action localization, and "CAS" means corresponding activation sequence of *CliffDiving*. The yellow boxes contain some cases that CO_2-Net fails to detect but can be successfully localized by our method. (Color figure online)

perspectives (one distant view and one close-up) in the area marked by the first yellow box, but CO_2-Net regards the two segments as the same continuous action, and with the help of RSC, our method correctly identifies the two segments and gives the truncation. In the area marked by the second yellow box, CO_2-Net incorrectly separates the celebratory scene into action instances of cliff diving, but our method correctly identifies these moments as backgrounds, and suppresses their activation values. This example reflects that CO_2-Net is likely to focus on the human body itself and ignore the correlation between actions, resulting in high CAS activation values in this scene. Our method provides the constraint for RGB features through Flow features (focusing on motion information), potentially weakening the excessive dependence of RGB features on the human body itself, so that the CAS activation values in this scene is basically maintained at zero. This also verifies our claim that background information disturbance in RGB features can be effectively reduced by applying RSC constraint between RGB and Flow features, allowing it to better support action localization and improve localization accuracy.

4.5 Ablation Studies

Impact about the number of regions. We can learn from the description of RSC in Sect. 3.3 that the number of regions is a key hyper-parameter that affects the consistency constraint, which has an effect on the final performance. It is worthwhile to investigate how to determine the number of regions. Therefore, to establish the optimal value, we employ all three baseline methods to conduct ablation studies about the number of regions on THUMOS14 [9] and ActivityNet1.2 [1] datasets. The quantitative results are shown in Table 3. The results show that no matter whatever baseline or dataset is used, the performance (in terms of average mAP) is the best when the number of regions is 4. At the same time, we can see that the number of regions should not be divided too much or too little, implying that while selecting the number of regions, we must take into account both the representation of features within the region and the variation of features across regions. Therefore, we can conclude that setting the number of regions to 4 is a reasonable value, which is exactly the number of regions that our method adopts.

Impact about the projection module. From Fig. 1, we can see that only RGB features are appended with the projection module f_{proj}. Here we also explore the remaining two cases on THUMOS14 [9] dataset: only the Flow features are appended with f_{proj} and all the two features are appended with f_{proj}. The quantitative results are shown in Table 4. The results show that no matter which baseline is used, the performance of appending f_{proj} after RGB features is the highest. Meanwhile, we can observe that appending f_{proj} after Flow features is the worst, which reflects that if RGB features are used as supervision to impose consistency constraint on Flow features, Flow features can not be enhanced. This also confirms our view that RGB features are easy to be disrupted by background, while Flow features are not easy to be disrupted by background, which can enhance the learning of RGB features, not vice versa. This also explains why we only append the projection module after the RGB features.

Table 3. Ablation studies about the number of regions on THUMOS14 and ActivityNet1.2 datasets. AVG is the average mAP under the thresholds 0.1:0.1:0.7 and 0.5:0.05:0.95 for THUMOS14 and ActivityNet1.2 datasets, respectively. The best results for each baseline method are bold.

Method	Regions	mAP@IoU (%)											
		THUMOS14								ActivityNet1.2			
		0.1	0.2	0.3	0.4	0.5	0.6	0.7	AVG	0.5	0.75	0.95	AVG
HAM-Net [10]	2	**67.0**	**60.3**	**51.0**	41.7	31.4	21.6	11.8	40.7	40.6	24.7	4.8	24.9
	4	66.9	60.2	**51.0**	42.0	31.7	**22.1**	**12.0**	**40.9**	**40.8**	**24.8**	**5.0**	**25.2**
	8	66.8	60.1	50.8	**42.1**	**31.8**	**22.1**	11.8	40.8	40.6	24.7	**5.0**	25.0
CoLA [32]	2	66.4	61.0	52.5	**44.2**	34.6	24.1	12.6	42.2	41.1	**27.9**	3.5	**26.7**
	4	**67.2**	**61.5**	**52.9**	43.9	**34.8**	**24.9**	**13.0**	**42.6**	41.0	**27.9**	**3.6**	**26.7**
	8	64.5	58.8	50.5	42.4	33.4	24.0	12.7	40.9	**41.2**	27.4	3.1	26.5
CO_2-Net [7]	2	69.3	62.9	54.2	45.7	37.5	25.0	12.2	43.8	**44.4**	26.7	5.4	26.8
	4	**70.6**	**64.2**	**55.9**	**47.7**	**38.9**	26.0	13.6	**45.3**	**44.4**	**26.8**	**5.7**	**27.1**
	8	69.0	63.0	54.1	45.5	38.3	**26.3**	**13.9**	44.3	44.0	26.1	4.8	26.4

Table 4. Ablation studies about the projection module on THUMOS14 dataset. AVG is the average mAP under the thresholds 0.1:0.1:0.7. The best results for each baseline method are bold.

Method	f_{proj}		mAP@IoU (%)							
	RGB	Flow	0.1	0.2	0.3	0.4	0.5	0.6	0.7	AVG
HAM-Net [10]	✓	–	**66.9**	**60.2**	**51.0**	**42.0**	**31.7**	**22.1**	**12.0**	**40.9**
	–	✓	56.9	49.8	40.2	29.6	20.3	11.7	5.9	30.6
	✓	✓	64.8	57.0	47.3	37.4	27.5	17.7	10.5	37.5
CoLA [32]	✓	–	**67.2**	**61.5**	**52.9**	**43.9**	**34.8**	**24.9**	**13.0**	**42.6**
	–	✓	54.4	48.6	40.0	31.0	21.0	12.7	5.6	30.5
	✓	✓	61.6	57.5	47.4	39.6	28.6	21.9	12.0	38.4
CO_2-Net [7]	✓	–	**70.6**	**64.2**	**55.9**	**47.7**	**38.9**	**26.0**	**13.6**	**45.3**
	–	✓	63.6	57.6	47.6	39.1	31.6	19.8	10.2	38.5
	✓	✓	67.1	61.1	51.4	42.6	35.1	23.4	11.7	41.8

5 Conclusion

In this paper, we propose a simple yet effective Regional Similarity Consistency (RSC) mechanism to suppress background information disturbance in RGB features by imposing a consistency constraint via Flow features, and we achieve the state-of-the-art Weakly-supervised Temporal Action Localization (WTAL) performance on the THUMOS14 [9] and ActivityNet1.2 [1] datasets. By explicitly applying the region consistency constraint across RGB and Flow modalities, the proposed method outperforms both the baseline and recent state-of-the-art methods in terms of average mAP on WTAL task. This plug-and-play approach is not only appropriate for WTAL task, but can also be applied to other tasks, such as action classification, action recognition, and so on, to aid improving feature extraction capabilities.

Acknowledgments. This work was supported by Shanghai Municipal Science and Technology Commission with Grant No. 22dz1204900.

References

1. Caba Heilbron, F., Escorcia, V., Ghanem, B., Carlos Niebles, J.: ActivityNet: a large-scale video benchmark for human activity understanding. In: IEEE Computer Vision and Pattern Recognition Conference, pp. 961–970 (2015)
2. Cao, M., Zhang, C., Chen, L., Shou, M.Z., Zou, Y.: Deep motion prior for weakly-supervised temporal action localization. IEEE Trans. Image Process. **31**, 5203–5213 (2022)
3. Carreira, J., Zisserman, A.: Quo vadis, action recognition? a new model and the kinetics dataset. In: IEEE Computer Vision and Pattern Recognition Conference, pp. 6299–6308 (2017)
4. Chao, Y.W., Vijayanarasimhan, S., Seybold, B., Ross, D.A., Deng, J., Sukthankar, R.: Rethinking the faster R-CNN architecture for temporal action localization. In: IEEE Computer Vision and Pattern Recognition Conference, pp. 1130–1139 (2018)

5. Gabeur, V., Sun, C., Alahari, K., Schmid, C.: Multi-modal transformer for video retrieval. In: European Conference on Computer Vision, pp. 214–229 (2020)
6. He, B., Yang, X., Kang, L., Cheng, Z., Zhou, X., Shrivastava, A.: ASM-LOC: action-aware segment modeling for weakly-supervised temporal action localization. In: IEEE Computer Vision and Pattern Recognition Conference, pp. 13925–13935 (2022)
7. Hong, F.T., Feng, J.C., Xu, D., Shan, Y., Zheng, W.S.: Cross-modal consensus network for weakly supervised temporal action localization. In: ACM International Conference on Multimedia, pp. 1591–1599 (2021)
8. Huang, L., Wang, L., Li, H.: Weakly supervised temporal action localization via representative snippet knowledge propagation. In: IEEE Computer Vision and Pattern Recognition Conference, pp. 3272–3281 (2022)
9. Idrees, H., et al.: The thumos challenge on action recognition for videos in the wild. Comput. Vis. Image Underst. **155**, 1–23 (2017)
10. Islam, A., Long, C., Radke, R.: A hybrid attention mechanism for weakly-supervised temporal action localization. In: AAAI Conference on Artificial Intelligence, vol. 35, pp. 1637–1645 (2021)
11. Ji, Y., Jia, X., Lu, H., Ruan, X.: Weakly-supervised temporal action localization via cross-stream collaborative learning. In: ACM International Conference on Multimedia, pp. 853–861 (2021)
12. Kingma, D.P., Ba, J.: Adam: a method for stochastic optimization. In: International Conference on Learning Representations (2015)
13. Lee, J.T., Jain, M., Park, H., Yun, S.: Cross-attentional audio-visual fusion for weakly-supervised action localization. In: International Conference on Learning Representations (2020)
14. Lee, P., Uh, Y., Byun, H.: Background suppression network for weakly-supervised temporal action localization. In: AAAI Conference on Artificial Intelligence, vol. 34, pp. 11320–11327 (2020)
15. Lerman, P.: Fitting segmented regression models by grid search. J. Royal Stat. Soc. **29**(1), 77–84 (1980)
16. Liu, D., Jiang, T., Wang, Y.: Completeness modeling and context separation for weakly supervised temporal action localization. In: IEEE Computer Vision and Pattern Recognition Conference, pp. 1298–1307 (2019)
17. Liu, Z., Wang, L., Zhang, Q., Tang, W., Yuan, J., Zheng, N., Hua, G.: ACSNet: action-context separation network for weakly supervised temporal action localization. In: AAAI Conference on Artificial Intelligence. vol. 35, pp. 2233–2241 (2021)
18. Long, F., Yao, T., Qiu, Z., Tian, X., Luo, J., Mei, T.: Gaussian temporal awareness networks for action localization. In: IEEE Computer Vision and Pattern Recognition Conference, pp. 344–353 (2019)
19. Luo, Z., et al.: Weakly-supervised action localization with expectation-maximization multi-instance learning. In: European Conference on Computer Vision, pp. 729–745 (2020)
20. Ma, F., et al.: SF-net: single-frame supervision for temporal action localization. In: European Conference on Computer Vision, pp. 420–437 (2020)
21. Narayan, S., Cholakkal, H., Khan, F.S., Shao, L.: 3C-Net: category count and center loss for weakly-supervised action localization. In: International Conference on Computer Vision, pp. 8679–8687 (2019)
22. Nguyen, P., Liu, T., Prasad, G., Han, B.: Weakly supervised action localization by sparse temporal pooling network. In: IEEE Computer Vision and Pattern Recognition Conference, pp. 6752–6761 (2018)

23. Paszke, A., et al.: PyTorch: an imperative style, high-performance deep learning library. In: Annual Conference on Neural Information Processing Systems, pp. 8026–8037 (2019)
24. Paul, S., Roy, S., Roy-Chowdhury, A.K.: W-TALC: weakly-supervised temporal activity localization and classification. In: European Conference on Computer Vision, pp. 563–579 (2018)
25. Shou, Z., Wang, D., Chang, S.F.: Temporal action localization in untrimmed videos via multi-stage CNNs. In: IEEE Computer Vision and Pattern Recognition Conference, pp. 1049–1058 (2016)
26. Snidaro, L., Micheloni, C., Chiavedale, C.: Video security for ambient intelligence. IEEE Trans. Syst. Man Cybern. **35**(1), 133–144 (2004)
27. Wang, L., Xiong, Y., Lin, D., Van Gool, L.: UntrimmedNets for weakly supervised action recognition and detection. In: IEEE Computer Vision and Pattern Recognition Conference, pp. 4325–4334 (2017)
28. Xu, Y., et al.: Segregated temporal assembly recurrent networks for weakly supervised multiple action detection. In: AAAI Conference on Artificial Intelligence, vol. 33, pp. 9070–9078 (2019)
29. Yang, W., Zhang, T., Yu, X., Qi, T., Zhang, Y., Wu, F.: Uncertainty guided collaborative training for weakly supervised temporal action detection. In: IEEE Computer Vision and Pattern Recognition Conference, pp. 53–63 (2021)
30. Zach, C., Pock, T., Bischof, H.: A duality based approach for realtime tv-l 1 optical flow. In: Joint Pattern Recognition Symposium, pp. 214–223 (2007)
31. Zhai, Y., Wang, L., Tang, W., Zhang, Q., Yuan, J., Hua, G.: Two-stream consensus network for weakly-supervised temporal action localization. In: European Conference on Computer Vision, pp. 37–54 (2020)
32. Zhang, C., Cao, M., Yang, D., Chen, J., Zou, Y.: CoLA: weakly-supervised temporal action localization with snippet contrastive learning. In: IEEE Computer Vision and Pattern Recognition Conference, pp. 16010–16019 (2021)
33. Zhang, C., et al.: Adversarial seeded sequence growing for weakly-supervised temporal action localization. In: ACM International Conference on Multimedia, pp. 738–746 (2019)
34. Zhao, Y., Xiong, Y., Wang, L., Wu, Z., Tang, X., Lin, D.: Temporal action detection with structured segment networks. In: International Conference on Computer Vision, pp. 2914–2923 (2017)

A Spatio-Temporal Identity Verification Method for Person-Action Instance Search in Movies

Yanrui Niu[1,2,3], Jingyao Yang[1,2,3], Chao Liang[1,2,3](✉), Baojin Huang[1,2,3], and Zhongyuan Wang[1,2,3]

[1] National Engineering Research Center for Multimedia Software (NERCMS), Wuhan, China
[2] Hubei Key Laboratory of Multimedia and Network Communication Engineering, Wuhan, China
[3] School of Computer Science, Wuhan University, Wuhan, China
cliang@whu.edu.cn

Abstract. As one of the challenging problems in video search, Person-Action Instance Search (P-A INS) aims to retrieve shots with a specific person carrying out a specific action from massive amounts of video shots. Most existing methods conduct person INS and action INS separately to compute the initial person and action ranking scores, which will be directly fused to generate the final ranking list. However, direct aggregation of two individual INS scores ignores spatial relationships of person and action, thus cannot guarantee their identity consistency and cause identity inconsistency problem (IIP). To address IIP, we propose a simple spatio-temporal identity verification method. Specifically, in the spatial dimension, we propose an identity consistency verification (ICV) step to revise the direct fusion score of person INS and action INS. Moreover, in the temporal dimension, we propose a double-temporal extension (DTE) operation to further improve P-A INS results. The proposed method is evaluated on the large-scale NIST TRECVID INS 2019–2021 tasks, and the experimental results show that it can effectively mitigate the IIP, and its performance surpasses that of the champion team in 2019 INS task and the second place teams in both 2020 and 2021 INS tasks.

Keywords: Person instance search · Action instance search · Double-temporal extension · Identity consistency verification

1 Introduction

With the rapid development of multimedia technology in recent years, videos have flooded our life. Finding specific targets from massive amounts of videos,

Y. Niu and J. Yang—These authors contribute equally to this work.

Supplementary Information The online version contains supplementary material available at https://doi.org/10.1007/978-3-031-27077-2_7.

D.-T. Dang-Nguyen et al. (Eds.): MMM 2023, LNCS 13833, pp. 82–94, 2023.
https://doi.org/10.1007/978-3-031-27077-2_7

i.e., video instance search (INS), is becoming increasingly important, and movies are suitable breeding grounds for it. Early INS research in movies mainly focuses on a single target, *i.e.*, single concept INS, such as finding a specific object [15,21], person [13,25,29], or action [7,17,26]. Recently, researchers started to investigate the more challenging combinatorial-semantic INS, which aims at retrieving specific instances with multiple attributes simultaneously. Representative works in this field include Person-Scene (P-S INS) [1,2,6] and Person-Action (P-A INS) [3–5]. The former aims at finding shots about the specific person in a specific scene, while the latter aims at finding shots about the specific person doing a specific action. In this paper we study the P-A INS in movies.

(a) "Bradley is carrying bag"

(b) "Pat is sitting on the couch"

Fig. 1. Examples of IIP in P-A INS. The blue and green boxes mark the target person and action, respectively. (Color figure online)

Existing methods [16,23,32] often adopt two different technical branches for person INS and action INS. Specifically, in the person INS branch, face detection and identification are conducted to compute ranking scores of video shots concerning the target person. In the action INS branch, the action recognition is conducted to compute ranking scores of video shots about the target action. Thereafter, two-branch INS scores are directly fused to generate the final ranking result. However, direct aggregation of scores cannot guarantee the identity consistency between person and action. For example, in Fig. 1(a), given "Bradley is standing" and "Danielle is carrying bag", the system [14] mistakes it as "Bradley is carrying bag" since the person "Bradley" and action "carrying bag" appear simultaneously; similar case happens in Fig. 1(b). We call it *identity inconsistency problem* (IIP).

To address the above problem, we propose a spatio-temporal identity verification method. In spatial dimension, we propose an identity consistency verification (ICV) scheme to compute the spatial consistency degree between face and action detection results. The higher spatial consistency degree means the larger overlapping area between the bounding boxes of face and action, thus the more likely that face and action belong to the same person. Furthermore, we find many face and action detection failures due to complex scenarios, such as non-frontal filming or object occlusion, hindering ICV from getting basic detection information. Considering the continuity of video frames in a shot and temporal continuity of some actions in adjacent shots, we propose a double-temporal extension (DTE) operation in the temporal dimension. The detection information of the interval frames is shared with intermediate frames through intra-shot DTE, and the fusion scores of adjacent shots are adjusted by inter-shot DTE.

The main contributions of this paper are as follows:

- We discover and study the IIP in the combinatorial P-A INS of movies. It shows that direct aggregation of single concept INS scores cannot always guarantee the identity consistency between person and action, which leads to the degraded performance of previous works.
- We propose a spatio-temporal identity verification method to address IIP, which uses ICV in the spatial dimension to check identity consistency between person and action, and DTE in the temporal dimension to share the detection information in successive frames in a shot and transferring the score information among adjacent shots.
- We verify the effectiveness of the proposed method on the large-scale TRECVID INS dataset. The performance surpasses the champion team in the 2019 INS task and the second place teams in both 2020 and 2021 INS tasks.

2 Related Work

2.1 Person INS

Person INS in videos aims to find shots containing a specific person from a video gallery, which is also termed as person re-identification. Most of the previous research working on person re-identification mainly focus on surveillance videos, where dresses rather than faces are more robust for identity discrimination [9, 31,35]. But in movies, due to massive amounts of close-up shots and frequent clothing changes, faces are more stable than dresses for person re-identification. Therefore, most of the existing works in movies mainly use face detection and face recognition algorithms for person INS [13,25,29].

2.2 Action INS

Existing research on action INS mainly relies on action recognition or detection technology [14,16,22,23,32]. The difference between them is that the former only recognizes the category of action, whereas the latter can provide the location bounding boxes of action, and we focus on action detection. According to different implementation strategies, action detection can be generally divided into image-based and video-based methods. The former is mainly designed for actions with obvious interactive objects but without rigorous temporal causality, *e.g.*, "holding glass" and "carrying bag". This corresponds to a specialized action detection task, *i.e.*, human-object interaction (HOI) detection [19,20,24], which aims to recognize the action (interaction) category, and meanwhile, locate human and object bounding boxes from images. The latter targets actions with rigorous temporal causality. *e.g.*, "open the door and enter" and "go up/down stairs". Hence, it usually works on successive multiple video frames, and representative methods are [27,28,30].

2.3 Fusion Strategy

For combinatorial-semantic P-A INS, the difficulty lies in how to combine the results of different branches. Most of the existing studies adopt a strategy of retrieving two instances separately and then aggregating individual scores in some ways [14,16,22,23,32]. For example, NII fuses scores of person INS and action INS by direct weighted summation [16]. Instead, WHU adopts a stepwise strategy of searching for the action based on a candidate person list. It first builds an initial candidate person shot list with person INS scores, then sorts the list according to scores of action INS [14,32]. PKU adopts a strategy of searching for the person based on a candidate action list [22,23]. However, direct aggregation of person INS and action INS results without checking their identity consistency may incur serious IIP. To solve this problem, Le *et al.* [18] raises a heuristic method. They calculate the distance between the target face and desired object, and assume that the shorter distance means a more positive relationship between person and action with the desired object. The method indirectly judges the identity consistency by the distance between the related object and the target face, but can not sufficiently prove the identity consistency of the target face and

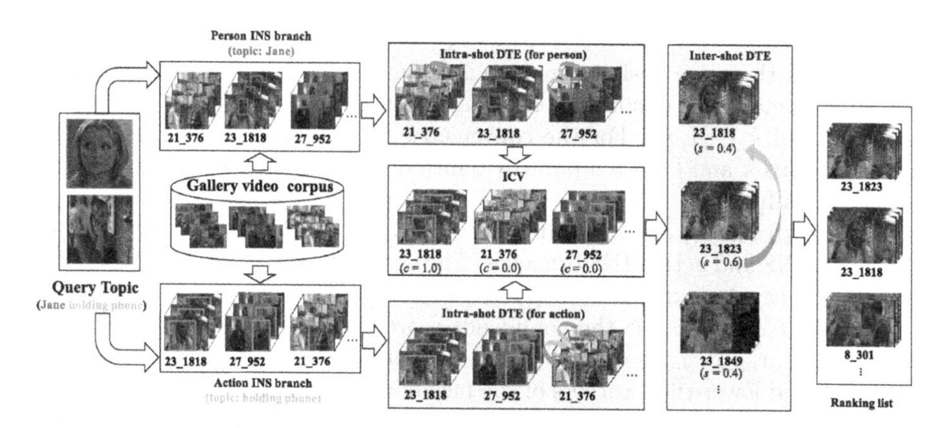

Fig. 2. The overall scheme of the spatio-temporal method for P-A INS. First, Person INS and action INS are conducted. Then intra-shot DTE recovers face and action detection information in the keyframes with detection failure. After that, ICV is conducted to filter out IIP shots. Then inter-shot DTE is used to adjust the final ranking of shots. At last, the ranking list is obtained by sorting all shots' scores. The yellow dotted boxes represents the recovered boxes, the orange arrows represent the directions of interpolation operation, c means the consistency degree of person and action, and s means shot score. (Color figure online)

specific action. Moreover, it works based on object detection, which means that it does not work for actions without obvious interactive objects, *e.g.*, "walking" and "standing".

Different from [18], we propose a spatio-temporal identity verification method for P-A INS, which can determine the identity consistency of the P-A pair without additional dependence on objects. Hence, it can be applied to both HOI and object-free actions.

3 Method

The overall scheme of our method is shown in Fig. 2. Given a topic and a video corpus, uniform sampling at an interval of 5 frames is first carried out to extract representative keyframes from shots of the video corpus. Then, person INS and action INS are conducted. Note that we apply detection in INS branches so we can obtain face/action detection scores as well as bounding boxes. Next, in the temporal dimension, intra-shot DTE is firstly conducted on failed detection shots, providing more detection information for ICV. Thereafter, in the spatial dimension, the ICV method is applied to check identity consistency between person and action, which filters out erroneous IIP shots. Finally, the maximum fusion score of all keyframes in a shot is taken as the INS score of the shot, and the inter-shot DTE is conducted to adjust the scores of shots, then the ranking list is obtained by sorting INS scores of all shots.

3.1 Preliminary

Assume that there are L shots in video gallery. For the l-th shot, K keyframes can be extracted. We denote the k-th keyframe in the shot l as $P^{(l,k)}$, where $l \in [1, L]$ and $k \in [1, K]$. For the convenience of the following discussion, the subscript signs k and l are temporarily omitted from all variables when they do not cause confusion.

For a keyframe P, assume that there are m faces and n actions detected in the person INS and action INS branches. The detection and identification results of i-th face can be expressed as $\langle ID_i, Conf_i, Box_i \rangle_{i=1}^{m}$, where ID_i represents the face id, $Conf_i$ records the confidence score of face identification, $Box_i = \langle x_{min_i}, y_{min_i}, x_{max_i}, y_{max_i} \rangle$ contains the horizontal and vertical coordinates of upper-left and lower-right corners of the face bounding box. Similarly, the result of j-th action can be expressed as $\langle ID_j, Conf_j, Box_j \rangle_{j=1}^{n}$, with similar notation definitions.

3.2 Identity Consistency Verification (ICV)

In order to address the IIP, we propose ICV to verify the identity consistency between person and action in the spatial dimension.

Specifically, for a keyframe P, we calculate spatial consistency degree matrix $\mathbf{C} = [c_{i,j}] \in \mathbb{R}^{m \times n}$ based on face and action bounding boxes obtained from person and action INS branches, in which $c_{i,j}$ is defined as:

$$c_{i,j} = \frac{\mathbf{Intersection}\left(Box_i^{\text{face}}, Box_j^{\text{action}}\right)}{\mathbf{Area}\left(Box_i^{\text{face}}\right)}, \tag{1}$$

where $\mathbf{Intersection}(\cdot,\cdot)$ is the function of computing the intersection area of two bounding boxes, $\mathbf{Area}(\cdot)$ is the function of computing the area of a bounding box.

Next, the proposed spatial consistency degree is applied to optimize the fusion score. Two representative fusion strategies are adopted.

- One simple strategy is the weighted fusion method ($Fusion_{wet}$) [14,16,32], which can be optimized as:

$$s_{i,j} = c_{i,j} \times \left[\alpha \times Conf_i^{\text{face}} + (1-\alpha) \times Conf_j^{\text{action}}\right], \tag{2}$$

 where $s_{i,j}$ stands for the fusion score of the i-th person and the j-th action, $\alpha \in [0,1]$ is the fusion coefficient, which is a hyperparameter.
- The other effective fusion strategy, *i.e.*, searching for the specific action based on a candidate person list ($Fusion_{thd}$), is widely used [14,22,23,32]. It can be improved by the proposed spatial consistency degree as:

$$s_{i,j} = c_{i,j} \times \left[\mathbf{F}_\delta\left(Conf_i^{\text{face}}\right) \times Conf_j^{\text{action}}\right], \tag{3}$$

 where $\mathbf{F}_\delta(\cdot)$ is a threshold function, δ is the threshold for face scores to determine whether the target person exists in the keyframes, *i.e.*, $\mathbf{F}_\delta(x) = 1$ if $x \geq \delta$, otherwise 0.

3.3 Double-Temporal Extension (DTE)

To address the detection failure problem caused by complex filming conditions, we propose DTE to transfer the information in the temporal dimension, which includes intra-shot DTE and inter-shot DTE.

intra-shot DTE shares the detection information among keyframes. We conduct the intra-shot DTE to recover face and action detection information in the keyframes with detection failure by linear interpolation. The shared detection information including confidence scores and coordinates of detection bounding boxes.

Inter-shot DTE shares the score information among shots. Because some actions have time continuity and can last more than one shot, the same query may appear in adjacent shots. Therefore, we adjust the final ranking of shots by diffusing the fusion scores of adjacent shots. The Gaussian curve is used to guide the score diffusion between shots with different distances:

$$\hat{s}_{i,j}^l = s_{i,j}^l + \sum_{-\gamma \leq d \leq \gamma} F_{dis}(d) \times \mathbf{max}\left(s_{i,j}^{l+d} - s_{i,j}^l, 0\right), \tag{4}$$

$$\boldsymbol{F}_{dis}(d) = \theta \cdot \frac{1}{\sqrt{2\pi}\sigma} exp\left(-\frac{d^2}{2\sigma^2}\right), \tag{5}$$

where $\hat{s}^l_{i,j}$ is the revised fusion score of i-th person conducting j-th action in the l-th shot after inter-shot DTE, $s^l_{i,j}$ is original fusion score, d is the distance between two shots, $\mathbf{max}(\cdot,\cdot)$ is used to limit diffusion direction, and $\boldsymbol{F}_{dis}(\cdot)$ is a distance based weight function, which decreases with the increase of shot distance. θ is used to adjust the contribution of distance, and σ is used to adjust the range of score diffusion, which determines the value of γ ($\gamma \approx 3 \cdot \sigma$).

3.4 Generating Ranking List

After obtaining fusion scores of all keyframes, the fusion score of the i-th person conducting the j-th action in l-th shot is the maximum score of keyframes in the shot:

$$s^l_{i,j} = \max_{k=1,\cdots,K} s^{(l,k)}_{i,j}. \tag{6}$$

Based on the fusion scores of all shots, we perform an inter-shot DTE in Sect. 3.3 to obtain the revised fusion scores. Then the ranking list concerning the topic of the i-th person conducting the j-th action is obtained by sorting the revised fusion scores of all shots. The complete flowchart of the proposed spatio-temporal identity verification method is presented in Algorithm D.1 in the supplementary material.

4 Experiments

4.1 Dataset and Evaluation Criteria

The TRECVID INS Dataset [5] comes from the 464-hour BBC soap opera "EastEnders", which is divided into 471,527 shots, containing about 7.84 million keyframes. NIST selects 70 topics based on it as representative samples for TRECVID 2019–2021 INS tasks [3–5]. The details of the dataset and topics are presented in Table A.1 and Table B.1-B.3 in the supplementary material.

According to the official evaluation criteria of TRECVID, Average Precision (AP) is adopted to evaluate the retrieval quality of each topic, and mean AP (mAP) is used to describe the overall performance among the given set of P-A INS topics. For each topic, only 1,000 shots at most can be evaluated.

4.2 Implementation Details

Person INS Branch. We adopt the RetinaFace detector [10] trained on the WIDER FACE [33] to obtain the face detection bounding boxes for each keyframe. and utilize the ArcFace [11] trained on the MS1Mv2 [11] to extracted 512-dimension features from normalized face images based on the detected face bounding boxes. Cosine similarity is used to calculate the face scores.

Action INS Branch. In the action INS branch, we especially apply two different action detection methods, *i.e.*, HOI detection on images and action detection on videos, according to different action characteristics. For topics with actions with obvious objects, we adopt PPDM [20] pre-trained on HICO-DET [8] (heatmap prediction network is DLA-34 [34]) to conduct HOI detection on images. For topics with actions lasting for a long time, we adopt ACAM [28] to conduct action detection on videos, which is trained on the AVA dataset [12].

Fusion Strategy. We test the effect of the parameters α and δ of two fusion methods, *i.e.*, $Fusion_{wet}$ and $Fusion_{thd}$, and compare the best performance of them. As shown in Figure C.1 in the supplementary material, $Fusion_{thd}$ is better than $Fusion_{wet}$, so we choose $Fusion_{thd}$ in the baseline model.

Double-Temporal Extension. We get the best parameters (refer to Figure C.2 in the supplementary material), where $\theta = 3$ and $\sigma = 5$.

4.3 Ablation Study

In this section, we evaluate the effectiveness of DTE and ICV on the NIST TRECVID 2019–2021 INS tasks.

Table 1. Ablation study results on NIST TRECVID 2019–2021 INS tasks. The black bold values mark the best value for each column (%).

Method			2019			2020			2021		
Base	*DTE*	*ICV*	P-A$_i$	P- A$_v$	P- A	P-A$_i$	P-A$_v$	P-A	P-A$_i$	P-A$_v$	P-A
✓			29.34	3.94	20.51	35.16	4.37	21.69	39.63	8.69	30.35
✓	✓		31.68	4.12	22.10	37.31	5.49	23.39	43.33	8.59	32.91
✓		✓	31.57	4.48	22.15	37.09	5.52	23.28	43.36	**9.54**	33.21
✓	✓	✓	**34.13**	**4.73**	**23.90**	**39.28**	**6.89**	**25.11**	**47.70**	9.27	**36.17**

Base. We construct a baseline model referred to as Base by eliminating all proposed methods. Specifically, in the Base model, the face and action scores are fused with $Fusion_{thd}$ to get scores of keyframes. Thereafter, the maximum score of keyframes is taken as the shot score. Finally, the ranking list is obtained by sorting the shot scores for each topic.

Then we add DTE and ICV gradually to Base. Note that we have two P-A INS combination methods since we adopt two action detection methods in the action INS branch, *i.e.*, image-based P-A INS (P-A$_\text{i}$ INS) and video-based P-A INS (P-A$_\text{v}$ INS). Table 1 shows ablation study results in 2019–2021 INS tasks. The mAP of topics corresponding to P-A$_\text{i}$ and P-A$_\text{v}$ columns are computed respectively, and the mAP of all topics is shown in the final P-A column.

Evaluation of DTE. We add DTE to Base, referred to as Base+DTE. In 2019 INS task, Base+DTE gains 1.59% (7.75% relative growth) improvement over the Base method. Similarly, in 2020 and 2021 INS tasks, the improvements are 1.70% (7.84% relative growth) and 2.56% (8.43% relative growth), which confirms the effectiveness of DTE.

Evaluation of ICV. We add ICV to Base, referred to as Base+ICV, which gains 1.64% (8.00% relative growth), 1.59% (7.33% relative growth), and 2.86% (9.42% relative growth) improvements over Base in 2019–2021 INS tasks respectively, confirming the effectiveness of ICV.

Evaluation of DTE and ICV. Furthermore, Base+DTE+ICV achieves the best performance in both experiments, which gains 3.39% (16.53% relative growth), 3.42% (15.77% relative growth), and 5.82% (19.18% relative growth) improvements over Base in 2019–2021 INS tasks. It can be seen that with the proposed method, the mAPs of P-A_i INS and P-A_v INS both improve, which proves the effectiveness of the proposed method is consistent, and the method works for both P-A INS branches on images and videos.

The visualization results of DTE and ICV are shown in Figure E.1 and E.2 in the supplementary material.

4.4 Comparison with Other Methods

We compare the proposed method with state-of-the-art methods on NIST TREC- VID 2019–2021 INS tasks. According to the official evaluation settings, each team is allowed to submit several runs for evaluation, and we select the best run of each top-3 team for comparison, whose details are shown in Section F in the supplementary material.

Figure 3 demonstrates the comparative results of our method and previous evaluation runs. As shown in Fig. 3(a), our method achieves the best performance on 10 topics and competitive performance on 9 topics. In Fig. 3(b), our method achieves 7 best and 4 competitive performance. And in Fig. 3(c), our method achieves 5 best and 11 competitive performance. The performance is relatively poor on other topics. Through observing the results of three-year INS tasks, we find that the reason for those relatively poor-performance topics is due to detection errors of some difficult action topics. For example, the actions in topics *9268, 9278, 9315* and *9316* are all "go up or down stairs", the actions in topics *9267, 9277, 9335* and *9336* are all "open the door and enter/leave", and the actions in topics *9306, 9307, 9337* and *9338* are all "holding cloth". It can be seen that the difficulty of action INS is an important factor limiting the performance of P-A INS. In general, we propose a simple INS method, compared with other methods with many tricks, our method still gets considerable performance. The mAP of our methods surpassed the state-of-the-art in 2019 INS task and the best runs of second place in 2020–2021 INS tasks.

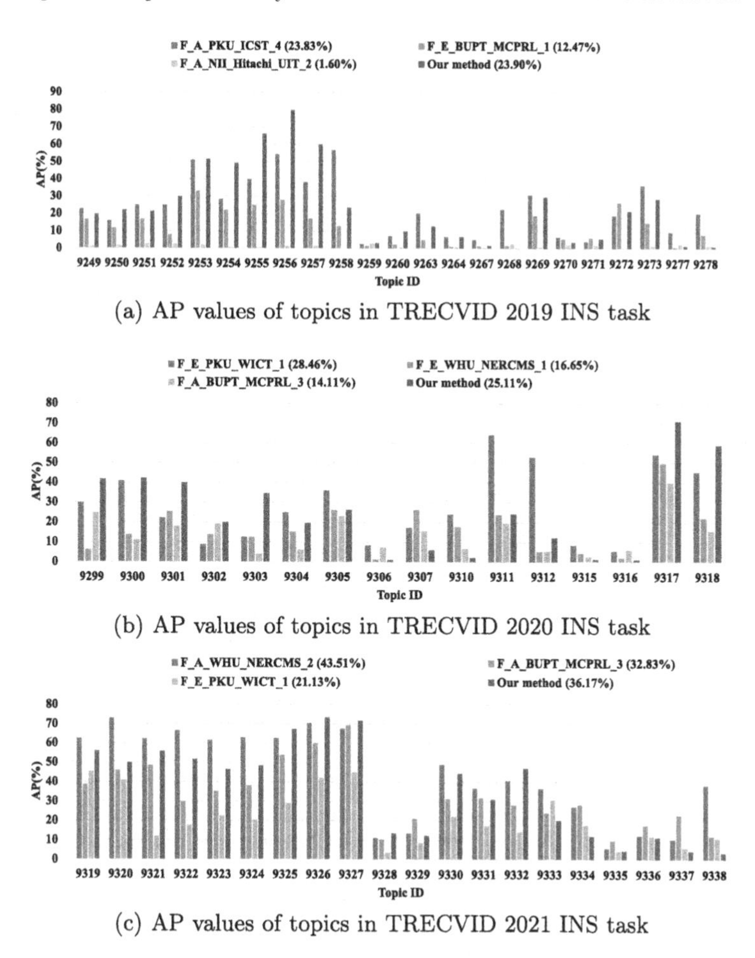

(a) AP values of topics in TRECVID 2019 INS task

(b) AP values of topics in TRECVID 2020 INS task

(c) AP values of topics in TRECVID 2021 INS task

Fig. 3. Comparisons with other P-A INS methods. The legend shows the mAP values. Blue, green and orange represents the best run of the first place, second place and third place team of INS 2019–2021 tasks, while red represents our method. (Color figure online)

5 Conclusion

We study the IIP between person and action in P-A INS in movies, and propose a simple but effective spatio-temporal identity verification method. The experimental results of our method on the large-scale TRECVID INS dataset verify its effectiveness and robustness. In the future, we will concentrate on improving the accuracy of identity verification by trying more methods, such as using other appearance-based features within the bounding boxes to infer identity consistency, or using human posture information to locate the face position in the action bounding boxes. And we will extend our method to more combinatorial-semantic INS tasks, e.g., the Person-Action-Scene INS.

Acknowledgement. This work is supported by the National Natural Science Foundation of China (No. U1903214, 61876135). The numerical calculations in this paper have been done on the supercomputing system in the Supercomputing Center of Wuhan University.

References

1. Awad, G., et al.: TRECVID 2018: Benchmarking video activity detection, video captioning and matching, video storytelling linking and video search. In: Proceedings of TRECVID 2018 (2018)
2. Awad, G., et al.: TRECVID 2017: evaluating ad-hoc and instance video search, events detection, video captioning, and hyperlinking. In: TREC Video Retrieval Evaluation (TRECVID) (2017)
3. Awad, G., et al.: TRECVID 2020: a comprehensive campaign for evaluating video retrieval tasks across multiple application domains. In: Proceedings of TRECVID 2020 (2020)
4. Awad, G., et al.: Evaluating multiple video understanding and retrieval tasks at TRECVID 2021. In: Proceedings of TRECVID 2021 (2021)
5. Awad, G., et al.: Trecvid 2019: an evaluation campaign to benchmark video activity detection, video captioning and matching, and video search retrieval. In: Proceedings of TRECVID 2019 (2019)
6. Awad, G., et al.: TRECVID 2016: evaluating video search, video event detection, localization, and hyperlinking. In: TREC Video Retrieval Evaluation (TRECVID) (2016)
7. Bojanowski, P., Bach, F., Laptev, I., Ponce, J., Schmid, C., Sivic, J.: Finding actors and actions in movies. In: Proceedings of the IEEE International Conference on Computer Vision (ICCV), pp. 2280–2287 (2013). https://doi.org/10.1109/ICCV.2013.283
8. Chao, Y.W., Liu, Y., Liu, X., Zeng, H., Deng, J.: Learning to detect human-object interactions. In: 2018 IEEE Winter Conference on Applications of Computer Vision (WACV), pp. 381–389 (2018). https://doi.org/10.1109/WACV.2018.00048
9. Chen, L., Yang, H., Xu, Q., Gao, Z.: Harmonious attention network for person re-identification via complementarity between groups and individuals. Neurocomputing **453**, 766–776 (2021). https://doi.org/10.1016/j.neucom.2020.07.118
10. Deng, J., Guo, J., Ververas, E., Kotsia, I., Zafeiriou, S.: Retinaface: single-shot multi-level face localisation in the wild. In: Proceedings of the IEEE/CVF Conference on Computer Vision and Pattern Recognition (CVPR) (2020). https://doi.org/10.1109/CVPR42600.2020.00525
11. Deng, J., Guo, J., Xue, N., Zafeiriou, S.: Arcface: additive angular margin loss for deep face recognition. In: Proceedings of the IEEE/CVF Conference on Computer Vision and Pattern Recognition (CVPR) (2019). https://doi.org/10.1109/CVPR.2019.00482
12. Gu, C., et al.: AVA: a video dataset of spatio-temporally localized atomic visual actions. In: Proceedings of the IEEE Conference on Computer Vision and Pattern Recognition (CVPR). pp. 6047–6056 (2018). https://doi.org/10.1109/CVPR.2018.00633
13. Haq, I.U., Muhammad, K., Ullah, A., Baik, S.W.: Deepstar: Detecting starring characters in movies. IEEE Access **7**, 9265–9272 (2019). https://doi.org/10.1109/ACCESS.2018.2890560

14. Jiang, L., et al.: Whu-nercms at trecvid 2019: Instance search task. In: Proceedings of TRECVID Workshop (2019). https://www-nlpir.nist.gov/projects/tvpubs/tv19.papers/whu_nercms.pdf
15. Jiang, W., Wu, Y., Jing, C., Yu, T., Jia, Y.: Unsupervised deep quantization for object instance search. Neurocomputing **362**, 60–71 (2019). https://doi.org/10.1016/j.neucom.2019.06.088
16. Klinkigt, M., et al.: Nii hitachi uit at trecvid 2019. In: Proceedings of TRECVID Workshop (2019). https://www-nlpir.nist.gov/projects/tvpubs/tv19.papers/nii_hitachi_uit.pdf
17. Laptev, I., Perez, P.: Retrieving actions in movies. In: 2007 IEEE 11th International Conference on Computer Vision (ICCV), pp. 1–8 (2007). https://doi.org/10.1109/ICCV.2007.4409105
18. Le, D.D., et al.: Nii_uit at trecvid 2020. In: Proceedings of TRECVID Workshop (2020). https://www-nlpir.nist.gov/projects/tvpubs/tv20.papers/nii_uit.pdf
19. Li, Y.L., et al.: Transferable interactiveness knowledge for human-object interaction detection. In: Proceedings of the IEEE/CVF Conference on Computer Vision and Pattern Recognition (CVPR), pp. 3585–3594 (2019). https://doi.org/10.1109/CVPR.2019.00370
20. Liao, Y., Liu, S., Wang, F., Chen, Y., Qian, C., Feng, J.: PPDM: Parallel point detection and matching for real-time human-object interaction detection. In: Proceedings of the IEEE/CVF Conference on Computer Vision and Pattern Recognition (CVPR), pp. 482–490 (2020). https://doi.org/10.1109/CVPR42600.2020.00056
21. Meng, J., Yuan, J., Yang, J., Wang, G., Tan, Y.P.: Object instance search in videos via spatio-temporal trajectory discovery. IEEE Trans. Multimedia **18**(1), 116–127 (2016). https://doi.org/10.1109/TMM.2015.2500734
22. Peng, Y., et al.: PKU-ICST at TRECVID 2019: Instance search task. In: Proceedings of TRECVID Workshop (2019). https://www-nlpir.nist.gov/projects/tvpubs/tv19.papers/pku-icst.pdf
23. Peng, Y., Ye, Z., Zhang, J., Sun, H.: PKU WICT at TRECVID 2020: Instance search task. In: Proceedings of TRECVID Workshop (2020). https://www-nlpir.nist.gov/projects/tvpubs/tv20.papers/pku-wict.pdf
24. Qi, S., Wang, W., Jia, B., Shen, J., Zhu, S.C.: Learning human-object interactions by graph parsing neural networks. In: Proceedings of the European Conference on Computer Vision (ECCV), pp. 401–417 (2018). https://doi.org/10.1007/978-3-030-01240-3_25
25. Kumar, N., Du, V., Doja, M.N., Shambharkar, P., Nimesh, U.K.: Automatic Face Recognition and Finding Occurrence of Actors in Movies. In: Ranganathan, G., Chen, J., Rocha, Álvaro. (eds.) Inventive Communication and Computational Technologies. LNNS, vol. 145, pp. 115–129. Springer, Singapore (2021). https://doi.org/10.1007/978-981-15-7345-3_10
26. Stoian, A., Ferecatu, M., Benois-Pineau, J., Crucianu, M.: Fast action localization in large-scale video archives. In: IEEE Trans. Cir. and Sys. for Video Technol. **26**(10), 1917–1930 (2016). https://doi.org/10.1109/TCSVT.2015.2475835
27. Tang, J., Xia, J., Mu, X., Pang, B., Lu, C.: Asynchronous Interaction Aggregation for Action Detection. In: Vedaldi, A., Bischof, H., Brox, T., Frahm, J.-M. (eds.) ECCV 2020. LNCS, vol. 12360, pp. 71–87. Springer, Cham (2020). https://doi.org/10.1007/978-3-030-58555-6_5

28. Ulutan, O., Rallapalli, S., Srivatsa, M., Torres, C., Manjunath, B.S.: Actor conditioned attention maps for video action detection. In: Proceedings of the IEEE/CVF Winter Conference on Applications of Computer Vision (WACV), pp. 527–536 (2020). https://doi.org/10.1109/WACV45572.2020.9093617
29. Wang, X., Liu, W., Chen, J., Wang, X., Yan, C., Mei, T.: Listen, look, and find the one: robust person search with multimodality index. ACM Trans. Multimedia Comput. Commun. Appl. 16(2) (2020). https://doi.org/10.1145/3380549
30. Wu, C.Y., Feichtenhofer, C., Fan, H., He, K., Krahenbuhl, P., Girshick, R.: Long-term feature banks for detailed video understanding. In: Proceedings of the IEEE/CVF Conference on Computer Vision and Pattern Recognition (CVPR), pp. 284–293 (2019). https://doi.org/10.1109/CVPR.2019.00037
31. Yang, F., Yan, K., Lu, S., Jia, H., Xie, D., Yu, Z., Guo, X., Huang, F., Gao, W.: Part-aware progressive unsupervised domain adaptation for person re-identification. IEEE Trans. Multimedia **23**, 1681–1695 (2021). https://doi.org/10.1109/TMM.2020.3001522
32. Yang, J., Kang'an Chen, Y.N., Fan, X., Liang, C.: WHU-NERCMS at TRECVID 2020: Instance search task. In: Proceedings of TRECVID Workshop (2020). https://www-nlpir.nist.gov/projects/tvpubs/tv20.papers/whu_nercms.pdf
33. Yang, S., Luo, P., Loy, C.C., Tang, X.: Wider face: A face detection benchmark. In: Proceedings of the IEEE Conference on Computer Vision and Pattern Recognition (CVPR), pp. 5525–5533 (2016). https://doi.org/10.1109/CVPR.2016.596
34. Yu, F., Wang, D., Shelhamer, E., Darrell, T.: Deep layer aggregation. In: Proceedings of the IEEE Conference on Computer Vision and Pattern Recognition (CVPR), pp. 2403–2412 (2018). https://doi.org/10.1109/CVPR.2018.00255
35. Zhang, W., Wei, Z., Huang, L., Xie, K., Qin, Q.: Adaptive attention-aware network for unsupervised person re-identification. Neurocomputing **411**, 20–31 (2020). https://doi.org/10.1016/j.neucom.2020.05.094

Binary Neural Network for Video Action Recognition

Hongfeng Han[1,3], Zhiwu Lu[2,3](✉), and Ji-Rong Wen[2,3]

[1] School of Information, Renmin University of China, Beijing, China
hanhongfeng@ruc.edu.cn
[2] Gaoling School of Artificial Intelligence, Renmin University of China, Beijing, China
{luzhiwu,jrwen}@ruc.edu.cn
[3] Beijing Key Laboratory of Big Data Management and Analysis Methods, Beijing, China

Abstract. In the typical video action classification scenario, it is critical to extract the temporal-spatial information in the videos with complex 3D convolution neural networks, which significantly expand both the computation cost and memory costs. In this paper, we propose a novel 1-bit 3D convolution block named 3D BitConv, capable of compressing the 3D convolutional networks while maintaining a high level of accuracy. Due to its high flexibility, the proposed 3D BitConv block can be directly embedded in the latest 3D convolutional backbone. Consequently, we binarize two representative 3D convolutional neural networks (C3D and ResNet3D) and validate the accuracy of action recognition tasks. The results of two widely used datasets demonstrate that the two proposed binary 3D networks achieve impressive performance at a lower cost. Furthermore, we carry out an extensive ablation analysis to test and verify the efficacy of the components in 3D BitConv.

Keywords: Binary neural network · Action recognition · 1-bit 3D convolution

1 Introduction

Video action recognition is a traditional challenge with deep networks, which aims to understand human actions and recognize the action of a subject in a short-cropped video. The development of the task is highly dependent on 3D CNNs whicht can capture spatio-temporal features. So, taking 3D CNN as a backbone achieves a good performance level of action recognition [7]. However, 3D CNN leads to significant computation and memory consumption for multi-layer network architectures and convolution operations. Hence, the action recognition task relies on computationally intensive hardware that can not be applied to resource-constrained mobile devices that heavily depend on high-performance hardware. It is a tremendous challenge with limited parameters and deploying

D.-T. Dang-Nguyen et al. (Eds.): MMM 2023, LNCS 13833, pp. 95–106, 2023.
https://doi.org/10.1007/978-3-031-27077-2_8

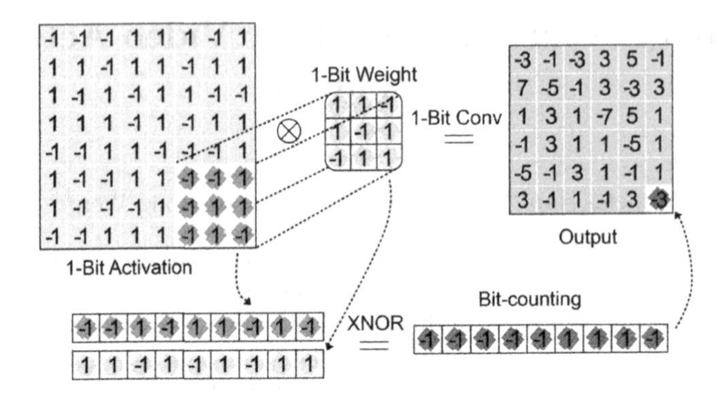

Fig. 1. The process of XNOR and Bit-counting from a 1-bit Convolutional Neural Network. Note that the convolution operations are simplified to XNOR and Bit-counting operations, which greatly reduces the memory footprint.

them in resource-limited environments [9,27]. To further deal with the problem, we use Binary Neural Network (BNN) to reduce computation and memory costs.

BNN [3] is first proposed to provide a model compression method and solve the challenge of too large parameter operation. The bit operation in the network replaces the multiplication and addition operations [21]. Figure 1 shows the whole process of 1-bit convolutional neural network. The Activation and Weight are also binarized to +1 or 1, and the convolution operations are simplified to XNOR and Bit-counting in 1-bit convolutional neural network. The 1-bit neural network retains the same structure of the original network design; all that it does is binarized the network's parameters and activation values. Specifically, 1-bit neural network achieves a theoretical 32 times storage space optimization (e.g. from float32 to 1bit) and 64 times speedup effect.

Recently, BNN has made continuous progress and achieved better performance [2,4,15]. Earlier researchers propose XNOR-Networks [21] that both the convolution layer's input and the convolution kernel's weights are binarized to +1 or -1. Bi-real Net [16] uses a shortcut and improves the 1-bit CNN training method. However, as far as we know, the current BNN is based on 1-bit 2D CNN for image classification tasks. For the first time, we propose BNN for action recognition and design a 1-bit 3D convolution model.

Therefore, the recent progress of BNN inspired us to employ binary networks to reduce the computation cost for the action recognition task. However, there are few binary works in 3D convolution networks because of the difficulty of preserving the accuracy of the binary 3D convolution operations. To alleviate the problem of high computing cost in 3D CNN and maintain high accuracy, we propose a novel 1-bit 3D convolution block (3D BitConv) to binarize CNNs. As we can see from Fig. 2, the bottom part shows that the proposed 3D BitConv consists of a designed binary 3D convolution, three learnable biases, and binary activation. This paper leverages the proposed 3D BitConv to binarize two representative 3D CNNs, C3D and ResNet3D, to validate the effectiveness of the 3D BitConv for video action recognition challenges.

As a result, our binary 3D CNNs can significantly improve the speed and computation cost for the action recognition task, which has essential application significance. As an example, we illustrate the binary C3D approach's architecture in Fig. 2 and binarized by the 3D BitConv. The input of the binary C3D is the clip sampled from the original video. Note that the first convolution module (Conv 1a) in binary C3D remains a common 3D convolution operation for initial feature extraction. We perform 1-bit 3D convolution from the second convolution module (Conv 2b) instead of binarizing all 3D convolution modules with corresponding 3D BitConv for better representation.

In conclusion, this work makes three primary contributions:

(1) We propose a novel binary neural network and design a 1-Bit 3D convolution (3D BitConv) model for the first time to alleviate the problem of high computing cost for video action recognition.
(2) We implement the binarization of two commonly used 3D CNNs: C3D and ResNet3D, and successfully employ them to complete the action recognition.
(3) The performance of extensive testing has demonstrated that the proposed 3D binary convolution greatly reduces computation costs while maintaining a high accuracy on two public action recognition benchmarks.

2 Related Work

2.1 Binary Neural Network

For a higher model compression rate and a faster processing speed, the binary neural network (BNN) has emerged as a prominent avenue of study in recent years. Many researchers have explored new network structures and training techniques. XNOR-NET [21] was proposed to binarize both weights and activations, further reducing quantization error [14], and an ApproxSign function was introduced [16] for gradient calculation. Notable results were achieved by applying knowledge distillation to the process of learning 8-bit networks utilizing full-precision models [25], and Li et al. [13] further designed a fully-quantized network. Unlike these methods that focused on the 2D CNNs, we first designed a novel 1-Bit 3D BitConv to quantize the 3D CNNs on the action recognition task.

2.2 BNN Application

BNN is utilized for image classification [10,19], object detection [18,28] and point cloud [20,26]. For the 1-Bit image classification application, the 20 kfps streaming camera with 1-Bit CNN in real-time streaming mode is to reduce 980x parameters [10]. For the 1-Bit object detection applications, Peng et al. [18] first proposed binary detection neural network (BDNN) with a low bit-width weight optimization method instead of large neural networks(e.g., Faster RCNN). For the 1-Bit point cloud application, Qin et al. [20] proposed BiPointNet, which is the first attempt at binarization on 3D point clouds to improve the 14.7 times processed speeds and reduce the 18.9 times memory footprint. In this paper, we attempt to apply 3D BitConv to the challenging application of video action recognition for 3D vision tasks.

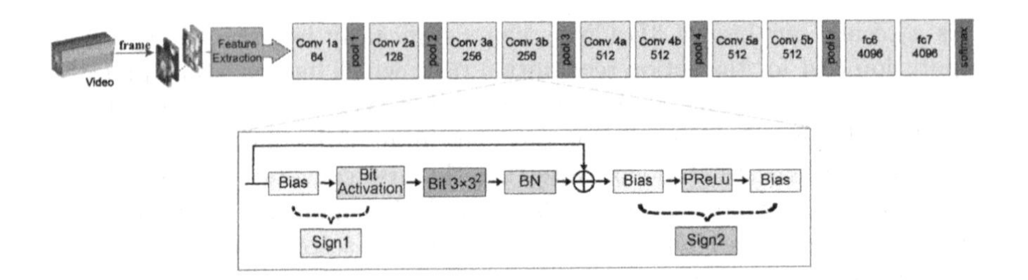

Fig. 2. The binary C3D architecture. The C3D network receives the extracted feature from the original video frames. In binary C3D, the first convolution module (Conv 1a) remains traditional 3D convolution for initial feature extraction. Then, we binarize all other convolution operations with the proposed 3D BitConv.

2.3 Video Action Recognition

More and more researchers have proposed different 3D CNN-based innovative approaches focusing on video-level features. Recently, Two-Stream ConvNet [22] was first proposed to be trained on dense optical flow, which consists of two CNNs streams to combine spatial and temporal networks for better spatial-temporal features. In this paper, we adopt another way that uses a binary neural network with 3D BitConv to complete the video action recognition for comparable performance on the public benchmark.

3 Methodology

3.1 Standard 1-Bit CNNs and Representation

With the development of 3D CNN, the limitations of hardware resources are gradually exposing defects. BNN is the most effective methods to produce compressed DNN, improving efficiency while preserving excellent performance. Therefore, 1-bit CNN is gradually applied in various 2D vision tasks. 1-Bit CNN refers to a CNN model that substitutes binary values for the actual values used in the weight parameters and activations. We use Sign functions for binary activations and weights, which helps to decrease the quantity of computation required. Memory is a precious resource, especially for mobile devices, and the 1-Bit representation of the weights significantly reduces its consumption. For the purpose of binarizing the weights and activations, CNNs are typically quantized to -1 or +1 using a non-linear sign function, as follows:

$$w_b = \operatorname{Sign}(w_r) = \begin{cases} +1 & \text{if } w_r \geq 0 \\ -1 & \text{otherwise} \end{cases} \tag{1}$$

where w_r is real weight, and w_b is the binary weight of -1 or +1.

$$a_b = \operatorname{Sign}(a_r) = \begin{cases} +1 \text{ if } a_r \geq 0 \\ -1 \text{ otherwise} \end{cases} \tag{2}$$

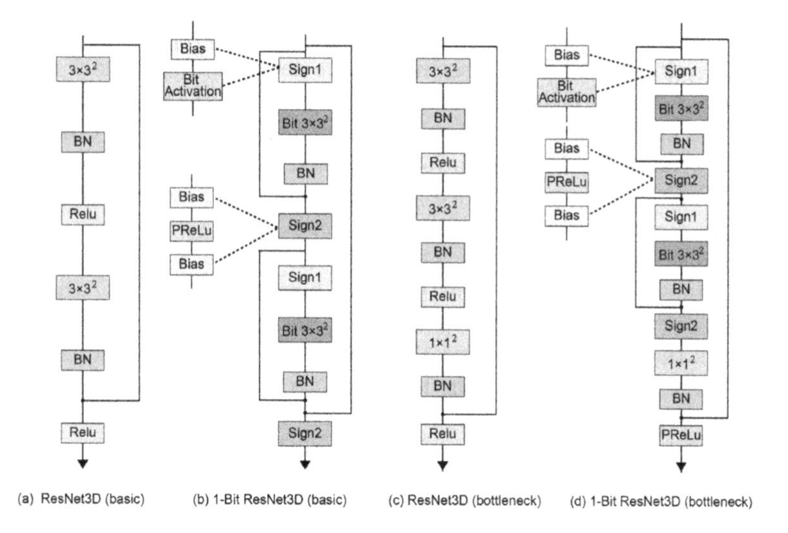

Fig. 3. Architecture of the original and binary 3D convolution blocks. (a) and (c) represent the original basic block and bottleneck block of ResNet3D, respectively. (b) and (d) indicate the corresponding binary basic block and bottleneck block. The Sign1 module contains a learnable bias and a binary activation, and the Sign2 module consists of two learnable bias and a PReLU activation.

where a_r is real activation, and a_b is also a binary activation. Note that a_r exists in training and inference processes.

Unlike other above mentioned approaches with 2D convolutional kernels, we focus on CNNs with 3D convolutional kernels, and discuss the implement of 3D BitConv for video action recognition tasks. In Fig. 2, we illustrate the architecture of a binarized C3D network embedded 1-Bit 3D convolution. The number of filters is reported inside each box. The C3D network receives video-level features from the original videos as input. In the real C3D backbone, it consists of eight 3D convolutions (from Conv1a to Conv5b), five max-pooling (from pool1 to pool5), and two fully connected (fc6 and fc7), then to a Softmax classifier. All of the 3D convolution kernels (from Conv1a to Conv5b) are $3 \times 3 \times 3$ and the strides are $1 \times 1 \times 1$ in temporal-spatial dimensions. The first 3D pooling kernel (pool1) and the stride are $1 \times 2 \times 2$, and other pooling kernels (from pool2 to pool5) are $2 \times 2 \times 2$ and the strides are $2 \times 2 \times 2$. Both of fully-connected (fc6 and fc7) have 4096 units.

In the process of binarizing the C3D network architecture, we adopt the following binarization scheme. The video-level feature is fed to the 1-Bit C3D, and the first convolution module (Conv1a) remains unchanged. From the second convolution module, we use 3D BitConv to replace the actual 3D convolution in Conv2a, conv3a, conv3b, conv4a, conv4b, conv5a, and conv5b, respectively. We illustrate the **3D conv3b block of C3D** in Fig. 2 for further detailed analysis. We design the 1-Bit 3D CNN module which contains the following blocks: Bias -> Bit Activation -> 1-Bit 3×3^2 convolution -> BatchNorm -> element-wise

Sum -> Bias -> PReLu -> Bias" with skip connection, to replace the 3D Conv3b of C3D network. For simpler representation, we define the Sign1 block and Sign2 block. The Sign1 block includes "Bias -> Activation" and the Sign2 block includes "Bias -> PReLu -> Bias". Therefore, the architecture of a 1-Bit C3D network is "Sign1 -> 1-Bit 3×3^2 convolution -> BatchNorm -> element-wise Sum -> Sign2". For the purpose of video action recognition, the trained 1-Bit C3D network can perform as a feature extractor.

To demonstrate the validation of the 3D BitConv even further, this paper extends it to the ResNet3D structure. We binarized ResNet3D (basic) and ResNet3D (bottleneck), and the specific refinement process is shown in Fig. 3, which shows the block of binarized Resnet3D network with embedded 3D BitConv. In Fig. 3(a), we illustrate the basic ResNet3D block, including the 3 x 3^2 convolution, BatchNorm, and ReLu operations. In more detail, the operations are connected in sequence as follows: "3 x 3^2 convolution -> BatchNorm -> ReLu -> 3 x 3^2 convolution -> BatchNorm -> ReLu", and there is a skip connection from the front of the 3D convolution operations to the rear of the BatchNorm operations. To binarize the basic ResNet3D architecture for action recognition, we apply the designed 3D BitConv to the basic ResNet3D block and form a 1-Bit ResNet3D (basic) architecture. The details are shown in Fig. 3(b). We design the 1-Bit 3D convolution block as "Sign1 -> Bit 3 x 3^2 convolution -> BatchNorm -> Sign2" with skip connection. In detail, the Sign1 block consists of "Bias -> Bit Activation", and the Sign2 block consists of "Bias -> PReLu -> Bias". So, there is double 1-Bit 3D convolution with skip connection in the 1-Bit ResNet3D (basic).

In Fig. 3(c), it is another ResNet3D (bottleneck), which is more complex than ResNet3D (basic). We also apply 3D BitConv as "Sign1 -> Bit 3 x 3^2 convolution -> BatchNorm -> Sign2" with skip connection to binarize ResNet3D (bottleneck). Figure 3(d) shows the block structure after binarization, including two concatenated "Sign1 -> Bit 3 x 3^2 convolution -> BatchNorm $->$ Sign2" with skip connection.

3.2 Training Strategy

The initialization is critical for the 1-Bit Training strategy. This paper uses the process of binarizing I3D to map the pre-training on ImageNet to the 3D through the flatten operation. For the binarization of C3D, we pre-train based on Sports-1M. When we complete the initialization process, we first add the Sign1 block and Sign2 block. The 3D convolution is not binarized at this time, and pre-training is performed first. After completing the previous pre-training stage, we use this result as the pre-training model. Then, we achieve 3D convolution binarization and train the model again to the result. The activation of ReLU has a non-negative value, whereas Sign either has a positive or negative value ($+1$ or -1). Given these considerations, if we use a real CNN with ReLU directly, which may not be suitable for the initial training of a 1-Bit CNN. Instead of ReLU, we use the clip (-1, x, 1) for pre-train CNN model with real-valued.

Table 1. Top-1 and Top-5 accuracies on UCF101 [23] and HMDB51 [12] datasets, where the views column indicates the number of input frames, the interval between two consecutive frames, and the number of input video clips.

Methods	Backbone	Pre-train	Views	UCF101 [23]		HMDB51 [12]	
				Top-1 Acc(%)	Top-5 Acc(%)	Top-1 Acc(%)	Top5-Acc(%)
C3D [24]	C3D [24]	Sports-1M [11]	$16 \times 1 \times 10$	83.27	95.90	74.35	91.13
1-Bit C3D	Bit-C3D			79.69	94.37	70.13	90.62
I3D [1]	ResNet3D-18 [8]	ImageNet [5]	$8 \times 4 \times 4$	73.12	90.48	70.49	88.30
	ResNet3D-50 [8]			74.52	91.09	72.75	89.32
1-Bit I3D	Bit-ResNet3D-18			69.50	88.13	68.82	88.95
	Bit-ResNet3D-50			70.02	88.37	70.71	89.68
SlowOnly [6]	ResNet3D-50 [8]	ImageNet [5]	$8 \times 2 \times 4$	81.95	95.43	77.25	91.28
1-Bit SlowOnly	Bit-ResNet3D-50			78.27	92.78	74.27	90.84

3.3 Implementation Details

The implementation framework is PyTorch [17]. In this paper, the 1-Bit C3D architecture uses an SGD optimizer, lr is 0.001, training epochs is 100, the learning rate warm_up takes the first ten rounds, and the batch size is 16. For another architecture, 1-Bit I3D also uses SGD; lr is 0.008, and the rest of the parameters are configured in the same way as before.

4 Experiments

4.1 Dataset and Evaluation Metrics

The proposed 3D BitConv block which is embedded in 1-Bit 3D CNN, is further evaluated on HMDB51 [12] and UCF101 [23]. HMDB51 consists of 51 categories with 6,766 clips, and there are a total of 13,320 videos included in UCF101, organized into 101 distinct action categories. For the UCF101 dataset, we adopt the official way of train-test splits. Due to the limitations of computing resources, it is unavailable to pre-train the model on Kinetics-400 [1]. Consequently, we used a random distribution method to partition the training and test sets into 4:1 for HMDB51. This paper reports the Top-1 and Top-5 accuracies (%). To provide a more accurate depiction of the BNN's memory capabilities, we further calculate the computation cost by the memory (MB) to evaluate the memory saving index after binarization. We reproduce other authors' models based on the divided and record the results for a fair comparison.

4.2 Experimental Results

Comparison to State-of-the-Art. We describe the details of the Top-1 accuracy (%) and Top-5 accuracy (%) of the binarized 1-Bit C3D and 1-Bit I3D on two datasets. Table 1 compares the performances produced by our 1-bit models and their corresponding state-of-the-art models.

Compare the proposed 1-Bit C3D models to the original C3D models, when we use the 1-Bit C3D model, its Top-1 accuracy (%) on the UCF101 dataset reduces from 83.27% to 79.69%, and its Top-5 accuracy (%) reduces from 95.90%

Table 2. Comparison of computaional cost between original network and the corresponding binary network testing on UCF101 dataset [23]. The Size column indicates the number of input frames and every frame's width and height.

Methods	Size	Backbone	Memory(Mb)	Memory saving	Top-1 Acc(%)
C3D [24]	16×112×112	C3D [24]	77.996	-	83.27
1-Bit C3D		Bit-C3D	51.225	1.52x	79.69
I3D [1]	8×224×224	ResNet3D-50 [8]	33.166	-	74.52
1-Bit I3D		Bit-ResNet3D-50	8.503	3.900x	70.02

to 94.37%. That is, it reduces the proportion by 4.3% and 1.6%, respectively. For another, on the HMDB51 dataset, the Top-1 accuracy (%) reduces from 74.35% to 70.13%, and its Top-5 accuracy (%) reduces from 91.13% to 90.62%. That is, it reduces the proportion by 5.7% and 0.6%, respectively. We also record detailed results in Table 1.

Particularly, for I3D and 1-Bit I3D models, we select ResNet3D-18 and ResNet3D-50 as the backbone to compare the performance. For the UCF101 dataset, when we use ResNet3D-18 backbone, the Top-1 accuracy (%) of 1-Bit I3D reduces from 73.12% to 69.50%, and its Top-5 accuracy (%) reduces from 90.48% to 88.13%. That is, it reduces the proportion by 5.0% and 2.6%, respectively. When we use a more complex ResNet3D-50 backbone, the Top-1 accuracy (%) of 1-Bit I3D reduces from 74.52% to 70.02%, and its Top-5 accuracy (%) reduces from 91.09% to 88.37%. That is, it reduces the proportion by 6.0% and 3.0%, respectively. For another dataset, HMDB51, when we use ResNet3D-18 backbone, the Top-1 accuracy (%) of 1-Bit I3D reduces from 70.49% to 68.82%, but its Top-5 accuracy (%) improves from 88.30% to 88.95%. In other words, it reduces the proportion by 2.4% with Top-1 accuracy (%) but improves by 0.7% Top-5 accuracy (%), respectively. When we replace to ResNet3D-50 backbone, the Top-1 accuracy (%) of 1-Bit I3D reduces from 72.75% to 70.71%, and its Top-5 accuracy (%) improves from 89.32% to 89.68%. The results indicate that it reduces the proportion by 2.8% and improves by 0.4%, respectively.

Moreover, we quantize the recent SlowOnly [6] model, which produces better performance as shown in Table 1. Specifically, for UCF101, we record the Top-1 accuracy (%) of 1-Bit SlowOnly reduces from 81.95% to 78.27%, and its Top-5 accuracy (%) reduces from 95.43 % to 92.78%. Thus, it reduces the proportion by 4.5% and 2.8%, respectively. Besides, for the HMDB51 dataset, the Top-1 accuracy (%) of 1-Bit SlowOnly reduces from 77.25% to 74.27%, and its Top-5 accuracy (%) reduces from 91.28% to 90.84%. That is to say, it reduces the proportion by 3.9% and 0.5%, respectively.

The Computational Cost. To further validate the performance with computational cost, we compared the relationship between the memory and performance parameters of the model. As we can see from Table 2, we can see that: when we use the C3D method, the Top-1 Acc(%) reaches 83.27%, and its memory footprint is 77.996Mbit. The 1-Bit C3D method achieves a Top-1 Acc(%) of 79.69%, and its memory usage is 51.225Mbit, so the memory saving after

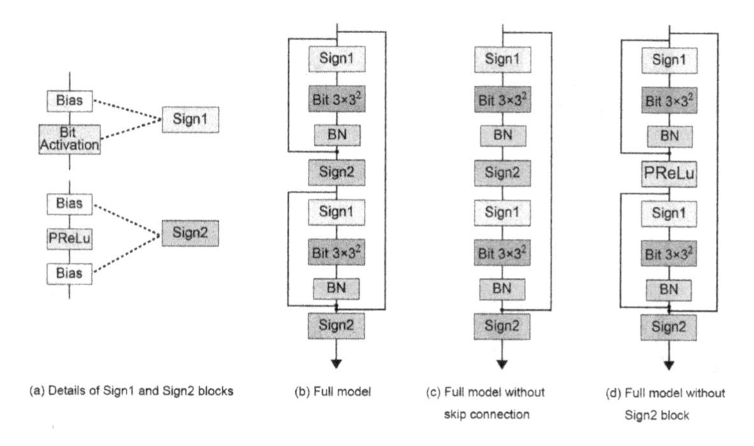

Fig. 4. Ablation study on different 1-Bit 3D convolution modules embedded in 1-Bit C3D and 1-Bit I3D-ResNet3D network architectures on UCF101 dataset. We define Sign1 and Sign2 blocks in (a), the Full model structure is defined in (b), the design of the Full model without skip connection is described in (c), and the design of the Full model without Sign2 block is described in (d).

Table 3. The effects of different components in 1-Bit C3D and 1-Bit I3D network for the final accuracy on UCF101 dataset. Note that we use 1-Bit ResNet3D-50 as the backbone of the 1-Bit I3D method.

Methods	Configuration	Top-1 Acc(%)	Top-5 Acc(%)
1-Bit C3D	No skip connection	79.06	93.75
	No Sign2 block	76.88	92.81
	Full model	79.69	94.37
1-Bit I3D	No skip connection	69.31	87.02
	No Sign2 block	66.69	86.47
	Full model	70.02	88.37

binarization reaches 1.52 times. In Table 2, the memory footprint of the I3D is 33.166Mbit, while the 1-Bit I3D is 8.503Mbit, which achieves 3.900 times memory saving compared to the I3D method. At the same time, the Top-1 Acc(%) only reduces from 74.52% to 70.02%.

To conclude, we can summarize that our designed 3D BitConv block embeds the typical 1-Bit 3D CNN networks, reducing the memory cost and being capable of learning results that match the original network.

4.3 Ablation Study

This paper performs the ablation research with different 3D BitConv blocks embedded in 1-Bit C3D and 1-Bit I3D-ResNet3D network on UCF101. In Fig. 4 (a) define the Sign1 and Sign2 blocks (b) illustrate the full model's structure (c) illustrate the design of the Full model without skip connection, and (d) illustrate the design of the Full model without Sign2 block. Table 3 records the effects of

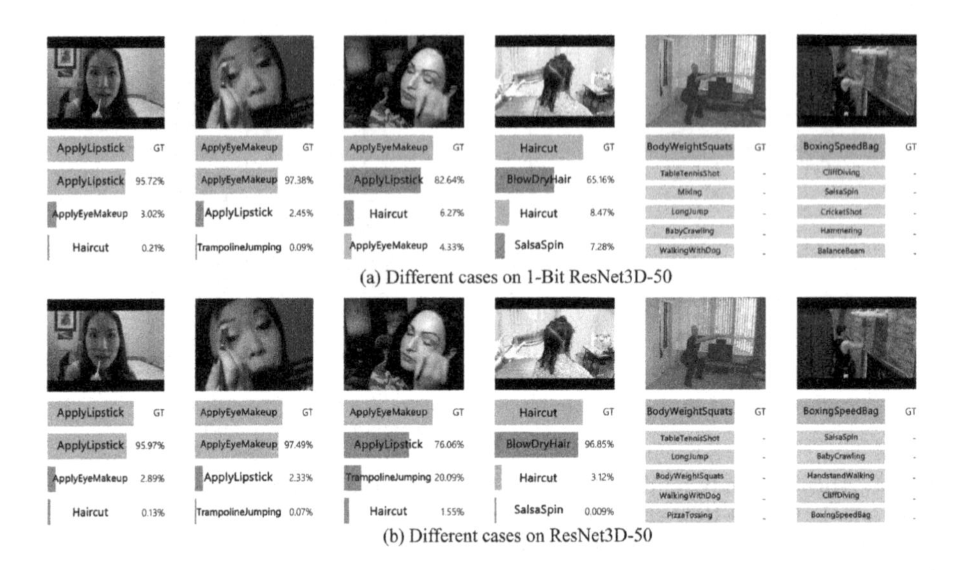

(a) Different cases on 1-Bit ResNet3D-50

(b) Different cases on ResNet3D-50

Fig. 5. Quantitative effect on UCF101. Figure (a) shows the different cases on 1-Bit ResNet3D-50, and Figure (b) shows the same cases as (a) on ResNet3D-50 for comparison. The first two columns in the figure are the visualized good cases, the middle columns are relatively good cases, and the last two are bad cases, respectively.

the ablation study. As for the 1-Bit C3D method, we show three different configurations. We use full model achieves 79.69% on Top-1 Acc(%) and 94.37% on Top-5 Acc(%) on UCF101 dataset. Then, we conduct the ablation study without skip connection, and the model achieves 79.06% on Top-1 Acc(%) and 93.75% on Top-5 Acc(%), while we conduct the ablation study without Sign2 block, and the model achieves 76.88% on Top-1 Acc(%) and 92.81% on Top-5 Acc(%). For the 1-Bit I3D method, we also show the same configurations. Top-1 Acc(%) and Top-5 Acc(%) are 70.02% and 88.37%, while both reduce to 69.31% and 87.02% without skip connection. It also reduces to 66.69% and 86.47% without Sign2 block, respectively.

Compared to 1-Bit C3D and 1-Bit I3D, we can analyze that the FULL model has the best results. Still, when the skip connection or the Sign2 block is not used, it can also achieve relatively good results, which shows that our model can still maintain good performance with reduced memory and almost learns the representation ability of the original model.

4.4 Quantitative Results

This section shows the quantitative results with different cases between 1-Bit ResNet3D-50 and ResNet3D-50 on the UCF101 dataset. From Fig. 5, the first two image columns visualize two good cases: ApplyLipstick and ApplyEyeMakeup. We can see that the binarized model 1-Bit ResNet3D-50 and the traditional model ResNet3D-50 get the highest possible level of performance on the

Top-1 Acc(%). Besides, the third and fourth columns visualize two relatively good cases. The Ground Truth of the two cases is ApplyEyeMakeup and Haircut. We can see neither model predicts correctly in Top-1, but the binarized model 1-Bit ResNet3D-50 is in the effect of Top-5, which is better than that of Top-1 and surpasses the traditional model in some cases. But in some bad cases (columns 5 and 6), neither our binarized model nor the traditional model can distinguish these confusingly indistinguishable actions, such as BodyWeightSquats and BoxingSpeedBag. Overall, our proposed 3D BitConv block embedded in 3D CNN can almost learn results that match the original network.

5 Conclusions

This paper proposes a novel 3D binary convolutional block, BitConv, to be flexibly embedded in various popular 3D CNNs. The proposed BitConv consists of a binary 3D convolution operation, three learnable biases, and a binary activation. Our new strategy of binary 3D convolution can not only ensure high accuracy but also greatly decrease the computation expense for 3D CNNs, which is critical in action recognition challenges. Meanwhile, we further binarize two representative 3D convolution-based methods (C3D, I3D-ResNet) to evaluate. The extensive experiment results indicate that BitConv has achieved impressive standards on two mainstream action recognition benchmarks.

Acknowledgements. This work was supported in part by National Natural Science Foundation of China (61976220 and 61832017), Beijing Outstanding Young Scientist Program (BJJWZYJH012019100020098), and the Research Seed Funds of School of Interdisciplinary Studies, Renmin University of China.

References

1. Carreira, J., Zisserman, A.: Quo vadis, action recognition? a new model and the kinetics dataset. In: CVPR, pp. 6299–6308 (2017)
2. Chen, T., Zhang, Z., Ouyang, X., Liu, Z., Shen, Z., Wang, Z.: BNN-BN=?: training binary neural networks without batch normalization. In: CVPR. pp. 4619–4629 (2021)
3. Courbariaux, M., Hubara, I., Soudry, D., El-Yaniv, R., Bengio, Y.: Binarized neural networks: training deep neural networks with weights and activations constrained to+ 1 or-1. arXiv preprint arXiv:1602.02830 (2016)
4. Darabi, S., Belbahri, M., Courbariaux, M., Nia, V.P.: BNN+: Improved binary network training (2018)
5. Deng, J., Dong, W., Socher, R., Li, L.J., Li, K., Fei-Fei, L.: ImageNet: a large-scale hierarchical image database. In: CVPR, pp. 248–255. IEEE (2009)
6. Feichtenhofer, C., Fan, H., Malik, J., He, K.: SlowFast networks for video recognition. In: ICCV, pp. 6202–6211 (2019)
7. Fernando, B., Gavves, E., Oramas, J.M., Ghodrati, A., Tuytelaars, T.: Modeling video evolution for action recognition. In: CVPR, pp. 5378–5387 (2015)
8. He, K., Zhang, X., Ren, S., Sun, J.: Deep residual learning for image recognition. In: CVPR, pp. 770–778 (2016)

9. Howard, A.G., et al.: MobileNets: efficient convolutional neural networks for mobile vision applications. arXiv preprint arXiv:1704.04861 (2017)
10. Jokic, P., Emery, S., Benini, L.: Binaryeye: A 20 KFPS streaming camera system on FPGA with real-time on-device image recognition using binary neural networks. In: SIES, pp. 1–7 (2018)
11. Karpathy, A., Toderici, G., Shetty, S., Leung, T., Sukthankar, R., Fei-Fei, L.: Large-scale video classification with convolutional neural networks. In: CVPR, pp. 1725–1732 (2014)
12. Kuehne, H., Jhuang, H., Garrote, E., Poggio, T., Serre, T.: HMDB: a large video database for human motion recognition. In: ICCV, pp. 2556–2563 (2011)
13. Li, R., Wang, Y., Liang, F., Qin, H., Yan, J., Fan, R.: Fully quantized network for object detection. In: CVPR, pp. 2810–2819 (2019)
14. Li, Z., Ni, B., Zhang, W., Yang, X., Wen, G.: Performance guaranteed network acceleration via high-order residual quantization. In: ICCV, pp. 2603-2611 (2017)
15. Lin, X., Zhao, C., Pan, W.: Towards accurate binary convolutional neural network. NIPS (2017)
16. Liu, Z., Wu, B., Luo, W., Yang, X., Liu, W., Cheng, K.T.: Bi-real net: Enhancing the performance of 1-bit CNNs with improved representational capability and advanced training algorithm. In: ECCV, pp. 722–737 (2018)
17. Paszke, A., et al.: PyTorch: an imperative style, high-performance deep learning library. NIPS (2019)
18. Peng, H., Chen, S.: BDNN: Binary convolution neural networks for fast object detection. PRL, pp. 91–97 (2019)
19. Phan, H., He, Y., Savvides, M., Shen, Z., et al.: MoBiNet: a mobile binary network for image classification. In: WACV, pp. 3453–3462 (2020)
20. Qin, H., Cai, Z., Zhang, M., Ding, Y., Zhao, H., Yi, S., Liu, X., Su, H.: Bipointnet: Binary neural network for point clouds. arXiv preprint arXiv:2010.05501 (2020)
21. Rastegari, M., Ordonez, V., Redmon, J., Farhadi, A.: XNOR-Net: ImageNet classification using binary convolutional neural networks. In: ECCV (2016)
22. Simonyan, K., Zisserman, A.: Two-stream convolutional networks for action recognition in videos. In: Adv. Neural Inf. Proc. Syst. (NeurIPS), pp. 568–576 (2014)
23. Soomro, K., Zamir, A.R., Shah, M.: Ucf101: A dataset of 101 human actions classes from videos in the wild. arXiv preprint arXiv:1212.0402 (2012)
24. Tran, D., Bourdev, L., Fergus, R., Torresani, L., Paluri, M.: Learning spatiotemporal features with 3D convolutional networks. In: ICCV, pp. 4489–4497 (2015)
25. Wei, Y., Pan, X., Qin, H., Ouyang, W., Yan, J.: Quantization mimic: towards very tiny CNN for object detection. In: ECCV, pp. 267–283 (2018)
26. Xu, S., Li, Y., Zhao, J., Zhang, B., Guo, G.: POEM: 1-bit point-wise operations based on expectation-maximization for efficient point cloud processing. arXiv preprint arXiv:2111.13386 (2021)
27. Zhang, X., Zhou, X., Lin, M., Sun, J.: Shufflenet: an extremely efficient convolutional neural network for mobile devices. In: CVPR, pp. 6848–6856 (2018)
28. Zhao, J., et al.: Data-adaptive binary neural networks for efficient object detection and recognition. PRL, pp. 239–245 (2022)

Image Quality Assessment and Enhancement

STN: Stochastic Triplet Neighboring Approach to Self-supervised Denoising from Limited Noisy Images

Bowen Wan, Daming Shi(✉), and Yukun Liu

College of Computer Science and Software Engineering, Shenzhen University, Shenzhen, China
dshi@szu.edu.cn

Abstract. With the rapid development of artificial intelligence in recent years, deep learning has shown great potential in the field of image denoising. However, most of the work is based on supervised learning, and the lack of clean images in the real application will limit neural network training. For this reason, self-supervised learning in the absence of clean images is getting more and more attentions. Nevertheless, since both the source and target in self-supervised training are the limited noisy image itself, such denoising methods suffer from overfitting. To this end, a stochastic triplet neighboring approach, thereafter referred to as STN, is proposed in this paper. Given an input noisy image, the source fed to the STN is the downsized sub-image via 4-neighbor sampling, whereas the target in STN training is a stochastic combination from the two neighbored sub-images. Such a mechanism is actually the augmentation of training data, which leads to the relief of the overfitting problem. Extensive experimental results show that our proposed STN approach outperforms the state-of-the-art image denoising methods.

Keywords: Image denoising · Self-supervised · Deep learning

1 Introduction

Recent data driven techniques outperform conventional model based methods and achieve the state of the art denoising quality [5,23,25]. A common disadvantage for the above methods is that all need to take a clean image as the target in training.Nevertheless, in medical and biological scenarios such as molecular imaging, there is often a lack or absence of clean images,leading to an infeasibility of the above methods. To this end, the noise-to-noise(N2N) method [12] is the first to reveal that deep neural networks (DNNs) can be trained with pairs of noisy-noisy images instead of noisy-clean images, in other words, training can be conducted with only two noisy images that are captured independently in the same scene. The N2N can be used in many tasks such as [3,4,6,20,24], since it creatively addresses the dependency on clean images. Unfortunately, pairs of

D.-T. Dang-Nguyen et al. (Eds.): MMM 2023, LNCS 13833, pp. 109–120, 2023.
https://doi.org/10.1007/978-3-031-27077-2_9

noisy images are still difficult to be obtained in some areas of practice, such as the dynamic imaging situation.

To further bring the N2N into practice, some research [1,9,10] concluded that it is still possible to train the network without using clean images if the noise between each region of the image is independent. Inspired by the above research, Neighbor2Neighbor (NB2NB) method [7] proposed a new sampling scheme to achieve better denoising effects with a single noisy image. Specifically, it used sampling algorithm to obtain two similar sub-images to replace pairs of noisy images for training. The advantage of this approach is that it does not need a prior noise model prior, like the recorrupted-to-recorrupted (R2R) [16], nor does it lose image information, like the Noise2Void (N2V) self-supervised methods N2V [9].

However, both the N2N and NB2NB still suffer from limitation in denoising the real world images. Our in-depth analysis, which is detailed in Sect. 2, shows that the limitation of above self-supervised learning methods approximating the supervised learning methods is caused by overfitting, moreover the overfitting is resulted from lack of clean images. As a matter of fact, the large requisite amount of training data is difficult to obtain in many practical scenarios.

To solve the overfitting problem of the self-supervised denoising methods, we propose a stochastic neighboring approach to augment the training data. The remainder of the paper is organized as follows. In Sect. 2, theoretical analysis shows how the overfitting problem is caused. To solve the overfitting problem, the details of our proposed STN is given in Sect. 3, followed by its implementation with deep learning in Sect. 4. The experiments are conducted in Sect. 5, with the conclusions drawn in Sect. 6.

2 Problem Statement

2.1 Unsupervised Image Denoising

The N2N [12] revealed that the noisy/true image pairs used to train the DNN can be replaced by noisy/noisy images pairs. The paired noisy images are represented by the mathematical equation: $y = x + n_1$, $z = x + n_2$, where n_1 and n_2 are uncorrelated. There are two main principles for N2N to successfully use the paired noisy images to train the network: the first is that the optimal solution obtained by network training is a mean value solution; the second is that the mean value of noise is zero. So the gradient generated by the network for a single noisy image is incorrect, but the gradient corresponding to the average of all noisy images is correct, which can be explained by the following equation:

$$\mathbb{E}\{z_i \mid y_i\} = \mathbb{E}\{x_i + n_i \mid y_i\} = \mathbb{E}\{x_i \mid y_i\} = x_i \tag{1}$$

where x, y, z stands for clean targets, noisy images and noisy targets, respectively. x_i denotes the i-th potentially clean image. n_i is independent on the noise contained in y_i, and the expectation of n_i is zero. Although the N2N alleviated the dependence on clean images, pairs of noisy images are still difficult to be obtained.

To eliminate N2N restrictions in dynamic scene denoising, AmbientGAN [2], a method of training generative adversarial networks also dispose the reliance on a clean target, and used a measurement function to generate a noise-corrupted image, which is then fed into the discriminator for comparison with the original noisy image. The Noiser2Noise (Nr2N) [14] forms pairs of noisy datasets by adding noise to the original noisy image. Then, R2R [16] overcomes the disadvantage of N2N [12] and generates paired noisy images by introducing a prior noise model. In particular, it used different noise levels to form pairs of noisy images with different damage degrees for a single noisy image. Although this approach is close to N2N [12] in denoising effect, it is limited in applications with real noisy images because the level and type of noise from real noisy images are difficult to determine.

2.2 Self-supervised Image Denoising

The N2V [9] offered a blind-spot network that can denoise using only a noisy image, and its main principle is to use the correlation among pixels to predict missing pixels. In this way, some pixel information is lost and the denoising effect is not ideal. The Self2Self [18] can realize image denoising by dropout in the case of only one noisy picture. Unlike N2V, it uses lost information as targets, but its training time is extremely long and a picture needs a training model, thus limiting its practical application.

The NB2NB [7] is the most recent a self-supervised denoising method, which can be viewed as an advanced version of the N2N [12] via a novel image sampling scheme to get rid of the requirement of paired noisy images. Specifically, it used single noisy images $\mathbf{Y}$ and sampler to estimate the optimal parameters θ of the regression problem of a network f_θ, which can be expressed by the following equation:

$$\begin{aligned} &\arg\min_\theta L_{reconstruction} + \gamma L_{regularization} = \\ &\arg\min_\theta \mathbb{E}_\mathbf{Y} \|f_\theta(y_1) - y_2\|_2^2 + \gamma \mathbb{E}_\mathbf{Y} \|f_\theta(y_1) - y_2 - r_1 + r_2\|_2^2 \end{aligned} \tag{2}$$

where $L_{reconstruction}$ is the measure of dissimilarity between the network output $f_\theta(y_1)$ and the adjacent noisy sub-images y_2, the regularization term $L_{regularization}$ is used to reduce the error due to using denoised adjacent sub-images r_1 ,r_2, and the γ is used to control the strength of the regularization term. According to Eq. 2, we can find that the target of the training network is noisy y_2. If the training data is limited, it is easy for the neural network to reproduce the noise of y_2, resulting in the overfitting problem.

In fact, many self-supervised denoising methods are less effective than supervised denoising methods when the datasets are limited [8,11,20,21,26]. In addition, we found that the error of DNN training due to using noisy images $\{y_i\}_{i=1}^N$ instead of clean images $\{x_i\}_{i=1}^N$ can be explained by the following equation:

$$\mathbb{E}_y\left[\frac{1}{N}\sum_i y_i - \frac{1}{N}\sum_i x_i\right]^2 = \mathbb{E}_y\left[\frac{1}{N}\sum_i n_i\right]^2 \tag{3}$$

Since we assume that the mean noise is zero, Eq. 3 can be represented by

$$\frac{1}{N^2}\sum_i \mathrm{Var}\left(n_i\right) \tag{4}$$

According to Eq. 4, we can find that the denoising performance gap between self-supervised methods and supervised methods lies in the number of the training noise samples and their variance. Therefore, the self-supervised denoising methods can approximate or even exceed the supervised denoising methods on the limited data set by implicitly increasing the number of noise samples in the training set.

3 Stochastic Triplet Neighbors

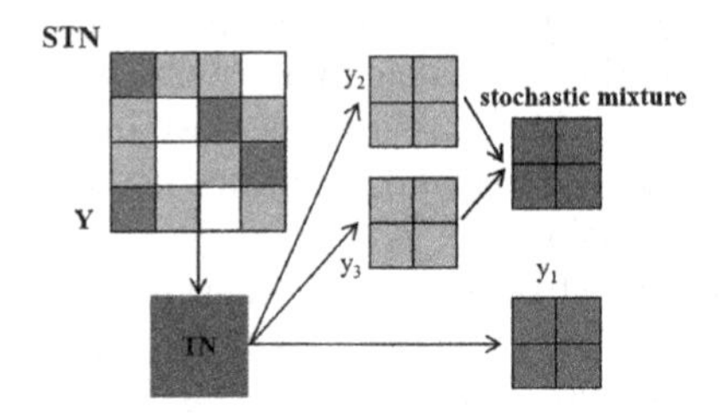

Fig. 1. The detail of stochastic triplet neighbors (STN). It contains triplet neighbors (TN) and stochastic mixing (SM). Three similar sub-images (y_1, y_2, y_3) can be obtained through single original noisy image Y and TN algorithm, then use Eq. 5 to process y_2 and y_3 to get stochastic as the network training target, and y_1 as the input for network training.

To relieve the overfitting problem with the self-supervised denoising methods, we improve NB2NB with a stochastic triplet neighboring approach, which contains two sub modules, namely, triplet neighbors sampler (TN) and stochastic mixing module (SM). The former module down-samples three similar sub-images from the input images, furthermore the latter module generates plenty of noisy samples by mixing two neighbor sub-images as follows:

$$sm = y_2 * \alpha + y_3 * (1 - \alpha) \tag{5}$$

where α is an image filled with random real numbers and has the same dimension as the noisy sub-images. Randomly changing of α while training epoch leads to

```
Input: a single noisy image Y
Output: two noisy sub-images y1,sm
1 for i = 0; i ≤ width(Y); i = i + 2 do
2     for j = 0; j ≤ height(Y); j = j + 2 do
3         Randomly select a pixel(A) from a sliding window and put it to y1;
4         Select the two pixels adjacent to A from this window and randomly put
          them to y2 and y3;
5         Put the three pixels obtained in step 2,3 into the three sub-images;
6     end
7 end
8 return y1, y2, y3;
```

Algorithm 1: The TN algorithm

a simulation of learning plenty of samples corrupted by unrelated noise. The randomly generated paired noisy sub-images can be expressed by the following equations:

$$\begin{aligned} y_1 &= x + n_1 \\ sm &= x + \varepsilon_{sm} + n_{sm} \\ n_{sm} &= n_2 * \alpha + n_3 * (1 - \alpha) \\ \varepsilon_{sm} &= \varepsilon_2 * \alpha + \varepsilon_3 * (1 - \alpha) \\ \varepsilon_i &= \mathbb{E}_{y_i|x}(y_i) - \mathbb{E}_{y_1|x}(y_1) \quad (i = 2, 3) \end{aligned} \tag{6}$$

In Eq. 6, n_1, n_2, and n_3 correspond to the noise contained in y_1, y_2 and y_3, ε_i is the difference between the potential clean subgraphs of y_1 and y_i. In addition, because of the randomness of α, the network generalization of training will also be enhanced. As for the number of noise samples generated by STN, if the batch size of DNN training is one, the number of noise sample is the product of the number of original noise sample and the number of iterative training networks.

On the other hand, it can be found from Eq. 6 that since α changes randomly with DNN training iteration, the noise contained in sm also changes randomly, but the potential clean image of sm remains. Therefore, the STN approach can prevent the network from learning the noise, so as to avoid overfitting.

4 Denoising with the STN

This section introduces how to train the denoising network with STN. Our method is first training the model with similar sub-images under the help of STN, then using the model directly denoise the large image, and designed a new loss which is as shown below:

$$\arg\min_{\theta} \mathbb{E}_{\mathbf{Y}} \left\| f_\theta(y_1) - sm \right\|_2^2 + \gamma \mathbb{E}_{\mathbf{Y}} \left\| f_\theta(y_1) - sm - r_1 + sm' \right\|_2^2 \tag{7}$$

where replace noisy sub picture and denoised sub-image in Eq. 2 with sm and sm', f_θ is the denoising network parameterized θ and $f_\theta(y_1)$ is denoised image. The first part of Eq. 7 is reconstruction term, and the latter part is regularization

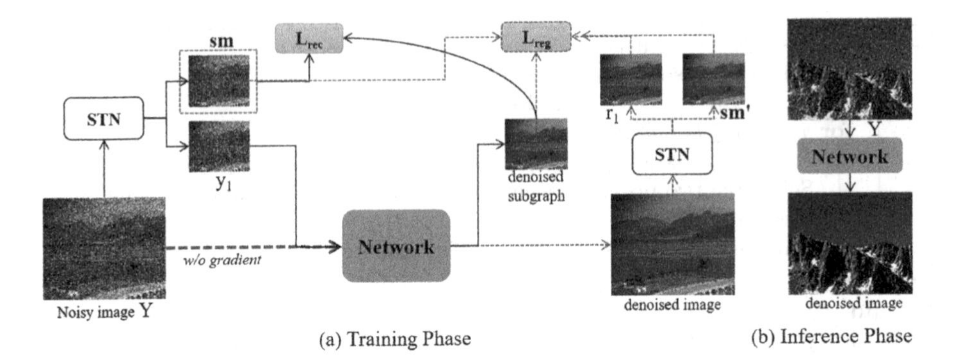

Fig. 2. The STN with denoising. (a) Overall training process. STN (stochastic triplet neighbors) obtained two similar noisy subgraphs (y_1, sm), and then carried out loss calculation after network training. The solid blue line was used to calculate reconstruction term L_{rec}, and the dotted blue line was used to calculate regularization term L_{reg}. The network is trained mainly with L_{rec} at the beginning, and the weight of L_{reg} is gradually increased as the network iterates. (b) Inference using the trained denoising model. The denoising network generates denoised images directly from the noisy images Y of the test set without additional operations. (Color figure online)

term. The hyperparameter γ is used to reduce the gap between the potential clean sub-images of adjacent sub-images. Figure 2 is a framework diagram of the entire self-supervised denoising.

Another point worth noting is that denoising by STN does not produce artifacts, and it also works for noisy images with high-frequency features, because we also introduce STN in the regular term. By subtracting r_1 and sm', we end up with mixed gap ε_{sm} once the network is trained, which is also the gap between the clean images corresponding to two noisy adjacent sub-images (y_1, sm).

5 Experiments

In this section, we demonstrate the effectiveness of our method with limited data sets through a large number of experiments.

We evaluate our method on real noisy datasets and synthesizing noisy images which add additive white Gaussian noise (AWGN). The evaluation metric adopted in this experiment is peak signal to noise ratio (PSNR) and structural similarity (SSIM). Then we proved our method to be able to solve the overfitting. To verify the robustness of our approach, we also tested the effect of different sample numbers on the image denoising effect. In addition, we performed ablation experiment on parameter α in Eq. 5 to analyze the improvement of denoising performance by its value range.

In the experiment, the Adam optimizer with an initial learning rate of 0.001 is used to train the model, and training epoch is 200. All experiments were done on the server with Python 3.8.5, PyTorch 1.6 [17] and Nvidia Tesla K80 GPUs. In addition, the hyperparameter of control regular loss is same as NB2NB.

5.1 Synthetic Experiments

Table 1. The quantitative experimental results of various methods at different noise levels on BSDS300.

Noise Level		10		25		25	
Metric		PSNR	SSIM	PSNR	SSIM	PSNR	SSIM
Supervised	N2C	35.67	0.969	31.67	0.928	28.79	0.872
Unsupervised	N2N	35.63	0.968	31.21	0.923	27.48	0.859
	Nr2N	32.40	0.944	29.38	0.889	25.89	0.794
	R2R	34.89	0.965	28.84	0.90	22.53	0.802
Self-Supervised	NB2NB	34.91	0.960	30.74	0.916	27.10	0.851
	STN	35.12	0.963	30.89	0.919	27.37	0.856

We used BSDS300 [13] dataset in the experiment of removing Gaussian synthetic noise. It contains 200 images for training and 100 images for test. Then we crop all the images to 128×128. The levels of AWGN = 10, 25, 50.

Our STN is compared with N2C, N2N [12], Nr2N [14], R2R [16] and NB2NB [7]. Among them, N2C needs clean images to train a network similar to UNet [19], N2N [12] need paired noisy images for training. Nr2N [14] and R2R [16] all need to know the noise model in advance. NB2NB [7] just needs a noisy picture of each scene to remove noise, the noisy picture it uses is formed by adding AWGN in the training process, that is to say, the noise sample number is equal to the product of the original sample number and the epoch of training network, but this is not consistent with reality, because we do not know the noise model. Therefore, the code adding noise when we test NB2NB [7] needs to be carried out before network training.

The results of the comparison method and our method are presented in Table 1. It is easy to see that our method is better than NB2NB in PSNR and SSIM at three different noise levels. With limited training samples, the denoising performance of N2N and NB2NB is greatly degraded compared with that of N2C, and it can be seen that the number of noise samples and noise variance are two key factors affecting whether self-supervised denoising is close to supervised denoising.

Table 2. The effect of different sample sizes on denoising results in BSDS300.

Datasets	Dataset-1		Dataset-2		Dataset-3	
Metric	PSNR	SSIM	PSNR	SSIM	PSNR	SSIM
N2C	31.67	0.928	30.58	0.914	30.15	0.898
N2N	31.21	0.923	30.43	0.912	29.94	0.894
NB2NB	30.74	0.916	30.04	0.904	29.55	0.883
STN	30.89	0.919	30.24	0.907	29.83	0.885

Next, the improvement of denoising performance of our method in the case of different sample numbers is shown. We used three BSDS300 datasets with different sample sizes. Dataset-1 was trained with 200 noisy images and tested with 100 noisy images, Dataset-2 was trained with 100 noisy images and tested with 25 noisy images, Dataset-3 was trained with 40 noisy images and tested with 10 noisy images, specific experimental results are shown in Table 2, and the experiment was carried out at a noise level of 25. It can be seen from Table 2 that the denoising performance of STN is improved more obviously for the dataset with fewer samples.

5.2 Real-world Experiments

Table 3. The quantitative experimental results of various methods for denoising real-world images from PolyU, Nam and CC.

Datasets	PolyU		Nam		CC	
Metric	PSNR	SSIM	PSNR	SSIM	PSNR	SSIM
N2C	37.62	0.973	39.95	0.991	34.14	0.961
NB2NB	37.30	0.972	37.65	0.987	34.01	0.956
STN	37.52	0.973	40.66	0.990	34.16	0.957

We verify the validity of STN through three real datasets, i.e. PloyU [22], Nam [15], CC [15]. For the convenience of training, the image sizes of these three datasets are uniformly cropped to 128×128. PolyU contains 40 scenes, which are indoor normal light scene and dark light scene, and outdoor normal light scene. Each scene was filmed 500 times in succession. Nam contains 11 scenes, mostly of similar objects and textures. A total of 500 JPEG images were taken for these 11 scenes. The CC dataset contains 15 noisy images. For these datasets, we divide the training set and test set in a ratio of 4:1. Since the actual noise model and paired noisy images are difficult to obtain, N2C and NB2NB are used in our comparison methods.

Table 3 contains quantitative comparison of three real noisy datasets. Firstly, we can see that the performance of NB2NB is degraded due to the small sample size of the real datasets. Then, after using our proposed STN, the denoising performance is greatly improved, especially in Nam dataset, our method even exceeds the supervised image denoising method.

5.3 The Experiment of Overfitting

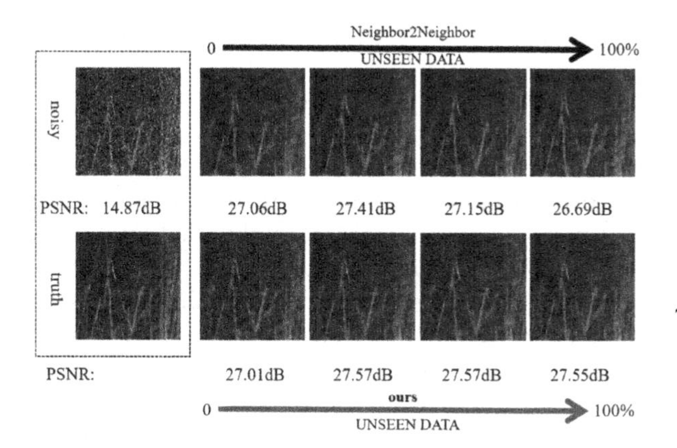

Fig. 3. Illustration of overfitting in Neighbor2Neighbor and the improvement effect of STN. The two images in the dotted box are noisy images and GT from the test set. The images on the right is the denoised image generated by the model generated by network training 50 times, 100 times, 150 times and 200 times.

In order to demonstrate the overfitting of NB2NB [7] in more detail and demonstrates the effectiveness of STN in solving the overfitting problem, we test it on BSDS300 [13], two hundred of which were used for training, 50 for testing, and we used to simulate real noisy pictures by introducing AWGN before training. According to Fig. 3, it can be seen from the 200 epochs of training that NB2NB [7] begin to overfitting around the 120th epoch.

On the other hand, thanks to the introduction of stochastic triplet neighbors, our denoiser achieve highest PSNR, the more detail is in Fig. 4.

5.4 Impact of Stochastic Mixing Parameter

In this section, the impact of the value of α on the performance of denoising network is explored. First of all, five small ranges of α are listed, namely [–3,–2], [–2,–1], [–1,0], [0,1], [1,2]. Then, the denoisers obtained in five small ranges were tested. The specific results are shown in Fig. 5 which is tested on the BSDS300.

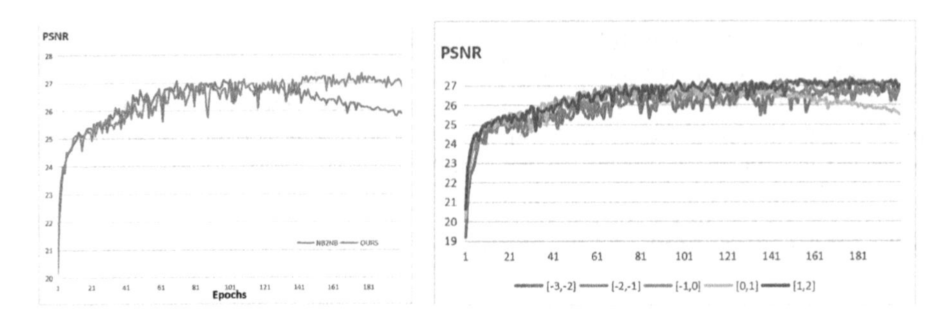

Fig. 4. Overfitting on a dataset of 200 single noisy images.

Fig. 5. Exploration of the value range of α.

Figure 5 shows [1,2] is the best in terms of denoising performance. As for why there is still weak overfitting in the range of 0 to 1, it can be explained by the following analysis: first, our training target is obtained by mixing y_2 and y_3 in Fig. 1, more specifically, by the pixels in the 2×2 window are mixed by the diagonal. When the average value of the mixing coefficient α is 0.5, it is easy to lead the random factors of the two diagonal pixels to be close to 0.5, and there are only two cases when the diagonal pixels are selected each time. In this way, the noise of the image sm does not change much in each training iteration; when the value range of α is [1, 2], the random factors of the two diagonal pixels are very different, and the mean values tend to be 1.5 and -0.5 respectively, resulting in the noise generated by each iteration is different. In short, the greater the noise changes in the target during network training, the more obvious the effect of removing overfitting is.

6 Conclusion

In this research, with limited training data, we propose to use stochastic triplet neighbors to obtain more supervisory information, leading to improve the denoising ability of self-supervised networks. The novel idea of the proposed STN is that a large number of paired noisy images can be generated by mixing adjacent sub-images with a random weight. The experimental results demonstrate that our method outperforms the state-of-the-art self-supervised methods. Compared with the supervised method, the STN deep network is capable of achieving the level of denoising performance, but our method enjoys the advantage of a much wider application areas, thanks to the non-requisite of paired noising target images or a prior noise model in the training phase. Our future work includes speeding up the convergence of the network training by optimizing the random weight systematically.

Acknowledgement. This work is supported by Ministry of Science and Technology China (MOST) Major Program on New Generation of Artificial Intelligence 2030 No. 2018AAA0102200. It is also supported by Natural Science Foundation China (NSFC) Major Project No. 61827814 and Shenzhen Science and Technology Innovation Commission (SZSTI) project No. JCYJ20190808153619413.

References

1. Batson, J., Royer, L.: Noise2self: Blind denoising by self-supervision. In: International Conference on Machine Learning, pp. 524–533. PMLR (2019)
2. Bora, A., Price, E., Dimakis, A.G.: Ambientgan: Generative models from lossy measurements. In: International Conference on Learning Representations (2018)
3. Buchholz, T.O., Jordan, M., Pigino, G., Jug, F.: Cryo-CARE: content-aware image restoration for cryo-transmission electron microscopy data. In: 2019 IEEE 16th International Symposium on Biomedical Imaging (ISBI 2019), pp. 502–506. IEEE (2019)
4. Ehret, T., Davy, A., Morel, J.M., Facciolo, G., Arias, P.: Model-blind video denoising via frame-to-frame training. In: Proceedings of the IEEE/CVF Conference on Computer Vision and Pattern Recognition, pp. 11369–11378 (2019)
5. Guo, S., Yan, Z., Zhang, K., Zuo, W., Zhang, L.: Toward convolutional blind denoising of real photographs. In: Proceedings of the IEEE Computer Society Conference on Computer Vision and Pattern Recognition, vol. 2019 (2019). https://doi.org/10.1109/CVPR.2019.00181
6. Hariharan, S.G., et al.: Learning-based x-ray image denoising utilizing model-based image simulations. In: Shen, D., et al. (eds.) MICCAI 2019. LNCS, vol. 11769, pp. 549–557. Springer, Cham (2019). https://doi.org/10.1007/978-3-030-32226-7_61
7. Huang, T., Li, S., Jia, X., Lu, H., Liu, J.: Neighbor2Neighbor: self-supervised denoising from single noisy images. In: Proceedings of the IEEE/CVF Conference on Computer Vision and Pattern Recognition, pp. 14781–14790 (2021)
8. Izadi, S., Mirikharaji, Z., Zhao, M., Hamarneh, G.: Whitenner-blind image denoising via noise whiteness priors. In: Proceedings of the IEEE/CVF International Conference on Computer Vision Workshops (2019)
9. Krull, A., Buchholz, T.O., Jug, F.: Noise2void-Learning denoising from single noisy images. In: Proceedings of the IEEE Computer Society Conference on Computer Vision and Pattern Recognition 2019, pp. 2124–2132 (2019). https://doi.org/10.1109/CVPR.2019.00223
10. Krull, A., Vičar, T., Prakash, M., Lalit, M., Jug, F.: Probabilistic noise2void: Unsupervised content-aware denoising. Front. Comput. Sci. **2**, 5 (2020)
11. Laine, S., Karras, T., Lehtinen, J., Aila, T.: High-quality self-supervised deep image denoising. Adv. Neural. Inf. Process. Syst. **32**, 6970–6980 (2019)
12. Lehtinen, J., et al.: Noise2Noise: learning Image Restoration without Clean Data. In: Dy, J., Krause, A. (eds.) Proceedings of the 35th International Conference on Machine Learning. Proceedings of Machine Learning Research, vol. 80, pp. 2965–2974. PMLR (2018)
13. Martin, D., Fowlkes, C., Tal, D., Malik, J.: A database of human segmented natural images and its application to evaluating segmentation algorithms and measuring ecological statistics. In: Proceedings 8th International Conference on Computer Vision, vol. 2, pp. 416–423 (2001)

14. Moran, N., Schmidt, D., Zhong, Y., Coady, P.: Noisier2noise: learning to denoise from unpaired noisy data. In: Proceedings of the IEEE/CVF Conference on Computer Vision and Pattern Recognition, pp. 12064–12072 (2020)
15. Nam, S., Hwang, Y., Matsushita, Y., Kim, S.J.: A holistic approach to cross-channel image noise modeling and its application to image denoising. In: Proceedings of the IEEE Conference on Computer Vision and Pattern Recognition, pp. 1683–1691 (2016)
16. Pang, T., Zheng, H., Quan, Y., Ji, H.: Recorrupted-to-Recorrupted: unsupervised deep learning for image denoising. In: Proceedings of the IEEE/CVF Conference on Computer Vision and Pattern Recognition, pp. 2043–2052 (2021)
17. Paszke, A., et al.: PyTorch: an imperative style, high-performance deep learning library. Adv. Neural. Inf. Process. Syst. **32**, 8026–8037 (2019)
18. Quan, Y., Chen, M., Pang, T., Ji, H.: Self2self with dropout: Learning self-supervised denoising from single image. Proceedings of the IEEE Computer Society Conference on Computer Vision and Pattern Recognition, pp. 1887–1895 (2020). https://doi.org/10.1109/CVPR42600.2020.00196
19. Ronneberger, O., Fischer, P., Brox, T.: U-Net: convolutional networks for biomedical image segmentation. In: Navab, N., Hornegger, J., Wells, W.M., Frangi, A.F. (eds.) MICCAI 2015. LNCS, vol. 9351, pp. 234–241. Springer, Cham (2015). https://doi.org/10.1007/978-3-319-24574-4_28
20. Wu, D., Gong, K., Kim, K., Li, X., Li, Q.: Consensus neural network for medical imaging denoising with only noisy training samples. In: Shen, D., et al. (eds.) MICCAI 2019. LNCS, vol. 11767, pp. 741–749. Springer, Cham (2019). https://doi.org/10.1007/978-3-030-32251-9_81
21. Xie, Y., Wang, Z., Ji, S.: Noise2same: optimizing a self-supervised bound for image denoising. arXiv preprint arXiv:2010.11971 (2020)
22. Xu, J., Li, H., Liang, Z., Zhang, D., Zhang, L.: Real-world noisy image denoising: a new benchmark. arXiv preprint arXiv:1804.02603 (2018)
23. Zhang, K., Zuo, W., Chen, Y., Meng, D., Zhang, L.: Beyond a gaussian denoiser: Residual learning of deep CNN for image denoising. IEEE Trans. Image Process. **26**(7), 3142–3155 (2017)
24. Zhang, Y., et al.: A poisson-gaussian denoising dataset with real fluorescence microscopy images. In: Proceedings of the IEEE/CVF Conference on Computer Vision and Pattern Recognition, pp. 11710–11718 (2019)
25. Zhang K., Z.W., Zhang, L.: FFDNet: toward a fast and flexible solution for CNN-based image denoising. IEEE Trans. Image Process. **27**(9), 4608–4622 (2018)
26. Zhussip, M., Soltanayev, S., Chun, S.Y.: Extending stein's unbiased risk estimator to train deep denoisers with correlated pairs of noisy images. Adv. Neural. Inf. Process. Syst. **32**, 1465–1475 (2019)

Fusion-Based Low-Light Image Enhancement

Haodian Wang, Yang Wang, Yang Cao, and Zheng-Jun Zha(✉)

University of Science and Technology of China, Hefei 230027, China
haodianwang@mail.ustc.edu.cn,
{ywang120,forrest,zhazj}@ustc.edu.cn

Abstract. Recently, deep learning-based methods have made remarkable progress in low-light image enhancement. In addition to poor contrast, the images captured under insufficient light suffer from severe noise and saturation distortion. Most existing unsupervised learning-based methods adopt the two-stage processing method to enhance contrast and denoise sequentially. However, the noise will be amplified in the contrast enhancement process, thus increasing the difficulty of denoising. Besides, the saturation distortion caused by insufficient illumination is not considered well in existing unsupervised low-light enhancement methods. To address the above problems, we propose a novel parallel framework, which includes a saturation adaptive adjustment branch, brightness adjustment branch, noise suppression branch, and fusion module for adjusting saturation, correcting brightness, denoise, and multi-branch fusion, respectively. Specifically, the saturation is corrected via global adjustment, the contrast is enhanced through curve mapping estimation, and we use BM3D to preliminary denoise. Further, the enhanced branches are fed to the fusion module for a trainable guided filter, which is optimized in an unsupervised training manner. Experiments on the LOL, MIT-Adobe 5k, and SICE datasets demonstrate that our method achieves better quantitation and qualification results than the state-of-the-art algorithms.

Keywords: Unsupervised · Low-light enhancement · Noise suppression · Saturation correction

1 Introduction

High-quality images are critical to computer vision tasks. However, due to the technical conditions and lighting limitations, images captured in insufficient light conditions inevitably appear with low contrast and unexpected noise and color shift, which will degrade both perceptual quality and downstream high-level vision tasks, such as object detection [16] and tracking [5]. Therefore, improving the quality of low-light images is urgently needed and has drawn significant attention in recent years.

D.-T. Dang-Nguyen et al. (Eds.): MMM 2023, LNCS 13833, pp. 121–133, 2023.
https://doi.org/10.1007/978-3-031-27077-2_10

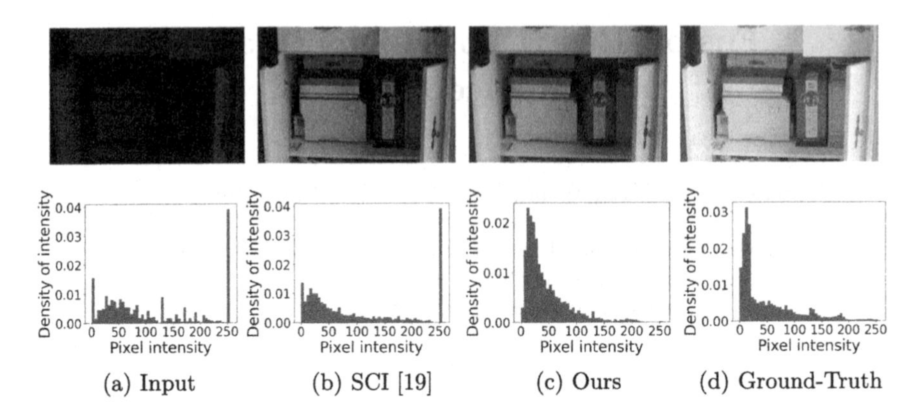

(a) Input (b) SCI [19] (c) Ours (d) Ground-Truth

Fig. 1. The statistical histograms of corresponding saturation channel by different enhancement methods. Our method achieves better saturation correction compared with SCI [19], which is the state-of-the-art LLIE method.

Recently, various supervised learning-based methods have been proposed for low-light image enhancement(LLIE) [8,12,17,23,25,27,28]. Nevertheless, due to training a deep model on paired data that may result in overfitting and limited generalization capability [15], unsupervised learning-based methods [4,8,10,12,17,19] are extensively used to perform LLIE. Some of the existing unsupervised LLIE algorithms adopt the end-to-end "one-stage method" [8,12,19] to enhance brightness, which does not sufficiently suppress the noise while enhancing the brightness. Therefore, some "two-stage methods" [4,10,17] are proposed, which in a "first enhancement then denoising" manner to enhance image contrast and suppress noise. However, this cascade processing method may lead to the accumulation of artifacts, and the noise will be amplified in the brightness enhancement process, thus increasing the difficulty of denoising. Furthermore, the existing unsupervised low-light enhancement algorithms [4,8,10,12,17,19] do not consider the saturation distortion caused by insufficient illumination, which leads to the incorrect color of the restored results, such as Fig. 1.

To address the above issues, we propose a parallel framework, which includes a saturation adaptive adjustment branch, brightness adjustment branch, noise suppression branch, and a fusion module for adjusting saturation, correcting brightness, denoise, and multi-branch fusion, respectively. Specifically, we propose a novel saturation adaptive adjustment method based on Gray World Algorithm [7] and Von Kries diagonal model [13] to adjust the saturation in HSV color space. As shown in Fig. 1, our method achieves better saturation correction than the state-of-the-art unsupervised method SCI [19]. Then we design a brightness adjustment branch, which uses the high-order mapping curve to adjust the original V channel at the pixel level in HSV color space. It takes the low-light image as the input and the parameter of the high-order mapping curve as the output. The estimated coefficients are used to adjust the dynamic range of the input through the high-order curve to obtain the corrected V channel. Meanwhile, we use BM3D [6] to initially denoise the input image in the noise suppression branch of the parallel framework. Finally, the output images of each branch of the parallel framework are fused through a fusion module composed of a trainable unsupervised guided filter network. The whole parallel framework is trained by unsupervised learning. It avoids the problem of insufficient denoise caused by the one-stage processing methods and intractable noise removal caused by the two-stage processing methods.

Our contributions are summarized as follows:

(1) We propose a parallel fusion framework to simultaneously perform contrast enhancement, saturation correction, and denoising. The whole framework is trained in an unsupervised manner.
(2) We propose a novel saturation adaptive adjustment method based on Gray World Algorithm and Von Kries diagonal model to correct saturation.
(3) We introduce a trainable unsupervised guided filter fusion module in multi-branch fusion to further suppress noise.
(4) Experiments on the LOL, MIT-Adobe 5k, and SICE datasets show that the proposed method achieves better saturation correction and noise suppression.

2 Related Work

2.1 Traditional LLIE Methods

Early LLIE methods are based on various image priors. For example, HE-based methods [21] focus on changing the image's dynamic range to improve the contrast, which may lead to insufficient or excessive results enhancement. Inspired by Retinex [14] theory, some methods decompose the image into pixel-level products of reflection and illumination map, and the enhanced results can be obtained through further processing. However, this method relies on intensive parameter adjustment, leading to inconsistent colors and noise in the enhanced results.

2.2 Deep Learning-Based LLIE Methods

In recent years, the method based on deep learning has shown surprising results on LLIE. High-quality normal light is usually used as the ground truth to guide low-light image enhancement. The first learning-based LLIE method, LL-Net [18], proposes a stacked automatic encoder for simultaneous denoising and enhancement. The work in [25] is based on Retinex theory and uses special subnetworks to enhance the illuminance and reflectance component, respectively. Zhang *et al.* [30] propose KinD, which uses three subnetworks for layer decomposition, reflectivity recovery, and illumination adjustment. The work in [28] proposes a recursive band network and trains it through a semi-supervised strategy. However, training a depth model on paired data may lead to overfitting and limit the generalization ability, so the unsupervised LLIE method has been widely used in recent years. EnGAN [12] proposed a generator and used unpaired data for training. ZeroDCE [8] designed a depth curve estimation network to adjust the dynamic range of low illumination images. Liu *et al.* [17] built a Retinex-inspired unrolling framework with architecture search. SCI [19] presents a lightweight enhancement network and achieves state-of-the-art.

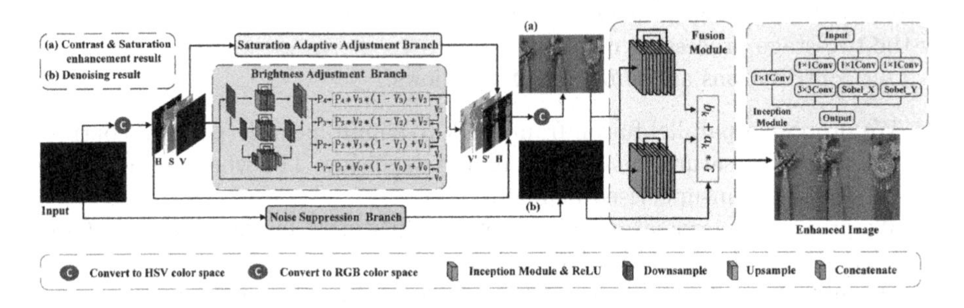

Fig. 2. Parallel fusion frame diagram, where the red box represents the brightness adjustment branch, and the green box represents the inception module. (Color figure online)

3 Proposed Method

The proposed framework is shown in Fig. 2. The framework consists of three main branches and a fusion module, and the input image is enhanced by brightness and saturation correction and denoising in parallel before fusion separately. The input image is converted from the RGB to HSV color space, and the S and V channels are corrected by the saturation and brightness correction branches, respectively. Meanwhile, the noise suppression branch performs preliminary denoise for input image parallelly on RGB color space. Finally, the output of each branch is fused through the fusion module. We explain the role of each branch and module in this section.

3.1 Saturation Adaptive Adjustment Branch

To solve the problem of saturation distortion, we propose a saturation adaptive adjustment method. It is assumed that the relative values of the three components of RGB remain unchanged after adjusting the saturation of an image. According to the Gray World Algorithm [7] and Von Kries diagonal model [13]:

$$\begin{cases} M'(x) = M(x) \cdot \dfrac{K}{\overline{M}(x)} \\ N'(x) = N(x) \cdot \dfrac{K}{\overline{N}(x)} \end{cases}, \tag{1}$$

where $M(x) = max\{R, G, B\}$, $N(x) = min\{R, G, B\}$ represent the maximum and minimum channels in the RGB color space, respectively. $\overline{M}(x)$ and $\overline{N}(x)$ represent the average value of the maximum value channel and the minimum value channel respectively. $M'(x)$ and $N'(x)$ represent the adjusted channel, and K is the gain coefficient. Corresponding to the HSV color space, the adjustment formula of the saturation channel can be derived:

$$\begin{aligned} S' &= \frac{M'(x) - N'(x)}{M'(x)} = 1 - \frac{N(x) \cdot \frac{K}{\overline{N}(x)}}{M(x) \cdot \frac{K}{\overline{M}(x)}} \\ &= 1 - (1 - S) \cdot \frac{\overline{M}(x)}{\overline{N}(x)} \end{aligned} \tag{2}$$

where S' is the saturation channel after adjustment. We adjust the saturation channel according to Eq. 2.

3.2 Brightness Adjustment Branch

Inspired by [8], we perform a parameters estimation network to estimate a set of best-fitting parameters of the pixel-wise light enhancement curve to adjust the brightness, as shown in the red box in Fig. 2. The module maps all pixels of the V channel by applying the curve parameters iteratively to obtain the final enhanced V channel. The iterative process is as follows:

$$I_n(x) = I_{n-1}(x) + P_n(x) \cdot I_{n-1}(x) \cdot (1 - I_{n-1}(x)), n = \{1, 2, 3, 4\}, \tag{3}$$

where x denotes pixel coordinates, n is the number of iterations, $P_n(x)$ is a parameter map with the same size as the given V channel. $I_n(x)$ is the result of each iteration, and $I_0(x)$ is the V channel of the original low-light image. We perform four iterations on the input V channel.

Unlike [8], we designed a multi-scale network to extract the input brightness information more effectively. Moreover, we only adjust the brightness in the V channel instead of in the RGB color channel. The parameter estimation network downsamples the input to 1/2, 1/4, and 1/8 size and then passes that through five layers of skip-connected inception module and ReLU, respectively. The features of the smallest size are upsampled and gradually fused with the features of larger size to obtain the parameters $P_n(x)$ of the curve. Finally, the iterative mapping is performed according to Eq. 3 to obtain the output. The inception module is shown in the green box of Fig. 2, which consists of parallel 1×1 convolution, 3×3 convolution, and horizontal and vertical Sobel filters.

3.3 Noise Suppression Branch

To obtain a relatively clean image and retain the edge and detail information of the input image, we perform preliminary denoise on the original image through the noise suppression branch. As shown in the lower branch of the Fig. 2, we use BM3D [6] as an initial denoise method. This branch can be replaced by any denoise method, and the denoise result will affect the final image quality.

3.4 Fusion Module

We design an unsupervised guided filtering network to fuse different branches in the parallel framework. We use the image with sharp edges obtained by the lower branch in the parallel framework to guide the images with correct brightness and saturation obtained by the upper branch for fusion, which removes noise and retains the proper brightness and saturation information.

Trainable Guided Filter. Traditional guided filtering [9] assumes that the guide and output images are local linearly correlated within the filter window. Suppose q is a linear transformation of guide image G in a window ω_k centered on pixel k:

$$q_i = a_k \cdot G_i + b_k \,, \forall i \in \omega_k \,, \tag{4}$$

where a_k, b_k are linear coefficients. As shown in the yellow box in Fig. 2, the guided filter fusion module takes the noisy image and the guide image as the input. It uses the five layers skip connected inception module to obtain the corresponding linear coefficients a_k and b_k, and calculate the output result of the fusion module according to Eq. 4.

Unsupervised Training Framework. The existing trainable guided filters [26] are almost all supervised. We split the input noise image according to the neighborhood of each 2×2 window and randomly select two adjacent pixels in each window as the split two images N_1 and N_2. So we get two images with independent conditions but similar contents. According to [11], the optimization problem is transformed into:

$$\arg\min_{\theta} \mathbb{E} \left\| f_\theta(N_1) - N_2 \right\|_2^2 \,, \tag{5}$$

where f_θ is the trainable guide filter denoising network parameterized by θ. Thus we implement unsupervised trainable guided filtering.

3.5 Loss Function

To train the **Brightness Adjustment Branch**, we use an exposure control loss L_{exp} to measure the distance between the average intensity value I_k of a local region to the well-exposedness level E:

$$L_{exp} = \frac{1}{R} \sum_{k=1}^{R} \left\| I_k - E \right\| \,, \tag{6}$$

where R represents the number of non-overlapping local regions of size 16×16, and we set E to 0.6 in our experiments. To preserve the monotonicity relations

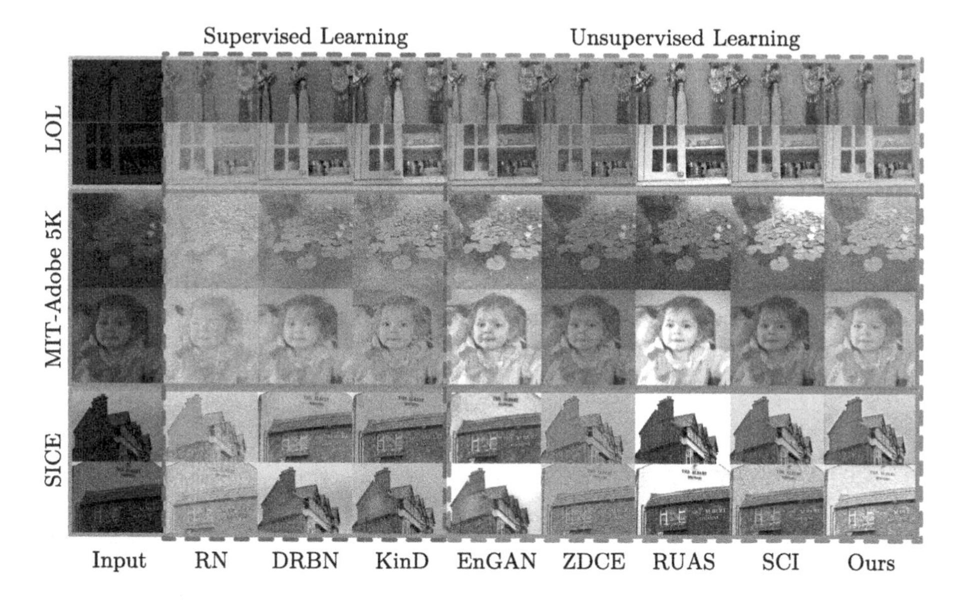

Fig. 3. Subjective results comparison on the LOL dataset [25], the MIT-Adobe 5K dataset [2] and the SICE dataset [3]. RN means RetinexNet, ZDCE means ZeroDCE. Compared with other methods, our method performs better in saturation and noise suppression.

between neighboring pixels, we add an illumination smoothness loss L_{tv} to each curve parameter map P_n:

$$L_{tv} = \frac{1}{T}\sum_{n=1}^{T}(\nabla_x P_n + \nabla_y P_n)^2 , \tag{7}$$

where T is the number of iterations, ∇_x and ∇_y represent the horizontal and vertical gradient operations, respectively. The total loss for training the parameter estimation network can be expressed as: $L_{total_1} = L_{exp} + \lambda_{tv} L_{tv}$, where λ_{tv} is the weight of the loss.

To train the **Fusion Module**, we use reconstruction loss L_{rec} to ensure structural similarity between the noisy input and output:

$$L_{rec} = \|f_\theta(N_1) - N_2\|_2^2 . \tag{8}$$

Since the ground-truth of N_1 and N_2 are different, directly applying reconstruction loss is inappropriate and leads to over-smoothing. So we add a regularization term loss L_{reg} [11]:

$$L_{reg} = \|f_\theta(N_1) - N_2 - (O_1 - O_2)\|_2^2 , \tag{9}$$

where O_1 and O_2 represent the two images split from the output. The total loss for training the trainable guided filter can be expressed as: $L_{total_2} = \lambda_{rec} L_{rec} + \lambda_{reg} L_{reg}$, where λ_{rec} and λ_{reg} are the weights of the losses.

4 Experiments

4.1 Experimental Setting

We conduct experiments on the LOL dataset [25], the MIT-Adobe 5K [2] dataset, and the SICE dataset [3]. We randomly sample 100 images from the MIT dataset for testing and others for training. The SICE dataset is a multi-exposure dataset consisting of 7 (or 9) pictures in various exposure levels for each scene. We select the first picture with the worst exposure in each scene in part II as the test images and select the third (or fourth) image as ground-truth.

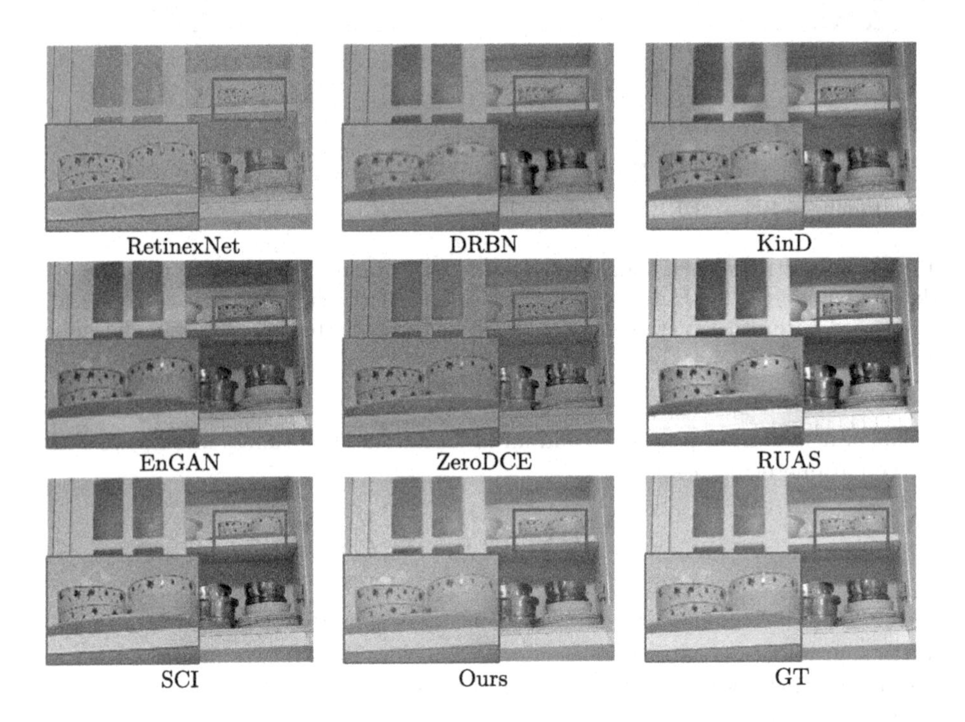

Fig. 4. Detailed comparison of existing methods on the LOL dataset.

4.2 Implementation Details

We implement our framework using PyTorch on an NVIDIA 3090 GPU and separately train the brightness adjustment module and the fusion module. We adopt the Adam optimizer with a learning rate of 0.0001. The cropped image and batch sizes are set to 512 and 8, respectively. λ_{tv} is set to 200, λ_{reg} and λ_{rec} are both set to 1.

Table 1. Quantitative results in terms of four full-reference metrics including PSNR, SSIM, LPIPS and MSE, and three no-reference metrics including LOE, NIQE, and EME on the LOL, MIT-Adobe 5K(MIT) and SICE datasets. The best result is shown in red, and the second-best result is blue.

Dataset	Metrics	Supervised Learning Methods			Unsupervised Learning Methods				
		RetinexNet	DRBN	KinD	EnGAN	ZeroDCE	RUAS	SCI	Ours
LOL	PSNR↑	17.5624	18.7139	20.0795	17.5713	16.8937	16.4268	18.2586	20.0607
	SSIM↑	0.6517	0.7344	0.8297	0.6537	0.5579	0.5039	0.5414	0.7132
	LPIPS↓	0.2730	0.1959	0.1776	0.2576	0.2701	0.2144	0.2892	0.2284
	MSE↓	1339.01	1871.47	1217.12	1978.54	1941.93	2543.65	1747.39	1222.38
	LOE↓	404.70	590.02	280.71	398.44	306.85	219.16	243.67	234.57
	NIQE↓	2.8993	3.6271	3.4088	3.2358	4.8792	4.1962	4.8968	3.1138
	EME↑	9.7901	10.1642	10.2616	15.0098	14.6715	14.6123	16.8015	16.8881
MIT	PSNR↑	15.7520	17.0130	18.7865	16.2985	18.0688	18.1560	19.3030	20.5714
	SSIM↑	0.6625	0.8013	0.8022	0.7953	0.7823	0.8022	0.8176	0.8046
	LPIPS↓	0.1763	0.2360	0.1495	0.1753	0.1547	0.1441	0.1312	0.1341
	MSE↓	1526.35	1159.68	1138.72	1195.00	1210.00	1368.95	1057.32	1025.54
	LOE↓	496.13	419.55	334.81	433.45	72.11	131.07	89.74	58.72
	NIQE↓	2.5585	2.8678	2.4399	2.3871	2.5323	3.8390	2.4482	2.4181
	EME↑	5.3471	4.5468	7.4647	6.5967	7.2509	5.2263	8.7124	8.7796
SICE	PSNR↑	17.2503	20.1136	21.3907	18.6267	19.9799	15.1072	20.7242	21.9516
	SSIM↑	0.7718	0.8183	0.8637	0.8476	0.8353	0.7295	0.8505	0.8618
	LPIPS↓	0.1974	0.1523	0.1162	0.1312	0.1408	0.2440	0.1416	0.1353
	MSE↓	1416.75	716.69	541.25	1123.68	756.43	2441.73	771.33	514.89
	LOE↓	645.69	515.43	409.92	527.92	436.33	485.80	408.06	405.83
	NIQE↓	2.7035	2.7721	2.5302	2.5556	2.6319	3.8231	2.5181	2.5152
	EME↑	8.2007	8.5509	10.6299	9.7228	9.4614	9.5468	12.5989	9.8908

4.3 Experimental Results

We compare our method with three advanced supervised learning methods, including RetinexNet [25], DRBN [28], and KinD [30], and four unsupervised learning methods, including EnGAN [12], ZeroDCE [8], RUAS [17], and SCI [19].

Qualitative Evaluation. We present the visual comparisons on typical low-light images in Fig. 3. Most of the previous methods cannot recover global illumination and structure well, such as RetinexNet [25] and RUAS [17], and uneven enhancement may occur in some areas of the image, such as KinD [30] and RUAS [17].Meanwhile, as shown in Fig. 4, most existing methods do not correct saturation and suppress noise well, such as DRBN [28], EnGAN [12], ZeroDCE [8], SCI [19] and RUAS [17]. Comparatively, our method achieves good perceptual visual quality, with proper illumination, saturation, as well as clean and sharp details.

Quantitative Evaluation. As shown in Table 1, We perform quantitative comparisons of different methods, and we use four full-reference metrics including PSNR, SSIM [24], LPIPS [29], and MSE, three no-reference metrics including

EME [1], LOE [22] and NIQE [20]. Our method achieves excellent results in most indicators compared with the existing methods. The results show that we have advantages in light restoration and structural restoration.

4.4 Ablation Study

In order to investigate the effectiveness of different components of our method, we conduct ablation experiments on several key components, including the proposed module and our proposed parallel framework.

Contribution of Each Branch. To verify the effectiveness of each branch, we conduct ablation studies on the LOL dataset [25]. Subjective experimental results and quantitative comparisons are shown in Fig. 5 and Table 2. The image will be very dark without the brightness adjustment branch. The overall image will be greenish without the saturation adaptive adjustment branch. Without the fusion module, the image will have severe noise.

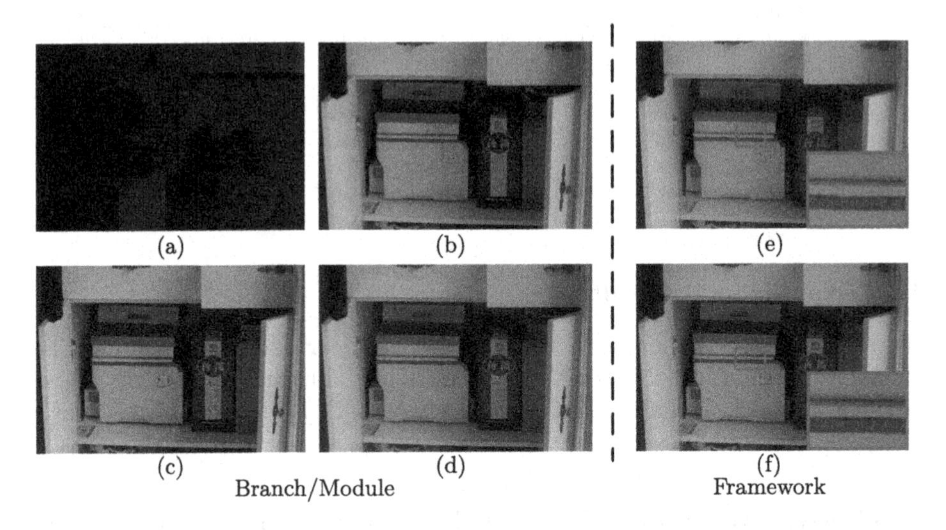

Fig. 5. Ablation on different branches/modules and frameworks. The first to sixth panels: (a) w/o brightness adjustment branch. (b) w/o saturation adaptive adjustment branch. (c) w/o fusion module. (d) full branches and module. (e) cascade framework. (f) our proposed parallel framework.

Effect of Parallel Framework. To verify the framework's effectiveness, we compare our proposed parallel framework with a "two-stage" cascade framework. We compare the experimental results of the cascade framework that removed the original noise suppression branch and changed it to self-guided filtering based on the output results of the original three branches. From Fig. 5 and Table 2, we can observe that our results are less noisy.

Table 2. Ablation on different branches/modules and frameworks. The best results are highlighted in bold. BAB means brightness adjustment branch, SAAB means saturation adaptive adjustment branch, FM means fusion module.

framework	w/o	LOE↓	NIQE↓	EME↑	PSNR↑	SSIM↑	LPIPS↓	MSE↓
parallel	BAB	308.93	4.7216	13.129	7.8010	0.2018	0.3533	12524.61
parallel	SAAB	245.828	3.6069	11.495	19.300	0.7091	0.2797	1449.928
parallel	FM	235.224	5.0333	12.12	19.1773	0.5458	0.3067	1347.22
cascade	\	246.522	4.992	11.253	19.6521	0.6755	0.2364	1289.655
parallel	\	**234.573**	**3.1138**	**16.888**	**20.0607**	**0.7132**	**0.2284**	**1222.377**

5 Conclusion

In this work, we propose a parallel unsupervised LLIE framework to improve brightness, correct saturation, and denoise, respectively. We designed a saturation correction branch based on the Gray World Algorithm and Von Kries model to correct saturation. In addition, we designed an unsupervised guided filtering module at the end of the parallel framework to fuse different branches. Various experiments show that our methods achieve better results quantitatively and qualitatively compared with the state-of-the-art unsupervised methods.

Acknowledgements. This work was supported by National Natural Science Foundation of China (No. 62206262).

References

1. Agaian, S.S., Silver, B., Panetta, K.A.: Transform coefficient histogram-based image enhancement algorithms using contrast entropy. IEEE Trans. Image Proc. **16**(3), 741–758 (2007)
2. Bychkovsky, V., Paris, S., Chan, E., Durand, F.: Learning photographic global tonal adjustment with a database of input/output image pairs. In: CVPR 2011, pp. 97–104. IEEE (2011)
3. Cai, J., Gu, S., Zhang, L.: Learning a deep single image contrast enhancer from multi-exposure images. IEEE Trans. Image Proc. **27**(4), 2049–2062 (2018)
4. Chang, R., Song, Q., Wang, Y.: RGNET: A two-stage low-light image enhancement network without paired supervision. In: 2021 International Joint Conference on Neural Networks (IJCNN), pp. 1–7. IEEE (2021)
5. Chu, Q., Ouyang, W., Li, H., Wang, X., Liu, B., Yu, N.: Online multi-object tracking using CNN-based single object tracker with spatial-temporal attention mechanism. In: Proceedings of the IEEE International Conference on Computer Vision, pp. 4836–4845 (2017)
6. Dabov, K., Foi, A., Katkovnik, V., Egiazarian, K.: Image denoising by sparse 3-d transform-domain collaborative filtering. IEEE Trans. Image proc. **16**(8), 2080–2095 (2007)
7. Gasparini, F., Schettini, R.: Color correction for digital photographs. In: 12th International Conference on Image Analysis and Processing, 2003. Proc, pp. 646–651. IEEE (2003)

8. Guo, C., Li, C., Guo, J., Loy, C.C., Hou, J., Kwong, S., Cong, R.: Zero-reference deep curve estimation for low-light image enhancement. In: Proceedings of the IEEE/CVF Conference on Computer Vision and Pattern Recognition, pp. 1780–1789 (2020)
9. He, K., Sun, J., Tang, X.: Guided image filtering. IEEE Trans. pattern Anal. Mach. Intell. **35**(6), 1397–1409 (2012)
10. Hu, J., Guo, X., Chen, J., Liang, G., Deng, F., Lam, T.L.: A two-stage unsupervised approach for low light image enhancement. IEEE Robot. Autom. Lett. **6**(4), 8363–8370 (2021). https://doi.org/10.1109/LRA.2020.3048667
11. Huang, T., Li, S., Jia, X., Lu, H., Liu, J.: Neighbor2neighbor: Self-supervised denoising from single noisy images. In: Proceedings of the IEEE/CVF conference on computer vision and pattern recognition, pp. 14781–14790 (2021)
12. Jiang, Y., Gong, X., Liu, D., Cheng, Y., Fang, C., Shen, X., Yang, J., Zhou, P., Wang, Z.: Enlightengan: Deep light enhancement without paired supervision. IEEE Trans. Image Proc. **30**, 2340–2349 (2021)
13. Kries, J.: Die gesichtsempfindungen. handbuch der physiologie des menschen von nagel, bd. 3 (1905)
14. Land, E.H.: The retinex theory of color vision. Sci. Am. **237**(6), 108–129 (1977)
15. Li, C., et al.: Low-light image and video enhancement using deep learning: A survey. In: IEEE Trans. Pattern Anal. Mach. Intell, pp. 1–1 (2021). https://doi.org/10.1109/TPAMI.2021.3126387
16. Lin, T.Y., Dollár, P., Girshick, R., He, K., Hariharan, B., Belongie, S.: Feature pyramid networks for object detection. In: Proceedings of the IEEE Conference on Computer vision and pattern Recognition, pp. 2117–2125 (2017)
17. Liu, R., Ma, L., Zhang, J., Fan, X., Luo, Z.: Retinex-inspired unrolling with cooperative prior architecture search for low-light image enhancement. In: Proceedings of the IEEE/CVF Conference on Computer Vision and Pattern Recognition, pp. 10561–10570 (2021)
18. Lore, K.G., Akintayo, A., Sarkar, S.: LLNet: A deep autoencoder approach to natural low-light image enhancement. Pattern Recognit. **61**, 650–662 (2017)
19. Ma, L., Ma, T., Liu, R., Fan, X., Luo, Z.: Toward fast, flexible, and robust low-light image enhancement. In: Proceedings of the IEEE/CVF Conference on Computer Vision and Pattern Recognition, pp. 5637–5646 (2022)
20. Mittal, A., Soundararajan, R., Bovik, A.C.: Making a "completely blind" image quality analyzer. IEEE Signal Proc. Lett. **20**(3), 209–212 (2012)
21. Pizer, S.M., Amburn, E.P., Austin, J.D., Cromartie, R., Geselowitz, A., Greer, T., ter Haar Romeny, B., Zimmerman, J.B., Zuiderveld, K.: Adaptive histogram equalization and its variations. Comput. Vis. Graph. Image Proc. **39**(3), 355–368 (1987)
22. Wang, S., Zheng, J., Hu, H.M., Li, B.: Naturalness preserved enhancement algorithm for non-uniform illumination images. IEEE Trans. Image Proc. **22**(9), 3538–3548 (2013)
23. Wang, Y., et al.: Progressive Retinex: Mutually reinforced illumination-noise perception network for low-light image enhancement. In: Proceedings of the 27th ACM international conference on multimedia, pp. 2015–2023 (2019)
24. Wang, Z., Bovik, A.C., Sheikh, H.R., Simoncelli, E.P.: Image quality assessment: from error visibility to structural similarity. IEEE Trans. Image proc. **13**(4), 600–612 (2004)
25. Wei, C., Wang, W., Yang, W., Liu, J.: Deep retinex decomposition for low-light enhancement. arXiv preprint arXiv:1808.04560 (2018)

26. Wu, H., Zheng, S., Zhang, J., Huang, K.: Fast end-to-end trainable guided filter. In: Proceedings of the IEEE Conference on Computer Vision and Pattern Recognition, . pp. 1838–1847 (2018)
27. Xu, K., Yang, X., Yin, B., Lau, R.W.: Learning to restore low-light images via decomposition-and-enhancement. In: Proceedings of the IEEE/CVF Conference on Computer Vision and Pattern Recognition, pp. 2281–2290 (2020)
28. Yang, W., Wang, S., Fang, Y., Wang, Y., Liu, J.: From fidelity to perceptual quality: a semi-supervised approach for low-light image enhancement. In: Proceedings of the IEEE/CVF conference on computer vision and pattern recognition, pp. 3063–3072 (2020)
29. Zhang, R., Isola, P., Efros, A.A., Shechtman, E., Wang, O.: The unreasonable effectiveness of deep features as a perceptual metric. In: Proceedings of the IEEE conference on computer vision and pattern recognition, pp. 586–595 (2018)
30. Zhang, Y., Guo, X., Ma, J., Liu, W., Zhang, J.: Beyond brightening low-light images. Int. J. Comput. Vis. **129**(4), 1013–1037 (2021)

Towards Interactive Facial Image Inpainting by Text or Exemplar Image

Ailin Li, Lei Zhao(✉), Zhiwen Zuo, Zhizhong Wang, Wei Xing, and Dongming Lu

College of Computer Sience and Technology, Zhejianng University, Hangzhou, China
{liailin,cszhl,zzwcs,endywon,wxing,ldm}@zju.edu.cn

Abstract. Facial image inpainting aims to fill visually realistic and semantically new pixels for masked or missing pixels in a face image. Although current methods have made progress in achieving high visual quality, the controllable diversity of face inpainting remains an open issue. This paper proposes a new facial image inpainting interaction mode, which enables filling semantic contents based on the texts or exemplar images provided by users. We use the powerful image-text representation abilities from the recently introduced Contrastive Language-Image Pre-training (CLIP) models to achieve this interactive face inpainting. We present two thoughts on our method. Specifically, we first explore a simple and effective optimization-based text-guided facial inpainting method in which a CLIP model is used as a loss network to modify the latent code iteratively in response to the text prompt. Next, we describe a multi-modal inpainting mapper network to map the input conditions (e.g., text or image) into corresponding latent code changes, supporting the guidance of different text prompts and exemplars within one model. We also introduce an exemplar-semantic similarity loss, which maps the inpainted facial image and the exemplar image into the CLIP's space to measure their similarity. This loss enables the generated image to include high-level semantic attributes from the exemplar image. Through extensive experiments, we demonstrate the effectiveness of our method in interactive facial inpainting based on the texts or exemplars.

Keywords: Multi-modal learning · Image inpainting · Vision-language pre-trained model

1 Introduction

Facial image inpainting is an interesting branch of image inpainting that targets generating visually realistic content in missing or masked regions while keeping semantic coherence in face images. It backs various applications, such as image extrapolation, portrait retouching, and so on. Compared to natural scenes, the face inpainting task is more challenging as it often requires generating semantic contents for the missing key components (e.g., eyes and mouths) that contain

D.-T. Dang-Nguyen et al. (Eds.): MMM 2023, LNCS 13833, pp. 134–148, 2023.
https://doi.org/10.1007/978-3-031-27077-2_11

Fig. 1. Examples of our face inpainting guided by text prompt and exemplar. The first row shows the reasonable outputs of text-driven latent optimization method and the second row shows both the text- or exemplar-guided results of inpainting mapper.

large appearance variations. In recent years, with the development of deep learning, many GAN-based image inpainting methods can generate high-quality visual content. Most of these approaches [7] generally attempt to generate one single result from a given image, ignoring many other plausible solutions.

However, image inpainting is intuitively not a deterministic problem. Usually, multiple solutions satisfy the semantically meaningful requirement of inpainting. Recently, several methods [24,25] focus on generating diverse image inpainting results from which users can select the desired one. A disadvantage of the above mentioned methods is that they cannot complete missing regions in a user-friendly way. Some methods attempt to add extra information such as edges [10], geometries [14,16] to guide the inpainting of structures and attributes. While these methods all enable a user to paint images using a limited vocabulary of semantic concepts, that is, the finite guidance information may lead to limited completed solutions. Nowadays, both text and image have become common mediums for daily communication. Inspired by these observations, there raises a question, *"Can we use an intuitive text prompt or an exemplary image to guide the face inpainting with user preferences?"*.

Benefiting from the development of cross-modal vision and language pre-training [12,13], text-guided image generation and manipulation has become possible. StyleCLIP [11] as one of the pioneers leverages the power of recently introduced Contrastive Language-Image Pre-training (CLIP) models [12] in order to achieve intuitive text-based semantic image manipulation. Facial image inpainting is a special case of image generation. Different from generating a completely new image, the requirement for text and exemplar-guided image inpainting is more stringent: the generated parts for missing regions are coherent with the remaining contents and semantically consistent with the text prompt and exemplar.

Inspired by the combination of cross-modal pre-training and powerful generative models, in this work, we attempt to achieve facial image inpainting via describing the desired visual concept with the user-provided text prompts and exemplars with the help of pre-trained models. To this end, we explore the interactive face inpainting method by leveraging the power of the CLIP model and propose two feasible ways, i.e., the optimization-based method and the multi-modal inpainting mapper method. We introduce the optimization-based approach for text-guided facial image inpainting. The multi-modal inpainting mapper considers both text and image conditions. These two approaches contribute to producing desired inpainted contents according to the text prompt and/or exemplar, as shown in Fig. 1, thus also indicating a promising way to strengthen the controllable diversity of facial image inpainting. Note that visual characteristics of the inpainted regions may be explicitly (smile, angry) or implicitly, indicating a real person (Taylor Swift). The main contributions of this paper are listed as follows:

1. A simple text-guided optimization-based method for facial image inpainting is proposed, resorting to the CLIP model as image-text similarity measurement.
2. We also design a multi-modal inpainting mapper network for mapping the input conditions (text or exemplar) into corresponding latent code changes. This solution can directly control the input conditions on the latent codes, and train one model for different texts and exemplars. Moreover, an exemplar-semantic similarity loss is designed to allow the inpainted facial image to contain similar high-level semantic attributes as those of the exemplar.
3. We conduct extensive experiments to demonstrate the effectiveness of our interactive facial image inpainting method with flexible control over masked images using either a text or an exemplar.

2 Related Work

2.1 Image Inpainting

Deep learning based image inpainting methods depend on deep neural networks and generative adversarial networks (GANs) [2] to produce semantic contents. A lot of efforts have been made to improve visual quality of the inpainted images, such as designing novel modules or architectures [6,17,19], adopting coarse-to-fine mechanism [21], improving semantic understanding [9]. However, these image inpainting methods produce only one result for each masked input.

Diverse Image Inpainting. Recently, diverse image inpainting has attracted much attention from many researchers. Pluralistic Image Completion (PIC) [25] proposes a two-path probabilistically principled framework, including two parallel training paths. The VAE-based reconstructive path utilizes the complete ground truth to get the prior distribution of missing regions and reconstruct the original image from this distribution while the generative path predicts the latent prior distribution for the missing regions conditioned on the visible pixels, from which can be sampled to generate random diverse results.

Semantic Image Inpainting. Moreover, by introducing auxiliary information, such as sketches [10], landmarks, geometries [14,16], the structures and attributes of the missing regions can be generated accordingly. However, these interactions are neither straightforward nor efficient. Recently, text as a flexible medium is explored to guide image inpainting systems. TDANet [22] designs a dual multimodal attention mechanism to extract explicit semantic information about the corrupted regions from the descriptive sentence. This may not make sense because some redundant words may introduce noisy pixels when generating predictions for the masked part. In contrast, we optimize or train our approaches resorting to the image-text representation ability in the CLIP space to achieve both text and exemplar image guidance.

2.2 Text-guided Image Generation and Manipulation

Text-guided image manipulation aims at editing images with given text descriptions. The development of text-to-image generation brings the possibility of manipulating images based on the text description. How to map text description to the image domain or learn a cross-modal language-vision representation space is the key to this task due to the domain gap between text and image.

A recent model, based on Contrastive Language-Image Pre-training (CLIP), learns a multi-modal embedding space, which may be used to estimate the semantic similarity between a given text and an image. Due to its powerful text-image representation learning capability, it can be integrated into multiple downstream multi-modal tasks. For instance, much research has attempted to combine StyleGAN and CLIP to achieve text-to-image generation and manipulation [1,11]. StyleCLIP [11] investigates three novel combination techniques to enable a wide variety of unique image manipulations. Inspired by StyleCLIP, we desire to utilize the CLIP to explore the facial image inpainting problem guided by a text or an exemplar prompt which is different from generating a completely new image.

3 Approach

3.1 Overview

In this work, we try to leverage image-text representation power of CLIP to achieve text and exemplar guided facial image inpainting. Two techniques are explored to combine the inpainting model and CLIP: the text-guided optimization-based method and the multi-modal inpainting mapper method. Both methods build upon the well pre-trained encoder and generator of Pluralistic Image Completion (PIC) [25]. Therefore, we first briefly introduce PIC and CLIP, respectively.

PIC. We have mentioned in Sect. 2.1 that PIC can obtain the latent prior distributions of missing regions in the VAE-based reconstructive path and predicts possible latent distributions conditioned on the visible pixels in the generative path. The reasons we choose pre-trained models from PIC are the followings: 1. The designing of the probability framework gives us the chance to directly obtain the learned distributions encoded from one given ground truth. 2. Taking a masked image as input, we can simply sample latent codes from its distribution from the well pre-trained image encoder and manipulate latent vectors without additional operations. 3. The manipulated latent vectors can then conveniently pass through the pre-trained generator to obtain the final inpainted results.

CLIP. is a multi-modality model that collects 400 million image-text pairs for training to learn transferable pre-trained knowledge, which has shown impressive zero-shot capabilities across multiple domains such as image classification and adaptation of generated images. It consists of two encoders, a text encoder and an image encoder. The goal of CLIP is to align the embedding spaces of language and visual during pre-training. It adopts the excellent contrastive learning framework, which minimizes the cosine distance between the encoded vectors of the paired image-text and maximizes the cosine distance of the unpaired ones. Resorting to large-scale pre-training, CLIP measures semantic similarity of a text and an image by aligning visual and language embedding spaces.

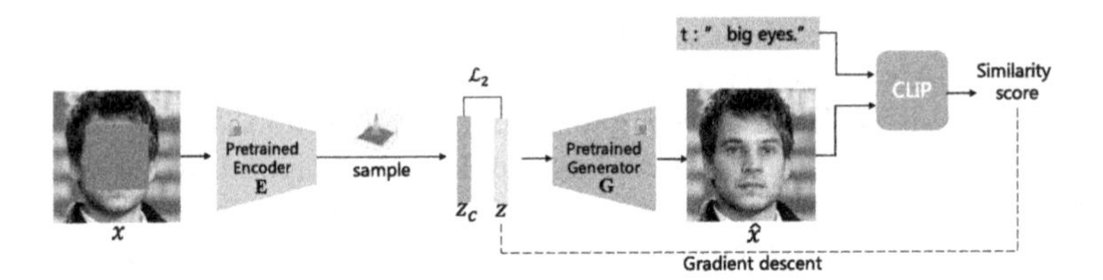

Fig. 2. The architecture of optimization-based method for text-guided face inpainting.

3.2 Text-guided Optimization Based Inpainting

To verify the text-image joint embedding ability of CLIP, we first attempt to leverage CLIP to introduce a simple and effective text-guided optimization method to guide face inpainting which optimizes latent vectors directly. Note that this method is only applied to text-driven face inpainting. We adopt two separately pre-trained networks. The first is a semantic similarity network $CLIP(t, \hat{x})$ that scores the semantic similarity between a provided text prompt t and an inpainted image $\hat{x}$. The second is the generative path of PIC $F = G(E(x))$ that is pre-trained to sample the optimal latent code $z_c = E(x)$ and synthesize an image $\hat{x} = G(z)$ given a masked image.

As shown in Fig. 2, the masked input image is fed into the pre-trained encoder E to obtain latent prior distribution for the missing regions conditioned on the visible pixels. We can sample a source optimal latent code z_c from this distribution. Then, we use CLIP to embed a user-provided text prompt and the generated image from the pre-trained generator G and measure the cosine similarity between the embeddings. The similarity is then reframed as a loss that we can use gradient descent to minimize. The optimization process is repeated to update the latent code z until the generated content gradually improves such that it semantically matches the target text prompt.

Specifically, given initial latent code z_c sampled from the missing region distribution encoded by the pre-trained encoder E and a text prompt t, we solve the following optimization problem:

$$\arg\min_{z} \lambda_{CLIP} D_{CLIP}(G(z), t) + \lambda_{L_2} ||z - z_c||, \tag{1}$$

where G is a pre-trained generator from the PIC network, and $D_{CLIP}(x, y)$ denotes the cosine similarity between the CLIP embeddings of x and y. In practice, we note that L_2 regularization improves the output's coherence and produces a better plausible image.

3.3 Multi-modal Inpainting Mapper

Although the optimization-based method can achieve facial image inpainting with a user-provided text prompt, it requires several minutes to optimize the latent code of a masked input image and infers an image at a time. It is also sensitive to the values of the hyper-parameters, which is difficult to control. In this section, we describe a new stable approach where a multi-modal inpainting mapper network is trained and requires a few seconds at reference, fully utilizing the text and image information extracted from CLIP space. This multi-modal inpainting mapper can both deal with text and image conditions, allowing text- and exemplar-guided face inpainting. As shown in Fig. 3, given the masked image to repair, we first use the pre-trained PIC encoder E to sample its optimal latent code z_c from the learned latent prior distribution. Meanwhile, given a text prompt t or an exemplar image x_e, we train an multi-modal inpainting mapper M_I to update the code as:

$$z = \mathcal{M}_{\mathcal{I}}(\mathcal{E}_t(t)) + z_c \quad or \quad z = \mathcal{M}_{\mathcal{I}}(\mathcal{E}_i(x_e)) + z_c \quad or \quad z = \mathcal{M}_{\mathcal{I}}(\mathcal{E}_t(t) + \mathcal{E}_i(x_e)) + z_c, \tag{2}$$

where $\mathcal{E}(\cdot)$ is the pre-trained CLIP encoder that projects the text or exemplar to the CLIP feature space. Given a masked face image and a text/exemplar prompt, we first obtain the source latent code z_c and text/image embedding (e_t and e_v) from the pre-trained image encoder of PIC and text/image encoder of CLIP, respectively. Then the multi-modal inpainting mapper learns to map the conditions into the corresponding latent code changes Δz based on z_c. Finally, the edited latent code z will be fed back into the pre-trained generator G of PIC to obtain the inpainting result. There are three ways of interactive face inpainting realizations, i.e., text guidance, exemplar guidance, and text+exemplar guidance:

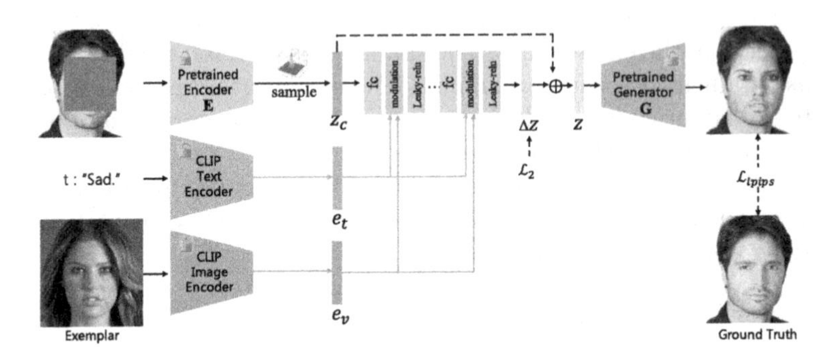

Fig. 3. The architecture of our proposed multi-modal inpainting mapper network, here we show an example with text and exemplar image as conditional inputs. This method supports to achieve face image inpainting according to the given text and exemplar, where text and exemplar are encoded by the text encoder and image encoder of CLIP to 512-dimensional vectors as conditional inputs for the fusion module, respectively. Note that only the mapper network is trainable.

Text Guidance. The inpainted contents are not only consistent with the textual description but coherent with the surrounding pixels.

Image Guidance. Since CLIP was trained in a contrastive manner, using textual descriptions, different images with the same high-level textual description are expected to receive a high similarity score, as their textual descriptions will be nearly identical. This allows the multi-modal inpainting mapper to enforce consistency based on the semantic properties of the image. The inpainted parts contain semantic features that constitute the high-level textual description of the exemplar image.

Text+Image Guidance. For the text+image conditions, because CLIP is well trained on large-scale image-text pairs, e_t and e_v both reside in the joint latent space of CLIP, we concatenate them as e. This naturally unifies the conditions from the text and image domains under one framework. The inpainted areas satisfy the above mentioned requirements conditioned on both text and image. Note that StyleCLIP requires retraining a separate model for each specific text description, but our approach not only supports multiple text conditions but image condition in a single model by fully exploiting the power of CLIP.

Semantic-preserved Modulation Module. The mapper network aims to preserve semantic information of the text prompt and exemplar by fusing text and image features. It consists of five blocks, as shown in Fig. 3. Each block consists of one fully connected (FC) layer, one newly designed semantic-preserved modulation module, and one non-linear activation layer (leaky relu). The modulation module is designed to directly control input conditions on latent codes. Instead of simply concatenating the condition embedding with the latent code, the semantic-preserved modulation module allows the model to have a multi-modal fusion process, which improves the manipulation ability of our method. We adopt two Multilayer Perceptrons (MLPs) to predict the channel-wise scaling parameters γ and shifting parameters β from condition embedding e, respectively:

$$\gamma = MLP_1(e), \quad \beta = MLP_2(e) \tag{3}$$

Given the intermediate output $\tilde{x}$ of the preceding f_c layer, the modulation module uses the condition embedding e to modulate $\tilde{x}$ as follows:

$$x' = (1+\gamma)\frac{\tilde{x} - \mu_{\tilde{x}}}{\sigma_{\tilde{x}}} + \beta, \tag{4}$$

where $\mu_{\tilde{x}}$ and $\sigma_{\tilde{x}}$ denote the mean and standard deviation of $\tilde{x}$ respectively.

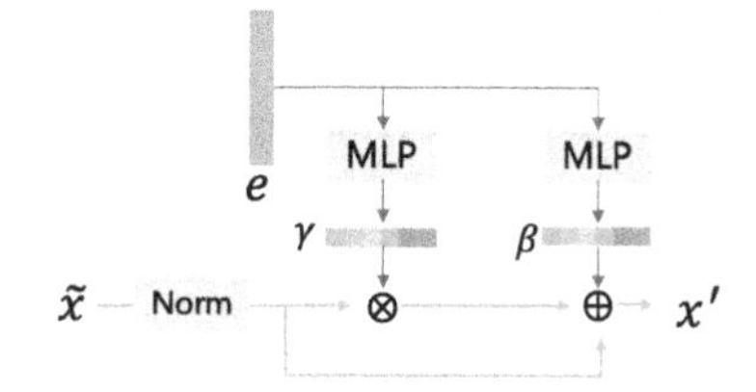

Fig. 4. The structure of the semantic-preserved modulation module.

Loss Functions. The key of this multi-modal inpainting mapper network is trained to map the input conditions into corresponding latent code changes. To better achieve the text and exemplar guidance, the following objectives are adopted:

CLIP Textual Loss. The CLIP textual loss, $\mathcal{L}^t_{CLIP}(z)$ guides the multi-modal inpainting mapper network to minimize the cosine distance between the generated image $G(z)$ and the text prompt t in the CLIP latent space:

$$\mathcal{L}^t_{CLIP}(z) = D_{CLIP}(G(z), t), \tag{5}$$

where G denotes again the pre-trained PIC generator.

Exemplar-semantic Similarity Loss. The image encoder $\mathcal{E}_i$ from CLIP, which, given an exemplar image x_e, provides a latent representation of its high-level textual description, $\mathcal{E}_i(x_e)$), in CLIP's latent space. We want the inpainted image $G(z)$ to possess similar high-level semantic attributes as those of the exemplar image. Exploiting the powerful potential of CLIP again, we encode them separately using CLIP's image encoder to measure their similarity in CLIP's latent space:

$$\mathcal{L}_{sim}(z) = 1 - cos(\mathcal{E}_i(G(z)), \mathcal{E}_i(x_e)), \tag{6}$$

where $cos(\cdot)$ means cosine similarity.

LPIPS Loss. To constrain the perceptual similarity between the inpainted image and the ground truth, we adopt the Learned Perceptual Image Patch Similarity (LPIPS) metric [23] to keep image quality. As demonstrated in [3], the LPIPS loss is observed better effect compared to the standard perceptual loss [4]. We found that this loss can improve the visual quality of the inpainted results with the conditions' guidance.

$$\mathcal{L}_{lpips} = ||\Phi(G(z)) - \Phi(G(z_s))||_2, \tag{7}$$

where Φ is the pretrained perceptual feature extractor and we adopt VGG [15] in our work.

Total Loss. To preserve the visual appearance of the ground truth image, we minimize the L_2 norm of the latent code changes step in the latent space. Finally, our total loss function is a weighted combination of these losses:

$$\mathcal{L}(z) = \lambda_{sim}\mathcal{L}_{sim}(z) + \lambda^t_{clip}\mathcal{L}^t_{CLIP}(z) + \lambda_{lpips}\mathcal{L}_{lpips} + \lambda_{L_2}||z - z_s||_2, \tag{8}$$

note that the total training losses depend on the input conditions.

4 Experiments

4.1 Evaluation Metrics

We perform evaluations by using widely adopted evaluation metrics: 1) Frechet Inception Score (FID) that evaluates the perceptual quality by measuring the distribution distance between the synthesized and real images; 2) mean l_1 error; 3) peak-signal-to-noise ratio (PSNR); and 4) structural similarity index (SSIM).

CLIP Similarity. Because we are interested in the interactive face inpainting guided by text prompt and exemplar image. The evaluation of this multi-modal face inpainting work is not a trivial problem. In this work, we design a CLIP similarity to measure the alignment between the text prompt and the according inpainted facial images. Denoting text encoder and image encoder of CLIP as $\mathcal{E}_t$ and $\mathcal{E}_i$, respectively, the similarity is computed as $Similarity = \frac{\mathcal{E}_t(t)\cdot\mathcal{E}_i(x)}{||\mathcal{E}_t(t)||_2\cdot||\mathcal{E}_i(x)||_2}$.

4.2 Implementation Details

We evaluate our methods on CelebA-HQ image dataset and randomly divide the 30,000 images into a training set of 27,000 images and a validation set of 3,000 images. Before using CLIP, we need to train PIC on the face datasets to obtain the pre-trained image encoder E and the corresponding generator G. The text-guided optimization is a training-free method which employs the Adam optimizer for 400–500 iterations with a learning rate of 0.01 to obtain one generated image at a time. The values of λ_{CLIP} and λ_{L_2} are set to 5 and 0.01 by default. For the multi-modal inpainting mapper network, the number of training iterations is set to 500,000 and the Adam [5] optimizer is used, with β_1 and β_2 set to 0.9 and 0.999, respectively. The learning rate is 0.0005 with a batch size of 1. The values of λ^t_{CLIP}, λ^v_{CLIP}, λ_{lpips} and λ_{L_2} are set to 2, 1, 0.5 and 1 by default. We employ the official provided pre-trained models or training codes for the comparative methods. All experiments are conducted on an NVIDIA GeForce GTX 3090 GPU.

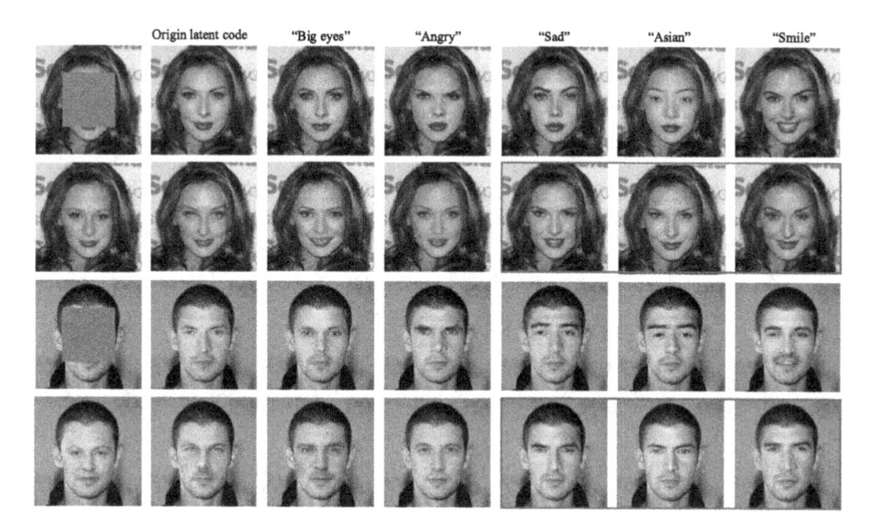

Fig. 5. Qualitative comparison results on two test images of CelebA-HQ. For each group, the first row shows the masked image and results of our approach given different text prompts. For the second row, from left to right, the images are: results of Edge Connect (EC) [10], deepfill v2 [18], MEDFE [8], WaveFill [20] and results of PIC [25] (with blue box).

4.3 Qualitative Results

Figure 5 contain results of our optimization-based method. We compare our method with the state-of-the-art inpainting method for visual quality. The single-solution inpainting methods such as edge connect [10] can only generate one result. Multiple-solution method PICNet produced various plausible results by sampling from the latent space conditional prior. Our results have competitive visual quality compared with these methods, but the inpainting of facial attributes cannot be controlled with these compared ones. For instance, PICNet generates images with random and subtle changes, while our method can generate text-controllable facial images.

Figure 6 shows the text- and exemplar-guided face inpainting results by our multi-modal inpainting mapper method. It can be seen that with the help of the designed inpainting mapper and CLIP loss, our approach can be individually or jointly based on the texts or exemplar images provided by users. Specifically, exemplar guidance can transfer the most globally notable semantic attributes from the exemplars. With the guidance of text and exemplar, the inpainting of facial attributes with our method can be controlled diversely.

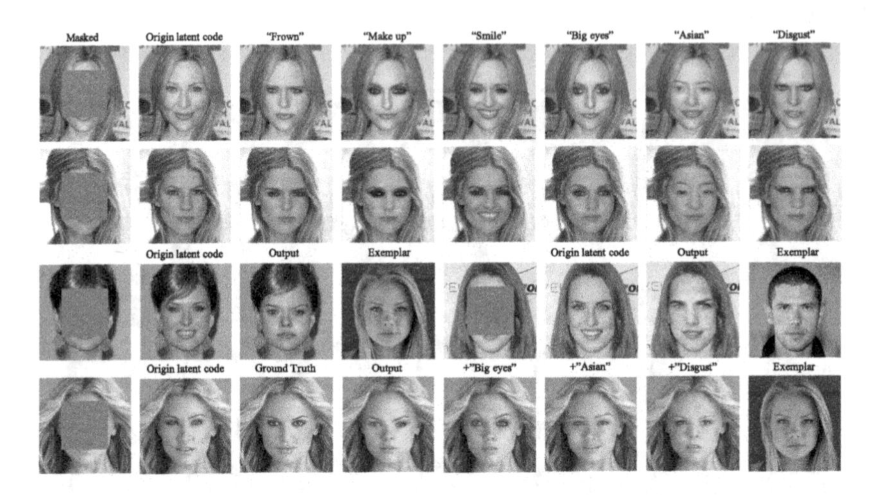

Fig. 6. Qualitative results of our multi-modal inpainting mapper method. The first two rows show the text-driven facial inpainted results, the third row shows the exemplar guided results, and the last row shows the results by both conditions.

We also attempt to complete a face region using a mask at a specified location. As shown in Fig. 7, we mask the surroundings of the eyes and achieve variations given different text prompts. This implementation is based on the pre-trained generator of the random mask from PIC. It can be seen the masked areas are inpainted well and also consistent with the text prompt.

Fig. 7. Qualitative results of a mask at a specified location.

4.4 Quantitative Results

For the image quality, we compare our method with EC [10], CA [17] and PIC [25]. Table 1 shows the performance comparisons, our method has the comparable visual results with these compared inpainting methods.

For CLIP similarity, we infer 50 inpainted images from multi-modal mapper inpainting method for each text prompt and compute the average similarity score as shown in Table 2. The similarity for a given text prompt is high, implying that the corresponding attributes are as consistent as one might expect.

Table 1. Quantitative comparison of different methods on the CelebA-HQ test set.

	EC [10]	CA [17]	PIC [25]	Ours
FID↓	8.16	37.61	11.74	11.70
SSIM↑	0.823	0.870	0.857	0.868
PSNR↑	21.75	23.65	23.93	24.56

Table 2. Average CLIP similarity score from different text prompts.

	Smile	Angry	Frown	Big eyes	Asian	Without makeup	Sad	Disgust
Similarity	0.74	0.80	0.79	0.77	0.73	0.72	0.75	0.72

4.5 Ablation Analysis

We mainly verify the CLIP losses and the inpainting mapper by alternately removing one of these components to retrain variants of the inpainting mapper method. Figure 8 shows the comparison results in the ablation study.

Importance of CLIP Textual Loss. Figure 8 (e) shows joint exemplar and text-driven ("big eyes") inpainted result. The CLIP textual loss $\mathcal{L}^t_{CLIP}$ guarantees the semantic similarity between the text prompt and inpainted facial contents in the CLIP space. Without $\mathcal{L}^t_{CLIP}$, Fig. 8 (f) shows that the result image cannot capture the "big eyes" attribute compared to Fig. 8 (e).

Importance of Exemplar-semantic Similarity Loss. This loss ensures that the inpainted face image is semantically similar to the exemplar in the latent image space of CLIP. We remove it to retrain the inpainting mapper to show its effectiveness. We can observe from Fig. 8 (g) that it is generated only driven by text prompt, without high-level semantics of the exemplar.

Importance of Semantic-preserved Modulation Module. To verify the modulation ability of the inpainting mapper, we remove the modulation module and simply concatenate the latent code with the input conditions and then feed them into the vanilla layernorm layer. Figure 8 (h) shows that it cannot capture the information of the input conditions well compared to Fig. 8 (e).

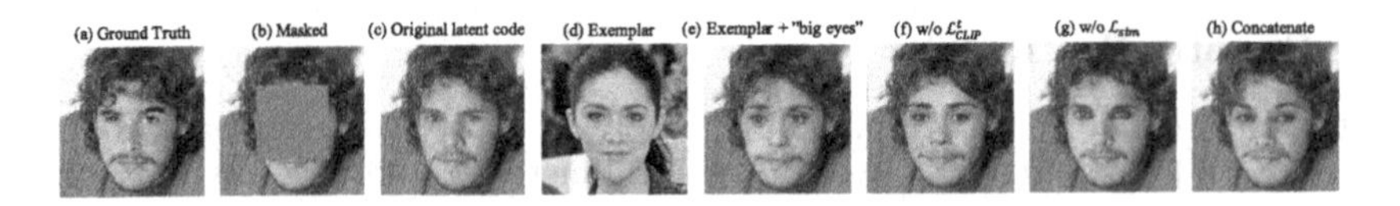

Fig. 8. Results generated by variants of the multi-modal inpainting mapper method.

Face Interpolation. We also demonstrated the smooth interpolation between source latent code z_A and edited latent code z_B. In detail, we combine the two latent codes by linear weighting to generate the intermediate latent code $z_I = \lambda z_A + (1 - \lambda) z_B$. Then, the inpainted image corresponding to the intermediate latent code z_I is produced. Images in Fig. 9 reflect the continuous changes.

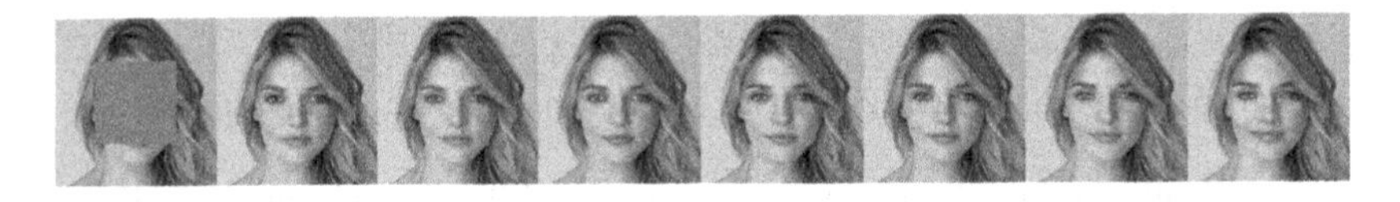

Fig. 9. Face interpolation results. Here by gradually increasing the parameter λ, we achieve text-driven face inpainting changing from unedited image to 'smile'.

5 Conclusion

In this work, we propose two ways for interactive facial image inpainting by leveraging CLIP's powerful text-image representation abilities. The first is based on per-target optimization to achieve text-guided face inpainting, while the second trains a multi-modal inpainting mapper network to perform both text- and exemplar-guided inpainting results within a few seconds at reference. This multi-modal interaction increases the flexibility and controllable diversity of facial image inpainting. Extensive experimental results have demonstrated the effectiveness of the proposed method.

References

1. Gal, R., Patashnik, O., Maron, H., Chechik, G., Cohen-Or, D.: StyleGAN-NADA: clip-guided domain adaptation of image generators. arXiv preprint arXiv:2108.00946 (2021)
2. Goodfellow, I., et al.: Generative adversarial nets. In: Advances in Neural Information Processing Systems 27 (2014)
3. Guan, S., Tai, Y., Ni, B., Zhu, F., Huang, F., Yang, X.: Collaborative learning for faster styleGAN embedding. arXiv preprint arXiv:2007.01758 (2020)
4. Johnson, J., Alahi, A., Fei-Fei, L.: Perceptual losses for real-time style transfer and super-resolution. In: Leibe, B., Matas, J., Sebe, N., Welling, M. (eds.) ECCV 2016. LNCS, vol. 9906, pp. 694–711. Springer, Cham (2016). https://doi.org/10.1007/978-3-319-46475-6_43

5. Kingma, D.P., Ba, J.: Adam: a method for stochastic optimization. arXiv preprint arXiv:1412.6980 (2014)
6. Lahiri, A., Jain, A.K., Agrawal, S., Mitra, P., Biswas, P.K.: Prior guided GAN based semantic inpainting. In: Proceedings of the IEEE/CVF Conference on Computer Vision and Pattern Recognition, pp. 13696–13705 (2020)
7. Liu, G., Reda, F.A., Shih, K.J., Wang, T.-C., Tao, A., Catanzaro, B.: Image inpainting for irregular holes using partial convolutions. In: Ferrari, V., Hebert, M., Sminchisescu, C., Weiss, Y. (eds.) ECCV 2018. LNCS, vol. 11215, pp. 89–105. Springer, Cham (2018). https://doi.org/10.1007/978-3-030-01252-6_6
8. Liu, H., Jiang, B., Song, Y., Huang, W., Yang, C.: Rethinking image inpainting via a mutual encoder-decoder with feature equalizations. In: Vedaldi, A., Bischof, H., Brox, T., Frahm, J.-M. (eds.) ECCV 2020. LNCS, vol. 12347, pp. 725–741. Springer, Cham (2020). https://doi.org/10.1007/978-3-030-58536-5_43
9. Liu, H., Jiang, B., Xiao, Y., Yang, C.: Coherent semantic attention for image inpainting. In: Proceedings of the IEEE/CVF International Conference on Computer Vision, pp. 4170–4179 (2019)
10. Nazeri, K., Ng, E., Joseph, T., Qureshi, F.Z., Ebrahimi, M.: Edgeconnect: Generative image inpainting with adversarial edge learning. arXiv preprint arXiv:1901.00212 (2019)
11. Patashnik, O., Wu, Z., Shechtman, E., Cohen-Or, D., Lischinski, D.: StyleCLIP: text-driven manipulation of styleGAN imagery. In: Proceedings of the IEEE/CVF International Conference on Computer Vision, pp. 2085–2094 (2021)
12. Radford, A., et al.: Learning transferable visual models from natural language supervision. In: International Conference on Machine Learning, pp. 8748–8763. PMLR (2021)
13. Ramesh, A., Dhariwal, P., Nichol, A., Chu, C., Chen, M.: Hierarchical text-conditional image generation with clip latents. arXiv preprint arXiv:2204.06125 (2022)
14. Ren, Y., Yu, X., Zhang, R., Li, T.H., Liu, S., Li, G.: Structureflow: image inpainting via structure-aware appearance flow. In: Proceedings of the IEEE/CVF International Conference on Computer Vision, pp. 181–190 (2019)
15. Simonyan, K., Zisserman, A.: Very deep convolutional networks for large-scale image recognition. arXiv preprint arXiv:1409.1556 (2014)
16. Xiong, W., et al.: Foreground-aware image inpainting. In: Proceedings of the IEEE/CVF Conference on Computer Vision and Pattern Recognition, pp. 5840–5848 (2019)
17. Yu, J., Lin, Z., Yang, J., Shen, X., Lu, X., Huang, T.S.: Generative image inpainting with contextual attention. In: Proceedings of the IEEE Conference on Computer Vision and Pattern Recognition, pp. 5505–5514 (2018)
18. Yu, J., Lin, Z., Yang, J., Shen, X., Lu, X., Huang, T.S.: Free-form image inpainting with gated convolution. In: Proceedings of the IEEE/CVF International Conference on Computer Vision, pp. 4471–4480 (2019)
19. Yu, T., et al.: Region normalization for image inpainting. In: Proceedings of the AAAI Conference on Artificial Intelligence, pp. 12733–12740 (2020)
20. Yu, Y., Zhan, F., Lu, S., Pan, J., Ma, F., Xie, X., Miao, C.: WaveFill: a wavelet-based generation network for image inpainting. In: Proceedings of the IEEE/CVF International Conference on Computer Vision, pp. 14114–14123 (2021)
21. Zeng, Y., Fu, J., Chao, H., Guo, B.: Learning pyramid-context encoder network for high-quality image inpainting. In: Proceedings of the IEEE/CVF Conference on Computer Vision and Pattern Recognition, pp. 1486–1494 (2019)

22. Zhang, L., Chen, Q., Hu, B., Jiang, S.: Text-guided neural image inpainting. In: Proceedings of the 28th ACM International Conference on Multimedia, pp. 1302–1310 (2020)
23. Zhang, R., Isola, P., Efros, A.A., Shechtman, E., Wang, O.: The unreasonable effectiveness of deep features as a perceptual metric. In: Proceedings of the IEEE Conference on Computer Vision and Pattern Recognition, pp. 586–595 (2018)
24. Zhao, L., et al.: UCTGAN: diverse image inpainting based on unsupervised cross-space translation. In: Proceedings of the IEEE/CVF Conference on Computer Vision and Pattern Recognition, pp. 5741–5750 (2020)
25. Zheng, C., Cham, T.J., Cai, J.: Pluralistic image completion (2019)

Dual-Feature Aggregation Network for No-Reference Image Quality Assessment

Yihua Chen, Zhiyuan Chen, Mengzhu Yu, and Zhenjun Tang(✉)

Guangxi Key Lab of Multi-Source Information Mining and Security, Guangxi Normal University, Guilin 541004, China
tangzj230@163.com

Abstract. We propose an effective Dual-Feature Aggregation Network (DFAN) for NR-IQA by using the Convolutional Neural Networks (CNN) called ResNet50 and Vision Transformer (ViT). The proposed DFAN-IQA method consists of three modules: the module of attention feature aggregation (MAFA), the module of semantic feature aggregation (MSFA), and the module of prediction (MP). The MAFA uses the pre-trained ViT to extract attention features of different levels and aggregates these attention features by a novel attention feature aggregation (AFA) block which consists of Full-Connected (FC) layers, Rectified Linear Activation Function (ReLU) and Layer normalization. The MSFA exploits the pre-trained ResNet50 to extract semantic features of different levels and aggregates these semantic features by a novel semantic feature aggregation (SFA) block which consists of convolution, Depth-Wise Convolution (DWConv), ReLU and Batch-Normalization (BN). The MP uses concatenation and FC layers to process the outputs of MAFA and MSFA for calculating quality score. Many experiments on open IQA datasets are done to test the proposed DFAN-IQA method. IQA performance comparisons illustrate that the proposed DFAN-IQA method outperforms some state-of-the-art NR-IQA methods in a whole. Evaluation experiment on across-datasets validates the generalization ability of the proposed DFAN-IQA method.

Keywords: Image quality assessment · Vision Transformer (ViT) · Convolutional Neural Networks (CNN) · Aggregation network

1 Introduction

As a crucial information medium, digital images play an essential role in the application fields of entertainment, communication and business, such as image sharing on social media, news presentations on the Internet, and advertising displays for companies. Image quality degradation is inevitable introduced during image formation, processing, and acquisition [1]. It is an important task of computer vision to design image quality assessment (IQA) method [2] for quantifying

D.-T. Dang-Nguyen et al. (Eds.): MMM 2023, LNCS 13833, pp. 149–161, 2023.
https://doi.org/10.1007/978-3-031-27077-2_12

image distortion. In this work, we propose a dual-feature aggregation network for No-Reference (NR) IQA.

Since humans are the perceivers of image information, it is most reliable to derive the mean opinion score (MOS) by subjective assessment of distorted images by humans. However, it requires more labor and time cost. Therefore, there is an urgent need to study objective assessment methods with high consistency with human visual perception. Generally, the objective assessment methods can be divided into three kinds: full-reference IQA (FR-IQA), reduce-reference IQA (RR-IQA) and NR-IQA. Among the three kinds of IQA methods, FR-IQA method can give more accurate quality scores due to the full access of reference image, and NR-IQA method usually provides relative inaccurate scores due to no use of reference image. In practice, the reference image is generally unavailable for many real applications. For example, it is difficult to find the reference image for an image downloaded from the Internet. Therefore, it is necessary to develop useful NR-IQA methods.

Many researchers exploited hand-crafted features to designed various NR-IQA methods [3–5]. These NR-IQA methods show good performances in evaluating some specific distortions, e.g., JPEG compression and noise. Recently, many researchers proposed to use deep learning techniques to perform the NR-IQA task. It is found that the performances of deep learning based NR-IQA methods generally outperform those of most hand-crafted features based NR-IQA methods. However, their performances are still not desirable yet. Aiming at this issue, we propose an effective Dual-Feature Aggregation Network (DFAN) for NR-IQA by using the Convolutional Neural Networks (CNN) called ResNet50 [6] and Vision Transformer (ViT) [7]. Specifically, the pre-trained ResNet50 is used to extract semantic features and the pre-trained ViT is used to extract attention features. The semantic features and the attention features are fed into the semantic feature aggregation block and the attention feature aggregation block, respectively. Finally, the outputs of the two blocks are used to predict quality score by concatenation and Full-Connected (FC) layers. Compared with the existing NR-IQA techniques, the main contributions are as follows.

(1) We design an attention feature aggregation module. This module uses the pre-trained ViT to extract attention features of different levels and aggregates these attention features by a novel attention feature aggregation (AFA) block. The AFA block consists of FC layers, linear rectification function (ReLU) and Layer normalization.
(2) We design a semantic aggregation fusion module. This module uses the pre-trained ResNet50 to extract semantic features of different levels and aggregates these semantic features by a novel semantic feature aggregation (SFA) block. The SFA block consists of convolution, Depth-Wise Convolution (DWConv), ReLU and Batch-Normalization (BN).
(3) We propose an effective DFAN-IQA method by combining the attention feature aggregation module and the semantic aggregation fusion module. The proposed DFAN-IQA method jointly exploits attention features and semantic features to predict quality score. Since the semantic features can cap-

ture semantic distortion and the attention features can indicate distortion in attention regions, the proposed DFAN-IQA method can effectively measure image distortion.

A lot of experiments on open datasets including LIVE [8], CSIQ [9], TID2013 [10], and KADID10K [11] are done. The results show that the proposed DFAN-IQA method reaches comparable performances compared to some state-of-the-art NR-IQA methods. The rest of this paper is structured as follows. Section 2 presents the related work. Section 3 explains details of the proposed DFAN-IQA method. Section 4 discusses experimental results and Sect. 5 presents the conclusions.

2 Related Work

In this section, we introduce NR-IQA from two aspects. Traditional NR-IQA methods are firstly introduced in Sect. 2.1, and the deep learning based NR-IQA methods are then presented in Sect. 2.2.

2.1 Traditional NR-IQA Methods

At the early stage of NR-IQA research, many researchers [12] designed NR-IQA methods for the distortions caused by specific operations. These methods can work well in predicting quality scores if the distortions of the evaluated images are introduced by their target operations. If the practical distortions are beyond the scope of the defined distortions, these methods are inefficient. Aiming at this problem, some researchers [3–5,13,14] proposed some NR-IQA methods for general distortion evaluation. These NR-IQA methods can accurately evaluate image distortions introduced by many operations. For example, Saad et al. [3] designed a fast single-stage method for NR-IQA. This method mainly relies on the statistical model of local DCT (discrete cosine transform) coefficients. In another work, Mittal et al. [4] designed a statistical model with locally normalized brightness coefficients in the spatial domain and used the model parameters to quantify the "naturalness" of image. A common of these NR-IQA methods is that they are all based on natural scene statistics (NSS). To address limitation of the NSS based NR-IQA methods, some researchers [15] used histogram features to calculate quality scores. Although many traditional NR-IQA methods show good IQA results, their performances are difficult to be significantly improved.

2.2 Deep Learning-based NR-IQA Methods

In recent years, the emergence of deep learning [16] has brought new ideas to the design of NR-IQA. The deep learning-based NR-IQA methods do not need to manually design feature extraction, and the image feature extraction can be obtained through an end-to-end learning. Some typical deep learning-based NR-IQA methods are briefly reviewed. Kang et al. [17] proposed a CNN-IQA method using a shallow layer. Yang et al. [18] used a bifurcated structure to design a CNN based network for NR-IQA. Recently, Zhang et al. [19] proposed a bilinear DB-CNN based network for NR-IQA. The above NR-IQA methods can directly learn the mapping between image and image quality scores. To design efficient methods that are closely matching human assessment, one builds mappings between the depth features of an image and image quality features. Researchers used the learned image quality features to regress quality score. For example, Su et al. [20] proposed the HyperIQA method which aggregates four features extracted from a pre-trained ResNet50 network. Recently, Golestaneh et al. [21] proposed the TReS method, which extracts features from the pre-trained Resnet50 and learns the attention between features using the encoder of ViT. It also adds the Self-Consistency loss and the Relative Ranking loss. In another work, Yang et al. [22] proposed the MANIQA method, which builds a mapping between the features of the pre-trained ViT network and the image quality features. In addition, it employs a two-branch prediction module to predict the image quality scores. The above-mentioned deep networks [20–22] are constructed by using complex network blocks to establish the relationship between depth and image quality features.

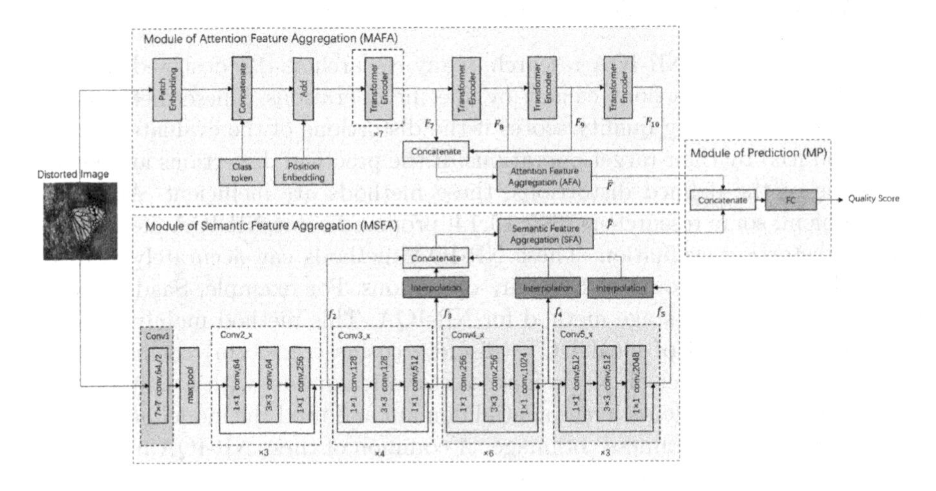

Fig. 1. Overview of the proposed DFAN-IQA method

3 Proposed Method

In this section, we first introduce overall pipeline of our proposed DFAN-IQA method. Then, we explain the module of attention feature aggregation and the module of semantic feature aggregation. Finally, we introduce the module of prediction .

3.1 Overall Pipeline

The framework of our proposed DFAN-IQA method is shown in Fig. 1. Our proposed DFAN-IQA method consists of three modules: the Module of Attention Feature Aggregation (MAFA), the Module of Semantic Feature Aggregation (MSFA), and the module of prediction (MP). The input of the proposed DFAN-IQA method is a distorted image and the output is the prediction score. The overviews of these modules are described in the following paragraphs.

The MAFA firstly exploits the ViT [7] pre-trained on ImageNet to extract attention features of different levels and then aggregates these attention features by a novel attention feature aggregation block. The input of the MAFA is an input image I and the output of the MAFA is the attention feature $\overline{F}$. Suppose that the input image $I \in \mathbb{R}^{H \times W \times 3}$, where H and W denote image height and image width. Thus, $\overline{F} \in \mathbb{R}^{1 \times N \times C}$ can be expressed as

$$\overline{F} = MAFA(I) \tag{1}$$

Note that N denotes the number of all tokens of an image, and C denotes the channel size.

The MSFA uses the ResNet50 [6] pre-trained on ImageNet to extract semantic features of different levels and aggregates these semantic features by a novel semantic feature aggregation block. The input of the MSFA is I and the output of the MSFA is the semantic feature $\widetilde{F}$. The semantic feature $\widetilde{F} \in \mathbb{R}^{\frac{H}{4} \times \frac{W}{4} \times 1}$ can be expressed as follows.

$$\widetilde{F} = MSFA(I) \tag{2}$$

The MP uses concatenation operation and FC layers to process the input data. The inputs of the MP are the outputs of the MAFA and the MSFA, i.e., $\overline{F}$ and $\widetilde{F}$, and the output of the MP is the quality score S. The quality score S can be expressed as follows.

$$S = MP\left(\overline{F}, \widetilde{F}\right) \tag{3}$$

Note that $S \in [0, 1]$ and a bigger S means better image quality.

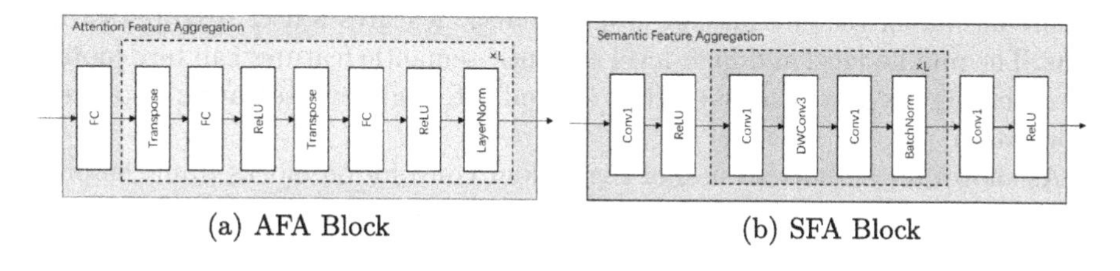

(a) AFA Block (b) SFA Block

Fig. 2. Structures of two feature aggregation blocks

3.2 Module of Attention Feature Aggregation

Attention feature plays an important role in IQA. Some existing techniques [23] show that attention feature can indicate the topic of an image which helps to evaluate quality. Here, we exploit the pre-trained ViT to extract attention features. As shown in the top half of Fig. 1, four Transformer encoders are exploited to find attention features of different stages. Specifically, the used ViT network has 12 layers, but only the outputs of four layers are used in our work. Let $F_i \in \mathbb{R}^{1\times N\times C}$ be the attention feature output from the i-th layer of ViT network, where $i \in \{7, 8, 9, 10\}$, C denotes the channel size, and N is the number of tokens. Thus, the attention features F_7, F_8, F_9, and F_{10} are used. Next, these attention features are concatenated to form a feature sequence $\widetilde{f} \in \mathbb{R}^{1\times N\times 4C}$ as follows.

$$\widetilde{f} = Concat\left(F_7, F_8, F_9, F_{10}\right) \tag{4}$$

To conduct feature aggregation, we design a novel Attention Feature Aggregation (AFA) block. As shown in Fig. 2 (a), the AFA block consists of a series of Fully Connected (FC) layer, Layer-Normalization (LayerNorm), Transposition (T) and Rectified Linear Activation Function (ReLU). The feature sequence $\widetilde{f}$ is sent to the AFA Block and then the output $\overline{F}_i$ can be obtained by the below equations.

$$\hat{f}' = ReLU\left(FC\left(T\left(FC\left(\widetilde{f}\right)\right)\right)\right) \tag{5}$$

$$\hat{f}'' = ReLU\left(FC\left(T\left(\hat{f}'\right)\right)\right) \tag{6}$$

$$\overline{F} = LayerNorm\left(\hat{f}''\right) \tag{7}$$

3.3 Module of Semantic Feature Aggregation

Semantic feature is important to the IQA method. In this work, we exploit the pre-trained ResNet50 [6] to extract semantic features at the different levels. These semantic features can indicate different-level distortions. Here, low-level semantic feature, middle-level semantic feature and high-level semantic semantic feature are all extracted. The low-level semantic features reflect texture distortions. The middle-level and high-level semantic semantic features can be denoted by combining low-level features. They can indicate high-level semantic distortion information.

As shown in the bottom half of Fig. 1, four convolution blocks from the pre-trained Resnet50 network are exploited to extract semantic features of different levels. Let $f_2 \in \mathbb{R}^{\frac{H}{4}\times\frac{W}{4}\times 4C'}$, $f_3 \in \mathbb{R}^{\frac{H}{8}\times\frac{W}{8}\times 8C'}$, $f_4 \in \mathbb{R}^{\frac{H}{16}\times\frac{W}{16}\times 16C'}$ and $f_5 \in \mathbb{R}^{\frac{H}{32}\times\frac{W}{32}\times 32C'}$ represent the outputs of the conv2_x, conv3_x, conv4_x and conv5_x blocks, respectively. Note that the C' is the dimension of the output of the conv1. Since the semantic features f_2, f_3, f_4 and f_5 have different sizes, we align them by bilinear interpolation (BI). Next, these semantic features are concatenated to form a sequence $f \in \mathbb{R}^{\frac{H}{4} \times \frac{W}{4} \times 60C'}$ according to the channel dimension. These operations can be expressed as follows.

$$f = Concat\left(f_2, BI\left(f_3\right), BI\left(f_4\right), BI\left(f_5\right)\right) \tag{8}$$

To perform feature aggregation, we design a novel semantic feature aggregation (SFA) block. As shown in Fig. 2 (b), the SFA block consists of a series of convolution, Depth-Wise Convolution (DWConv), Rectified Linear Activation Function (ReLU) and Batch-Normalization (BN). The feature sequence f is sent to the SFA block and then the output $\widetilde{F}_i$ can be obtained by the below equations.

$$f' = ReLU\left(Conv1\left(f\right)\right) \tag{9}$$

$$f'' = BN\left(Conv1\left(DWConv3\left(Conv1\left(f'\right)\right)\right)\right) \tag{10}$$

$$\widetilde{F} = ReLU\left(Conv1\left(f''\right)\right) \tag{11}$$

where DWConv3 is the depth-wise convolution with 3×3 filters, and Conv1 indicates the convolution with 1×1 filters.

3.4 Module of Prediction

The outputs of the MAFA and the MSFA, i.e., $\widetilde{F}$ and $\overline{F}$, are fed to the MP. Before data connection operation, we need to extract the quality token of $\widetilde{F}$ and flatten $\overline{F}$. Then, the quality token of $\widetilde{F}$ and the flattened $\overline{F}$ are concatenated to construct a vector. The constructed vector is sent to the FC layers with two layers to regress the image quality score S which is calculated by the below equation.

$$S = FC\left(Concat\left(QT\left(\overline{F}\right), Flatten\left(\widetilde{F}\right)\right)\right) \tag{12}$$

where $QT(\cdot)$ represents the quality token extraction.

4 Experimental Results

Many experiments are done to demonstrate performances of the proposed DFAN-IQA method. Section 4.1 describes the used datasets, and Sect. 4.2 presents details of our implementation. Section 4.3 and Sect. 4.4 discuss performance comparison and ablation study, respectively.

4.1 Datasets

We use four open datasets to train and validate the performance of the proposed DFAN-IQA method. These used datasets are the LIVE [8], CSIQ [9], TID2013 [10], and KADID10K [11]. The LIVE has 779 images, CSIQ has 866 images, TID2013 has 3000 images and KADID10K has 10206 images.

We evaluate performance of the proposed DFAN-IQA method by conducting two kinds of experiments. The experiments of the first kind are done on the same dataset. In other words, the training set, validation set and test set are all taken from one dataset. Following the previous techniques [20–22], we randomly divide each dataset into the training set (60% images), validation set (20% images) and test set (20% images). The experiments of the second kind are conducted on cross datasets. Specifically, the proposed DFAN-IQA method is trained on the KADID10K dataset and tested on the LIVE, CSIQ, and TID2013 datasets.

4.2 Implementation Details

Our experiments are done on an NVIDIA TESLA V100S with PyTorch 1.8.2 and CUDA 10.2. The pre-trained ViT and ResNet50 network have size requirements for input images. Since the sizes of images of the used datasets do not match the required size, we randomly select an image patch sized 224×224 from each image and use it as input image. During network training, we randomly flip the image patch horizontally with probability 0.5 to augment data, and the selected batch size of images is 8. We use the cosine annealing mode, set the learning rate to 9×10^{-6}, and use an ADAM optimizer with a weight decay of 1×10^{-6}. For the cosine annealing function, $T_{\max}$ and $eta_{\min}$ are set to 50 and 0. For the SFA and AFA blocks, the L values are both 1. We select the mean square error (MSE) as our objective function and train our network with 300 epochs.

During network testing, we randomly select 20 image patches from each test image. Since each image patch inherits the distortion of the corresponding test image, the quality score of image patch can reflect the score of the test image. To make accurate score, we calculate the average value of the predicted scores of 20 image patches and use the average value to represent the final quality score of the test image.

Note that we train and test our network 3 times with different random seeds to find average result of the proposed DFAN-IQA method. In other words, the calculated result listed in the below sections is the mean of 3 results.

4.3 Performance Comparison

In this section, we conduct experiments on the same dataset and the across datasets, respectively. The purpose of the experiments on the same dataset is to verify the prediction accuracy of the IQA method. The purpose of the experiments on the across datasets is to validate the generalization ability of the IQA method. Following the existing work, two famous metrics called PLCC [24] and

SRCC [25] are used to assess IQA performance. The PLCC indicates the linear correlation between the MOS and the predicted quality score, while SRCC describes the level of monotonic correlation. The bigger the PLCC/SRCC value, the better the IQA performance.

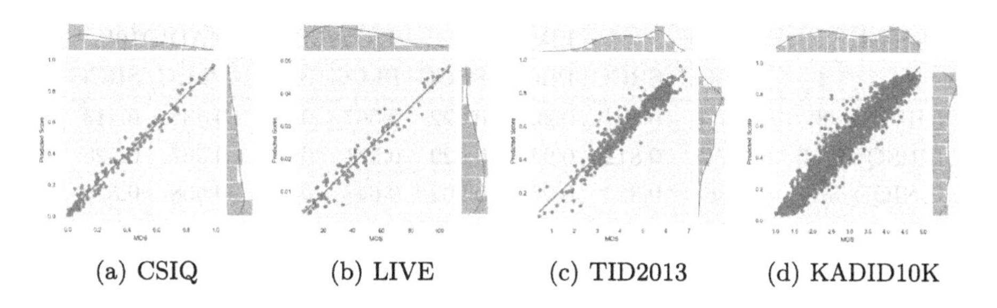

(a) CSIQ (b) LIVE (c) TID2013 (d) KADID10K

Fig. 3. Scatter plots of the MOS values and predicted scores

Evaluation on the Same Dataset. We compare the proposed DFAN-IQA method with some state-of-the-art methods on different datasets. Table 1 presents the IQA performance results of different methods. In Table 1, the best results are shown in bold and the second best results are shown in blue font. Note that the PLCC and SRCC values of the compared methods are reported in their corresponding papers [22]. From Table 1, it can be seen that the PLCC and SRCC values of the proposed DFAN-IQA method are bigger than those of the compared methods on CSIQ, TID2013 and KADID10K datasets. On the LIVE dataset, the MANIQA method reaches the biggest PLCC and SRCC values, and our PLCC and SRCC values are the second biggest PLCC and SRCC values. Therefore, the proposed DFAN-IQA method outperforms the compared methods in a whole.

To visually show IQA performance of the proposed DFAN-IQA method, the scatter plots of the MOS values and predicted scores are presented in Fig. 3. In the Fig. 3, the x-axis is the predicted score of the proposed DFAN-IQA method and the y-axis is the MOS value. It is observed that most points are around the line. This indicates that our predicted scores and the MOS values are highly correlated.

Evaluation on the Cross-Datasets. To verify the generalization of the proposed DFAN-IQA method, We train it on the KADID10K dataset. The KADID10K dataset has 10206 images, which is bigger than those of other three datasets. In the testing phase, we use all images of the CSIQ, LIVE and TID2013 datasets to validate performance on these databases. Table 2 presents our IQA performance on the cross-datasets. On CSIQ dataset, our PLCC and SRCC values are both bigger than 0.83. On LIVE dataset, our PLCC and SRCC values are both bigger than 0.91. On TID2013 dataset, our PLCC and SRCC values are both bigger than 0.76. High PLCC and SRCC values indicate that the proposed DFAN-IQA method has good generalization ability.

Table 1. IQA Performance comparisons on four open datasets

Method	CSIQ		LIVE		TID2013		KADID10K	
	PLCC	SRCC	PLCC	SRCC	PLCC	SRCC	PLCC	SRCC
DIIVINE [3]	0.776	0.804	0.908	0.892	0.567	0.643	0.435	0.413
BRISQUE [4]	0.748	0.812	0.944	0.929	0.571	0.626	0.567	0.528
ILNIQE [5]	0.865	0.822	0.906	0.902	0.648	0.521	0.558	0.528
BIECON [13]	0.823	0.815	0.961	0.958	0.762	0.717	0.648	0.623
MEON [14]	0.864	0.852	0.955	0.951	0.824	0.808	0.691	0.604
WaDIQaM [26]	0.844	0.852	0.955	0.960	0.855	0.835	0.752	0.739
DBCNN [19]	0.959	0.946	0.971	0.968	0.865	0.816	0.856	0.851
TIQA [27]	0.838	0.825	0.965	0.949	0.858	0.846	0.855	0.850
MetaIQA [28]	0.908	0.899	0.959	0.960	0.868	0.856	0.775	0.762
P2P-BM [29]	0.902	0.899	0.958	0.959	0.856	0.862	0.849	0.840
HyperIQA [20]	0.942	0.923	0.966	0.962	0.858	0.840	0.845	0.852
TReS [21]	0.942	0.922	0.968	0.969	0.883	0.863	0.858	0.915
MANIQA [22]	0.968	0.961	**0.983**	**0.982**	0.943	0.937	0.946	0.944
Our DFAN-IQA	**0.982**	**0.979**	0.976	0.971	**0.965**	**0.957**	**0.969**	**0.967**

Table 2. Performance trained on KADID10K

Test on	CSIQ		LIVE		TID2013	
	PLCC	SRCC	PLCC	SRCC	PLCC	SRCC
DFAN-IQA	0.861	0.836	0.911	0.932	0.775	0.766

4.4 Ablation Study

To verify the effectiveness of our designed blocks, some ablation experiments on the LIVE dataset are done. Note that the MAFA can be viewed as a combination of attention feature block and AFA block. Similarly, the MSFA can be viewed as a combination of semantic feature block and SFA block. We test the IQA performances of the proposed DFAN-IQA method with different block combinations. The results are shown in Table 3, where the best results are highlighted in bold face. Clearly, the proposed DFAN-IQA method using all blocks outperforms the proposed DFAN-IQA method using one block or two blocks. This experimentally proves effectiveness of our designed blocks.

Table 3. Ablation experiments on LIVE dataset

No.	MSFA		MAFA		PLCC	SRCC
	Semantic features	SFA	Attention features	AFA		
1	✓				0.899	0.875
2	✓	✓			0.957	0.953
3			✓		0.970	0.965
4			✓	✓	0.973	0.968
5	✓		✓		0.970	0.966
6	✓	✓	✓	✓	**0.976**	**0.971**

5 Conclusions

We have proposed a new DFAN-IQA method in this paper. The proposed DFAN-IQA method exploits the pre-trained ViT and ResNet50 to extract attention features and semantic features from a distorted image, respectively. In addition, we have designed the AFA block and the SFA block. The designed AFA block and SFA block are used to aggregate the attention features and semantic features, respectively. The outputs of the AFA block and the SFA block are fed to the MP. The MP uses the concatenation operation and FC layers to calculate the quality score. Many experiments have been done on open datasets to validate the performance of the proposed DFAN-IQA method. The IQA performance comparisons have illustrated that the proposed DFAN-IQA method outperforms some state-of-the-art NR-IQA methods in a whole. The evaluation experiment on the across-datasets has validated the generalization ability of the proposed DFAN-IQA method. The experiment of ablation study has shown the effectiveness of our designed blocks.

Acknowledgments. This work is partially supported by the Guangxi Natural Science Foundation (2022GXNSFAA035506), the National Natural Science Foundation of China (62272111, 61962008), Guangxi "Bagui Scholar" Team for Innovation and Research, Guangxi Talent Highland Project of Big Data Intelligence and Application, Guangxi Collaborative Innovation Center of Multi-source Information Integration and Intelligent Processing, and the Innovation Project of Guangxi Graduate Education (YCSW2022177). Dr. Zhenjun Tang is the corresponding author.

References

1. Zheng, H., et al.: Learning conditional knowledge distillation for degraded-reference image quality assessment. In: Proceedings of the IEEE/CVF International Conference on Computer Vision (ICCV), pp. 10222–10231 (2021)

2. Yu, M., et al.: Perceptual hashing with complementary color wavelet transform and compressed sensing for reduced-reference image quality assessment. IEEE Trans. Circuits Syst. Video Technol. **32**(11), 7559–7574 (2022)
3. Saad, M.A., et al.: Blind image quality assessment: a natural scene statistics approach in the DCT domain. IEEE Trans. Image Process. **21**(8), 3339–3352 (2012)
4. Mittal, A., et al.: No-reference image quality assessment in the spatial domain. IEEE Trans. Image Process. **21**(12), 4695–4708 (2012)
5. Zhang, L., et al.: A feature-enriched completely blind image quality evaluator. IEEE Trans. Image Process. **24**(8), 2579–2591 (2015)
6. He, K., et al.: Deep residual learning for image recognition. In: Proceedings of the IEEE Conference on Computer Vision and Pattern Recognition (CVPR), pp. 770–778 (2016)
7. Dosovitskiy, A., et al.: An image is worth 16x16 words: transformers for image recognition at scale. arXiv preprint arXiv:2010.11929 (2020)
8. Sheikh, H., et al.: A statistical evaluation of recent full reference image quality assessment algorithms. IEEE Trans. Image Process. **15**(11), 3440–3451 (2006)
9. Larson, E., Chandler, D.: Most apparent distortion: full-reference image quality assessment and the role of strategy. J. Electronic Imaging **19**, 011006 (2010)
10. Ponomarenko, N., et al.: Color image database tid2013: peculiarities and preliminary results. In: Proceedings of the European Workshop on Visual Information Processing (EUVIP), pp. 106–111 (2013)
11. Lin, H., et al.: KADID-10k: a large-scale artificially distorted iqa database. In: Proceedings of the Eleventh International Conference on Quality of Multimedia Experience (QoMEX), pp. 1–3 (2019)
12. Ferzli, R., Karam, L.J.: A no-reference objective image sharpness metric based on the notion of just noticeable blur (jnb). IEEE Trans. Image Process. **18**(4), 717–728 (2009)
13. Kim, J., Lee, S.: Fully deep blind image quality predictor. IEEE J. Select. Top. Signal Process. **11**(1), 206–220 (2017)
14. Ma, K., et al.: End-to-end blind image quality assessment using deep neural networks. IEEE Trans. Image Process. **27**(3), 1202–1213 (2018)
15. Li, Q., et al.: Blind image quality assessment using statistical structural and luminance features. IEEE Trans. Multimedia **18**(12), 2457–2469 (2016)
16. Feng, P., Tang, Z.: A survey of visual neural networks: current trends, challenges and opportunities. Multimedia Systems, pp. 1–32 (2022)
17. Kang, L., et al.: Convolutional neural networks for no-reference image quality assessment. In: Proceedings of the IEEE Conference on Computer Vision and Pattern Recognition, pp. 1733–1740 (2014)
18. Yang, W., et al.: Deep learning for single image super-resolution: a brief review. IEEE Trans. Multimedia **21**(12), 3106–3121 (2019)
19. Zhang, W., et al.: Blind image quality assessment using a deep bilinear convolutional neural network. IEEE Trans. Circuits Syst. Video Technol. **30**(1), 36–47 (2020)
20. Su, S., et al.: Blindly assess image quality in the wild guided by a self-adaptive hyper network. In: Proceedings of the IEEE/CVF Conference on Computer Vision and Pattern Recognition (CVPR), pp. 3664–3673 (2020)
21. Golestaneh, S.A., et al.: No-reference image quality assessment via transformers, relative ranking, and self-consistency. In: Proceedings of the IEEE/CVF Winter Conference on Applications of Computer Vision (WACV), pp. 3989–3999 (2022)
22. Yang, S., et al.: MANIQA: multi-dimension attention network for no-reference image quality assessment. arXiv preprint arXiv:2204.08958 (2022)

23. Chefer, H., et al.: Transformer interpretability beyond attention visualization. In: Proceedings of the IEEE/CVF Conference on Computer Vision and Pattern Recognition (CVPR), pp. 782–791 (2021)
24. Tang, Z., et al.: Perceptual image hashing with weighted dwt features for reduced-reference image quality assessment. Computer J. **61**, 1695–1709 (2018)
25. Zar, J.H.: Spearman rank correlation. Encyclopedia of biostatistics 7 (2005)
26. Bosse, S., et al.: Deep neural networks for no-reference and full-reference image quality assessment. IEEE Trans. Image Process. **27**(1), 206–219 (2018)
27. You, J., Korhonen, J.: Transformer for image quality assessment. In: Proceedings of the IEEE International Conference on Image Processing (ICIP), pp. 1389–1393 (2021)
28. Zhu, H., et al.: MetaIQA: deep meta-learning for no-reference image quality assessment. In: Proceedings of the IEEE/CVF Conference on Computer Vision and Pattern Recognition (CVPR), pp. 14131–14140 (2020)
29. Ying, Z., et al.: From patches to pictures (PAQ-2-PIQ): Mapping the perceptual space of picture quality. In: Proceedings of the IEEE/CVF Conference on Computer Vision and Pattern Recognition (CVPR), pp. 3572–3582 (2020)

Multimedia Analytics Application

Single Cross-domain Semantic Guidance Network for Multimodal Unsupervised Image Translation

Jiaying Lan[1], Lianglun Cheng[1], Guoheng Huang[1(✉)], Chi-Man Pun[2], Xiaochen Yuan[3], Shangyu Lai[4], HongRui Liu[5], and Wing-Kuen Ling[1]

[1] Guangdong University of Technology, Guangzhou, China
kevinwong@gdut.edu.cn
[2] University of Macau, Macau, China
[3] Macao Polytechnic University, Macau, China
[4] University of Maryland College Park, Maryland, MD 20742, USA
[5] San José State University, San José, CA 95192, USA

Abstract. Multimodal image-to-image translation has received great attention due to its flexibility and practicality. The existing methods lack the generality of effective style representation, and cannot capture different levels of stylistic semantic information from cross-domain images. Besides, they ignore the parallelism for cross-domain image generation, and their generator can only be responsible for specific domains. To address these issues, we propose a novel Single Cross-domain Semantic Guidance Network (SCSG-Net) for coarse-to-fine semantically controllable multimodal image translation. Images from different domains are mapped to a unified visual semantic latent space by a dual sparse feature pyramid encoder, and then the generative module generates the result images by extracting semantic style representation from the input images in a self-supervised manner guided by adaptive discrimination. Especially, our SCSG-Net meets the needs of users in different styles as well as diverse scenarios. Extensive experiments on different benchmark datasets show that our method can outperform other state-of-the-art methods both quantitatively and qualitatively.

Keywords: Multimodal image translation · Semantic guidance · Unsupervised learning

1 Introduction

The extensive research of image translation can be applied both in computer vision and in image processing problems, such as style transfer [2,28], image inpainting [21,25], image synthesis [8,12,19,23,27]. The task of image translation is to transform an image from one domain to another while preserving the content

D.-T. Dang-Nguyen et al. (Eds.): MMM 2023, LNCS 13833, pp. 165–177, 2023.
https://doi.org/10.1007/978-3-031-27077-2_13

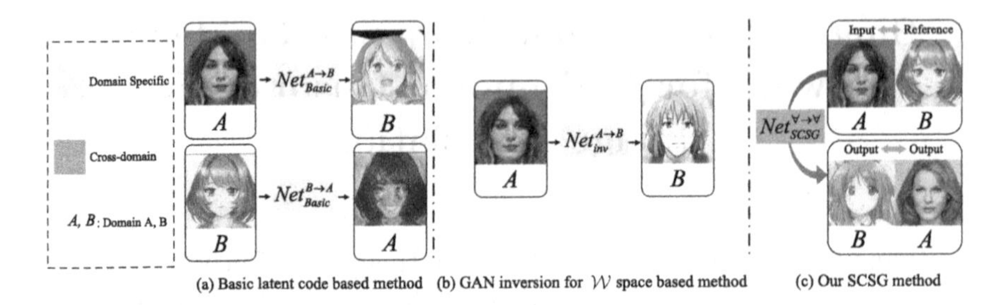

Fig. 1. The framework of various image translation pipelines. (a) shows the latent code-based methods [8,12] where the multiple domain-specific encoders and generators are used for multimodal image translation. (b) shows the $\mathcal{W}$ space-based methods [21,25,27] where a pre-trained StyleGAN model is used to design a style encoder for image translation, and the results of these methods are domain-specific. (c) shows our proposed SCSG-Net method that maps cross-domain images for multimodal image translation, and it is built upon our unified visual semantic latent space for disentangled representation learning.

of the image. In this case, domain means that a set of images can be divided into categories with different shapes and textures, such as a human face and a cat face, and a human face and a cartoon face, which contain two different domains respectively. Many solutions for this task, however, are only able to translate a certain domain to another by a domain-specific generator, which means they are not cross-domain.

To translate images to another, we need to address the generality of effective style representation. One basic idea of translating an image is to separate the codes into two parts, one for content and the other for style, as shown in Fig. 1(a). Then a style generator can translate the content of one image using the style of another. The MUNIT [8] is based on this idea. Yet some limitations exist for this method. First, the generators are usually domain-specific. In other words, one generator can only translate an image to certain styles. The ideal multimodal image translation captures the stylistic representation of the different semantic layers of each modality, as it allows us to perform image translation in a fine-grained manner. At the same time, the parallelism of generation is maintained, which means that the model does not require multiple domain-specific generators and discriminators.

Some other approaches based on StyleGAN inversion have also been proposed for image-to-image translation. However, they require not only a domain-specific encoder but also a domain-specific pre-trained StyleGAN generator, as shown in Fig. 1(b). In other words, they are limited in crossing different domains.

To solve the problems mentioned above, we propose a Single Cross-domain Semantic Guidance Network (SCSG-Net) in this paper. A Cross-domain network takes any input and any reference, as shown in Fig. 1(c), and the result should remain the shape of the input image and the style texture of the reference image. Our SCSG-Net requires neither a specific encoder nor a generator for each domain image. Also, no pre-trained domain-specific StyleGAN generator is needed. Its encoder turns the input combined with the reference into a unified visual semantic latent space and then puts it into the generator. Besides, our SCSG-Net reduces redundant modules. It does not require domain-specific encoders to sample the image data distribution for each domain, and the proposed encoders and generators have better generalizability. The training process is unsupervised and does not rely on paired data, and the discriminator performs adaptive discrimination of semantic features based on the current network output. From the perspective of GANs design, our purposed Cross-domain Network samples cross-domain image data as an overall distribution. It utilizes non-domain specific encoders and generators to obtain the stylistic representations of different semantic levels after the inputs are convoluted into the unified visual semantic space, and it uses the latent code of this space to inject into the generator to generate target domain images. Specifically, we also propose the Semantic-Aware Adaptive Discrimination for adaptive discrimination of the output image.

The main contributions of this paper are summarized as follows: **1)** We propose a novel SCSG-Net to effectively generate diverse user-desirable cross-domain images with better texture details while maintaining the source domain shape in a unified visual semantic latent space. **2)** We design a novel Dual Sparse Feature Pyramid Encoder based on the sparse feature pyramid. It aims at mapping the cross-domain images into a unified hierarchical semantic latent space, which consists of the shape code and texture code. **3)** We propose a Semantic Patch Confidence Evaluation Module (SPCEM) as adaptive discrimination. It can make better use of the hierarchical semantic features and can better select patch regions for self-supervised semantic guidance discrimination.

2 Related Work

2.1 Image-to-image Translation

Image-to-image translation techniques [4,8,9,15,17,20,26,30] have been widely explored via GANs. For image translation with paired data, Isola et al. proposed the first image-to-image network Pix2Pix [9]. For image translation with unpaired data, Zhu et al. [30] proposed a cycle-consistency loss. For image translation in multi-modal, Huang et al. presented MUNIT [8], a multi-modal extension of UNIT [17] that decomposes images into content and style latent space. Choi et al. [4] presented both latent-guided and reference-guided synthesis for image translation utilizing both latent codes and reference images. The autoencoder structure was used by Park et al. [20] to swap the style and content of two provided images, and it could also be considered a reference-guided synthesis approach. And these methods require domain-specific encoders, generators, and discriminators for each target domain, as shown in Fig. 1(a).

2.2 Image Encoder

Self-Supervised Learning is one of the unsupervised learning, which is aimed to learn generic feature representation for downstream tasks, focused on different pre-training tasks [13]. He et al. proposed an asymmetric encoder-decoder architecture called MAE [5] via self-supervised learning. Autoencoding [7] is also one of the unsupervised learning, which is a classical study of representational learning. Richardson et al. [21] and Tov et al. [25] produce high-quality image embedding which is manipulated by self-supervised reconstruction. However, these latent spaces and image embedding are domain-specific and the construction required domain-specific pre-trained StyleGAN models. Our SCSG-Net is an unsupervised training image translation network trained on unpaired data. Specifically, we design a Dual Sparse Feature Pyramid Encoder to encode cross-domain images to obtain the corresponding unified visual semantic latent codes, and then decode and reconstruct them by self-supervised learning. The encoder learns a generic feature representation for subsequent multimodal image translation tasks.

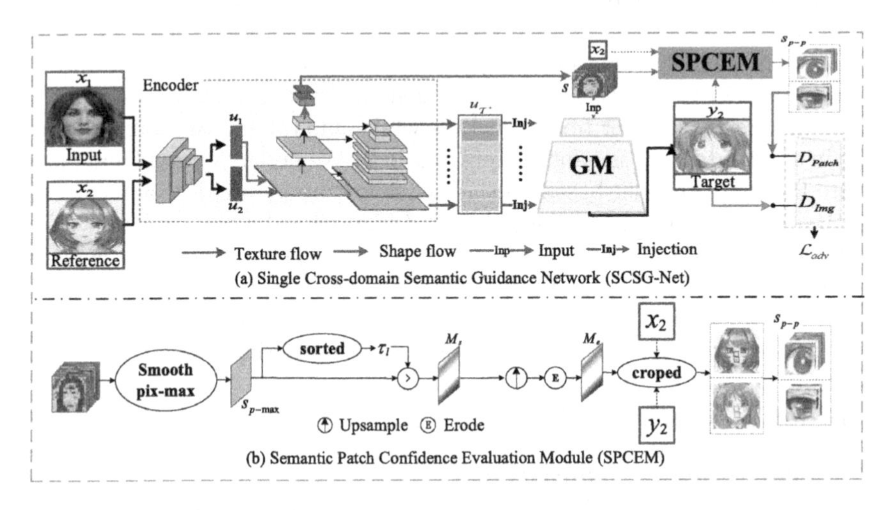

Fig. 2. Overview of the proposed SCSG-Net. (a) shows the SCSG-Net network architecture. Encoder E encodes the shape code s of any source image and the texture code u_T^+ is extracted from any reference image. Generative Module (GM) generates the target images y_2 from a combination of the sampled inputs s and u_T^+. SPCEM updates the appropriate patch pair s_{p-p} provided to D_{Patch} using the self-supervised semantic features. And the detailed structure of SPCEM is in (b).

3 Proposed Method

3.1 Overview

Many current methods translate specific domains of images into limited genres of styles. In other words, one model is merely useful in specific cases. To make a more general model in image translation, we propose a novel method via self-supervised visual semantic features and apply coarse-to-fine style control from cross-domain image data. To apply self-supervised semantic features and coarse-to-fine style control in cross-domain image translation, we propose the SCSG-Net in this paper, as shown in Fig. 2.

Our SCSG-Net (Fig. 2(a)) aims to generate diverse user-desirable images with more desirable texture details in the texture space while maintaining the source domain shape. To deeply exploit how self-supervised hierarchical semantic features in a unified visual semantic latent space can help subsequent adaptive discrimination, SCSG-Net adds the SPCEM (Fig. 2(b)) as an advanced strategy. The evaluation is performed under the assumption that the $\mathcal{S}$ space is known to have semantic information from the arguments above. The SPCEM is proposed to make better use of these hierarchical semantic features as well as to better select patch regions.

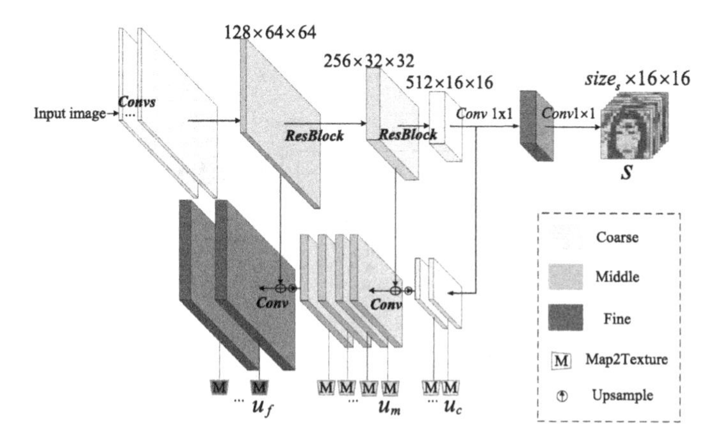

Fig. 3. The architecture of our proposed Dual Sparse Feature Pyramid Encoder. For each of the eight output textures code, a tiny mapping network Map2Texture is trained to map the features to texture codes from the corresponding feature layers. Map2texture is a tiny fully convolutional network that uses a sequence of convolutions and LeakyReLU activations to gradually reduce the spatial size of the input features. And $size_s$ is the output channel of shape code s.

3.2 Dual Sparse Feature Pyramid Encoder

A straightforward method for embedding the target domain is to immediately encode a particular input image into a latent space that uses a single uniform vector derived from the encoder in the final layer of the network. As a result, it learns all style vectors simultaneously [10]. However, such a design creates a significant bottleneck that prevents it from fully reflecting the hierarchical fine features of the input image.

Karras et al. [11] indicated that style inputs at different feature levels equate to varying degrees of detail, which are essentially grouped into coarse, medium, and fine in StyleGAN. We design a novel Dual Sparse Feature Pyramid Encoder as a result of this insight. One branch captures deep shape features from the input images, and another branch captures different levels of semantic texture for domain-free image manipulation. It is not only able to compress the image to obtain the shape code, but also generates three different levels of texture codes extracted by the mapping network Map2Texture, as shown in Fig. 3. Also, we note the existing PSP encoder [21], but we differ from it in that 1) our encoder not only represents the style but also captures the shape features of the input images and encodes them, 2) our encoder is sparse.

We define our Encoder E as the following equation:

$$E(X) = (s, u_{\mathcal{T}^+}), \tag{1}$$

where $E(X)$ is a mapping of X combining the shape code s and the texture code $u_{\mathcal{T}^+}$. And $u_{\mathcal{T}^+}$ also can be grouped as $u_{\mathcal{T}^+} = \{u_c, u_m, u_f\}$, where u_c, u_c and u_f represent the texture codes of coarse (0–1), middle (2–4), and fine (5–6) of the encoder output in Fig. 3, respectively.

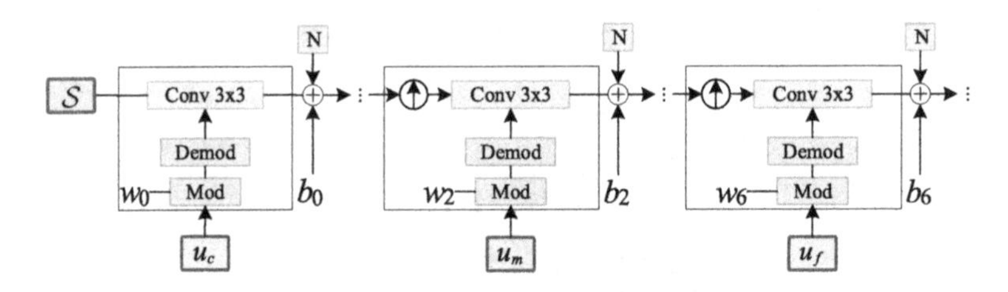

Fig. 4. Our Generative Module (GM).

3.3 Generative Module

We are inspired by the model of StyleGAN2 [11], where the generative module is similar to its modulation module. We first use the shape code s with geometric shape representations as the initial input to the GM. Then, the coarse-to-fine hierarchical texture features $u_{\mathcal{T}^+}$ are injected layer by layer into the style texture features as the modulation module. The subsequent generation is finally performed. More details of the GM are shown in Fig. 4.

Our separable code extraction follows the framework of a classic auto-encoder. First, we define x_1 as the input image and x_{recon} as the reconstructed image, and the equation is shown below.

$$x_{recon} = GM(E(x_1)), \tag{2}$$

where $GM(E(x_1))$ is the result of our Generative Module.

Then, the $\mathcal{L}1$ loss function will be used by us as the reconstruction loss $\mathcal{L}_{recon}$ of the encoder.

$$\mathcal{L}_{recon}(E, GM) = \mathbb{E}_{x_1 \sim X_1}\left[\|x_{recon} - x_1\|_1\right]. \tag{3}$$

We can also perform fine-grained texture stylization control at different semantic levels by modifying the injected u_{T^+} to get $u_{T\mathcal{M}}$. It could be formulated in Eq. 4.

$$u_{T\mathcal{M}} = \sum_{n=0}^{N}(1-\beta)u_{T_1} + \beta u_{T_2}, \tag{4}$$

where u_{T_1} and u_{T_2} represent two different input texture codes, respectively. And $\beta \in [0,1]$ is a vector and the different values of β determine the degree of blending of the two u_T. This blending operation does not need to be done on every layer in a practical test. In Sect. 4.4, we show the results of blending.

3.4 Semantic Patch Confidence Evaluation Module

When our encoder learns a factored representation using the aforementioned restrictions, the resultant representation will not be intuitive for conducting coarse-to-fine image translation. Here, it is not guaranteed that s and u_{T^+} actually represent the shape and the texture, respectively.

To overcome this problem, we assume that any generated target image based on the same u_{T^+} as the injected texture code should maintain consistent texture features. We define the loss function for local joint patches termed discriminator $\mathcal{L}_{\text{Patch}}$ as the following equation:

$$\begin{aligned}\mathcal{L}_{\text{Patch}}(E, GM, D_{Patch}) = \\ \mathbb{E}_{x_2\sim X_2}\left[-\log\left(D_{\text{patch}}\left(crop\left(GM\left(s_1, u_2^{T^+}\right)\right), crop\left(x_2\right)\right)\right)\right],\end{aligned} \tag{5}$$

where the D_{Patch} represents a patch discriminator [20], and *crop* is an operation to get updated patches. It promotes the output images to retain key texture from the source images. And s_1 and $u_2^{T^+}$ are the shape code and the texture code of the input image x_1, respectively.

The objective of the generator is to generate an output image $G(s, u_{T^+})$ in which any patches from the output cannot be differentiated from a set of patches from input x_2. In addition, the operation of *crop* is related to the module SPCEM. The module SPCEM takes the latent code s as an input. It enters a smooth Pix-max operation to get the corresponding positions s_{p-max}, and filters these positions according to the threshold τ_l to get the mask M_s with more response after noise reduction. Then upsampling and erosion operations are performed on M_s to optimize the positioning of the patch location and obtain mask M_e. Ultimately, the patch location is selected based on M_e, and the appropriate patch pair s_{p-p} is updated. Additionally, it also means that each area can be randomized to produce a more relevant distribution of style samples instead of considering a small area of the style patch. Figure 2(b) illustrates the series of steps described above. We suppose that there are hybrid image x_1 and reference image x_2. The number of patches extracted from x_1 and x_2 do not match. In order to learn the texture feature of x_2, patches are extracted in the vicinity of x_2 for each s_{p-p} of x_1. This method guides the GM to extract the texture feature.

Then we define the adversarial loss $\mathcal{L}_{\text{adv}}^{'}$ as the following equation:

$$\mathcal{L}_{\text{adv}}^{'}(E, GM, D_{Patch}, D_{Img}) = \mathcal{L}_{\text{adv}}(GM, D_{Img}) + \mathcal{L}_{\text{Patch}}(E, GM, D_{Patch}). \tag{6}$$

Fig. 5. Visual comparisons of various image-to-image translation methods on the Face2Anime dataset.

4 Experiments

4.1 Datasets and Settings

Datasets. Three popular benchmarks of multimodal unsupervised image the translation was conducted by us. Face2Anime [16] is larger and has a wider range of anime styles (for example, face postures, drawing styles, strokes, haircuts, eye shapes, and facial features) than selfie2anime, and it has 17,796 photos in total, with 8,898 anime-faces and genuine photo-faces. The photo2portrait and cat2dog data [14] is composed of 6452 real-face images, 814 western painting images, 871 cat images, and 1364 dog images, We randomly select 100 images from each of the categories as test data.

Evaluation Metrics. Fréchet Inception Distance (FID) [6] and Kernel Inception Distance (KID) [1] are calculated. FID is an upgrade of IS [22], and KID is an improved measure of quality, realism, and GANs convergence. Lower KID and FID scores indicate that the target and generated images are visually comparable, and demonstrate the better quality of the generated images.

Optimization. The Adam optimizer is applied to train our SCSG-Net with momentum 0.7, weight decay 3×10^{-4}, batch size 4×1 T V100 GPUs.

4.2 Qualitative Comparison

In Fig. 5, the reality of the image generated by the previous methods [8,12,18,30] is not very well. In the second row of Fig. 5, StarGANv2 cannot retain the long hair feature of the source face and generates a short hair result that is

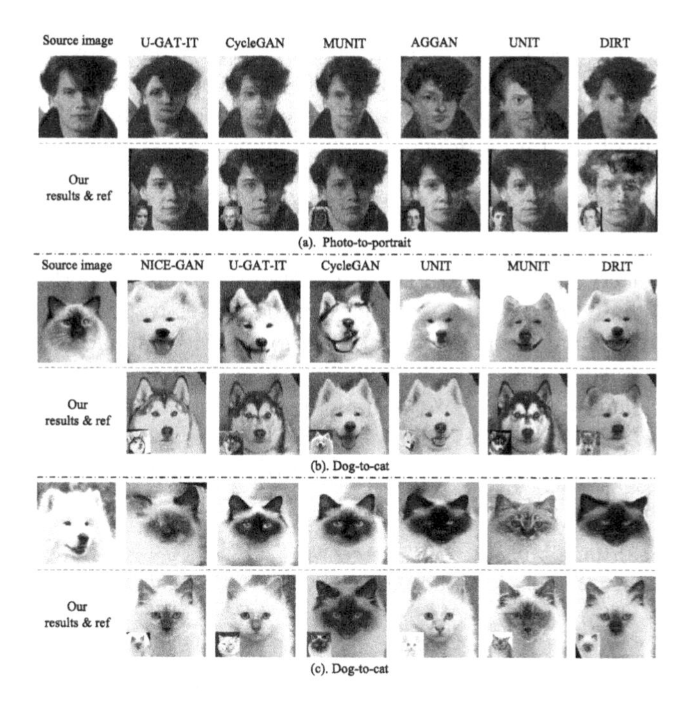

Fig. 6. (a): Visual comparisons on the photo2portrait dataset. (b), (c): Visual comparisons on the cat $\leftrightarrow$ dog dataset.

Table 1. Quantitative evaluation results of Face2Anime (f2a), photo-to-portrait (p2p), dog-to-cat (d2c) and cat-to-dog (c2d) translation. Kernel Inception Distance $\times 100 \pm std \times 100$. The best results are highlighted in boldface.

Model	f2a	Model	p2p	Model	d2c		c2d	
	FID↓		KID↓		KID↓	FID↓	KID↓	FID↓
CycleGAN [30]	50.09	CycleGAN [30]	1.84	CycleGAN [30]	4.93	119.32	6.93	125.30
U-GAT-IT [12]	42.84	U-GAT-IT [12]	1.79	U-GAT-IT [12]	3.22	80.75	2.49	64.36
MUNIT [8]	43.75	MUNIT [8]	4.75	NICE-GAN [3]	1.58	48.79	1.20	44.67
FUNIT [18]	56.81	UNIT [17]	1.20	MUNIT [29]	1.26	53.25	2.42	60.84
StarGAN-v2 [4]	39.71	AGGAN [24]	2.33	UNIT [17]	1.94	59.56	1.94	63.78
AniGAN [16]	38.45	DRIT [14]	5.85	DRIT [14]	5.23	94.50	4.53	79.57
Ours	**35.27**	**Ours**	**1.12**	**Ours**	**0.38**	**34.10**	**0.35**	**23.97**

more similar to the reference anime face. And the AniGAN constructs an anime face without the background information of the source. The photo-to-portrait translation is shown in Fig. 6. The facial features of our results are more defined than in the baseline model and Western-style brushstrokes and shading are more realistic than in the baseline models. The dog $\leftrightarrow$ cat translation in Fig. 6 shows that the image generated by our SCSG-Net has more detailed features of the cat while retaining the general pose of the source image. For example, the cat whiskers are a detailed feature of cats that are easily ignored by other models, but they can still be generated by our SCSG-Net realistically. In summary, the images generated by SCSG-Net guarantee diverse outputs while maintaining the texture realism of the target domain.

4.3 Quantitative Comparison

Table 1 shows that the KID and FID scores of our method outperform all baseline models. There are two main reasons for the better results achieved by our SCSG-Net. On one hand, the presence of fine geometric shape features preserves more identity features, thus enabling the images generated by the GM to better preserve the identity features of the input images as well. On the other hand, discriminator D_{Patch} promotes translated images to retain the crucial texture of the reference image. Table 1 shows that the KID and FID scores of our method outperform all baseline models.

Table 2. Ablation study of our SCSG-Net on Face2Anime.

Model	Setting	FID↓
B1	SCSG-Net	**35.27**
B2	SCSG-Net $-\mathcal{L}_{\text{recon}}$	36.42
B3	SCSG-Net $-\mathcal{L}_{\text{adv}}$	37.56
B4	SCSG-Net $-\mathcal{L}_{\text{Patch}}$	38.03
B5	SCSG-Net - SPCEM	36.97

4.4 Ablation Study

Effect of the Different Loss Terms. We undertake a comprehensive ablation experiment on the Face2Anime dataset to validate each component of the proposed SCSG-Net. SCSG-Net has 5 baselines as shown in Table 2. **1)** B1 is our SCSG-Net baseline. **2)** B2 is our SCSG-Net without $\mathcal{L}_{\text{recon}}$. **3)** B3 is our SCSG-Net without $\mathcal{L}_{\text{adv}}$. **4)** B4 is our SCSG-Net without $\mathcal{L}_{\text{Patch}}$. **5)** B5 is our SCSG-Net without SPCEM. By the way, we adopt the strategy of randomly cropping patches to compensate for D_{patch} production patches.

Table 2 shows the results of the ablation experiment. We can see that both B1 and B2 achieves better FID score than B3, which substantiates the importance of $\mathcal{L}_{\text{adv}}$ as a discriminative constraint of image levels. B1 outperforms B2 showing that adding $\mathcal{L}_{\text{recon}}$ loss term will further enlarge the accuracy of style representation. The score of B4 is lower than others, indicating that $\mathcal{L}_{\text{Patch}}$ is the most effective constraint of SCSG-Net for multimodal image translation, which can better maintain the reference image style and texture features. B5 achieves a better FID score than B4, demonstrating the effectiveness of the SPCEM.

Effect of Parameters in Coarse-to-fine Style Image Translation. Figure 7 shows an example of visual semantic manipulation of two photos, guided by four reference photos, in the direction of coarse-to-fine from the bottom to the top. We constrain the shift parameter β for each layer. When β takes the values of 0.25, 0.5, and 0.75, it corresponds to the results of coarse constraint, medium constraint, and fine constraint, respectively.

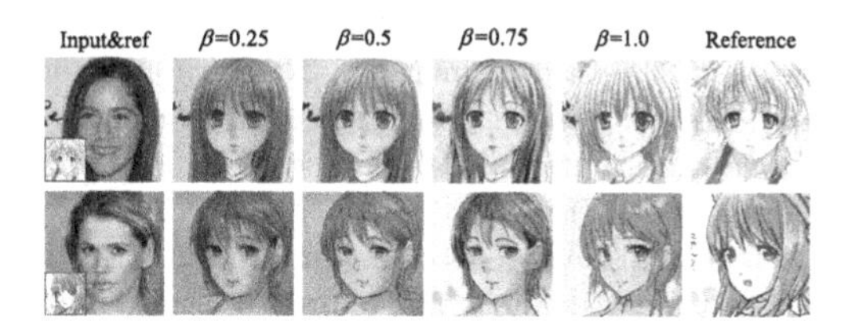

Fig. 7. Effect of parameters in coarse-to-fine style image translation.

5 Conclusion

In this paper, we propose a novel SCSG-Net to address the task of multimodal unsupervised image translation, which aims to perform image translation with a single pair of encoder and generator on a source domain image based on a cross-domain reference image. Our proposed Dual Sparse Feature Pyramid Encoder effectively maps the multimodal images to a unified visual semantic latent space. Based on the StyleGAN2 generative module, we evaluate the unified visual semantic latent space that is semantically controllable. We also evaluate the SPCEM module which could make better use of these hierarchical semantic features as well as better select patch regions. We verify the effectiveness of our SCSG-Net both quantitatively and qualitatively.

Acknowledgements. This work was supported in part by the Science and technology research in key areas in Foshan under Grant 2020001006832, the Key-Area Research and Development Program of Guangdong Province under Grant 2018B010109007 and 2019B010153002, and the Guangzhou R&D Programme in Key Areas of Science and Technology Projects under Grant 202007040006, and the Guangdong Provincial Key Laboratory of Cyber-Physical System under Grant 2020B1212060069, and the Program of Marine Economy Development (Six Marine Industries) Special Foundation of Department of Natural Resources of Guangdong Province under Grant GDNRC [2020]056.

References

1. Bińkowski, M., Sutherland, D., Arbel, M., Gretton, A.: Demystifying MMD GANs. In: ICLR (2018)

2. Chen, H., et al.: Artistic style transfer with internal-external learning and contrastive learning. In: NIPS 34 (2021)
3. Chen, R., Huang, W., Huang, B., Sun, F., Fang, B.: Reusing discriminators for encoding: towards unsupervised image-to-image translation. In: CVPR, pp. 8168–8177 (2020)
4. Choi, Y., Uh, Y., Yoo, J., Ha, J.W.: StarGAN v2: diverse image synthesis for multiple domains. In: CVPR, pp. 8188–8197 (2020)
5. He, K., Chen, X., Xie, S., Li, Y., Dollár, P., Girshick, R.: Masked autoencoders are scalable vision learners. In: CVPR, pp. 16000–16009 (2022)
6. Heusel, M., Ramsauer, H., Unterthiner, T., Nessler, B., Hochreiter, S.: GANs trained by a two time-scale update rule converge to a local nash equilibrium. In: Proceedings of the 31st International Conference on Neural Information Processing Systems, pp. 6629–6640 (2017)
7. Hinton, G.E., Zemel, R.: Autoencoders, minimum description length and Helmholtz free energy. In: Advances in Neural Information Processing Systems 6 (1993)
8. Huang, X., Liu, M.-Y., Belongie, S., Kautz, J.: Multimodal unsupervised image-to-image translation. In: Ferrari, V., Hebert, M., Sminchisescu, C., Weiss, Y. (eds.) ECCV 2018. LNCS, vol. 11207, pp. 179–196. Springer, Cham (2018). https://doi.org/10.1007/978-3-030-01219-9_11
9. Isola, P., Zhu, J.Y., Zhou, T., Efros, A.A.: Image-to-image translation with conditional adversarial networks. In: CVPR, pp. 1125–1134 (2017)
10. Karras, T., Laine, S., Aila, T.: A style-based generator architecture for generative adversarial networks. In: CVPR, pp. 4401–4410 (2019)
11. Karras, T., Laine, S., Aittala, M., Hellsten, J., Lehtinen, J., Aila, T.: Analyzing and improving the image quality of styleGAN. In: CVPR, pp. 8110–8119 (2020)
12. Kim, J., Kim, M., Kang, H., Lee, K.H.: U-GAT-IT: unsupervised generative attentional networks with adaptive layer-instance normalization for image-to-image translation. In: ICLR (2019)
13. Komodakis, N., Gidaris, S.: Unsupervised representation learning by predicting image rotations. In: ICLR (2018)
14. Lee, H.-Y., Tseng, H.-Y., Huang, J.-B., Singh, M., Yang, M.-H.: Diverse image-to-image translation via disentangled representations. In: Ferrari, V., Hebert, M., Sminchisescu, C., Weiss, Y. (eds.) ECCV 2018. LNCS, vol. 11205, pp. 36–52. Springer, Cham (2018). https://doi.org/10.1007/978-3-030-01246-5_3
15. Lee, H.Y., et al.: Drit++: diverse image-to-image translation via disentangled representations. Int. J. Comput. Vision **128**(10), 2402–2417 (2020)
16. Li, B., Zhu, Y., Wang, Y., Lin, C.W., Ghanem, B., Shen, L.: AniGAN: style-guided generative adversarial networks for unsupervised anime face generation. IEEE Trans. Multimedia (2021)
17. Liu, M.Y., Breuel, T., Kautz, J.: Unsupervised image-to-image translation networks. In: NIPS, pp. 700–708 (2017)
18. Liu, M.Y., Huang, X., Mallya, A., Karras, T., Aila, T., Lehtinen, J., Kautz, J.: Few-shot unsupervised image-to-image translation. In: ICCV, pp. 10551–10560 (2019)
19. Long, J., Lu, H.: Multi-level gate feature aggregation with spatially adaptive batch-instance normalization for semantic image synthesis. In: Lokoč, J., et al. (eds.) MMM 2021. LNCS, vol. 12572, pp. 378–390. Springer, Cham (2021). https://doi.org/10.1007/978-3-030-67832-6_31
20. Park, T., Zhu, J.Y., Wang, O., Lu, J., Shechtman, E., Efros, A., Zhang, R.: Swapping autoencoder for deep image manipulation. In: NIPS 33, pp. 7198–7211 (2020)

21. Richardson, E., Alaluf, Y., Patashnik, O., Nitzan, Y., Azar, Y., Shapiro, S., Cohen-Or, D.: Encoding in style: a styleGAN encoder for image-to-image translation. In: CVPR, pp. 2287–2296 (2021)
22. Salimans, T., Goodfellow, I., Zaremba, W., Cheung, V., Radford, A., Chen, X.: Improved techniques for training GANs. In: NIPS, pp. 2234–2242 (2016)
23. Sun, Yanbei, Lu, Yao, Lu, Haowei, Zhao, Qingjie, Wang, Shunzhou: Multimodal unsupervised image-to-image translation without independent style encoder. In: Jónsson, B., et al. (eds.) MMM 2022. LNCS, vol. 13141, pp. 624–636. Springer, Cham (2022). https://doi.org/10.1007/978-3-030-98358-1_49
24. Tang, H., Xu, D., Sebe, N., Yan, Y.: Attention-guided generative adversarial networks for unsupervised image-to-image translation. In: 2019 International Joint Conference on Neural Networks (IJCNN), pp. 1–8. IEEE (2019)
25. Tov, O., Alaluf, Y., Nitzan, Y., Patashnik, O., Cohen-Or, D.: Designing an encoder for styleGAN image manipulation. TOG **40**(4), 1–14 (2021)
26. Wang, T.C., Liu, M.Y., Zhu, J.Y., Tao, A., Kautz, J., Catanzaro, B.: High-resolution image synthesis and semantic manipulation with conditional GANs. In: CVPR, pp. 8798–8807 (2018)
27. Yang, S., Jiang, L., Liu, Z., Loy, C.C.: Pastiche master: exemplar-based high-resolution portrait style transfer. In: CVPR (2022)
28. Zhai, S., Hu, X., Chen, X., Ni, B., Zhang, W.: Resolution booster: global structure preserving stitching method for ultra-high resolution image translation. In: Ro, Y.M., et al. (eds.) MMM 2020. LNCS, vol. 11961, pp. 702–713. Springer, Cham (2020). https://doi.org/10.1007/978-3-030-37731-1_57
29. Zhao, Y., Wu, R., Dong, H.: Unpaired image-to-image translation using adversarial consistency loss. In: Vedaldi, A., Bischof, H., Brox, T., Frahm, J.-M. (eds.) ECCV 2020. LNCS, vol. 12354, pp. 800–815. Springer, Cham (2020). https://doi.org/10.1007/978-3-030-58545-7_46
30. Zhu, J.Y., Park, T., Isola, P., Efros, A.A.: Unpaired image-to-image translation using cycle-consistent adversarial networks. In: ICCV, pp. 2223–2232 (2017)

Towards Captioning an Image Collection from a Combined Scene Graph Representation Approach

Itthisak Phueaksri[1(✉)], Marc A. Kastner[2], Yasutomo Kawanishi[1,3], Takahiro Komamizu[1], and Ichiro Ide[1]

[1] Nagoya University, Nagoya, Aichi, Japan
phueaksrii@cs.i.nagoya-u.ac.jp
[2] Kyoto University, Kyoto, Japan
[3] RIKEN, Seika, Kyoto, Japan

Abstract. Most content summarization models from the field of natural language processing summarize the textual contents of a collection of documents or paragraphs. In contrast, summarizing the visual contents of a collection of images has not been researched to this extent. In this paper, we present a framework for summarizing the visual contents of an image collection. The key idea is to collect the scene graphs for all images in the image collection, create a combined representation, and then generate a visually summarizing caption using a scene-graph captioning model. Note that this aims to summarize common contents across all images in a single caption rather than describing each image individually. After aggregating all the scene graphs of an image collection into a single scene graph, we normalize it by using an additional concept generalization component. This component selects the common concept in each sub-graph with ConceptNet based on word embedding techniques. Lastly, we refine the captioning results by replacing a specific noun phrase with a common concept from the concept generalization component to improve the captioning results. We construct a dataset for this task based on the MS-COCO dataset using techniques from image classification and image-caption retrieval. An evaluation of the proposed method on this dataset shows promising performance.

Keywords: Multiple-image summarization · Image captioning · Scene graph captioning

1 Introduction

With an increasing number of images on the Web and on Social Media, it has become a challenge to describe and understand these images. Describing a collection of images with a short description is often easier to grasp overall contexts than describing them individually. For a single image, image captioning is a popular task that generates an image description in the form of a sentence. However,

D.-T. Dang-Nguyen et al. (Eds.): MMM 2023, LNCS 13833, pp. 178–190, 2023.
https://doi.org/10.1007/978-3-031-27077-2_14

Fig. 1. Example of image summarization compared with image collection captioning: The left side shows the existing approach, in which the description is limited to words or noun phrases. The right side shows the proposed approach, in which an image collection is described in a single and refined sentence.

it cannot be easily adjusted to describe multiple images simultaneously. Image collection summarization [22,24,32] is a challenging new task which aims to generate a shared caption for all images in an image collection. However, existing approaches are limited to summarizing an image collection only in the form of concept words or tag words. A recent work [24] presents a method for summarizing the texture, style, and material of similar objects in an image collection. Other works [22,32] summarize an image collection as a set of keywords. To be more informative in describing an image collection in semantic contexts, we propose image collection captioning to describe an image collection in the form of a single sentence. Figure 1 shows the proposed task compared to the existing image summarization task.

Our approach aims to understand an image collection based on scene graph representations generated from the images in the collection. A scene graph is a popular means of describing a region-based image context by detecting objects in the image and their relationships. It has also been leveraged as a bottom-up mechanism for image captioning tasks [12]. A scene graph is a structured list of triplets consisting of a subject, a predicate, and an object. In the proposed method, with multiple scene graphs from images in the collection, we combine all scene graphs into a single scene graph representation. We then estimate all nodes and relations to find the most prominent combined context and generate a summarized scene graph for the captioning model to generate a phrase.

The challenge of the image collection captioning task is to generate a caption that can simultaneously describe all images in the image collection. Inspired by abstractive text summarization [7], we propose two components to generalize specific words to more general word choices with ConceptNet [23]. The first component is *Sub-Graph Concept Generation* that processes all image scene graphs in response to expanding the word concept following the ConceptNet. Incorporating the idea of word communities [1], we find the representative words in each community to be the word choices for the captioning. For example, when constructing a word community for "bird" and "bear," we find a representative word "animal" after expanding the concept. The second component is *Sentence Refinement*, which integrates the result of sub-graph concept generation and the captioning result to rebuild the caption, focusing on the noun phrase. When the captioning result is, e.g., "a polar bear standing in the snow near the water," the refined result becomes "animal standing in the snow near the water." Here, it replaces the phrase "a polar bear" with a more general word "animal," which is the knowledge gathered from the *Sub-Graph Concept Generation* component.

Due to this task being novel, there is no existing dataset available. Thus, we build an image collection dataset from the popular image caption dataset, MS-COCO [17]. However, the number of captions in the MS-COCO dataset is restricted to five captions per image. Correspondingly with the dataset limitation and the idea of summarizing scene graphs, we build a scene graph captioning model trained by a single image. Then, we transfer the model to our framework. To evaluate the proposed method, we compare it with text summarization methods, which are most similar to our work. These methods are evaluated on several automatic evaluation metrics designed to evaluate text generation and text summarization tasks. This work consists of the following:

- First, we propose a new challenging task, image collection captioning, which aims to describe an image collection in the form of a single sentence.
- Second, we propose a baseline for the novel task based on a combined scene graph captioning approach. We build a combined scene graph representing all images in the image collection and then generate a caption based on it.
- Last, we construct a dataset for this task based on the MS-COCO dataset by incorporating image classification and image-caption retrieval tasks.

2 Related Work

Image Captioning. Image captioning [12] is an image-to-text translation task that aims to describe the scene, location, objects, and interactions in an image in the form of a sentence. BEiT-3 [27] is the current state-of-the-art in vision-language task with multi-way transformers. Other methods [19,33] introduce a scene graph into their captioning models to improve the performance. In our work, we build upon this approach by extending this idea to scene graphs generated from multiple images and using their combined representation.

Multiple Image Summarization. Multiple image summarization was introduced recently, which aims to generate common keywords for an image collection. Samani et al. [22] propose a method to find the semantic concept of an image collection. They generalize word concepts in a specific domain and aim to find semantic similarities between them. Zhang et al. [32] present a method to generate a visual summary and a textual topic of an image collection by mapping and discovering the textual topics and corresponding images. Trieu et al. [24] propose extending the transformer-based architecture of a single image captioning model to generate a caption for an image collection, which aims to describe the texture, style, and material of the image collection in the form of a noun phrase. Due to the dataset limitation of this task, they also introduce a dataset construction method. By gathering images from Web pages, they collect 2.1 million image collections containing at least five images each. In our work, we propose a scene graph captioning model with a combined scene graph that can describe the interaction between the objects in the form of a sentence, which is more informative than implementing transformer-based methods.

Scene Graph Generation. Scene graph generation [9] is a method of describing object contexts in an image. Many scene graph generation methods start by finding object regions using Fast R-CNN [6] as an object detector. They then find the relationship between objects in both local and global contexts. Neural Motifs [29] is a model constructed from a stacked Motif Network (MotifNet), which is strongly predictive of relation labels on the Visual Genome Dataset and further evaluated on the MS-COCO Dataset. Their method includes three main predicting stages bounding regions, labels for regions, and relationships. RelDN [30] is a recent novel scene graph generation method focusing on improving the accuracy of relationship classification, raising entity confusion and loss over predicate classes. In our work, we make use of the Neural Motif network to detect the relationship between objects in an image using ResNet [11] as an object detector.

Text Summarization. Text summarization is a text-to-text generation task that aims to generate a short description from multiple documents. Text summarization tasks can be divided into two main paradigms: extractive summarization and abstractive summarization. Extractive summarization aims to identify the salient information that appears in documents. T5 [21] is a strong baseline of the supervised summarization model, which is pre-trained by the Wikipedia dataset[1]. Moreover, it is also introduced using unsupervised learning techniques. SUPERT [5] is an unsupervised multiple-document summarization model that evaluates the sentences in documents and selects one of them to be a topic. However, it is restricted to the document content. Meanwhile, abstractive summarization aims to rewrite the summarized sentence from sentences in the document by finding the semantic representation or generating a common word from a word corpus. XL-Sum [10] introduces the BBC news dataset and a multi-lingual abstractive summarization method that fine-tunes with the T5 model with their dataset. Our work is inspired by abstractive text summarization, as we aim to summarize an image collection into a generalized caption.

3 Proposed Method: Image Collection Captioning

Our proposed method starts with a collection of images. First, we generate a scene graph for each image. Next, all of the sub-graphs in the collection are combined. Word communities are also constructed from the sub-graphs to find common words. Lastly, we generate a caption from the summarized graph and refine the caption with the common words.

Following this idea, the proposed method consists of five components which are shown in Fig. 2 and discussed in detail in the following sections. The first

[1] https://www.tensorflow.org/datasets/catalog/wikipedia/ (accessed Sept. 9, 2022)

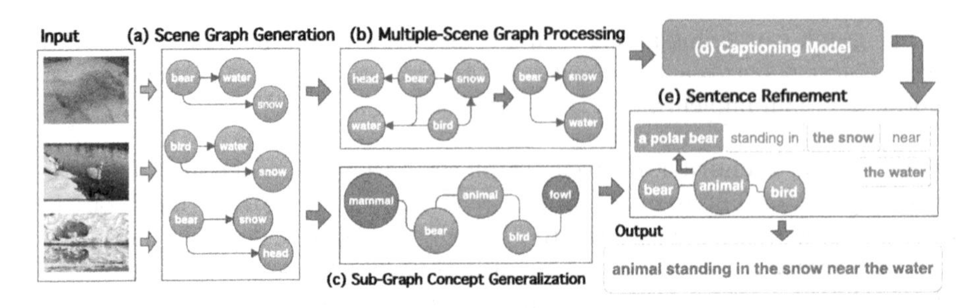

Fig. 2. Overview of the proposed method, consisting of five components: (a) *Scene Graph Generation* extracts a scene graph for each image. (b) *Multiple-Scene Graph Processing* combines all scene graphs and finds a representative graph. (c) *Sub-Graph Concept Generalization* finds word communities from all scene graphs and generates a common word. (d) *Captioning Model* generates the initial caption for the representative graph. (e) *Sentence Refinement* rephrases the caption with the common words.

one is *Scene Graph Generation* which extracts image features and generates a scene graph for each image (sub-graph). Next, all the sub-graphs are parallelly passed into two components: *Multiple Scene Graph Processing* to merge and select part of the combined graph, and *Sub-Graph Concept Generalization* to find general concept words through the word communities. Then, the *Captioning Model* generates a sentence based on the representative scene graph. The generated sentence from the captioning model and the community word graphs are finally passed to the *Sentence Refinement* to output the final caption.

3.1 Scene Graph Generation

Following the existing work on image captioning model leveraging scene graphs, we use the current state-of-the-art scene graph generation method with ResNet101 [11] + Neural Motif [29] as a scene graph parser in the proposed method. This model is retrained on the Visual Genome dataset [15], a popular practice for scene graph captioning. A recent image captioning work [2] shows that manually cleaning up duplicate labels of the Visual Genome dataset from 2,500/1,000/500 to 1,600/400/20 of objects/attributes/relations can improve image captioning performance. We also follow this idea. The result of scene graph generation is represented in a directed graph, which includes subjects, predicates, and objects.

3.2 Multiple-Scene Graph Processing

We build the multiple-scene graph processing module as a feature selection of all sub-graphs from the *Scene Graph Generation Component*. All the sub-graphs are merged into a single directed graph, as shown in Eq. 1, in which G is the merged graph and g_i is a directed graph represented as a set of triplets. In the merging process, we count the occurrence of each feature and the number of edges to be the weight in the selection step.

$$G = \bigcup_{i=1}^{n} g_i \tag{1}$$

Next, we select the top-n nodes and the top-m relations from the sub-graphs (with $n = 36$ and $m = 100$ used in the following experiments, which are feature numbers of our captioning model). In preliminary experiments, we found that implementing betweenness centrality, which refers to the summarization of the fraction of the shortest path in finding the center nodes, is the most efficient method compared with other aspects, which is represented as:

$$g(v) = \sum_{s \neq v \neq t} \frac{\sigma_{st}(v)}{\sigma_{st}}, \tag{2}$$

where $\sigma_{st}(v)$ is the number of paths from s to t passing through v, and σ_{st} is the total number of the shortest paths from s to t.

3.3 Sub-graph Concept Generalization

Next, we discuss how to generalize specific contexts within the sub-graphs. For example, given an image collection of animals in which each image contains "bear" or "bird," our idea is to generalize these words as a common word: "animal." We build the *Sub-Graph Concept Generalization Component* to find a common concept. A popular text-based semantic network named ConceptNet [23] is employed to extend the word relations of specific words and then select a general word to represent them.

Inspired by text analysis based on word synonym relationships, we build a community of words and find the representative word in each word community. The process is shown in Fig. 3. First, all the object words that appear in the sub-graph are lemmatized. We then incorporate ConceptNet to expand the word relationships of each node based on *synonym* and *isA* relations. In a preliminary experiment, we found that finding a concept word from more different numbers of expandable relations results in generating a more general word. We hence limit the maximum number of each relation to ten.

After expanding the concept, a word graph community is generated by joining all the expanded sub-graphs, and non-degree nodes are dropped. To estimate the representativeness of the node to be the common concept of each sub-graph, we encode all nodes using GloVe word embedding [20] and then calculate the similarity between each node as the distance. When calculating the weight, we select a word by calculating the highest node degree using cosine similarity as a weight to find each sub-graph word concept. We determine the representative in each word community by considering the average shortest distance node over all nodes. We implement the improved closeness centrality, which can estimate the graph with many connections [28] as:

$$C(u) = \frac{n-1}{N-1} \frac{n-1}{\sum_{v=1}^{n-1} d(u,v)}, \tag{3}$$

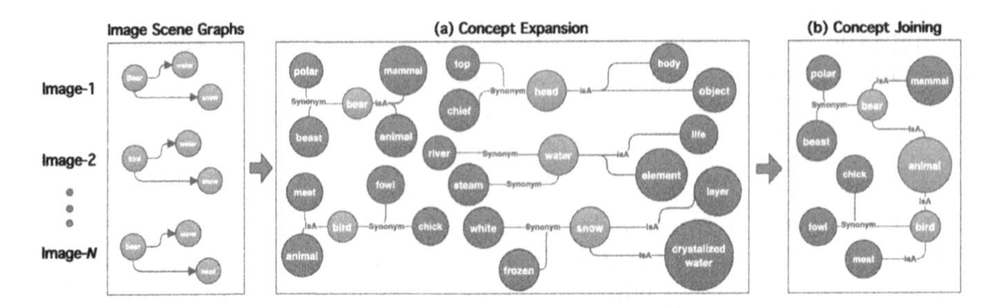

Fig. 3. Example of constructing word communities. (a) *Concept Expansion* expands word concepts of each word in sub-graphs by incorporating with ConceptNet [23]. (b) *Concept Joining* joins all the same words together and selects the central word node as the representative.

where $C(u)$ is the closeness centrality of node u, n is the number of all reachable nodes, N is the number of nodes in the graph, $d(u, v)$ is the distance between nodes u and v.

3.4 Captioning Model

The captioning model consists of a Graph Convolutional Network (GCN) [19] and the Attention-based LSTM model [2]. We build the GCN to process the triplet of *subject*, *predicate*, and *object* features. Each feature is extracted from the scene graph generation process, which consists of 36 subject features, 36 object features, and 100 predicate features. Each feature dimension is 1,024, which is the feature size of the *Scene Graph Generation* output. Next, we build the attention-based LSTM model following the top-down LSTM captioning with two layers of attention-based LSTM, both layers with a size of 512.

3.5 Sentence Refinement

To improve the caption, we modify the beam search of sentence generation to generalize the caption, mainly focusing on general noun words of the caption result. First, word tokens are extracted from the caption and labeled with NLTK POS tagging [18]. Next, a noun phrase is found and labeled with its object component. Finally, the object component of each noun phrase is mapped with the *Sub-Graph Concept Generation Component* to replace the word with the representative of the word community in the sub-graph concept. In the following experiment, we select a beam size of five for generating the final caption.

4 Dataset Construction

The proposed method aims to generate a caption for an image collection, i.e., a sentence describing multiple images. Due to this task being novel, there is no existing dataset for this. The MS-COCO dataset [17] is a popular image captioning dataset which is closest to this task. However, it is typically used only for the single-image captioning task in which each image is captioned with one or more sentences.

In our work, we build upon the MS-COCO dataset by estimating the semantic contents of images and captions and use this to augment the dataset towards image collection captioning. The dataset contains numerous similar images, and the annotated labels are not distinct [26]. It is thus straightforward to use it to generate image collections. To construct a dataset for the proposed task, we implement and compare two approaches based on image classification and image-caption retrieval to estimate the semantic contents of the 5K testing.

4.1 Image Classification Approach

This approach refers to the common concepts gathered from scene graphs. We classify image classes using an image classification model and use that knowledge to collect similar images to construct image collections. In our experiment, we use ResNet101 [11] pre-trained with the ImageNet dataset [3]. First, the top-five classes of each image are predicted. Then, the intersection of classes between the images is found. The prediction scores of each class in each image collection are ranked, and the top-five prediction scores are selected, thus limiting the number of images for each collection. Each of the 880 classes forms the ImageNet concept classes, in which each class contains at least 5 images. The ground-truth captions for the image collection are the concatenated set of all captions of its composite images. Thus, we end up with 15 and 25 sentences from the original description, which are used for evaluation.

4.2 Image-Caption Retrieval Approach

This approach considers both semantic image contents and semantics of the captions annotated to each image. In the following experiment, VSE++ [4] can query the top K images in the embedding space by estimating their visual-semantic embedding. We generate 5K collections for our testing set and limit the query number to five, which results in each image collection in our testing set containing six images. The ground-truth captions for the image collection are the concatenated set of all captions of its composite images. Thus, we end up with 30 sentences from the original description, which are used for the evaluation.

5 Evaluation

5.1 Captioning Model Pre-training Strategy

Due to the limits of the dataset, we first train and evaluate our captioning model for single images using the MS-COCO dataset [17]. Afterwards, the single image captioning model is integrated with the proposed image collection captioning

Fig. 4. Examples of the proposed method on image collections built with the image classification approach. The results show that the proposed method can extract important features from the image collections and generate a general caption describing the most occurred image contents in an image collection. Images with a border indicate that they fit the image collection caption, while those without indicate outliers.

framework, as discussed in Sect. 3. In our training phase, we follow the Karpathy split [13] in which the sizes of the training image set is 118K, the validation image set is 5K, and the testing image set is 5K images. We implement a learning rate decay of 0.8 for every eight epochs, the initial learning rate of 0.0008, dropout 0.5, employing Adam optimization [14], cross-entropy loss, and multi-label margin loss. The best checkpoint of a single caption with CIDEr [25] evaluation is used as the captioning model in the image collection captioning framework.

5.2 Evaluation Metrics

Various evaluation metrics are introduced in the image captioning and text summarization field. Due to our method aiming to perform image captioning over summarization of multiple images, we use ROUGE-1 (R-1), ROUGE-2 (R-2), and ROUGE-L (R-L) [16] as text summarization evaluation metrics. We further implement CIDErBtw [26], a similarity evaluation between sentences based on the CIDEr metric. However, these evaluation metrics are limited in evaluating abstract contexts. Thus, we use BERTScore [31], which is a text generation metric based on calculating word similarity score, and WEEM4TS [8], which is a metric introduced to evaluate abstractive summarization by evaluating the similarity between the summary and the ground-truth using word embedding.

5.3 Results

We evaluate the proposed method on 5K images from each testing dataset created by the two dataset construction approaches. A caption is generated for

Table 1. Evaluation of the results of image collections built with the image classification approach compared to SUPERT [5], T5 [21], and XL-Sum [10].

Method		R-1 ↑	R-2 ↑	R-L ↑	BERTScore ↑	CIDErBtw ↑	WEEM4TS ↑
SUPERT [5]		0.3116	0.0823	0.2848	0.5889	0.4095	0.0746
T5 [21]		0.2938	0.0728	0.2601	0.5710	0.3002	0.0573
XL-Sum [10]		0.1873	0.0284	0.1632	0.4630	0.1004	0.0616
Proposed	w/o CG	**0.3254**	**0.0912**	**0.2958**	**0.5999**	**0.5308**	0.1088
	w/ CG	0.3077	0.0823	0.2777	0.5895	0.4755	**0.1132**

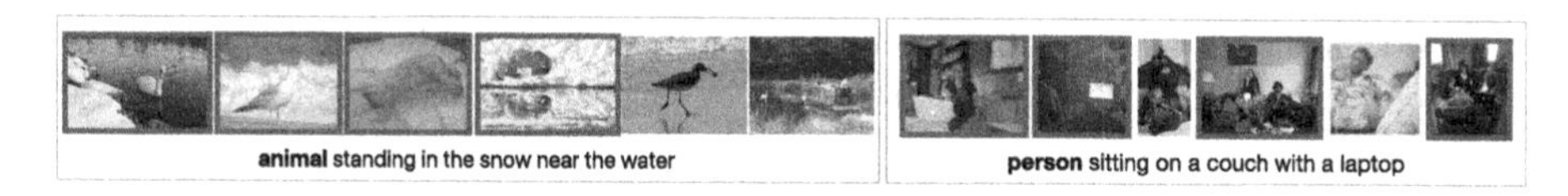

Fig. 5. Examples of the proposed method for image collections built with the image-caption retrieval approach. The results show that the proposed method can extract important features from the image collection and normalize the word concepts in captioning results. Images with a border indicate that they fit the image collection caption, while those without indicate outliers.

each image collection and is compared with the ground-truth caption set of each image collection using automatic evaluation metrics by averaging the scores. We evaluate the proposed method by comparing it to extractive and abstractive text summarization models. However, we realize that automatic evaluation is limited in abstractive summarization. To make the evaluation clear, we ablate two methods, *with* and *without* the *Sub-Graph Concept Generalization (CG)* component, and compare their results.

Image Classification Dataset. We first show results on the dataset constructed with the image classification approach, which consists of 880 image collections, in Fig. 4. It shows that the proposed method detects the most frequently occurring contents, ignoring less appearing contents. Then, it generates a general caption to describe most image contents in an image collection. The evaluation results are shown in Table 1, which shows that the proposed method achieved the best result on overall automatic evaluations.

Image-Caption Retrieval Dataset. We next show results on the dataset constructed with the image-caption retrieval approach in Fig. 5. It shows that the prediction relates to the most frequent content in the image collection, and unrelated contents are ignored. They also keep the main specific content if it can describe the image collection. Table 2 shows the evaluation results, which show that the proposed method beats text summarization methods in this novel task. The experiments above show a novel image collection captioning baseline compared with text summarization methods. However, we also found a limitation of automatic evaluation metrics when the captioning results are refined. The

Table 2. Evaluation of the results of image collections built with the image-caption retrieval approach compared to SUPERT [5], T5 [21], and XL-Sum [10].

Method		R-1 ↑	R-2 ↑	R-L ↑	BERTScore ↑	CIDErBtw ↑	WEEM4TS ↑
SUPERT [5]		0.3756	0.1105	0.3231	0.6166	0.7016	0.1083
T5 [21]		0.3441	0.1037	0.3025	0.6057	0.5524	0.1031
XL-Sum [10]		0.2148	0.0367	0.1833	0.4678	0.1023	0.0860
Proposed	w/o CG	**0.3782**	**0.1265**	**0.3409**	**0.6270**	**0.7955**	0.1062
	w/ CG	0.3517	0.1105	0.3140	0.6093	0.7156	**0.1096**

proposed method without refining the captioning results, achieved better evaluation scores in ROUGE, BERTScore, and CIDerBtw, due to the nature of these metrics focusing on text similarity. However, WEEM4TS is a novel metric for abstractive summarization evaluation and shows a promising direction for the image collection captioning task being more suitable for our task of generalizing across images.

6 Conclusion

We introduced a challenging new task which aims to produce a single fitting caption for a collection of images. The key idea was to generate a scene graph for each image in the collection, combine them to generate a generalized combined representation, and then generate a caption. The proposed method showed potential for transferring a single-image captioning model to image collection captioning. Inspired by text summarization methods generating a summary, the proposed method improves the abstractiveness of the summarized image collection caption by finding generalized words using graph theory and word communities. We additionally introduced the prospect of using an augmented version of the MS-COCO dataset, a popular image captioning dataset, in the image collection captioning task. The results are promising and pioneering steps toward captioning an image collection with a shared description. In the future, we plan to work also on a more challenging dataset and also improve the captioning model focusing on estimating the overall semantic context of an image collection incorporating external knowledge. Our project can be found at https://www.cs.is.i.nagoya-u.ac.jp/opensource/nu-icc/.

Acknowledgements. Parts of this work were supported by JSPS Grant-in-aid for Scientific Research (21H03519) and a joint research project with National Institute of Informatics.

References

1. Alrasheed, H.: Word synonym relationships for text analysis: a graph-based approach. PLoS ONE **16**(7), e0255127 (2021)
2. Anderson, P., et al.: Bottom-up and top-down attention for image captioning and visual question answering. In: 2018 IEEE Conference on Computer Vision and Pattern Recognition, pp. 6077–6086 (2018)
3. Deng, J., et al.: ImageNet: a large-scale hierarchical image database. In: 2009 IEEE Computer Society Conference on Computer Vision and Pattern Recognition, pp. 248–255 (2009)
4. Faghri, F., et al.: VSE++: improving visual-semantic embeddings with hard negatives. In: 29th British Machine Vision Conference (2018)
5. Gao, Y., et al.: SUPERT: Towards new frontiers in unsupervised evaluation metrics for multi-document summarization. In: 58th Annual Meeting of the Association for Computational Linguistics, pp. 1347–1354 (2020)

6. Girshick, R.: Fast R-CNN. In: 16th IEEE International Conference on Computer Vision, pp. 1440–1448 (2015)
7. Gupta, S., et al.: Abstractive summarization: an overview of the state of the art. Expert Syst. Appl. **121**, 49–65 (2019)
8. Hailu, T.T., et al.: A framework for word embedding based automatic text summarization and evaluation. Information **11**(2), 78–100 (2020)
9. Han, X., et al.: Image scene graph generation (SGG) benchmark. Comput. Res. Reposit. arXiv preprint arXiv:2107.12604 (2021)
10. Hasan, T., et al.: XL-Sum: large-scale multilingual abstractive summarization for 44 languages. In: Findings of the Association for Computational Linguistics: ACL-IJCNLP 2021, pp. 4693–4703 (2021)
11. He, K., et al.: Deep residual learning for image recognition. In: 2016 IEEE Conference on Computer Vision and Pattern Recognition, pp. 770–778 (2016)
12. Hossain, M.Z., et al.: A comprehensive survey of deep learning for image captioning. ACM Comput. Survey **51**(6), 1–36 (2019)
13. Karpathy, A., et al.: Deep visual-semantic alignments for generating image descriptions. In: 2015 IEEE Conference on Computer Vision and Pattern Recognition, pp. 3128–3137 (2015)
14. Kingma, D.P., et al.: Adam: a method for stochastic optimization. In: 3rd International Conference on Learning Representations (2014)
15. Krishna, R., et al.: Visual genome: connecting language and vision using crowdsourced dense image annotations. Int. J. Comput. Vis. **123**(1), 32–73 (2017)
16. Lin, C.Y.: ROUGE: a package for automatic evaluation of summaries. In: ACL-04 Workshop on Text Summarization Branches Out, pp. 74–81 (2004)
17. Lin, T.-Y., et al.: Microsoft COCO: common objects in context. In: Fleet, D., Pajdla, T., Schiele, B., Tuytelaars, T. (eds.) ECCV 2014. LNCS, vol. 8693, pp. 740–755. Springer, Cham (2014). https://doi.org/10.1007/978-3-319-10602-1_48
18. Loper, E., et al.: NLTK: the natural language toolkit. In: 42nd Annual Meeting of the Association for Computational Linguistics, vol. 1, pp. 63–70 (2002)
19. Milewski, V., et al.: Are scene graphs good enough to improve image captioning? In: Joint Conference 59th Annual Meeting of the Association for Computational Linguistics and 11th International Conference on Natural Language Processing (2020)
20. Pennington, J., et al.: GloVe: global vectors for word representation. In: 2014 Conference on Empirical Methods in Natural Language Processing, pp. 1532–1543 (2014)
21. Raffel, C., et al.: Exploring the limits of transfer learning with a unified text-to-text transformer. J. Mach. Learn. Res. **21**(140), 1–67 (2020)
22. Samani, Z.R., et al.: A knowledge-based semantic approach for image collection summarization. Multimed. Tools Appl. **76**(9), 11917–11939 (2017)
23. Speer, R., et al.: ConceptNet 5.5: an open multilingual graph of general knowledge. In: 31st AAAI Conference on Artificial Intelligence, pp. 4444–4451 (2017)
24. Trieu, N., et al.: Multi-image summarization: textual summary from a set of cohesive images. Comput. Res. Reposit. arXiv preprint arXiv:2006.08686 (2020)
25. Vedantam, R., et al.: CIDEr: consensus-based image description evaluation. In: 2015 IEEE Conference on Computer Vision and Pattern Recognition, pp. 4566–4575 (2015)
26. Wang, J., Xu, W., Wang, Q., Chan, A.B.: Compare and reweight: distinctive image captioning using similar images sets. In: Vedaldi, A., Bischof, H., Brox, T., Frahm, J.-M. (eds.) ECCV 2020. LNCS, vol. 12346, pp. 370–386. Springer, Cham (2020). https://doi.org/10.1007/978-3-030-58452-8_22

27. Wang, W., et al.: Image as a foreign language: BEiT pretraining for all vision and vision-language tasks. Compt. Res. Reposit. arXiv preprint arXiv:2208.10442 (2022)
28. Wasserman, S., et al.: Social Network Analysis: Methods and Applications, vol. 8. Cambridge University Press, Cambridge (1994)
29. Zellers, R., et al.: Neural motifs: scene graph parsing with global context. In: 2018 IEEE Conference on Computer Vision and Pattern Recognition, pp. 5831–5840 (2018)
30. Zhang, J., et al.: Graphical contrastive losses for scene graph parsing. In: 2019 IEEE Conference on Computer Vision and Pattern Recognition, pp. 11535–11543 (2019)
31. Zhang, T., et al.: BERTScore: evaluating text generation with BERT. In: 9th International Conference on Learning Representations (2020)
32. Zhang, W., et al.: Joint optimisation convex-negative matrix factorisation for multi-modal image collection summarisation based on images and tags. IET Comput. Vis. **13**(2), 125–130 (2019)
33. Zhong, Y., Wang, L., Chen, J., Yu, D., Li, Y.: Comprehensive image captioning via scene graph decomposition. In: Vedaldi, A., Bischof, H., Brox, T., Frahm, J.-M. (eds.) ECCV 2020. LNCS, vol. 12359, pp. 211–229. Springer, Cham (2020). https://doi.org/10.1007/978-3-030-58568-6_13

Health-Oriented Multimodal Food Question Answering

Jianghai Wang[1], Menghao Hu[2], Yaguang Song[2,3,4], and Xiaoshan Yang[2,3,4](✉)

[1] Zhengzhou University, Zhengzhou, China
[2] Peng Cheng Laboratory, Shenzhen, China
humh01@pcl.ac.cn
[3] Institute of Automation, Chinese Academy of Sciences (CASIA), Beijing, China
songyaguang2019@ia.ac.cn, xiaoshan.yang@nlpr.ia.ac.cn
[4] University of Chinese Academy of Sciences (UCAS), Beijing, China

Abstract. Health-oriented food analysis has become a research hotspot in recent years because it can help people keep away from unhealthy diets. Significant progress has been made in recipe retrieval, food recommendation, nutrition and calorie estimation. However, existing works still cannot well balance the individual preference and the health. Multimodal food question and answering (Q&A) has a great potential in practical applications, but it is still not well studied. In this paper, we build a health-oriented multimodal food Q&A dataset (MFQA) with 9K question and answer pairs based on a multimodal food knowledge graph collected from a food-sharing website. In addition, we propose a knowledge-based multimodal food Q&A framework, which consists of three important parts: encoder module, retrieval module, and answer module. Extensive experimental results on the MFQA dataset demonstrate the effectiveness of our method. The code and dataset are available at https://github.com/Wjianghai/HMFQA.

Keywords: Multimodel Q&A · Food analysis · Health

1 Introduction

Food is the indispensable material basis for human survival and development. With the advanced technologies of agriculture and biology, the quantity and types of food have significantly increased over the past few decades. Whereas, people also suffer from various diseases (e.g., diabetes and cardiovascular diseases) due to unhealthy diets (e.g., excessive intake of high-fat and high-sugar foods) in a material-rich life. About 1.9 billion adults worldwide are at high risk of these diseases due to being overweight. People with different health states have different food needs from a dietary perspective. For example, obese people need low-sugar foods, and people with diabetes need some foods that can lower blood glucose. However, people are always overwhelmed when preparing food at

J. Wang and M. Hu—Equal contribution

D.-T. Dang-Nguyen et al. (Eds.): MMM 2023, LNCS 13833, pp. 191–203, 2023.
https://doi.org/10.1007/978-3-031-27077-2_15

home or ordering food from restaurants, mainly because they lack the relevant knowledge and experience to judge whether a particular food is appropriate for them.

To help people eat healthy, researchers have used a large amount of food-related data and applied advanced techniques to study tasks such as food recognition [2], cooking activity recognition [6], food detection [1], dietary recommendation [3], dietary intervention [22], and consumption patterns analysis [20]. Significant progress has been made in recipe retrieval, food recommendation, nutrition and calorie estimation. For example, Chen et al. [3] proposed a personalized food recommendation method, which utilizes attentive memory networks to achieve bidirectional interaction between text queries and knowledge bases. However, existing works still cannot well balance the individual preference and the health. Multimodal food Q&A has a great potential in practical applications, because it can find the appropriate food to meet the user's multimodal query (described by a pair of sentence and image) in a user-friendly manner based on multimodal data including the cooking process, ingredients, nutritional content, visual information of different foods. However, this task is still not well studied due to the lack of relevant datasets.

In this paper, we present a health-oriented multimodal food Q&A dataset, MFQA, which has 9K question and answer pairs generated based on a multimodal food knowledge graph with over 171K entities and 1.9M triples collected from a food-sharing website Meishijie. Based on the multimodal food dataset, we propose a knowledge-based multimodal food Q&A framework, which consists of three parts. The first part is responsible for encoding input texts, images and knowledge. In the second part, the model uses multimodal information and performs a preliminary screening of knowledge based on attention mechanisms. In the third part, the model conducts various interactions between the question and knowledge and finally infers the correct answer.

In summary, the contributions of this work are two-fold. **(1)** We propose a health-oriented multimodal food Q&A dataset with 9K question and answer pairs. To build the dataset, we also collect a multimodal food knowledge graph, which has the largest number of entities. **(2)** We propose a knowledge-based multimodal food Q&A framework to comprehensively analyze the multimodal information in the query and knowledge base to infer the correct answer. This method is the first health-oriented multimodal Q&A work in food computing.

2 Related Work

2.1 Health-Oriented Food Analysis

Recently, health-oriented food analysis has become a research hotspot. Okamoto et al. [19] propose a calorie estimation system that alerts people to high-calorie food based on images. Forde et al. [15] study how visual and odor information affects people's food choices. Song et al. [21] use the user's dynamic preference for food calories to recommend favorite foods to the user. Mejova et al. [16] use Internet images to study food consumption for public health analysis.

Datasets are significant for research and development in food computing community. Recently, there are also health-related food datasets, e.g., Market2Dish [25], Foodkg [8], and PFoodReq [3], that are generally derived from the logs of health applications and the sharing of professional food websites. These datasets contain calories, fat, and nutritional content and can be used to analyze food balance and calorie estimation tasks. These datasets are not suitable for multimodal food Q&A in terms of health due to data volume and type limitations. More work related to food computing can be found in the survey [17].

2.2 Knowledge-Based Question Answering

Knowledge-based Question Answering (KBQA) aims to answer natural language questions through external knowledge bases. There are two mainstream solutions: Semantic Parsing (SP) [11] and Information Retrieval (IR) [4]. The SP method first converts the natural language question into symbolic form and then queries the answer in the knowledge base. The IR method first retrieves a question-related sub-graph from the knowledge base and then sorts the entities in the sub-graph according to their relevance to the question. Dai et al. [5] conduct sub-graph pruning and sub-graph selection through entity detection with Bi-GRU and linear chain CRF. Zhao et al. [28] propose an n-gram based method for sub-graph selection. Luo et al. [14] proposed a BERT-based with relation-aware attention single-relation Q&A method, which can perform entity detection and entity linking. Most existing KBQA methods only consider using single-modal information as input. Different from these methods, this paper focuses on a multimodal food Q&A task that requires the model to handle multimodal information.

Knowledge-based visual Q&A (KBVQA) is also related to our work. In recent years, external knowledge has been widely used to solve the VQA task because it can improve the accuracy and interpretability. Wang et al. [24] retrieve triples related to image from the knowledge graph as candidate answers. Wu et al. [26] fuse automatically generated image descriptions with external knowledge bases to predict the correct answer. Narasimhan et al. [18] use graph convolutional networks to build an end-to-end reasoning system for VQA with a knowledge base. Zhu et al. [29] represent an image as a multimodal heterogeneous graph, which contains information from fact graph to solve the VQA task. Zhang et al. [27] propose an enhanced VQA framework based on rich visual knowledge. Although exiting KBVQA methods have achieved significant progress, they mainly consider the image dataset with general objects and scenes. Therefore, they cannot be directly applied to solve the health-oriented food Q&A task in this paper.

3 Methodology

3.1 Multimodal Food Q&A Dataset

We crawl a Chinese food dataset from food sharing website Meishijie. To build the multimodal knowledge graph, we deem recipe, ingredient, and type label as graph entities. And we use the recipe image, nutrient information, and cooking process as attributes of the recipe entity. Finally, we obtain a multimodal food KG (MFKG) with 171K entities and 1.9M triples. Figure 1 provides an example of a recipe in the MFKG. We build the multimodal food Q&A (MFQA) dataset by creating question and answer pairs from the MFKG with manually defined question templates. We construct the questions using three disease labels: hyperglycemia, hypertension, hyperlipidemia, and one function label: weight loss. Finally, we end up with a dataset with over 9K question and answer pairs, where each pair is a tuple of <question, image, triplet, answer>.

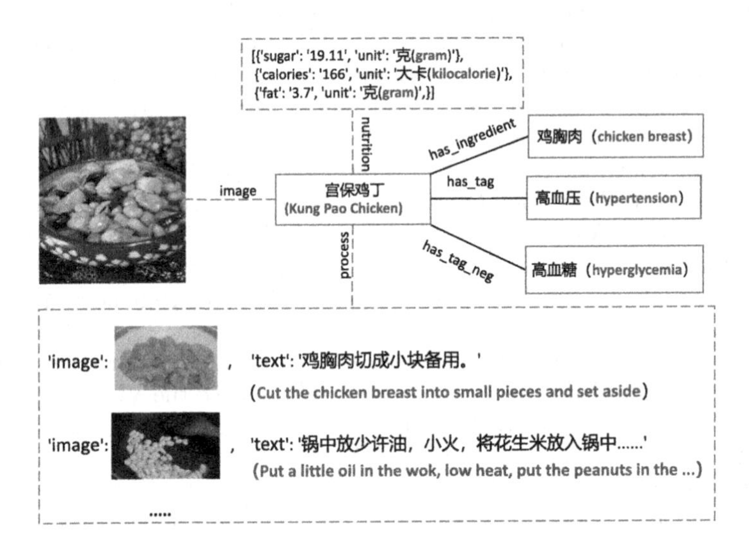

Fig. 1. An example of the multimodal food knowledge graph.

3.2 Problem Definition

We assume a multimodal food knowledge graph G and a multimodal food Q&A dataset $\mathcal{D}$ are given. The multimodal food knowledge graph $G=\{\mathcal{E}, \mathcal{R}, \mathcal{F}\}$, where $\mathcal{E}$, $\mathcal{R}$ and $\mathcal{F}$ are sets of entities, relations and facts respectively. Each fact f is denoted as a triplet $(h, r, t) \in \mathcal{F}$, where h, r and t are the head entity, the relationship and the tail entity. We denote the multimodal food Q&A dataset as $\mathcal{D} = \{\mathcal{I}, \mathcal{Q}, \mathcal{W}, \mathcal{A}\}$, where $\mathcal{I}$, $\mathcal{Q}$, $\mathcal{W}$ and $\mathcal{A}$ are the sets of images, questions, knowledge and answers respectively. Each data sample $d \in \mathcal{D}$ is defined as a quad (i, q, w, a), where i represents an input food image, q represents a natural language question, $w = \{f_i\}_{i=1}^{m}$ represents knowledge capable of answering questions, and a represents the answer. The goal of our task is to train a model based on G and $\mathcal{D}$, so that the learned model can accept the question q and image i as input, retrieve top k knowledge $G_k = \{g_i'\}_{i=1}^{k}$ from the food knowledge base G, and output the correct answer a. Figure 2 gives an overview of our model, which is divided into three main modules: Encoder Module, Retrieval Module, Answer Module. In Encoder Module, we extract the features of the input data and get the multimodal query representation m_q. The knowledge were screened preliminary and fed into the encoder to obtain n feature vectors $G_n = \{g_i\}_{i=1}^{n}$. In Retrieval Module, we select k candidate knowledge G_k that are most relevant to answer the question ($n > k$). Finally, we feed the top-k knowledge feature G_k and multimodal query representation m_q into the Answer Module to get the answer a.

3.3 Encoder Module

Query Encoder. We use pre-trained BERT and ResNet to encode text and image data. Specifically, for a Chinese sentence question q, we use a BERT pre-trained on Chinese corpus as the encoder and take the output value of the last output layer as the sentence feature vector $t \in \mathbb{R}^{d_h}$ (d_h=768). For each image i, we use ResNet18 model pre-trained on ImageNet and get the image feature vector $v \in \mathbb{R}^{d_v}$ (d_v=512) from the last fully connected layer. Then, we feed v into a fully connected layer ($\mathbb{R}^{d_v} \mapsto \mathbb{R}^{d_h}$), and concatenate it with t to get a joint representation $m_q \in \mathbb{R}^{2 \times d_h}$: $m_q = [t, \mathrm{FC}(v)]$.

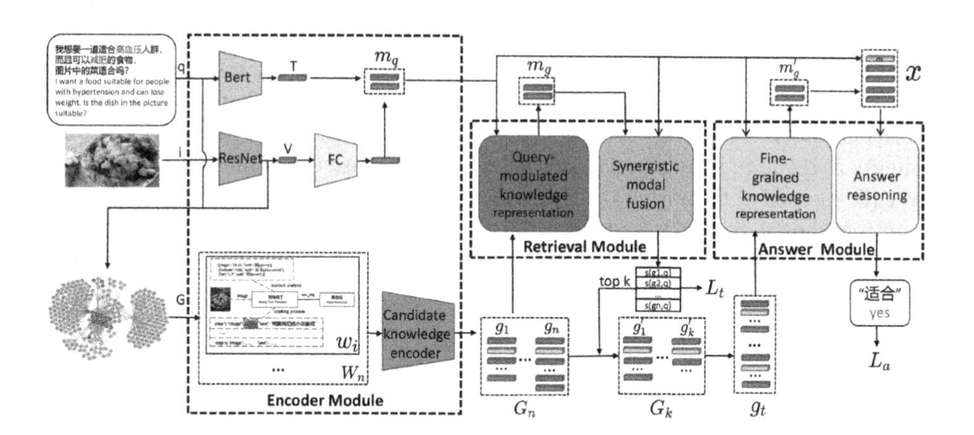

Fig. 2. Overview of our knowledge-based multimodal food Q&A framework.

Candidate Knowledge Generation. Generally, we can utilize all the food KG and the input query for reasoning, leading to a huge computational cost. Therefore, it is necessary to conduct a preliminary screening of the KG before conducting the knowledge-based reasoning. Specifically, each recipe in the G has a corresponding image. We utilize the same ResNet18 used in query encoder to encode recipe images and calculate the distances to the input image feature v. By taking the top k_h most similar recipe entities, we get the set of head entities $\{h_i\}_{i=1}^{k_h}$. We then use the string matching algorithm to get the set of tail entities $\{t_i\}_{i=1}^{k_t}$ that match the question text q. We collect all paths $W_n = \{w_i\}_{i=1}^{n}$ between $\{h_i\}_{i=1}^{k_h}$ and $\{t_i\}_{i=1}^{k_t}$ that exist in G as the candidate knowledge.

Candidate Knowledge Encoder. We extract three kinds of features for each candidate knowledge w: structural information w^s, type information w^t, and context information w^c, which will be illustrated as follows.

The structure information represents the entity-relation-entity structure of the knowledge. We use a sentence with the order of head entity, relation and tail entity, and input this sentence into the pre-trained BERT to get a feature vector $w^s \in \mathbb{R}^{d_h}$.

We also consider the type information as a knowledge feature. Specifically, for each knowledge $w = \{f_i\}_{i=1}^{m}$, we extract its type information (f_1^{hid}, f_1^{rid}, f_1^{tid}..., f_m^{hid}, f_m^{rid}, f_m^{tid}), and input them into a type embedding layer before into a bidirectional LSTM [10] to get a feature vector $w^t \in \mathbb{R}^{d_h}$. Note that *hid*, *rid* and *tid* are the type ids of head entity, relation and tail entity respectively.

The context information represents the attribute information of the candidate's knowledge, including food images, nutrition analysis and cooking process. We encode the text and image in the context information using the pre-trained BERT and ResNet18 respectively and get text feature vector $w_t^c \in \mathbb{R}^{attr_{tnum} \times d_h}$ and visual feature vector $w_v^c \in \mathbb{R}^{attr_{vnum} \times d_v}$. The $attr_{tnum}$ and $attr_{vnum}$ are the number of context sentences and context images. Note that we change the nutrition information into a sentence by simply concatenating the nutrient components and theirs values in a fixed order. We then use the same fully connected layer ($\mathbb{R}^{d_v} \mapsto \mathbb{R}^{d_h}$) to map w_v^c to the same dimension as w_t^c, and then concatenate them together to form context information feature $w^c \in \mathbb{R}^{(attr_{tnum}+attr_{vnum}) \times d_h}$. Finally, we concatenate these three kinds of knowledge features and get the final representation vector $g \in \mathbb{R}^{attr_{num} \times d_h}$, where $attr_{num} = attr_{tnum} + attr_{vnum} + 2$.

3.4 Retrieval Module

The Retrieval Module takes multimodal query representation m_q and candidate knowledge G_n as input and retrieves a knowledge set G_k that can help answer the question. Specifically, for the feature g_i of each knowledge in G_n, we calculate a score $s(g_i, m_q)$ and finally get the score set $S = \{s(g_i, m_q)\}_{i=1}^{n}$. Then we select top k candidate knowledge G_k as the output according to the scores. The similarity score $s(g_i, m_q)$ is calculated through query-modulated knowledge representation and synergistic modal fusion, which will illustrated as follows.

Query-Modulated Knowledge Representation. The knowledge feature g contains three kinds of features extracted from text and visual data. Each of them has different contribution to answering the question. Here we design a query-modulated knowledge representation (QMKR) mechanism to focus on important components of knowledge g with the consideration of the multimodal query feature m_q. It consists of a series of Self-Attention blocks [23]. Specifically, we deem the multimodal query m_q and the knowledge g as a query and key-value pair and input them to a Self-Attention block which consists of scaled dot product attention, feed-forward, and add&norm operations. The query-modulated knowledge feature $m_g \in \mathbb{R}^{2\times d_h}$ is calculated as follows:

$$Q_q = m_q W_{Q_q}, K_g = gW_{K_g}, V_g = gW_{V_g}, \tag{1}$$

$$m_g = \text{Attention}(Q_q, K_g, V_g) = \text{softmax}\left(\frac{Q_q K_g^\top}{\sqrt{d_h}}\right) V_g. \tag{2}$$

Here, all projection matrices $W_{[\cdot]} \in \mathbb{R}^{d_h \times d_h}$ are learnable parameters.

Synergistic Modal Fusion. Both the multimodal query representation m_q and the query-modulated knowledge representation m_g have the features of text and visual data. The synergistic modal fusion module explores bidirectional attention between m_q and m_g and fuses the features from different modalities. It consists of a series of cross-modal Self-Attention blocks. Specifically, we deem the m_q as a query and the m_g as a key-value pair and input them into a cross-attention block to compute the enhanced multimodal query $m'_q \in \mathbb{R}^{2\times d_h}$:

$$Q'_q = m_q W_{Q'_q}, K'_g = m_g W_{K'_g}, V'_g = m_g W_{V'_g}, \tag{3}$$

$$m'_q = \text{Attention}\left(Q'_q, K'_g, V'_g\right) = \text{softmax}\left(\frac{Q'_q {K'_g}^\top}{\sqrt{d_h}}\right) V'_g. \tag{4}$$

Similarly, we deem the m_g as a query and m_q as a key-value pair to compute enhanced query-modulated knowledge $m'_g \in \mathbb{R}^{2\times d_h}$:

$$Q'_g = m_g W_{Q'_g}, K'_q = m_q W_{K'_q}, V'_q = m_q W_{V'_q}, \tag{5}$$

$$m'_g = \text{Attention}\left(Q'_g, K'_q, V'_q\right) = \text{softmax}\left(\frac{Q'_g {K'_q}^\top}{\sqrt{d_h}}\right) V'_q. \tag{6}$$

Here, all projection matrices $W_{[\cdot]} \in \mathbb{R}^{d_h \times d_h}$ are learnable parameters.

Finally, we extract the feature vectors corresponding to the first tokens in m'_q and m'_g as the unified features $s_q \in \mathbb{R}^{d_h}$ and $s_g \in \mathbb{R}^{d_h}$. Then, we calculate the dot product of s_g and s_q to get the similarity score $s(g, m_q)$ for the knowledge feature g and the query m_q. Based on the score set $\{s(g_i, m_q)\}_{i=1}^n$, we extract the top k knowledge features $G_k = \{g'_i\}_{i=1}^k$ as the input of the Answer Module.

3.5 Answer Module

Fine-Grained Knowledge Representation. After fusing the text and visual information, we need to further find the knowledge that is most relevant to the input query. We first concatenate knowledge features in G_k as an integrated knowledge representation g_t. Here, $g_t \in \mathbb{R}^{\sum_{i=0}^{k} attr_{num_i} \times d_h}$ and $attr_{numi}$ represents the number of attributes of the ith knowledge feature g'_i. Then, we feed the multimodal query m_q and the integrated knowledge feature g_t into the fine-grained knowledge representation (FGKR) module and get the fine-grained knowledge feature $m'_g \in \mathbb{R}^{2\times d_h}$: $m'_g = f_{FGKR}(g_t, m_q)$. Note that FGKR used here has a similar network structure as the QMKR of the Retrieval Module, but with different learnable parameters.

Answer Reasoning. To comprehensively combine the information from the query and the knowledge, we construct a CLS token represented by a zero vector *cls* with d_h dimension. By concatenating the CLS token with m_q and m'_g, we get a feature $x \in \mathbb{R}^{5\times d_h}$. Then, we feed x into a series of Self-Attention blocks and take the feature vector $x' \in \mathbb{R}^{d_h}$ which corresponds to the output of the CLS token:

$$Q_x = xW_{Q_x}, K_x = xW_{K_x}, V_x = xW_{V_x}, \tag{7}$$

$$\text{Attention}\left(Q_x, K_x, V_x\right) = \text{softmax}\left(\frac{Q_x K_x^T}{\sqrt{d_h}}\right) V_x. \tag{8}$$

Here, all projection matrices $W_{[\cdot]} \in \mathbb{R}^{d_h \times d_h}$ are learnable parameters.

Finally we feed the x' into a classifier $f_{classifier(\cdot)}$: $\mathbb{R}^{d_h} \mapsto \mathbb{R}^{N_c}$, which consists of a linear projection layer followed by Softmax. N_c is the number of answers.

3.6 Optimization

The overall framework is optimized by the following loss function: $\mathcal{L} = \alpha_t L_t + \alpha_a L_a$, where L_t and L_a are used to constrain the Retrieval Module and the Answer Module, respectively. α_t and α_a are balance weights.

The L_t is used to train the Retrieval Module, which is defined as a multi-label margin loss function:

$$L_t(S, gt_r) = \sum_{i,j \notin gt_r} \frac{\max(0, 1 - (S[gt_r[i]] - S[j]))}{|S|}, \tag{9}$$

where S represents the predicted score, gt_r represents the index of the groundtruth, $i \in \{1, \cdots, |gt_r|\}$, $1 \le gt_r[i] \le |S|$, $j \in \{1, \cdots, |S|\}$.

The L_a is defined as a Cross-Entropy of the prediction and ground-truth gt_a: $L_a = CE(f_{classifier}(x'), gt_a)$.

4 Experiments

4.1 Implementation Details

In our method, the Retrieval Module and the Answer Module consist of a series of attention blocks. We set the number of attention blocks in synergistic modal fusion module, query-modulated knowledge representation, answer reasoning module and fine-grained knowledge representation module as 2, 2, 8, and 8 respectively. We set the number of head entities (i.e., k_h) in candidate knowledge generation as 15 and the number of selected knowledge features (i.e., k) in Retrieval Module as 16. We set the two balance weights (i.e., α_t and α_a) of the overall loss function as 0.5 and 0.5, respectively. The overall framework is trained with optimizer SGD. We use the learning strategy of stepLR. The batch size is 16 and the number of epochs is 50. The initial learning rate is set to 1×10^{-3}, which decays once every 4 epochs. The decay factor is set to 0.68. An early stop will occur if the learning rate does not decline within 8 epochs. For all experiments, we train and test the model on a Tesla V100 GPU with PyTorch. We divide the MFQA dataset into training, validation and test set with a ratio of 80%:10%:10%. We use F1-Score and Accuracy metrics to evaluate the results.

4.2 Comparison with State-of-the-Art Methods

According to the usage of KG, there are two types of KGQA models. The first type uses entities in KG as the answers and defines the Q&A task as computing the probability of each entity as the answer to a query, e.g., BAMnet [4]. The second type has answers from Q&A datasets and needs to calculate the probability of each candidate answer, e.g., ConceptBert [7], HAN [12], Hypergraph Transformer [9], BAN [13]. Since these models have different feature preprocessing methods, we unify them by replacing the image, text and KG feature processing schemes with the Encoder Module in this paper.

BAMnet is a state-of-the-art KBQA method that retrieves answers directly from KG. We use the Encoder Module in our method to replace the Input Module and Memory Module (feature extraction for Query and KG, respectively) of BAMnet. After knowledge retrieval and representation, the answer reasoning of our work is adopted to conduct the answer prediction.

Table 1. Comparison of our model and existing methods on the MFQA dataset.

Method	F1-Score	Accuracy
BAMnet [4]	0.681	0.546
ConceptBert [7]	0.664	0.662
HAN [12]	0.668	0.664
BAN [13]	0.676	0.671
Hypergraph Transformer [9]	0.693	0.642
Ours	**0.710**	**0.682**

ConceptBert [7]: consists of four parts: the vision-language module that fuses image and text information, the concept-language module that fuses the text and KG information, the aggregator that implements feature aggregation and the classifier that conducts answer prediction. Here, we use the Encoder Module in our model to replace the feature extraction process of ConceptBert for image, question and KG (i.e. faster-RCNN for image feature, BERT for word embedding and GCN for KG representation).

BAN [13], HAN [12], Hypergraph Transformer [9]: For each of these three methods, we directly use the Encoder Module of our work to replace the feature processing part and retain the attention Q&A part to conduct the answer prediction.

Table 1 shows the evaluation results of our method and the compared baselines. Our method achieves the accuracy of 0.682 and the F1-score of 0.710, which are better than all the compared methods. Although the BAN method has the second best accuracy result, it has a relatively low F1 score. Because this method cannot achieve a good balance between precision and recall. These results demonstrate the effectiveness of our method.

4.3 Ablation Study

We conduct an ablation study to systematically investigate the impact of different components in our method. To verify the necessity of encoding three kinds of information (i.e., structure information, type information, context information) in Knowledge Encoder, we conduct the ablation of each information. *w/o Ws in Knowledge Encoder* means that we take away structural information to verify the importance of structural information of knowledge. Similarly, *w/o Wt in Knowledge Encoder* and *w/o Wc in Knowledge Encoder* mean that we take away the type information and context information respectively. Then, we design variant models for the Retrieval Module. *w/o QMKR in retrieval* is the model after removing the query-modulated knowledge representation in the Retrieval Module. *w/o SMF in retrieval* is the model after removing the synergistic modal fusion in the Retrieval Module. *w/o retrieval* is the variant model after completely removing the Retrieval Module. We also design two variant models for the Answer Module. *w/o FGKR in answer* is the model after removing the fine-grained knowledge representation in the Answer Module. *w/o AR in answer* is the model after removing the answer reasoning.

Table 2 shows the experimental results of the ablation experiment. The ablation results for the Knowledge Encoder show that these three kinds of information are all important in helping answer the questions, among which the *type information* is the most important according to the F1 score. The structure and context also contain important features for answering questions. However, these two kinds of information are more detailed, which may also contain noise and thus may interfere with answering questions. Retrieval Module can filter useless knowledge for the answer prediction and increase the relevance of the knowledge representation, thereby improving the Q&A performance. The results show that the query-modulated knowledge representation and synergistic modal fusion

Table 2. Ablation study results.

Method	F1-Score	Accuracy
w/o Ws in Knowledge Encoder	0.674	0.605
w/o Wt in Knowledge Encoder	0.654	0.629
w/o Wc in Knowledge Encoder	0.689	0.664
w/o QMKR in retrieval	0.700	0.681
w/o SMF in retrieval	0.707	0.677
w/o retrieval	0.680	0.653
w/o FGKR in answer	0.674	0.617
w/o AR in answer	0.660	0.653
Ours(full model)	**0.710**	**0.682**

both are both important in the Retrieval Module,because the former can find the knowledge that is more relevant to the input query while the latter can fuse the information from different modalities by bidirectional attention.

After removing the Retrieval Module, the result becomes worse, which shows the importance of the Retrieval Module. The answer reasoning plays a more important role than the fine-grained knowledge representation in the Answer Module. Interestingly, the FGKR has the same network structure as the QMKR but it contributes more to the answer prediction. These results demonstrate the effectiveness of each component in our framework.

4.4 Parameter Analysis

Impact of the Number of Attention Blocks. Our framework contains multiple attention blocks. We denote the number of attention blocks in the synergistic modal fusion, query-modulated knowledge representation, answer reasoning, and fine-grained knowledge representation as J, K, M, and N, respectively. We assume that the number of attention blocks in each module (J, K, M, or N) are set to 2, 4 or 8. Figure 3(a)-(b) show the Accuracy and F1 result when changing two of the four hyperparameters. As shown, the best performance is obtained with combination of $J = 2$, $K = 2$, $M = 8$, $N = 8$.

Impact of Balance Weights. For the two balance weights α_t and α_a illustrated in Sect. 3.6, we assume that $\alpha_t + \alpha_a = 1$. Figure 3(c) shows the result when changing one of the balance weights from 0.1 to 0.9 with the interval of 0.2. As shown, the best performance is obtained when $\alpha_t = 0.5$, $\alpha_a = 0.5$. This demonstrates that the Retrival Module and Answer Module are equally important in our method.

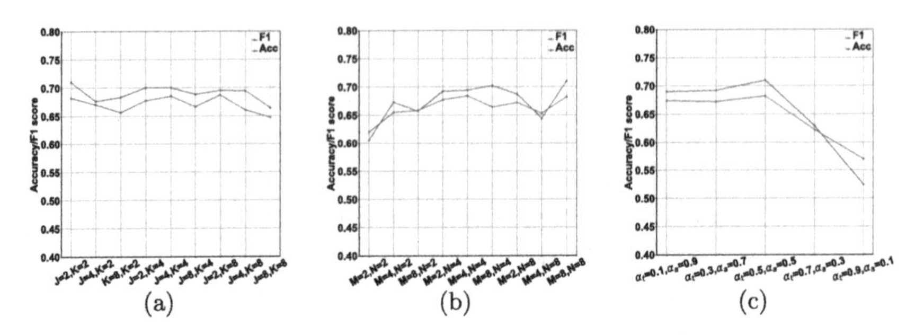

Fig. 3. Impact of the number of attention blocks (a-b) and balance weights (c).

5 Conclusion

In this paper, we propose a multimodal food Q&A dataset MFQA, which is created based on a newly collected multimodal food knowledge graph MFKG. MFQA and MFKG are of great importance for the food computing community, providing new research data for many tasks, such as cooking guidance, food recommendation, and food Q&A. Moreover, we propose a new health-oriented multimodal food Q&A framework, which consists of three elaborately designed modules, i.e., encoder module, retrieval module and answer module, which can comprehensively analyze the multimodal information in the user query and knowledge graph to predict the correct answer. Experimental results demonstrate the effectiveness of the proposed method.

Acknowledgments. This work was supported by National Key Research and Development Program of China (No. 2018AAA0100604), National Natural Science Foundation of China (No. 61720106006, 61721004, 62072455, U1836220, U1705262, 61872424), Key Research Program of Frontier Sciences of CAS (QYZDJ-SSW-JSC039), and Beijing Natural Science Foundation (L201001).

References

1. Aguilar, E., Remeseiro, B., Bolaños, M., Radeva, P.: Grab, pay, and eat: Semantic food detection for smart restaurants TMM, pp. 3266–3275 (2018)
2. Chen, X., Zhou, H., Diao, L.: Chinesefoodnet: A large-scale image dataset for chinese food recognition. CoRR abs/1705.02743 (2017)
3. Chen, Y., Subburathinam, A., Chen, C., Zaki, M.J.: Personalized food recommendation as constrained question answering over a large-scale food knowledge graph. In: WSDM, pp. 544–552 (2021)
4. Chen, Y., Wu, L., Zaki, M.J.: Bidirectional attentive memory networks for question answering over knowledge bases. In: NAACL-HLT, pp. 2913–2923 (2019)
5. Dai, Z., Li, L., Xu, W.: CFO: conditional focused neural question answering with large-scale knowledge bases. In: ACL (2016)
6. Damen, D., et al.: Scaling egocentric vision: The EPIC-KITCHENS dataset. CoRR abs/1804.02748 (2018)

7. Gardères, F., Ziaeefard, M., Abeloos, B., Lécué, F.: Conceptbert: Concept-aware representation for visual question answering. In: EMNLP, pp. 489–498 (2020)
8. Haussmann, S., et al.: Foodkg: A semantics-driven knowledge graph for food recommendation. In: ISWC, pp. 146–162 (2019)
9. Heo, Y., Kim, E., Choi, W.S., Zhang, B.: Hypergraph transformer: Weakly-supervised multi-hop reasoning for knowledge-based visual question answering. In: ACL, pp. 373–390 (2022)
10. Hochreiter, S., Schmidhuber, J.: Long short-term memory. Neural computation, pp. 1735–1780 (1997)
11. Hu, S., Zou, L., Yu, J.X., Wang, H., Zhao, D.: Answering natural language questions by subgraph matching over knowledge graphs. TKDE, pp. 824–837 (2018)
12. Kim, E., Kang, W., On, K., Heo, Y., Zhang, B.: Hypergraph attention networks for multimodal learning. In: CVPR, pp. 14569–14578 (2020)
13. Kim, J., Jun, J., Zhang, B.: Bilinear attention networks. In: NeurIPS, pp. 1571–1581 (2018)
14. Luo, D., Su, J., Yu, S.: A bert-based approach with relation-aware attention for knowledge base question answering. In: IJCNN, pp. 1–8 (2020)
15. McCrickerd, K., Forde, C.: Sensory influences on food intake control: moving beyond palatability. Obesity Reviews, pp. 18–29 (2016)
16. Mejova, Y., Haddadi, H., Noulas, A., Weber, I.: #foodporn: Obesity patterns in culinary interactions. In: e-Health, pp. 51–58 (2015)
17. Min, W., Jiang, S., Liu, L., Rui, Y., Jain, R.C.: A survey on food computing. ACM Comput. Surv. **52**(5), 92:1–92:36 (2019)
18. Narasimhan, M., Lazebnik, S., Schwing, A.G.: Out of the box: Reasoning with graph convolution nets for factual visual question answering. In: NeurIPS, pp. 2659–2670 (2018)
19. Okamoto, K., Yanai, K.: An automatic calorie estimation system of food images on a smartphone. In: MM, pp. 63–70 (2016)
20. Phan, T., Perez, D.G.: Healthy #fondue #dinner: analysis and inference of food and drink consumption patterns on instagram. In: MUM, pp. 327–338 (2017)
21. Song, Y., Yang, X., Xu, C.: Self-supervised calorie-aware heterogeneous graph networks for food recommendation. TOMM (2022)
22. Sonnenberg, L., Gelsomin, E., Levy, D.E., Riis, J., Barraclough, S., Thorndike, A.N.: A traffic light food labeling intervention increases consumer awareness of health and healthy choices at the point-of-purchase. ACPM, pp. 253–257 (2013)
23. Vaswani, A., et al.: Attention is all you need. In: NeurIPS. pp. 5998–6008 (2017)
24. Wang, P., Wu, Q., Shen, C., Dick, A.R., van den Hengel, A.: Explicit knowledge-based reasoning for visual question answering. In: IJCAI, pp. 1290–1296 (2017)
25. Wang, W., Duan, L., Jiang, H., Jing, P., Song, X., Nie, L.: Market2dish: Health-aware food recommendation. TOMCCAP, pp. 33:1–33:19 (2021)
26. Wu, Q., Wang, P., Shen, C., Dick, A.R., van den Hengel, A.: Ask me anything: Free-form visual question answering based on knowledge from external sources. In: CVPR, pp. 4622–4630 (2016)
27. Zhang, L., et al.: Rich visual knowledge-based augmentation network for visual question answering. TNN pp. 4362–4373 (2021)
28. Zhao, W., Chung, T., Goyal, A.K., Metallinou, A.: Simple question answering with subgraph ranking and joint-scoring. In: NAACL-HLT, pp. 324–334 (2019)
29. Zhu, Z., Yu, J., Wang, Y., Sun, Y., Hu, Y., Wu, Q.: Mucko: Multi-layer cross-modal knowledge reasoning for fact-based visual question answering. In: IJCAI, pp. 1097–1103 (2020)

MM-Locate-News: Multimodal Focus Location Estimation in News

Golsa Tahmasebzadeh[1,2](✉), Eric Müller-Budack[1,2], Sherzod Hakimov[3], and Ralph Ewerth[1,2]

[1] TIB–Leibniz Information Centre for Science and Technology, Hannover, Germany
{golsa.tahmasebzadeh,eric.mueller,ralph.ewerth}@tib.eu
[2] L3S Research Center, Leibniz University Hannover, Hannover, Germany
[3] Computational Linguistics, University of Postdam, Potsdam, Germany
sherzod.hakimov@uni-potsdam.de

Abstract. The consumption of news has changed significantly as the Web has become the most influential medium for information. To analyze and contextualize the large amount of news published every day, the geographic focus of an article is an important aspect in order to enable content-based news retrieval. There are methods and datasets for geolocation estimation from text or photos, but they are typically considered as separate tasks. However, the photo might lack geographical cues and text can include multiple locations, making it challenging to recognize the focus location using a single modality. In this paper, a novel dataset called Multimodal Focus Location of News (MM-Locate-News) is introduced. We evaluate state-of-the-art methods on the new benchmark dataset and suggest novel models to predict the focus location of news using both textual and image content. The experimental results show that the multimodal model outperforms unimodal models.

Keywords: Multimodal geolocation estimation · News analytics · Computer vision · Natural language processing

1 Introduction

With the rapidly growing amount of information on the Web and social media, news articles are increasingly conveyed in multimodal form; apart from text, images and videos have become inseparable from news to report on events. Besides, there is an increasing research interest in the geographical content of news in fields like information retrieval or Web science. Thus, it is important to determine the *focus* location of the main story, i.e., a piece of metadata which indicates the geographic focus of a news article [7]. The focus locations are vital, for instance, to identify the location of a natural disaster for crisis management or to analyse political or social events across the world. Even if the news article focuses on one particular event, the text usually mentions various locations which are not the focus locations (see Fig. 1b). On the other hand, a news image

D.-T. Dang-Nguyen et al. (Eds.): MMM 2023, LNCS 13833, pp. 204–216, 2023.
https://doi.org/10.1007/978-3-031-27077-2_16

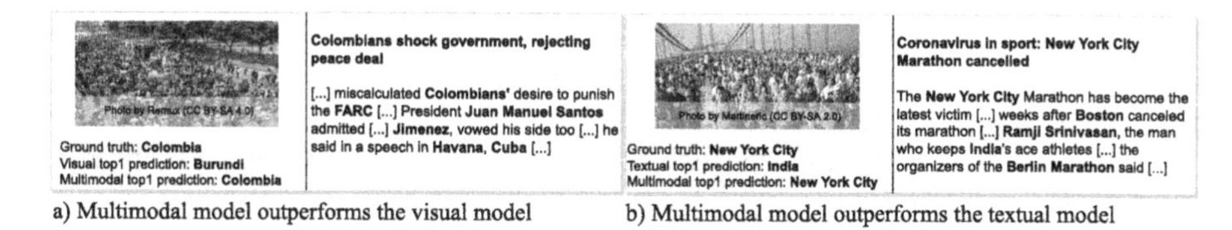

Photo by Remux (CC BY-SA 4.0)
Ground truth: Colombia
Visual top1 prediction: Burundi
Multimodal top1 prediction: Colombia
Colombians shock government, rejecting peace deal
[...] miscalculated Colombians' desire to punish the FARC [...] President Juan Manuel Santos admitted [...] Jimenez, vowed his side too [...] he said in a speech in Havana, Cuba [...]
a) Multimodal model outperforms the visual model
Photo by Martinefic (CC BY-SA 2.0)
Ground truth: New York City
Textual top1 prediction: India
Multimodal top1 prediction: New York City
Coronavirus in sport: New York City Marathon cancelled
The New York City Marathon has become the latest victim [...] weeks after Boston canceled its marathon [...] Ramji Srinivasan, the man who keeps India's ace athletes [...] the organizers of the Berlin Marathon said [...]
b) Multimodal model outperforms the textual model

Fig. 1. Samples from the MM-Locate-News dataset. (a) The visual model fails due to lack of visual geo-representative clues. (b) The textual model fails due to various locations in the text. In both cases, a multimodal model helps specify the correct location. Images are replaced with similar ones due to license restrictions.eps

alone usually does not include sufficient geo-representative features to identify the focus location reliably (see Fig. 1a). Therefore, approaches for multimodal geolocation estimation are required to bridge these gaps.

Most of previous approaches for geolocation estimation depend on either text [7,10,14,20] or image [15,16,24]. Text-based approaches assign geo-coordinates to the focus location at the document level [1,7] or based on specific events [10, 14]. On the other hand, the majority of image-based methods focus on locating photos from certain environments such as cities [8,16]. In recent years, several deep learning approaches have been introduced for geolocation estimation at global scale without any prior assumptions [23,24,28,33]. However, information from image and text are required for a more robust focus location estimation in multimodal news, as the examples in Fig. 1 illustrate. But, so far, only a few methods have considered multiple modalities for geolocation estimation [17,18]. Overall, these methods suffer from two major limitations: (1) They do not make use of recent multimodal transformer models [26] or other unimodal state-of-the-art approaches to extract rich textual (e.g., [6]) and visual information (e.g., [24]); (2) Although there are various large-scale datasets for text-based geolocation estimation of news at document level (e.g., Local-Global Lexicon (LGL) [22], GeoVirus [9]), or based on events (e.g., Atrocity Event Data [30], political news from New York Times [14]), they do not provide the accompanying images and thus do not allow for multimodal focus location.

Some multimodal datasets, such as the *MediaEval Placing Task* benchmark datasets [21], WikiSatNet [31], or *Multiple Languages and Modalities (MLM)* [2], contain image-text pairs with the corresponding geo-coordinates, but do not cover news articles and consequently do not address the related challenges such as multiple entities in the text or the lack of geo-representative features in the images. To the best of our knowledge, *BreakingNews* [27] is the biggest dataset of multimodal news articles with corresponding location information. However, the locations are primarily extracted based on the *Rich Site Summary* (RSS) and, if not available, on heuristics such as the publisher location or story text [27]. As a result, each document is provided with a number of potential locations that do not necessarily correspond to the focus location, are inaccurate, or even wrong in some cases. In summary, there is a clear need for multimodal datasets of news articles with image and text and high-quality focus location labels. In addition, a multimodal approach is required which benefits from state-of-the-art approaches to provide rich visual and textual features for news documents.

In this paper, we address the task of focus location estimation; specifically, we target the gap between lack of visual geo-representative clues in news images and multiple location mentions in the body text of news. Our contributions are summarized as follows. (1) We present the *MM-Locate-News (Multimodal Focus Location of News)* dataset that consists of 6395 news articles covering 237 cities and 152 countries across all continents as well as multiple domains such as *health*, *environment*, and *politics*. The dataset is collected in a weakly-supervised manner, and multiple data cleaning steps are applied to remove articles with potential inaccurate geolocation information. The acquired dataset addresses drawbacks of other datasets such as BreakingNews [27] as it considers multimodal content of news to label the corresponding location. Furthermore, we provide a test set of 591 manually annotated news articles. (2) We propose several neural network models that, unlike previous work [4,17,18,27,29], exploit state-of-the-art image-based [24,34], text-based [6], and multimodal approaches [26] to extract embeddings for multimodal geolocation estimation of news. (3) Experimental results and comparisons to existing solutions on the proposed dataset and on BreakingNews [27] demonstrate the feasibility of the proposed approach.

The remainder of the paper is structured as follows. Section 2 describes the dataset acquisition, while the proposed model for multimodal geolocation estimation is presented in Sect. 3. We discuss the experimental results in Sect. 4. Section 5 concludes the paper and outlines potential directions for future work.

2 MM-Locate-News Dataset

In this section, we present our novel dataset called *Multimodal Focus Location of News (MM-Locate-News)*[1]. As discussed in Sect. 1, existing datasets for geolocation estimation are either not related to news [21,31,32], do not provide images along with news texts [9,22], or contain unreliable labels for locations extracted from the news feed [27]. In contrast, MM-Locate-News includes image-text pairs of news and the location focus at different geospatial resolutions, such as city, country, and continent. In the sequel, data collection and cleaning steps (Fig. 2) as well as annotation process and dataset statistics are presented.

Data Collection: The dataset has been collected in a weakly-supervised fashion. To cover a variety of locations from all six continents, we extract all countries, capitals, highly populated cities (minimum population of 500 000; minimum area of 200 km^2), and medium populated cities (population between 20 000 and 500 000; area between 100 km^2 and 200 km^2) from Wikidata [32]. For each location, we query *EventRegistry*[2] for events between 2016 and 2020 from the following categories: *sports*, *business*, *environment*, *society*, *health*, and *politics*.

[1] Source code & dataset: https://github.com/TIBHannover/mm-locate-news.

[2] http://eventregistry.org/.

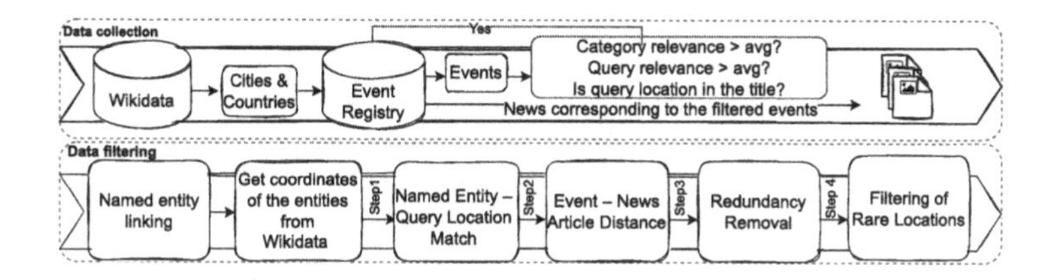

Fig. 2. MM-Locate-News data collection and filtering steps.

Note that *EventRegistry* automatically clusters news articles reporting on the same (or similar) events and that the news title of the cluster centroid represents the event name. To ensure the quality, we filter out events that do not include the location in their name or when their *category relevance* and *query relevance* scores, provided by *EventRegistry*, are below the average scores of all events per query location. The intuition behind this step is that an event with a location mentioned in its name more likely provides news articles focusing on the queried location. Finally, we extract all news articles from the remaining event clusters.

Data Filtering: We apply the following steps to remove irrelevant samples.
1) Named Entity – Query Location Match: We assume that an article is related to a query location if it is geographically close to at least one named entity. Following related work [25], we extract the named entities using *spaCy* [13] and use *Wikifier* [3] to link them to *Wikidata* for disambiguation. We extract *coordinate location* (*Wikidata Property P625*), which is available primarily for locations (e.g., landmarks, cities, or countries). For persons, we extract the *place of birth* (*Wikidata Property P19*) as they likely act in the respective country (or even city). We compute the Great Circle Distance (GCD) between the geographical coordinates of the query location and the extracted entity locations. We keep news articles that include at least one named entity whose GCD from the query location is smaller than $\sqrt[k]{a}$ where a is the area (*Wikidata Property P2046*) of the query location, and k is a hyperparameter as defined in Sect. 4.
2) Event – News Article Distance: Each news article in *EventRegistry* is assigned a similarity measure that represents the closeness to an event. We discard articles with a lower similarity than the average similarity of all articles of the same cluster to keep the news articles that are most related to the respective event.
3) Redundancy Removal: We compute the similarity between news articles using TF-IDF vectors (Term Frequency; Inverse Document Frequency) and discard one of the articles when the similarity is higher than 0.5 to remove redundancy.
4) Filtering of Rare Locations: After applying the filtering step 1–3, we remove rare locations (and corresponding articles) with less than five articles as they contain too few articles for training.

Dataset Statistics: In total, we queried 853 locations and extracted 13 143 news articles. After the data cleaning steps, we end up with 6395 news articles for 389 locations (237 cities and 152 countries). We divided the *MM-Locate-News*

Table 1. Distribution of train, validation and test samples in MM-Locate-News among continents (AF: Africa, SA: South America, EU: Europe, AS: Asia, NA: North America, OC: Oceania)

	AF	SA	EU	AS	NA	OC	Total
Train	854	216	1604	1842	589	161	5266
Val	93	24	147	160	88	23	535
Test	84	29	179	202	71	26	591

Table 2. Manual annotation criteria (C) for the MM-Locate-News test set variants (T). Answers "yes", "no" or "unsure" are denoted as "–", while "u" and "✓" denote "unsure" and "yes".

	T1	T2	T3
C1: Image depicts query location	–	✓	u
C2: Text focuses on query location	✓	✓	✓
C3: Image and text conceptually related	–	–	✓
Number of samples	591	65	154

dataset into train, validation and test data splits by equally distributing news articles among locations as given in Table 1, yielding approximately 80:10:10 splits (see Fig. 1 for samples from the dataset).

Data Annotation: The test split of the dataset is manually annotated. Users annotated a given news article along with its image and the query location to provide "yes", "no", or "unsure" labels to three criteria (*C1* - *C3*) given in Table 2. Different criteria depending on the answers are turned into a different variant of the test data to evaluate the geolocation estimation models. In the *T1* version, the text focuses on the query location, and in the *T2* both image and text represent the query location. Since it was difficult to find images where the query location is shown, we made the *T3* version where the annotators were not certain about whether the image shows the location. Thus, in cases where the text focuses on the location and the image and text are related, we assume that the image also shows the location.

Annotator Agreement: A total of three users annotated the test set, where two users annotated each sample. The inter-coder agreement for the criteria *C1*, *C2*, and *C3* is 0.44, 0.38, and 0.55, respectively (according to Krippendorff's alpha [19]). Despite relatively moderate agreement scores, we noticed that the agreement in percent is quite high: *C2* and *C3* are 80% and *C1* is 66.6%. This is caused by the annnotators' bias towards the answer "yes" for all criteria.

3 Multimodal Focus Location Estimation

We define focus location estimation as a classification problem where for each article, i.e., image-text pair, the total number of n query locations (country or city) are considered as target classes. The n-dimensional one-hot encoded ground truth vector $\mathbf{y} = \langle y_1, y_2, \ldots, y_n \rangle \in \{0, 1\}^n$ represents the query location. We extract textual and visual features as follows.

Visual Features: The visual *Scene* descriptor (representing a place in a general sense) is based on ResNet-152 model [12] to recognize 365 places (pre-trained on the Places365 dataset [34]). The *Location* descriptor, $base(M, f^*)$, is taken from a state-of-the-art photo geolocation estimation approach [24]. The *Object* descriptor utilizes the ResNet-152 model [12] pre-trained on the ImageNet dataset [5]. The $CLIP_i$ descriptor is extracted using CLIP (Contrastive Language-Image Pretraining) [26] image encoder.

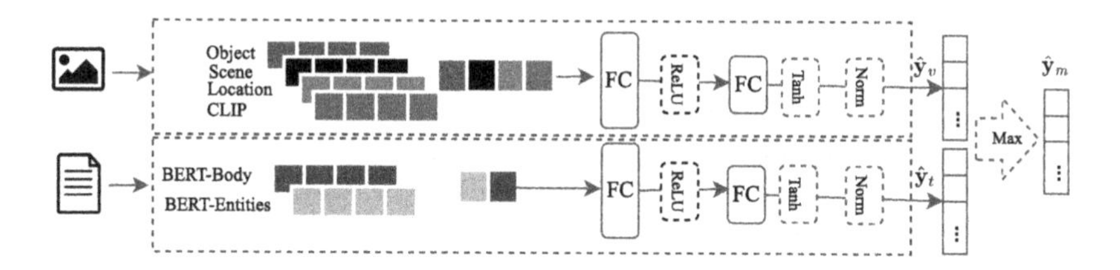

Fig. 3. The model architecture for multimodal focus location estimation in news.

Textual Features: A pre-trained BERT (Bidirectional Encoder Representations from Transformers) [6] model is used to extract two distinct feature vectors. The first one represents the whole body text (BERT-Body), while the other refers to mentions of named entities and averages the individual word embeddings of all named entities (BERT-Entities).

Multimodal Architecture: As shown in Fig. 3, the textual and visual embeddings are concatenated and passed to the fully connected (FC) layers with an output size of 1024, followed by a *ReLU* layer. Next, the outputs are fed to another *FC* layer followed by a $\tanh(u_i^l)$ layer. The norm function with clamp $\min = 10^{-12}$ is applied to extract the visual $\hat{\mathbf{y}}_v$ and textual vector $\hat{\mathbf{y}}_t$ with size of $n = 389$ (number of locations). We obtain the multimodal output $\hat{\mathbf{y}}_m$ using the maximum probabilities of the visual $\hat{\mathbf{y}}_v$ and textual $\hat{\mathbf{y}}_t$ outputs.

Network Optimization: During training, the models are optimized as follows. For unimodal models, the cross-entropy loss between the ground truth one-hot encoded vector $\mathbf{y}$ to one ($\hat{\mathbf{y}}_v$ or $\hat{\mathbf{y}}_t$) is optimized. For multimodal models, the same procedure is applied, but on both vectors $\hat{\mathbf{y}}_v$ and $\hat{\mathbf{y}}_t$. We freeze the weights of the visual models during training.

4 Experimental Setup and Results

In this section, the evaluation setup (Sect. 4.1) and experimental results including a comparison with state-of-the-art approaches on the *MM-Locate-News* (Sect. 4.2) and *BreakingNews* [27] (Sect. 4.3) datasets are reported.

4.1 Evaluation Setup

Evaluation Metrics: We use the geographical location of the class with highest probability in the network output ($\hat{\mathbf{y}}_v$, $\hat{\mathbf{y}}_t$, or $\hat{\mathbf{y}}_m$) as the prediction and measure the Great Circle Distance (GCD) to the ground truth location. As suggested by Hays and Efros [11], the models are evaluated with respect to the percentage of samples predicted within different accuracy levels: *city*, *region*, *country*, and *continent* with maximum tolerable GCD to the ground location of 25, 200, 750, and 2500 kilometers, respectively. For the evaluation on BreakingNews (Sect. 4.3), we use mean and median GCD metrics as suggested by Ramisa et al. [27].

Table 3. Accuracy [%] of focus location estimates for baselines and our models on the test variants of MM-Locate-News (best results per modality and GCD accuracy level are highlighted). Text features: BERT-Body (B-Bd), BERT-Entities (B-Et). Visual Features: Location (Lo), Object (Ob), Scene (Sc), CLIP$_i$. GCD accuracy levels: City (CI, max. 25 km GCD to the ground truth location), Region (RE, max. 200 km), Country (CR, max. 750 km), Continent (CT, max. 2500 km)

Approach	Modality	T1				T2				T3			
		CI	RE	CR	CT	CI	RE	CR	CT	CI	RE	CR	CT
Mordecai [10] (only country-level)	Textual	–	–	**74.1**	82.9	–	–	72.3	84.6	–	–	**72.7**	81.8
Cliff-clavin [7]	Textual	36.9	53.1	71.2	**86.5**	38.5	56.9	66.2	87.7	33.1	**48.7**	**72.7**	**85.1**
B-Bd	Textual	22.8	27.4	41.1	68.4	33.8	35.4	49.2	70.8	19.5	23.4	40.9	63.6
B-Et	Textual	**48.1**	**53.5**	66.3	79.5	58.5	**64.6**	**75.4**	81.5	**42.9**	46.8	61.0	77.9
B-Bd + B-Et	Textual	42.5	47.9	60.4	78.5	**60.0**	**64.6**	73.8	**89.2**	37.0	41.6	53.2	71.4
ISNs [24]	Visual	2.5	4.4	12.0	31.0	**16.9**	**26.2**	**40.0**	**55.4**	0.6	1.9	7.8	29.9
CLIP$_i$	Visual	**4.6**	**6.3**	**15.4**	**41.1**	4.6	4.6	13.8	41.5	5.2	**8.4**	**19.5**	42.2
Lo + Ob	Visual	3.2	3.7	8.5	27.7	10.8	10.8	13.8	27.7	3.9	4.5	9.7	27.3
Sc + Ob	Visual	1.2	1.5	6.4	20.6	3.1	4.6	9.2	21.5	1.9	2.6	9.1	26.0
Lo + Sc	Visual	2.4	3.0	9.0	27.1	6.2	6.2	12.3	33.8	1.9	3.2	9.1	26.0
Lo + Sc + Ob	Visual	2.5	3.4	9.1	31.0	7.7	7.7	13.8	30.8	3.2	5.2	11.7	33.8
Lo + CLIP$_i$	Visual	3.7	5.6	12.9	36.9	4.6	7.7	13.8	41.5	5.2	7.1	16.2	37.7
Sc + CLIP$_i$	Visual	3.9	5.6	12.7	36.4	3.1	4.6	13.8	44.6	**5.8**	**8.4**	16.9	40.3
Lo + Sc + CLIP$_i$	Visual	2.5	3.7	10.5	33.8	4.6	4.6	10.8	33.8	3.2	5.2	11.7	**43.5**
CLIP$_i$ + B-Bd + B-Et	Textual+Visual	63.6	68.9	78.5	86.6	69.2	**76.9**	**84.6**	**90.8**	61.0	64.9	74.7	81.8
CLIP$_i$ + Lo + B-Bd + B-Et	Textual+Visual	61.4	66.0	76.8	86.3	66.2	70.8	81.5	**90.8**	61.0	64.9	74.0	81.8
CLIP$_i$ + Sc + B-Bd + B-Et	Textual+Visual	63.1	68.0	78.0	86.0	63.1	67.7	76.9	86.2	**63.6**	68.8	77.3	83.8
Lo + Sc + B-Bd + B-Et	Textual+Visual	65.1	69.5	78.7	84.8	70.8	75.4	81.5	86.2	**63.6**	68.2	77.9	80.5
CLIP$_i$ + Lo + Sc + B-Bd + B-Et	Textual+Visual	**65.5**	**70.6**	**81.2**	**88.7**	**72.3**	**76.9**	83.1	**90.8**	**63.6**	**69.5**	**81.2**	**85.7**

Hyperparameters: We use the Adam optimizer, learning rate of 1×10^{-4}, and batch size of 128 for optimization. We train the models for 500 epochs and perform validation after every epoch. The model with the highest mean accuracy in city and region GCD thresholds on the validation set is used for testing. We set the parameter for data filtering to $k = 6$ (Sect. 2) based on an empirical evaluation of a small subset (150 samples) of the dataset.

Compared Systems: We evaluate different combinations of the proposed model based on the feature modalities. We also compare against two popular text-based methods *Cliff-clavin* [7], *Mordecai* [10] and one image-based state-of-the-art model (*ISNs: Individual Scene Networks* [24]).

4.2 Experimental Results on MM-Locate-News

The results are reported in Table 3 and discussed below.

Textual Models: For smaller GCD thresholds specifically city and region, in *T2*, the combination *B-Et* + *B-Bd* improves the performance, and in *T1* and *T3* the *B-Et* model provides the best results. When used separately, *B-Et* has a more substantial impact than *B-Bd*, indicating that named entities and their frequency play a vital role to predict the focus location in the news. While *Mordecai* and *Cliff-clavin* achieve the best results at country and continent level for *T1* and *T3*, respectively, these baselines are either not applicable (*Mordecai*) or achieve worse results (*Cliff-clavin*) compared to our models on more fine-grained levels.

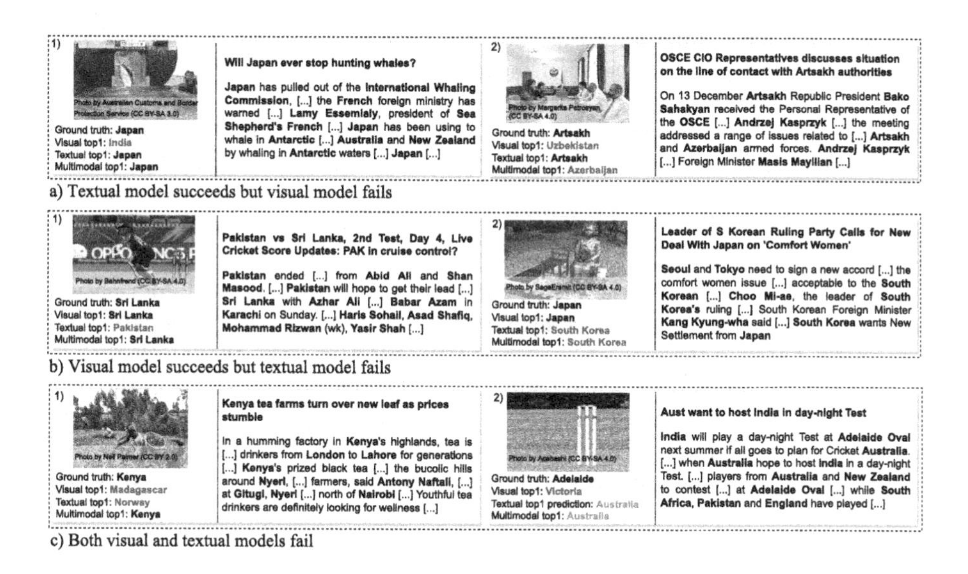

Fig. 4. Sample outputs of the MM-Locate-News dataset. In the left column the multimodal model predicts the correct location but in the right column it fails. Images are replaced with similar ones due to license restrictions.

Visual Models: The results show that $CLIP_i$ performs well in all test variants providing the best results on *T1* and *T3*, and that combinations with scene (Sc + $CLIP_i$) and location features ($Lo + Sc + CLIP_i$) can further improve the results. *ISNs* specifically trained for photo geolocalization achieve superior results on *T2* where images depict the query location and provide enough geographical cues. Unlike $CLIP_i$, ISNs do not generalize well on other test variants.

Multimodal Models: The combination of $CLIP_i$ with multimodal information drastically improves the results compared to unimodal models in all test data variants and distance thresholds. Even though our visual models do not outperform *ISNs* in *T2*, they considerably improve the results when combined with textual features (*Lo + Sc + B-Bd + B-Et*). These results suggest that a multimodal architecture is beneficial for focus location estimation in news.

Qualitative Results: Figure 4 presents the sample outputs for different models. As expected, it is easier for the textual model to predict the correct focus location when there are multiple mentions of the ground truth location in the text (Fig. 4-a.1), or person names related to the focus location such as Fig. 4-a.2, which mentions *Masis Mayilian*. However, the textual model fails when there are multiple locations (Fig. 4-b.2) or person names (Fig. 4-b.1) which are irrelevant to the focus location. It is very challenging for the visual model to predict the correct focus location due to lack of visual geo-representative features in news images (Fig. 4-a.1, 4-c.2) specifically when the image is indoor (Fig. 4-a.2). In contrast, the image sometimes provides rich content as displayed in Fig. 4-b.1, which shows the sport competition mentioned in the text, or Fig. 4-b.2 which illustrates the *Comfort Women statue*, referring to *Japan*. The multimodal model fails when there are multiple mentions of irrelevant entities and the image lacks geo-representative features such as illustrated in Fig. 4-a.2. In Fig. 4-c.1, both unimodal models fail but the combination of visual and textual information succeed since the text is about the event shown in the picture which is related to the focus location *Kenya*. Another example is Fig. 4-c.2, where the image contains scene information (Cricket field) but it does not provide visual information about the event. Therefore, the multimodal model fails to predict the city *Adelaide*, but it predicts the correct focus country *Australia*. Even though the image provides rich visual features when combined with irrelevant locations in the text the model predicts the wrong location. This demonstrates how much entities in text are important for multimodal focus location estimation. To sum up, the qualitative examples confirm the need for both image and text to predict the focus location of news.

Table 4. Mean and median GCD (km) using the best models per modality on the BreakingNews test set (mean and median values are divided by 1000).

Approach	Mean	Median
$CLIP_i$	2.84	1.3
B-Bd + B-Et	1.93	0.88
$CLIP_i$ + Lo + Sc + B-Bd + B-Et	**1.88**	**0.83**
Places [27]	3.40	**0.68**
W2V matrix [27]	1.92	0.90
VGG19 + Places + W2V [27]	**1.91**	0.88

4.3 Experimental Results on BreakingNews

We also evaluate our proposed architectures for focus location estimation on BreakingNews [27] for comparison. We divide the dataset into 33 376, 11 210, and 10 581 samples for train, validation, and test splits. However, please note that the quality of ground truth labels in BreakingNews is much lower compared to our proposed MM-Locate-News dataset as the test data are not annotated manually for focus locations. For a better comparison to the approaches presented by Ramisa et al. [27], we train a regression model by minimizing the GCD between the ground truth (the geo-coordinates of the first location as mentioned in the original paper [27]) and the predicted geographical coordinates based on the various types of state-of-the-art visual and textual features presented in Sect. 3.

Table 4 shows the comparison of our models to the regression models provided by Ramisa et al. [27] using Mean and Median GCD values. From our proposed models, as expected the multimodal model $CLIP_i$ *+ Lo + Sc + B-Bd + B-Et* achieves the best performance. In comparison with *VGG19 + Places + W2V matrix* from Ramisa et al. [27], our model achieves better results with 30 km and 50 km lower mean and median GCD, respectively. Unimodal models based on text (*B-Bd + B-Et*) outperform the visual models ($CLIP_i$). For textual models, *B-Bd + B-Et* achieves slightly better results (difference of 20 km) than *W2V matrix* [27] with respect to the median value. Regarding the visual models, $CLIP_i$ outperforms *Places* [27] in terms of mean value by 560 km but is worse with a margin of 620 km for the median value. This may be explained by more (bad) outlier predictions of the *Places* model. Overall, the results suggest that a multimodal architecture is beneficial for focus location estimation in news.

5 Conclusion

In this paper, we have introduced a novel dataset called *MM-Locate-News*, which is a multimodal collection of image-text pairs for focus location estimation of news articles. A weakly-supervised method has been suggested to collect news with focus locations. We manually annotated the test data to acquire three different test data variants. We have proposed various baselines and multimodal approaches using state-of-the-art transformer-based architectures for the task of focus location estimation. The experimental results have shown that the combination of textual and visual features for geolocation estimation outperforms the compared approaches that rely on features from a single modality.

In future work, we will extend the dataset by considering additional languages and locations. We will also investigate multimodal geolocation estimation on different news domains and apply further noise removal to the images.

Acknowledgements. This work was partially funded by the EU Horizon 2020 research and innovation program under the Marie Skłodowska-Curie grant agreement no. 812997 (CLEOPATRA ITN), and by the Ministry of Lower Saxony for Science and Culture (Responsible AI in digital society, project no. 51171145).

References

1. Andogah, G., Bouma, G., Nerbonne, J.: Every document has a geographical scope. Data Knowl. Eng. **81**, 1–20 (2012)
2. Armitage, J., Kacupaj, E., Tahmasebzadeh, G., Swati Maleshkova, M., Ewerth, R., Lehmann, J.: MLM: a benchmark dataset for multitask learning with multiple languages and modalities. In: International Conference on Information and Knowledge Management (CIKM), pp. 2967–2974 (2020). https://doi.org/10.1145/3340531.3412783
3. Brank, J., Leban, G., Grobelnik, M.: Semantic annotation of documents based on wikipedia concepts. Informatica (Slovenia) **42**(1), 23–32 (2018). http://www.informatica.si/index.php/informatica/article/view/2228
4. Crandall, D.J., Backstrom, L., Huttenlocher, D.P., Kleinberg, J.M.: Mapping the world's photos. In: International Conference on World Wide Web (WWW), pp. 761–770 (2009). https://doi.org/10.1145/1526709.1526812
5. Deng, J., Dong, W., Socher, R., Li, L.J., Li, K., Fei-Fei, L.: ImageNet: a large-scale hierarchical image database. In: IEEE Conference on Computer Vision and Pattern Recognition, pp. 248–255 (2009)
6. Devlin, J., Chang, M., Lee, K., Toutanova, K.: BERT: pre-training of deep bidirectional transformers for language understanding. In: Conference of the North American Chapter of the Association for Computational Linguistics: Human Language Technologies (NAACL-HLT), pp. 4171–4186 (2019). https://doi.org/10.18653/v1/n19-1423
7. D'Ignazio, C., Bhargava, R., Zuckerman, E., Beck, L.: CLIFF-CLAVIN: determining geographic focus for news articles. In: NewsKDD: Data Science for News Publishing Workshop co-located with ACM SIGKDD Conference on Knowledge Discovery and Data Mining (2014)
8. Gordo, A., Almazán, J., Revaud, J., Larlus, D.: Deep image retrieval: learning global representations for image search. In: European Conference on Computer Vision (ECCV), pp. 241–257 (2016). https://doi.org/10.1007/978-3-319-46466-4_15
9. Gritta, M., Pilehvar, M.T., Collier, N.: Which melbourne? augmenting geocoding with maps. In: 56th Annual Meeting of the Association for Computational Linguistics (ACL), pp. 1285–1296 (2018). https://aclanthology.org/P18-1119/
10. Halterman, A.: Mordecai: Full text geoparsing and event geocoding. J. Open Source Softw. **2**(9), 91 (2017). https://doi.org/10.21105/joss.00091
11. Hays, J., Efros, A.A.: IM2GPS: estimating geographic information from a single image. In: Conference on Computer Vision and Pattern Recognition (CVPR) (2008)
12. He, K., Zhang, X., Ren, S., Sun, J.: Identity mappings in deep residual networks. In: European Conference on Computer Vision (ECCV), pp. 630–645 (2016). https://doi.org/10.1007/978-3-319-46493-0_38
13. Honnibal, M., Montani, I.: spaCy 2: natural language understanding with bloom embeddings, convolutional neural networks and incremental parsing (2017). https://spacy.io
14. Imani, M.B., Chandra, S., Ma, S., Khan, L., Thuraisingham, B.M.: Focus location extraction from political news reports with bias correction. In: International Conference on Big Data (BigData), pp. 1956–1964 (2017). https://doi.org/10.1109/BigData.2017.8258141

15. Izbicki, M., Papalexakis, E.E., Tsotras, V.J.: Exploiting the earth's spherical geometry to geolocate images. In: European Conference on Machine Learning and Knowledge Discovery in Databases (ECML PKDD), pp. 3–19 (2019). https://doi.org/10.1007/978-3-030-46147-8_1
16. Kim, H.J., Dunn, E., Frahm, J.: Learned contextual feature reweighting for image geo-localization. In: Conference on Computer Vision and Pattern Recognition (CVPR), pp. 3251–3260 (2017). https://doi.org/10.1109/CVPR.2017.346
17. Kordopatis-Zilos, G., Papadopoulos, S., Kompatsiaris, I.: Geotagging text content with language models and feature mining. Proc. IEEE **105**(10), 1971–1986 (2017). https://doi.org/10.1109/JPROC.2017.2688799
18. Kordopatis-Zilos, G., Popescu, A., Papadopoulos, S., Kompatsiaris, Y.: Placing images with refined language models and similarity search with PCA-reduced VGG features. In: MediaEval 2016 Workshop. vol. 1739 (2016). http://ceur-ws.org/Vol-1739/MediaEval_2016_paper_13.pdf
19. Krippendorff, K.: Computing krippendorff's alpha-reliability (2011). https://repository.upenn.edu/asc_papers/43
20. Kulkarni, S., Jain, S., Hosseini, M.J., Baldridge, J., Ie, E., Zhang, L.: Multi-level gazetteer-free geocoding. In: International Workshop on Spatial Language Understanding and Grounded Communication for Robotics, pp. 79–88 (2021)
21. Larson, M.A., Soleymani, M., Gravier, G., Ionescu, B., Jones, G.J.F.: The benchmarking initiative for multimedia evaluation: Mediaeval 2016. IEEE MultiMedia **24**(1), 93–96 (2017). https://doi.org/10.1109/MMUL.2017.9
22. Lieberman, M.D., Samet, H., Sankaranarayanan, J.: Geotagging with local lexicons to build indexes for textually-specified spatial data. In: International Conference on Data Engineering (ICDE), pp. 201–212 (2010). https://doi.org/10.1109/ICDE.2010.5447903
23. Lin, T., Belongie, S.J., Hays, J.: Cross-view image geolocalization. In: Conference on Computer Vision and Pattern Recognition (CVPR), pp. 891–898 (2013). https://doi.org/10.1109/CVPR.2013.120
24. Müller-Budack, E., Pustu-Iren, K., Ewerth, R.: Geolocation estimation of photos using a hierarchical model and scene classification. In: European Conference on Computer Vision (ECCV), pp. 575–592 (2018). https://doi.org/10.1007/978-3-030-01258-8_35
25. Müller-Budack, E., Theiner, J., Diering, S., Idahl, M., Hakimov, S., Ewerth, R.: Multimodal news analytics using measures of cross-modal entity and context consistency. Int. J. Multimedia Inf. Retrieval **10**(2), 111–125 (2021). https://doi.org/10.1007/s13735-021-00207-4
26. Radford, A., et al.: Learning transferable visual models from natural language supervision. In: International Conference on Machine Learning (ICML), pp. 8748–8763 (2021). http://proceedings.mlr.press/v139/radford21a.html
27. Ramisa, A., Yan, F., Moreno-Noguer, F., Mikolajczyk, K.: BreakingNews: article annotation by image and text processing. IEEE Trans. Pattern Anal. Mach. Intell. **40**(5), 1072–1085 (2018). https://doi.org/10.1109/TPAMI.2017.2721945
28. Seo, P.H., Weyand, T., Sim, J., Han, B.: CPlaNet: enhancing image geolocalization by combinatorial partitioning of maps. In: European Conference on Computer Vision (ECCV), pp. 544–560 (2018). https://doi.org/10.1007/978-3-030-01249-6_33
29. Serdyukov, P., Murdock, V., van Zwol, R.: Placing flickr photos on a map. In: SIGIR Conference on Research and Development in Information Retrieval (SIGIR), pp. 484–491 (2009). https://doi.org/10.1145/1571941.1572025

30. Ulfelder, J., Schrodt, P.: Political Instability Task Force Worldwide Atrocities Event Data Collection Codebook. version 1.0 b2 (2009)
31. Uzkent, B., et al.: Learning to interpret satellite images using wikipedia. In: International Joint Conference on Artificial Intelligence (IJCAI), pp. 3620–3626 (2019). https://doi.org/10.24963/ijcai.2019/502
32. Vrandecic, D., Krötzsch, M.: Wikidata: a free collaborative knowledgebase. Commun. ACM **57**(10), 78–85 (2014). https://doi.org/10.1145/2629489
33. Weyand, T., Araujo, A., Cao, B., Sim, J.: Google landmarks dataset v2 - A large-scale benchmark for instance-level recognition and retrieval. In: Conference on Computer Vision and Pattern Recognition (CVPR), pp. 2572–2581 (2020)
34. Zhou, B., Lapedriza, À., Khosla, A., Oliva, A., Torralba, A.: Places: a 10 million image database for scene recognition. IEEE Trans. Pattern Anal. Mach. Intell. **40**(6), 1452–1464 (2018). https://doi.org/10.1109/TPAMI.2017.2723009

Multimedia Content Generation

C-GZS: Controllable Person Image Synthesis Based on Group-Supervised Zero-Shot Learning

Jiyun Li[1], Yuan Gao[1], Chen Qian[1](✉), Jiachen Lu[2], and Zhongqin Chen[1]

[1] School of Computer Science and Technology, Donghua University, Shanghai, China
{jyli,chen.qian}@dhu.edu.cn, 2212653@mail.dhu.edu.cn
[2] Culver Academies, Culver, IN 46511, USA
jiachen.lu@culver.org

Abstract. The objective of person image synthesis is to generate an image of a person that is perceptually indistinguishable from an actual one. However, the technical challenges that occur in pose transfer, background swapping, and so forth ordinarily lead to an uncontrollable and unpredictable result. This paper proposes a zero-shot synthesis method based on group-supervised learning. The underlying model is a twofold auto-encoder, which first decomposes the latent feature of a target image into a disentangled representation of swappable components and then extracts and recombines the factors therein to synthesize a new person image. Finally, we demonstrate the superiority of our work through both qualitative and quantitative experiments.

Keywords: Person image synthesis · Zero-shot synthesis · Group-supervised learning

1 Introduction

Person image synthesis is a trending topic in the field of computer vision. It aims to produce a person image that is perceptually indistinguishable from an actual one. However, the uncontrollability of human attributes, e.g., the diversity of appearances and poses, is a tough problem. For example, pose-guided methods that synthesize person images based on pose transfer easily outcome an unrealistic result due to the deformed textures. In order to tackle the challenge, virtual try-on networks, such as VITON [4] and its extensions [6,17] are proposed. Nevertheless, such methods cannot transfer a person's characteristics from multiple sources and edit the person's image at a component level, e.g., modifying clothes or background. Attribute-decomposed GAN [14] makes

This work has been supported by the National Key R&D Program of China under Grant 2019YFE0190500, the Fundamental Research Funds for the Central Universities of Ministry of Education of China (Grant No.2232021D-22), and the Initial Research Funds for Young Teachers of Donghua University.

D.-T. Dang-Nguyen et al. (Eds.): MMM 2023, LNCS 13833, pp. 219–230, 2023.
https://doi.org/10.1007/978-3-031-27077-2_17

progress by converting the source person image to code so the reconstruction can be regarded as re-coding. Although the flexible and continuous control of attributes is someway attained, it still lacks the comprehensive disentanglement constraints, let alone manipulating the specific attribute-related dimensions in the disentangled representation.

In this paper, we present a controllable person image synthesis method based on group-supervised zero-shot learning (C-GZS), which is an evolution of the group-supervised zero-shot synthesis (GZS) [3] network. Notably, our approach provides high controllability in creating person images based on attribute disentanglement. The architecture of the C-GZS network consists of two independent paths, one for control factor extraction and the other for zero-shot synthesis. The former maps the samples onto a disentangled latent representation space, which gains more global control over the person's image. While the latter reconstructs the desired feature attributes by adding control factors in a specific order. Together they make a giant benefit for person image synthesis. The main contributions in this paper are summarized as follows:

1. We proposed the C-GZS approach that can control the synthesis procedure of person images with specific human attributes relying solely on a collection of unprocessed dressing images.
2. The C-GZS network is specially designed for attribute-oriented data mining by adding the control factors, which not only apply the comprehensive disentanglement constraints but also grant the user more flexible and continuous control of human attributes.
3. Unlike existing methods utilizing a human parser to extract component layouts, we use an encoder-decoder for pose transfer instead of human semantic segmentation. The experiments prove that our technique performs better in the fashion domain.

2 Related Work

GAN models have been widely used in image synthesis [1,20] and face editing [16,19]. For example, in comparison to the pix2pix framework [7], pix2pixHD [18] improves the image rate and the output diversity. CycleGAN [22] defines a mapping relationship between a source image and a target instead of pairing samples. StyleGAN [8] proposes a generator for image synthesis controlled by adaptive instance normalization, while StyleGAN2 [9] ameliorates by adjusting the weights of the convolution kernel to control the channel variance. We hereby categorize the existing methods into two groups in terms of the control of human attributes.

The first group is the uncontrollable methods, which have a thumping majority yet cannot realize the delicate image synthesis. For instance, β-VAE [5] applies an adjustable hyperparameter β on the VAE framework [10] to balance latent channel capacity. However, it fails to obtain the controllable disentanglement because the latent factors of variation in unsupervised data cannot map onto the user-controllable attributes. PG^2 network [12] has an ability limited to synthesize the source and target poses into an arbitrary pose image, which seems less available in practice. Hence, our work is beyond the scope of pose-guided image generation, i.e., our research objective is person image synthesis with the aid of controlling component-level human attributes.

The second group is the controllable methods that have increasingly attracted attention in recent years. For example, FashionAI [15] implements two encoders using the LSTM network and U-Net, respectively, which fuse the features for further image synthesis via a decoder. But, it can control only one single attribute at a time. StarGAN [2] can disentangle different attributes of a face for transferring through exchanging specific parts of the latent code across images. Zhang et al. [21] introduce spatially-adaptive warped normalization (SAWN), integrating a learned flow-field to warp modulation parameters to align the style and pose features. Attribute-decomposed GAN [14] introduces a new generator architecture that embeds the source person image into the latent space as a series of decomposed component codes and recombines these codes in a specific order to construct the complete style code. But, it uses a multilayer perceptron to learn the styles, which may result in losing information about styles in a character's appearance. GZS [3] network uses disentangled representation learning in a group-supervised fashion to simulate human imagination and accomplish zero-shot image synthesis. However, it requires manipulation of specific attribute-related dimensions captured in the decomposition representation, which makes the image synthesis rigid and insufficient, especially not adaptable for practical domains. Hence, our method adds control factors for different dimensions to ensure convenient and flexible image synthesis specializing in the fashion domain.

3 Our Approach

In this paper, we propose a novel method called C-GZS as the abbreviation of controllable person image synthesis based on group-supervised zero-shot learning. The framework of our approach is demonstrated in Fig. 1. Since the background information and person's poses are usually abstract and indescribable in the clothing dataset, we need to perform a data augmentation by disentangling the human attributes. Then, during the training process, we investigate the relationships among attributes by the multiple graph structure represented by the group-supervised learning. Subsequently, We can explicitly spot the positions of the extracted control factors through iterative training with swapping attributes. Remarkably, the control factors added to the model are the key to controllable person image synthesis. In the rest of this section, we will elaborate on every module shown in Fig. 1.

3.1 Data Augmentation

The dataset collected from the fashion domain consists typically of images in clear and straightforward styles, which are beneficial to express the visual effect of the clothes but do not align well with our research context. Therefore, at the

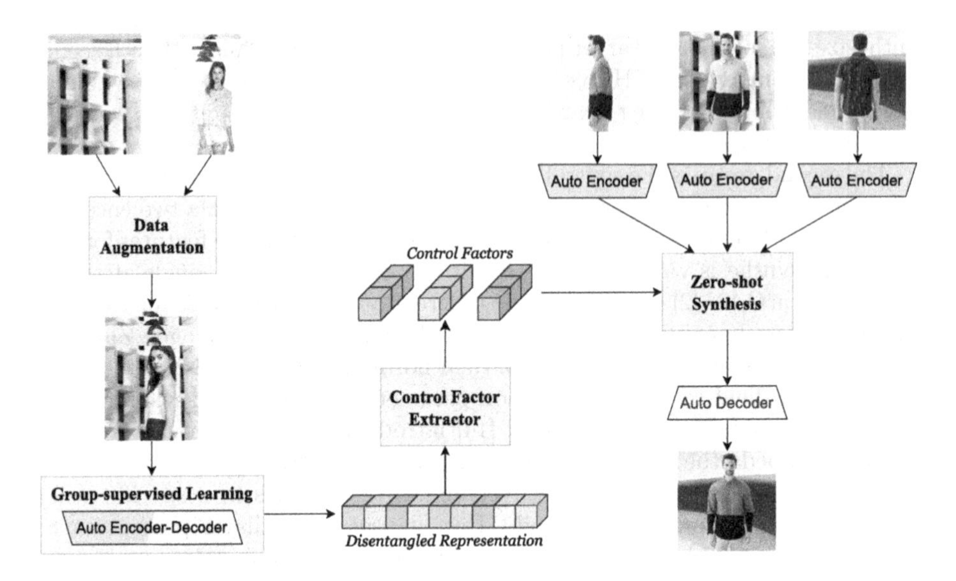

Fig. 1. An overview of the C-GZS approach.

beginning of our approach, we need to augment the data using an attribute-oriented disentanglement method. As a result, we can obtain more images with sufficient attributes. For example, if we use three landscape backgrounds to replace the monochromatic backgrounds of three fashion photographs, nine resulting images are eventually generated.

The data augmentation is executed based on DeepLabv3+, which is a semantic segmentation model with atrous separable convolution for semantic image segmentation. In addition, its architecture contains a spatial pyramid pooling module to capture more contextual information by pooling features of different resolutions and an encoder-decoder to draw more explicit object boundaries. Figure 2 demonstrates the workflow of our data augmentation.

3.2 Group-Supervised Learning

Establishing the relationship between different images' attributes is the prerequisite for control factor extraction. We hereby use a multigraph to depict the commonalities and variabilities in terms of the attributes in the dataset. Notably, two vertices may be connected by more than one edge in a multigraph. In our approach, each vertex indicates an image, while each edge denotes an attribute. Thus, we can puzzle out what an image is made of and, furthermore, what an attribute can be replaced by. It is significantly helpful for zero-shot synthesis. For example, as shown in Fig. 3, in the multigraph of the person image dataset, the label of an edge denotes the name of an attribute which are the same, shared by two samples. S_1 is a set of images that contains enough attributes to generate a new image. S_2 is another set, but we cannot generate a new image directly

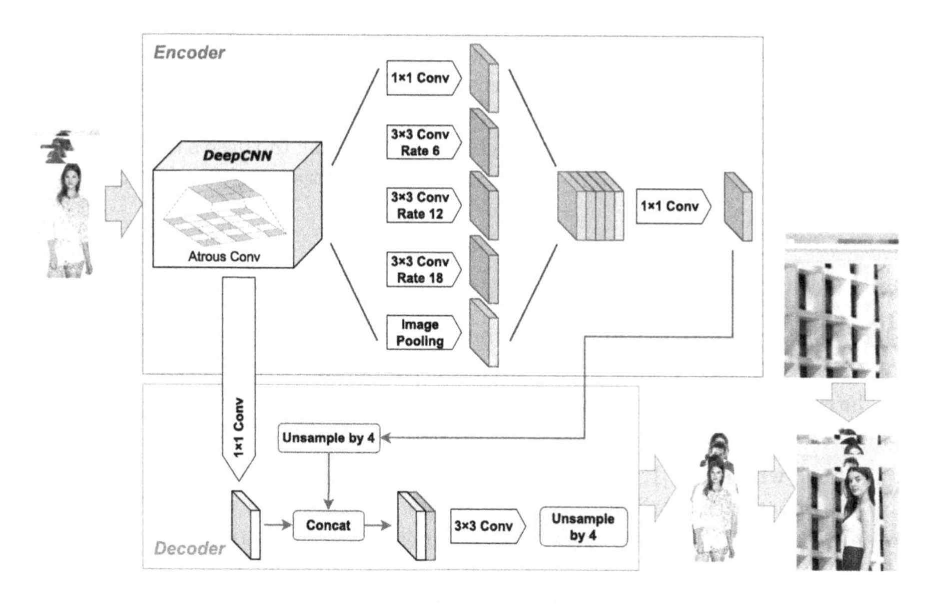

Fig. 2. Data augmentation by attribute-oriented disentanglement.

from it due to the lack of a specific attribute. However, the missing attribute can be found in an image, namely I, from S_1. So, we can yield this attribute by information extracted from I's related images belonging to S_2.

For every image, we execute a self-reconstruction method realized by a pair of encoder and decoder. The encoder is responsible for mapping the samples onto the latent vectors, whereas the decoder maps the latent vectors onto the samples. Notably, unlike the conventional encoder-decoder, we use the space of sample pairwise relationships in the encoding part. As a natural result, the output image is indeed identical to the input one, but we can successfully separate the underlying attributes during the procedure, which can be formally described as follows:

$$\begin{aligned} E &: X_1 \times M_1 \rightarrow R^d \times M_1 \\ D &: R^d \times M_1 \rightarrow X_1 \end{aligned} \tag{1}$$

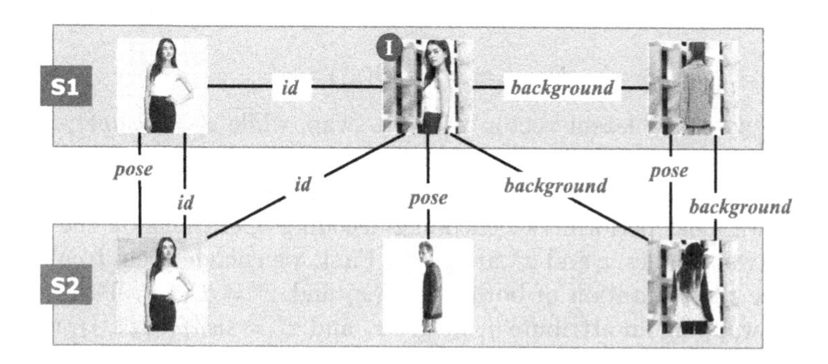

Fig. 3. Multigraph of the person image dataset.

where M_1 denotes the space of samples' pairwise relationships added to the decoder. C-GZS implements a decoder that satisfies the $(X, M) \in (X_1, M_1)$, X denotes a batch of training samples, and M denotes a subgraph of the pairwise relationship of training samples. We achieve the training of these sample pairs by swapping the selection of latent vectors and finally extracting the control factors we need.

3.3 Control Factor Extractor

The C-GZS network implements attribute swaps across images by swapping the corresponding positions on the latent representations of the images. We define the latent space as $Z = E(X, M)$, where $z^{(1)} = [g_1^{(1)}, g_2^{(1)}, g_3^{(1)}, \ldots, g_m^{(1)}]$ is the latent vector. It connects three row vectors $g_j^{(1)} \in R^{d_j}$, j is the number of attributes, where $j \in [1, m]$, and $d = \sum_{j=1}^{m} d_j$. They are hyperparameters.

To extract each attribute's control factors, we use the swap disentanglement training: $R^d \times R^d \times [1, m] \rightarrow R^d \times R^d$. We choose the position of the swap by adding the number of positions. The position of the swap is the latent vector for which an attribute is swapped. We introduce the formula for swapping two latent vectors:

$$\begin{aligned} swap(z^{(1)}, z^{(2)}) &= swap([g_1^{(1)}, g_2^{(1)}, g_3^{(1)}, \ldots, g_m^{(1)}], [g_1^{(2)}, g_2^{(2)}, g_3^{(2)}, \ldots, g_m^{(2)}], 2) \\ &= [g_1^{(1)}, g_2^{(2)}, g_3^{(1)}, \ldots, g_m^{(1)}], [g_1^{(2)}, g_2^{(2)}, g_3^{(2)}, \ldots, g_m^{(2)}] \end{aligned} \quad (2)$$

We define two strategies of swapping attributes in our approach, viz., one-overlap and cycle attribute swap, as illustrated in Fig. 4. During the extraction process of control factors, we choose ad-hoc strategy under different circumstances.

On the one hand, if the two sample images share some same attributes, a one-overlap attribute swap should be adopted. We use the attribute connection in the constructed multigraph to find two samples where a particular attribute is identical (x', x), with $(|M(x', x)| = 1)$. The swap does not affect other attributes. First encoded x' and x, the latent space is then expressed as $z = E(x)$ and $z' = E(x')$. We want to swap two samples with the same attributes (e.g., attribute j). After getting the swapped potential space representation, we can retrieve an image by decoding it. The formulas for a one-overlap attribute swap are as below:

$$D(z_s) \approx x \qquad D(z'_s) \approx x' \quad (3)$$

where z_s denotes the latent vector after the swap, while $z'_s = swap(z, z', j)$.

On the other hand, if two sample images have no attribute in common, a cycle attribute swap should be utilized. Unlike the one-overlap attribute swap approach, we must perform two encoding-decoding operations on the samples. Assuming two samples x and x^* are given. First, we encode them to obtain the latent space representation of both $z = E(x)$ and $z^* = E(x^*)$. Then, through randomly swapping an attribute j, we get z_s and $z_s^* = swap(z, z^*, j)$, which are further decoded to get $x_2 = D(z_s)$ and $x_2^* = D(z_s^*)$, respectively. After that, we

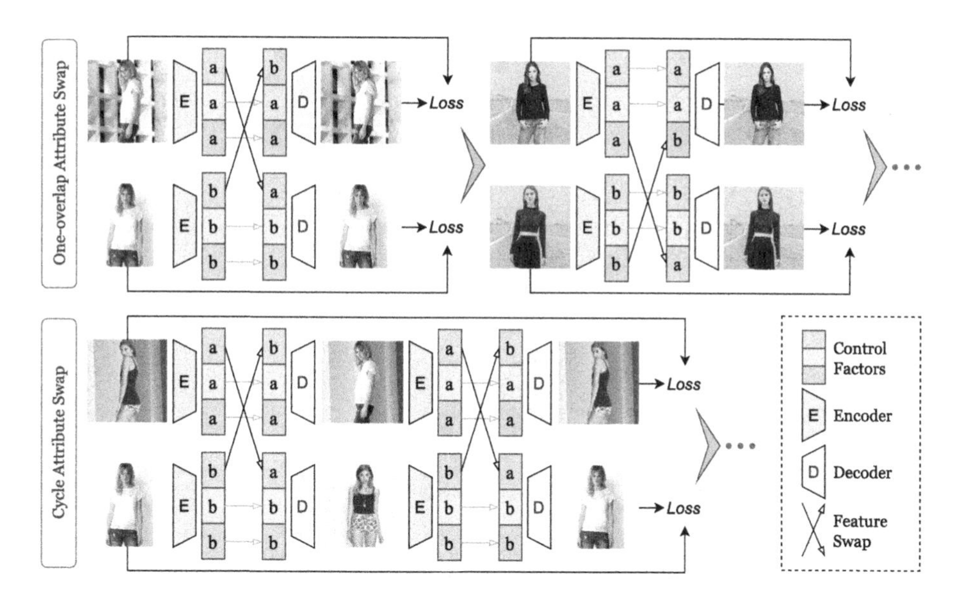

Fig. 4. The two strategies of swapping attributes.

conduct a similar operation to encode x_2 and x_2^*, which results in $z_2 = E(x_2)$ and $z_2^* = E(x_2^*)$, respectively. Likewise, we repeat the encoding process, until the last swapped attribute z_{s2} and $z_{s2}^* = swap(z_2, z_2^*, j)$. Finally, we extract z_{s2} and z_{s2}^* as control factors to provide the possibility of subsequent control of zero-shot synthesis. The result is obtained as follows:

$$D(z_{s2}) \approx x \qquad D(z_{s2}^*) \approx x^* \tag{4}$$

which suggest that swapping one attribute will not affect others. Thereby, we can extract the latent vector of attributes by swapping them and eventually harvest the expected control factors.

3.4 Zero-Shot Synthesis

After the control factors are extracted, we are ready to perform a person image synthesis. Firstly, we pick three images as the source images, which should never have been used in the previous stages. Secondly, we disentangle their attributes in order to yield the latent space representations, respectively. Thirdly, we compare the newly generated representations with the control factors to identify the commonalities, which are subsequently combined for new latent space representations. Finally, we decode the combined representations to reconstruct the image according to the target feature attributes. The C-GZS network is specially designed for attribute-oriented data mining by adding the control factors, which not only apply the comprehensive disentanglement constraints but also grant the user more flexible and continuous control of human attributes. Figure 5 depicts the workflow of person image synthesis.

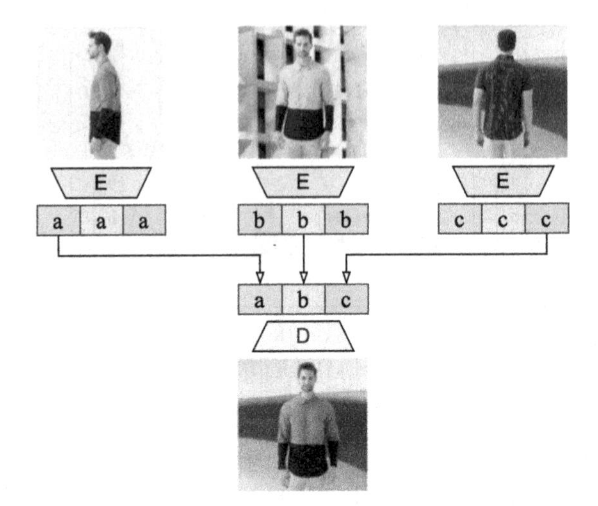

Fig. 5. Zero-shot synthesis of a person image.

4 Experiments

In this section, we verify our proposed method by training the C-GZS network to synthesize person images in a controllable manner. The evaluation is conducted through qualitative and quantitative experiments, respectively.

We use the DeepFashion dataset [11] to train our model for the experiments. The images from DeepFashion contain three attributes: person identity, pose, and background. There are three types of poses, i.e., viewing from the front, side, and back, and 11 categories of backgrounds. The image resolution is set to 256256.

4.1 Quantitative Experiments

We calculate the model-based confusion matrix between image attributes to evaluate the degree of disentanglement. Ideally, the result should be a unit matrix. We compare our C-GZS network with the GZS network, the latent disentanglement representation of the auto-encoder, and the direct use of attribute labels to supervise the auto-encoder (AE+DS). AE+DS completes the classification task using supervised learning to retain the discriminative information. However, it is not sufficient to synthesize vivid images. According to the comparison result, our method outperforms others, as displayed in Table 1.

Table 1. A comparison based on the disentanglement degree.

	Our C-GZS			GZS			Auto-encoder			AE+DS		
	I	P	B	I	P	B	I	P	B	I	P	B
I	**0.9**	0.2	0.4	0.7	0.6	0.3	0.4	0.3	0.6	0.6	0.7	0.4
P	0.7	**1.0**	0.3	0.3	0.9	0.2	0.3	0.7	0.8	0.6	**1.0**	0.3
B	0.3	0.2	**1.0**	0.7	0.8	0.5	0.5	0.5	0.5	0.4	0.3	0.6

Note: I denotes ID, P denotes pose, and B denotes background.

We also evaluate the image quality using the peak signal-to-noise ratio (PSNR), which measures the pixel-level error. Table 2 shows the comparison result.

Table 2. A comparison based on the PSNR

Model	PSNR
Auto-encoder	15.21
AE+DS	16.44
GZS	23.45
Our C-GZS	**29.45**

Moreover, we use the contextual score (CX) [13] to measure the quality of texture detail synthesis of the person on the non-aligned task. Table 3 shows the comparison result.

Table 3. A comparison based on the CX

Model	CX
PG^2	2.795
Attribute-decomposed GAN	2.474
StarGAN	2.673
Our C-GZS	**2.421**

In summary, the benchmarks of the quantitative experiments are the consistency of the appearance of the source image and the target image. Apparently, our method scores highest in PSNR and lowest in CX. The former indicates that the proposed C-GZS created person images with less distortion, while the latter proves that our method generates images that are more similar to the input ones.

4.2 Qualitative Experiments

Figure 6 demonstrates the results of person image synthesis, which are produced by three methods at the cutting edge and ours. It is worth noting that these three method are pose-guided. In other word, we hereby synthesize person image using two attributes, e.g., target persons and their poses. By comparison, the proposed C-GZS yields a more realistic image with more clear-cut details in both the overall structure and garment textures.

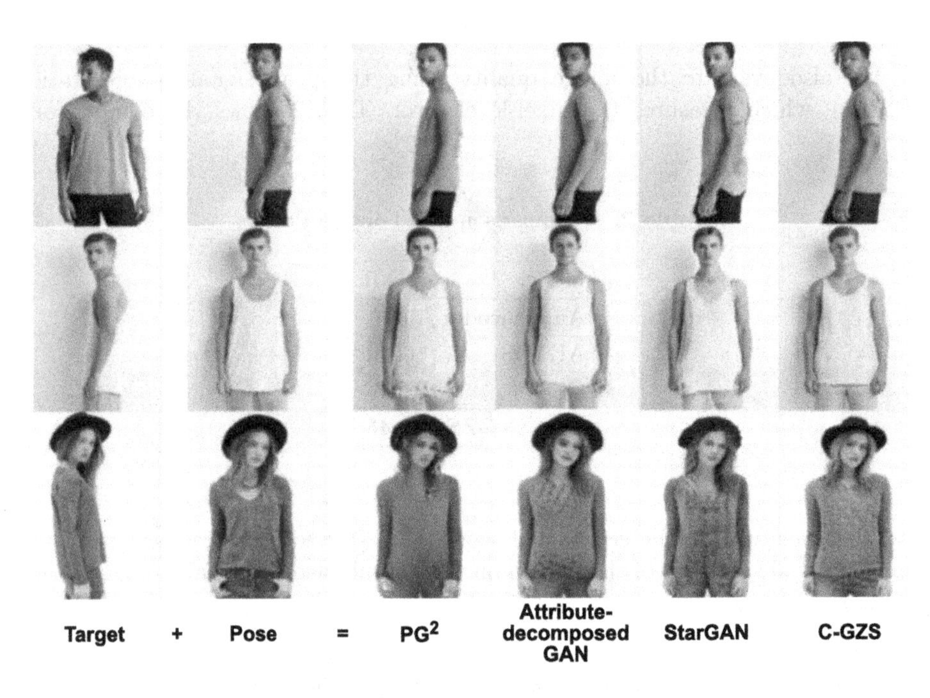

Fig. 6. Qualitative comparison of methods using two attributes.

Fig. 7. Qualitative comparison of methods using three attributes.

Finally, we compare our C-GZS method with the original GZS. The experiment uses images with three attributes in the dataset. Figure 7 exhibits a part of the results. Undoubtedly, the proposed C-GZS can significantly improve image quality with the aid of control factors.

5 Conclusion

This paper presents a novel person image synthesis method called C-GZS. Enlightened by the GZS network, we disentangle the feature attributes by group-supervised learning and synthesize images by zero-shot learning. In addition, we extract the control factors from the disentanglement representation in order to precise control which characteristics should be reformed to generate the image. Finally, we conducted a series of experiments to evaluate our method's effectiveness. In comparison with existing methods, the proposed C-GZS shows a more powerful ability to synthesize images combining person, pose, and background in both quantitative and qualitative manners. In the future, we will consider adapting our method for broader domains, which requires upgrading the built-in algorithms to extract more abstract and specific attributes and generate more realistic images.

References

1. Brock, A., Donahue, J., Simonyan, K.: Large scale GAN training for high fidelity natural image synthesis. In: International Conference on Learning Representations (2018). poster
2. Choi, Y., Choi, M., Kim, M., Ha, J.W., Kim, S., Choo, J.: StarGAN: unified generative adversarial networks for multi-domain image-to-image translation. In: Proceedings of the IEEE Conference on Computer Vision and Pattern Recognition, pp. 8789–8797 (2018)
3. Ge, Y., Abu-El-Haija, S., Xin, G., Itti, L.: Zero-shot synthesis with group-supervised learning. In: Proceedings of the International Conference on Learning Representations (2021). poster
4. Han, X., Wu, Z., Wu, Z., Yu, R., Davis, L.S.: VITON: an image-based virtual try-on network. In: Proceedings of the IEEE Conference on Computer Vision and Pattern Recognition, pp. 7543–7552 (2018)
5. Higgins, I., et al.: beta-VAE: learning basic visual concepts with a constrained variational framework. In: International Conference on Learning Representations (2017). poster
6. Honda, S.: VITON-GAN: virtual try-on image generator trained with adversarial loss. arXiv preprint arXiv:1911.07926 (2019)
7. Isola, P., Zhu, J.Y., Zhou, T., Efros, A.A.: Image-to-image translation with conditional adversarial networks. In: Proceedings of the IEEE Conference on Computer Vision and Pattern Recognition, pp. 1125–1134 (2017)
8. Karras, T., Laine, S., Aila, T.: A style-based generator architecture for generative adversarial networks. In: Proceedings of the IEEE/CVF Conference on Computer Vision and Pattern Recognition, pp. 4401–4410 (2019)

9. Karras, T., Laine, S., Aittala, M., Hellsten, J., Lehtinen, J., Aila, T.: Analyzing and improving the image quality of StyleGAN. In: Proceedings of the IEEE/CVF Conference on Computer Vision and Pattern Recognition, pp. 8110–8119 (2020)
10. Kingma, D.P., Welling, M.: Auto-encoding variational bayes. arXiv preprint arXiv:1312.6114 (2013)
11. Liu, Z., Luo, P., Qiu, S., Wang, X., Tang, X.: DeepFashion: powering robust clothes recognition and retrieval with rich annotations. In: Proceedings of the IEEE Conference on Computer Vision and Pattern Recognition, pp. 1096–1104 (2016)
12. Ma, L., Jia, X., Sun, Q., Schiele, B., Tuytelaars, T., Van Gool, L.: Pose guided person image generation. In: Guyon, I., (eds.), Advances in Neural Information Processing Systems. vol. 30. Curran Associates, Inc. (2017). https://proceedings.neurips.cc/paper/2017/file/34ed066df378efacc9b924ec161e7639-Paper.pdf
13. Mechrez, R., Talmi, I., Zelnik-Manor, L.: The contextual loss for image transformation with non-aligned data. In: Proceedings of the European Conference on Computer Vision (ECCV), pp. 768–783 (2018)
14. Men, Y., Mao, Y., Jiang, Y., Ma, W.Y., Lian, Z.: Controllable person image synthesis with attribute-decomposed GAN. In: Proceedings of the IEEE/CVF Conference on Computer Vision and Pattern Recognition, pp. 5084–5093 (2020)
15. Sun, W., Bappy, J.H., Yang, S., Xu, Y., Wu, T., Zhou, H.: Pose guided fashion image synthesis using deep generative model. arXiv preprint arXiv:1906.07251 (2019)
16. Tran, L., Yin, X., Liu, X.: Disentangled representation learning GAN for pose-invariant face recognition. In: Proceedings of the IEEE Conference on Computer Vision and Pattern Recognition, pp. 1415–1424 (2017)
17. Wang, B., Zheng, H., Liang, X., Chen, Y., Lin, L., Yang, M.: Toward characteristic-preserving image-based virtual try-on network. In: Proceedings of the European Conference on Computer Vision (ECCV), pp. 589–604 (2018)
18. Wang, T.C., Liu, M.Y., Zhu, J.Y., Tao, A., Kautz, J., Catanzaro, B.: High-resolution image synthesis and semantic manipulation with conditional GANs. In: Proceedings of the IEEE Conference on Computer Vision and Pattern Recognition, pp. 8798–8807 (2018)
19. Yadav, N.K., Singh, S.K., Dubey, S.R.: CSA-GAN: cyclic synthesized attention guided generative adversarial network for face synthesis. Appl. Intell. 1–20 (2022)
20. Zhan, F., Zhu, H., Lu, S.: Spatial fusion GAN for image synthesis. In: Proceedings of the IEEE/CVF Conference on Computer Vision and Pattern Recognition, pp. 3653–3662 (2019)
21. Zhang, J., et al.: Controllable person image synthesis with spatially-adaptive warped normalization. arXiv preprint arXiv:2105.14739 (2021)
22. Zhu, J.Y., Park, T., Isola, P., Efros, A.A.: Unpaired image-to-image translation using cycle-consistent adversarial networks. In: Proceedings of the IEEE International Conference on Computer Vision, pp. 2223–2232 (2017)

DiffMotion: Speech-Driven Gesture Synthesis Using Denoising Diffusion Model

Fan Zhang[1,2], Naye Ji[2](✉), Fuxing Gao[2], and Yongping Li[1,3]

[1] Faculty of Humanities and Arts, Macau University of Science and Technology, Macau, China

[2] College of Media Engineering, Communication University of Zhejiang, Hangzhou, China

{fanzhang,jinaye,fuxing}@cuz.edu.cn

[3] College of Digital Technology and Engineering, Ningbo University of Finance and Economics, Ningbo, China

liyongping@nbufe.edu.cn

Abstract. Speech-driven gesture synthesis is a field of growing interest in virtual human creation. However, a critical challenge is the inherent intricate one-to-many mapping between speech and gestures. Previous studies have explored and achieved significant progress with generative models. Notwithstanding, most synthetic gestures are still vastly less natural. This paper presents *DiffMotion*, a novel speech-driven gesture synthesis architecture based on diffusion models. The model comprises an autoregressive temporal encoder and a denoising diffusion probability Module. The encoder extracts the temporal context of the speech input and historical gestures. The diffusion module learns a parameterized Markov chain to gradually convert a simple distribution into a complex distribution and generates the gestures according to the accompanied speech. Compared with baselines, objective and subjective evaluations confirm that our approach can produce natural and diverse gesticulation and demonstrate the benefits of diffusion-based models on speech-driven gesture synthesis. Project page: https://github.com/zf223669/DiffMotion.

Keywords: Gesture generation · Gesture synthesis · Cross-modal · Speech-driven · Diffusion model

1 Introduction

Recently, 3D virtual human technology has become increasingly popular with the rise of the metaverse. To provide natural and relatable characters, one main task is to implement non-verbal(co-speech) gestures that look natural and match the verbal ones that communicate like humans. To date, despite motion capture

Supplementary Information The online version contains supplementary material available at https://doi.org/10.1007/978-3-031-27077-2_18.

D.-T. Dang-Nguyen et al. (Eds.): MMM 2023, LNCS 13833, pp. 231–242, 2023.
https://doi.org/10.1007/978-3-031-27077-2_18

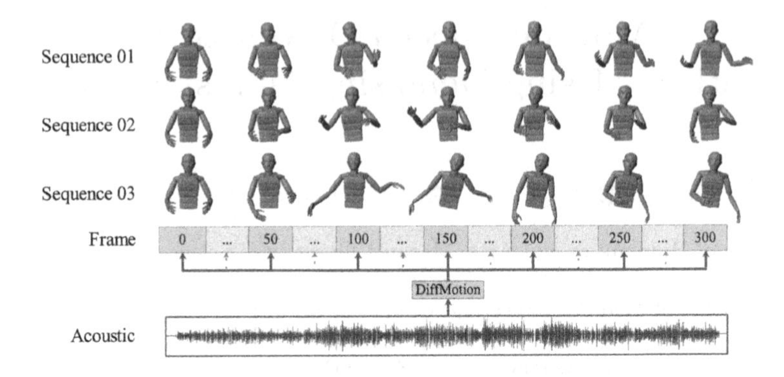

Fig. 1. Random samples from DiffMotion can give many distinct yet natural output gestures within and between sequences, even if the input speech audio is the same.

systems meeting this task, they require particular hardware, space, and actors, which can be expensive. Automatic generation is the cheapest way to generate gestures, which does not require human effort at production time. Speech-driven gesture generation is one considerable potential solution. However, the primary challenge for generating relevant and well-timed gestures from input speech is the inherent cross-modal one-to-many mapping between speech and gesture. That is, it is questioning to model the connection, which is that the same utterance is often accompanied by significantly different gestures at different times, even for the same or other speakers [16].

Under the assumption of one-to-one mapping, previous rule-based or deterministic deep learning methods fail to achieve this task. The former limited to the provided gesture units results in repetitive movements, and the latter, trained by minimizing a mean square error, is prone to mean pose. Thus, the present research has shifted to probabilistic generative models(such as GANs, VAEs, and Normalizing Flow). Despite that, most synthetic gestures are still significantly less natural compared with the original motion-capture dataset [30]. Diffusion models which can generate high-quality and diverse samples have shown tremendously impressive results on various generation tasks. Nevertheless, diffusion models have yet gained little attention in speech-driven gesture synthesis tasks.

This paper proposes **DiffMotion**, a novel diffusion-based probabilistic architecture for speech-driven gesture generation. The model learns on sizable sets of unstructured gesture data with zero manual annotation. Furthermore, as shown in Fig. 1, our method can estimate natural, various gestures, even those not present in the dataset. Our contributions are as follows:

(1) We propose DiffMotion, the first instance of the Diffusion-based generative model, to solve the cross-modal speech-driven gesture synthesis.
(2) We innovatively integrated an Autoregressive Temporal Encoder and a Denoising Diffusion Probabilistic Module, which can learn the complex regularities between gestures and the accompanying speech and generate realistic, various motions which match the rhythm of the speech.
(3) Experiments show that DiffMotion outperforms state-of-the-art baselines, objectively and subjectively.

2 Related Work

Due to the research now shifting from *deterministic* to *generative* model, we only discuss the novel approaches for speech-driven gesture generation briefly.

Ylva et al. [28] introduced GANs [10] for converting speech to 3D gesture motion using multiple discriminators. Though this approach improves significantly concerning standard regression loss training, the dataset needs to be hand-annotated. Further, they admitted the generated motions lacked realism and discontinuity because they assumed it was a gesture phase classification task. Impressively, owning to the probabilistic generative models, called *Normalizing Flows* [5] [6], which can tackle the exact log-likelihood and latent-variable inference, Alexanderson et al. [23]constructed a Glow-based [7,18] network derived from normalizing flows and RNN that successfully modeled the conditional probability distribution of the gestures given speech as input and obtained various motions given the same speech signal. Further on, Taylor et al. [22] extended normalizing flows by combining variational autoencoder, demonstrating that the approach can produce expressive body motion close to the ground truth using a fraction of the trainable parameters. Though normalizing flows are powerful enough to capture high-dimensional complexity yet still trainable, these methods require imposing topological constraints on the transformation [5,6,31].

Diffusion models(A survey in [26]), the more flexible architectures, use parameterized Markov chain to convert as simple distribution into complex data distribution gradually and can be efficiently trained by optimizing the Variational Lower Bound. After overtaking GAN on image synthesis [4,9], the diffusion model has shown remarkable impressive results on various generation tasks, such as computer vision [15], and natural language processing [1]. In particular, the diffusion model has demonstrated impressive results in multi-modal modeling [2,32], and time series forecasting [21], which owns the enormous prospect in cross-model sequence-to-sequence modeling and generation. However, it has yet earned little attention in speech-driven gesture synthesis tasks. Inspired by the discussions above, we design to construct a diffusion-based architecture to explore the capability of the diffusion model in speech-driven gesture generation.

3 Our Approach

We present **DiffMotion** for cross-modal speech-driven gesture synthesis. The model aims to generate gesticulations conditioned on speech features and historical gestures. In this section, we first formulate the problem in Sect. 3.1 and then elaborate on the DiffMotion architecture in Sect. 3.2. Finally, the training and inference process is described in Sect. 3.3 and Sect. 3.4.

3.1 Problem Formulation

We denote the gesture features and the acoustic signal as $x_t^0 \in [x_1^0, ..., x_t^0, ..., x_T^0]$ and $c_t \in [c_1, ..., c_t, ..., c_T]$, where $x_t^0 = \mathbb{R}^D$ is 3D skeleton joints angle at frame

t, and D indicates the number of channels of the skeleton joints. c_t is the current acoustic subsequence signal constructed by the sub-sequence excerpted from speech acoustic feature sequence $[a_1, ..., a_T]$, and T is the sequence length. Let $p_\theta(\cdot)$ denote the Probability Density Function(PDF), which aims to approximate the actual gesture data distribution $p(\cdot)$ and allows for easy sampling. Given t' as the past frame before current frame t, and τ as the past window frames and the initial motion $x^0_{1:\tau}$, we are tasked with generating the pose $x^0_t \sim p_\theta(\cdot)$ frame by frame according to its conditional probability distribution given historical poses $x^0_{t'-\tau:t'-1}$ and acoustic signal c_t as covariate:

$$\begin{aligned} x_t^0 \sim p_\theta\left(x_t^0 | x_{t-\tau:t-1}^0, c_t\right) \approx p(\cdot) := p\left(x_t^0 | x_{t-\tau:t-1}^0, c_t\right) \\ := p\left(x_{1:\tau}^0\right) \cdot \prod_{t'=\tau+1}^{t} p\left(x_{t'}^0 | x_{t'-\tau:t'-1}^0, c_{t'}\right) \end{aligned} \tag{1}$$

The autoregressive temporal encoder extracts the conditional information, and the $p_\theta(\cdot)$ aims to approximate $p(\cdot)$ that is trained by denoising diffusion module. We discuss these two modules in detail in Sect. 3.2.

3.2 DiffMotion Architecture

DiffMotion architecture consists of two modules: Autoregressive Temporal Encoder (AT-Encoder) and Denoising Diffusion Probabilistic Module(DDPM). The whole architecture is shown in Fig. 2.

Autoregressive Temporal Encoder(AT-Encoder). Multi-layer LSTM is adopted for AT-Encoder to encode the temporal context of speech acoustic features and past poses up to frame $t-1$ via updated hidden state h_{t-1} :

$$h_{t-1} = g\left(x_{t-\tau:t-1}^0, c_t, h_{t-2}\right) = \mathrm{LSTM}_\theta\left(\text{concatenate}\left(\mathrm{x}_{\mathrm{t}-\tau:\mathrm{t}-1}^0, \mathrm{c}_\mathrm{t}\right), \mathrm{h}_{\mathrm{t}-2}\right), \tag{2}$$

where h_t represents the LSTM hidden state evolution. LSTM_θ is parameterized by sharing weights θ and $h_0 = 0$. Then, we use a sequence of neutral(mean) poses for the initial motion $x_{1:\tau}$. Thus we can approximate Eq. (1) by the *Motion Diffusion* model, and its negative log-likelihood(NLL) is :

$$p_\theta\left(x_{1:\tau}^0\right) \cdot \prod_{t'=\tau+1}^{t} p_\theta\left(x_{t'}^0 | h_{t'-1}\right), \quad NLL := \Sigma_{t'=\tau+1}^{t} - \log p_\theta\left(x_{t'}^0 | h_{t'-1}\right) \tag{3}$$

Denoising Diffusion Probabilistic Module (DDPM). The DDPM is a latent variable model [24] of the form $p_\theta := \int p_\theta\left(x^{0:N}\right) dx^{1:N}$, where $x^1, ..., x^N$ are latent of the same dimensionality as the data x^n at the n-th diffusion time stage. The module contains two processes, namely *diffusion process* and *generation process*. At training time, the diffusion process gradually converts the original data(x^0) to white noise(x^N) by optimizing a variational bound on the data likelihood. At inference time, the generation process recovers the data by reversing this noising process through the Markov chain using Langevin sampling [19]. The gesture can be generated by sampling from the conditional data distribution at each frame and then fed back to the AT-Encoder for producing the next frame. The Markov chains in the diffusion process and the generation process are:

$$p\left(x^n|x^0\right) = \mathcal{N}\left(x^n; \sqrt{\overline{\alpha}^n}x^0, \left(1-\overline{\alpha}^n\right)I\right) \quad and$$
$$p_\theta\left(x^{n-1}|x^n, x^0\right) = \mathcal{N}\left(x^{n-1}; \tilde{\mu}^n\left(x^n, x^0\right), \tilde{\beta}^n I\right), \tag{4}$$

where $\alpha^n := 1 - \beta^n$ and $\overline{\alpha}^n := \prod_{i=1}^{n} \alpha^i$. As shown by [9], β^n is a increasing variance schedule $\beta^1, ..., \beta^N$ with $\beta^n \in (0,1)$, and $\tilde{\beta}^n := \frac{1-\overline{\alpha}^{n-1}}{1-\overline{\alpha}^n}\beta^n$.

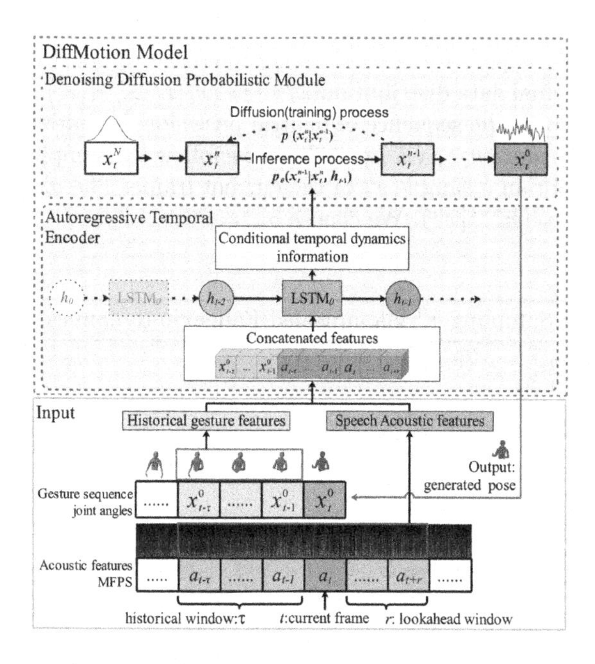

Fig. 2. DiffMotion schematic. The model consists of an autoregressive temporal encoder and a denoising diffusion probabilistic module.

3.3 Training

The training objective is to optimize the parameters θ that minimize the NLL via Mean Squared Error(MSE) loss between the true noise $\epsilon \sim \mathcal{N}(0, I)$ and the predicted noise ϵ_θ:

$$\mathbb{E}_{x_t^0,\epsilon,n}[||\epsilon - \epsilon_\theta\left(\sqrt{\overline{\alpha}^n x_t^0} + \sqrt{1-\overline{\alpha}^n}\epsilon, h_{t-1}, n\right)||^2], \tag{5}$$

Here ϵ_θ is a neural network, which uses input x_t^0 , h_{t-1} and n that to predict the ϵ, and contains the similar architecture employed in [21]. The complete training procedure is outlined in Algorithm 1.

Algorithm 1: Training for each frame $t \in [\tau+1, T]$

Input: data $x_t^0 \sim p\left(x_t^0 | x_{t-\tau:t-1}^0, c_t\right)$ and LSTM state h_{t-1}
repeat
  Initialize $n \sim \text{Uniform}(1, ..., N)$ and $\epsilon \sim \mathcal{N}(0, I)$
  Take gradient step on

$$\nabla_\theta || \epsilon - \epsilon_\theta \left(\sqrt{\overline{\alpha}_n} x_t^0 + \sqrt{1 - \overline{\alpha}_n} \epsilon, h_{t-1}, n\right) ||^2$$

until *converged*;

3.4 Inference

After training, we expect to use variational inference to generate new gestures matching the original data distribution($x_t^0 \sim p_\theta\left(x_t^0 | x_{t-\tau:t-1}^0, c_t\right)$). Firstly, we run the AT-Encoder over the sequence of neutral poses for the initial motion $x_{1:\tau}$ to obtain the hidden state h_{t-1} via Eq. 2. Then we follow the sampling procedure in Algorithm 2 to obtain a sample x_t^0 of the current frame. The σ_θ is the standard deviation of the $p_\theta\left(x^{n-1} | x^n\right)$. We choose $\sigma_\theta := \tilde{\beta}^n$.

Algorithm 2: Sampling x_t^0 via annealed Langevin dynamics

Input: noise $x_t^N \sim \mathcal{N}(0, I)$ and state h_{t-1}
for $n = N$ **to** 1 **do**
  if $n > 1$ **then**
    $z \sim \mathcal{N}(0, I)$
  else
    $z = 0$
  end if
  $x_t^{n-1} = \frac{1}{\sqrt{\alpha^n}}\left(x_t^n - \frac{\beta^n}{\sqrt{1-\overline{\alpha}^n}} \epsilon_\theta\left(x_t^n, h_{t-1}, n\right)\right) + \sqrt{\sigma_\theta} z$
end for
Return: x_t^0

During inferencing, the past poses $[x_{t-\tau-1}^0, ..., x_{t-1}^0]$ and acoustic features $[a_{t-\tau}, ...a_{t+r}]$ are concatenated and sent to the AT-Encoder for extract the context, then as a conditional information to Diffusion Module to generate current gesture(x_t^0). DiffMotion outputs one gesture in each frame and is then fed back to the generated gesture sequence. At the same time, the past pose window slides from $[t-\tau, t-1]$ to $[t-\tau+1, t]$, and the acoustic window also moved forward by one frame for the next gesture(x_{t+1}^0) generation, as shown in Fig. 2.

4 Experiments

We compare DiffMotion(**DM**) with previous baselines objectively and subjectively. For fair and comparable, we select baselines followed by: 1) using Trinity

Gesture Dataset [29] recommended by GENEA Workshop [13]; 2) Skeleton structure is consistent with DM; 3) Joint angles represented by exponential map [8]; 4) Open source provided. A flow-based method, called StyleGestures(**SG**) [23], meets the requirement above. Meanwhile, the Audio2Getsture method (**AG**) [11] was introduced for only subjective evaluation since its output format is somewhat different from **DM** and **SG**. All experiments include the ground truth(**GT**). We focus on 3D upper body beat gesture, which makes up more than 50% of all co-speech gestures and is rhythmically connected to the accompanying speech [17].

4.1 Training-Data Processing

Trinity Gesture Dataset we train on includes 23 takes, totaling 244 min of motion capture and audio of a male native English speaker producing spontaneous speech on different topics. The actor's motion was captured with 20 Vicon cameras at 59.94 frames per second(fps), and the skeleton includes 69 joints.

The upper-body skeleton was selected from the first spine joint to the hands and head and excluded the finger motion, keeping only 15 upper-body joints. We followed the data process method presented by [23] and obtained 20,665×2 samples. Each with 80×27 speech features(80 frames(4s) with 27-channel Mel-frequency power spectrograms, MFPS) as input and 80×45 joint angle features as output. The frame per second is downsampled from 60 fps to 20 fps. Each joint angle was represented by an exponential map to avoid discontinuities.

4.2 Model Settings

The LSTM in AT-Encoder consists of 2 layers with hidden state $h_t \in \mathbb{R}^{512}$. The similar network set, proposed by [21], is employed for ϵ_θ. The quaternary variance schedule starts from $\beta_1 = 1 \times 10^{-4}$ till $\beta^N = 0.1$. We set quantile=0.5 at the inference stage for catching the median value.

In truth, the gestures must be prepared in advance [3,12]. We let the control inputs c_t at time instance t contain not only the current acoustic features a_t but also the window of surrounding $c_t = a_{t-\tau:t+r}$, where the lookahead $r = 20$ is set so that a sufficient amount of future information can be taken into account, shown in Fig. 2. To avoid animation jitter, we apply Savitzky-Golay Smoothing Filter [20] to filter all joint channels of the sequence along the frame and set the window length to 31 and poly order to 4.

The model is built on TorchLightning framework using a batch size of 80, Adam optimizer with a learning rate of 1.5×10^{-3}. All experiments run on an Intel i9 processor and a single NVIDIA GTX 3090 GPU.

4.3 Objective Evaluation

We quantitatively compare DiffMotion **DM** with **SG** and the **GT** in the team of realism, Time consistency, and diversity: 1) **Realism**: We adopt L_1 distance of joint position [11] and the Percentage of Correct 3D Keypoints(PCK) [27] to evaluate the realism of the generated motion. 2) **Time consistency**: A Beat Consistency Score(BCS) metric [11] is introduced to measure the motion-speech beat correlation(time consistency). 3) **Diversity**: The diversity metric evaluates the variations among generated gestures and in a sequence. We synthesize multiple motion sequences sampled N times for the same speech input and then split each sequence into equal-length clips without overlap. Finally, we calculate the averaged L_1 distance of the whole clips.

The results are listed in Table 1. The quantitative results show that our method outperforms **SG** on realism and time consistency. On the Diversity metric, our model achieves higher score(6.25 on average) than **SG**(3.3) and **GT**(2.6). It indicates that both **DM** and **SG** have the capability to generate novel gestures that are not presented in **GT**, and our model obtains a better result than **SG**, due to the Diffusion Model can sample a wider range of samples [21].

Table 1. Quantitative results. ↑ means higher is better, ↑ means lower is better. We perform 20 tests and report their average and best scores(in parentheses).

Method	L_1↓	PCK↑	BCS↑	Diverisity↑
DM(Ours)	**11.6(10.26)**	**0.61(0.35)**	0.79(0.80)	**6.25(6.31)**
SG	16.35(15.22)	0.41(0.50)	0.68(0.69)	3.3(3.6)
GT	0	0	**0.93**	2.6

Table 2. Parameter counts, training time, and average synthesis time per frame with 95% confidence intervals.

Method	Param.Count↓	Train.time↓	Synth.time↓
DM(Ours)	**10.4 M**	**8.02 min**	1.60 ± 0.06 s
SG	109.34 M	6.2 H	$\mathbf{0.08 \pm 0.06}$ **s**

We also report parameter counts, training time, and average synthesis time per frame in Table 2. The results show that the parameter count for **DM**(10.4M) is much less than **SG**(109.34M). Our method achieves less training time(8.02 min.) than **SG**(6.2 hour). However, with the inherent characteristics of diffusion models [26], the generation phase in **DM**(1.60 ± 0.06) takes a longer time than SG(0.08 ± 0.06).

4.4 Subjective Evaluation

The ultimate goal of speech-driven gesture generation is to produce natural, convincing motions. Considering that *objective evaluation of gesture synthesis is generally tricky and does not always translate into superior subjective quality for human observes* [23, 25], we perform subjective human perception. A question set

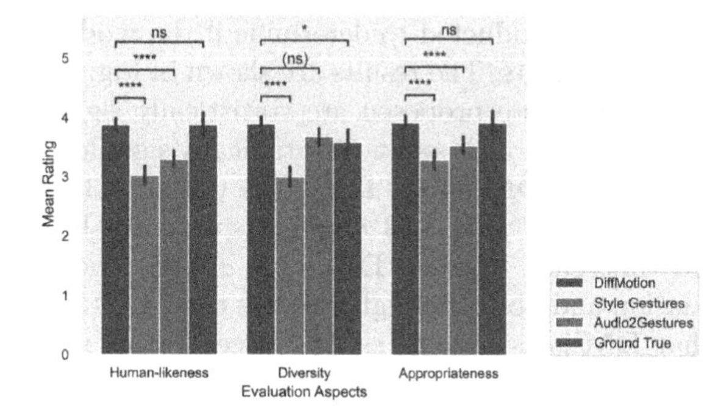

Fig. 3. Mean ratings with 95% confidence intervals. Asterisks indicated significant effects(*: $p < 0.05$, ****: $p <$ 1.00e-04, ns: no significant difference).

Table 3. The mean perceptual rating score.

Model	Human-likeness↑	Diverse↑	Appropriateness↑
DM	$\mathbf{3.89 \pm 0.95}$	$\mathbf{3.91 \pm 0.96}$	$\mathbf{3.93 \pm 0.93}$
SG	3.03 ± 1.34	3.01 ± 1.43	3.30 ± 1.36
AG	3.61 ± 1.12	$\mathbf{3.89 \pm 0.92}$	3.55 ± 1.12
GT	$\mathbf{3.89 \pm 0.93}$	3.60 ± 1.02	$\mathbf{3.93 \pm 0.92}$

consisting of three evaluation aspects with a 5-point Likert scale to subjectively evaluate baselines(**SG**, **AG**), **DM**, and **GT**. Three aspects are *human-likeness*, *diversity*, and *appropriateness*, respectively: 1) **Human-likeness**: whether the generated gestures are natural and look like the motion of an actual human, without accounting for the speech; 2) **Diversity**: which motion has more gesture patterns; 3) **Appropriateness**: the time consistency, that is, whether the generated gestures match the rhythm of the speech.

First, we trained each model and generated 20 clips with the same speech audio as input. Each clip lasts for 18 s. Next, we randomly selected 3 clips generated by each model for valuation. Then, we built up the video by GENEA_visualizer [14] for user study.

30 volunteer participants were recruited, including 16 males and 14 females, aged 19–23. All of them(20 from China, 10 international students from USA, UK, etc.) are good at English. They were asked to rate the scale for the evaluation aspects. The scores were assigned from 1 to 5, representing the worst to best.

Firstly, we introduced the method to all participants and showed them some example clips which not in the valuation set. After the participants fully understood the process, we started the formal experiment. All participants were instructed to wear headphones and sit in front of a computer screen. The environment was quiet and had no interference. Participants were unaware of which method each video belonged to. The order of videos was random, but each video was guaranteed to appear three times, presented and scored by the participants.

One-way ANOVA was conducted to determine if the models' scores differed on the three evaluation aspects. The results are shown in Fig. 3 and Table 3. The mean rating scores of **DM** we proposed are statistically significantly different from the other two baseline models and not statistically significant with **GT**. The score of human-likeness for **DM** and **GT** is 3.89 ± 0.95 and 3.89 ± 0.93, and the appropriateness is 3.93 ± 0.93 and 3.93 ± 0.92, respectively. Interestingly, there was no significant difference between **DM** (3.91 ± 0.96) and **AG**(3.89 ± 0.92) on diversity evaluation and obtained higher scores than **GT**(3.60 ± 1.02). These results suggest that **DM** is as capable as **AG** of generating more generous gestures that are not present in ground-truth. The results reveal that the proposed method outperforms previous SOTA methods and demonstrate the diffusion-based method benefits speech-driven gesture generation tasks.

4.5 Ablation Study

We found that the length of diffusion step N is a crucial hyperparameter that can affect the quality and effectiveness of gesture generation. Despite larger N allowing x^N to be approximately Gaussian [24], it results in more time in the inference process. For making a trade-off between generation efficiency and effectiveness, we tune the number of epochs for early stopping(50 epochs) and evaluated $N = 1, 50, 100, 200, 500, ..., 1000, ..., 2500$ while keeping all other hyperparameters unchanged. The results are listed in Table 4. Faster generation is achieved but causes jittery when $N < 100$. A continuous motion can be achieved when $N \geq 100$(especially $N = 100$, we reach the best result). However, as N increases, the time consumed increases substantially. Nevertheless $N > 1000$, the model occasionally produces bizarre poses due to the diffusion step destroying the detail of the raw information [31].

Table 4. Evaluation for the number of diffusion step N.

Metric	1	100	300	500	600	800	1000	1500	2000	2500
Synth. time per frame(sec.)	0.004 ± 0.001	0.34 ± 0.02	0.97 ± 0.04	1.60 ± 0.06	1.91 ± 0.06	2.55 ± 0.08	3.23 ± 0.10	4.81 ± 0.14	6.43 ± 0.19	8.02 ± 0.23
Training Time (min.)	8.63 ± 0.06	8.63 ± 0.07	8.62 ± 0.10	8.51 ± 0.08	8.55 ± 0.09	8.52 ± 0.08	8.93 ± 0.08	8.84 ± 0.08	8.74 ± 0.07	8.67 ± 0.08
Train Loss	0.99	0.32	0.20	0.17	0.16	0.15	0.13	0.12	0.10	0.09
Val Loss	0.99	0.34	0.22	0.19	0.18	0.16	0.15	0.13	0.11	0.10

5 Conclusion

In this paper, we propose a novel framework **DiffMotion** for automatically co-speech gesture synthesis. The framework consists of an Autoregressive Temporal Encoder and a Denoising Diffusion Probability Module. The architecture can learn to obtain the complex one-many mapping between gestures and the accompanying speech and can generate 3D co-speech gesticulation that is natural, diverse, and well-timed. Experiments confirm that our system outperforms previous baselines, quantitatively and qualitatively. Still, there are some limitations. For example, the model tends to generate redundant movements that lack relaxation, and it experiences slow inference due to the inherent characteristics of DDPM. For future work, we may plan the following attempts: 1) considering the breathing space for relaxation; 2) introducing multimodal features such as semantic and affective expressions; 3) investigating a real-time system enabling users to interact with virtual human interfaces.

Acknowledgements. This work was supported by the Key Program and development projects of Zhejiang Province of China (No.2021C03137), the Public Welfare Technology Application Research Project of Zhejiang Province, China (No.LGF22F020008), and the Key Lab of Film and TV Media Technology of Zhejiang Province (No.2020E10015).

References

1. Austin, J., Johnson, D.D., Ho, J., Tarlow, D., van den Berg, R.: Structured denoising diffusion models in discrete state-spaces. Adv. Neural Inf. Process. Syst. **34**, 17981–17993 (2021)
2. Avrahami, O., Lischinski, D., Fried, O.: Blended diffusion for text-driven editing of natural images. In: Proceedings of the IEEE/CVF Conference on Computer Vision and Pattern Recognition, pp. 18208–18218 (2022)
3. David, M.: Gesture and Thought. University of Chicago press, Chicago (2008)
4. Dhariwal, P., Nichol, A.: Diffusion models beat GANs on image synthesis. Adv. Neural Inf. Process. Syst. **34**, 8780–8794 (2021)
5. Dinh, L., Krueger, D., Bengio, Y.: NICE: non-linear independent components estimation. arXiv preprint arXiv:1410.8516 (2014)
6. Dinh, L., Sohl-Dickstein, J., Bengio, S.: Density estimation using real NVP. arXiv preprint arXiv:1605.08803 (2016)
7. Eje, H.G., Simon, A., Jonas, B.: MoGlow: probabilistic and controllable motion synthesis using normalising flows. ACM Trans. Graph. **39**(6), 1–14 (2020)
8. Grassia, F.: Sebastian: Practical parameterization of rotations using the exponential map. J. Graph. Tool. **3**(3), 29–48 (1998)
9. Ho, J., Jain, A., Abbeel, P.: Denoising diffusion probabilistic models. Adv. Neural Inf. Process. Syst. **33**, 6840–6851 (2020)
10. Ian, G., et al.: Generative adversarial nets. In: Advances in Neural Information Processing Systems **27** (2014)
11. Jing, L., et al.: Audio2Gestures: generating diverse gestures from speech audio with conditional variational autoencoders. In: Proceedings of the IEEE/CVF International Conference on Computer Vision, pp. 11293–11302 (2021)
12. Kendon, A.: Gesticulation and speech: two aspects of the process of utterance. Relat. verbal Nonverbal Commun. **25**(1980), 207–227 (1980)
13. Kucherenko, T., Jonell, P., Yoon, Y., Wolfert, P., Henter, G.E.: The GENEA challenge 2020: benchmarking gesture-generation systems on common data. In: International Workshop on Generation and Evaluation of Non-Verbal Behaviour for Embodied Agents (GENEA workshop) 2020 (2020)

14. Kucherenko, T., Jonell, P., Yoon, Y., Wolfert, P., Henter, G.E.: A large, crowd-sourced evaluation of gesture generation systems on common data: the GENEA challenge 2020. In: 26th International Conference on Intelligent User Interfaces, pp. 11–21 (2021)
15. Li, H., et al.: SRDiff: single image super-resolution with diffusion probabilistic models. Neurocomputing **479**, 47–59 (2022)
16. Matthew, B.: Voice puppetry. In: Proceedings of the 26th Annual Conference on Computer Graphics and Interactive Techniques, pp. 21–28 (1999)
17. McNeill, D.: Hand and mind: what gestures reveal about thought. In: Advances in Visual Semiotics, p. 351 (1992)
18. P., K.D., Prafulla, D.: Glow: generative flow with invertible 1x1 convolutions. arXiv preprint arXiv:1807.03039 (2018)
19. Paul, L.: sur la théorie du mouvement brownien. C. R. Acad. Sci. **65**(11), 146, 530–533 (1908), publisher: American Association of Physics Teachers
20. Press, W.H., Teukolsky, S.A.: Savitzky-golay smoothing filters. Comput. Phys. **4**(6), 669–672 (1990)
21. Rasul, K., Seward, C., Schuster, I., Vollgraf, R.: Autoregressive denoising diffusion models for multivariate probabilistic time series forecasting. In: International Conference on Machine Learning, pp. 8857–8868 (2021)
22. Sarah, T., Jonathan, W., David, G., Iain, M.: Speech-driven conversational agents using conditional flow-VAEs. In: European Conference on Visual Media Production, pp. 1–9 (2021)
23. Simon, A., Eje, H.G., Taras, K., Jonas, B.: Style-controllable speech-driven gesture synthesis using normalising flows. In: Computer Graphics Forum. vol. 39, no. 2, pp. 487–496. Wiley Online Library (2020)
24. Sohl-Dickstein, J., Weiss, E., Maheswaranathan, N., Ganguli, S.: Deep unsupervised learning using nonequilibrium thermodynamics. In: International Conference on Machine Learning, pp. 2256–2265. PMLR (2015)
25. Wolfert, P., Robinson, N., Belpaeme, T.: A review of evaluation practices of gesture generation in embodied conversational agents. IEEE Trans. Human Mach. Syst. **52**(3), 379–389 (2022)
26. Yang, L., Zhang, Z., Hong, S., Zhang, W., Cui, B.: Diffusion models: A comprehensive survey of methods and applications (Sep 2022)
27. Yi, Y., Deva, R.: Articulated human detection with flexible mixtures of parts. IEEE Trans. Pattern Anal. Mach. Intell. **35**(12), 2878–2890 (2012)
28. Ylva, F., Michael, N., Rachel, M.: Multi-objective adversarial gesture generation. In: Motion, Interaction and Games, pp. 1–10. ACM, Newcastle upon Tyne United Kingdom (2019)
29. Ylva, F., Rachel, M.: Investigating the use of recurrent motion modelling for speech gesture generation. In: Proceedings of the 18th International Conference on Intelligent Virtual Agents, pp. 93–98 (2018)
30. Yoon, Y., et al.: The GENEA challenge 2022: A large evaluation of data-driven co-speech gesture generation (2022)
31. Zhang, Q., Chen, Y.: Diffusion normalizing flow. In: Advances in Neural Information Processing Systems. vol. 34 (2021)
32. Zhu, Y., Wu, Y., Olszewski, K., Ren, J., Tulyakov, S., Yan, Y.: Discrete contrastive diffusion for cross-modal and conditional generation (2022)

TG-Dance: TransGAN-Based Intelligent Dance Generation with Music

Dongjin Huang(✉), Yue Zhang, Zhenyan Li, and Jinhua Liu

Shanghai Film Academy, Shanghai University, Shanghai, China
{djhuang,20723208}@shu.edu.cn

Abstract. Intelligent choreographic from music is a popular field of study currently. Many works use fragment splicing to generate new motions, which lacks motion diversity. When the input is only music, the frame-by-frame generation methods lead to similar motions generated by the same music. Some works improve this problem by adding motions as one of the inputs, but requires a high number of frames. In this paper, a new transformer-based neural network, TG-dance, is proposed for predicting high-quality 3D dance motions that follow the musical rhythms. We propose a new idea of multi-level expansion of motion sequences and design a new motion encoder, using a multi-level transformer-upsampling layer. The multi-head attention in the transformer allows better access to contextual information. The upsampling can greatly reduce motion frames input, and is memory friendly. We use generative adversarial network to effectively improve the quality of generated motions. We designed experiments on the publicly available large dataset AIST++. The experimental results show that TG-dance network outperforms the latest models in quantitative and qualitative. Our model inputs fewer frames of motion sequences and audio features to predict high-quality 3D dance motion sequences that follow the rhythm of the music.

Keywords: Dance motion generation · Multimodal fusion · Upsampling · Transformer · Multi-head attention

1 Introduction

With the rise of the Internet, dance has now become an art form that can be seen everywhere. Dance performance requires a certain level of creativity, and choreography experts are usually better at creating great choreography than novices, but it takes a lot of time and effort. Therefore, how to use artificial intelligence methods instead of human creativity to generate dance motions from music has become a hot research topic in recent years. For choreographers, this task can

This work was supported by the Shanghai Natural Science Foundation of China under Grant No.19ZR1419100 and the Shanghai talent development funding of China under Grant No.2021016.

D.-T. Dang-Nguyen et al. (Eds.): MMM 2023, LNCS 13833, pp. 243–254, 2023.
https://doi.org/10.1007/978-3-031-27077-2_19

help provide creative reference and increase their efficiency. In entertainment, it has been applied in scenarios such as game character interaction [1] and virtual idol performance, and is very promising research for application.

The classical method used in dance generation tasks is fragment splicing, which generates new motion sequences by reassembling the motion fragments of the dataset. This method solves the motion generation problem by relying on the selection of motion fragments of the dataset and has difficulties in generating innovative dance. Therefore the frame-by-frame generation method becomes the choice for a large amount of works. In recent years deep learning has a wide range of applications. Convolutional neural network is proved to be able to solve the sequence problem and gradually becomes the main method for motion generation tasks. However, the problem of gradient disappearance or explosion tends to produce frozen motion sequences. Some recent works have experimented with transformer, which is widely used in NLP tasks. The attention mechanism provides better access to contextual information than convolutional neural network, and is able to generate longer sequences of unfrozen motions.

Although some progress has been made in recent years on the task of dance generation, some challenges remain. First, the fragment splicing method can generate higher quality motion sequences, but it is difficult to create new motions. Secondly, although the frame-by-frame generation method can create some new motions based on music compared to combining pre-cut motion clips, most of the current works only use music as input, resulting in similarity of motions generated from the same music. Third, recent tasks have added motion sequences as one of the inputs to improve the similarity problem, but generating high-quality motion sequences requires more frames.

To address these challenges, we propose a transformer-based TG-dance network. We use frame-by-frame prediction to ensure the diversity of generated motions, and input music and motion together to solve the problem of motion consistency. In addition, to reduce input frames for motion sequences, we use multi-level upsampling to extend motion sequences and add the discriminator to improve the quality of the predicted motions. Experiments have shown that our results produce higher quality and richer 3D dance sequences compared to the results of the latest methods.

To summarize, our contributions are as follows:

1. We propose a new frame-by-frame generation idea to learn to generate long time motion sequences. The use of multi-level encoding and upsampling to increase the sequence length can reduce the motion sequence input length and is memory friendly.
2. We propose a transformer-based network TG-dance, where the transformers use multi-head self-attention to encode the input sequences and an adversarial training method to improve the quality of the generated motions.
3. Experiments on the dataset AIST++ prove the superiority of our method, which can generate higher quality and richer 3D dance sequences compared to other advanced methods.

2 Related Work

Earlier works used statistical models [2,3] to generate motions. Motion fragment splicing [1,4–6] is the classical solution for dance motion generation. ChoreoMaster [4] was based on earlier motion graphs [7–9]. The work [9] was the first attempt to add constraints of motion beats and rhythms. ChoreoNet [5] created dance motion units (CAUs) and predicted CAUs in the generation module. DanceIt [6] designed a cross-modal alignment module to learn the correlations of motion and music. The fragment splicing method transforms the motion generation problem into a fragment selection problem and is able to generate higher quality motions sequences, but it lacks motion diversity.

With the development of neural network, motion generation tasks are also gradually adopting neural network as the mainstream method, such as improved RNN [10–12], one-dimensional convolutional neural network (CNN) [13], and graph convolutional network (GCN) [14]. Several works improved dance motions by adding unique modules. DanceNet [15] added gated activation unit to improve the interaction between different features and model complexity. Duan et al. [1] used a co-ascent mechanism on unlabeled dataset. In addition, an improved dilated convolution text-to-speech model with an autoregressive network [16] was shown to solve the motion-music generation problem as well.

In contrast to convolutional neural network, the recent new structure transformer [17] completely abandons recursion and convolution and uses only attention mechanisms. Works in different domains [18,19] demonstrated that the transformer plays a role in solving temporal problems and is able to fuse better for different modalities [20,21]. Cai et al. [22] encoded the joint trajectories and then input them to the transformer. ST-Transformer [23] used a decoupled temporal and spatial self-attention mechanism. Huang et al. [24] used RNN as the encoder of the transformer. TSMT [25] proposed a novel two-stream motion transformer generative model that handles motion and music sequences separately. Danceformer [26] used a transformer-like network, DanTrans, that first generated key motion frames and then predictd the motions between frames by music features. FACT [27] was a full-attention cross-modal transformer network, capable of predicting motions from a long motion sequence and music. Transformer has higher parallelizability than the widely used convolutional neural network and can obtain better contextual information, which solves the problem of generating frozen motions to some extent.

To increase the diversity of generated motions, we use a small number of motion frames as one of the inputs and expand the sequences with an upsampling layer to generate long time motion sequences. Considering the advantages of transformer for sequence tasks, we use transformer as the basic block of our network to encode motions and learn the correlation between motion and music.

3 Our Approach

In this section, we propose the TG-dance model for generating 3D dance motions from music, as shown in Fig. 1. TG-dance uses a transformer-based adversarial

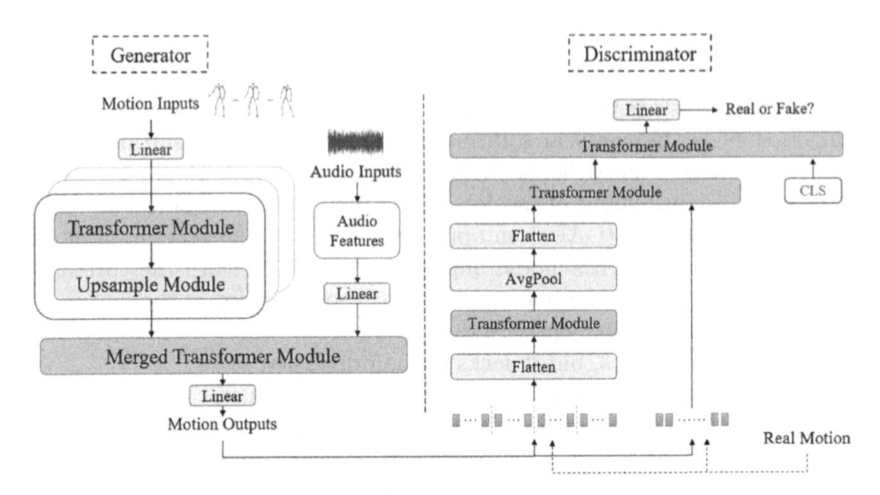

Fig. 1. The structure of the TG-dance network. It contains a generator **G** and a discriminator **D**. The input of the generator module **G** includes motion sequences and music feature sequences, which are aligned in the first frame. The generator is used to predict the future motion sequences. Discriminator module **D** discriminates the predicted motion sequences as true or false.

network, consisting of the generator **G** and the discriminator **D**. The motion encoder in the generator encodes and upsamples the motion sequences. The hybrid encoder combines the motion features and music features to predict motion sequences. The discriminator has three transformers, respectively encoding the combination of motion features and token, and finally obtaining the token to determine the authenticity of the motion sequences.

3.1 Generator Module

Transformer has been widely used in the field of natural language processing and performs well for sequence tasks. We use transformer as the basic block of the network. Each transformer encoding layer contains a position encoding layer, layernorm layer, multi-head self-attention layer and feed-forward MLP with GELU non-linearity, as shown in Fig. 2(a). Multi-head self-attention mechanism can capture richer features than self-attention and the equation is as follows:

$$\begin{aligned} MultiHead(Q,K,V) &= Concat\left(head_1,\ldots,head_h\right)W^O, \\ where\ head_i &= Attention\left(QW_i^Q,KW_i^K,VW_i^V\right). \end{aligned} \tag{1}$$

where Q,K,V are matrices composed of word vectors and W^O are dot product weights. W_i^Q,W_i^K,W_i^V are the weight parameters of matrices Q,K,V. The attention layer is calculated as follows, and d_k is the dimension of the input:

$$Attention(Q, K, V) = softmax\left(\frac{QK^T}{\sqrt{d_k}}\right) \cdot V \tag{2}$$

Due to the complexity of human motions, we first encode the motion sequences and subsequently add music to learn the correlation between the two. Our generator module consists of the motion encoder and the hybrid encoder. One of the inputs to the network is music, which is extracted by the tool library librosa [28] to obtain music features.

Another input to the network, the motion sequence first passes through the linear layer and the output size is $l_m \times C$, where l_m is the number of frames of the input sequence and C is the hidden size. Inspired by Transgan [29], we design the motion encoder as a combination of the transformer layer and the upsampling layer. The transformer layer with the multi-head attention mechanism encodes the sequence and gets the motion features. The upsampling layer takes a similar approach to image upsampling to add the length of motion sequence. Specifically, we define l_s by analogy with the pixel values of the image length $\boldsymbol{H}$ and width $\boldsymbol{W}$ such that $l_m = l_s \times l_s$. After one transformer-upsampling layer, a sequence of $2l_s \times 2l_s \times C/4$ is obtained. The second stage is the same as the first stage, and the encoder finally outputs a sequence of $4l_s \times 4l_s \times C/16$. Compared to the initial input, a 16 times longer motion sequence is obtained after two transformer-upsampling layers, and the multi-level upsampling is able to be memory friendly during the training process.

The hybrid encoder encodes audio features and motion features obtained by the motion encoder together. The encoder is also based on the transformer, which can fuse the two modalities to learn the correlation between them and output long motion sequences.

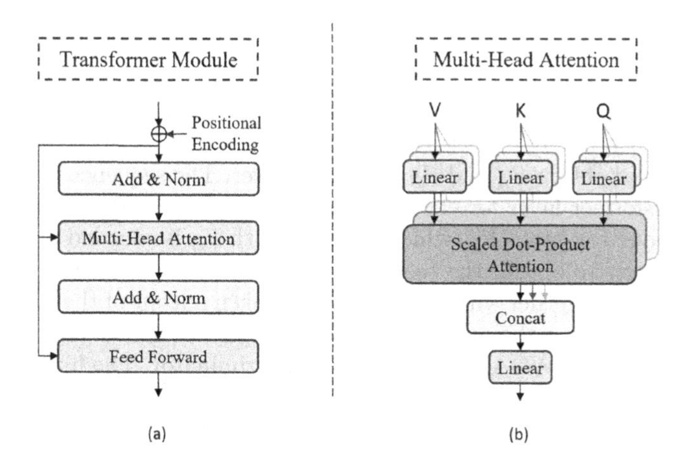

Fig. 2. Transformer details. (a) Structure of the transformer in the TG-dance network. (b) The multi-head self-attention structure used in transformer.

The generator loss uses a combination of L_1 loss, MSE loss and BCE loss, with parameters of 1:1:0.01. L_1 loss and MSE loss calculate the distance between the target motion sequence and the predicted motion sequence, making the predicted motion sequence closer to the target sequence. The equations for L_1 loss and MSE loss are as follows:

$$loss_{L_1}(x, y) = \frac{1}{n}\sum_{i=1}^{n} \mid y_i - f(x_i) \mid \tag{3}$$

$$loss_{MSE}(x, y) = \frac{1}{n}\sum_{i=1}^{n} (y_i - f(x_i))^2 \tag{4}$$

where x is the input value, $f(x_i)$ is model predicted value of x, and y_i is the real sample value.

BCE loss is one of the frequently used losses in classification tasks. The generated motion sequences and the real motion sequences go through the discriminator to get the classification results, and we use BCE loss to calculate the distance between the two:

$$loss_{BCE}(p, t) = -w\left[tlog(p) + (1-t)\, log(1-p)\right] \tag{5}$$

where p is the model predicted value, t is the real label value, and w is the weight.

3.2 Discriminator Module

The generator module predicts the future dance motion sequences with the input of motion sequences and music features. The discriminator module judges the truth and falsity of the generated motion sequences and the real ones. The generator is able to improve the quality of the generated motions by constantly fighting against the discriminator.

The discriminator has three transformer layers. We first generate motion sequences of 3/4 and 1/4 the length of the input sequence by two-dimensional convolution, and obtain S_A and S_B by flatten layer. The sequence S_A is input to the first transformer layer for encoding. Then it goes through the pooling layer and flatten layer to obtain S'_A, which together with S_B is encoded by the second transformer layer to obtain the sequence S_C. Finally, it goes through the third transformer layer together with the token. The token output of the discriminator is the result of judging the authenticity of the input sequence.

We use the loss of WGAN-GP [30] in the discriminator. The lipschitz restriction of WGAN is weight clipping. In order to solve the problem of difficult training and slow convergence, WGAN-GP adds a regular term - gradient penalty, which allows weights to be distributed more uniformly. The equation is as follows:

$$loss = \mathop{\mathbb{E}}_{\tilde{x}\sim\mathbb{P}_g}\left[D(\tilde{x})\right] - \mathop{\mathbb{E}}_{x\sim\mathbb{P}_r}\left[D(x)\right] + \lambda \mathop{\mathbb{E}}_{\hat{x}\sim\mathbb{P}_{\hat{x}}}\left[(\| \nabla_{\hat{x}} D(\hat{x}) \|_2 - 1)^2\right] \tag{6}$$

where λ is the parameter of the gradient penalty term. $\boldsymbol{x}$ is the real sample value, $\widetilde{\boldsymbol{x}}$ is the model predicted value. $\hat{\boldsymbol{x}}$ is interpolated from $\widetilde{\boldsymbol{x}}$ and $\boldsymbol{x}$. If gradient norm moves away from its target norm value 1, then the higher the penalty of the loss.

Multiple transformers are used in the generator and discriminator, and different transformers have different depths. The generator predicts the future motion sequences for a long time. Since the more forward part of the future motion sequence has higher correlation with the input, our discriminator supervises only the first N frames.

4 Experimental Results and Analysis

Our experiments are conducted using Ubuntu 20.04.2 Linux OS, Intel(R) Xeon(R) CPU E5-2620 v4, Nvidia TITAN Xp, and we use pytorch 1.11.0 as the deep learning framework.

4.1 Dataset and Experimental Settings

We use the currently publicly available 3D human dance dataset AIST++ [27] for our experiments, which has rich variety and a large amount of unit street dance motion data. The dataset is divided into training set and test set, which do not overlap. The training set contains nearly one thousand combinations of dance and music. The motion data in AIST++ is 60 frames per second. This high frame rate setting makes the motion smoother, but requires more motion frames for the same motion change target. To be able to achieve more motion changes with fewer motion frames, we first extracted the motion data of AIST++ to contain 30 frames per second. Each frame having 75 dimensions, containing 3-dimensional position data of the root node and 3-dimensional rotation data of the 24 nodes. We use the audio feature processing tool library librosa [28] for music feature extraction, including MFCC, chroma, spectral and beat. During the training period, we sliced the motion-music combinations by window and we use max-min normalization for them to unify the input data before putting them into the network.

During the training, the input to our model consists of a 16-frame motion sequence and a 64-frame audio feature sequence, and they are aligned in the first frame. The motion encoder in the generator encodes and upsamples the input motion sequence twice. The two phases generate motion sequences of 64 and 256 frames respectively, but we only supervise the first 16 frames. Each transformer has a different encoding role and therefore has a different number of depths and hidden layer size. The learning rate of both the generator and the discriminator takes a linear decrease, starting from 1e-4 and decreasing to 0.8 of the previous learning rate after every 50 steps.

4.2 Quantitative Evaluation

There are currently no uniform evaluation metrics for dance generation tasks. We use the dance quality evaluation method used in FACT [27] to compare our method with the baseline methods. FID is one of the common evaluation metrics used in various fields. The distributed distances between the real and generated motion sequences in the joint position space $\mathbb{R}^{T\times N\times 3}$ are used to calculate the FID, which is used to evaluate the quality of the generated motions. Dist is calculated by the features of the motion changes and is used to evaluate the diversity of motion changes. Beat Alignment Score (BeatAlign) is used to evaluate the correlation between dance and input music. The equation is as follows:

$$BeatAlign = \frac{1}{m}\sum_{i=1}^{m} exp\left(-\frac{min_{\forall t_j^y \in B^y} \parallel t_i^x - t_j^y \parallel^2}{2\sigma^2}\right) \tag{7}$$

where t_i is the motion beat, t_j is the music beat, and σ is the normalized sequence parameter.

We used the test set of the AIST++ dataset for quantitative evaluation, including 40 generated and real motion sequences, each generated motion sequence having 20 s. In Table 1, the data for the four baselines are obtained from the FACT [27] article. Compared to the four baselines, our method has the lowest FID value 34.48 and the highest BeatAlign value 0.247, as well as the second highest Dist value 6.69, so our model has the best overall performance. FACT [27] is the most recent model, and can generate motion sequences similar to real dance motions. However, since real dance motions are mainly composed of unit choreographic motions, it leads to a lack of some motion diversity. Li et al. [25] reached the highest Dist value 6.85 for the five models, but this was due to the discontinuity of the generated motions and obtained the highest FID value 86.43. DanceNet [15] and DanceRevolution [24] were able to generate more continuous motions, but the rigid motions resulted in motion quality evaluation value FID lower than FACT [27] of 35.35. Besides, it is indicated in FACT [27] that its full training requires 300k steps, but our network only requires 2.6k steps to converge and generate motions with higher quality and diversity. Figure 3 provides the visualization results of TG-dance generated dances, and the dance motions are rich and follow the music.

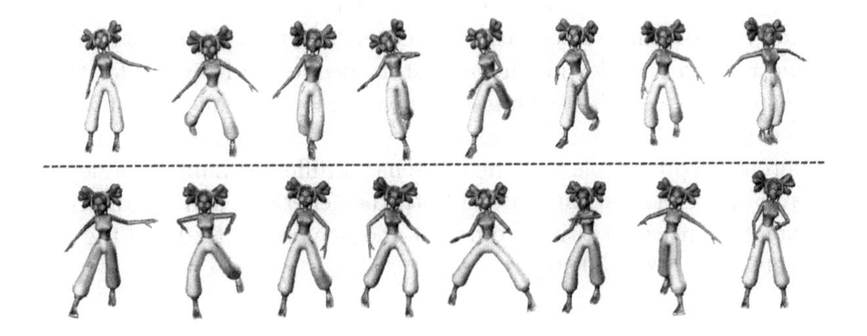

Fig. 3. Visualization results of the motions generated by TG-dance.

Table 1. Quantitative results of TG-dance and other baselines on AIST++ dataset. ↑ means that the larger the value of the corresponding objective index, the better the enhancement result.

Method	FID↓	Dist↑	BeatAlign↑
Li et al. [25]	86.43	**6.85**	0.232
DanceNet [15]	69.18	2.86	0.232
DanceRevolution [24]	73.42	3.52	0.220
FACT [27]	35.35	5.94	0.241
TG-dance(ours)	**34.48**	6.69	**0.247**

4.3 Ablation Study

To demonstrate the effectiveness of our model, we conduct ablation experiments on the use of the discriminator and the motion encoder in the structure. (1) We set up no discriminator in TG-dance and obtain the prediction results on the same steps of network training. The results in Table 2 show that the generated dance motions have higher scores for motion diversity without discriminator because of the discontinuity of the generated motions. (2) We set the depth of transformers of the motion encoder to 1, and adjust the depth of the hybrid encoder at the same time to ensure that the sum of the depths of the transformers is the same.

Compared to the two ablation models, the results in Table 2 show that our model obtains the best overall results. When using only the generator, the model generates discontinuous motions and thus obtains the highest Dist value 10.69. The results of the ablation study are visualized in Fig. 4. When the discriminator is not used, richer dance sequences can be generated, but the motions are not smooth enough. When the depth of transformer block of the motion encoder is 1, the generated motions freeze quickly.

Table 2. Quantitative results of the models that remove the discriminator and use low motion encoder depth.

Method	FID↓	Dist↑	BeatAlign↑
Only Generator	66.09	**10.69**	0.246
Low Motion Encoder Depth	56.41	4.13	0.194
TG-dance(ours)	**34.48**	6.69	**0.247**

4.4 User Study

We compare the results using our method and FACT [27]. The participants in our investigation include 5 professional dancers and 20 non-professional dancers. We

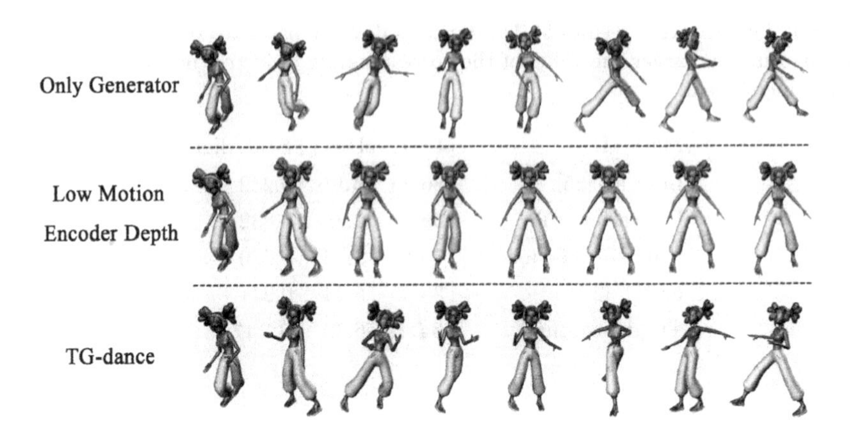

Fig. 4. Visualization results of TG-dance ablation study.

chose the AIST++ dataset for our experiments and provide the subjects with the videos of results generated by FACT and our method. Each set of videos has the same input motions and music. Each subject is required to watch 10 randomly selected sets of videos from the 40 sets and then score two results from each set of videos on three questions: (1) The level of dance quality in the video without considering the music. (2) In the video, the level of dancing to the music. (3) The level of the creativity of the dance in the video. Each question can be scored from one of five scores (very bad-1, bad-2, average-3, good-4, excellent-5).

We count the feedback and the scoring results for both groups from subjects. In the results of non-professional dancers' scores for the three questions, we scored 73.7%, 70.6% and 76.7% of the full scores, while the FACT [27] results obtain 61.9%, 65.3% and 54.8%. Professional dancers score our dances 68.4%, 65.6% and 84.8%, while FACT [27] generated dances scored 67.2%, 64.8% and 61.6%. Figure 5 shows the results, which illustrate the higher quality and diversity of our dances and the higher correlation with music.

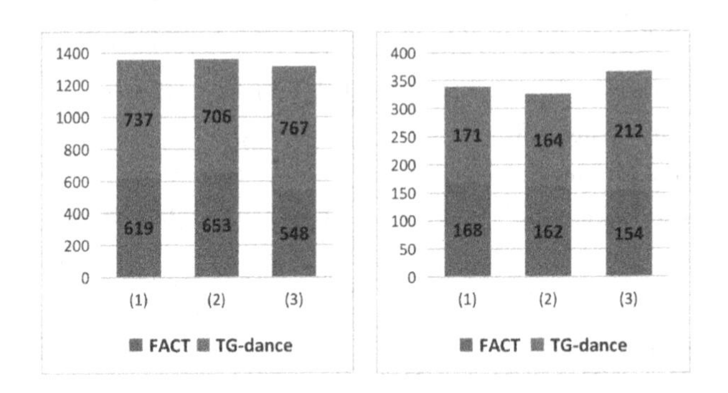

Fig. 5. User study results. Left: Scoring results of 20 non-professional dancers. Right: Scoring results of 5 professional dancers.

5 Conclusion

In this paper, we propose a new transformer-based network TG-dance, which is able to generate long 3D dance sequences associated with music by using fewer motion sequences as inputs. The motion sequences have rich dance diversity, and our model has high learning efficiency. It has the best overall performance in the quantitative comparison results and the highest score in the subjective evaluation. In future work, we have yet to experiment in more music styles, such as classical dance and house dance. We will explore how to better learn the correlation information between dance and music, and other methods to evaluate the correlation between the two other than music rhythm.

References

1. Duan, Y., et al.: Semi-supervised learning for in-game expert-level music-to-dance translation. arXiv preprint arXiv:2009.12763 (2020)
2. Bowden, R.: Learning statistical models of human motion. In: IEEE Workshop on Human Modeling, Analysis and Synthesis (CVPR), vol. 2000 (2000)
3. Pullen, K., Bregler, C.: Animating by multi-level sampling. In: Proceedings Computer Animation 2000, pp. 36–42. IEEE (2000)
4. Chen, K., et al.: ChoreoMaster: choreography-oriented music-driven dance synthesis. ACM Trans. Graph. (TOG) **40**(4), 1–13 (2021)
5. Ye, Z., et al.: ChoreoNet: towards music to dance synthesis with choreographic action unit. In: Proceedings of the 28th ACM International Conference on Multimedia, pp. 744–752 (2020)
6. Guo, X., Zhao, Y., Li, J.: DanceIt: music-inspired dancing video synthesis. IEEE Trans. Image Process. **30**, 5559–5572 (2021)
7. Arikan, O., Forsyth, D.A.: Interactive motion generation from examples. ACM Trans. Graph. (TOG) **21**(3), 483–490 (2002)
8. Kovar, L., Gleicher, M., Pighin, F.: Motion graphs. In: ACM SIGGRAPH 2008 classes, pp. 1–10 (2008)
9. Kim, T.H., Park, S.I., Shin, S.Y.: Rhythmic-motion synthesis based on motion-beat analysis. ACM Trans. Graph. (TOG) **22**(3), 392–401 (2003)
10. Jain, A., Zamir, A. R., Savarese, S., Saxena, A.: Structural-RNN: deep learning on spatio-temporal graphs. In: Proceedings of the IEEE Conference on Computer Vision and Pattern Recognition, pp. 5308–5317 (2016)
11. Ghosh, P., Song, J., Aksan, E., Hilliges, O.: Learning human motion models for long-term predictions. In: 2017 International Conference on 3D Vision (3DV), pp. 458–466. IEEE (2017)
12. Wallace, B., Martin, C. P., Torresen, J., Nymoen, K.: Towards movement generation with audio features. arXiv preprint arXiv:2011.13453 (2020)
13. Ahn, H., Kim, J., Kim, K., Oh, S.: Generative autoregressive networks for 3d dancing move synthesis from music. IEEE Robot. Autom. Lett. **5**(2), 3501–3508 (2020)
14. Ferreira, J.P., et al.: Learning to dance: a graph convolutional adversarial network to generate realistic dance motions from audio. Comput. Graph. **94**, 11–21 (2021)
15. Zhuang, W., Wang, C., Chai, J., Wang, Y., Shao, M., Xia, S.: Music2Dance: DanceNet for music-driven dance generation. ACM Trans. Multimed. Comput. Commun. Appl. (TOMM) **18**(2), 1–21 (2022)

16. Lee, J., Kim, S., Lee, K.: Listen to dance: music-driven choreography generation using autoregressive encoder-decoder network. arXiv preprint arXiv:1811.00818 (2018)
17. Vaswani, A., et al.: Attention is all you need. In: Advances in Neural Information Processing Systems, pp. 5998–6008 (2017)
18. Zheng, C., Zhu, S., Mendieta, M., Yang, T., Chen, C., Ding, Z.: 3D human pose estimation with spatial and temporal transformers. In: Proceedings of the IEEE/CVF International Conference on Computer Vision, pp. 11656–11665 (2021)
19. Huang, C. Z. A., et al.: Music transformer. arXiv preprint arXiv:1809.04281 (2018)
20. Bhattacharya, U., Rewkowski, N., Banerjee, A., Guhan, P., Bera, A., Manocha, D.: Text2Gestures: a transformer-based network for generating emotive body gestures for virtual agents. In: 2021 IEEE Virtual Reality and 3D User Interfaces (VR), pp. 1–10. IEEE (2021)
21. Gan, C., Huang, D., Chen, P., Tenenbaum, J.B., Torralba, A.: Foley music: learning to generate music from videos. In: Vedaldi, A., Bischof, H., Brox, T., Frahm, J.-M. (eds.) ECCV 2020. LNCS, vol. 12356, pp. 758–775. Springer, Cham (2020). https://doi.org/10.1007/978-3-030-58621-8_44
22. Cai, Y., et al.: Learning progressive joint propagation for human motion prediction. In: Vedaldi, A., Bischof, H., Brox, T., Frahm, J.-M. (eds.) ECCV 2020. LNCS, vol. 12352, pp. 226–242. Springer, Cham (2020). https://doi.org/10.1007/978-3-030-58571-6_14
23. Aksan, E., Kaufmann, M., Cao, P., Hilliges, O.: A spatio-temporal transformer for 3D human motion prediction. In: 2021 International Conference on 3D Vision (3DV), pp. 565–574. IEEE (2021)
24. Huang, R., Hu, H., Wu, W., Sawada, K., Zhang, M., Jiang, D.: Dance revolution: long-term dance generation with music via curriculum learning. arXiv preprint arXiv:2006.06119 (2020)
25. Li, J., et al.: Learning to generate diverse dance motions with transformer. arXiv preprint arXiv:2008.08171 (2020)
26. Li, B., Zhao, Y., Zhelun, S., Sheng, L.: DanceFormer: music conditioned 3D dance generation with parametric motion transformer. In: Proceedings of the AAAI Conference on Artificial Intelligence. vol. 36, no. 2, pp. 1272–1279 (2022)
27. Li, R., Yang, S., Ross, D. A., Kanazawa, A.: Ai choreographer: music conditioned 3D dance generation with AIST++. In: Proceedings of the IEEE/CVF International Conference on Computer Vision, pp. 13401–13412 (2021)
28. McFee, B., et al.: librosa: audio and music signal analysis in python. In: Proceedings of the 14th Python in Science Conference. vol. 8, pp. 18–25 (2015)
29. Jiang, Y., Chang, S., Wang, Z.: TransGAN: two transformers can make one strong GAN. arXiv preprint arXiv:2102.07074 1(3) (2021)
30. Gulrajani, I., Ahmed, F., Arjovsky, M., Dumoulin, V., Courville, A.C.: Improved training of wasserstein GANs. In: Advances in Neural Information Processing Systems. vol. 30 (2017)

Visual Question Generation Under Multi-granularity Cross-Modal Interaction

Zi Chai[1,2], Xiaojun Wan[1,2](✉), Soyeon Caren Han[3], and Josiah Poon[3]

[1] Wangxuan Institute of Computer Technology, Peking University, Beijing, China
{chaizi,wanxiaojun}@pku.edu.cn
[2] The MOE Key Laboratory of Computational Linguistics, Peking University, Beijing, China
[3] School of Computer Science, The University of Sydney, Camperdown, Australia
{caren.han,josiah.poon}@sydney.edu.au

Abstract. Visual question generation (VQG) aims to ask human-like questions automatically from input images targeting on given answers. A key issue of VQG is performing effective cross-modal interaction, i.e., dynamically focus on answer-related regions during question. In this paper, we propose a novel framework based on multi-granularity cross-modal interaction for VQG containing both object-level and relation-level interaction. For object-level interaction, we leverage both semantic and visual features under a contrastive learning scenario. We further illustrate the importance of high-level relations (e.g., spatial, semantic) between regions and answers for generating deeper questions. Since such information were somewhat ignored by prior VQG studies, we propose relation-level interaction based on graph neural networks. We perform experiments on VQA2.0 and Visual7w datasets under automatic and human evaluations and our model outperforms all baseline models.

Keywords: Visual question generation · Multi-modal understanding

1 Introduction

Visual question generation (VQG) aims at teaching machines to ask fluent, logical and reasonable questions based on a given image. It is an emerging topic in both natural language processing and computer vision fields with wide applications. As it can be regarded as the dual task of visual question answering (VQA), jointly dealing with the two tasks is a natural idea [17,32]. VQG models can also alleviate the dataset bias problem in VQA by enriching the diversity of questions through data augmentation [33]. In visual dialog systems, asking questions can enhance the interactiveness and persistence of human-machine interaction [36]. VQG models are useful in generating questions for children education [15] and facilitating clinical diagnosis based on radiology images [26].

Existing VQG studies can be categorized into two categories. In *answer-based VQG*, generated questions are targeted on certain input answers. In *answer-free VQG*, answers are not considered as prior conditions. Since the lack of strong

D.-T. Dang-Nguyen et al. (Eds.): MMM 2023, LNCS 13833, pp. 255–266, 2023.
https://doi.org/10.1007/978-3-031-27077-2_20

constraints in answer-free VQG, such models often suffer from generating safe generic questions, e.g., "what is in this picture", which are not useful in practice [14,33]. Besides, in real-world cases like data augmentation and educational material generation, question-answer pairs are required instead of pure questions. To this end, we focus on answer-based VQG, i.e., both answer and image are provided as inputs, and each generated question should be targeted on the given answer based on the input image.

One of the key issues of answer-based VQG is performing effective cross-modal interaction [20]. Since the given answer is often a short phrase or even a single word, it is less informative compared with the input image. To this end, VQG models are required to dynamically focus on important answer-related regions during question generation. Based on this observation, prior works adopted object-level interaction scenarios which dynamically focus on answer-related regions during question generation. [20] encoded the input answer, image into text, visual representations respectively and further applied multi-modal attention. As there is no explicit regional bounding box during the attention process, it is an *implicit* answer-related region selection process. To perform *explicit* region selection, [29,33] performed object detection and generated semantic labels for salient objects. By comparing the semantics between target answer and object labels, they picked up answer-related regions during the decoding process. It turned out that comparing target answers with explicit bounding boxes for answer-related region selection always get better performances. However, only semantic features were used in prior works for answer-related region selection. This may lead to potential bias when comparing object labels with object-unrelated answers (e.g., numbers, relations, attributes, etc.) and zero-shot new objects. To this end, we propose a novel scenario by adopting contrastive learning to distill semantic knowledge from object labels into region representations. Compared with prior works using single semantic distances, we leverage both semantic and visual features for region selection and get better experimental results.

As illustrated by image captioning studies [4,31], high-level relations (e.g., spatial, semantic) between visual objects is important for scene understanding. However, most prior VQG works somewhat ignored such relations, which is essential for generating "deeper" questions, e.g., "what color is the railing *behind* the boy", "what is the man *on the left riding*", etc. To this end, we propose a relation-level interaction scenario that explicitly considers multiple types of relations among regions (objects) and answer. We connect regions and answer by different kinds of relation-edges and use GCN layers to capture interactions between them. Above all, the main contributions of this paper are:

- We propose a novel multi-granularity cross-modal interaction framework for VQG, by considering both object-level interaction and relation-level interaction between regions and target answer.
- We propose a contrastive scenario for object-level interaction to leverage multi-modal features for answer-related region selection.
- We perform experiments on VQA2.0 and Visual7w datasets and verify the effectiveness of our model under automatic and human evaluations.

2 Related Work

VQG has a natural connection with image captioning since questions can be regarded as a special type of captions. Some early studies performed VQG by modifying image captions into questions [24]. Others directly applied image captioning architectures on VQG datasets [21]. Under these scenarios, questions are generated directly from given images without any constraints, i.e., they perform *unconditional VQG*. However, these models suffered from generating safe generic questions that are not useful in practice [14,33]. Besides, different from image captioning focusing on global image understanding, human tends to ask questions based on a certain local region [12,33]. As a result, most recent studies focus on *conditional VQG* generating questions, especially answer-based VQG. Under this scenario, the VQG model should selectively and dynamically focus on different image regions to make generated questions strongly related with prior constraints, which is rather different with image captioning.

As mentioned above, object-level interactions were adopted to dynamically focus on answer-related regions during question generation. For implicit region selection, [20] designed dynamic multi-modal attention and [36] adopted reinforcement learning techniques. [29,33] further illustrated that explicit region selection based on object detection always lead to better results. These models compare semantic distances between target answers and semantic labels of salient object for region selection and further adopted GCN during question generation. However, using single semantic information may lead to potential bias. We use both semantic and visual features under our contrastive learning scenario for answe-related object-level interaction.

3 Model

The framework of our model is shown in Fig. 1. Based on the object detector, we get visual and semantic representations respectively for salient regions. For object-level interaction, the cross-modal fusion unit performs answer-related region selection under the guidance of our contrastive scenario. For relation-level interaction, we connect regions and answers into a graph using multiple types of relation-edges discovered by our relation extractor. We further leverage GCN layers to capture high-level relation information before question decoding.

3.1 Feature Extraction

Before cross-modal interaction, we first perform feature extraction. For the target answer A, we get $\boldsymbol{a}_0 \in \mathbb{R}^{768}$ as its global representation. For the input image I, we select n salient regions and get their visual, semantic representations as $\boldsymbol{r}^v, \boldsymbol{r}^s \in \mathbb{R}^{n \times 2048}$ respectively. These features are fixed during the training process.

Answer Representations. For answer A composed by m words $\{w_i\}_{i=1}^m$, we add a special token *CLS* denoted by w_0 at its beginning. Using a pre-trained BERT encoder [7] on $\{w_i\}_{i=0}^m$, we get a sequence of 768-dimensional vectors $\boldsymbol{a} = \{\boldsymbol{a}_i\}_{i=0}^m \in \mathbb{R}^{(m+1) \times 768}$ and regard $\boldsymbol{a}_0$ as the global answer representation.

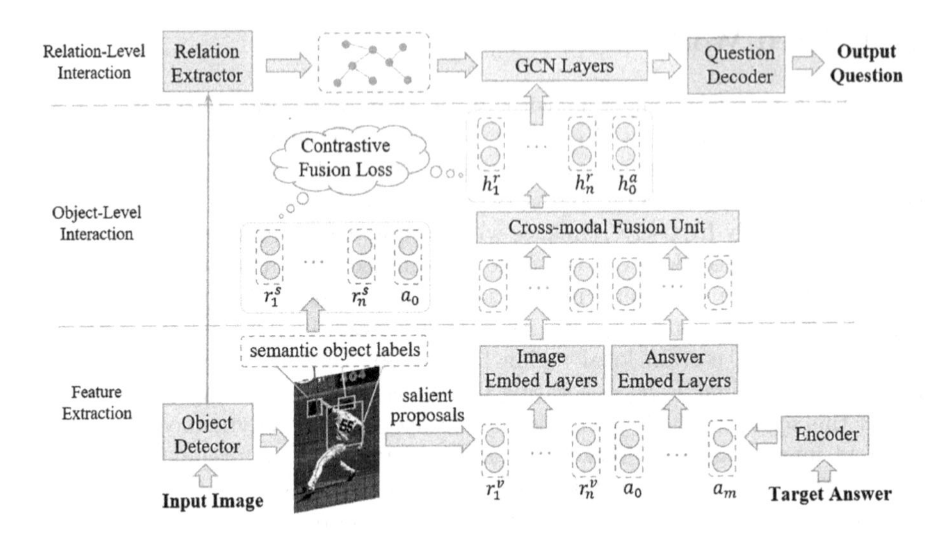

Fig. 1. Architecture of our multi-granularity cross-modal interaction. Red components denote image-domain, blue components denote text-domain and green components show cross-modal interaction. (Color figure online)

Region Representations. For image I, we utilize a pre-trained object detector to discover silent candidate object regions. Based on the observation that only salient objects are used to ask questions in most cases [12,29], we consider the top-n proposals with the highest object classification scores. According to the left border of their bounding boxes, we place these regions from left to right into a sequence. Their corresponding proposals are a sequence containing n 2048-dimensional vectors $\boldsymbol{r}^v = \{\boldsymbol{r}_i^v\}_{i=1}^n \in \mathbb{R}^{n\times 2048}$. Based on the sequence of proposals, we generate semantic labels $o = \{o_i\}_{i=1}^n$ for all regions. For fair comparison with [29,33], we use the same method based on Faster R-CNN [25] for object detection and label generation, where each label is a short region (object) caption containing nouns (e.g., object category) and attributes (e.g., color or size). Similar with getting global answer representations, we use the above BERT encoder on each semantic label to get $\boldsymbol{r}^s = \{\boldsymbol{r}_i^s\}_{i=1}^n \in \mathbb{R}^{n\times 768}$. For n salient regions, $\boldsymbol{r}^v, \boldsymbol{r}^s$ provide visual and semantic representations respectively.

Embedding Layers. Since visual and text representations are in different dimensions, we project them into a common hidden space for further cross-modal interaction. For visual representations $\boldsymbol{r}^v \in \mathbb{R}^{n\times 2048}$ of the n regions, we use a linear projection layer with *ReLU* activation [1] to map them into $\boldsymbol{x}^r \in \mathbb{R}^{n\times 1024}$. Leveraging the same structure with different parameters, answer representations $\boldsymbol{a} \in \mathbb{R}^{(m+1)\times 768}$ are mapped into the same hidden space $\boldsymbol{x}^a = \{\boldsymbol{x}_i^a\}_{i=0}^m \in \mathbb{R}^{(m+1)\times 1024}$, where $\boldsymbol{x}_0^a$ is the global representation.

3.2 Object-Level Cross-Modal Interaction

We perform object-level interaction using a cross-modal fusion unit under the guidance of a contrastive fusion loss. Our cross-modal fusion unit is based on the idea of multi-head self-attention. In the traditional attention mechanism, we have (queries, keys, values) represented by (Q, K, V). For simplicity, we set the dimension of each (query, key, value) as d, and the calculation is performed as:

$$Attn(Q, K, V) = softmax(\frac{QK^T}{\sqrt{d}})V \tag{1}$$

By setting $Q = K = V$, Eq. 1 becomes self-attention. [27] further proposed multi-head self-attention leveraging h self-attention heads to focus on different aspects. Since the attention process ignores position information which is important for capturing context dependency between words and relative position between regions, they also add positional encoding on the input embeddings. We denote the whole process as $Y = SA(X)$ where X is a sequence of vectors containing positional information and Y is the output sequence.

Cross-Modal Fusion Unit. As shown on the right of Fig. 1, our cross-modal fusion unit captures the relations between region-representations $\boldsymbol{x}^r \in \mathbb{R}^{n\times 1024}$ and answer-representations $\boldsymbol{x}^a \in \mathbb{R}^{(m+1)\times 1024}$. It is similar with the encoder of Transformer, except that the self-attention is performed hierarchically:

$$[\boldsymbol{h}^r; \boldsymbol{h}^a] = FF(SA[SA(\boldsymbol{x}^r); SA(\boldsymbol{x}^a)]) \tag{2}$$

where FF is a single-layer feed forward network, SA denotes multi-head self attention with positional encoding, $\boldsymbol{h}^r \in \mathbb{R}^{n\times 1024}$, $\boldsymbol{h}^a \in \mathbb{R}^{(m+1)\times 1024}$ correspond with $\boldsymbol{x}^r, \boldsymbol{x}^a$ respectively, and we ignore the residual connection with layer normalization layers denoted by "Add & Norm" blocks in Fig. 1 for simplicity. These blocks are useful to stabilize the training process. In other words, we first perform self-attention on vision domain and text domain respectively and then concatenate their results for cross-modal self-attention.

Contrastive Fusion Loss. After the cross-modal fusion unit, region representations are $\boldsymbol{h}^r \in \mathbb{R}^{n\times 1024}$ and the answer embedding is $\boldsymbol{h}_0^a \in \mathbb{R}^{1024}$. Recall that in Sect. 3.1, region (object) labels and the target answer are mapped into a semantic hidden space denoted by $\boldsymbol{r}^s \in \mathbb{R}^{n\times 768}$, $\boldsymbol{a}_0 \in \mathbb{R}^{768}$ receptively using a pre-trained BERT encoder. Note that $[\boldsymbol{r}^s; \boldsymbol{a}_0]$ is not adopted by our cross-modal fusion unit, and thus it provides a different semantic view of regions and answer compared with $[\boldsymbol{h}^r; \boldsymbol{h}_0^a]$ in visual view. As mentioned above, prior works [29,33] only used semantic view for explicit answer-related region selection, and we find it is better to consider both semantic and visual features.

To leverage multi-modal features for answer-realted region selection, we propose a bootstrapping contrastive scenario regarding $[\boldsymbol{r}^s; \boldsymbol{a}_0]$, $[\boldsymbol{h}^r; \boldsymbol{h}_0^a]$ as a positive pair. By contrasting the two views of representations, we expect the outputs of our cross-model fusion unit contain more semantic information. More specifically, traditional contrastive learning methods sample and compare positive and negative instances [5,11] during the training process. With BYOL [10] pioneered in performing contrastive learning without negative samplings, using bootstrapping contrastive learning has becoming a new trend [3,6]. Following their ideas, we first project $[\boldsymbol{h}^r; \boldsymbol{h}^a_0]$ into a dense vector $\boldsymbol{h}^c \in \mathbb{R}^n$ by:

$$\boldsymbol{h}^c = softmax(\tilde{\boldsymbol{h}}^c) \;\; \text{where } \tilde{\boldsymbol{h}}^c_i = f_h(\boldsymbol{h}^r_i, \boldsymbol{h}^a_0) \; (i = 1, ..., n) \tag{3}$$

where f_h is a similarity function (we directly set f_h as dot-product). Similarly, we project $[\boldsymbol{r}^s; \boldsymbol{a}_0]$ into another dense vector $\boldsymbol{r}^c \in \mathbb{R}^n$. Based on the two views of projections $\boldsymbol{h}^c, \boldsymbol{r}^c \in \mathbb{R}^n$, we set a contrastive loss based on their KL-distance:

$$L_c = -\sum_{i=1}^{n} \boldsymbol{r}^c_i \log \boldsymbol{h}^c_i \tag{4}$$

Based on this bootstrapping contrastive scenario, we expect our model to generate better region and answer representations for region selection.

We can also regard the contrastive loss as distilling semantic information from $\boldsymbol{r}^c$ into cross-modal region representations $\boldsymbol{h}^c$. Under this interpretation, the BERT model in Sect. 3.1 can be regarded as a teacher and $\boldsymbol{r}^c$ are soft labels generated for our model for learning better representations.

3.3 Relation-Level Cross-Modal Interaction

After object-level cross-modal interaction, salient regions and the target answer are represented by $\boldsymbol{h}^r \in \mathbb{R}^{n \times 1024}$, $\boldsymbol{h}^a_0 \in \mathbb{R}^{1024}$ respectively. Inspired by image captioning studies leveraging high-level relation information, we connect regions and answer into a graph using multiple kinds of relations and leverage GCN layers to capture high-level relations.

Graph Construction. To connect the n salient regions, we first perform relation extraction. For fair comparison with prior works based on image captioning, we use the same MOTIFS relationship detector [35] after the object detector. We extract multiple kinds of geometric (e.g., above, behind), possessive (e.g., has, part of), semantic (e.g., carrying, eating) and misc (e.g., from, made of) relationships. Since the long-tail distribution for different kinds of relations, we only consider the top-k most frequent ones. Corresponding with the k kinds of relations, we adopt k types of edges to connect n regions into a graph. We further regard the target answer as a special node connecting with all region nodes by a special type of edges. In this way, we build a graph $\mathcal{G}$ containing n region-nodes, one answer-node and $k+1$ types of edges. We initialize n region-nodes and the answer-node based on $\boldsymbol{h}^r \in \mathbb{R}^{n \times 1024}$, $\boldsymbol{h}^a_0 \in \mathbb{R}^{1024}$ respectively. In Sect. 3.1, we order different regions into a sequence according to their bounding boxes from left to right. Based on this sequence, we assign each region-node a position label from 1 to n, and we assign label 0 to the target answer. Similar with [8], we further add position embeddings into $\boldsymbol{h}^r, \boldsymbol{h}^a_0$ and the results are final initial node representations denoted by $\boldsymbol{v}^{(0)} = \{\boldsymbol{v}^{(0)}_i\}_{i=0}^{n} \in \mathbb{R}^{(n+1) \times 1024}$.

Graph Convolutional Network. Based on $\mathcal{G}$ containing $n+1$ nodes and $k+1$ types of edges, we leverage GCN layers to update node representations from $\boldsymbol{v}^{(0)}$ recurrently. At time-step t, each node first computes a hidden state $\boldsymbol{s}_i^{(t)} \in \mathbb{R}^{1024}$ by gathering information from neighbors according to the types of edges:

$$\begin{aligned} \boldsymbol{s}_i^{(t)} &= \sum_{v_j \in \mathcal{N}(v_i)} \boldsymbol{W}_l \ \boldsymbol{v}_j^{(t-1)} + \boldsymbol{b}_l \\ \text{where } l &= \text{type-ID}(e_{i,j}) \in \{0, 1, ..., k\} \end{aligned} \tag{5}$$

$\mathcal{N}(v_i)$ contains all neighbors of v_i and $e_{i,j}$ is the edge between v_i, v_j. Since the graph $\mathcal{G}$ contains $k+1$ kinds of edges, we map each kind of edges into a unique type-ID from 0 to k. We further use a function type-ID$(e_{i,j}) \in \{0, 1, .., k\}$ to denote the exact type-ID of $e_{i,j}$. In other words, there are $k+1$ candidates $\{\boldsymbol{W}_l, \boldsymbol{b}_l\}_{l=0}^{k}$ and we chose one of them as $\boldsymbol{W}_l, \boldsymbol{b}_l$ in Eq. 5 according to the exact edge-type of $e_{i,j}$. Based on $\boldsymbol{s}_i^{(t)}$, each node further updates its representation under a gated mechanism:

$$\begin{aligned} \boldsymbol{v}_i^{(t)} &= \boldsymbol{z}_i^{(t)} \odot \boldsymbol{v}_i^{(t-1)} + (\boldsymbol{1} - \boldsymbol{z}_i^{(t)}) \odot \tilde{\boldsymbol{v}_i}^{(t)} \\ \tilde{\boldsymbol{v}_i}^{(t)} &= tanh(\boldsymbol{W}_v[\boldsymbol{s}_i^{(t)}; \boldsymbol{y}_i^{(t)} \odot \boldsymbol{v}_i^{(t-1)}]) \\ \boldsymbol{y}_i^{(t)} &= \sigma(\boldsymbol{W}_y[\boldsymbol{s}_i^{(t)}; \boldsymbol{v}_i^{(t-1)}]) \\ \boldsymbol{z}_i^{(t)} &= \sigma(\boldsymbol{W}_z[\boldsymbol{s}_i^{(t)}; \boldsymbol{v}_i^{(t-1)}]) \end{aligned} \tag{6}$$

where $\boldsymbol{y}, \boldsymbol{z}$ are update and reset gates, $\boldsymbol{W}_v, \boldsymbol{W}_y, \boldsymbol{W}_z$ are trainable parameters, σ are sigmoid function and $\odot$ is element-wise multiplication. We iteratively repeat the process for T times to get a sequence $[\boldsymbol{v}_i^{(0)}, ..., \boldsymbol{v}_i^{(T)}]$ and use the average value of all these vectors as the final output $\boldsymbol{u}_i$ $(i = 0, ..., n) \in \mathbb{R}^{1024}$ for each node.

3.4 Decoder

Our question decoder is a standard Transformer-decoder containing a masked multi-head self-attention layer and a multi-head attention layer based on final region and answer representations $\boldsymbol{u} = \{\boldsymbol{u}_i\}_{i=0}^{n} \in \mathbb{R}^{(n+1)\times 1024}$. The output layers recurrently predict each word of generated question, and our final loss function is their combination, i.e., $L = \lambda L_{mle} + (1 - \lambda) L_c$.

4 Experiment

4.1 Dataset

We perform experiments on VQA2.0 and Visual7w datasets which have been widely used in recent works. The VQA2.0 dataset [9] contains 82783, 40504, 81434 images for training, validation and testing collected from MS COCO [19]. Three question-answer pairs are collected for each image. Since the answers in test dataset are not available, we adopt the commonly used off-line data split in previous works [12,14,29,33] containing 5k images for validation and testing, respectively. The Visual7w dataset [37] contains 47300 images. Following [33], we use its telling QA subset where each image is associated with five questions on average. There are 69817, 28020 and 42031 questions for training, validation and testing respectively.

4.2 Baseline Models and Evaluation Metrics

We compare our model with three image captioning models, two joint learning models and four answer-based VQG models. For the three image captioning models, *LSTM* [21] is a neural seq2seq model containing a CNN encoder and an LSTM decoder for image captioning. *LSTM-AN* [21] is an improved version of *LSTM*, which feeds the concatenated features of images and target answers at each time step during text generation. *SAT* [30] is a typical attention-based captioning method. We set the initial state of its decoder LSTM as the joint embedding of input images and answers. For the two joint learning models, *iQAN* [17] adopts the MUTAN architecture [2] for joint learning, and *DL-VQG* [32] leverages reinforcement learning. For the four answer-based VQG models, *IVQA* [20] leverages dynamic cross-modal attention for answer-related region selection. *IM-VQG* [14] leverages the VAE framework for generating diverse questions by maximizing the mutual information between questions, imput images and answer-categories. In our experiment, we consider each answer as a unique category in the dataset to avoid extra humman annotation. *Radial-GCN* [33] performs explicit object-level cross-modal interaction by selecting a core answer area and building an answer-related GCN graph structure. *MOAG* [29] adopts co-attention to focus on multiple answer-related features and uses the attention weights to perform soft region selection on the GCN interaction layers. Since their code has not been released to public yet, we report the performances on Visual7w under our reproduction model.

We adopt automatic and human evaluation metrics. Following the conventions, we adopt the BLEU_4 [22], ROUGE_L [18], METEOR [16] and CIDEr [28] as automatic evaluation metrics. We perform human evaluation in three aspects: Fluency (F) measures if a question is grammatically correct and is fluent to read. It takes values from 0, 1, 2 where higher values reflect better quality; Image-relevance (R-I), answer-relevance (R-A) are binary 0/1 values measuring if the generated question is related to the input image, the given answer respectively. Since performing human evaluation is rather expensive, we only compare our model with strong baselines RadialGCN and MOAG on 150 random sampled image-question pairs from Visual7w.

To make the training process more stable, we initialize the parameters of the question decoder by performing auto-encoding on all questions from training set. Besides, we borrow ideas from curriculum learning, i.e., we first train our model on simple questions and then we feed more difficult questions. We classify different questions using a pre-trained VQA model [13]. A question is simpler if it is correctly answered with higher confidence, while it is more difficult if it is wrong answered with higher confidence.

Table 1. Automatic evaluation results. B_4, R_L stands for $BLEU_4$ and $ROUGE_L$. We underline the performance of strongest baselines while set the best result of each metric in bold. Numbers in $*$ are our reproduction results.

Method	VQA2.0				Visual7w			
	B_4	R_L	METEOR	CIDEr	B_4	R_L	METEOR	CIDEr
LSTM [21]	15.2	47.1	19.8	1.32	20.2	46.8	19.2	1.13
LSTM-AN [21]	22.8	52.6	24.3	1.62	21.9	50.1	22.9	1.34
SAT [30]	23.1	53.4	24.4	1.65	22.3	50.3	23.4	1.34
iQAN [17]	27.1	56.8	26.8	2.09	23.1	52.0	25.1	1.44
DL-VQG [32]	24.4	55.9	26.1	1.88	22.9	51.2	24.3	1.38
IVQA [20]	23.9	55.3	25.7	1.84	22.7	50.8	23.7	1.36
IM-VQG [14]	24.8	56.3	26.3	1.94	23.0	51.7	24.6	1.41
Radial-GCN [33]	27.9	57.2	27.1	2.10	<u>23.6</u>	<u>52.7</u>	25.9	<u>1.52</u>
MOAG [29]	<u>28.1</u>	<u>60.4</u>	<u>27.8</u>	<u>2.39</u>	23.2*	52.3*	<u>26.1</u>*	1.48*
Ours	**28.7**	**61.2**	**28.5**	**2.51**	**24.2**	**53.6**	**26.5**	**1.63**

4.3 Results

The automatic evaluation results are illustrated in Table 1. Our model reaches the best performance on both datasets, illustrating the efficacy of using multi-granularity cross-modal interaction on both object-level and relation-level. The models regarding VQG as image captioning get rather poor performance compared with VQG baselines, indicating that answer-based VQG is rather different with image captioning. When we compare VQG baselines, the strong baselines Radial-GCN and MOAG outperform others since they perform explicit answer-related region selection. Compared with Radial-GCN focusing on a core answer region, MOAG gets better results by focusing on multiple objects. Besides, models on VQA2.0 always reach better results. Compared with Visual7w, it contains more training samples with longer but informative questions.

Table 2. Human evaluation results.

Method	F	R-Img	R-Ans
RadialGCN [33]	1.62	0.79	<u>0.72</u>
MOAG [29]	<u>1.68</u>	<u>0.81</u>	0.71
Ours	**1.73**	**0.82**	**0.75**

In Table 2, we compare our model with strong baselines under human evaluation and it gets best performances under each metric. This shows that our model generates most fluent and input-related questions. When it comes to fluency and R-Img, all models get competitive results. However, performances under R-Ans are significantly lower, indicating that making the generated questions answer-related is more difficult than image-related. We think this is because target answers often contain less information than input images.

5 Conclusion and Future Work

In this paper, we focus on answer-based VQG task aiming to ask human-like questions from given image targeting on certain answers. To perform cross-modal interaction which is one of the key issues of VQG, we propose the multi-granularity cross-modal interaction framework. We first perform object-level interaction to capture the relations between target answers and salient image regions. Different from prior works using semantic distances for explicit region selection, we leverage both semantic and visual features. We further propose relation-level interaction which captures high-level relations between regions and target answer, which is helpful for generating deeper questions.

In the future, a promising idea is leveraging the Transformer architecture to introduce more granularity of cross-modal interaction. Especially performing large-scale cross-modal pre-training based on ViT [23] as well as using modality-shared ViT encoders [34] on insufficient VQG datasets. Besides, more powerful object-level and relation-level interaction scenario is also worth exploring. Finally, it is interesting to perform question generation across more modalities, e.g. from audio and video.

Acknowledgements. This work was supported by National Key R&D Program of China (2021YFF090 1502), National Science Foundation of China (No. 62161160339), State Key Laboratory of Media Convergence Production Technology and Systems and Key Laboratory of Science, Technology and Standard in Press Industry (Key Laboratory of Intelligent Press Media Technology). We appreciate the anonymous reviewers for their helpful comments. Xiaojun Wan is the corresponding author.

References

1. Agarap, A.F.: Deep learning using rectified linear units (ReLU). arXiv preprint arXiv:1803.08375 (2018)
2. Ben-Younes, H., Cadene, R., Cord, M., Thome, N.: MUTAN: multimodal tucker fusion for visual question answering. In: Proceedings of the IEEE International Conference on Computer Vision, pp. 2612–2620 (2017)
3. Caron, M., et al.: Emerging properties in self-supervised vision transformers. In: Proceedings of the IEEE/CVF International Conference on Computer Vision, pp. 9650–9660 (2021)
4. Chen, S., Jin, Q., Wang, P., Wu, Q.: Say as you wish: fine-grained control of image caption generation with abstract scene graphs. In: Proceedings of the IEEE/CVF Conference on Computer Vision and Pattern Recognition, pp. 9962–9971 (2020)
5. Chen, T., Kornblith, S., Norouzi, M., Hinton, G.: A simple framework for contrastive learning of visual representations. In: International Conference on Machine Learning, pp. 1597–1607. PMLR (2020)

6. Chen, X., He, K.: Exploring simple siamese representation learning. In: Proceedings of the IEEE/CVF Conference on Computer Vision and Pattern Recognition, pp. 15750–15758 (2021)
7. Devlin, J., Chang, M.W., Lee, K., Toutanova, K.: BERT: pre-training of deep bidirectional transformers for language understanding. arXiv preprint arXiv:1810.04805 (2018)
8. Dosovitskiy, A., et al.: An image is worth 16×16 words: transformers for image recognition at scale. arXiv preprint arXiv:2010.11929 (2020)
9. Goyal, Y., Khot, T., Summers-Stay, D., Batra, D., Parikh, D.: Making the V in VQA matter: elevating the role of image understanding in visual question answering. In: Proceedings of the IEEE Conference on Computer Vision and Pattern Recognition, pp. 6904–6913 (2017)
10. Grill, J.B., et al.: Bootstrap your own latent-a new approach to self-supervised learning. Adv. Neural Inf. Process. Syst. **33**, 21271–21284 (2020)
11. He, K., Fan, H., Wu, Y., Xie, S., Girshick, R.: Momentum contrast for unsupervised visual representation learning. In: Proceedings of the IEEE/CVF Conference on Computer Vision and Pattern Recognition, pp. 9729–9738 (2020)
12. Huang, Q., et al.: Aligned dual channel graph convolutional network for visual question answering. In: Proceedings of the 58th Annual Meeting of the Association for Computational Linguistics, pp. 7166–7176 (2020)
13. Kim, J.H., Jun, J., Zhang, B.T.: Bilinear attention networks. In: Advances in Neural Information Processing Systems. vol. 31 (2018)
14. Krishna, R., Bernstein, M., Fei-Fei, L.: Information maximizing visual question generation. In: Proceedings of the IEEE/CVF Conference on Computer Vision and Pattern Recognition, pp. 2008–2018 (2019)
15. Kunichika, H., Katayama, T., Hirashima, T., Takeuchi, A.: Automated question generation methods for intelligent english learning systems and its evaluation. In: Proceedings of ICCE (2004)
16. Lavie, A., Agarwal, A.: METEOR: An automatic metric for MT evaluation with high levels of correlation with human judgments. In: Proceedings of the Second Workshop on Statistical Machine Translation, pp. 228–231. Association for Computational Linguistics (2007)
17. Li, Y., et al.: Visual question generation as dual task of visual question answering. In: Proceedings of the IEEE Conference on Computer Vision and Pattern Recognition, pp. 6116–6124 (2018)
18. Lin, C.Y.: Rouge: A package for automatic evaluation of summaries. Text Summarization Branches Out (2004)
19. Lin, T.-Y., et al.: Microsoft COCO: common objects in context. In: Fleet, D., Pajdla, T., Schiele, B., Tuytelaars, T. (eds.) ECCV 2014. LNCS, vol. 8693, pp. 740–755. Springer, Cham (2014). https://doi.org/10.1007/978-3-319-10602-1_48
20. Liu, F., Xiang, T., Hospedales, T.M., Yang, W., Sun, C.: iVQA: inverse visual question answering. In: Proceedings of the IEEE Conference on Computer Vision and Pattern Recognition, pp. 8611–8619 (2018)
21. Mostafazadeh, N., Misra, I., Devlin, J., Mitchell, M., He, X., Vanderwende, L.: Generating natural questions about an image. arXiv preprint arXiv:1506.00278 (2016)
22. Papineni, K., Roukos, S., Ward, T., Zhu, W.J.: Bleu: a method for automatic evaluation of machine translation. In: Proceedings of the 40th Annual Meeting on Association for Computational Linguistics, pp. 311–318. Association for Computational Linguistics (2002)

23. Radford, A., et al.: Learning transferable visual models from natural language supervision. In: International Conference on Machine Learning, pp. 8748–8763. PMLR (2021)
24. Ren, M., Kiros, R., Zemel, R.: Exploring models and data for image question answering. In: Advances in Neural Information Processing Systems. vol. 28 (2015)
25. Ren, S., He, K., Girshick, R., Sun, J.: Faster R-CNN: towards real-time object detection with region proposal networks. In: Advances in Neural Information Processing Systems. vol. 28 (2015)
26. Sarrouti, M., Abacha, A.B., Demner-Fushman, D.: Visual question generation from radiology images. In: Proceedings of the First Workshop on Advances in Language and Vision Research, pp. 12–18 (2020)
27. Vaswani, A., et al.: Attention is all you need. In: Advances in Neural Information Processing Systems. vol. 30 (2017)
28. Vedantam, R., Lawrence Zitnick, C., Parikh, D.: CIDEr: consensus-based image description evaluation. In: Proceedings of the IEEE Conference on Computer Vision and Pattern Recognition, pp. 4566–4575 (2015)
29. Xie, J., Cai, Y., Huang, Q., Wang, T.: Multiple objects-aware visual question generation. In: Proceedings of the 29th ACM International Conference on Multimedia, pp. 4546–4554 (2021)
30. Xu, K., et al.: Show, attend and tell: neural image caption generation with visual attention. In: International Conference on Machine Learning, pp. 2048–2057. PMLR (2015)
31. Xu, N., Liu, A.A., Liu, J., Nie, W., Su, Y.: Scene graph captioner: Image captioning based on structural visual representation. J. Vis. Commun. Image Represent. **58**, 477–485 (2019)
32. Xu, X., Song, J., Lu, H., He, L., Yang, Y., Shen, F.: Dual learning for visual question generation. In: 2018 IEEE International Conference on Multimedia and Expo (ICME), pp. 1–6. IEEE (2018)
33. Xu, X., Wang, T., Yang, Y., Hanjalic, A., Shen, H.T.: Radial graph convolutional network for visual question generation. IEEE Trans. Neural Netw. Learn. Syst. **32**(4), 1654–1667 (2020)
34. You, H., et al.: Ma-clip: towards modality-agnostic contrastive language-image pre-training (2021)
35. Zellers, R., Yatskar, M., Thomson, S., Choi, Y.: Neural Motifs: scene graph parsing with global context. In: Proceedings of the IEEE Conference on Computer Vision and Pattern Recognition, pp. 5831–5840 (2018)
36. Zhang, J., Wu, Q., Shen, C., Zhang, J., Lu, J., Van Den Hengel, A.: Goal-oriented visual question generation via intermediate rewards. In: Proceedings of the European Conference on Computer Vision (ECCV), pp. 186–201 (2018)
37. Zhu, Y., Groth, O., Bernstein, M., Fei-Fei, L.: Visual7W: grounded question answering in images. In: Proceedings of the IEEE Conference on Computer Vision and Pattern Recognition, pp. 4995–5004 (2016)

Multimodal and Multidimensional Imaging Application

Optimizing Local Feature Representations of 3D Point Clouds with Anisotropic Edge Modeling

Haoyi Xiu[1,2], Xin Liu[1(✉)], Weimin Wang[3(✉)], Kyoung-Sook Kim[1], Takayuki Shinohara[4], Qiong Chang[2], and Masashi Matsuoka[2]

[1] Artificial Intelligence Research Center, Tokyo, Japan
{hiroki-shuu,xin.liu,ks.kim}@aist.go.jp
[2] Tokyo Institute of Technology, Tokyo, Japan
q.chang@c.titech.ac.jp, {xiu.h.aa,matsuoka.m.ab}@m.titech.ac.jp
[3] Dalian University of Technology, Dalian, China
wangweimin@dlut.edu.cn
[4] Pasco Corporation, Tokyo, Japan
taarkh6651@pasco.co.jp

Abstract. An edge between two points describes rich information about the underlying surface. However, recent works merely use edge information as an ad hoc feature, which may undermine its effectiveness. In this study, we propose the Anisotropic Edge Modeling (AEM) block by which edges are modeled adaptively. As a result, the local feature representation is optimized where edges (e.g., object boundaries defined by ground truth) are appropriately enhanced. By stacking AEM blocks, AEM-Nets are constructed to tackle various point cloud understanding tasks. Extensive experiments demonstrate that AEM-Nets compare favorably to recent strong networks. In particular, AEM-Nets achieve state-of-the-art performance in object classification on ScanObjectNN, object segmentation on ShapeNet Part, and scene segmentation on S3DIS. Moreover, it is verified that AEM-Net outperforms the strong transformer-based method with significantly fewer parameters and FLOPs, achieving efficient learning. Qualitatively, the intuitive visualization of learned features successfully validates the effect of the AEM block.

Keywords: Edge modeling · Anisotropic diffusion · 3D point clouds · Semantic segmentation

1 Introduction

A 3D point cloud is a basic shape representation in which the scanned surface is represented as a set of points in the 3D space. With the advent of

---

Supplementary Information The online version contains supplementary material available at https://doi.org/10.1007/978-3-031-27077-2_21.

D.-T. Dang-Nguyen et al. (Eds.): MMM 2023, LNCS 13833, pp. 269–281, 2023.
https://doi.org/10.1007/978-3-031-27077-2_21

cost-effective sensors, an increasing number of large-scale point cloud datasets have been released to researchers, facilitating deep learning–based point cloud understanding. Typical applications include autonomous driving and augmented reality.

A point cloud is naturally unstructured, hindering the application of convolutional neural networks (CNNs) for regular grid data. Therefore, many prior studies have focused on applying CNNs by projecting point clouds to regular grids [8,19]. However, there is information loss caused by the projection. To remedy this issue, PointNet applies shared multi-layer perceptrons (MLPs) and symmetric functions to raw 3D points, consuming the point cloud in a lossless manner [15]. PointNet++ subsequently applies PointNets to local subsets of points, following the CNN-like design [16]. Recently, various convolution methods for point clouds have been developed [11,22,27]. Despite these efforts, learning point clouds remains challenging due to the difficulty in inferring the underlying continuous surface from discrete point samples.

Recently, it is found that exploring the edge information benefits the modeling of the local surface. For instance, researchers use edges as auxiliary information [25,26,34] that describes local structures. In these works, edges are simply treated as an additional feature or converted to spatial weights. Consequently, edge information successfully improves performance; however, we believe that using edges without careful treatment may undermine their usefulness since edges contain rich structural information. Moreover, some studies explicitly supervise the model with edge information [5,7,20]. However, these methods require additional per-point and clean labels, which are costly to create in practice.

In this study, instead of treating edges as merely an auxiliary feature describing local structures, we propose the Anisotropic Edge Modeling (AEM) block dedicated to the modeling of edges without additional supervision. The AEM block is inspired by the classic anisotropic diffusion [14] which performs edge-preserving filtering. Specifically, an AEM block learns to optimize local feature representations through the adaptive sharpening of edges (e.g., object boundaries). A visualization of the effect of AEM is shown in Fig. 1. This adaptability makes the AEM block applicable to different stages of modern hierarchical learning in which relationships between different concepts are modeled. Using the AEM block as the basic building block, a deep neural network (DNN) termed AEM-Nets are constructed and applied to various point cloud recognition tasks. Extensive experiments show that AEM-Nets achieve superior performance in comparison with recent strong networks.

The key contributions are highlighted as follows:

- We propose the AEM block that optimizes the local feature representation by the adaptive sharpening of edges.
- A DNN termed AEM-Nets are constructed by stacking AEM blocks to tackle various point cloud recognition tasks.
- Extensive experiments verify the effectiveness and design choices of the AEM-Nets. In particular, AEM-Nets achieve state-of-the-art performance in object classification on ScanObjectNN, object segmentation on ShapeNet Part, and scene segmentation on S3DIS.

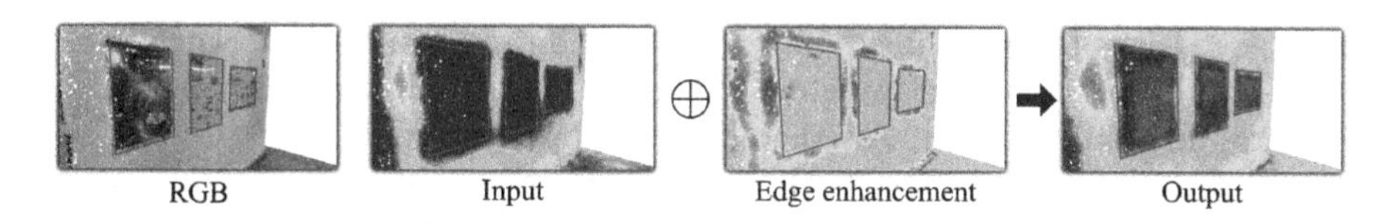

Fig. 1. The effect of Anisotropic edge modeling (AEM). AEM produces more discriminative features through edge enhancement.

2 Related Work

2.1 Deep Learning for 3D Point Clouds

Projection-Based Methods. Projection-based methods project point clouds to 2D [8,19] and 3D [3,35] regular grids to make regular convolution applicable to point clouds. However, they lose fine-grained details through projections.

MLP-Based Methods. Pioneered by PointNet [15], MLP-based methods prevent information loss by operating directly on the raw points. Subsequently, PointNet++ [16] and others [10,33] have made major steps toward the convolution-like operation by applying shared MLPs to local subsets of points.

Convolution-Based Methods. Various point convolutions have been realized by defining convolution operations on unstructured point clouds. Some studies construct pseudo grids on which the points are projected, enabling the regular convolution [13,22]; the others dynamically generate convolution filters based on positional features [11,27]. Though effective, convolution-based methods do not explicitly model edges that provide more accurate representations of the surface.

Transformer-Based Methods. Motivated by the success of Transformers [24], increasing efforts are put into its application on point cloud processing. PAT [30] applies original self-attention (SA) directly. Point Transformer [34] applies local vector attention to better model the local structure. Point-BERT [32] enhances the performance by masked point modeling. Stratified Transformer [9] achieves efficient modeling of long-range dependencies with the stratified strategy.

2.2 Edge-Aware Methods

Edge-aware methods integrate edge information explicitly into the network design. Pioneered by EdgeConv [26], some methods (e.g., [28,29]) use edges as auxiliary features to model the local geometric structure. Although performance improves, edge features are exploited by concatenation and MLPs without careful treatments, which may undermine their usefulness. On the other hand, some methods regard edge information as a similarity measure. Such similarity is often used for describing connectivity among neighboring points [25]. In other cases, edges are often combined with the attention mechanism [34], which converts edges into normalized weights. However, the forced conversion may lose rich structural information. Another type of approach guides the network by edge-related supervision. Specifically, networks are taught to maintain spatial consistency [7], perform edge detection [5] or be aware of boundaries [20]. Such methods involve dedicated loss functions or models, and often require clean and per-point ground truth, thereby making their practical applications challenging.

In contrast, the proposed AEM block makes full use of structural information contained in edges by concentrating on the modeling of edges. Moreover, AEM models edges without additional supervision.

3 Method

In this section, we design the AEM block and explain the rationale of its design. We first present an in-depth analysis of classic anisotropic diffusion on which our method is based. Then, we present the definition of AEM and the AEM block. Lastly, we describe network architectures constructed by stacking AEM blocks.

3.1 Revisiting Anisotropic Diffusion

Classic anisotropic diffusion [14] is a method that performs edge-preserving smoothing. Let $\{\mathbf{u}_i\}_{i=1}^N \in \mathbb{R}^{N\times d}$ denote the features of a point cloud from an intermediate layer. N and d denote the number of points in a point cloud and the dimension of each point feature, respectively. Further, let $\mathcal{N}(i)$ denote the spatial neighborhood of point i. Anisotropic diffusion is defined as:

$$\mathbf{u}_i' = \mathbf{u}_i + \frac{\lambda}{|\mathcal{N}(i)|}\sum_{j\in\mathcal{N}(i)} g(||\mathbf{u}_j - \mathbf{u}_i||)(\mathbf{u}_j - \mathbf{u}_i), \tag{1}$$

where $g : \mathbb{R} \rightarrow \mathbb{R}$ is the "edge-stopping" function, λ is a constant that determines the rate of diffusion, $|\mathcal{N}(i)|$ represents the number of neighbors around point i, and $\mathbf{u}_i'$ denotes the updated feature. The core of Eq. (1) is the "edge-stopping" function g. g is a function of the edge magnitude, thereby making the equation adaptive to the local structure. The exact form of g can be chosen based on practical requirements. In most cases, the behavior of g is governed by handcrafted thresholds which allow smoothing when edges are small and preserving otherwise. Consequently, Eq. (1) preserves salient edges while smoothing out small discontinuities. Therefore, when it is applied to DNNs-based point cloud analysis, it would likely be beneficial to the learning of local representations.

Analysis of g. However, designing g for a specific application is non-trivial as it requires domain-specific knowledge and numerous trial and error. Concretely, a suitable function and handcrafted thresholds must be determined carefully so that the significant edges are preserved and others are smoothed out. This becomes particularly problematic when one desires to incorporate Eq. (1) in

DNNs since DNNs typically have multiple layers with each layer modeling different concepts. Furthermore, g is typically chosen to obey the maximum principle for stability [14]. In other words, Eq. (1) cannot sharpen the feature beyond the global maximum of the initial data. On the contrary, we believe that it would be desirable to emphasize certain features beyond the restriction if doing so benefits the learning. Therefore, to enable the adaptive modeling of edges in DNNs, one must design a g whose form is flexible without handcrafted parameters and restrictions while being adaptive to the local structure.

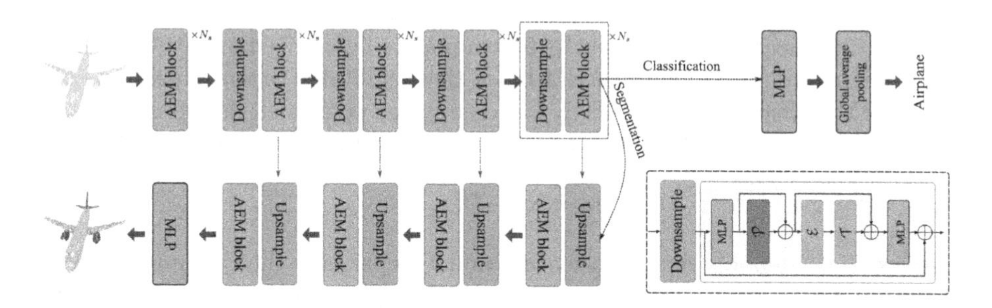

Fig. 2. Structure of the AEM block (bottom right) and network architectures.

3.2 Anisotropic Edge Modeling (AEM)

Definition. To resolve the aforementioned issues of g upon applying Eq. (1) to DNNs, we propose Anisotropic Edge Modeling (AEM) that allows the explicit modeling of edges to optimize the local representation. We define AEM as follows:

$$\mathbf{u}_i' = \mathbf{u}_i + \mathcal{T}\left(\frac{1}{|\mathcal{N}(i)|}\sum_{j\in\mathcal{N}(i)}\mathcal{E}\left(\mathbf{u}_j - \mathbf{u}_i\right)\right), \tag{2}$$

where $\mathcal{E}:\mathbb{R}^d \to \mathbb{R}^d$ is the edge modeling function and $\mathcal{T}:\mathbb{R}^d \to \mathbb{R}^d$ denotes a nonlinear transformation.

Similar to g in Eq. (1), we design $\mathcal{E}$ as a function of edge; therefore, AEM is adaptive to the local structure. However, instead of using a fixed function, we design $\mathcal{E}$ as a trainable shared-MLP (or 1D convolution with nonlinearities) so that the form of the function is determined in a data-dependent manner. Consequently, AEM frees users from choosing appropriate functions and thresholds by trial and error. Meanwhile, since $\mathcal{E}$ is no longer bounded by the maximum principle, Eq. (2) can perform sharpening/smoothing without restriction. Moreover, notice also that $g:\mathbb{R}\to\mathbb{R}$ in Eq. (1) produces a coefficient for the entire edge vector. Although applying a common coefficient for all channels ensures a stable behavior, it inevitably results in reduced adaptability since each channel may encode information of varying importance. In contrast, $\mathcal{E}$ produces refined edge vectors with each channel being a nonlinear combination of others.

In other words, enhanced adaptability is obtained by applying different coefficients to each channel. We additionally apply $\mathcal{T}$ to facilitate the training. We set $\mathcal{T} = \texttt{ReLU} \circ \texttt{BatchNorm}$ for regularization [6] and to encourage sparsity [2].

In this form, AEM inherits the merits of classic anisotropic diffusion while being more expressive. As a result, AEM explicitly models edges such that the local feature representation is optimized adaptively.

3.3 AEM Block

Since AEM deals primarily with the relationship between points, individual point features need to be sufficiently discriminative before feeding into the AEM layer. Therefore, we design the AEM block that facilitates the optimization and enhances the effectiveness of AEM. Concretely, we construct the point embedding (PE) layer that embeds point features in preparation for the AEM layer:

$$\mathbf{u}_i = \mathbf{x}_i + \mathcal{P}(\mathbf{x}_i), \tag{3}$$

where $\mathbf{x}_i \in \mathbb{R}^d$ denotes the input feature and $\mathcal{P}$ represents the PE function. We design the PE layer as a residual mapping to ease optimization [4]. In particular, we choose KPConv [22] with the depthwise separable scheme [18] enabled as the instantiation for $\mathcal{P}$ owing to its popularity and effectiveness [5,9].

Further, we construct the AEM block by incorporating the PE and AEM layers into the bottleneck residual structure [4] to reduce the overall computation without compromising performance. As a result, the AEM block can effectively optimize the local feature representation and serve as a basic building block for constructing DNNs. The illustration of the AEM block is shown in Fig. 2.

3.4 AEM-Nets

We construct network architectures based on AEM blocks to tackle various point cloud recognition tasks. The overview of the architectures is shown in Fig. 2.

The five-stage encoder is constructed by stacking multiple AEM and downsampling blocks. Specifically, the input points are downsampled at the beginning of each stage (excluding the first stage). Subsequently, N_s AEM blocks are applied sequentially following the downsampling block where s denotes the stage. Each downsampling block doubles the feature dimension whereas other blocks maintain it. The output of the encoder is a sparse version of the input with high-dimensional features. As a consequence, hierarchical feature learning can be achieved efficiently and features gradually become aware of global structures.

The classification head is appended to the encoder for producing object classifications. Specifically, a shared-MLP is applied to the output of the encoder followed by global average pooling to produce the final object score.

For segmentation, the U-Net–like decoder is constructed in a similar way to that of the encoder. AEM blocks are appended to an upsampling block in each stage. Point coordinates and features are upsampled until the point cloud recovers its input resolution. Finally, an MLP is applied to obtain point scores.

4 Experiments

In this section, we report the result of experiments performed on several highly competitive benchmarks. We implement all experiments using PyTorch. Experiments related to object classification and segmentation are performed using one NVIDIA Tesla V100 GPU. Experiments related to scene segmentation are conducted using four NVIDIA Tesla V100 GPUs. Other details and settings are provided in the supplementary materials due to the limited space.

Table 1. Comparisons with recent strong methods on object classification, object part segmentation, and scene segmentation tasks.

Method	Year	ScanObjectNN	ShapeNet Part	S3DIS
CurveNet [28]	2021	–	<u>86.8</u>	–
Point Transformer [34]	2021	–	86.6	<u>70.4</u>
Point-BERT [32]	2022	83.1	85.6	–
PointMLP [12]	2022	85.7	86.1	–
RepSurf [17]	2022	<u>86.0</u>	–	68.9
Stratified Transformer [9]	2022	–	86.6	**72.0**
Ours	–	**86.9**	**87.1**	**72.0**

4.1 Object Classification

ScanObjectNN [23] is used to validate the effectiveness of the AEM-Net for classification. The dataset consists of real-world 3D scans where 15k objects are included. Each object is labeled as one of the 15 classes. Each point cloud includes some measurement errors and background points. We use the hardest set of the dataset and adopt the official train-test split. Following previous works [12,17, 23,32], we use 1,024 points as input and adopt overall accuracy as the evaluation metric. The input is augmented with random rotation, scaling, and translation.

The result is listed in Table 1. The AEM-Net achieves the best performance among the recent networks. In particular, it outperforms the strong transformer-based [32], MLP-based [12], and explicit surface representation-based [17] methods by 3.8%, 1.2%, and 0.9%, respectively, demonstrating its effectiveness. We speculate that optimizing local representations leads to more discriminative global representations, resulting in superior performance.

4.2 Object Segmentation

We evaluate the AEM-Net on the ShapeNet Part dataset [31] for object segmentation. In total, 16,880 models are included in the dataset, where 14,006 models are used for training and 2,874 for testing. It includes 16 object classes and 50 object parts, each of which is annotated into two to six parts. We use the data provided by [16] for benchmarking. We use all available points with their surface normals as input. The input is augmented with random scaling and translation. We use Instance-wise IoU as the evaluation metric [16].

The result is listed in Table 1. Note that a gap of 0.2 or more in scores can be considered significant as the benchmark is highly competitive and the metric is rigid. The AEM-Net outperforms CurveNet [28] by 0.3, validating the effectiveness of edge modeling in object segmentation. We believe that the ability to adaptively sharpen edges leads to the refined local representation. The verification of this insight is provided in Sect. 5.4.

4.3 Scene Segmentation

We evaluate the AEM-Net on large-scale scene segmentation using the S3DIS [1] dataset. Six indoor environments containing 272 rooms are included. Each point is assigned one of the 13 categories. Similar to [21], we advocate using Area five for testing and others for training. Following the common practice [9,22,34], we downsample the rooms with the voxel size of 0.04cm for training. During training, each mini-batch contains at most 24,000 points. During testing, every point is evaluated and each mini-batch contains at most 80,000 points. We adopt mIoU as the evaluation metric. The input is augmented by random rotation, jittering, color auto-contrasting, and color dropping.

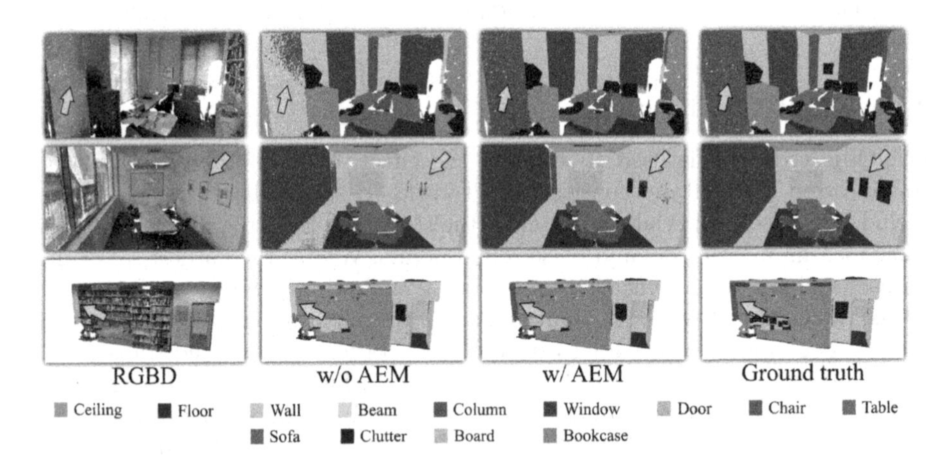

Fig. 3. The qualitative result of scene segmentation on S3DIS.

The result is listed in Table 1. The AEM-Net performs on par with the current state of the art Stratified Transformer [9], demonstrating its effectiveness in scene understanding. To better grasp the role of AEM, qualitative results are shown in Fig. 3. As can be seen, the network equipped with AEM obtains much smoother predictions compared with its plain counterpart (w/o AEM). Moreover, the impact of AEM is especially evident on the boundary of objects. For instance, objects that the plain network can barely detect are detected by AEM-Net (the second row).

5 Ablation Study and Analysis

In this section, the design of the AEM block and its efficiency are validated. Subsequently, we provide useful insights about what is learned by the AEM block through visualizing activations.

5.1 Ablation Studies of the AEM Block

As shown in Table 2, removing the AEM layer from the AEM block significantly reduces the performance by 1.9%; hence, the AEM layer plays a key role in achieving high performance. Furthermore, replacing AEM with EdgeConv reduces the performance by 2.1%, showing the superior effectiveness of AEM compared with the popular edge-aware method.

Table 2. Ablation studies of the AEM block on S3DIS.

	mIoU
AEM block	72.0
w/o AEM	70.1 (−1.9)
Replace AEM with EdgeConv [26]	69.9 (−2.1)

Table 3. Ablation studies of AEM components on S3DIS.

	mIoU
Full	72.0
w/o $\mathcal{T}$	69.8 (−2.2)
w/o $\mathcal{E}$ & $\mathcal{T}$	67.6 (−4.4)

Table 4. Analysis of computational complexity on S3DIS. C denotes the output channel width of the first stage.

Method	Params. (M)	FLOPs (G)	mIoU
Point Transformer (C = 32) [34]	7.8	9.0	70.4
Ours (C = 32)	1.4	2.6	71.5
Ours (C = 80)	8.7	12.6	72.0

5.2 Ablation Studies of AEM Components

The result is reported in Table 3. The performance drops significantly by 2.2% when $\mathcal{T}$ is removed. We conjecture that the optimization becomes difficult without $\mathcal{T}$ since the network is deep. As expected, removing both $\mathcal{E}$ and $\mathcal{T}$ substantially reduces performance due to the loss of adaptability in the AEM block.

5.3 Computational Efficiency

To analyze computational efficiency, we construct a smaller model (AEM-Net (C=32)) to examine the cost-performance trade-offs. As shown in Table 4, AEM-Net can achieve better performance with significantly fewer computations compared to Point Transformer [34]. Furthermore, increasing C from 32 to 80 further improves the performance from 71.5 to 72.0, achieving the state of the art. Therefore, we believe that AEM-Net can achieve the efficient use of available resources based on the practical requirements at hand.

5.4 Analysis of AEM

In this section, we investigate the adaptability of the AEM block. To this end, we visualize the activation of $\mathbf{u}_i' - \mathbf{u}_i$ in the last layer to apprehend the working mechanism of the AEM block. As can be seen in Fig. 4, strong activations are observed on the part boundaries of objects. In this case, AEM-Net simply locates and enhances the boundary. On the other hand, we observe a more sophisticated behavior for scene segmentation. As shown in Fig. 5, AEM-Net learns to enhance the boundaries for some classes and suppress them for the other classes. As a result, the cross-boundary differences, i.e., edges, are enhanced.

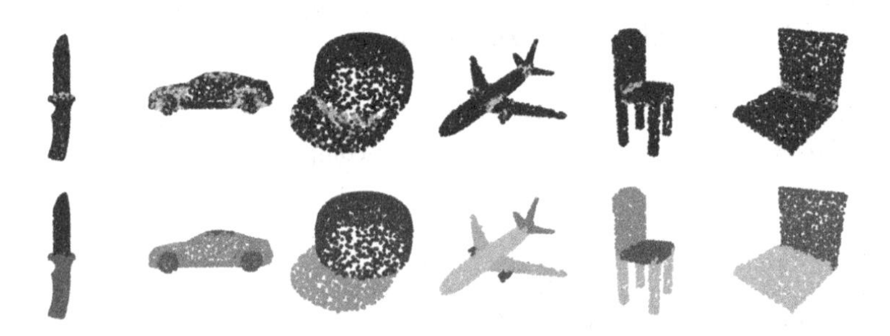

Fig. 4. Effect of the AEM block using ShapeNet. Top: activations of $\mathbf{u}_i'$ - $\mathbf{u}_i$. Bottom: ground truth. The AEM block precisely locates and enhances the part boundaries.

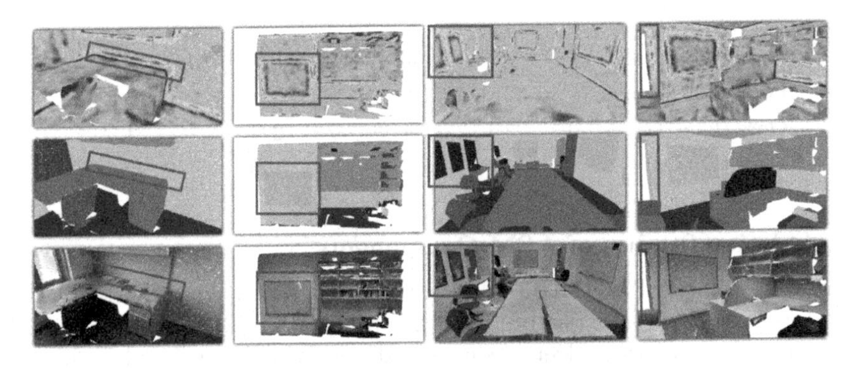

Fig. 5. Effect of the AEM block using S3DIS. Top: activations of $\mathbf{u}_i'$ - $\mathbf{u}_i$. Middle: ground truth. Bottom: color. The AEM block sharpens the boundaries by simultaneously enhancing one side of the boundaries and suppressing the other.

6 Conclusion

In this study, we propose the AEM block that optimizes the local feature representation by the adaptive sharpening of edges. AEM-Nets are constructed by stacking AEM blocks and applied to various point cloud understanding tasks.

Extensive experiments show that AEM-Nets achieve superior performance compared to recent strong methods. Furthermore, we verified that the AEM block sharpens the edges that correspond well to the object boundaries. We believe that it would be interesting to perform a principled analysis of the AEM block and investigate its potential through model scaling in the future.

Acknowledgement. This work was partially supported by the projects commissioned by the New Energy and Industrial Technology Development Organization (JPNP18010 and JPNP20006), JSPS Grant-in-Aid for Scientific Research (21K12042), and Fundamental Research Funds for the Central Universities (DUT21RC(3)028).

References

1. Armeni, I., et al.: 3D semantic parsing of large-scale indoor spaces. In: Proceedings of the IEEE Conference on Computer Vision and Pattern Recognition, pp. 1534–1543 (2016)
2. Glorot, X., Bordes, A., Bengio, Y.: Deep sparse rectifier neural networks. In: Proceedings of the Fourteenth International Conference on Artificial Intelligence and Statistics, pp. 315–323. JMLR Workshop and Conference Proceedings (2011)
3. Graham, B., Engelcke, M., Van Der Maaten, L.: 3D semantic segmentation with submanifold sparse convolutional networks. In: Proceedings of the IEEE Conference on Computer Vision and Pattern Recognition, pp. 9224–9232 (2018)
4. He, K., Zhang, X., Ren, S., Sun, J.: Deep residual learning for image recognition. In: Proceedings of the IEEE Conference on Computer Vision and Pattern Recognition, pp. 770–778 (2016)
5. Hu, Z., Zhen, M., Bai, X., Fu, H., Tai, C.: JSENet: joint semantic segmentation and edge detection network for 3D Point clouds. In: Vedaldi, A., Bischof, H., Brox, T., Frahm, J.-M. (eds.) ECCV 2020. LNCS, vol. 12365, pp. 222–239. Springer, Cham (2020). https://doi.org/10.1007/978-3-030-58565-5_14
6. Ioffe, S., Szegedy, C.: Batch normalization: accelerating deep network training by reducing internal covariate shift. In: International Conference on Machine Learning, pp. 448–456. PMLR (2015)
7. Jiang, L., Zhao, H., Liu, S., Shen, X., Fu, C.W., Jia, J.: Hierarchical point-edge interaction network for point cloud semantic segmentation. In: Proceedings of the IEEE/CVF International Conference on Computer Vision, pp. 10433–10441 (2019)
8. Kanezaki, A., Matsushita, Y., Nishida, Y.: RotationNet: joint object categorization and pose estimation using multiviews from unsupervised viewpoints. In: Proceedings of the IEEE Conference on Computer Vision and Pattern Recognition, pp. 5010–5019 (2018)
9. Lai, X., et al.: Stratified transformer for 3D point cloud segmentation. In: Proceedings of the IEEE/CVF Conference on Computer Vision and Pattern Recognition, pp. 8500–8509 (2022)
10. Lan, S., Yu, R., Yu, G., Davis, L.S.: Modeling local geometric structure of 3D point clouds using Geo-CNN. In: Proceedings of the IEEE/CVF Conference on Computer Vision and Pattern Recognition, pp. 998–1008 (2019)
11. Li, Y., Bu, R., Sun, M., Wu, W., Di, X., Chen, B.: PointCNN: Convolution on χ-transformed points. In: Proceedings of the 32nd International Conference on Neural Information Processing Systems, pp. 828–838 (2018)

12. Ma, X., Qin, C., You, H., Ran, H., Fu, Y.: Rethinking network design and local geometry in point cloud: a simple residual MLP framework. arXiv preprint arXiv:2202.07123 (2022)
13. Mao, J., Wang, X., Li, H.: Interpolated convolutional networks for 3D point cloud understanding. In: Proceedings of the IEEE/CVF International Conference on Computer Vision, pp. 1578–1587 (2019)
14. Perona, P., Malik, J.: Scale-space and edge detection using anisotropic diffusion. IEEE Trans. Pattern Anal. Mach. Intell. **12**(7), 629–639 (1990)
15. Qi, C.R., Su, H., Mo, K., Guibas, L.J.: PointNet: deep learning on point sets for 3D classification and segmentation. In: Proceedings of the IEEE Conference on Computer Vision and Pattern Recognition, pp. 652–660 (2017)
16. Qi, C.R., Yi, L., Su, H., Guibas, L.J.: PointNet++: deep hierarchical feature learning on point sets in a metric space. In: Advances in Neural Information Processing Systems. vol. 30 (2017)
17. Ran, H., Liu, J., Wang, C.: Surface representation for point clouds. In: Proceedings of the IEEE/CVF Conference on Computer Vision and Pattern Recognition, pp. 18942–18952 (2022)
18. Sifre, L., Mallat, S.: Rigid-motion scattering for texture classification. arXiv preprint arXiv:1403.1687 (2014)
19. Su, H., Maji, S., Kalogerakis, E., Learned-Miller, E.: Multi-view convolutional neural networks for 3D shape recognition. In: Proceedings of the IEEE International Conference on Computer Vision, pp. 945–953 (2015)
20. Tang, L., Zhan, Y., Chen, Z., Yu, B., Tao, D.: Contrastive boundary learning for point cloud segmentation. In: Proceedings of the IEEE/CVF Conference on Computer Vision and Pattern Recognition, pp. 8489–8499 (2022)
21. Tchapmi, L., Choy, C., Armeni, I., Gwak, J., Savarese, S.: SEGCloud: semantic segmentation of 3D point clouds. In: 2017 International Conference on 3D Vision (3DV), pp. 537–547. IEEE (2017)
22. Thomas, H., Qi, C.R., Deschaud, J.E., Marcotegui, B., Goulette, F., Guibas, L.J.: KPConv: flexible and deformable convolution for point clouds. In: Proceedings of the IEEE/CVF International Conference on Computer Vision, pp. 6411–6420 (2019)
23. Uy, M.A., Pham, Q.H., Hua, B.S., Nguyen, T., Yeung, S.K.: Revisiting point cloud classification: a new benchmark dataset and classification model on real-world data. In: Proceedings of the IEEE/CVF International Conference on Computer Vision, pp. 1588–1597 (2019)
24. Vaswani, A., et al.: Attention is all you need. arXiv preprint arXiv:1706.03762 (2017)
25. Wang, L., Huang, Y., Hou, Y., Zhang, S., Shan, J.: Graph attention convolution for point cloud semantic segmentation. In: Proceedings of the IEEE/CVF Conference on Computer Vision and Pattern Recognition, pp. 10296–10305 (2019)
26. Wang, Y., Sun, Y., Liu, Z., Sarma, S.E., Bronstein, M.M., Solomon, J.M.: Dynamic graph CNN for learning on point clouds. ACM Trans. Graph. (ToG) **38**(5), 1–12 (2019)
27. Wu, W., Qi, Z., Fuxin, L.: PointConv: deep convolutional networks on 3D point clouds. In: Proceedings of the IEEE/CVF Conference on Computer Vision and Pattern Recognition, pp. 9621–9630 (2019)
28. Xiang, T., Zhang, C., Song, Y., Yu, J., Cai, W.: Walk in the cloud: learning curves for point clouds shape analysis. In: Proceedings of the IEEE/CVF International Conference on Computer Vision (ICCV) (October 2021)

29. Xu, M., Ding, R., Zhao, H., Qi, X.: PAConv: position adaptive convolution with dynamic kernel assembling on point clouds. arXiv preprint arXiv:2103.14635 (2021)
30. Yang, J., et al.: Modeling point clouds with self-attention and gumbel subset sampling. In: Proceedings of the IEEE/CVF Conference on Computer Vision and Pattern Recognition, pp. 3323–3332 (2019)
31. Yi, L., et al.: A scalable active framework for region annotation in 3D shape collections. ACM Trans. Graph. (ToG) **35**(6), 1–12 (2016)
32. Yu, X., Tang, L., Rao, Y., Huang, T., Zhou, J., Lu, J.: Point-BERT: pre-training 3D point cloud transformers with masked point modeling. In: Proceedings of the IEEE/CVF Conference on Computer Vision and Pattern Recognition, pp. 19313–19322 (2022)
33. Zhang, Z., Hua, B.S., Yeung, S.K.: ShellNet: efficient point cloud convolutional neural networks using concentric shells statistics. In: Proceedings of the IEEE/CVF International Conference on Computer Vision, pp. 1607–1616 (2019)
34. Zhao, H., Jiang, L., Jia, J., Torr, P.H., Koltun, V.: Point transformer. In: Proceedings of the IEEE/CVF International Conference on Computer Vision, pp. 16259–16268 (2021)
35. Zhou, Y., Tuzel, O.: VoxelNet: end-to-end learning for point cloud based 3D object detection. In: Proceedings of the IEEE Conference on Computer Vision and Pattern Recognition, pp. 4490–4499 (2018)

Floor Plan Analysis and Vectorization with Multimodal Information

Tao Wen[1,2,3], Chao Liang[1,2,3(✉)], You-Ming Fu[1,2,3], Chun-Xia Xiao[3], and Hai-Ming Xiang[4]

[1] National Engineering Research Center for Multimedia Software (NERCMS), Wuhan University, Wuhan, China
[2] Hubei Key Laboratory of Multimedia and Network Communication Engineering, Wuhan University, Wuhan, China
[3] School of Computer Science, Wuhan University, Wuhan, China
cliang@whu.edu.cn
[4] VIVECO, Wuhan, China

Abstract. Floor plan analysis and vectorization are of practical importance in real estate and interior design fields. The analysis usually serves as a preliminary to the vectorization by extracting structural elements and room layouts. However, existing analysis methods mainly focus on the visual modality, which is insufficient for identifying rooms due to the lack of semantic clues about room types. On the other hand, standard floor plan images have rich textual annotations that provide semantic guidance of room layouts. Motivated by this fact, we propose a multimodal segmentation network (OCR)^2 that exploits additional textual information for the analysis of floor plan images. Specifically, we extract texts that indicate the room layouts with optical character recognition (OCR) and fuse them with visual features by a cross-attention mechanism. Thereafter, we further optimize the state-of-the-art vectorization method in efficiency by (1) replacing the gradient-descent steps with the fast principle components analysis (PCA) to convert doors and windows, and (2) removing the unnecessary iterative steps when extracting room contours. Both quantitative and qualitative experiments validate the effectiveness and efficiency of our proposed method.

Keywords: Floor plan analysis · Semantic segmentation · Image vectorization

1 Introduction

Floor plans describe architectural structure and layout, which are of critical importance for real estate and interior design. They are scaled drawings of indoor spaces that usually consist of vector graphics and textual annotations. While professionals edit floor plans using software such as AutoCAD in a vector-graphics representation, the drawings are eventually visualized, which are rendered into

D.-T. Dang-Nguyen et al. (Eds.): MMM 2023, LNCS 13833, pp. 282–293, 2023.
https://doi.org/10.1007/978-3-031-27077-2_22

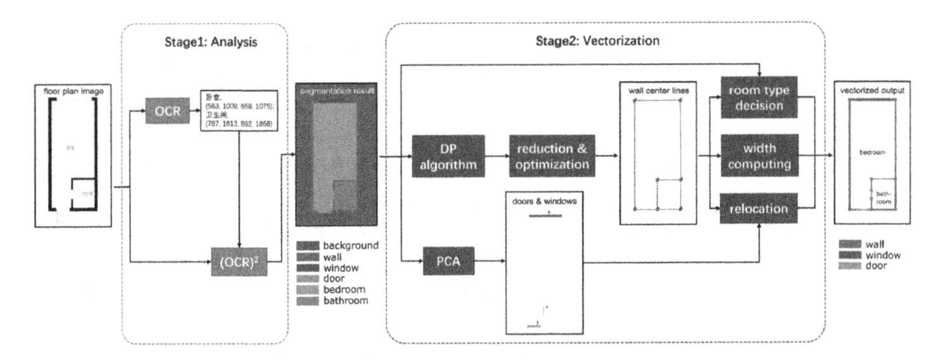

Fig. 1. The flowchart of our floor plan analysis and vectorization method. A floor plan image is fed into the pipeline that yields a vectorized representation of the original floor plan.

pixel images with rasterization at the cost of vectorized information. The vectorization of floor plan images is a problem with practical meaning to recover the lost vectorized information needed for further applications, such as modification, numerical computation, and scene synthesis.

The state-of-the-art vectorization methods usually require a preliminary analysis of the floor plan images to reconstruct their vector-graphics representations [9,11], where the structural elements (e.g., walls, doors, and windows) and room layouts are extracted. Therefore, we adopt a two-stage approach that performs floor plan analysis in the first stage and then completes the vectorization as shown in Fig. 1.

Traditional methods of floor plan analysis relies on a sequence of low-level image processing heuristics. However, their level of performance falls far behind that of human annotators or what is required for production [9]. Currently, semantic segmentation has been the standard technique for floor plan analysis that produces robust pixel-wise predictions of various elements in a data-driven approach. However, most of these methods focus on the visual modality and are insufficient for identifying rooms. One of the reasons is that floor plan images highlight only the frameworks of rooms and lack necessary semantic clues about the room types. Moreover, the current floor plan datasets are small and inadequate for training a powerful deep segmentation network. Nevertheless, it is worth noticing that standard floor plans have rich textual annotations that provide semantic guidance of room layouts, and we believe that incorporating the textual modality will improve the effectiveness of floor plan analysis. Inspired by the state-of-the-art segmentation network OCRNet [18], we propose a multimodal segmentation network that accepts results from optical character recognition (OCR) and yields enhanced segmentation results with a cross-attention mechanism, and we call this multimodal segmentation network $(OCR)^2$. Although some floor plan analysis methods also exploit textual information to identify rooms [3,11], our method takes advantage of multimodal information in a single deep segmentation model. To the best of our knowledge, we are the first to introduce OCR in a semantic segmentation problem setting.

A few works present their vectorization methods following the semantic analysis of floor plan images. While their performance is encouraging, we find that the state-of-the-art vectorization method [11] suffers from inefficiency and redundancy, which is featured by intense and iterative optimization-solving in the gradient-descent style. To address this problem, we further optimize the vectorization method by replacing the steps involving complex differentiable rasterization [10] with the fast principle components analysis (PCA) to convert doors and windows into rectangles. Additionally, we remove the unnecessary iterative steps that produce approximated polygons of room contours.

Our contributions are summarized as follows:

- We propose a multimodal segmentation network $(\text{OCR})^2$ for floor plan analysis, which yields enhanced results by exploiting additional textual modality.
- We further optimize the state-of-the-art floor plan vectorization method in efficiency by replacing time-consuming steps and removing redundant steps.
- We conduct both quantitative and qualitative experiments to evaluate our method's effectiveness and efficiency.

2 Related Works

2.1 Floor Plan Analysis

The problem of floor plan analysis has attracted researchers for decades. Traditional methods rely on low-level image features and image-processing heuristics and are weak in generalization. Walls are fundamental for a floor plan and are a priority in the analysis. Some of the earliest works identified walls by directly finding connected components in pixel images [6,13]. Advanced techniques like Hough transform and image vectorization were used to detect the lines of walls [1,3,8]. Various heuristics like convex hull approximation, edge-linking, or color analysis [1,12] were applied to overcome the malformed structure of walls. Doors and windows were located either by looking at the openings on walls [1] or by feature descriptors. [2,12]. Finally, rooms were extracted by recursively partitioning image regions [12] or finding the components of the connected pixels surrounded by the walls [8].

As deep learning technology develops, significant progress in generalization has been seen in floor plan analysis. Segmentation networks such as Fully convolutional network (FCN), UNet, and DeepLabv3+ are utilized to extract semantic areas of walls, doors, windows, and rooms in floor plan images [5,9,11,16]. Zeng *et al.* [19] introduced a room-boundary guided attention mechanism, improving the model's performance on floor plan images. Besides, object detection methods such as Fast R-CNN and YOLO series are used to find local objects such as doors, windows, and furniture icons [5,11,21]. However, textual information of room types in floor plan images has received little attention. Dodge *et al.* [5]

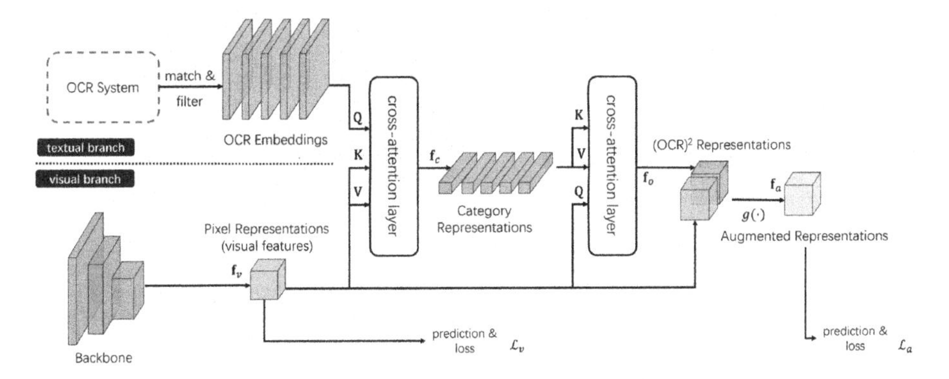

Fig. 2. The network architecture of $(\text{OCR})^2$. The components in dotted boxes could be generic and are not essential for $(\text{OCR})^2$.

employed Google Vision API to recognize text only to estimate the actual dimension of rooms. Lv *et al.* [11] implemented text detection functions to support the decision of room types. However, their system disjoints visual and textual modalities as the textual information is exploited only in the vectorization stage.

As far as we are concerned, we are the first to introduce OCR system in a semantic segmentation problem setting for floor plan analysis.

2.2 Floor Plan Vectorization

Liu *et al.* [9] used a deep regression network to predict junctions in a floor plan image (e.g., wall corners or door/window end-points), followed by an integer programming (IP) step to construct its vectorized representation. However, most segmentation-based methods produce pixel-level prediction and do not fit vectorized representation. Lv *et al.* [11] presented a vectorization algorithm for segmentation results, which suffers from inefficiency and redundancy because of the intense optimization-solving for even the most simple vector shapes.

We further optimize the vectorization method in [11] by removing unnecessary steps and replacing time-consuming ones with efficient ones.

3 Method

We present a two-stage approach that produces vectorized representaions of the input floor plan images. The complete flowchart of our method is shown in Fig. 1. In the analysis stage, we propose a multimodal segmentation network $(\text{OCR})^2$ that exploits additional textual modality. And in the vectorization stage, we implement the optimized floor plan vectorization method.

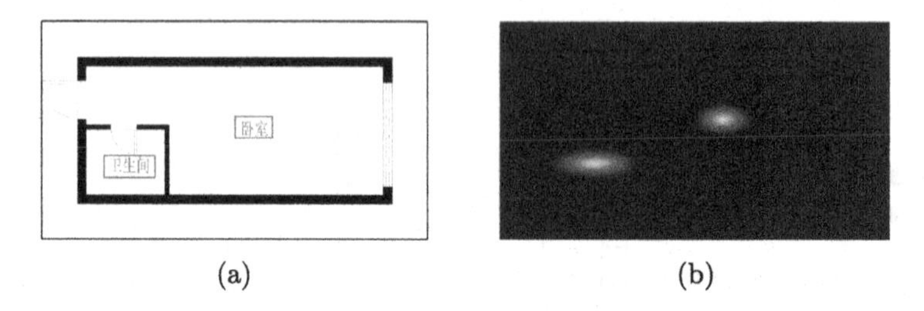

Fig. 3. The illustration of OCR embedding. (a) Texts that indicate room types in the floor plan image are marked by red boxes. (b) The category channels of the OCR embedding are merged for visualization (Color figure online)

3.1 (OCR)^2

Inspired by OCRNet [18], we propose a multimodal segmentation network that exploits both visual and texutal modalities as shown in Fig. 2.

In the beginning, two branches extract textual information and visual features of floor plan images respectively. In the textual branch, a generic OCR extract texts indicating room types, which are later embedded. In the visual branch, a backbone network for segmentation is used.

OCR Embeddings. Although there are different methods for OCR embedding, we choose to embed all the matched texts as heatmaps that reflect their position and categories. As illustrated in Fig. 3, a gaussian distribution is generated for each matched text to cover its bounding box area. Later, the OCR embedding is created as an aggregation of all the distributions according to their room categories. Assuming that the floor plan image is of size $w \times h$, the created OCR embedding E is a 3D tensor of size $w \times h \times C$ whose number of channels C equals the total number of room categories.

Cross-Attention Layers. We use a cross-attention mechanism to fuse the separated textual and visual modalities. The attention [17] is computed as follows. Three inputs are required: a set of N_q queries $\mathbf{Q} = \{\mathbf{q}_i \in \mathbb{R}^d | 1 \leq i \leq N_q\}$, a set of N_p keys $\mathbf{K} = \{\mathbf{k}_j \in \mathbb{R}^d | 1 \leq j \leq N_p\}$ and values $\mathbf{V} = \{\mathbf{v}_j \in \mathbb{R}^d | 1 \leq j \leq N_p\}$. For each query $\mathbf{q}_i$, the attention output is the linear aggregation of values:

$$\mathrm{Attn}(\mathbf{q}_i, \mathbf{K}, \mathbf{V}) = \sum_{j=1}^{N_p} \alpha_{ij} \mathbf{v}_j. \tag{1}$$

The weight α_{ij} is computed as softmax normalization of dot-product between the query $\mathbf{q}_i$ and the key $\mathbf{k}_j$:

$$\alpha_{ij} = \frac{e^{\frac{1}{\sqrt{d}} \mathbf{q}_i^T \mathbf{k}_j}}{Z_i} \text{ where } Z_i = \sum_{j=1}^{N_p} \alpha_{ij} \mathbf{v}_j. \tag{2}$$

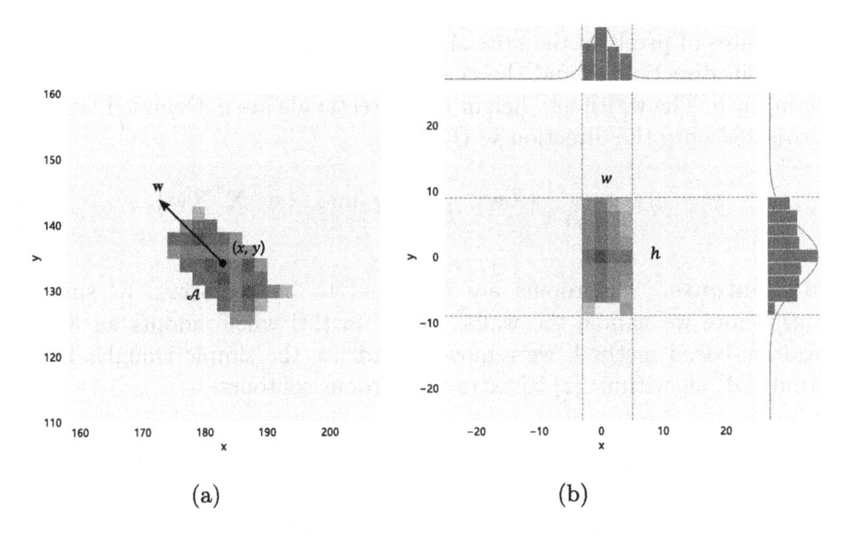

(a) (b)

Fig. 4. The illustration of vectorizing doors and windows with PCA.

The first cross-attention layer outputs representations $\mathbf{f}_c$ for each room category. The queries are flattened slices of each category from the OCR embedding, and the keys and values are the visual features $\mathbf{f}_v$ s from the backbone. The second cross-attention layer generates $(\text{OCR})^2$ representations $\mathbf{f}_o$ at each pixel. The queries are the representations at each position on the image. The keys and values are the category representations $\mathbf{f}_c$ from the first attention layer.

Prediction and Losses. Instead of directly using the $(\text{OCR})^2$ representations for prediction, we concatenate them with the original pixel representations from the backbone. As shown in Fig. 2, the final augmented representations $\mathbf{f}_a$ are updated by the aforementioned two parts of features:

$$\mathbf{f}_a = g([\mathbf{f}_v, \mathbf{f}_o]), \tag{3}$$

where $g(\cdot)$ is a 1×1 conv transformation.

Supervised by the same ground-truth annotations, the original pixel representations and the augmented representations are supposed to yield identical predictions. Given this insight, we derive two losses $\mathcal{L}_v$ and $\mathcal{L}_a$ from the $(\text{OCR})^2$ network and compute the total loss as the sum of the two:

$$\mathcal{L}_{\text{total}} = \mathcal{L}_v + \mathcal{L}_a, \tag{4}$$

where the standard pixel-wise cross-entropy loss are used for each item.

3.2 Vectorization

Doors and Windows. We convert pixels of doors and windows into rectangles with principal component analysis (PCA). As ilustrated in Fig. 4a, by treating the coordinates of pixels in the area $\mathcal{A}$ as a set of 2D data $\mathbf{X}$, we apply PCA to find the main direction $\mathbf{w}$ and the centroid (x, y) of $\mathbf{X}$ with the optimization-solving in Eq. 5. The width and height of the rectangle are determined later after $\mathbf{X}$ is projected onto the direction $\mathbf{w}$ (Fig. 4b).

$$\mathbf{w} = \arg\max_{\|\mathbf{w}\|=1} \left\{\|\mathbf{X}\mathbf{w}\|^2\right\} = \arg\max_{\|\mathbf{w}\|=1} \left\{\mathbf{w}^\mathsf{T}\mathbf{X}^\mathsf{T}\mathbf{X}\mathbf{w}\right\}. \tag{5}$$

Room Contours. The rooms are supposed to be vectorized as simplified polygons before we handle the walls. Unlike in [11] which adopts an iterative optimization-based method, we remove it and use the simple Douglas-Peucker algorithm (DP algorithm) [7] to extract the room contours.

Wall Ceter Lines and Widths. We follow the steps in [11] to vectorize walls. Wall center lines are represented as an undirected graph $\mathcal{G} = (\mathcal{V}, \mathcal{E})$ which is updated by performing veritces reduction and coordinates optimization iteratively.

The width of each wall is decided subsequently by maximizing the IoU metric between the area of the wall and the pixel area of the room boundary.

Last Steps. After the walls are vectorized, the doors and windows should be relocated on the walls by projecting the rectangles onto the nearest edge of the graph $\mathcal{G}$. In addition, because the segmentation results and the detected texts may specify different probability distributions over the room categories, each room type should be determined by a combination of the two.

4 Experiments

4.1 Experimental Setting

Datasets. R2V was released along with the vectorization method of the same name, containing 815 images [9]. We followed the train-test split of the datasets in [19] for a fair comparison. Besides, we collected floor plan drawings on shejijia website and created a dataset with 1050 images (50 images are included in test split). We plan to release this dataset in the future.

Implementation Details. We called BaiduOCR API offline to imitate the generic OCR system of our $(\text{OCR})^2$ method. We implemented the segmentation network based on Pytorch [14] and adopted OCRNet [18] as the backbone, initializing it with the weights pre-trained on ImageNet. The model was trained with a SGD optimizer (*momentum* = 0.9 and *weight_decay* = 0.0005), and the global dropout was set to 0.1. The learning rate was initially set to 0.01, and a polynomial learning rate policy with factor $(1-(\frac{iter}{iter_{max}})^{0.9})$ was performed. We implemented the vectorization algorithm described in [11] on our own to compare it with ours. The PCA used to vectorize doors and windows was provided by scikit-learn library [15].

Table 1. Comparison with Raster-to-Vector (R2V), Deep Floor Plan Recognition (DFPR), and general segmentation networks.

	$(\mathrm{OCR})^2$ (ours)	R2V	DFPR	DeepLabV3+	PSPNet
overall_acc.	**0.92**	0.84	0.90	0.88	0.88
mIoU	**0.80**	–	0.76	0.69	0.70

Table 2. Ablation study of $(\mathrm{OCR})^2$ network architecture.

	w/o text.	w/o category feat.	w/ visual pred.	full
overall_acc.	0.95	0.97	0.96	0.97
mIoU	0.74	0.78	0.76	**0.82**

All the procedures mentioned above were run and estimated with NVIDIA A5000 accelerator, 32-core Intel Xeon E5-2620 v4 processors, and 128GB RAM.

4.2 Comparison with State-of-the-Art

We compare our $(\mathrm{OCR})^2$ method with Raster-to-Vector (R2V) [9] and Deep Floor Plan Recognition (DFPR) [19] on R2V dataset. In addition, DeepLabV3+ [4] and PSPNet [20] are also included to evaluate the performance of general segmentation networks on floor plan images.

The overall pixel accuracy (*overall_acc.*) and mean intersection over union (*mIoU*) are reported in Table 1 for quantitative evaluation. We take the experimental results from [19], and the evaluation metrics of our method are averaged over multiple runs.

4.3 Ablation Study

To verify the effectiveness of our $(\mathrm{OCR})^2$ network architecture, we conduct ablation studies on shejijia dataset. The variants of our method are listed in the following:

- **w/o text.**: the textual branch is removed, and the backbone is used directly to make predictions;
- **w/o category feat.**: the first cross-attention layer is removed, and the OCR embeddings are treated as the queries of the other cross-attention layer;
- **w/ visual pred.**: the pixel representations in the visual branch are replaced with pixel-wise predictions;
- **full**: our full implementation of $(\mathrm{OCR})^2$.

Table 2 shows the comparison between the defective schemes and the complete method. The reported overall pixel accuracy (*overall_acc.*) and mean intersection over union (*mIoU*) are averaged over multiple runs. We can see that

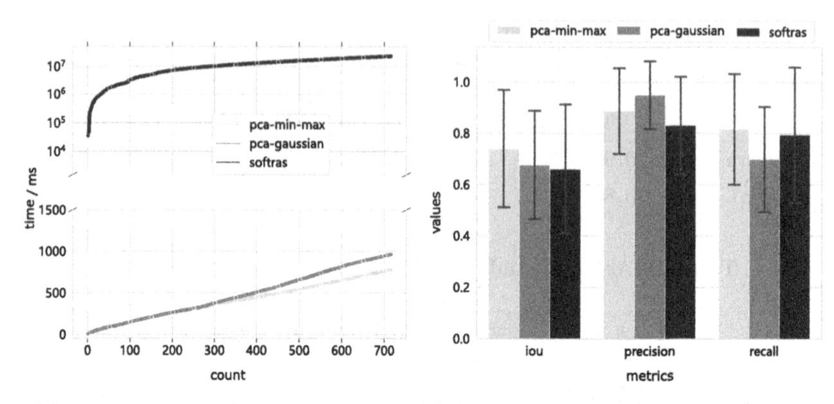

(a) The accumulated running time. (b) The metrics of fitting rectangles.

Fig. 5. The experimental results of vectorization methods of doors and windows.

Table 3. The accuracy of room contour extraction methods.

	IoU	Precision	Recall
w/ optimization	0.96 ± 0.02	0.99 ± 0.01	0.97 ± 0.01
w/o optimization	0.97 ± 0.02	0.99 ± 0.01	0.98 ± 0.02

w/o text. scores the lowest amongst all the variants (underlined in Table 2). It is reasonable to conclude that textual information boosts the performance of the segmentation network. The result of **w/o category feat.** shows that removing the first attention layer leads to a decrease in performance. Besides, replacing the pixel representation with sparse pixel-wise prediction also brings damage to the performance.

4.4 PCA Vectorization

We implement the vectorization method of doors and windows in [11] and compare it with our PCA-based method in efficiency and accuracy. **softras** is the vectorization method described in [11] based on differentiable rasterization [10]. **PCA-min-max** is the PCA-based method proposed in Sect. 3.2, and the sizes of rectangle are determined by the mininum and maximum coornidnates along x, y axis. **PCA-Gaussian** is a similar PCA-based method, but the sizes are computed as 3σ after gaussian fitting.

We collected 50 well-formed segmentation results of floor plan images and applied the vectorization methods above to evaluate their efficiency and accuracy. The iou, precision, and recall are reported to evaluate how well the vector graphics fit the pixel areas. Figure 5 shows that our proposed PCA-based methods achieve a better fit with significantly less time.

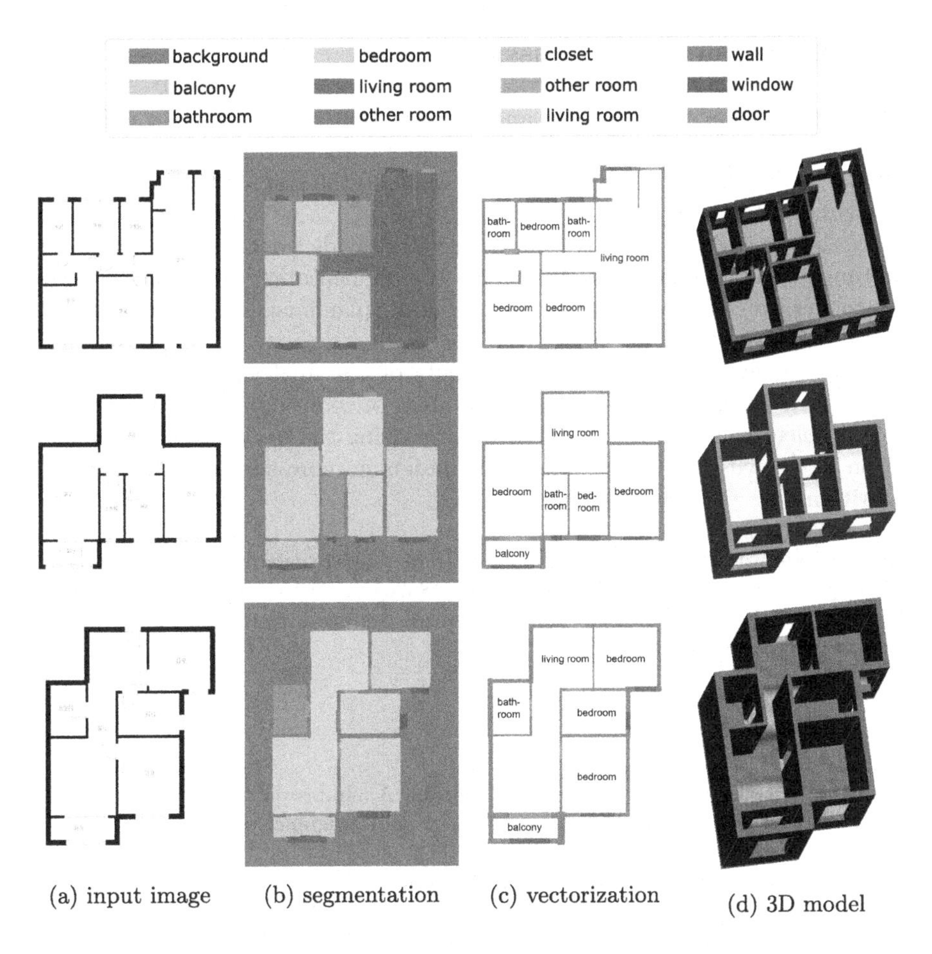

Fig. 6. The qualitative results of floor plan analysis and vectorization.

4.5 Room Contour Extraction

We remove the iterative optimization steps in [11] when extracting room contours, and the experimental results on 50 well-formed segmentation results justify our modification to this routine.

Both **w/ optimization** and **w/ optimization** schemes are tested. The IoU, precision, and recall are reported to evaluate how well the extracted polygons fit the room pixel areas. Table 3 shows that the optimization steps do not improve the accuracy of approximated polygons and thus are unnecessary.

4.6 Qualitative Results

Figure 6 gives the qualitative results of our floor plan analysis and vectorization method. The 3D models in the last column are created from the vectorized representation of input floor plan images.

5 Conclusion and Discussion

We present a floor plan analysis and vectorization method that exploits both visual and textual modalities. A multimodal segmentation network called $(OCR)^2$ is proposed, and optimization is made to the state-of-the-art floor plan vectorization method.

We believe that the $(OCR)^2$ network will provide further insights into how multimodal information is used to boost the segmentation performance. Moreover, our optimized vectorization method of floor plan images has practical meaning for industrial applications.

However, we also find that the outputs of our analysis and vectorization method are corrupted when the segmentation of room structure is malformed. This problem could be coped by either improving the performance of segmentation model with expanded dataset, or designing more advanced vectorization algorithm.

Acknowledgements. This work is supported by the National Natural Science Foundation of China (No. U1903214, 61876135, 61862015). The numerical calculations in this paper have been done on the supercomputing system in the Supercomputing Center of Wuhan University.

References

1. Ahmed, S., Liwicki, M., Weber, M., Dengel, A.: Improved automatic analysis of architectural floor plans. In: 2011 International Conference on Document Analysis and Recognition, pp. 864–869. IEEE (2011)
2. Ahmed, S., Liwicki, M., Weber, M., Dengel, A.: Automatic room detection and room labeling from architectural floor plans. In: 2012 10th IAPR International Workshop on Document Analysis Systems, pp. 339–343. IEEE (2012)
3. Ahmed, S., Weber, M., Liwicki, M., Dengel, A.: Text/graphics segmentation in architectural floor plans. In: 2011 International Conference on Document Analysis and Recognition, pp. 734–738. IEEE (2011)
4. Chen, L.-C., Zhu, Y., Papandreou, G., Schroff, F., Adam, H.: Encoder-decoder with atrous separable convolution for semantic image segmentation. In: Ferrari, V., Hebert, M., Sminchisescu, C., Weiss, Y. (eds.) ECCV 2018. LNCS, vol. 11211, pp. 833–851. Springer, Cham (2018). https://doi.org/10.1007/978-3-030-01234-2_49
5. Dodge, S., Xu, J., Stenger, B.: Parsing floor plan images. In: 2017 Fifteenth IAPR International Conference on Machine Vision Applications (MVA), pp. 358–361. IEEE (2017)
6. Dosch, P., Tombre, K., Ah-Soon, C., Masini, G.: A complete system for the analysis of architectural drawings. Int. J. Doc. Anal. Recogn. **3**(2), 102–116 (2000)
7. Douglas, D.H., Peucker, T.K.: Algorithms for the reduction of the number of points required to represent a digitized line or its caricature. Cartographica Int. J. Geographic Inf. Geovisual. **10**(2), 112–122 (1973)
8. de las Heras, L.P., Ahmed, S., Liwicki, M., Valveny, E., Sánchez, G.: Statistical segmentation and structural recognition for floor plan interpretation. Int. J. Doc. Anal. Recogn. (IJDAR) **17**(3), 221–237 (2014)

9. Liu, C., Wu, J., Kohli, P., Furukawa, Y.: Raster-to-vector: revisiting floorplan transformation. In: Proceedings of the IEEE International Conference on Computer Vision, pp. 2195–2203 (2017)
10. Liu, S., Li, T., Chen, W., Li, H.: Soft rasterizer: a differentiable renderer for image-based 3d reasoning. In: Proceedings of the IEEE/CVF International Conference on Computer Vision, pp. 7708–7717 (2019)
11. Lv, X., Zhao, S., Yu, X., Zhao, B.: Residential floor plan recognition and reconstruction. In: Proceedings of the IEEE/CVF Conference on Computer Vision and Pattern Recognition, pp. 16717–16726 (2021)
12. Macé, S., Locteau, H., Valveny, E., Tabbone, S.: A system to detect rooms in architectural floor plan images. In: Proceedings of the 9th IAPR International Workshop on Document Analysis Systems, pp. 167–174 (2010)
13. Or, S.H., Wong, K.H., Yu, Y.K., Chang, M.M.V., Kong, H.: Highly automatic approach to architectural floorplan image understanding & model generation. Pattern Recogn. pp. 25–32 (2005)
14. Paszke, A., et al.: Pytorch: an imperative style, high-performance deep learning library. In: Wallach, H., Larochelle, H., Beygelzimer, A., d'Alché-Buc, F., Fox, E., Garnett, R. (eds.) Advances in Neural Information Processing Systems, vol. 32, pp. 8024–8035. Curran Associates, Inc. (2019), https://papers.neurips.cc/paper/9015-pytorch-an-imperative-style-high-performance-deep-learning-library.pdf
15. Pedregosa, F., et al.: Scikit-learn: machine learning in Python. J. Mach. Learn. Res. **12**, 2825–2830 (2011)
16. Surikov, I.Y., Nakhatovich, M.A., Belyaev, S.Y., Savchuk, D.A.: Floor plan recognition and vectorization using combination UNet, Faster-RCNN, statistical component analysis and Ramer-Douglas-Peucker. In: Chaubey, N., Parikh, S., Amin, K. (eds.) COMS2 2020. CCIS, vol. 1235, pp. 16–28. Springer, Singapore (2020). https://doi.org/10.1007/978-981-15-6648-6_2
17. Vaswani, A., et al.: Attention is all you need. In: Advances in Neural Information Processing Systems, vol. 30 (2017)
18. Yuan, Y., Chen, X., Wang, J.: Object-contextual representations for semantic segmentation. In: Vedaldi, A., Bischof, H., Brox, T., Frahm, J.-M. (eds.) ECCV 2020. LNCS, vol. 12351, pp. 173–190. Springer, Cham (2020). https://doi.org/10.1007/978-3-030-58539-6_11
19. Zeng, Z., Li, X., Yu, Y.K., Fu, C.W.: Deep floor plan recognition using a multi-task network with room-boundary-guided attention. In: Proceedings of the IEEE/CVF International Conference on Computer Vision, pp. 9096–9104 (2019)
20. Zhao, H., Shi, J., Qi, X., Wang, X., Jia, J.: Pyramid scene parsing network. In: Proceedings of the IEEE Conference on Computer Vision and Pattern Recognition, pp. 2881–2890 (2017)
21. Ziran, Z., Marinai, S.: Object detection in floor plan images. In: Pancioni, L., Schwenker, F., Trentin, E. (eds.) ANNPR 2018. LNCS (LNAI), vol. 11081, pp. 383–394. Springer, Cham (2018). https://doi.org/10.1007/978-3-319-99978-4_30

Safe Contrastive Clustering

Pengwei Tang[1], Huayi Tang[1], Wei Wang[2], and Yong Liu[1(✉)]

[1] Gaoling School of Artificial Intelligence, Renmin University of China, Beijing, China
tangpwei@163.com, tangh4681@gmail.com, liuyonggsai@ruc.edu.cn
[2] China Unicom Research Institute, Beijing, China
wangw200@chinaunicom.cn

Abstract. Contrastive clustering is an effective deep clustering approach, which learns both instance-level consistency and cluster-level consistency in a contrastive learning fashion. However, the strategies of data augmentation used by contrastive clustering is an important prior knowledge such that inappropriate strategies may severely cause performance degradation. By converting the different strategies of data augmentations into a multi-view problem, we propose a safe contrastive clustering method which is guaranteed to alleviate the reliance on prior knowledge of data augmentations. The proposed method can maximize the complementary information between these different views and minimize the noise caused by the inferior views. Such a method addresses the safeness that contrastive clustering with multiple data augmentation strategies is no worse than that with one of those strategies. Moreover, we provide the theoretical guarantee that the proposed method can achieve empirical safeness. Extensive experiments demonstrate that our method can reach safe contrastive clustering on popular benchmark datasets.

Keywords: Contrastive clustering · Multi-view · Unsupervised learning

1 Introduction

Recently, contrastive learning has drawn much attention in unsupervised learning, which can extract all-purpose representations so that it obtains great success in many downstream tasks [1,2]. Contrastive learning can maximize the agreement between positive pairs and minimize the agreement between negative pairs. Contrastive clustering introduces the idea of contrastive learning to clustering, which achieves promising performance [3–5]. SCAN [3] pretrains a feature extraction model by contrastive learning and then utilizes the model to train its cluster assignment module by exploiting the agreement between samples and their nearest neighbors. CC [4] utilizes both the instance-level consistency and cluster-level consistency to perform contrastive clustering. GCC [5] constructs a KNN graph to mine graph information of adjacent samples by contrastive clustering.

D.-T. Dang-Nguyen et al. (Eds.): MMM 2023, LNCS 13833, pp. 294–305, 2023.
https://doi.org/10.1007/978-3-031-27077-2_23

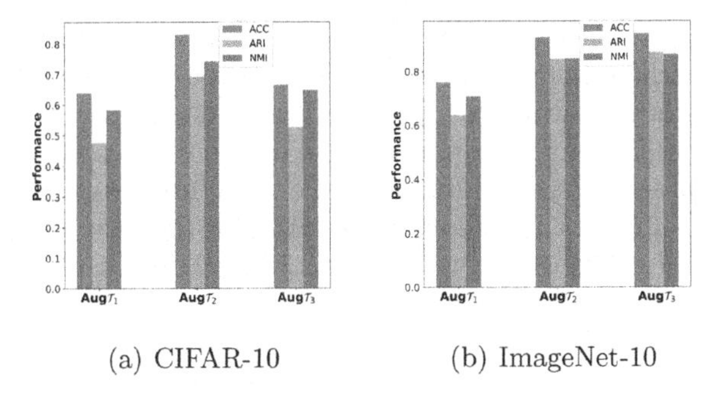

(a) CIFAR-10 (b) ImageNet-10

Fig. 1. Clustering Performance of different data augmentation strategies $\mathcal{T}_1$, $\mathcal{T}_2$ and $\mathcal{T}_3$ on CIFAR-10 and ImageNet-10 datasets. The strategy $\mathcal{T}_1$ is {resized cropping, horizontal flipping, color jitter}, the strategy $\mathcal{T}_2$ is {resized cropping, horizontal flipping, color jitter, grayscale} and the strategy $\mathcal{T}_3$ is {resized cropping, horizontal flipping, color jitter, grayscale, gaussian blur}.

In contrastive clustering, inappropriate strategies of data augmentations may cause severe performance [4]. The data augmentation methods used by contrastive clustering usually include cropping, horizontal flipping, color distortion, gaussian blur, and so on [2,4,5]. The strategies of data augmentation are the combinations of these data augmentation methods. Different strategies of data augmentation can provide different kinds of or different numbers of the view of the samples. In contrastive clustering, the two-way data augmentations even lead to more differences between view pairs. Such difference may cause the performance of the contrastive clustering model to be different. For instance, in Fig. 1, three different strategies of data augmentation are adopted to perform the GCC model, respectively. We can observe that the data augmentation strategy $\mathcal{T}_2$ performs best in the CIFAR-10 dataset while in the ImageNet-10 dataset, the data augmentation strategies $\mathcal{T}_3$ achieve the best clustering performance.

As observed, prior knowledge of data augmentations heavily decide the clustering performance in contrastive clustering. However, in the real world, there are no labels to evaluate the performance of the clustering tasks. Thus, it is hard to know which types of combinations of data augmentation strategies are effective for contrastive clustering models. Such a disadvantage motivates us to develop a method to achieve safe data augmentations for contrastive clustering.

One natural assumption is that different data augmentation strategies may cause the contrastive clustering model to extract representations with different high-level semantics. Another natural conjecture is that such representations with different high-level semantics are distinct views of the complete invariant high-level semantic information. These representations are similar to the meaning of the "view" of multi-view learning. **In this paper, one kind of view representation is corresponding to one kind of strategy of data augmentations**.

Based on the above discussions, we propose a safe contrastive clustering model to achieve a safe data augmentation mechanism. Our method is a two-stage method, which learns the different view representations by classic contrastive clustering in the first stage and utilizes these different views to achieve a safe data augmentation mechanism in the second stage. In the second stage, from the standpoint of representation learning, our proposed model is supposed to take advantage of the complementary information from these different views. Furthermore, from the perspective of safeness, our proposed model is required to effectively search for the applicable coefficients of different views to avoid the noise caused by inferior views. To solve these two issues, we develop a bi-level optimization framework in the second stage.

Our main contributions are summarized as follows.

(1). We first propose a two-stage method to achieve a safe data augmentation mechanism in contrastive clustering. Our proposed method can guarantee that contrastive clustering with multiple data augmentation strategies is no worse than that with one of those strategies.

(2). We give a formal and complete definition of *safe contrastive clustering*. We provide the theoretical analysis that our proposed model can achieve the defined empirical safe contrastive clustering.

(3). We conduct experiments on six challenging datasets to verify that our proposed model can achieve safe data augmentation in contrastive clustering.

2 Related Work

2.1 Contrastive Clustering

Deep clustering methods achieve promising clustering performance due to the powerful ability of representations. Existing work can be categorized into autoencoder based methods [6,7] and label feature based methods [8–10]. The former methods adopt an auto-encoder to learn latent features by reconstruction loss. The latter methods utilize a fully connected layer that can predict the cluster assignments and regard cluster assignments as label features to calculate the loss function. However, these two categories of methods only exploit the similarity of the positive pairs, causing clusters not to be separable. Recently, based on contrastive learning [1,2], new deep clustering methods could enforce the positive pairs to be closer and push negative pairs to be more separable. Compared to contrastive learning, contrastive clustering methods [3–5] additionally introduce a clustering head to encourage the samples and their augmentations to share similar cluster assignment distribution. Nevertheless, an observation is that the clustering performance of contrastive clustering is heavily dependent on the strategies of data augmentations. To address this problem, we propose a method that achieves safe data augmentation for contrastive clustering. Futhermore, our key conjecture is that representations learning through distinct data augmentation strategies are different views of the complete invariant high-level semantic information, which is inspired by the multi-view clustering [16–18].

2.2 Safeness in Machine Learning

The meaning of safeness in machine learning is to avoid the performance degradation caused by inferior learners or noisy features. For classic semi-supervised learning, the work in [12] develops safe semi-supervised support vector machines by taking advantage of multiple low-density separators to approximate the ground-truth decision boundary, and the method proposed by [13] learns a safe prediction from multiple semi-supervised regressors. For weakly supervised learning, the paper [19] presents a unified ensemble learning scheme to achieve safeness by integrating multiple weakly supervised learners. Recently, safeness in deep learning has also attracted increasing attention. DS^{3}L [14] learns a neural network to reweight the regularization loss of unlabeled samples, alleviating the performance degradation resulting from class distribution mismatch in deep semi-supervised learning. Different from the above work, our work is to guarantee that contrastive clustering with multiple data augmentation strategies is no worse than that with one of those strategies.

3 Preliminaries

Our proposed method, namely safe contrastive clustering (**SCC**), is a two-stage method: in the first stage, SCC adopts a classic contrastive clustering to train the encoders with multiple different data augmentation strategies; in the second stage, based on these aforementioned trained encoders, SCC can reach multiple views of the complete invariant semantic information, and then employ these views to achieve safeness mechanism of data augmentation by a bi-level optimization framework.

3.1 Notations

$\mathcal{D} = \{x_i\}_{i=1}^n$ denotes a dataset. $\binom{m}{r}$ denotes the combination number $\frac{m!}{(m-r)!r!}$. The clusters of $\mathcal{D}$ is denote as K. The set of strategies of stochastic data augmentation is denoted as $\{\mathcal{T}_i\}_{i=1}^O$. $\{f_i\}_{i=1}^O$ denote the encoders obtained from the classic contrastive clustering, where the encoder f_i is associated with the strategy of data augmentations $\mathcal{T}_i$. We call $f_i(x)$ one view representation of x.

3.2 Contrastive Clustering Approach

In the first stage, we use a classic contrastive clustering method to reach encoders $\{f_i\}_{i=1}^O$ with multiple data augmentation strategies $\{\mathcal{T}_i\}_{i=1}^O$. Sample x are first transformed by strategy $\mathcal{T}$, then fed to a encoder f such as ResNet-18 [15] and finally used to compute the consistency loss by two projection heads. Our work use the current published state-of-the-art method GCC [5] in the first stage.

4 Proposed Method for Safe Contrastive Clustering

In this section, we introduce the key technique to achieve safe contrastive clustering, *i.e.*, the method in the second stage. We first give the assumption and definition of the safe contrastive clustering, then introduce the safeness mechanism of the contrastive clustering, and finally give a theoretical analysis.

4.1 Assumption and Definition

The multilayer perceptron (MLP) that mines the cluster-level consistency in the first stage, also called the cluster projection head, can predict the cluster assignments. However, it is hard to achieve safeness by using the cluster projection head. The reason is that the projection head has been trained so long that it is only focused on one view representation. Thus, it is impractical to train the cluster assignment module initialized as the projection head. Therefore, to take advantage of all the information from different views of the complete semantics, our proposed method trains the cluster assignment module from scratch. We boldly introduce the following assumption.

Assumption 1. *The clustering performance of the cluster projection head can be recovered by learning from the view representation obtained by the trained encoders* $\{f_i\}_{i=1}^{O}$ *in the first stage.*

The safe contrastive clustering hopes that the clustering performance with multiple strategies of data augmentations is not worse than the clustering performance with only one of those data augmentation strategies, which helps reduce the reliance on prior knowledge. However, it is hard to evaluate clustering performance without ground-truth labels. As a result, we evaluate the performance based on the empirical clustering loss. Based on Assumption 1, the definition of safe contrastive clustering is presented as follows.

Definition 1. *(**Empirical Safe Contrastive Clustering**) If the empirical clustering risk of a model learning with multiple view representations is no higher than that of the model learning with only one of those view representations, this model is said to achieve **empirical safe contrastive clustering**.*

4.2 Safeness Mechanism of Contrastive Clustering

For contrastive clustering with different data augmentation strategies, the encoders could extract different view representations of the complete invariant information. In other words, $\{f_i(x)\}_{i=1}^{O}$ could be different view representations of the sample x. On the basis of the encoders trained in the first stage, our proposed method aims to achieve safeness mechanism of the data augmentation strategies in the second stage.

To satisfy Assumption 1, a powerful deep divergence-based clustering model [16–18], including alignment module $\mathcal{M}$, safe module $\mathcal{S}$ and cluster assignment module $\mathcal{A}$, is introduced to exploit the view representations. Next, we will detail these three modules and a powerful deep divergence-based clustering loss in order.

The alignment module $\mathcal{M}$ uses a set of neural networks with the same architecture to map distinct view representations $\{f_i(x)\}_{i=1}^{O}$ into the same feature space, which is formulated as:

$$\mathcal{M}(x) = \{g_i(f_i(x))\}_{i=1}^{O}, \tag{1}$$

where g_i denotes the mapping neural networks for $f_i(x)$.

To achieve safe contrastive clustering with less prior knowledge about strategies of data augmentations, we introduce a safe module $\mathcal{S}$ to obtain a fusion from the aligned view representations $\mathcal{M}(x) = \{g_i(f_i(x))\}_{i=1}^{O}$:

$$\mathcal{S}(\mathcal{M}(x)) = \sum_{i=1}^{O} \lambda_i g_i(f_i(x)), \tag{2}$$

where $\{\lambda_i\}_{i=1}^{O}$ are the safe coefficients which are the learnable parameters, satisfying two constraints, namely, $\lambda_i \in [0,1]$ and $\sum_{i=1}^{O} \lambda_i = 1$. The proposed SCC model first learns the safe coefficients without any constraints and then utilizes softmax normalization to achieve these constraints.

The proposed model finally obtains the cluster assignments via a cluster assignment module $\mathcal{A}$. The cluster assignment module $\mathcal{A}$ is also implemented by neural networks. Let $\mathcal{L}$ denote the loss function in the second stage. The cluster assignment module $\mathcal{A}$ includes two sub-components, *i.e.*, one encoding network with multiple fully connected layers and another fully connected layer for prediction. The fusion obtained by the safe module $\mathcal{S}$ is first fed to the multiple fully connected layers to extract the clustering-friendly hidden features $\mathbf{h}$. On top of the hidden features, the final fully connected layer predicts the cluster assignments. A softmax operation is adopted to normalize the probabilities.

To extract complementary information from different views $\{f_i(x)\}_{i=1}^{O}$, the proposed model adopts deep divergence-based clustering loss, which exhibits its huge advantages in the multi-view clustering task [16–18]. Let $\mathbf{C} \in \mathbb{R}^{n \times K}$ denote the cluster assignment matrix by the module $\mathcal{A}$. Let $\mathbf{K} \in \mathbb{R}^{n \times n}$ denote the similarity matrix of n samples and its element is computed by $\mathbf{K}_{ij} = \exp\left(-\|\mathbf{h}_i - \mathbf{h}_j\|^2\right) / \left(2\sigma^2\right)$. Let $\mathbf{e}_j \in \mathbb{R}^K$ represent the standard simplex. $\mathbf{D} \in \mathbb{R}^{n \times K}$ measures how similar the cluster assignments to the standard simplex via $\mathbf{S}_{ij} = \exp\left(-\|\mathbf{C}_{i,:} - \mathbf{e}_j\|^2\right)$. According to the deep divergence-based clustering loss, the objective function is formulated as:

$$\begin{aligned}\mathcal{L}(\Lambda, \Theta) = & \frac{1}{\binom{K}{2}} \sum_{p=1}^{K-1} \sum_{r>p} \frac{\mathbf{C}_{:,p}^{\top} \mathbf{K} \mathbf{C}_{:,r}}{\sqrt{\mathbf{C}_{:,p}^{\top} \mathbf{K} \mathbf{C}_{:,p} \mathbf{C}_{:,r}^{\top} \mathbf{K} \mathbf{C}_{:,r}}} + \frac{1}{\binom{n}{2}} \sum_{i=1}^{n} \sum_{j>i} \mathbf{C}_{i,:}^{\top} \mathbf{C}_{j,:} \\ & + \frac{1}{\binom{K}{2}} \sum_{p=1}^{K-1} \sum_{r>p} \frac{\mathbf{S}_{:,p}^{\top} \mathbf{K} \mathbf{S}_{:,r}}{\sqrt{\mathbf{S}_{:,p}^{\top} \mathbf{K} \mathbf{S}_{:,p} \mathbf{S}_{:,r}^{\top} \mathbf{K} \mathbf{S}_{:,r}}},\end{aligned} \tag{3}$$

where Λ includes $\{\lambda_i\}_{i=1}^{O}$ and Θ includes the network parameters of the mapping networks $\{g_i\}_{i=1}^{O}$ and the cluster assignment module $\mathcal{A}$. The first item increases the compactness of one cluster and the separability between distinct clusters. The second term leads to orthogonal assignments for different samples. The last term aims at obtaining assignments close to the standard simplex.

To optimize Λ and Θ, a bi-level optimization framework is formulated as:

$$\begin{aligned} \Lambda^* &= \arg\min_{\Lambda} \frac{1}{n}\sum_{i=1}^{n} \mathcal{L}\left(\mathcal{A}\left(\mathcal{S}\left(\mathcal{M}\left(x_i\right)\right)\right); \Lambda, \Theta^*\right), \\ \text{where} \quad \Theta^* &= \arg\min_{\Theta} \frac{1}{n}\sum_{i=1}^{n} \mathcal{L}\left(\mathcal{A}\left(\mathcal{S}\left(\mathcal{M}\left(x_i\right)\right)\right); \Lambda, \Theta\right). \end{aligned} \tag{4}$$

We propose a bi-level optimization [11] approach to optimize the divergence-based loss. However, it is challenging to solve a bi-level optimization. To trade off accuracy and efficiency, we utilize a simplified optimization method to alternately update the network parameters Θ and safe coefficients Λ by mini-batch gradient descent. The overall training procedure is summarized in Algorithm 1.

4.3 Theoretical Analysis

Theorem 1. *(Empirical Safeness.) Let the empirical clustering risk of the model training with multiple view representations* $\{f_i\}_{i=1}^{O}$ *be*

$$\hat{\mathcal{L}}_n = \frac{1}{n}\sum_{i=1}^{n} \mathcal{L}\left(\mathcal{A}\left(\mathcal{S}\left(\mathcal{M}\left(x_i\right)\right)\right)\right). \tag{5}$$

Let the empirical clustering risk of the model training with the view representation f_o *be*

$$\hat{\mathcal{L}}_n^o = \frac{1}{n}\sum_{i=1}^{n} \mathcal{L}\left(\mathcal{A}\left(g_o\left(x_i\right)\right)\right). \tag{6}$$

The optimal solution of $\hat{\mathcal{L}}_n$ *is denoted as* $\hat{\mathcal{L}}_n^*$. *We can prove that the following inequality holds:*

$$\hat{\mathcal{L}}_n^* \leq \min\left\{\left(\hat{\mathcal{L}}_n^1\right), \left(\hat{\mathcal{L}}_n^2\right), \ldots, \left(\hat{\mathcal{L}}_n^O\right)\right\}. \tag{7}$$

The proof of Theorem 1 is as follows.

Proof. Based on the definitions of the module $\mathcal{S}$ and $\mathcal{M}$, we have

$$\hat{\mathcal{L}}_n|_{\lambda_o=1,\lambda_p=0,p\neq o} = \hat{\mathcal{L}}_n^o. \tag{8}$$

The optimal solution of $\hat{\mathcal{L}}_n$ is denoted as $\hat{\mathcal{L}}_n^*$. And the $\hat{L}_n^o$ is a special case of $\hat{L}_n$. Thus, we have

$$\hat{\mathcal{L}}_n^* \leq \min\left\{\left(\hat{\mathcal{L}}_n^1\right), \left(\hat{\mathcal{L}}_n^2\right), \ldots, \left(\hat{\mathcal{L}}_n^O\right)\right\}.$$

The proof of Theorem 1 is completed.

According to Assumption 1 and Definition 1, Theorem 1 reveals that the proposed bi-level optimization framework can achieve **empirical safe contrastive clustering** in the second stage. That is to say, the empirical clustering risk of a model learning with multiple view representations is no higher than that of the model learning with only one of those view representations.

Algorithm 1: Safe contrastive Clustering

Input: datasets $\mathcal{D}$, cluster number K, data augmentation strategies $\{\mathcal{T}_1, \mathcal{T}_2, \ldots, \mathcal{T}_O\}$, the number of epochs $T_1, T_2, T_\Theta, T_\Lambda$

Output: A safe contrastive clustering model

```
1  // Stage One:
2  for i to O do
3      for k to T_1 do
4          Initialize the contrastive clustering model;
5          Use 𝒯_i to make data augmentations;
6          Update the model by minimizing the contrastive clustering loss;
7      end
8      Obtain the embeddings {f_i(x_j)}_{j=1}^n by the trained encoder f_i;
9  end
10 // Stage Two:
11 for i to T_2 do
12     for j to T_Θ do
13         Fixing Λ, update Θ by optimizing loss ℒ;
14     end
15     for j to T_Λ do
16         Fixing Θ, update Λ by optimizing loss ℒ;
17     end
18 end
```

5 Experiments

5.1 Datasets and Metrics

We evaluate our proposed method on six datasets, including CIFAR-10, CIFAR-100-20, STL-10, ImageNet-10, ImageNet-Dogs and Tiny-ImageNet. Three standard metrics including Accuracy (**ACC**), Adjusted Rand Index (**ARI**) and Normalized Mutual Information (**NMI**) are adopted to evaluate the clustering performance. Higher values of all three metrics mean better clustering performance.

5.2 Implementation Details

In the first stage, we adopt the same setting as GCC [10] but with three different data augmentation strategies. The three strategies contain $\mathcal{T}_1$({resized cropping, horizontal flipping, color jitter}), $\mathcal{T}_2$({resized cropping, horizontal flipping, color jitter, grayscale}) and $\mathcal{T}_3$({resized cropping, horizontal flipping, color jitter, grayscale, gaussian blur}). In the second stage, we adopt two Adam optimizers with learing rate 0.00001 and 0.01, respectively, to update the network parameters Θ and the safe coefficients Λ. T_2 of the CIFAR-10, CIFAR-100-20, and Tiny-ImageNet datasets is set to 5, and that of other datasets with fewer images is set to 10. Let T_Θ and T_Λ both be 10. We initialize λ_1, λ_2 and λ_3 as $\frac{1}{3}$. Let the batch size be 1000. The hyperparameter σ used in the similarity matrix $\boldsymbol{K}$ was set to 15% of the median pairwise distance between hidden

Table 1. Clustering performance comparisons of GCC and SCC method. The "$\mathcal{T}_i$" in the brackets shows the data augmentation strategies.

Datasets	CIFAR-10			CIFAR-100			STL-10		
Methods	ACC	NMI	ARI	ACC	NMI	ARI	ACC	NMI	ARI
GCC($\mathcal{T}_1$)	0.640	0.584	0.477	0.383	0.394	0.238	0.576	0.507	0.409
GCC($\mathcal{T}_2$)	0.830	0.743	0.693	0.467	0.485	0.316	0.770	0.680	0.628
GCC($\mathcal{T}_3$)	0.666	0.649	0.528	0.449	0.464	0.293	**0.798**	0.688	0.638
SCC($\mathcal{T}_1$)	0.653	0.604	0.503	0.403	0.423	0.265	0.560	0.499	0.390
SCC($\mathcal{T}_2$)	**0.868**	0.788	0.750	0.472	0.506	0.323	0.770	0.682	0.615
SCC($\mathcal{T}_3$)	0.829	0.781	0.711	0.460	0.502	0.313	0.791	0.687	0.617
SCC($\mathcal{T}_1, \mathcal{T}_2, \mathcal{T}_3$)	**0.868**	**0.794**	**0.755**	**0.494**	**0.540**	**0.352**	0.780	**0.699**	**0.641**

Datasets	ImageNet-10			ImageNet-Dogs			Tiny-ImageNet		
Methods	ACC	NMI	ARI	ACC	NMI	ARI	ACC	NMI	ARI
GCC($\mathcal{T}_1$)	0.761	0.709	0.637	0.438	0.414	0.270	0.096	0.266	0.044
GCC($\mathcal{T}_2$)	0.927	0.847	0.849	0.496	0.480	0.329	**0.127**	0.326	**0.059**
GCC($\mathcal{T}_3$)	0.940	0.871	0.864	**0.564**	0.520	**0.388**	0.111	0.307	0.053
SCC($\mathcal{T}_1$)	0.793	0.732	0.665	0.466	0.465	0.292	0.091	0.279	0.035
SCC($\mathcal{T}_2$)	0.942	0.876	0.875	0.494	0.514	0.333	0.121	0.345	0.046
SCC($\mathcal{T}_3$)	0.941	0.867	0.875	0.534	0.526	0.361	0.102	0.324	0.041
SCC($\mathcal{T}_1, \mathcal{T}_2, \mathcal{T}_3$)	**0.946**	**0.882**	**0.884**	0.527	**0.531**	0.374	**0.127**	**0.353**	0.047

features within each mini-batch [17]. The architecture of one mapping neural network g of alignment module $\mathcal{M}$ is implemented by fully connected layers with dimensions of 512 – ReLU – 512 – ReLU – 512 – ReLU. The architecture of assignment module $\mathcal{A}$ is implemented by fully connected layers with dimensions of 100 – ReLU – BatchNorm – K – Softmax. Following [16,18], we select a clustering result with the lowest divergence-based by running the experiments 20 times. We report the average of the 5 such selected clustering results.

Next, we introduce the hyperparameters of data augmentations methods. The size, scale and probability of resized cropping are set to the size of input images, $[0.2, 1]$ and 1, respectively. The probability of horizontal flipping is set to 0.5. The brightness, contrast, saturation, hue and probability of color jitter are set to 0.4, 0.4, 0.4, 0.1 and 0.8, respectively. The probability of grayscale is set to 0.5. The kernel size and probability of gaussian blur are set to 23 and 1, respectively. The standard deviation σ of gaussian blur is selected uniformly at random to lie in $[0, 1, 2.0]$.

5.3 Experimental Results and Analysis

In Table 1, we report the clustering results of SCC, and we report the clustering results of GCC and SCC with only one strategy of data augmentations.

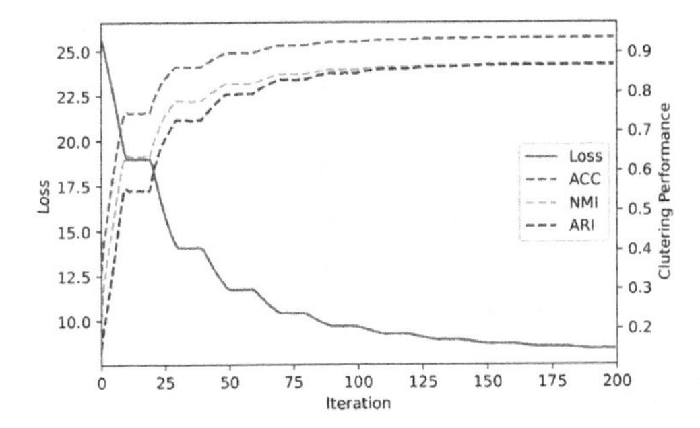

Fig. 2. The evolution of loss function and clustering performance of the proposed SCC across training process on ImageNet-10 dataset.

The Analysis of Assumption. Assumption 1 is an important premise of the theoretical guarantee provided by us. However, such an assumption seems a little bold. As observed, the clustering results of our SCC model learning from one view representations are comparable to the clustering results of the corresponding GCC model, indicating that our divergence-based model can recover the clustering performance of the GCC model. This observation demonstrates that our SCC model used in the second stage can satisfy Assumption 1.

The Analysis of Safeness. Based on Assumption 1, the definition of safeness mechanism can be transferred to the second stage, namely Definition 1. As we can see in Table 1, the clustering results of SCC($\mathcal{T}_1, \mathcal{T}_2, \mathcal{T}_3$) is comparable to the best of that of the SCC($\mathcal{T}_i$), $i = 1, 2, 3$. This provides the experimental guarantee for Theorem 1, showing that our SCC model learning with multiple view representations performs no worse than that learning with one view representation.

To sum up, the experimental results demonstrate the superiority of the safeness mechanism achieved by our proposed SCC model.

5.4 Convergence Analysis

In Fig. 2, the loss value curves and clustering results are shown to further analyze the convergence properties of our proposed SCC model in the second stage. We can see that the loss value first decreases across the training process and then finally converges. Accordingly, the clustering performance first increases across the training process and then finally becomes convergent. The trend of loss value reveal the character of the bi-level optimization framework.

5.5 The Analysis of Safe Coefficients

Optimal Values of the Safe Coefficients. We report the optimal values of the safe coefficients in Table 2. Combining Table 1 and 2, we can see that approximately, the better clustering performance that the view representation f_i achieves, the higher value λ_i. The observation is reasonable, which is consistent with intuition.

Table 2. Safe coefficients of SCC on all datasets.

	λ_1	λ_2	λ_3
CIFAR-10	0.235	0.422	0.343
CIFAR-100-20	0.215	0.425	0.360
STL-10	0.193	0.496	0.311
ImageNet-10	0.276	0.344	0.380
ImageNet-dogs	0.224	0.391	0.385
Tiny-ImageNet	0.294	0.386	0.320

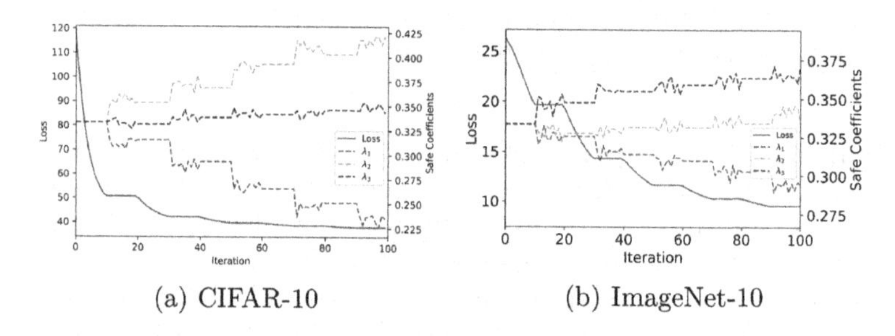

(a) CIFAR-10 (b) ImageNet-10

Fig. 3. The evolution of loss function and safe coefficients of the proposed SCC across training process on CIFAR-10 and ImageNet-10.

Optimization Process of the Safe Coefficients. To further study the optimization of safe coefficients, we plot the evolution of safe coefficients Λ with increasing iteration on CIFAR-10 and ImagNet-10 datasets, as shown in Fig. 3. First, Λ alternately is updated and remains unchanged, showing the characteristic of the bi-level optimization. Then, as the increase of iterations, the λ_i of superior view representation f_i becomes larger and λ_i of inferior view representation f_i becomes smaller gradually. Finally, the Λ converges.

The above two analyses demonstrate that our SCC can automatically search for applicable coefficients.

6 Conclusion

In this work, we develop a two-stage method to achieve safeness mechanism for contrastive clustering. We regard the feature representations learning through distinct data augmentation strategies as different views of the complete invariant information. Our SCC can maximize the complementary information from multiple views and minimize the noise caused by inferior views by a bi-level optimization in the second stage. Furthermore, we prove the empirical safeness of SCC in theory. Extensive experiments verify the safeness of SCC. One limitation of SCC is not an end-to-end method, which brings about the disadvantage that the optimization tends to be stuck in the local optimum. We plan to develop an end-to-end safe contrastive clustering method in the future.

Acknowledgments. This work is supported by the Fundamental Research Funds for the Central Universities, and the Research Funds of Renmin University of China (2021030199).

References

1. He, K., Fan, H., Wu, Y., Xie, S., Girshick, R.B.: Momentum contrast for unsupervised visual representation learning. In: CVPR (2020)
2. Chen, T., Kornblith, S., Norouzi, M., Hinton, G.E.: A simple framework for contrastive learning of visual representations. In: ICML (2020)
3. Van Gansbeke, W., Vandenhende, S., Georgoulis, S., Proesmans, M., Van Gool, L.: SCAN: learning to classify images without labels. In: Vedaldi, A., Bischof, H., Brox, T., Frahm, J.-M. (eds.) ECCV 2020. LNCS, vol. 12355, pp. 268–285. Springer, Cham (2020). https://doi.org/10.1007/978-3-030-58607-2_16
4. Li, Y., et al.: Contrastive clustering. In: AAAI (2021)
5. Zhong, H., et al.: Graph contrastive clustering. In: ICCV (2021)
6. Dizaji, K.G., Herandi, A., Deng, C., (Tom) Cai, W., Huang, H.: Deep clustering via joint convolutional autoencoder embedding and relative entropy minimization. In: ICCV (2017)
7. Ji, P., Zhang, T., Li, H., Salzmann, M., Reid, I.D.: Deep subspace clustering networks. In: NeurIPS (2017)
8. Caron, M., Bojanowski, P., Joulin, A., Douze, M.: Deep clustering for unsupervised learning of visual features. In: Ferrari, V., Hebert, M., Sminchisescu, C., Weiss, Y. (eds.) Computer Vision – ECCV 2018. LNCS, vol. 11218, pp. 139–156. Springer, Cham (2018). https://doi.org/10.1007/978-3-030-01264-9_9
9. Hu, W., Miyato, T., Tokui, S., Matsumoto, E., Sugiyama, M.: Learning discrete representations via information maximizing self-augmented training. In: ICML (2017)
10. Ji, X., Vedaldi, A., Henriques, J.F.: Invariant information clustering for unsupervised image classification and segmentation. In: CVPR (2019)
11. Franceschi, L., Frasconi, P., Salzo, S., Grazzi, R., Pontil, M.: Bilevel programming for hyperparameter optimization and meta-learning. In: ICML (2018)
12. Li, Y.-F., Zhou, Z.-H.: Towards making unlabeled data never hurt. IEEE TPAMI **37**, 175–188 (2015)
13. Li, Y.-F., Zha, H., Zhou, Z.-H.: Learning safe prediction for semi-supervised regression. In: AAAI (2017)
14. Guo, L.-Z., Zhang, Z.-Y., Jiang, Y., Li, Y.-F., Zhou, Z.H.: Safe deep semi-supervised learning for unseen-class unlabeled data. In: ICML (2016)
15. He, K., Zhang, X., Ren, S., Sun, J.: Deep residual learning for image recognition. In: CVPR (2016)
16. Trosten, D.J., Løkse, S., Jenssen, R., Kampffmeyer, M.C.: Reconsidering representation alignment for multi-view clustering. In: CVPR (2021)
17. Kampffmeyer, M.C., Løkse, S., Bianchi, F.M., Livi, L.F., Salberg, A.-B., Jenssen, R.: Deep divergence-based approach to clustering. Neural Netw. **13**, 91–101 (2019)
18. Tang, H., Liu, Y.: Deep safe multi-view clustering: reducing the risk of clustering performance degradation caused by view increase. In: CVPR (2022)
19. Li, Y.-F., Guo, L.-Z., Zhou, Z.-H.: Towards safe weakly supervised learning. IEEE TPAMI **43**, 334–346 (2021)

SRes-NeRF: Improved Neural Radiance Fields for Realism and Accuracy of Specular Reflections

Shufan Dai[1], Yangjie Cao[1](✉), Pengsong Duan[1], and Xianfu Chen[2]

[1] Zhengzhou University, Zhengzhou, China
{caoyj,duanps}@zzu.edu.cn
[2] Technical Research Centre of Finland Ltd., Espoo, Finland
xianfu.chen@ieee.org

Abstract. The Neural Radiance Fields (NeRF) is a popular view synthesis technique that represents a scene using a multilayer perceptron (MLP) combined with classic volume rendering and uses positional encoding techniques to increase image resolution. Although it can effectively represent the appearance of a scene, they often fail to accurately capture and reproduce the specular details of surfaces and require a lengthy training time ranging from hours to days for a single scene. We address this limitation by introducing a representation consisting of a density voxel grid and an enhanced MLP for a complex view-dependent appearance and model acceleration. Modeling with explicit and discretized volume representations is not new, but we propose Swish Residual MLP (SResMLP). Compared with the standard MLP+ReLU network, the introduction of layer scale module allows the shallow information of the network to be transmitted to the deep layer more accurately, maintaining the consistency of features. Introduce affine layers to stabilize training, accelerate convergence and use the Swish activation function instead of ReLU. Finally, an evaluation of four inward-facing benchmarks shows that our method surpasses NeRF's quality, it only takes about 18 min to train from scratch for a new scene and accuracy capture the specular details of the scene surface. Excellent performance even without positional encoding.

Keywords: Scene representation · View synthesis · Image-based rendering · Volume rendering · 3D deep learning · Spectral bias

1 Introduction

Novel view synthesis is a long-standing problem in computer vision and graphics. The ability to leverage deep learning to reason about 3D scenes with sparse image sets has broad applications in entertainment, virtual and augmented reality, and many other applications. Emerging neural rendering techniques have recently enabled photorealistic image quality for these tasks. One of the most prominent

D.-T. Dang-Nguyen et al. (Eds.): MMM 2023, LNCS 13833, pp. 306–317, 2023.
https://doi.org/10.1007/978-3-031-27077-2_24

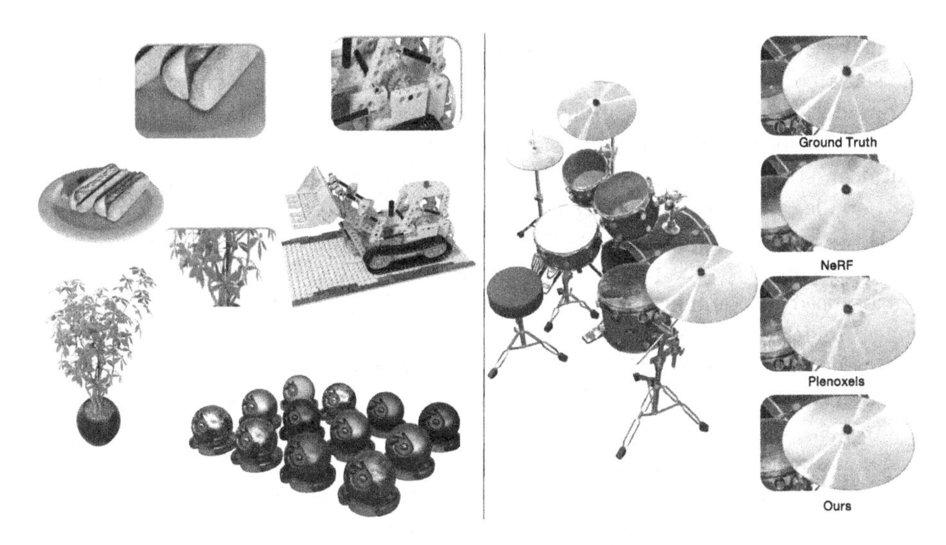

Fig. 1. We present a method to represent complex signals such as specular reflections. Our method is able to match the expressiveness of coordinate-based MLPs while retaining reconstruction and rendering speed of voxel grids.

recent advances in neural rendering is Neural Radiance Fields (NeRF) [25] which, given a handful of images of a static scene, learns an implicit volumetric representation of the scene that can be rendered from novel viewpoints. Although the current neural rendering technology has achieved leading image rendering quality, it still does not perform well in terms of model acceleration and image highlight detail.

By sampling the 3D coordinates in the scene, and using the MLP to infer the density of the location and the view-dependent color value, NeRF renders compelling photorealistic images of 3D scenes from novel viewpoints using a neural volumetric scene representation. Due to the extreme sampling requirements and costly neural network queries, rendering a NeRF is agonizingly slow. To address this limitation, we use the dense voxel grid to output the density and view-dependent color features of the scene using a linear interpolation strategy, which is similar to the DirectVoxGo [37] model.

Although NeRF employs positional encoding that maps the inputs to a higher dimensional space using high-frequency functions to improve renderings that perform poorly at representing high-frequency variation in color and geometry, it still renders poorly on specular surfaces. Figure 1 shows that NeRF and its variants rendering quality is still not ideal on specular objects. It is observed that the rendering results are very rough on the drums category. These rough artifacts are main caused by spectral bias [29]. Variants [10,12,20,21,31,33,48] of NeRF work well in the direction of acceleration, but there is little work that combines acceleration with image quality, especially for scene highlight details. They almost use a deeper MLP combined with the ReLU activation function to implicitly represent the 3D scene, and that position encoding [29] is applied to the sampled point coordinates and the viewpoint vector before they are input to the model, mapping them to a high-dimensional space and allowing the MLP to fit the high-frequency function more effectively. This MLP+ReLU implicit representation structure has not been effectively improved in the subsequent NeRFs method, leading to a long-term limitation in the image rendering quality of NeRFs, where the model cannot effectively learn the high-frequency details of the image. To address this problem, we propose the SResMLP architecture, which can be used in combination with a learnable 3D Grid to significantly improve image rendering quality and satisfactory model training speed. The MLP architecture used by NeRFs [25] can cause their feature vectors to vary with depth, feature space inconsistency, and problems such as gradient loss and exploding. Therefore, we refer to the Residual Multi-Layer Perceptrons (ResMLP) [41] and Vision Transformer (ViT) [6], which have an excellent performance in the field of image classification, and we adapt them to improve them. NeRFs use the combination of standard MLP+ReLU. On this basis, our method introduces skip-connection and layerscale to ensure the consistency of shallow network features and deep network features, which plays an important role in effectively fitting high and low-frequency components. We use Affine layers to stabilize training and accelerate convergence. Using the Swish activation function instead of the ReLU activation function, experiments have shown that Swish performs the best on the peak signal-to-noise ratio. Although this is a small change, it significantly improves the expressiveness of the model, and the ability to capture image details further improve.

To summarize, we make the following contributions:

1. We improve a tradeoff between rendering speed and image quality.
2. Significantly improves the neural radiance fields ability to represent the specular details of scene appearance.
3. Compared with NeRF, the average peak signal-to-noise ratio (PSNR) on Synthetic-NeRF dataset is increased by 1.51 dB, Synthetic-NSVF dataset by 5 dB, BlendedMVS dataset by 4.35 dB, TanksTemples dataset by 3.04 dB. Effectively improve the PSNR, SSIM, and LPIPS quantitative metric of the model on public datasets.

2 Related Work

Neural Radiance Fields. Recently, NeRF has caused a new boom in new view synthesis tasks. By simply inputting images of the sparse angles of the scene and the corresponding camera parameters, images of the new view can be obtained. Compared to traditional explicit and discrete volume representations such as voxel lattices and MPI, NeRF performs extremely well in the novel view synthesis task, using a coordinate-based MLP as an implicit and continuous volume representation. NeRF achieves appealing quality and has good flexibility with many follow-up extensions to various setups, e.g., relighting [2,3,36,49], deformation [8,26,27,27,42], self-calibration [15,19,24,44,47], meta-learning [38], dynamic scene modeling [9,18,23,28,45], and generative modeling [4,17,33]. However, NeRF also has some limitations, such as the model training time being very slow and cannot effectively fit the rich highlight details of the scene objects.

Enhanced Standard MLP. As a classic neural network, MLP is applied to various tasks of deep learning. Transformers [43] built by MLP shine in natural language processing, image classification, and recognition tasks. MLP-Mixer [40] uses Mixer's MLP to replace ViT's Transformer [6], which reduces the degree of freedom of feature extraction and can cleverly exchange information between patches and information within patches alternately. From the results, pure MLP seems to be feasible, and The complex structure of Transformer is omitted, and it becomes more concise. Recently, Facebook AI Lab proposed ResMLP [41] for tasks such as image classification, a purely MLP-based architecture that uses residual operations to update projection features, and finally average pools all block features classification later. It is more stable than Transformer training and more concise than MLP-Mixer. Inspired by ResNLP, we adapt it to represent neural radiation fields with surprising performance. On the NeRF public dataset, the peak signal-to-noise ratio is improved by an overall average of 1.5dB.

Spectral Bias. Recent works [25,29,34,39] have shown that a standard MLP with ReLU [11] shows limited performance in representing high-frequency textures. Researchers call this phenomenon spectral bias. Its presence leads to some limitations of the coordinate-based MLP to implicitly represent 3D scenes, such as the inability to fit high-frequency details of object surfaces. Similar limitations exist in the hyper-division domain. Various methods have been proposed to alleviate this problem. For example, researchers have proposed the SIREN [34] periodic activation function to replace the ReLU activation function, which can achieve accelerated convergence as well as improved image quality. Other approaches [25,39] are to map input coordinates into high-dimensional Fourier space by using position encoding or Fourier feature mapping before passing an MLP. This is also the scheme used by NeRFs, but we found that the images rendered by NeRFs still have problems with highlight details being difficult to capture. We consider that the ReLU activation function is still not the optimal choice, so we use the Swish [30] activation function to replace ReLU, and the introduction of the layerscale module in MLP can ensure the consistency of network features and significantly improve the network's ability to fit high-frequency details (Fig. 2).

3 Method

In this section, first, we will briefly describe the learnable density voxel grid, which mainly acts as an acceleration in the model, which is not our main innovation. Second, we describe how SResMLP structures utilize skip-connection

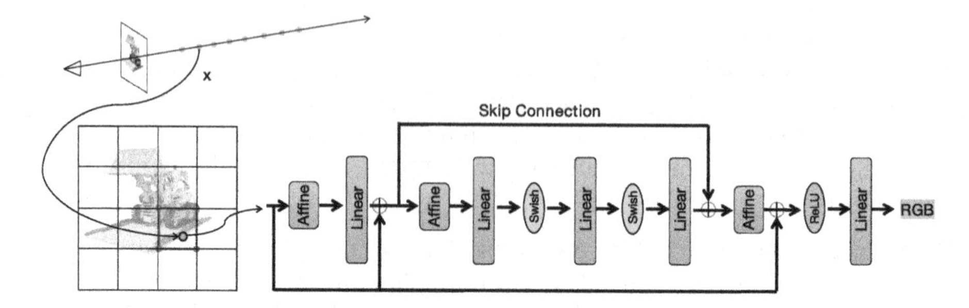

Fig. 2. The SRes-NeRF architecture. The coordinates $\boldsymbol{x}$ of the ray sampling points are linearly interpolated by the learnable 3D Grid and then stitched with the viewpoint vector $\boldsymbol{d}$. Finally, they are fed into the SResMLP network to predict the color values. The SResMLP network is stably trained and accelerates convergence using an affine transformation layer. Layerscale and Skip Connection are used to enhance the network's ability to fit high-frequency information.

and layerscale to enhance the neural network's ability to fit high and low-frequency components. The Affine layer is introduced to stabilize training, accelerate convergence, and replace ReLU with the Swish activation function to further improve the peak signal-to-noise ratio. By explicitly using these components in standard MLP, SResMLP can accurately reproduce the appearance of specular highlights and reflections. In addition, with the positional encoding removed, our model still has amazing performance in image rendering quality.

3.1 Direct Voxel Grid Optimization

DirectVoxGo adopt a representation consisting of a density voxel grid for scene geometry and a feature voxel grid with a shallow MLP for complex view-dependent appearance. In order to balance the model training speed and image rendering quality, we use a 3D voxel grid similar to DirectVoxGo to represent the geometric structure and color features of the 3D scene:

$$\boldsymbol{\sigma} = \operatorname{interp}\left(\boldsymbol{x}, \boldsymbol{V}^{(\text{density })}\right) \tag{1a}$$

$$\boldsymbol{c}^{(\text{feat})} = \operatorname{interp}\left(\boldsymbol{x}, \boldsymbol{V}^{(\text{feat })}\right) \tag{1b}$$

We use a higher-resolution density voxel grid to represent the geometry of the object:

$$\boldsymbol{V}^{(\text{density})} \in \mathbb{R}^{1 \times N_x \times N_y \times N_z} \tag{2a}$$

$$\boldsymbol{V}^{(\text{feat})} \in \mathbb{R}^{D \times N_x \times N_y \times N_z} \tag{2b}$$

and the sampled coordinate points are linearly interpolated to obtain the $\boldsymbol{\sigma}$ value. Using the same approach, we can obtain the color feature values. First, search the coarse geometry of a scene. Then reconstruct the fine detail including view-dependent effects.

3.2 Swish Residual MLP

NeRF and its variants still has difficulty fitting image highlight details using standard MLP, mainly due to deep networks being biased towards learning lower frequency functions. This phenomenon is called spectral bias. To improve this phenomenon, the specific approach in NeRF is to convert the coordinates of the sampled points into Fourier space before inputting them into the MLP for mapping, so that the high-frequency sine cosine function will distinguish the adjacent coordinate points, and in such a high-dimensional space, the adjacent coordinate points will have the opportunity to be in different positions, and therefore the MLP can better fit the data containing high-frequency variations. In short, by mapping the input data to a high-dimensional space, the MLP can classify the sampled points more effectively. However, this still cannot effectively fit the highlight details of the object surface, and because the MLP layers of NeRF are deep, there are problems such as gradient disappearance and gradient explosion, which lead to unstable training.

We propose the SResMLP for representing neural radiance fields. First, since the input data are the coordinates of the sampled points $\boldsymbol{x}$ and the viewpoint vector $\boldsymbol{d}$, we use a linear layer to map the input data to higher dimensions, followed by Affine layer, which is very simple and has the role of a Layer Normalization layer:

$$\mathrm{Aff}_{\boldsymbol{\alpha},\boldsymbol{\beta}}(\mathbf{x},\mathbf{d}) = \mathrm{Diag}(\boldsymbol{\alpha}) \cdot \mathbf{cat}(\mathbf{x},\mathbf{d}) + \boldsymbol{\beta} \tag{3}$$

where $\boldsymbol{\alpha}$ and $\boldsymbol{\beta}$ are learnable weight vectors. The Affine layer simply rescales and shifts the input elements. This operation has several advantages over other normalization operations: first, it has no cost in inference compared to Layer Normalization because it can be absorbed in the neighboring linear layers. Second, the Affine operator does not rely on batch statistics as opposed to Batch-Norm [13] and Layer Normalization:

$$\boldsymbol{Z} = \mathrm{cat}(\boldsymbol{x},\boldsymbol{d}) + \mathrm{Linear}(\mathrm{Aff}(\boldsymbol{x},\boldsymbol{d})) \tag{4a}$$

$$\boldsymbol{F} = \mathrm{Aff}(\boldsymbol{Z} + \mathrm{Aff}(\mathrm{Swish}(\mathrm{Linear}(\mathrm{Aff}(\boldsymbol{Z}))))) \tag{4b}$$

$$\boldsymbol{C} = \mathrm{Linear}(\mathrm{ReLU}(\boldsymbol{F})) + \mathrm{cat}(\boldsymbol{x},\boldsymbol{d})) \tag{4c}$$

where $\boldsymbol{Z}$ denote by $\mathrm{Aff}(\boldsymbol{Z})$ the Affine operation applied independently to each column of the matrix $\boldsymbol{Z}$. However, this approach brings another problem: the low-frequency classification cannot be fitted well, which leads to image rendering results with more noise, and defects. To be able to fit the high and low-frequency functions in a balanced way, we concatenate the signal data before the input to the neural network, the residuals to the last layer of the network. The reason for this is that we want the MLP to be able to constantly fit the high-frequency information while also taking into account the fitting of the low-frequency information. To maintain the consistency of the shallow features and deep features of the neural network, the layerscale layer is introduced. Experiments with neural tangent kernel (NTK) [14] show that layerscale can significantly improve the ability of standard MLP to fit high frequency components. AutoInt [20] has been shown to represent neural radiance fields in ReLU, SIREN [34] and Swish activation functions, where Swish activation performs best in terms of PSNR.

Table 1. Quantitative comparisons on the Synthetic-NeRF, Synthetic-NSVF, BlendedMVS and Tanks&Temples datasets. Golden yellow represents the first, orange represents the second, and light yellow represents the third. Our method excels in rendering quality and is better than the original NeRF [25] and the improved DirectVoxGo [37] on the four datasets under all metrics. We also show better results than most of the recent methods.

	Synthetic-NeRF				Synthetic-NSVF			
Methods	PSNR↑	SSIM↑	LPIPS↓(Vgg)	LPIPS↓(Alex)	PSNR↑	SSIM↑	LPIPS↓(Vgg)	LPIPS↓(Alex)
SRN [35]	22.26	0.846	0.170	-	24.33	0.882	-	0.141
NV [22]	26.05	0.893	0.160	-	25.83	0.892	-	0.125
NeRF [25]	31.01	0.947	0.081	-	30.81	0.952	-	0.043
JaxNeRF [5]	31.69	0.953	0.068	-	-	-	-	-
Mip-NeRF [1]	33.09	0.961	0.043	-	-	-	-	-
AutoInt [20]	25.55	0.911	-	-	-	-	-	-
FastNeRF [10]	29.90	0.937	-	-	-	-	-	-
SNeRG [12]	30.38	0.950	-	-	-	-	-	-
NSVF [21]	31.74	0.953	-	0.047	35.13	0.979	-	0.015
PlenOctrees [48]	31.71	0.958	0.053	-	-	-	-	-
Plenoxels [7]	31.71	0.958	0.049	-	-	-	-	-
DirectVoxGo [37]	31.93	0.956	0.053	0.035	35.08	0.975	0.033	0.019
KiloNeRF [32]	31.00	0.950	-	-	33.37	0.970	-	-
Ours	32.52	0.959	0.048	0.032	35.83	0.978	0.029	0.015

	BlendedMVS				Tanks&Temples			
Methods	PSNR↑	SSIM↑	LPIPS↓(Vgg)	LPIPS↓(Alex)	PSNR↑	SSIM↑	LPIPS↓(Vgg)	LPIPS↓(Alex)
SRN [35]	20.51	0.770	-	0.294	24.09	0.847	-	0.251
NV [22]	23.03	0.793	-	0.243	23.70	0.834	-	0.260
NeRF [25]	24.15	0.828	-	0.192	25.78	0.864	-	0.198
JaxNeRF [5]	-	-	-	-	27.94	0.904	0.168	-
NSVF [21]	26.90	0.898	-	0.113	28.40	0.900	-	0.153
PlenOctrees [48]	-	-	-	-	27.99	0.917	0.131	-
Plenoxels [7]	-	-	-	-	-	-	-	-
DirectVoxGo [37]	28.02	0.922	0.101	0.075	28.41	0.911	0.155	0.148
KiloNeRF [32]	27.39	0.920	-	-	28.41	0.910	-	-
Ours	28.50	0.929	0.095	0.069	28.82	0.920	0.147	0.137

4 Experimental

4.1 Datasets

We used four types of publicly available data sets for comparison: Synthetic-NeRF [25], Synthetic-NSVF [21], BlendedMVS [46], Tanks&Temples [16]. Synthetic-NeRF is the standard Blender dataset from the original NeRF paper. Synthetic-NeRF and Synthetic-NSVF contain eight objects. We set the image resolution to 800 × 800 pixels and let each scene have 100 views for training and 200 views for testing. BlendedMVS and Tanks&Temples are also similar datasets.

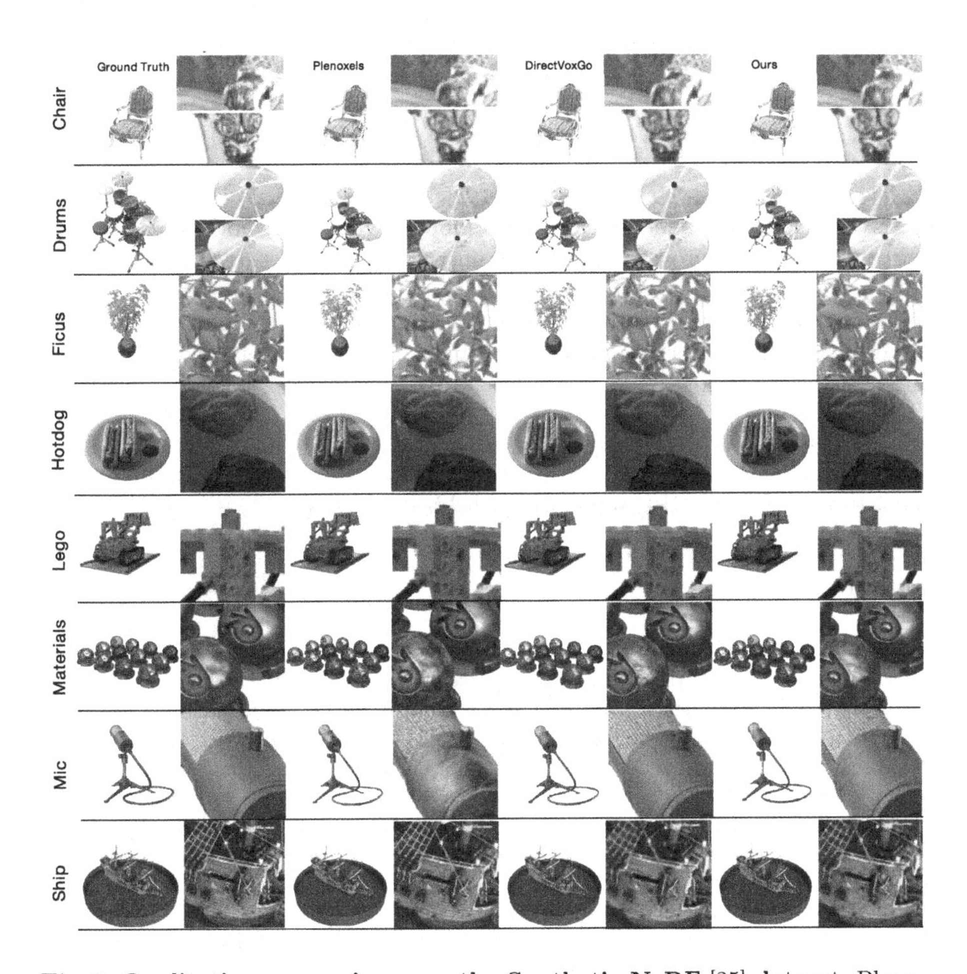

Fig. 3. Qualitative comparisons on the Synthetic-NeRF [25] dataset. Plenoxels [7] and DirectVoxGo [37] fail reconstruct the highlight details on the surface of an object, such as the drums category. Our model reconstructs the highlight details on the surface of drums almost perfectly. Also, our model reconstructs smoother surface of Hotdog.

4.2 Comparisons

Our rendering quality is better than the original NeRF [25] and the improved DirectVoxGo [37] on the four datasets under all metrics. Table 1 and Fig. 3 also show leading results for most of the recent methods. Our model performs very well in all metrics. Although the rendering quality performance of Mip-NeRF [1] is also good, it has a long training time. When compared with other accelerated models, although PlenOctree [48], Plenoxels [7], and DirectVoxGo made significant improvements in acceleration, they did not take into account the rendering quality of images. Table 1 and Fig. 3 shows the superiority of our method in rendering quality, and the training time required for the model is tens of minutes. Our method balances model training speed with image rendering quality.

5 Conclusion

Our MLP network designed for NeRFs combined with direct optimization of 3D Grid significantly improves image rendering quality. Compared with the existing NeRFs methods, our method achieves better performance in terms of image rendering quality, especially in specular reflection and image highlight detail. In the future, reducing memory overheads and speeding up network inference is worth further exploring.

Acknowledgements. The authors would like to thank the School of Cyber Science and Engineering, Zhengzhou University, the Zhengzhou City Collaborative Innovation Major Project, and the GPU support provided by the 3DCV lab.

References

1. Barron, J.T., Mildenhall, B., Tancik, M., Hedman, P., Martin-Brualla, R., Srinivasan, P.P.: Mip-NeRF: a multiscale representation for anti-aliasing neural radiance fields. In: Proceedings of the IEEE/CVF International Conference on Computer Vision, pp. 5855–5864 (2021)
2. Bi, S., et al.: Neural reflectance fields for appearance acquisition. arXiv preprint arXiv:2008.03824 (2020)
3. Boss, M., Braun, R., Jampani, V., Barron, J.T., Liu, C., Lensch, H.: Nerd: neural reflectance decomposition from image collections. In: Proceedings of the IEEE/CVF International Conference on Computer Vision, pp. 12684–12694 (2021)
4. Chan, E.R., Monteiro, M., Kellnhofer, P., Wu, J., Wetzstein, G.: pi-GAN: periodic implicit generative adversarial networks for 3d-aware image synthesis. In: Proceedings of the IEEE/CVF Conference on Computer Vision and Pattern Recognition, pp. 5799–5809 (2021)
5. Deng, B., Barron, J.T., Srinivasan, P.P.: JaxNeRF: an efficient JAX implementation of nerf (2020). https://github.com/google-research/google-research/tree/master/jaxnerf
6. Dosovitskiy, A., et al.: An image is worth 16x16 words: Transformers for image recognition at scale. arXiv preprint arXiv:2010.11929 (2020)
7. Fridovich-Keil, S., Yu, A., Tancik, M., Chen, Q., Recht, B., Kanazawa, A.: Plenoxels: Radiance fields without neural networks. In: Proceedings of the IEEE/CVF Conference on Computer Vision and Pattern Recognition, pp. 5501–5510 (2022)
8. Gafni, G., Thies, J., Zollhofer, M., Nießner, M.: Dynamic neural radiance fields for monocular 4D facial avatar reconstruction. In: Proceedings of the IEEE/CVF Conference on Computer Vision and Pattern Recognition, pp. 8649–8658 (2021)

9. Gao, C., Saraf, A., Kopf, J., Huang, J.B.: Dynamic view synthesis from dynamic monocular video. In: Proceedings of the IEEE/CVF International Conference on Computer Vision, pp. 5712–5721 (2021)
10. Garbin, S.J., Kowalski, M., Johnson, M., Shotton, J., Valentin, J.: FastNeRF: High-fidelity neural rendering at 200fps. In: Proceedings of the IEEE/CVF International Conference on Computer Vision, pp. 14346–14355 (2021)
11. Glorot, X., Bordes, A., Bengio, Y.: Deep sparse rectifier neural networks. In: Proceedings of the Fourteenth International Conference on Artificial Intelligence and Statistics, pp. 315–323. JMLR Workshop and Conference Proceedings (2011)
12. Hedman, P., Srinivasan, P.P., Mildenhall, B., Barron, J.T., Debevec, P.: Baking neural radiance fields for real-time view synthesis. In: Proceedings of the IEEE/CVF International Conference on Computer Vision, pp. 5875–5884 (2021)
13. Ioffe, S., Szegedy, C.: Batch normalization: accelerating deep network training by reducing internal covariate shift. In: International Conference on Machine Learning, pp. 448–456. PMLR (2015)
14. Jacot, A., Gabriel, F., Hongler, C.: Neural tangent kernel: convergence and generalization in neural networks. In: Advances in Neural Information Processing Systems, vol. 31 (2018)
15. Jeong, Y., Ahn, S., Choy, C., Anandkumar, A., Cho, M., Park, J.: Self-calibrating neural radiance fields. In: Proceedings of the IEEE/CVF International Conference on Computer Vision, pp. 5846–5854 (2021)
16. Knapitsch, A., Park, J., Zhou, Q.Y., Koltun, V.: Tanks and temples: benchmarking large-scale scene reconstruction. ACM Trans. Graphics (ToG) **36**(4), 1–13 (2017)
17. Kosiorek, A.R., et al.: NeRF-VAE: a geometry aware 3D scene generative model. In: International Conference on Machine Learning, pp. 5742–5752. PMLR (2021)
18. Li, Z., Niklaus, S., Snavely, N., Wang, O.: Neural scene flow fields for space-time view synthesis of dynamic scenes. In: Proceedings of the IEEE/CVF Conference on Computer Vision and Pattern Recognition, pp. 6498–6508 (2021)
19. Lin, C.H., Ma, W.C., Torralba, A., Lucey, S.: BARF: bundle-adjusting neural radiance fields. In: Proceedings of the IEEE/CVF International Conference on Computer Vision, pp. 5741–5751 (2021)
20. Lindell, D.B., Martel, J.N., Wetzstein, G.: Autoint: automatic integration for fast neural volume rendering. In: Proceedings of the IEEE/CVF Conference on Computer Vision and Pattern Recognition, pp. 14556–14565 (2021)
21. Liu, L., Gu, J., Zaw Lin, K., Chua, T.S., Theobalt, C.: Neural sparse voxel fields. Adv. Neural. Inf. Process. Syst. **33**, 15651–15663 (2020)
22. Lombardi, S., Simon, T., Saragih, J., Schwartz, G., Lehrmann, A., Sheikh, Y.: Neural volumes: learning dynamic renerable volumes from images. ACM Trans. Graphics **38**(4), 1–14 (2019). https://doi.org/10.1145/3306346.3323020
23. Martin-Brualla, R., Radwan, N., Sajjadi, M.S., Barron, J.T., Dosovitskiy, A., Duckworth, D.: NeRF in the wild: neural radiance fields for unconstrained photo collections. In: Proceedings of the IEEE/CVF Conference on Computer Vision and Pattern Recognition, pp. 7210–7219 (2021)
24. Meng, Q., et al.: GNeRF: GAN-based neural radiance field without posed camera. In: Proceedings of the IEEE/CVF International Conference on Computer Vision, pp. 6351–6361 (2021)
25. Mildenhall, B., Srinivasan, P.P., Tancik, M., Barron, J.T., Ramamoorthi, R., Ng, R.: NeRF: representing scenes as neural radiance fields for view synthesis. In: Vedaldi, A., Bischof, H., Brox, T., Frahm, J.-M. (eds.) ECCV 2020. LNCS, vol. 12346, pp. 405–421. Springer, Cham (2020). https://doi.org/10.1007/978-3-030-58452-8_24

26. Noguchi, A., Sun, X., Lin, S., Harada, T.: Neural articulated radiance field. In: Proceedings of the IEEE/CVF International Conference on Computer Vision, pp. 5762–5772 (2021)
27. Park, K., et al.: Deformable neural radiance fields (2020)
28. Pumarola, A., Corona, E., Pons-Moll, G., Moreno-Noguer, F.: D-NeRF: neural radiance fields for dynamic scenes. In: Proceedings of the IEEE/CVF Conference on Computer Vision and Pattern Recognition, pp. 10318–10327 (2021)
29. Rahaman, N., et al.: On the spectral bias of neural networks. In: International Conference on Machine Learning, pp. 5301–5310. PMLR (2019)
30. Ramachandran, P., Zoph, B., Le, Q.V.: Searching for activation functions. arXiv preprint arXiv:1710.05941 (2017)
31. Rebain, D., Jiang, W., Yazdani, S., Li, K., Yi, K.M., Tagliasacchi, A.: DeRF: decomposed radiance fields. In: Proceedings of the IEEE/CVF Conference on Computer Vision and Pattern Recognition, pp. 14153–14161 (2021)
32. Reiser, C., Peng, S., Liao, Y., Geiger, A.: KiloNeRF: speeding up neural radiance fields with thousands of tiny MLPs. In: Proceedings of the IEEE/CVF International Conference on Computer Vision, pp. 14335–14345 (2021)
33. Schwarz, K., Liao, Y., Niemeyer, M., Geiger, A.: GRAF: generative radiance fields for 3d-aware image synthesis. Adv. Neural. Inf. Process. Syst. **33**, 20154–20166 (2020)
34. Sitzmann, V., Martel, J., Bergman, A., Lindell, D., Wetzstein, G.: Implicit neural representations with periodic activation functions. Adv. Neural. Inf. Process. Syst. **33**, 7462–7473 (2020)
35. Sitzmann, V., Zollhöfer, M., Wetzstein, G.: Scene representation networks: Continuous 3d-structure-aware neural scene representations. In: Advances in Neural Information Processing Systems, vol. 32 (2019)
36. Srinivasan, P.P., Deng, B., Zhang, X., Tancik, M., Mildenhall, B., Barron, J.T.: NeRV: neural reflectance and visibility fields for relighting and view synthesis. In: Proceedings of the IEEE/CVF Conference on Computer Vision and Pattern Recognition, pp. 7495–7504 (2021)
37. Sun, C., Sun, M., Chen, H.T.: Direct voxel grid optimization: super-fast convergence for radiance fields reconstruction. In: Proceedings of the IEEE/CVF Conference on Computer Vision and Pattern Recognition, pp. 5459–5469 (2022)
38. Tancik, M., et al.: Learned initializations for optimizing coordinate-based neural representations. In: Proceedings of the IEEE/CVF Conference on Computer Vision and Pattern Recognition, pp. 2846–2855 (2021)
39. Tancik, M., et al.: Fourier features let networks learn high frequency functions in low dimensional domains. Adv. Neural. Inf. Process. Syst. **33**, 7537–7547 (2020)
40. Tolstikhin, I.O., et al.: MLP-mixer: an all-MLP architecture for vision. Adv. Neural. Inf. Process. Syst. **34**, 24261–24272 (2021)
41. Touvron, H., et al.: ResMLP: feedforward networks for image classification with data-efficient training. arXiv preprint arXiv:2105.03404 (2021)
42. Tretschk, E., Tewari, A., Golyanik, V., Zollhöfer, M., Lassner, C., Theobalt, C.: Non-rigid neural radiance fields: reconstruction and novel view synthesis of a dynamic scene from monocular video. In: Proceedings of the IEEE/CVF International Conference on Computer Vision, pp. 12959–12970 (2021)
43. Vaswani, A., et al.: Attention is all you need. In: Advances in Neural Information Processing Systems, vol. 30 (2017)
44. Wang, Z., Wu, S., Xie, W., Chen, M., Prisacariu, V.A.: NeRF-: neural radiance fields without known camera parameters. arXiv preprint arXiv:2102.07064 (2021)

45. Xian, W., Huang, J.B., Kopf, J., Kim, C.: Space-time neural irradiance fields for free-viewpoint video. In: Proceedings of the IEEE/CVF Conference on Computer Vision and Pattern Recognition, pp. 9421–9431 (2021)
46. Yao, Y., et al.: BlendedMVS: a large-scale dataset for generalized multi-view stereo networks. In: Proceedings of the IEEE/CVF Conference on Computer Vision and Pattern Recognition, pp. 1790–1799 (2020)
47. Yen-Chen, L., Florence, P., Barron, J.T., Rodriguez, A., Isola, P., Lin, T.Y.: INeRF: inverting neural radiance fields for pose estimation. In: 2021 IEEE/RSJ International Conference on Intelligent Robots and Systems (IROS), pp. 1323–1330. IEEE (2021)
48. Yu, A., Li, R., Tancik, M., Li, H., Ng, R., Kanazawa, A.: PlenOctrees for real-time rendering of neural radiance fields. In: Proceedings of the IEEE/CVF International Conference on Computer Vision, pp. 5752–5761 (2021)
49. Zhang, X., Srinivasan, P.P., Deng, B., Debevec, P., Freeman, W.T., Barron, J.T.: NeRFactor: neural factorization of shape and reflectance under an unknown illumination. ACM Trans. Graphics (TOG) **40**(6), 1–18 (2021)

Real-Time and Interactive Application

LiteHandNet: A Lightweight Hand Pose Estimation Network via Structural Feature Enhancement

Zhi-Yong Huang[1,2], Song-Lu Chen[1,2], Qi Liu[1,2], Chong-Jian Zhang[1,2], Feng Chen[2,3], and Xu-Cheng Yin[1,2](✉)

[1] University of Science and Technology Beijing, Beijing 100083, China
{huang.zhiyong,qiliu7,chongjianzhang}@xs.ustb.edu.cn,
{songluchen,xuchengyin}@ustb.edu.cn
[2] USTB-EEasyTech Joint Lab of Artificial Intelligence, Beijing 100083, China
[3] EEasy Technology Company Ltd., Zhuhai 519000, China
cfeng@eeasytech.com

Abstract. This paper presents a real-time lightweight network, LiteHandNet, for 2D hand pose estimation from monocular color images. In recent years, keypoint heatmap representation is dominant in pose estimation due to its high accuracy. Nevertheless, keypoint heatmaps require high-resolution representation to extract accurate spatial features, which commonly means high computational costs, e.g., high delay and tremendous model parameters. Therefore, the existing heatmap-based methods are not suitable for the scenes with computation-limited resources and high real-time requirements. We find that high-resolution representation can obtain more clear structural features of a hand, e.g., contours and key regions, which can provide high-quality spatial features to the keypoint heatmap, thus improving the robustness and accuracy of a model. To fully extract the structural features without introducing unnecessary computational costs, we propose a lightweight module, which consists of two parts: a multi-scale feature block (MSFB) and a spatial channel attention block (SCAB). MSFB can extract structural features from hands using multi-scale information, while SCAB can further screen out high-quality structural features and suppress low-quality features. Comprehensive experimental results verify that our model is state-of-the-art in terms of the tradeoff between accuracy, speed, and parameters.

Keywords: Hand pose estimation · Liteweight · Structural feature

1 Introduction

2D hand pose estimation has a wide range of practical application scenarios, such as gesture recognition, virtual reality (VR), augmented reality (AR), human-computer interaction, etc. In recent years, pose estimation based on convolutional neural networks has been developed rapidly. 2D human pose estimation [17,23,28] aims in determining the position of body keypoints, which is a

D.-T. Dang-Nguyen et al. (Eds.): MMM 2023, LNCS 13833, pp. 321–333, 2023.
https://doi.org/10.1007/978-3-031-27077-2_25

mature research field in pose estimation and greatly promotes the development of 2D hand pose estimation. Structural features inherent in the hands are significant in hand pose estimation and are used to improve model performance [1,10,11]. In addition, some methods [2,22,30] have verified the significance of high-resolution representation, which can provide accurate spatial information for pose estimation to achieve SOTA performance. However, existing models are hard to be applied to real-world scenarios with computation-limited resources or high real-time requirements due to the expensive computational costs incurred for processing high-resolution representation.

LiteHRNet [30] is a SOTA lightweight high-resolution network with fewer model parameters and FLOPs than other lightweight pose estimation methods [4,6,16,19]. However, the multi-branch network architecture of LiteHRNet leads to high memory access cost (MAC) [16], which makes the running speed too slow to satisfy the scenarios requiring real-time calculation. The lightweight network [20] also suffers from slow speed due to the complexity of self-attention and dense connection. SRHandNet [26] is a real-time 2D hand pose estimation model and achieves SOTA on the OneHand10K [25] and RHD [34] datasets, but tremendous computational costs limit its application scenarios.

To efficiently extract spatial information with few unnecessary computational costs, we propose MSFB and SCAB. Unlike most lightweight modules [4,6,16,19,30] that focus on reducing redundant costs of convolution operations, our lightweight modules focus on providing high-quality structural features (see Fig. 1) for keypoint heatmaps. MSFB can extract structural features with rich multi-scale information, while SCAB can further screen out high-quality structural features and suppress low-quality features by channel attention.

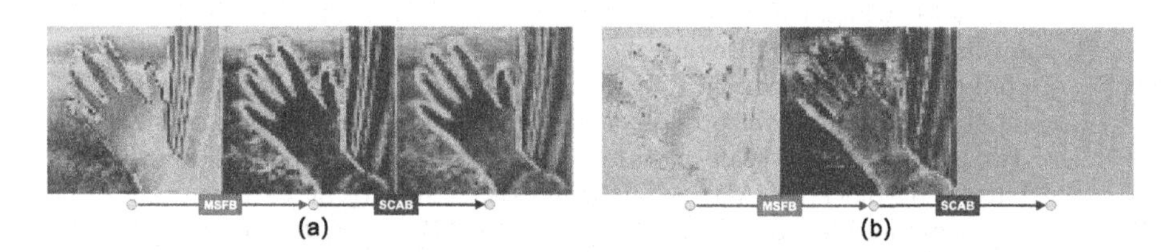

Fig. 1. Illustration of feature maps extracted by MSFB and SCAB. (a) illustrates the process of enhancing high-quality structural features. (b) illustrates the process of supressing low-quality structural features.

Our main contributions are as follows.

- We propose an lightweight 2D hand pose estimation network that focuses on reducing the computational costs of high-resolution representation.
- We introduce the multi-scale feature block and the spatial channel attention block to extract high-quality structural features and enhance network robustness.
- We achieve a state-of-the-art performance in terms of the tradeoff between accuracy, speed, and parameters on challenging hand datasets, including Panoptic, OneHand10K, FreiHand, and RHD.

2 Related Work

2.1 Hand Pose Estimation

There are two main keypoint representations for 2D hand pose estimation, i.e., 2D heatmap [2,17,22,30] and pose regression [3,12,20,23]. Compared with pose regression, 2D heatmap representation generally has higher performance due to retaining spatial information. The 2D heatmap representation, however, requires high-resolution representation to achieve high performance, which makes it have a higher computational complexity. In recent years, some regression-based methods [3,12,27] can achieve performance close to the heatmap-based methods but require complicated train tricks, such as well-designed supervision and longer training epoches. Other keypoint representations, DSNT [18] and SimDR [15], which implicitly learns the 2D heatmap and retains spatial information of the 2D heatmap, can reduce dependence on high-resolution representation. However, there is still a performance gap between other methods and the heatmap-based for high-resolution representation. Given the high-performance potential of 2D heatmap, we propose a lightweight network to effectively extract clear structural features required for 2D heatmap while reducing the computational cost of processing high-resolution representation.

2.2 Multi-scale Feature Extraction

High-to-low and low-to-high resolution processing [22] are widely used in recent 2D pose estimation, which is conducive to extracting multi-scale features to solve the problem of scale variation. The processing is generally found in two representative network architectures: the encoder-decoder network [17] and the multi-branch network [22]. Commonly, the multi-brach network architecture can reach better performance by retaining the high-resolution representation, which contains accurate spatial information. However, high memory access cost (MAC) [16] of multi-brach networks leads to high time delay. Li et al. [13] propose a multi-scale residual block, which introduces convolution kernels of different sizes to enhance multi-scale detection, but the convolution with big kernel sizes introduces expensive computational costs. Therefore, we propose a lightweight module, MSFB, and put it on the input and output layers of an encoder-decoder network, which can effectively extract multi-scale features without introducing heavy computational costs.

2.3 Channel Attention Network

Some methods [7,14,29,33] introduce attention mechanisms into the field of computer vision. Instead of the costly 1×1 convolution, LiteHRNet [30] proposes a channel attention module to fuse multi-scale features. Although channel attention networks can enhance channel-wise features with a few computational costs, the channel features extracted by global average pooling or global max pooling lose a lot of spatial information. Therefore, we propose SCAB, which can retain more spatial information to screen out high-quality structural features by channel weighting.

3 Method

We adopt the encoder-decoder network with asymmetric input sizes to extract multi-scale features with less computational complexity (see Fig. 2). To efficiently extract clear structural features, we use two techniques as follows. First, we enhance the input and output layers of the encoder-decoder network using MSFB and SCAB, which are key parts of the network for extracting rich multi-scale features. Second, we add many residual connections between the encoder and decoder to pass the shallow features directly to the deep features, which can retain the spatial information of the high-resolution representation.

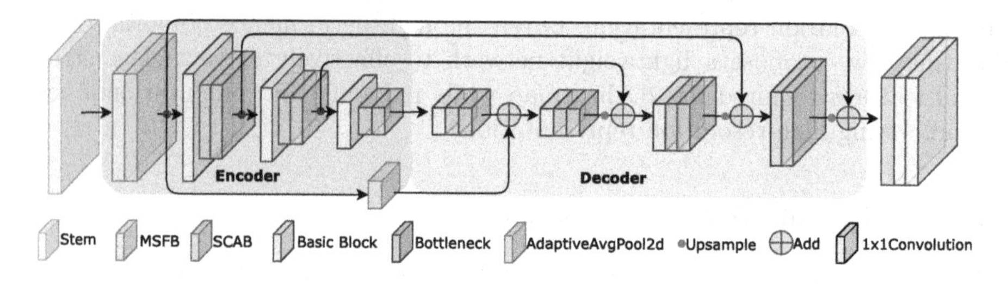

Fig. 2. Illustration of the LiteHandNet architecture.

Our model is designed in a lightweight manner with the following principles. First, we apply depthwise convolutions [6] only for the high-resolution representation and regular convolutions for the rest low-resolution representation, which reduces computational cost while retaining the powerful channel-wise information exchange capability of the encoder-decoder network. Second, the input and output of the convolution modules are maintained at 128 channels instead of increasing the number of channels after downsampling operations. The above strategies have two main advantages. One is to reduce the computational costs without having a significant impact on performance. The other is to overcome the drawback of cross-channel information being blocked in group convolutions [8] and depthwise convolutions. In addition, accurate spatial information is the key to 2D hand pose estimation. Therefore, we propose MSFB and SCAB to extract spatial information, e.g. structural features, without introducing too much redundant computational costs. Table 1 demonstrates that MSFB and SCAB can effectively reduce the computational complexity and model parameters. w and h are the spatial sizes of the feature map. The number of input channels and output channels is C_i and C_o. g is the group numbers of convolution. The bias of a convolution layer is not included in our computation. FLOPs and #Params are calculated under the arguments: $w = h = 64$, $C_i = C_o = g = 128$.

Table 1. Computational comlexity and parameters of modules. Our lightweight modules can efficiently reduce the computational costs with the same input and output channels.

Module	Theory complexity	FLOPs	Theory parameters	#Params
1×1 convolution	whC_oC_i	67.109M	C_oC_i	0.016M
3×3 convolution	$17whC_oC_i$	1140.851M	$9C_oC_i$	0.147M
DPBlock	$whC_i(C_i + C_o + 34)$	152.044M	$C_i(C_i + C_o + 18)$	0.035M
MSFB	$whC_i(9C_i/4 + C_o + 68)$	253.755M	$C_i(9C_i/4 + C_o + 36)$	0.058M
SCAB	$17C_i + C_iC_o + C_o$	0.019M	$9C_i + C_o(C_i + 1)$	0.018M

The illustration of our network and subnetworks are shown in Fig. 2 and Fig. 3, respectively. The pipeline of our network is as follows. First, we use a simple stem module (see Fig. 3(a)) to extract the 4× downsampled feature map, and the stem module is inspired by PeleeNet [24], which can extract rich shallow features with a little computational complexity. Second, MSFB and SCAB of the encoder-decoder network extract high-quality structural features from shallow features and deep features. Finally, the 2D keypoint heatmap is estimated through one bottleneck and two 1×1 convolutions.

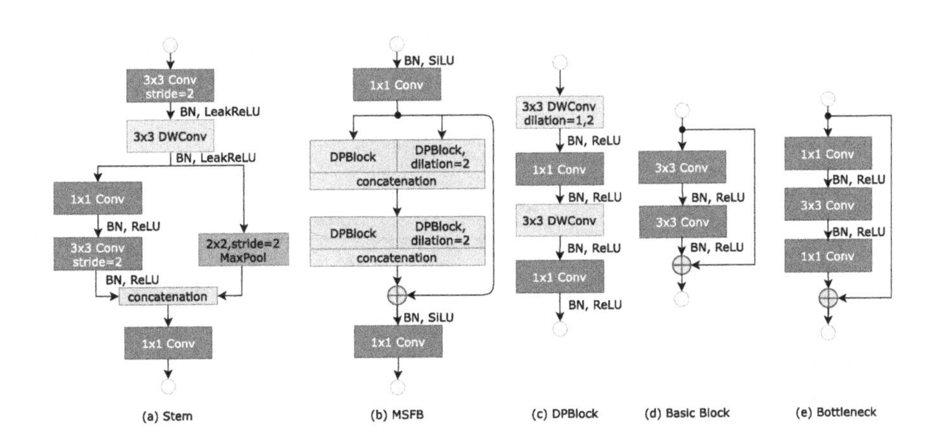

Fig. 3. The modules of LiteHandNet.

3.1 Multi-scale Feature Blcok (MSFB)

The architecture of MSFB is shown in Fig. 3. On the high-resolution representation of the encoder-decoder network, we use MSFB to effectively extract rich structural features. To reduce the computational complexity, we first use a 1×1 convolution to reduce the number of channels C to C/2. There are two groups of DPBlock and two DPBlock in the same group with different receptive fields by adopting a dilation rate of 2. The 1×1 convolution at the end of each DPBlock is for changing the channel dimensions. The channel dimensions of DPBlock in the first group are C/2, C/2, C/2, and C/4, while they are all C/2 in the second group. Finally, we use a residual connection and a 1×1 convolution to fuse multi-scale information. In our experiment, we set channel C as 128.

3.2 Spatial Channel Attention Block (SCAB)

The illustration of SCAB is shown in Fig. 4. SCAB can efficiently screen out high-quality spatial features, which depends on retaining more spatial information when compressing channel features. Therefore, unlike the previous channel attention modules [7,29], we do not use the adaptive average pooling to directly extract a global value of each channel, which will lose too much spatial information. We use the adaptive average pooling to obtain a 3×3 feature matrix for each channel, which compresses the spatial information on the original feature map into nine grids. The spatial information of the matrix will be extracted through a 3×3 depthwise convolution to generate a global value of each channel. Finally, the channel features with more accurate spatial information are fused by a full connection layer to generate channel attention.

The SCAB performs a matrix-vector multiplication at each position.

$$Y = \mathbf{A} \otimes X \tag{1}$$

where X and Y are input and output maps, and $\mathbf{A}$ is the channel attention vector.

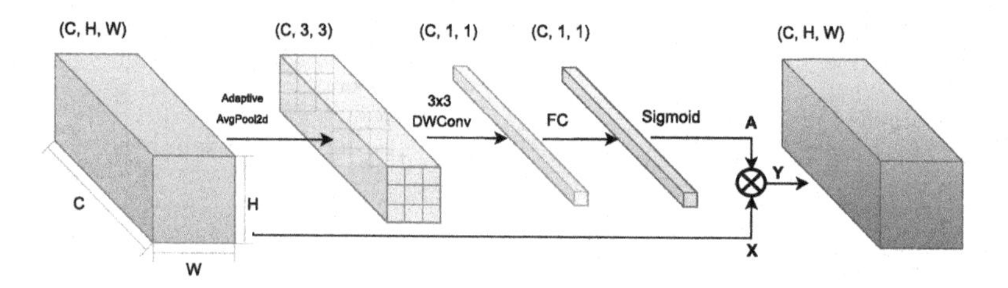

Fig. 4. The illustration of spatial channel attention block. DWConv means depthwise convolution. FC indicates a fully connected layer. The shape of the features maps is a triple (channels C, height H, and width W).

3.3 Loss Function

We discover that MSE loss on the Gaussian heatmap can not solve the problem of imbalance between positive and negative samples. Therefore, we use the MSE loss with balance factors to supervise the model training. Our loss function $\mathcal{L}$ is as follows.

$$\mathcal{L} = \lambda \mathcal{L}_1 + \mathcal{L}_2 \tag{2}$$

where:

$$\begin{cases} \mathcal{L}_1 = \frac{N}{N_p} \times \sum MSE(p, \hat{p}), \hat{p} > \rho \\ \mathcal{L}_2 = \frac{N}{N_n} \times \sum MSE(p, \hat{p}), \hat{p} \leq \rho \end{cases} \tag{3}$$

If a pixel value $\hat{p}$ of the ground truth heatmap is greater than the threshold ρ, we view it as a positive sample, otherwise a negative sample. p is a pixel value of the predicted heatmap with the same position as $\hat{p}$. N is the number of pixels on the heatmap. N_p is the number of positive samples, and N_n is the number of negative samples. In our experiments, we set the loss balance term λ to 0.1 and the threshold ρ to 0.5.

4 Experiments

4.1 Experimental Settings

Datasets. We conduct experiments on the Panoptic [21], OneHand10K [25], FreiHand [35], and RHD [34] datasets. Panoptic is a well-known hand pose estimation dataset, which contains hands from the wild and synthetic data. OneHand10K is an in-the-wild 2D hand pose dataset with 10k RGB images, and each image contains a hand and its annotations, including the segmentation mask and 21 labeled keypoints. FreiHand is a large-scale, multi-view hand dataset with annotations of 3D hand pose and mask. RHD is a synthetic dataset and provides 21 labeled keypoints, an RGB image, a depth map, and a segmentation mask for each hand sample. The overview of these four datasets is shown in Table 2.

Table 2. The overview of datasets. #Kpt means the number of keypoints. Wild denotes whether the dataset is collected in the wild.

Dataset	Size	#Kpt	Wild	Images		
				Train	Test	Total
FreiHand [35]	224 × 224	21		104192	13024	117216
RHD [34]	256 × 256	21		41255	2727	43982
Panoptic [21]	256 × 256	21		16729	846	17575
OneHand10K [25]	256 × 256	21	✓	10000	1703	11703

Training. We use 4 NVIDIA Titan RTX GPUs for training with mini-batch size 16 per GPU. We used Adam [9] optimizer with an initial learning rate of $5e^{-4}$. The image resolution is 224 × 224 for the FreiHand dataset and 256 × 256 for the other datasets. In addition, we adopt some common data augmentation, such as random scaling, random flipping, random image center shift, etc.

Testing. All of our results do not use test time augmentation (TTA), such as the flip test and multi-scale testing. These operations will seriously affect the running speed, which makes TTA rarely used in actual scenarios. We adopt DARK [31] post-processing to reduce the quantization error when taking keypoint coordinates from the heatmap. We calculate the average running speed of 10,000 images on a GPU (NVIDIA Titan RTX) and a CPU (Intel Xeon e5-2680 V4), respectively.

Evaluation. There are several metrics to evaluate the model performance, such as PCK (probability of correct keypoint), PCKh (head-normalized probability of correct keypoint), and AUC (area under the curve). All datasets use PCK as the default metric, except for Panoptic which uses PCKh. To evaluate the overall performance, we use a uniform metric mAUC, which is the average of the AUCs of the four datasets. In addition, we use frames per second (FPS) to measure the running speed. The calculation formulas are as follows.

$$\mathrm{PCK}^*_{\sigma,\mathrm{S}} = \frac{\sum_{i=1}^{N}\left(\frac{\|p_i - p_i^*\|_2}{S} \leq \sigma\right)\delta(v_i > 0)}{\sum_{i=1}^{N}\delta(v_i > 0)} \tag{4}$$

$\delta(v_i > 0)$ is a Boolean function that determines whether the i-th keypoint is visible or not. N is the number of keypoints. P_i and P_i^* represent the predicted coordinate and a ground truth for the i-th keypoint, respectively. S is a scale-normalized factor.

$$PCK = PCK^*_{\sigma=0.2,S=\max(w,h)} \tag{5}$$

σ is a threshold. w and h are the size of the hand.

$$PCKh = PCK^*_{\sigma=0.5,S=h} \tag{6}$$

h is a head-normalized factor.

$$AUC = \frac{1}{20}\sum_{i=1}^{20} PCK^*_{\sigma=0.05*(i-1),S=30} \tag{7}$$

4.2 Ablation Study

We perform ablation experiments on the FreiHand dataset, as shown in Table 3, which verified the effectiveness of the MSFB and SCBA modules. MSFB increases by 4.5 AUC in comparison to a module with one Basic Block and two Bottleneck, but the parameters and GFLOPs decrease by 20.3% and 55.5%, respectively. Meanwhile, SCAB increases the AUC by 3.0–12.6% compared with SEBlock [7], CBAM [29], 1×1 convolution, and Identity. In addition, SCAB only increases the parameters by 1.8%, while GFLOPs barely increases. We discover that adding SEBlock after MSFB will lead to the degradation of model performance because SEBlock only focuses on the global information, resulting in weak generalization ability. Compared with Identity, the AUC of using CBAM and

SCAB modules can be increased by 0.6% and 3.6%, respectively, which means the detailed spatial information is conducive to enhancing the effectiveness of channel attention. Compared with CBAM, SCAB achieves higher performance and lower computational costs.

Table 3. Ablation study on FreiHand dataset. HRM denotes high-resolution module. CEM denotes channel enhancement module. Conv denotes convolution. *Block**is a module consisting of one Basic Block and two Bottleneck. Identity indicates no additional operation.

Model	HRM	CEM	GFLOPs	#Params	PCK@0.2	AUC
LiteHandNet	Blocks*	SCAB	2.56	2.76M	95.5	77.8
LiteHandNet	MSFB	Identity	1.14	2.20M	96.4	78.7
LiteHandNet	MSFB	1×1Conv	1.22	2.24M	96.4	78.6
LiteHandNet	MSFB	SEBlock [7]	1.14	2.21M	92.7	73.1
LiteHandNet	MSFB	CBAM [29]	2.74	2.83M	96.7	79.3
LiteHandNet	MSFB	SCAB	1.14	2.24M	**97.8**	**82.3**

4.3 Comparative Results

We compare our method with other state-of-the-art methods in Table 4. LiteHandNet achieves the best mAUC of 71.08%, an improvement of 1.11% over LiteHRNet30, indicating that our method achieves the best overall performance. MobileNetV2 [19] and ResNet18 [5] are the backbones of SimpleBaseline [32]. Our method achieves the best AUC on Panoptic and OneHand10K, 65.20% and 51.39%, respectively. On the FreiHand and RHD datasets, our model can achieve competitive AUC compared with SRHandNet [26] and LiteHRNet30 [30].

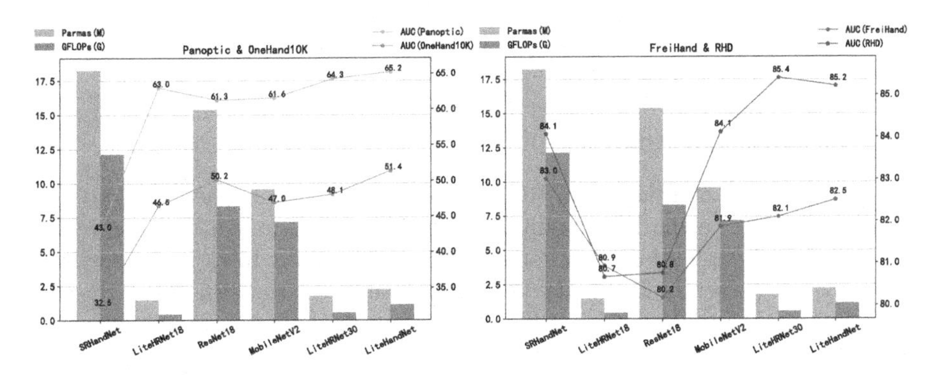

Fig. 5. Comparison of the computational costs and AUC on OneHand10K, Panoptic, FreiHand, and RHD.

The comparison of inference speed is shown in Table 5. Compared with LiteHRNet30, which is ranked second in mAUC, our model has improved the FPS on GPU and CPU by 574% and 470%, respectively. ResNet18 [5], MobileNetV2 [19], and SRHandNet [26] run remarkably fast on the GPU due to their similar network architecture. However, the above methods have higher computational costs and lower model performance compared to our method. Figure 5 illustrates that our model can remarkably reduce computational costs while retaining high performance. In conclusion, our model achieves the best trade-off between performance and speed.

Table 4. Comparative experiments. All of the models are trained with the same training setting, and our method can achieve the best mAUC with competitive model parameters.

model	GFLOPs	#Params	Panoptic		OneHand10K		FreiHand		RHD		mAUC
			PCKh@0.5	AUC	PCK@0.2	AUC	PCK@0.2	AUC	PCK@0.2	AUC	
MobileNetV2	7.15	9.59M	98.82	61.58	95.52	46.98	97.62	81.86	94.59	84.12	68.64
ResNet18	8.31	15.38M	98.76	61.30	97.05	50.18	96.90	80.17	92.44	80.76	68.10
SRHandNet	12.14	18.26M	97.59	42.97	84.56	32.48	98.03	**83.01**	94.25	84.86	60.83
LiteHRNet18	0.42	1.48M	98.76	63.02	95.63	46.50	97.32	80.93	92.47	80.67	67.78
LiteHRNet30	0.56	1.77M	98.87	64.26	95.78	48.12	97.93	82.10	95.62	**85.41**	69.97
LiteHandNet(ours)	1.14	2.24M	98.73	**65.20**	97.64	**51.39**	97.78	82.51	95.02	85.21	**71.08**

Table 5. Comparison of inference speed. Our method can run in real time on the GPU and achieve a comparable speed to other methods on the CPU.

Model	GFLOPs	#Params	Input size	GPU			CPU		
				FPS	mean (ms)	std (ms)	FPS	mean (ms)	std (ms)
MobileNetV2	7.15	9.59M	256×256	130.7	7.7	0.3	16.3	61.5	20.3
ResNet18	8.31	15.38M	256×256	239.7	4.2	0.2	20.1	49.7	16.0
SRHandNet	12.14	18.26M	256×256	125.5	8.0	0.3	10.5	95.0	25.2
LiteHRNet18	0.42	1.48M	256×256	18.2	54.8	2.1	3.9	257.8	49.4
LiteHRNet30	0.56	1.77M	256×256	12.7	78.5	6.6	3.0	336.8	57.2
LiteHandNet(ours)	1.14	2.24M	256×256	72.9	13.7	0.6	14.1	71.1	20.1

4.4 Visualization Analysis

The illumination condition of an RGB image have a great impact on the detection results. Due to the high-quality structural features, our model has higher robustness to illumination variations, which is significant to estimate accurate keypoints. As shown in Fig. 6, the lack of structural information under the dim illumination condition makes it difficult for other methods to estimate accurate keypoints.

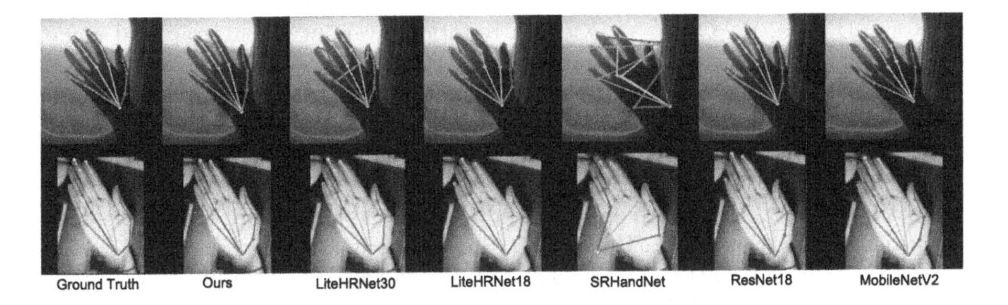

Fig. 6. Qualitative results under different illumination conditions. Compared with other methods, our method is considerably robust in dim illumination.

5 Conclusion

In this paper, we propose a lightweight 2D hand pose estimation network that can efficiently extract high-quality structural features from high-resolution representations. Comprehensive experiments verify that our method can achieve a state-of-the-art performance in terms of the tradeoff between accuracy, speed, and parameters. Our method can be further applied to related studies in the future, such as whole-body pose estimation and multi-hand pose estimation.

Acknowledgements. The research is supported by National Key Research and Development Program of China (2020AAA0109701), National Natural Science Foundation of China (62076024, 62006018).

References

1. Chen, Y., et al.: Nonparametric structure regularization machine for 2D hand pose estimation. In: WACV, pp. 370–379 (2020)
2. Cheng, B., Xiao, B., Wang, J., Shi, H., Huang, T.S., Zhang, L.: Higherhrnet: scale-aware representation learning for bottom-up human pose estimation. In: CVPR, pp. 5385–5394 (2020)
3. Gu, K., Yang, L., Yao, A.: Removing the bias of integral pose regression. In: ICCV, pp. 11047–11056 (2021)
4. Han, K., Wang, Y., Tian, Q., Guo, J., Xu, C., Xu, C.: Ghostnet: more features from cheap operations. In: CVPR, pp. 1577–1586 (2020)
5. He, K., Zhang, X., Ren, S., Sun, J.: Deep residual learning for image recognition. In: CVPR, pp. 770–778 (2016)
6. Howard, A.G., et al.: Mobilenets: efficient convolutional neural networks for mobile vision applications. arXiv (2017)
7. Hu, J., Shen, L., Sun, G.: Squeeze-and-excitation networks. In: CVPR, pp. 7132–7141 (2018)
8. Ioannou, Y., Robertson, D.P., Cipolla, R., Criminisi, A.: Deep roots: improving CNN efficiency with hierarchical filter groups. In: CVPR, pp. 5977–5986 (2017)
9. Kingma, D.P., Ba, J.: Adam: a method for stochastic optimization. In: ICLR (2015)

10. Kong, D., Ma, H., Chen, Y., Xie, X.: Rotation-invariant mixed graphical model network for 2d hand pose estimation. In: WACV, pp. 1535–1544 (2020)
11. Kong, D., Ma, H., Xie, X.: SIA-GCN: a spatial information aware graph neural network with 2D convolutions for hand pose estimation. In: BMVC (2020)
12. Li, J., et al.: Human pose regression with residual log-likelihood estimation. In: ICCV, pp. 11005–11014 (2021)
13. Li, J., Fang, F., Mei, K., Zhang, G.: Multi-scale residual network for image super-resolution. In: Ferrari, V., Hebert, M., Sminchisescu, C., Weiss, Y. (eds.) ECCV 2018. LNCS, vol. 11212, pp. 527–542. Springer, Cham (2018). https://doi.org/10.1007/978-3-030-01237-3_32
14. Li, X., Wang, W., Hu, X., Yang, J.: Selective kernel networks. In: CVPR, pp. 510–519 (2019)
15. Li, Y., et al.: Is 2D heatmap representation even necessary for human pose estimation? arXiv (2021)
16. Ma, N., Zhang, X., Zheng, H.-T., Sun, J.: ShuffleNet V2: practical guidelines for efficient CNN architecture design. In: Ferrari, V., Hebert, M., Sminchisescu, C., Weiss, Y. (eds.) Computer Vision – ECCV 2018. LNCS, vol. 11218, pp. 122–138. Springer, Cham (2018). https://doi.org/10.1007/978-3-030-01264-9_8
17. Newell, A., Yang, K., Deng, J.: Stacked hourglass networks for human pose estimation. In: Leibe, B., Matas, J., Sebe, N., Welling, M. (eds.) ECCV 2016. LNCS, vol. 9912, pp. 483–499. Springer, Cham (2016). https://doi.org/10.1007/978-3-319-46484-8_29
18. Nibali, A., He, Z., Morgan, S., Prendergast, L.A.: Numerical coordinate regression with convolutional neural networks. arXiv (2018)
19. Sandler, M., Howard, A.G., Zhu, M., Zhmoginov, A., Chen, L.: Mobilenetv 2: inverted residuals and linear bottlenecks. In: CVPR, pp. 4510–4520 (2018)
20. Santavas, N., Kansizoglou, I., Bampis, L., Karakasis, E.G., Gasteratos, A.: Attention! A lightweight 2D hand pose estimation approach. arXiv (2020)
21. Simon, T., Joo, H., Matthews, I.A., Sheikh, Y.: Hand keypoint detection in single images using multiview bootstrapping. In: CVPR, pp. 4645–4653 (2017)
22. Sun, K., Xiao, B., Liu, D., Wang, J.: Deep high-resolution representation learning for human pose estimation. In: CVPR, pp. 5693–5703 (2019)
23. Toshev, A., Szegedy, C.: Deeppose: human pose estimation via deep neural networks. In: CVPR, pp. 1653–1660 (2014)
24. Wang, R.J., Li, X., Ao, S., Ling, C.X.: PELEE: a real-time object detection system on mobile devices. In: ICLR (2018)
25. Wang, Y., Peng, C., Liu, Y.: Mask-pose cascaded CNN for 2D hand pose estimation from single color image. IEEE Trans. Circuits Syst. Video Technol. **29**, 3258–3268 (2019)
26. Wang, Y., Zhang, B., Peng, C.: SrhandNet: real-time 2D hand pose estimation with simultaneous region localization. IEEE Trans. Image Process. **29**, 2977–2986 (2020)
27. Wang, Z., Nie, X., Qu, X., Chen, Y., Liu, S.: Distribution-aware single-stage models for multi-person 3d pose estimation. arXiv (2022)
28. Wei, S., Ramakrishna, V., Kanade, T., Sheikh, Y.: Convolutional pose machines. In: CVPR, pp. 4724–4732 (2016)
29. Woo, S., Park, J., Lee, J.-Y., Kweon, I.S.: CBAM: convolutional block attention module. In: Ferrari, V., Hebert, M., Sminchisescu, C., Weiss, Y. (eds.) ECCV 2018. LNCS, vol. 11211, pp. 3–19. Springer, Cham (2018). https://doi.org/10.1007/978-3-030-01234-2_1

30. Yu, C., et al.: Lite-hrNet: a lightweight high-resolution network. In: CVPR, pp. 10440–10450 (2021)
31. Zhang, F., Zhu, X., Dai, H., Ye, M., Zhu, C.: Distribution-aware coordinate representation for human pose estimation. In: CVPR, pp. 7091–7100 (2020)
32. Zhang, Z., Tang, J., Wu, G.: Simple and lightweight human pose estimation. arXiv (2019)
33. Zhao, H., et al.: PSANet: point-wise spatial attention network for scene parsing. In: Ferrari, V., Hebert, M., Sminchisescu, C., Weiss, Y. (eds.) ECCV 2018. LNCS, vol. 11213, pp. 270–286. Springer, Cham (2018). https://doi.org/10.1007/978-3-030-01240-3_17
34. Zimmermann, C., Brox, T.: Learning to estimate 3D hand pose from single RGB images. In: ICCV, pp. 4913–4921 (2017)
35. Zimmermann, C., Ceylan, D., Yang, J., Russell, B.C., Argus, M.J., Brox, T.: Freihand: a dataset for markerless capture of hand pose and shape from single RGB images. In: ICCV, pp. 813–822 (2019)

DilatedSegNet: A Deep Dilated Segmentation Network for Polyp Segmentation

Nikhil Kumar Tomar, Debesh Jha(✉), and Ulas Bagci

Machine and Hybrid Intelligence Lab, Department of Radiology,
Northwestern University, Chennai, India
debesh.jha@northwestern.edu

Abstract. Colorectal cancer (CRC) is the second leading cause of cancer-related death worldwide. Excision of polyps during colonoscopy helps reduce mortality and morbidity for CRC. Powered by deep learning, computer-aided diagnosis (CAD) systems can detect regions in the colon overlooked by physicians during colonoscopy. Lacking high accuracy and real-time speed are the essential obstacles to be overcome for successful clinical integration of such systems. While literature is focused on improving accuracy, the speed parameter is often ignored. Toward this critical need, we intend to develop a novel real-time deep learning-based architecture, DilatedSegNet, to perform polyp segmentation on the fly. DilatedSegNet is an encoder-decoder network that uses pre-trained ResNet50 as the encoder from which we extract four levels of feature maps. Each of these feature maps is passed through a dilated convolution pooling (DCP) block. The outputs from the DCP blocks are concatenated and passed through a series of four decoder blocks that predicts the segmentation mask. The proposed method achieves a real-time operation speed of 33.68 frames per second with an average dice coefficient (DSC) of 0.90 and mIoU of 0.83. Additionally, we also provide heatmap along with the qualitative results that shows the explanation for the polyp location, which increases the trustworthiness of the method. The results on the publicly available Kvasir-SEG and BKAI-IGH datasets suggest that DilatedSegNet can give real-time feedback while retaining a high DSC, indicating high potential for using such models in real clinical settings in the near future. The GitHub link of the source code can be found here: https://github.com/nikhilroxtomar/DilatedSegNet.

Keywords: Deep learning · Polyp segmentation · Colonoscopy · Residual network · Generalization · Real-time segmentation

1 Introduction

Missed polyp during routine colonoscopy examination is the primary source of interval colorectal cancer (CRC). The polyps that are not recognized within the colonoscope are the major source contributor to this problem. Colonoscopy is

D.-T. Dang-Nguyen et al. (Eds.): MMM 2023, LNCS 13833, pp. 334–344, 2023.
https://doi.org/10.1007/978-3-031-27077-2_26

considered the gold standard for colon cancer diagnosis and follow-up. However, 22–28% of polyps are missed during a routine examination [12]. Some of these polyps can cause post-colonoscopy colorectal cancer (CRC). One of the reasons for the polyp miss-rate is either the polyp was not visible during the examination or was not recognized despite being in the visual field because of the faster colonoscope withdrawal time. Deep learning based algorithms can highlight the presence of pre-cancerous tissue in the colon and have the potential to improve the diagnostic performance of endoscopists. Improving the polyp detection rate as well as its accurate segmentation is an unmet clinical need. In practice, precise polyp segmentation provides important information in the early detection of colorectal cancer via their shape, texture, and location information.

Tomar et al. [17] proposed a feedback attention network for biomedical image segmentation where they utilized the previous epoch mask with the current training epoch in an iterative fashion to further improve the performance. Fan et al. [3] used Res2Net-based [4] backbone where they used a parallel partial decoder and parallel reverse attention mechanism for the accurate polyp segmentation. Jha et al. [9] proposed an efficient architecture where they utilized the strength of the residual block, atrous spatial pyramidal pooling, with squeeze and excitation block for polyp segmentation. Shen et al. [15] proposed a hard region enhancement network (HRENet) that consists of an informative context enhancement (ICE) module and trained the model on edge and structure consistency aware loss (ESCLoss) to improve the polyp segmentation on the precise edge. Zhao et al. [21] proposed a multi-scale subtraction network (MSNet) for automatic polyp segmentation. Despite of several architectures proposed in the literature, most existing methods often neglect the encoder and tend to focus more on the decoder part of the network, which led to the loss of significant features from the encoder part. In our proposed method, we focus more on the encoder part of the network by utilizing different scales features which are passed through multiple dilated convolutions to capture more enlarged features, leading to improved polyp segmentation. Unlike other decoders, the design of our decoder is straightforward. It utilizes simple sequences of layers such as an upsampling layer, concatenation, residual block and an attention layer. We introduce the novel deep learning architecture, DilatedSegNet, to address the critical need for clinical integration of polyp segmentation routine, which is real-time and retains high accuracy. The main contribution of the study are as follows:

1. We introduce a novel network named DilatedSegNet for polyp segmentation. The architecture begins with a pre-trained ResNet50 [5] and utilizes dilated convolution [19] pooling block to increase the receptive field for capturing more diverse and reliable features for a better delineation.
2. DilatedSegNet showed outstanding performance by outperforming nine standard benchmarking methods with two widely used publicly available polyp segmentation datasets.

3. Extensive experimental results and cross-dataset test results on two unseen datasets showed the better generalization capability of the DilateSegNet. Explored deep features showed via heatmaps that the proposed network model is focusing on the target polyp regions and their boundaries, proving visual interpretability of the model.

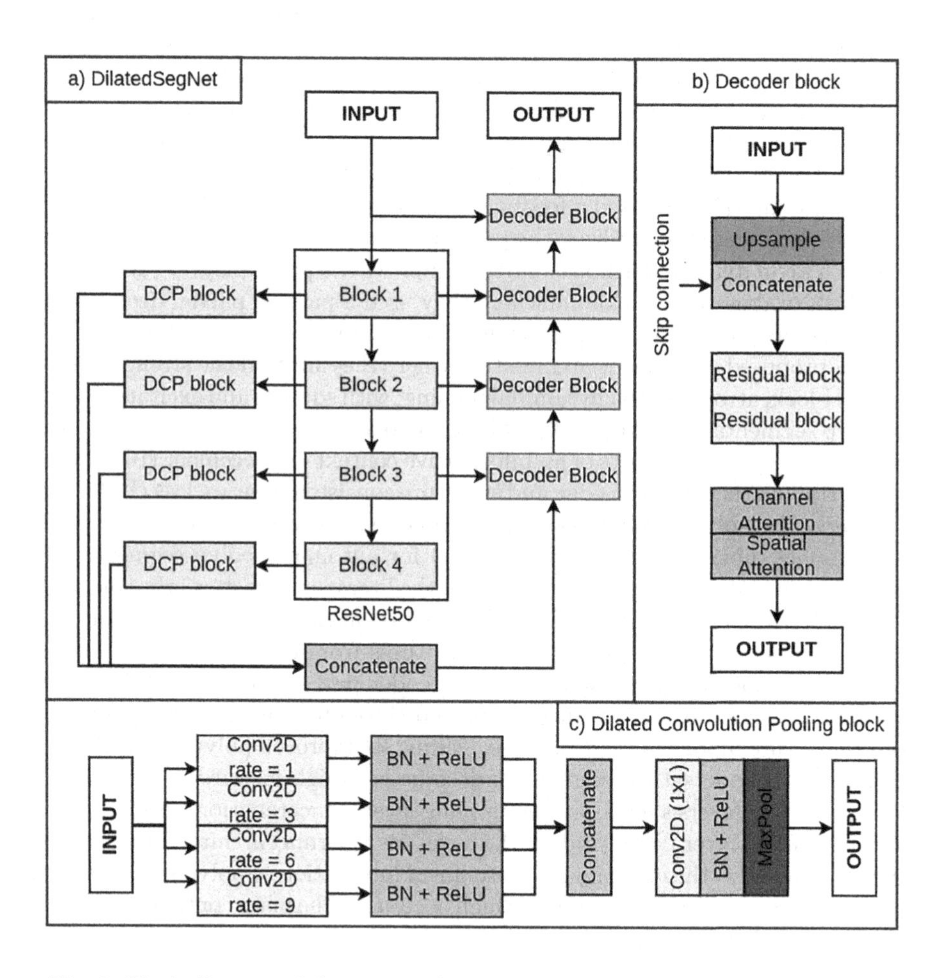

Fig. 1. Block diagram of the proposed DilatedSegNet along with its components.

2 Method

Figure 1 shows the block diagram of the proposed DilatedSegNet along with its core components. It follows an encoder-decoder scheme much like the U-Net [14], consisting of a pre-trained ResNet50 [5] as an encoder. The input image I with a resolution of $[h \times w \times 3]$ is fed to the pre-trained encoder from which we extract four levels of features maps $\{f_i : i = 1, 2, 3, 4\}$ with varying resolution of $[h/2^k \times w/2^k : k = 1, 2, 3, 4]$. Each of these feature maps is then passed through a Dilated Convolution Pooling (DCP) block, where four parallel dilated convolutions with the rate $1, 3, 6, 9$ are applied to enhance the field of view. The output from all the DCP blocks is concatenated and passed to the first decoder block, where the feature map is upsampled and concatenated with a skip connection from the pre-trained encoder. Next, it is passed through some residual block and then a Convolutional Block Attention Module (CBAM) [18]. The output of the CBAM is passed to the next decoder for further transformation. Finally, the output from the last decoder block is passed to a 1×1 convolution followed by a sigmoid activation function.

2.1 Dilated Convolution Pooling (DCP) Block

The DCP block begins with four parallel 3×3 convolution layers having a dilation rate of 1, 3, 6 and 9. The dilated convolution increases the receptive field of the 3×3 kernel, which helps it to cover more area over the input feature maps. Thus, by increasing dilation rate, we get better feature maps with each layer. The output from each convolutional layer is followed by batch normalization and a ReLU activation function. Next, we combine the output from each ReLU activation function to form a concatenated feature map, which is followed by a 1×1 convolutional layer to reduce the number of feature channels. The 1×1 convolutional layer is further followed by batch normalization and a ReLU activation function. The output of the ReLU activation function is passed through a max-pooling layer to reduce its spatial dimensions.

2.2 Decoder Block

The decoder block begins with a bilinear upsampling where the feature map spatial dimensions (height and width) are increased by a factor of two. After that, we concatenate the upsampled feature map with the feature map from the pre-trained encoder through the skip connections. These skip connections fetch the necessary features directly from the encoder to the decoder which is sometimes lost due to the depth of the network. The concatenated feature maps are passed through a set of two residual blocks which helps to learn more meaningful semantic features from the input. These features are further refined by using an attention mechanism called CBAM [18]. CBAM consists of channel attention followed by spatial attention to highlight the more significant features and suppress the irrelevant ones.

3 Experimental Setup

In this section, we present the datasets, evaluation metrics and the implementation details.

3.1 Datasets and Evaluation Metrics

We have selected the Kvasir-SEG [8] and BKAI-IGH [11] datasets to evaluate the performance of the proposed DilatedSegNet. Both of the datasets are

Table 1. Complexity of the models with image size of 256 × 256.

Method	Publication venue	Backbone	Parameters (Millions)	Flops (GMac)	FPS
U-Net [14]	MICCAI'15	–	31.04	54.75	160.27
ResU-Net [20]	GRSL'18	–	8.22	45.42	197.94
U-Net++ [22]	DLMIAW'18	–	9.16	34.65	123.45
DeepLabV3+ [2]	ECCV'18	ResNet50	39.76	43.31	99.16
ResU-Net++ [9]	ISM'19	–	4.06	15.81	55.86
DDANet [16]	ICPRW'20	–	3.36	18.2	86.46
PraNet [3]	MICCAI'20	Res2Net	32.55	6.93	36.21
ColonSegNet [7]	IEEE Access'21	–	5.01	62.16	122.42
HarDNet-MSEG [6]	Arxiv'21	–	33.34	6.02	41.20
FANet [17]	IEEE TNNLS'22	–	7.72	94.75	65.53
CaraNet [13]	SPIE MI'22	Res2Net	46.64	11.48	20.13
DilatedSegNet (Ours)	–	ResNet50	18.11	27.1	33.68

publicly available and can be easily accessible. Kvasir-SEG [8] consists of 1000 polyp images, their corresponding masks, and the bounding box information. Similarly, BKAI-IGH [11] consists of 1000 polyp images in the training dataset, and separate 200 images in the test dataset. However, the ground truth of the test dataset is not made publicly available by the dataset provided. So, we only experiment on the training dataset. Additionally, each of the polyp in dataset is categorized neo-plastic (potential to become cancerous) and non-neoplastic (non cancerous). However, we treat the dataset as a binary class problem (i.e. polyp and normal tissue). We have used standard segmentation metrics such as DSC, mean intersection over Union (mIoU), precision, recall, F2-score and frame per second (FPS) to benchmark the performance of our proposed model.

3.2 Implementation Details

In this study, we have implemented our proposed DilatedSegNet and all the other benchmark models using the PyTorch framework and trained on a RTX 3090 GPU. We have used the same hyperparameters for all the models for a fair comparison. The images and masks were first split into training, validation and testing datasets. For Kvasir-SEG, we have utilized the official split of 880/120, where 880 images and masks were used for training and the rest of the 120 were used for validation and testing. For the BKAI dataset, we followed an $80:10:10$ split, where 80% images and masks were used for training, 10% was used for validation and the remaining 10% was used for the testing. All the images and masks were resized to 256 × 256 pixels. To make the model more robust, we have used an online data augmentation strategy with random rotation, horizontal flipping, vertical flipping and coarse dropout. All the models were trained by an Adam optimizer [10] with a learning rate of $1e^{-4}$ and a batch size of 16. We have used a combination of dice loss and binary cross-entropy as the loss function. ReduceLROnPlateau was used while training to reduce the learning rate for better performance, while early stopping was used to stop the training when the model stopped improving.

Table 2. Results on the Kvasir-SEG [8] and BKAI-IGH [11] datasets.

Method	Publication	DSC	mIoU	Recall	Precision	F2
Train and test data: Kvasir-SEG [8]						
U-Net [14]	MICCAI'15	0.8264	0.7472	0.8504	0.8703	0.8353
ResU-Net [20]	GRSL'18	0.7642	0.6634	0.8025	0.8200	0.7740
U-Net++ [22]	DLMIAW'18	0.8228	0.7419	0.8437	0.8607	0.8295
DeepLabV3+ [2]	ECCV'18	0.8837	0.8173	0.9014	0.9028	0.8904
ResU-Net++ [9]	ISM'19	0.6453	0.5341	0.6964	0.7080	0.6575
DDANet [16]	ICPRW'20	0.7415	0.6448	0.7953	0.7670	0.7640
PraNet [3]	MICCAI'20	0.8942	0.8296	0.9060	**0.9126**	0.8976
ColonSegNet [7]	IEEE Access'21	0.7920	0.6980	0.8193	0.8432	0.7999
HarDNet-MSEG [6]	Arxiv'21	0.8260	0.7459	0.8485	0.8652	0.8358
FANet [17]	IEEE TNNLS'22	0.7844	0.6975	0.8503	0.8165	0.8054
CaraNet [13]	SPIE MI'22	0.8742	0.8001	**0.9289**	0.8614	0.8996
DilatedSegNet (Ours)	–	**0.8957**	**0.8336**	0.9169	0.9096	**0.9034**
Train and test data: BKAI-IGH [11]						
U-Net [14]	MICCAI'15	0.8286	0.7599	0.8295	0.8999	0.8264
ResU-Net [20]	GRSL'18	0.7433	0.6580	0.7447	0.8711	0.7387
U-Net++ [22]	DLMIAW'18	0.8275	0.7563	0.8388	0.8942	0.8308
DeepLabV3+ [2]	ECCV'18	0.8937	0.8314	0.8870	**0.9333**	0.8882
ResU-Net++ [9]	ISM'19	0.7130	0.6280	0.7240	0.8578	0.7132
DDANet [16]	ICPRW'20	0.7269	0.6507	0.7454	0.7575	0.7335
PraNet [3]	MICCAI'20	0.8904	0.8264	0.8901	0.9247	0.8885
ColonSegNet [7]	IEEE Access'21	0.7748	0.6881	0.7852	0.8711	0.7746
HarDNet-MSEG [6]	Arxiv'21	0.7627	0.6734	0.7532	0.8344	0.7528
FANet [17]	IEEE TNNLS'22	0.8305	0.7578	0.8285	0.9169	0.8243
CaraNet [13]	SPIE MI'22	0.8948	0.8309	0.8907	0.9280	0.8911
DilatedSegNet (Ours)	–	**0.8950**	**0.8315**	**0.9082**	0.9111	**0.8991**

4 Results

We present quantitative and qualitative results along with the heatmaps for model interpretability.

4.1 Performance Test on Same Dataset

Table 2 shows the result of the DilatedSegNet on the Kvasir-SEG [8] and BKAI-IGH [11] datasets, respectively. DilatedSegNet obtains an DSC score of 0.8957 and mIoU of 0.8336 with Kvasir-SEG and an DSC-score of 0.8950 and mIoU of 0.8315 on the BKAI-IGH dataset, outperforming nine state-of-the-art benchmarks. The most competitive network to our network was PraNet [3] which obtained DSC and mIoU 0.8942 and 0.8296, respectively, for the Kvasir-SEG. DeepLabv3+ obtained the most competitive results with BKAI-IGH: a DSC of 0.8937 and mIoU of 0.8314. DilatedSegNet achieved a real-time operation speed of real-time speed of 33.68 FPS. The number of parameters used in DeepLabv3+ was 39.76 million and the number of flops utilized was 43.31 GMac. However, our

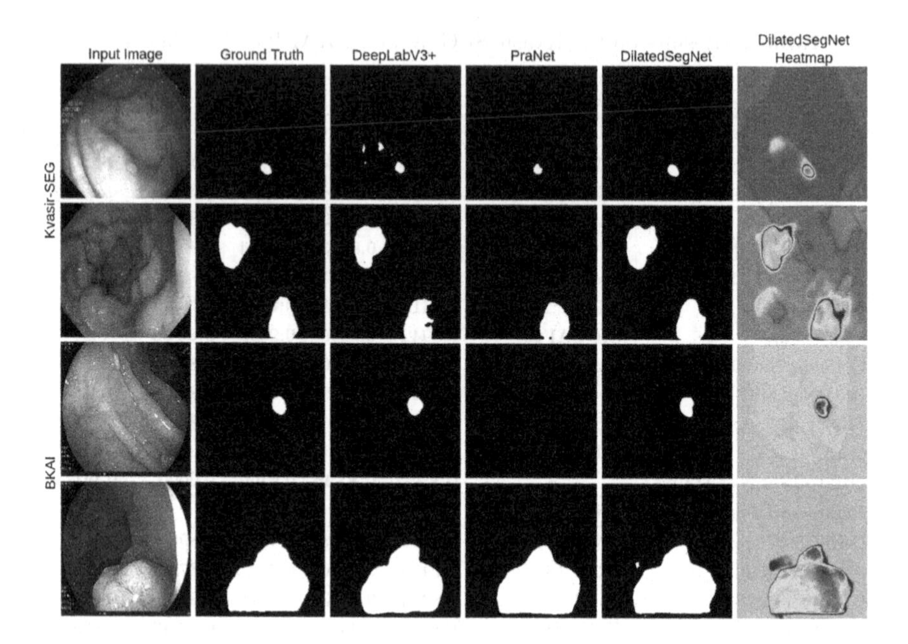

Fig. 2. The figure shows qualitative results comparison of the three best methods. The heatmaps are obtained with respect to the convolutional layer at the bottleneck. The produced heatmap shows both important and unimportant pixels. Here, the heatmap shows that DilatedSegNet utilized correct pixels from the input image while making predictions for polyp and non-polyps. The qualitative comparison between the ground truth and the heatmap produced by DilatedSegNet shows that the heatmap is precise. This show that the prediction made by the proposed model is trustworthy. (Color figure online)

proposed architecture has only 18.11 million parameters and 27.1 GMac flops (refer Table 1), substantially better performance by lowering the parameters and flops, thanks to our lightweight architectural design allowing for real-time processing (Fig. 3).

Figure 2 shows the qualitative results of DilatedSegNet and two state-of-the-art networks (i.e., PraNet [3] and DeepLabv3+ [2]). The qualitative result shows that DilatedSegNet can correctly segment smaller and medium-sized polyps that are commonly missed during routine colonoscopy examinations due to their size. For diminutive polyps, DeepLabv3+ shows over-segmentation and PraNet shows under-segmentation. Similarly, PraNet misses challenging and flat polyps for two cases, and DeepLabv3+ shows under segmentation (for the second example). The visual results comparison shows that DilatedSegNet has a better ability to capture regular and flat polyps. Thus, both the qualitative and quantitative results exhibit the high overall performance of DilatedSegNet. Additionally, we determined the heatmap results of the DilatedSegNet. The heatmap results show the relevance of the individual polyp and non-polyp pixels. The heatmaps can be useful to understand the convolutional neural network and helps towards

Table 3. Cross dataset results of models trained on Kvasir-SEG [8] and tested on independent CVC-ClinicDB [1] and BKAI-IGH [11].

Model	DSC	mIoU	Recall	Precision	F2
Train: Kvasir-SEG [8], **Test: CVC-ClinicDB** [1]					
U-Net [14]	0.6336	0.5433	0.6982	0.7891	0.6563
ResU-Net [20]	0.5970	0.4967	0.6210	0.8005	0.5991
U-Net++ [22]	0.6350	0.5475	0.6933	0.7967	0.6556
DeepLabV3+ [2]	0.8142	0.7388	0.8331	0.8735	0.8198
ResU-Net++ [9]	0.4642	0.3585	0.5880	0.5770	0.5084
DDANet [16]	0.5234	0.4183	0.6502	0.5935	0.5718
PraNet [3]	0.8046	0.7286	0.8188	**0.8968**	0.8077
ColonSegNet [7]	0.6126	0.5090	0.6564	0.7521	0.6246
HarDNet-MSEG [6]	0.6960	0.6058	0.7173	0.8528	0.7010
FANet [17]	0.6524	0.5579	0.7560	0.7243	0.6872
CaraNet [13]	0.8254	0.7450	**0.8568**	0.8696	**0.8389**
DilatedSegNet (Ours)	**0.8278**	**0.7545**	0.8462	0.8921	0.8336
Train: Kvasir-SEG [8], **Test: BKAI-IGH** [11]					
U-Net [14]	0.6347	0.5686	0.6986	0.7882	0.6591
ResU-Net [20]	0.5836	0.4931	0.6716	0.6549	0.6177
U-Net++ [22]	0.6269	0.5592	0.6900	0.7968	0.6493
DeepLabV3+ [2]	0.7286	0.6589	0.7919	0.8123	0.7493
ResU-Net++ [9]	0.4166	0.3204	0.6979	0.3922	0.5019
DDANet [16]	0.5006	0.4115	0.6612	0.4825	0.5592
PraNet [3]	0.7298	0.6609	0.8007	0.824	0.7484
ColonSegNet [7]	0.5765	0.4910	0.7191	0.6644	0.6225
HarDNet-MSEG [6]	0.6502	0.5711	0.7420	0.7469	0.6830
FANet [17]	0.5153	0.4412	**0.8395**	0.5505	0.5913
CaraNet [13]	0.7470	0.6749	0.8234	0.8102	**0.7742**
DilatedSegNet (Ours)	**0.7545**	**0.6906**	0.7886	**0.8750**	0.7649

model interpretability. Here, "red" and "yellow" denote the important regions the models learns as polyp, whereas "blue" color shows that the model considers those regions as less significant areas.

4.2 Performance Test on Completely Unseen Datasets

Table 3 shows the cross-dataset results. In the experimental setting #1, we train the dataset on Kvasir-SEG and test it on CVC-ClinicDB (a completely unseen dataset). The proposed method obtains a high DSC score of 0.8278 and mIoU of 0.7545 and outperforms the best performing DeepLabv3+ by 1.36% in DSC and 1.57% in mIoU. Similarly, in setting #2 when the model is trained on Kvasir-SEG and tested on BKAI-IGH data, the DilatedSegNet surpass the best performing PraNet [3] and obtains 2.47% more in DSC and 2.97% in mIoU.

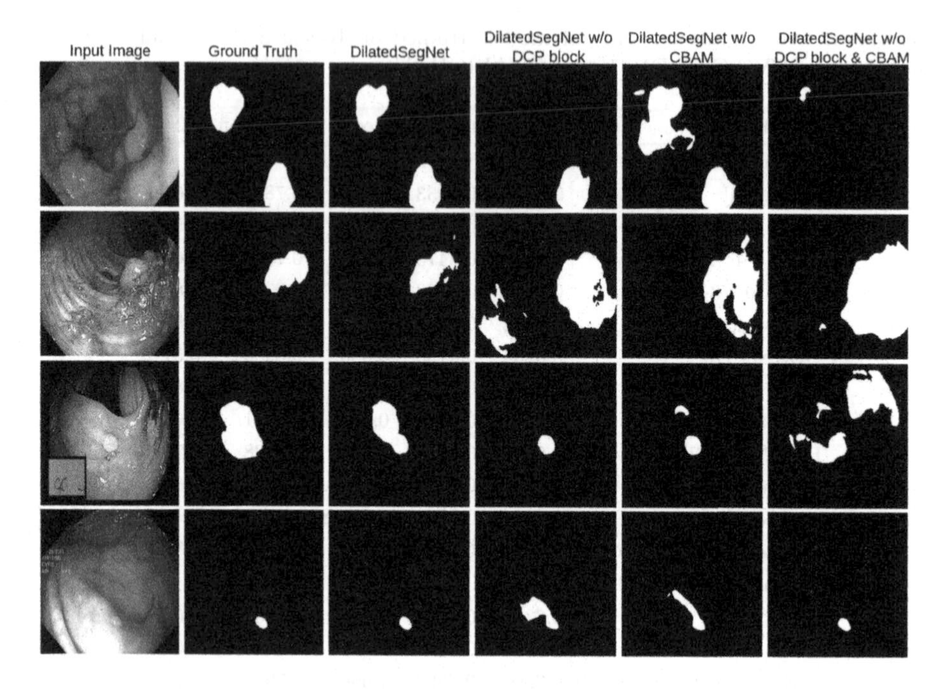

Fig. 3. The Figure shows examples of qualitative results comparison of the ablation study from the Kvasir-SEG dataset. The leftmost column shows the input image and the other column next to it shows the ground truth indicating the area covered by polyp and non-polyp. The name of the network used for training during the ablation study is indicated at the top. The qualitative examples show that the proposed network is the best. Eliminating DCP or attention block or both affect the quality of prediction. This is evidenced by the over-segmentation or under-segmentation results produced under the same setting without incorporating the individual or both of the blocks.

Table 4. Ablation study of the proposed DilatedSegNet on the Kvasir-SEG [8].

No.	Method	DSC	mIoU	Recall	Precision
#1	DilatedSegNet w/o DCP block	0.8725	0.8067	0.8917	0.9025
#2	DilatedSegNet w/o Attention	0.8832	0.8135	0.9076	0.8966
#3	DilatedSegNet w/o DCP block & Attention (CBAM)	0.8627	0.7946	0.8871	0.8947
#4	DilatedSegNet	**0.8957**	**0.8336**	**0.9169**	**0.9096**

5 Ablation Study

In the Table 4, we present the results of the ablation study to verify the effectiveness and the importance of each blocks. Here, we test the DilatedSegNet without DCP block (setting #1), without attention (# 2), and without DCP block & attention (#3). The proposed architecture has an improvement of 3.3% in DSC and 3.9% in mIoU, 2.98% recall and 1.49% in precision as compared to the setting #3. Therefore, we showed that the proposed method had performance improvement with the utilization of DCP and attention block.

6 Conclusion

In this work, we proposed the DilatedSegNet architecture that utilizes a dilated convolution pooling (DCP) block and CBAM to accurately segment polyps with high performance and real-time speed, which has never been addressed before. The experimental results on the same dataset testing and completely unseen dataset testing results showed that DilateSegNet achieves a high DSC and outperforms the state-of-the-art polyp segmentation models. The design of the architecture was supported by the ablation study. The qualitative, quantitative and heatmap suggest that DilatedSegNet can be a strong benchmark for building early polyp detection in clinics. Additionally, the presented heatmap was effective in discriminating different polyp and non-polyp (normal tissue) pixels from the colonoscopy image. In the future, we plan to explore DilatedSegNet with the multi-centre dataset, evaluate its robustness, and explore the results on the federated learning settings.

Acknowledgement. This project is supported by the NIH funding: R01-CA246704 and R01-CA240639.

References

1. Bernal, J., Sánchez, F.J., Fernández-Esparrach, G., Gil, D., Rodríguez, C., Vilariño, F.: WM-DOVA maps for accurate polyp highlighting in colonoscopy: validation vs saliency maps from physicians. Comput. Med. Imaging Graph. **43**, 99–111 (2015)
2. Chen, L.-C., Zhu, Y., Papandreou, G., Schroff, F., Adam, H.: Encoder-decoder with atrous separable convolution for semantic image segmentation. In: Ferrari, V., Hebert, M., Sminchisescu, C., Weiss, Y. (eds.) ECCV 2018. LNCS, vol. 11211, pp. 833–851. Springer, Cham (2018). https://doi.org/10.1007/978-3-030-01234-2_49
3. Fan, D.-P., et al.: PraNet: parallel reverse attention network for polyp segmentation. In: Martel, A.L., et al. (eds.) MICCAI 2020. LNCS, vol. 12266, pp. 263–273. Springer, Cham (2020). https://doi.org/10.1007/978-3-030-59725-2_26
4. Gao, S.H., Cheng, M.M., Zhao, K., Zhang, X.Y., Yang, M.H., Torr, P.: Res2net: a new multi-scale backbone architecture. IEEE Trans. Pattern Anal. Mach. Intell. **43**(2), 652–662 (2019)
5. He, K., Zhang, X., Ren, S., Sun, J.: Deep residual learning for image recognition. In: Proceedings of the IEEE Conference on Computer Vision and Pattern Recognition (CVPR), pp. 770–778 (2016)
6. Huang, C.H., Wu, H.Y., Lin, Y.L.: HarDNet-MSEG A Simple Encoder-Decoder Polyp Segmentation Neural Network that Achieves over 0.9 Mean Dice and 86 FPS. arXiv preprint arXiv:2101.07172 (2021)
7. Jha, D., et al.: Real-time polyp detection, localization and segmentation in colonoscopy using deep learning. IEEE Access **9**, 40496–40510 (2021)

8. Jha, D., et al.: Kvasir-SEG: a segmented polyp dataset. In: Proceedings of the International Conference on Multimedia Modeling (MMM), pp. 451–462 (2020)
9. Jha, D., et al.: ResUNet++: an advanced architecture for medical image segmentation. In: Proceedings of the International Symposium on Multimedia (ISM), pp. 225–2255 (2019)
10. Kingma, D.P., Ba, J.: Adam: a method for stochastic optimization. arXiv preprint arXiv:1412.6980 (2014)
11. Lan, P.N., et al.: NeoUNet: towards accurate colon polyp segmentation and neoplasm detection. arXiv preprint arXiv:2107.05023 (2021)
12. Leufkens, A., Van Oijen, M., Vleggaar, F., Siersema, P.: Factors influencing the miss rate of polyps in a back-to-back colonoscopy study. Endoscopy **44**(05), 470–475 (2012)
13. Lou, A., Guan, S., Ko, H., Loew, M.H.: Caranet: context axial reverse attention network for segmentation of small medical objects. In: Medical Imaging 2022: Image Processing, vol. 12032, pp. 81–92 (2022)
14. Ronneberger, O., Fischer, P., Brox, T.: U-Net: convolutional networks for biomedical image segmentation. In: Navab, N., Hornegger, J., Wells, W.M., Frangi, A.F. (eds.) MICCAI 2015. LNCS, vol. 9351, pp. 234–241. Springer, Cham (2015). https://doi.org/10.1007/978-3-319-24574-4_28
15. Shen, Y., Jia, X., Meng, M.Q.-H.: HRENet: a hard region enhancement network for polyp segmentation. In: de Bruijne, M., et al. (eds.) MICCAI 2021. LNCS, vol. 12901, pp. 559–568. Springer, Cham (2021). https://doi.org/10.1007/978-3-030-87193-2_53
16. Tomar, N.K., : DDANet: dual decoder attention network for automatic polyp segmentation. In: Proceedings of the International Conference on Pattern Recognition Workshop, pp. 307–314 (2021)
17. Tomar, N.K., et al.: FaNet: a feedback attention network for improved biomedical image segmentation. IEEE Trans. Neural Networks Learn. Syst. (2022)
18. Woo, S., Park, J., Lee, J.-Y., Kweon, I.S.: CBAM: convolutional block attention module. In: Ferrari, V., Hebert, M., Sminchisescu, C., Weiss, Y. (eds.) ECCV 2018. LNCS, vol. 11211, pp. 3–19. Springer, Cham (2018). https://doi.org/10.1007/978-3-030-01234-2_1
19. Yu, F., Koltun, V.: Multi-scale context aggregation by dilated convolutions. arXiv preprint arXiv:1511.07122 (2015)
20. Zhang, Z., Liu, Q., Wang, Y.: Road extraction by deep residual U-Net. IEEE Geosci. Remote Sens. Lett. **15**(5), 749–753 (2018)
21. Zhao, X., Zhang, L., Lu, H.: Automatic polyp segmentation via multi-scale subtraction network. In: International Conference on Medical Image Computing and Computer-Assisted Intervention, pp. 120–130 (2021)
22. Zhou, Z., Rahman Siddiquee, M.M., Tajbakhsh, N., Liang, J.: UNet++: a nested u-net architecture for medical image segmentation. In: Deep Learning in Medical Image Analysis and Multimodal Learning for Clinical Decision Support, pp. 3–11 (2018)

Music Instrument Classification Reprogrammed

Hsin-Hung Chen(✉) and Alexander Lerch

Music Informatics Group, Georgia Institute of Technology, Atlanta, USA
hchen605@gatech.edu

Abstract. The performance of approaches to Music Instrument Classification, a popular task in Music Information Retrieval, is often impacted and limited by the lack of availability of annotated data for training. We propose to address this issue with "reprogramming," a technique that utilizes pre-trained deep and complex neural networks originally targeting a different task by modifying and mapping both the input and output of the pre-trained model. We demonstrate that reprogramming can effectively leverage the power of the representation learned for a different task and that the resulting reprogrammed system can perform on par or even outperform state-of-the-art systems at a fraction of training parameters. Our results, therefore, indicate that reprogramming is a promising technique potentially applicable to other tasks impeded by data scarcity.

Keywords: Reprogramming · Instrument classification

1 Introduction

The task of Music Instrument Classification (MIC) aims at automatically recognizing the musical instruments playing in a music recording. MIC can provide important information for a variety of applications such as music recommendation, music discovery, and automatic mixing. In recent years, Deep Learning (DL) models have shown superior performance in practically all Music Information Retrieval (MIR) tasks including MIC. However, the lack of large-scale annotated data remains a major problem for data-hungry supervised machine learning algorithms in this field [7,16,21]. Beyond small-scale expert-annotated datasets, larger datasets are often collected by crowd-sourcing annotations, which leads to noisy and sometimes incomplete labels. For example, the majority of labels in the OpenMIC dataset [21] —a popular dataset for polyphonic MIC— are missing; this data scarcity can negatively impact the training of complex classifiers for this multi-label task.

One established approach to address this data challenge is transfer learning. In this approach, the knowledge of a source domain with sufficient training data is transferred to a related but different target domain with insufficient training data. This knowledge transfer is often achieved by either directly using a pre-learned representation as classifier input or by fine-tuning a pre-trained source-domain model with the target-domain data. For instance, the VGGish

D.-T. Dang-Nguyen et al. (Eds.): MMM 2023, LNCS 13833, pp. 345–357, 2023.
https://doi.org/10.1007/978-3-031-27077-2_27

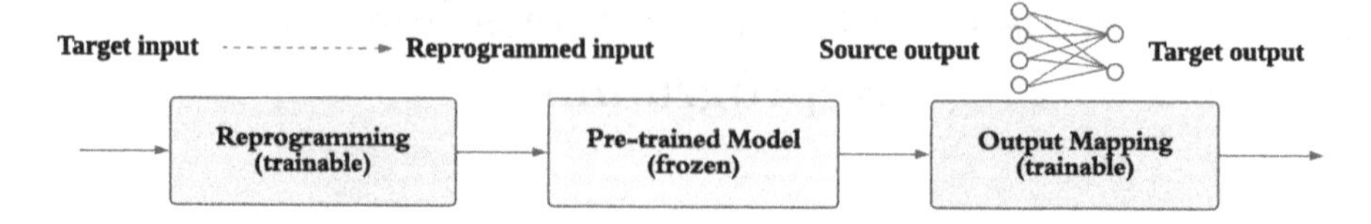

Fig. 1. The concept of model reprogramming.

representation [19], trained on a wide variety of audio data, has been successfully utilized for MIC [16]. *Model reprogramming* aims at expanding the transfer learning paradigm by treating the source-domain models as unmodified black-box machine learning models extended only by input pre-processing and output post-processing. Model reprogramming was first introduced in 2018 by Elsayed et al. [8], who showed that a trainable universal input transformation function can reprogram a pre-trained ImageNet model (without changing the model weights) to solve the MNIST/CIFAR-10 image classification task with high accuracy. Figure 1 illustrates the basic concept of model reprogramming: a trainable model for input reprogramming modifies the input data to be fed into the frozen black-box model pre-trained on the source task, followed by an output transformation that maps the outputs of the pre-trained model to the target categories. Thus, the reprogramming layer serves to "reprogram" the pre-trained model to work with a new target task with different input data and different target classes. Since the complexity of the input and output transformation can be low, model reprogramming can combine the advantage of leveraging a powerful deep pre-trained representation with the advantages of (i) reduced training complexity and (ii) reduced data requirements. Reprogramming methods have been successfully applied to various tasks such as medical image classification [33], time-series classification [35], and language processing [28]. Results show that reprogramming methods can perform on par or better than state of the art methods, thus demonstrating the feasibility of reprogramming pre-trained models and showing the potential of this method for improving the performance on other tasks with low amounts of data.

In this work, we investigate reprogramming for the task of MIC with the incompletely labeled OpenMIC dataset. We choose a pre-trained state-of-the-art audio classification model and extend it by input pre-processing and label-mapping. We provide extensive results on various input processing configurations. The main contributions of the paper are

(i) the presentation of a system with low complexity that is able to outperform state-of-the-art MIC systems, and
(ii) the introduction of the reprogramming paradigm to the field of MIR with a multitude of applications with insufficient training data.

The remainder of the paper is structured as follows. The following Sect. 2 presents a brief overview of relevant work. The pre-trained model and the proposed reprogramming methods are introduced in Sect. 3. The evaluation and analysis are presented in Sect. 4. We conclude with a brief summary and directions of future work.

2 Related Work

This related work is structured into three main parts, an overview of MIC, a short survey of transfer learning, and recent work on reprogramming.

While earlier research on MIC has focused on the detection of instruments from audio only containing one instrument [1,9,10,26] or on the detection of the pre-dominant instrument in a mixture [4], current research has focused on recognizing instruments in polyphonic and poly-timbral audio recordings containing multiple instruments playing multiple voices simultaneously. Similar to other audio classification tasks, earlier systems tended to use traditional machine learning approaches with low-level audio features at the input [9,27] while modern approaches are dominated by neural networks. Li et al. [25] proposed to learn features for instrument recognition with a Convolutional Neural Network (CNN) using the MedleyDB dataset [3]. Hung et al. proposed to detect instrument activity at a high time-resolution and showed the advantage of pitch-conditioning on instrument recognition performance [22]. Gururani et al. [16] introduced training attention-based models to the task for enhanced accuracy and implemented a partial binary cross-entropy loss to ignore missing labels in the OpenMIC dataset [21]. Gururani and Lerch showed that a semi-supervised approach based on consistency loss to adapt and leverage the data with missing labels outperforms other systems [15]. Despite previous efforts on data curation and annotation, the access to fully annotated data on a large scale remains a challenge. The IRMAS dataset [4] is a polyphonic dataset with mixed genres, however, it targets predominant instrument recognition and is therefore not suitable for multi-instrument classification. The MedleyDB [3] and Mixing Secrets [14] datasets are both multi-track datasets with strong annotations of instrument activity. However, only a few hundred of songs are available, creating potential issues not only with respect to the dataset size itself but also regarding data distribution and diversity. The OpenMIC dataset for polyphonic instrument recognition, published by Humphrey et al. [21], presents a reasonably large sample size across various genres. Unfortunately, a considerable number of labels are missing as not all clips are labeled with all 20 instruments due to the crowd-sourced annotation process. Slakh2100 is another large music dataset containing mixed tracks of 34 instrument categories and with perfect annotation. However, it is synthesized and rendered from MIDI files instead of real recordings. Transfer learning is an important tool in machine learning when facing the fundamental problem of insufficient training data. It aims to transfer the knowledge from a source domain to the target domain by relaxing the assumption that the training data and the test data must be independent and identically distributed [32]. It is based on the idea that a powerful representation, learned for a source task with large datasets, can be adapted to a related but not identical target task that lacks training data. For example, Qiuqiang et al. [24] present the adaptation of pre-trained models such as ResNet [18] and MobileNet [20] to audio tagging tasks, showing the generalizability of systems pre-training on large-scale datasets to audio pattern recognition. Choi et al. [5] show that representations pre-trained on the music tagging task can be successfully transferred to various music classification tasks and can lead to competitive results. Jordi et al. introduced the "musicnn" representations, featuring a set of CNNs pre-trained on the music audio tagging task [30]. In the context of MIC, Gururani et al. [16] successfully adopt the VGGish pre-trained representation [19] as the input features for their attention-based model.

The promising results of transfer learning outlined above led to Tsai et al. framing two new research questions [33]: (i) "is finetuning a pre-trained model necessary for learning a new task?" and (ii) "can transfer learning be expanded where nothing but only the input-output model responses are observable?" The attempt to answer these questions inspired by ideas of adversarial approaches led to the idea of "reprogramming." Adversarial machine learning aims at manipulating the prediction of a well-trained deep learning model by designing and learning perturbations to the data inputs without changing the target model [2,8,33]. The success of these approaches shows that the classifier output can be changed just by modifying its input, and thus suggests that such methods might be applicable in a non-adversarial context by modifying the input of a pre-trained model to "adapt" the model to a target task. This leads to the concept of model reprogramming, also referred to as adversarial reprogramming [8], which is a technique that aims at leveraging the knowledge of a model pre-trained for a different task by pre-processing the input and post-processing the output. Elsayed et al. showed that pre-trained ImageNet models can be reprogrammed to classifying MNIST and CIFAR-10 by adding learnable parameters around the input image [8]. Tsai et al. demonstrate the advantage of reprogramming on label-limited data such as biomedical image classification [33], and combine reprogramming with multi-label mapping. Model reprogramming has also been used in tasks other than image classification such as natural language processing. For example, Hambardzumyan et al. propose Word-level Adversarial ReProgramming (WARP) for language understanding by adding learnable tokens to the input sequence [17]. The evaluation shows that WARP outperforms all models with less parameters. Neekhara et al. demonstrate re-purposing character-level classification tasks to sentiment classification tasks by implementing a trainable adversarial sequence with surrounding input tokens [28]. Yang et al. [35] applied reprogramming to acoustic models for time-series classification on the UCR Archive benchmark [6]. The input audio is treated as time-series and is padded to be reprogrammed. The model achieves state-of-the-art accuracy on 20 out of 30 datasets with considerably fewer trainable parameters than the pre-trained models.

The presented reprogramming methods achieve promising results with simple learnable reprogramming operations. Therefore, we can conclude that reprogramming is an effective new transfer learning approach inspired by adversarial methods that could address data insufficiency problems by requiring less training complexity.

3 Proposed Method

3.1 Pre-trained Model

The criteria for choosing the pre-trained model to be used as black-box model in our reprogramming setup were that the model (i) offers state-of-the-art performance in audio classification, (ii) is trained on a comparable but different task, (iii) is preferably an attention-based model to make it better suited to work on the weakly labeled OpenMIC dataset (see below), (iv) is of sufficiently high complexity to learn a powerful representation, and (v) has been trained on a large number of data points.

Given these criteria, the Audio Spectrogram Transformer (AST) [12,13] was selected. AST is a convolution-free, purely attention-based model with an audio spectrogram input, achieving state-of-the-art results on AudioSet [11], ESC-50 [29], and Speech Commands V2 [34]. Choosing the AST model trained on AudioSet provides us with a pre-trained model that should be suitable for music audio. The input audio is pre-processed into 128-dimensional Log-Mel spectrogram features computed with a 25 ms von-Hann window every 10 ms. The spectrogram is split into a sequence of 16×16 patches with overlap, and then linearly projected and added to a learnable positional embedding as the input of the transformer encoder. For the AST pre-trained on AudioSet, the number of output classes is 527.

3.2 Reprogramming

The overall structure follows the flow-chart presented in Fig. 1 with the reprogramming stage, the pre-trained model, and the output label mapping stage. To explore the potential and performance of reprogramming, we investigate various forms of input and output reprogramming in this work.

Input Reprogramming. The input reprogramming step aims to find a trainable input modification that can be applied to inputs universally to transform them into an input representation useful for the pre-trained model. Previously proposed methods include the superposition of noise to the input also known as adversarial reprogramming [8,33]. We investigate this approach, but we also propose a novel method to extend this simple superposition by transforming the input spectrogram by means of a neural network. In this way, a well-crafted perturbation of the input might improve "compatibility" with the pre-trained model.

Noise Reprogramming: Previous reprogramming methods add a learnable noise component to input to translate the target data to the source domain of the pre-trained model. This is similar to what many approaches to adversarial attacks do [8,33]. Unlike previous methods, we choose to add the noise to the spectrogram and not the time domain signal. We hypothesize that a spectrogram is a more fitting representation, because of the higher complexity of music audio compared to other signals reprogramming has been applied to. The operation can be formulated as $\hat{X} = X + N$. X is the input spectrogram with size $T \times F$ where T is the number of time bins, and F is the number of mel-band filters. $\hat{X}$ is the input of the pre-trained model. The learnable noise N is universal to all target data and independent of X. The dimension of N is identical to X, the size of the input spectrogram.

CNN Reprogramming: The application of CNN to audio spectrograms is considered a standard baseline in many audio classification tasks. Therefore, we propose to use trainable CNN layers as an input transformation of the input data. These layers replace the noise superposition as input processing. The transformation can be formally described by $\hat{X} = F(X)$, in which F represents the CNN consisting of two 2D convolutional layers with a receptive field of 3×3, a stride of 1×1, and a padding size of 1×1. Note that in this special case, no max-pooling is applied as the input dimension matches the input dimension of the pre-trained model. The difference between Noise Reprogramming and CNN Reprogramming is that CNN layers apply learnable transformation to the input itself instead of simply adding learnable noise that is independent of the input. Due to the idea of weight sharing in CNN, the training parameters needed are the CNN kernels; hence, the amount of parameters is considerable less than Noise Reprogramming.

U-Net Reprogramming: The idea behind using CNN Reprogramming is that the input audio spectrogram can be transformed into a suitable and compatible spectrogram for the pre-trained model. In order to provide more flexibility for the reprogramming learning, it is fair to consider the features at different time resolutions. Therefore, in addition to CNN reprogramming, we further propose U-Net reprogramming for MIC. U-Net is a CNN structure first developed for biomedical image segmentation [31] and has become popular in both speech and music separation tasks. The architecture consists of a contraction path to capture context and a symmetric expansion path to reconstruct the extracted features back to input resolution [31]. With convolutional layers and skip connections, U-Net is able to represent the input in both high-level features on coarser time-frequency scales and detailed features in deep CNN layers. These features are combined using bilinear upsampling blocks, yielding multi-scale features with both high-level and deep representations [23]. The U-Net's success in music source separation tasks might also imply that U-Net processing on a spectrogram is effective to differentiate instrument content. To the best of our knowledge, this is the first time a U-Net structure has been proposed for reprogramming. The proposed U-Net structure for reprogramming is shown in Fig. 2. The formulation is identical to the CNN reprogramming mentioned above as the transform function is applied to the input. It consists of three convolutional layers in both contraction path and expansion path, and each convolutional layer is followed by a batch normalization layer and a ReLU activation. A 2×2 max-pooling is applied for the CNN layer in the contraction path while upsampling is applied for that in expansion path.

Output Reprogramming. The output categories of the pre-trained system obviously do not match the music instrument classes to be classified. Therefore, the outputs of the pre-trained model have to be mapped to the target labels. For example, Yang et al. propose to use a many-to-one label mapping [33,35].

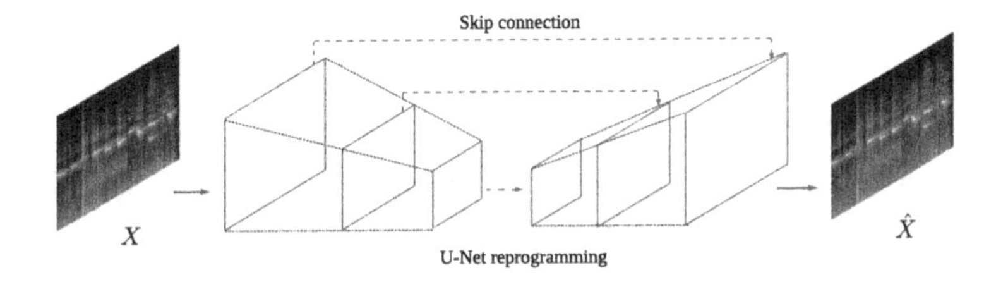

Fig. 2. U-Net structure for input reprogramming.

For each target label, its class prediction will be the averaged class predictions over the set of source labels assigned to it. We investigate this approach, but also propose a new output mapping utilizing fully-connected (FC) layers to fit the targets. Here, the output probabilities are mapped to the target labels by the FC layer (FCL) and a sigmoid activation function. The mapping can thus be learned during the training phase. Note that the original last layer activation in AST is removed in our model.

4 Experimental Setup

To evaluate the impact of different input and output reprogramming options, the results for the following systems will be reported:

- *Noise Reprogramming (AST-NRP)*: pre-trained model with added noise at the input and with output FCL label mapping,
- *CNN Reprogramming (AST-CNNRP)*: pre-trained model with CNN input processing and with output FCL label mapping, and
- *U-Net Reprogramming (AST-URP)*: pre-trained model with U-Net input processing and with output FCL label mapping.

In addition to the three methods introduced above (AST-NRP, AST-CNNRP, AST-URP), we also add the following systems for comparison: (i) the pre-trained AST (AST-BS) without input reprogramming as a baseline to evaluate the power of the AST representation for MIC, (ii) a CNN baseline (CNN-BS) with roughly the same number of training parameters as our proposed input transformation methods AST-CNNRP and AST-URP, (iii) a transfer learning approach by fine-tuning the AST system with the target data (AST-TL), (iv) the previous state-of-the-art Mean Teacher (MT) model by Gururani and Lerch [15], and (v) the Random Forest (RF) baseline released with the OpenMic dataset [21].

The implementation of the proposed methods is publicly available.[1]

4.1 Dataset

We use the OpenMIC dataset for the experiments in this paper. OpenMIC is the first open multi-instrument music dataset with a comparably diverse set of

[1] github.com/hchen605/ast_inst_cls, last accessed: Nov 9, 2022.

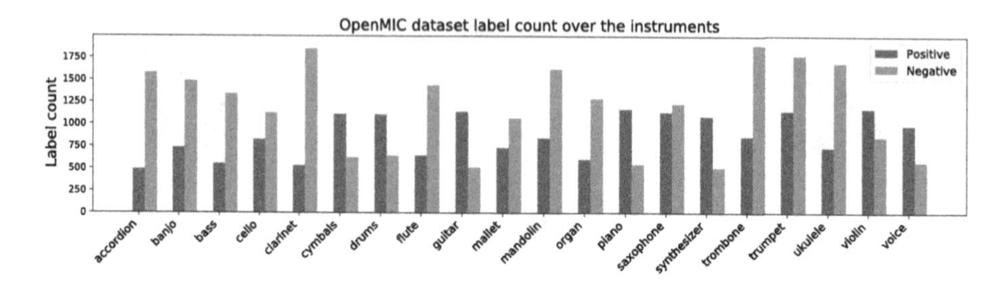

Fig. 3. OpenMIC: positive vs. negative labels.

musical instruments and genres, addressing issues in other previously existing datasets for MIC [21]. It consists of 20,000 audio clips, each of 10 s length. Every clip is labeled with the presence (positive) or absence (negative) of least one of 20 musical instruments, and each instrument class has at least 500 confirmed positives and at least totally 1500 confirmed labels. Note that if the dataset were fully labeled, it would come with 20000 clips$\cdot$20 instruments = 400000 labels, however, the actual number of labels in the dataset is 41, 268, meaning that approximately 90% of the labels are missing. Moreover, each 10 s clip has instrument presence or absence tags without specifying onset and offset times, also referred to as weak labels. Therefore, models cannot be trained using fine-grained instrument activity annotation. Figure 3 visualizes the overall label distribution of OpenMIC. We can observe the unbalanced nature; commonly seen instruments, such as piano, voice, and violin, generally have more positive labels than the others. Another possible reason that we can see more positive labels for these common instruments might be that crowd-sourced annotators were more familiar with these common timbres.

The models investigated in this study are trained to identify the presence or absence of musical instruments in OpenMIC dataset. The publicly available data splits are used for training and testing: approx. 25% of data are used for testing, and 15% of the training data is sampled randomly to form the validation set.

4.2 Training Procedure and Evaluation Metrics

The reprogramming model is trained with a batch size of 8, and the Adam optimizer with binary cross-entropy loss is used. We apply an initial learning rate of 5e-5 for the first 10 epochs, and then the learning rate is cut into half every 5 epochs until reaching 50 epochs.

We are interested in both the classification performance and the model complexity of each investigated system. To be comparable to previously reported results on the MIC task, the macro F1 score, calculated by purely averaging the per-instrument F1 scores and ignoring the weight of per-class data amount, is reported. The final results for each setup will be reported by averaging the macro F1 score of 10 experiments. In addition to this classification metric, the number of training parameters of each of the evaluated systems will be reported.

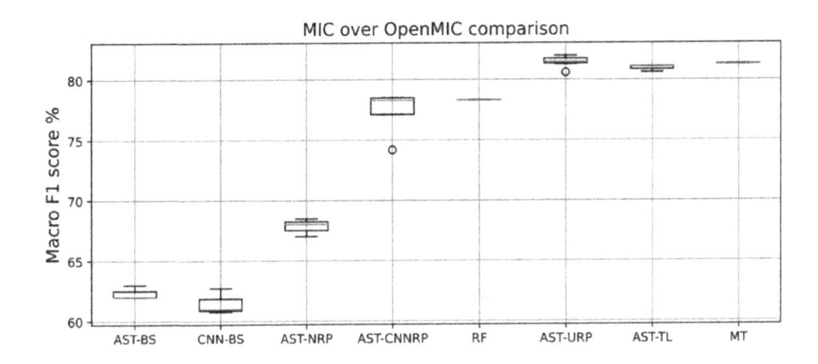

Fig. 4. Macro F1 scores of the evaluated methods.

4.3 Results and Discussion

Input Reprogramming. The overall average macro F1 scores are shown in Fig. 4. We note that the previous state-of-the-art MT [15] is reported with the result 81.3% as shown in the paper, and the RF baseline [21] is reported with the result 78.3% calculated from the released Python Notebook. We can make the following observations. First, simply using the pre-trained AST without input processing does not lead to convincing results, although at around 62% the result is considerably higher than guessing (50%). The performance of the un-tuned system roughly matches the performance of the simple CNN-BS trained on data for the task, indicating that AST has learned a powerful and useful representation. Second, a trained noise signal that is simply added to the input can improve this baseline performance by more than 6%, but the resulting system remains far from the performance of state-of-the-art systems. This implies that while traditional reprogramming approaches can work to a certain degree, the complexity and variability of music signals requires an input-adaptive transform as opposed to the addition a constant signal. Third, the results for AST-TL show the effectiveness of transfer learning showing performance roughly on par with the state-of-the-art. This result emphasizes that transfer learning is a powerful tool with competitive results for weakly-labeled data. Fourth, both the AST-CNNRP and AST-URP pre-processing steps dramatically improve classification performance over the AST-NRP with the U-Net-based reprogramming performing about 4% better than the CNN. We also note that transfer learning with fine-tuning results in only limited improvement over the reprogramming methods, implying the AST representation without fine-tuning is suitable for this task. Fifth, AST-URP slightly outperforms all presented system and

Table 1. Comparison of model training parameters.

Method	AST-BS	CNN-BS	AST-TL	AST-NRP	AST-CNNRP	AST-URP	MT
#Param.(M)	0.017	0.017	87.873	0.148	0.017	0.018	0.111

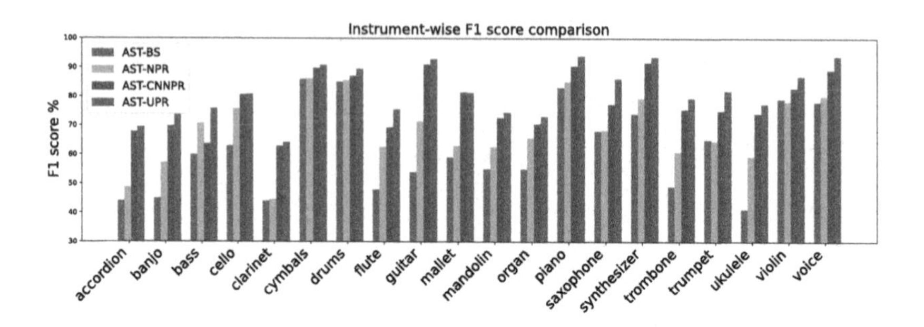

Fig. 5. F1 score comparison between the reprogramming methods.

even the state-of-the-art MT system. Diving deeper into the detailed results, Fig. 5 displays the instrument-wise F1 scores. We can observe a lot of variation in classification performance over instruments. These variations are related to the amount of positive labels as shown in Fig. 3, a clear indicator that the number of (positive) labels directly influences the classification performance. In fact, the correlation coefficient of the amount of instrument-wise positive labels and AST-BS F1 score is 0.75, showing the high correlation. We can see that the U-Net reprogramming model AST-URP outperforms all other models for all instruments, and we observe considerable performance gains especially for the instrument classes with few positive labels. For example, accordion has the lowest number of positive labels with around 500 and the AST-URP improvement is over 25%. This demonstrates the capability of the proposed reprogramming method to solve the data scarcity issue by leveraging powerful other representations.

The results support our assumption that an input transformation utilizing both high-level and low-level features benefits reprogramming. Overall, we can see that both AST-CNNRP and AST-URP are effective approaches to adapt a pre-trained model to a new task. It encourages future research on reprogramming pre-trained models with other methods of transformation.

Complexity Analysis. One of the main advantages of reprogramming is reduced training complexity as the pre-trained model remains unchanged. As an indicator of model complexity, Table 1 reports the number of training parameters along with the F1 score previously visualized in Fig. 4. In terms of complexity, it is clear that the training complexity of reprogramming is very low. We can observe that MT, the previous state of the art system using a Mean Teacher approach with consistency loss has a higher complexity (one order of magnitude) than either our CNN-based approach or the high-performing U-Net approach, despite them using the VGGish representation as input to their system. Compared to the number of AST training parameters, the proposed AST-URP model has only a fraction of training parameters (0.02%) but is still the best-performing system.

5 Conclusion

In this work, we propose to apply model reprogramming to the task of music instrument classification. We extend the existing reprogramming approaches by utilizing a novel U-Net-based input reprogramming method. By leveraging the power of pre-trained audio spectrogram transformer model, we show that our model can achieve state-of-the-art performance at a fraction of the training complexity of other models. We provide detailed results on the impact of different input and output reprogramming approaches. Without applying data augmentation and advanced model architectures like semi-supervised MT learning, reprogramming a pre-trained model is a low-complexity approach to achieving state-of-the-art performance. Although the workload during the inference stage remains unchanged, the low training complexity of this new transfer learning approach in combination with our promising results opens up a multitude of possible use cases in tasks in MIR and other fields with insufficient data.

In future work, we plan to explore variations of the input reprogramming stages beyond CNN or U-Net structures. Furthermore, we plan to test the reprogramming algorithm with various other pre-trained models from other audio and non-audio tasks. We believe that reprogramming is a promising approach with the potential to be used in many other MIR tasks.

References

1. Benetos, E., Kotti, M., Kotropoulos, C.: Musical instrument classification using non-negative matrix factorization algorithms and subset feature selection. In: ICASSP (2006)
2. Biggio, B., Roli, F.: Wild patterns: ten years after the rise of adversarial machine learning. Pattern Recognit. **84**, 317–331 (2018)
3. Bittner, R.M., Salamon, J., Tierney, M., Mauch, M., Cannam, C., Bello, J.P.: Medleydb: a multitrack dataset for annotation-intensive MIR research. In: ISMIR (2014)
4. Bosch, J.J., Janer, J., Fuhrmann, F., Herrera, P.: A comparison of sound segregation techniques for predominant instrument recognition in musical audio signals. In: ISMIR (2012)
5. Choi, K., Fazekas, G., Sandler, M.B., Cho, K.: Transfer learning for music classification and regression tasks. In: ISMIR (2017)
6. Dau, H.A., et al.: The UCR time series archive. IEEE/CAA J. Automatica Sinica **6**(6), 1293–1305 (2019)
7. Defferrard, M., Benzi, K., Vandergheynst, P., Bresson, X.: FMA: a dataset for music analysis. In: ISMIR (2017)
8. Elsayed, G.F., Goodfellow, I., Sohl-Dickstein, J.: Adversarial reprogramming of neural networks. In: ICLR (2019)
9. Eronen, A., Klapuri, A.: Musical instrument recognition using cepstral coefficients and temporal features. In: ICASSP (2000)
10. Essid, S., Richard, G., David, B.: Hierarchical classification of musical instruments on solo recordings. In: ICASSP (2006)

11. Gemmeke, J.F., et al.: Audio set: an ontology and human-labeled dataset for audio events. In: ICASSP (2017)
12. Gong, Y., Chung, Y.A., Glass, J.R.: AST: Audio spectrogram transformer. In: Interspeech (2021)
13. Gong, Y., Lai, C.I., Chung, Y.A., Glass, J.R.: SSAST: self-supervised audio spectrogram transformer. In: AAAI (2021)
14. Gururani, S., Lerch, A.: Mixing secrets: a multi-track dataset for instrument recognition in polyphonic music. In: ISMIR (2017)
15. Gururani, S., Lerch, A.: Semi-supervised audio classification with partially labeled data. In: IEEE ISM (2021)
16. Gururani, S., Sharma, M., Lerch, A.: An attention mechanism for musical instrument recognition. In: ISMIR (2019)
17. Hambardzumyan, K., Khachatrian, H., May, J.: WARP: word-level adversarial ReProgramming. In: IJCNLP (2021)
18. He, K., Zhang, X., Ren, S., Sun, J.: Deep residual learning for image recognition. In: CVPR (2016)
19. Hershey, S., et al.: CNN architectures for large-scale audio classification. In: ICASSP (2017)
20. Howard, A.G., et al.: Mobilenets: efficient convolutional neural networks for mobile vision applications. In: CVPR (2017)
21. Humphrey, E., Durand, S., McFee, B.: Openmic-2018: an open data-set for multiple instrument recognition. In: ISMIR (2018)
22. Hung, Y., Yang, Y.: Frame-level instrument recognition by timbre and pitch. In: ISMIR (2018)
23. Jansson, A., Humphrey, E.J., Montecchio, N., Bittner, R.M., Kumar, A., Weyde, T.: Singing voice separation with deep u-net convolutional networks. In: ISMIR (2017)
24. Kong, Q., Cao, Y., Iqbal, T., Wang, Y., Wang, W., Plumbley, M.D.: PANNs: large-scale pretrained audio neural networks for audio pattern recognition. In: IEEE/ACM Transactions on Audio, Speech, and Language Processing (2020)
25. Li, P.Q., Qian, J., Wang, T.: Automatic instrument recognition in polyphonic music using convolutional neural networks. CoRR abs/1511.05520 (2015)
26. Lostanlen, V., Cella, C.E.: Deep convolutional networks on the pitch spiral for music instrument recognition. In: ISMIR (2016)
27. Nagawade, M.S., Ratnaparkhe, V.R.: Musical instrument identification using MFCC. In: RTEIC) (2017)
28. Neekhara, P., Hussain, S., Dubnov, S., Koushanfar, F.: Adversarial reprogramming of text classification neural networks. In: EMNLP-IJCNLP (2019)
29. Piczak, K.J.: ESC: Dataset for Environmental Sound Classification. In: ACM MM, pp. 1015–1018. ACM (2015)
30. Pons, J., Serra, X.: MusiCNN: pre-trained convolutional neural networks for music audio tagging. In: ISMIR (2019)
31. Ronneberger, O., Fischer, P., Brox, T.: U-Net: convolutional networks for biomedical image segmentation. In: Navab, N., Hornegger, J., Wells, W.M., Frangi, A.F. (eds.) MICCAI 2015. LNCS, vol. 9351, pp. 234–241. Springer, Cham (2015). https://doi.org/10.1007/978-3-319-24574-4_28
32. Tan, C., Sun, F., Kong, T., Zhang, W., Yang, C., Liu, C.: A survey on deep transfer learning. In: Kůrková, V., Manolopoulos, Y., Hammer, B., Iliadis, L., Maglogiannis, I. (eds.) ICANN 2018. LNCS, vol. 11141, pp. 270–279. Springer, Cham (2018). https://doi.org/10.1007/978-3-030-01424-7_27

33. Tsai, Y.Y., Chen, P.Y., Ho, T.Y.: Transfer learning without knowing: reprogramming black-box machine learning models with scarce data and limited resources. In: ICML. PMLR (2020)
34. Warden, P.: Speech commands: a dataset for limited-vocabulary speech recognition. CoRR abs/1804.03209 (2018)
35. Yang, C.H.H., Tsai, Y.Y., Chen, P.Y.: Voice2series: reprogramming acoustic models for time series classification. In: ICML. PMLR (2021)

Cascading CNNs with S-DQN: A Parameter-Parsimonious Strategy for 3D Hand Pose Estimation

Mingqi Chen, Shaodong Li(✉), Feng Shuang, and Kai Luo

Guangxi Key Laboratory of Intelligent Control and Maintenance of Power Equipment, School of Electrical Engineering, Guangxi University, Nanning 530004, China
mqchen0916@outlook.com, lishaodongyx@126.com, fshuang@gxu.edu.cn, luokaijiayou@163.com

Abstract. This paper proposes a cascaded parameter-parsimonious 3D hand pose estimation strategy to improve real-time performance without sacrificing accuracy. The estimation process is first decomposed into feature extraction and feature exploitation. The feature extraction is seen as a dimension reduction process, where convolutional neural networks (CNNs) are used to ensure accuracy. Feature exploitation is considered as a policy optimization process, and a shallow reinforcement learning (RL)-based feature exploitation module is proposed to improve running rapidity. Ablation studies and experiments are carried out on NYU and ICVL datasets to evaluate the performance of the strategy, and multiple baselines are used to evaluate generalization. The results show that the improvement on testing time reaches 8.1% and 14.6% by the proposed strategy. Note that the overall accuracy also reaches state-of-the-art, which further shows the effectiveness of the proposed strategy.

Keywords: Hand pose estimation · Feature exploitation · Reinforcement learning · Real-time performance

1 Introduction

3D hand pose estimation remains a key issue for vision-based human-robot interaction [22]. Deep learning methods are used to regress [2,5,8] or detect [7,17] 3D hand joint locations, and good accuracy has been achieved. However, the real-time performance of 3D hand pose estimation frameworks is usually less discussed. Actually, to overcome some open challenges in 3D hand pose estimation like self-occlusion and self-similarity, complex frameworks and multi-modal methods need to be introduced to enhance accuracy, while causing bad real-time performance [8,14,18]. Meanwhile, real-time performance is vital to achieve

This work was supported in part by the funding of basic ability promotion project for young and middle-aged teachers in Guangxi's colleges and universities (Grant No. 2022KY0008), in part by special fund of Guangxi Bagui Scholars, and in part by National Natural Science Foundation of China (Grant No. 61720106009).

D.-T. Dang-Nguyen et al. (Eds.): MMM 2023, LNCS 13833, pp. 358–369, 2023.
https://doi.org/10.1007/978-3-031-27077-2_28

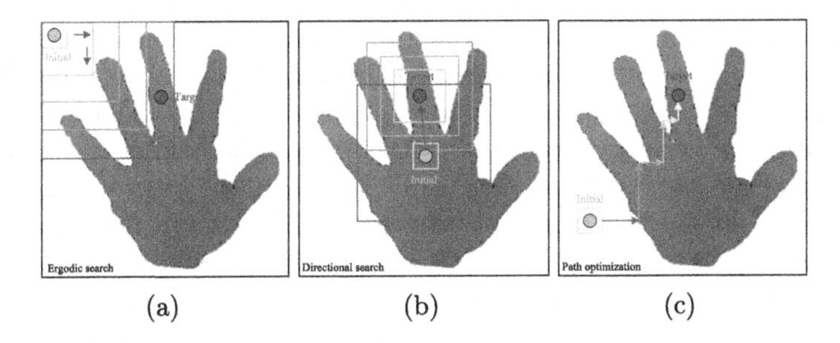

Fig. 1. Illustration of the proposed feature exploitation module. Traditional exploitation frameworks in 3D hand pose estimation follows bottom-up (a) or top-down (b) strategy, which need exhaustive search on the feature regions. In this paper, the feature exploitation is considered as a path optimization issue (c), where the features are not ergodic.

immersive human-robot interaction. Thus, it is also important to improve real-time performance of the existing 3D hand pose estimation frameworks without sacrificing accuracy.

3D hand pose estimation aims at establishing a mapping from the images to 3D joint locations, which can be achieved by a two-step process [6,15,23]. The first step is reducing the dimension of input images to low-dimensional feature vectors. The second step is refining the vectors to estimate 3D hand joint locations. Thus, the overall process of 3D hand pose estimation can be achieved by a novel strategy containing a feature extraction module and a feature exploitation module. The feature extraction module first performs dimension reduction from numerous high-dimensional data. The feature exploitation module then refines the feature vectors to the corresponding estimations. Both modules can maintain accuracy, while the feature exploitation module takes more effect on running rapidity. Our aim in this paper is improving real-time performance without loosing overall accuracy. Thus, CNNs are used in feature extraction to ensure accuracy, and RL is used in feature exploitation to improve real-time performance.

Existing feature extraction methods follow bottom-up strategy to obtain features from numerous data [19,24]. Feature exploitation modules also follow bottom-up strategy to predict 3D joint locations, where ergodic search is performed on all regions of the feature map, as shown in Fig. 1a. The strategy leads to redundancy on region searching and is time-consuming [20]. Recently, top-down strategy is introduced to exploit features in pixel-wise detection issues. The strategy finds optimal directions sequentially to search the corresponding feature region [1]. Top-down strategy achieves better running rapidity than bottom-up strategy, while exhaustive search is also performed in each step to obtain the optimal region proposal, as shown in Fig. 1b. In fact, both exploitation strategies needs CNNs to further reduce dimensions of the features. However, in 3D hand pose estimation, the extracted features can be low-dimensional, and the exploitation process can be considered as sequential point-wise translations of hand joints in pixel or camera coordinates, as shown in Fig. 1c. Thus, the feature exploitation process of 3D hand pose estimation can be further considered as a path optimization issue without exhaustive searching the features.

RL is excellent to explore an optimized path by training an agent under the guidance of a reward function, which has been widely used in image and video processing [3,11,12,20], etc. Although the advantage of RL-based modules in training is usually not obvious, the rapidity of testing improves comparing to other strategies, which further enhances real-time performance. However, to the best of our knowledge, RL has not been used in 3D hand pose estimation.

This paper addresses the issue of real-time performance in 3D hand pose estimation based on a novel strategy. The strategy decomposes the estimation process into feature extraction and feature exploitation, and a framework combining deep learning (DL) and shallow RL to perform rapid estimation. DL is used to extracted features, and shallow RL is used to obtain an optimized strategy to locate hand joints. To perform feature exploitation in a parameter-parsimonious manner, a shallow representation network-based DQN (S-DQN) is proposed to achieve sequential point-wise translations. To the best of our knowledge, the strategy is firstly proposed in 3D hand pose estimation, and RL is firstly introduced to improve real-time performance. Experiments are carried out to evaluate the performance of the proposed feature exploitation module. The results show the proposed module successfully increases testing speed of 3D hand pose estimation without weakening the overall accuracy. The contribution of this paper can be summarized as follows:

- A parameter-parsimonious strategy of 3D hand pose estimation is proposed containing a feature extraction module and a feature exploitation module. CNNs are used in the feature extraction module to ensure accuracy, and shallow RL is used in the feature exploitation module to improve real-time performance.
- S-DQN is introduced to achieve sequential point-wise translations. The continuous process of feature exploitation in 3D hand pose estimation can then be achieved using the proposed framework using the discretized representation of actions.
- Experiments are carried out to evaluate the performance of the proposed strategy using two different baselines on ICVL and NYU datasets. The running time on each baseline is reduced by 5.6% and 14.6%, which reaches state-of-the-art. The overall accuracy also reaches state-of-the-art, which further shows the effectiveness of the proposed strategy.

The rest of the paper is organized as follows: Sect. 2 introduces the proposed 3D hand pose estimation strategy in detail, especially the feature exploitation module based on S-DQN. Section 3 performs experiments to validate the performance of the proposed strategy, and Sect. 4 concludes the paper.

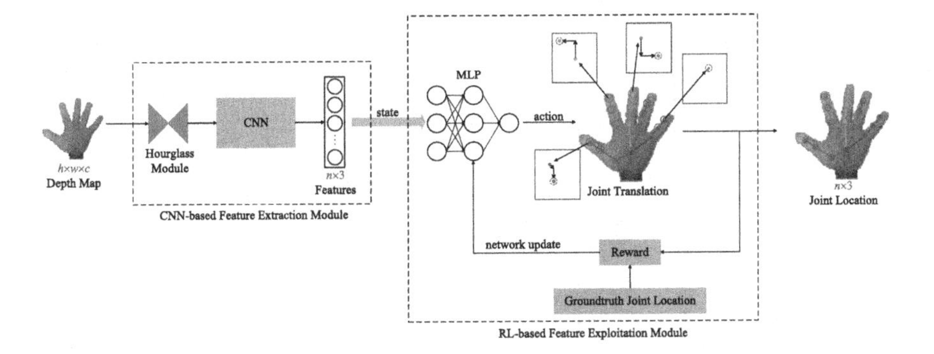

Fig. 2. Overview of the proposed 3D hand pose estimation strategy.

2 The Proposed Method

2.1 Overview

In this paper, a parameter-parsimonious framework for 3D hand pose estimation is proposed to improve real-time performance based on the proposed strategy, which is shown in Fig. 2. The framework consists a feature extraction module based on CNNs, and a feature exploitation module based on S-DQN. The feature extraction module takes an $h \times w \times c$ depth image as input, and outputs an $n \times 3$-dimensional flattened feature vector, where h, w, c represents the dimensions of the depth map, and n represents number of hand joints. The feature exploitation module takes the feature vector as input and finally outputs an $n \times 3$ matrix denoting accurate 3D hand joint locations by translation in pixel or camera coordinates. The overall accuracy is mainly ensured in the feature extraction module, while the running speed is improved in the feature exploitation module. Thus, two representative frameworks are used to extract features to ensure accuracy, and S-DQN is used to improve real-time performance of feature exploitation.

2.2 Feature Extraction Module

The aim of the proposed framework is improving real-time performance of estimation without sacrificing accuracy, while the feature extraction module is not the priority. Thus, two representative frameworks with state-of-the-art accuracy are used to extract features to ensure the overall accuracy, namely HandFoldingNet [5] and JGR-P2O [7]. Both modules take the depth maps as input, and output $n \times 3$-dimensional feature vector for the subsequent feature exploitation.

HandFoldingNet is a 3D CNN-based hand pose regression framework, where point clouds are transformed from depth maps and used as input. The network uses an encoder-decoder format to extract features using global and multiple local folding blocks. JGR-P2O is a 2D CNN-based hand pose detection framework. The framework extracts and decodes features using a graph convolutional

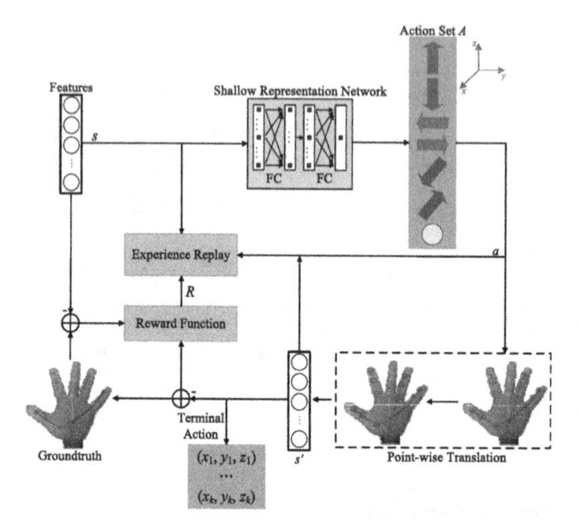

Fig. 3. Working procedure of the S-DQN-based feature exploitation module.

network, and heatmaps are used to obtain features in both pixel and camera coordinates.

Note that HandFoldingNet contains multiple local folding blocks, and JGR-P2O introduces multi-stage framework to refine the features. The iteration of a single block or a stage limits real-time performance. Thus, in the proposed framework, HandFoldingNet is used with a single local folding block, and one-stage JGR-P2O is used to extract feature. Moreover, the original baselines are used in ablation studies to evaluate the performance of the proposed strategy.

2.3 The Proposed Feature Exploitation Module

The feature exploitation module takes the extracted $n \times 3$-dimensional feature vector as input, and trains an agent to obtain the estimation results. As mentioned above, the exploitation process can be considered as a path optimization process containing sequential point-wise translations of hand joints in pixel or camera coordinates. DQN [13] is a classical framework in deep RL, which achieves good performance in low-dimensional and discrete action space. Therefore, DQN is introduced in the feature exploitation module to improve real-time performance, which uses shallow representation networks to tackle the continuance and correlations between input states, and obtains an optimized path of point-wise translations in a parameter-parsimonious manner. We denote the proposed module as S-DQN, whose process is shown in Fig. 3.

Process Modeling. The feature exploitation process in the proposed strategy is considered as a Markov Decision Process (MDP) to meet the requirements of S-DQN. An MDP can be represented using a four-tuple $\langle S, A, P, R\rangle$, where S represents the set of the states, A is the set of actions, P is the state transition probabilities, and R is a discrete reward function. The elements in the MDP can be specified as follows:

(1) State Set S. The state set S is taken as the input of the feature exploitation module, which is defined according to the output of the feature extraction module. In the feature exploitation module, the 3D camera coordinates is used to describe the state space. In fact, 2D pixel coordinates can also be used to describe the state space. The input state of the module is the flattened representation of the aforementioned $n \times 3$ matrix. Meanwhile, the initial states are the extracted features of the previous module.

(2) Action Set A. The action set A denotes the direction of translation in camera coordinates, namely 6 basic types of actions $X+, X-, Y+, Y-, Z+, Z-$. Moreover, a *stop* action is also defined as the terminal action, where the joint keeps its previous location. When the variation of estimation error reaches a certain threshold, a *done* notation is received, and the agent outputs a *stop* action to finish the exploitation process.

(3) Reward Function R. In depth-based 3D hand pose estimation, the Euclidean distance between the estimated joint and the corresponding groundtruth, namely the 3D joint error, is the most frequently used evaluation metric. Thus, to achieve accurate exploitation, R is set according to the variation of the 3D joint errors. When the error after translation is bigger than the previous error, a penalty of -1 is given, and the variation between the new error and the original error is computed. When the variation between the new error and the original error, the translation is considered to be finished, and a reward of 1 is given. The reward function can be written as

$$R = \begin{cases} -1, & e_{old} - e_{new} < 0 \\ 1, & e_{ori} - e_{new} > e_{threshold} \\ 0, & \text{else} \end{cases} \tag{1}$$

where the reward is 0 under other circumstances. e_{old}, e_{new}, and e_{ori} denote the estimation errors measured from the input, output, and the original, respectively. $e_{threshold}$ is a predefined threshold of the error variation. Once the variation reaches $e_{threshold}$, the exploitation is finished, a *done* notation is also returned with a *stop* action.

2.4 Network Training

In this paper, the feature extraction module and the feature exploitation module are trained sequentially. Thus, loss functions are defined separately for each module. The loss of HandFoldingNet is as

$$\mathcal{L}_{HandFold} = \sum_{j=1}^{J} L1_{\text{smooth}}\left(\mathbf{j}_j^0 - \mathbf{j}_j^*\right) + \sum_{j=1}^{J} L1_{smooth}\left(\mathbf{j}_j^1 - \mathbf{j}^*\right) \tag{2}$$

where $\mathbf{j}_j^0$ and $\mathbf{j}_j^1$ represent the estimated locations of the jth joint of each folding block in the network, and $\mathbf{j}_j^*$ denotes the groundtruth location of the jth joint. Meanwhile, the loss of JGR-P2O is

$$\mathcal{L}_{JGR} = \mathcal{L}_{\text{coordinate}} + \beta \mathcal{L}_{\text{offset}} \tag{3}$$

where $\mathcal{L}_{\text{coordinate}}$ represents the Huber loss function between the estimated joints in pixel coordinates and the groundtruth, $\mathcal{L}_{\text{offset}}$ represents the loss when generating offset maps in each axis, and $\beta = 0.0001$ is a balancing weight factor. The loss of the feature extraction module can be further referenced in [5] and [7].

The loss of the proposed S-DQN-based feature exploitation module follows the original loss function of DQN [13] as

$$\mathcal{L}_{S-DQN} = \begin{cases} r & \text{if } done \\ \|r + \gamma \max_a Q(s', a; \theta^-) - Q(s, a; \theta)\|_2 & \text{else} \end{cases} \tag{4}$$

where θ and θ^- represent the parameters of the evaluation network and the target network, respectively.

Adam optimizer is introduced to train the feature extraction module, and stochastic gradient descent (SGD) is used to train the feature exploitation module. In the feature exploitation module, the discount factor γ is set to 0.9, and ϵ-greedy is used to balance exploration and exploitation, where ϵ is set to 0.9.

3 Experiments

3.1 Datasets and Settings

NYU [17] and ICVL [16] are used to evaluate the performance of the framework. The NYU hand pose dataset contains 72757 training and 8252 testing frames, where 3 different views are captured, and 36 joints are annotated. Following most previous works, one single view and 14 annotated joints are used to perform hand pose estimation. The ICVL hand pose dataset contains 22084 training and 1596 testing frames, with 16 joints being annotated.

The experiments are performed on a single NVIDIA TITAN V GPU using PyTorch. For each feature extraction module, the preprocessing procedure and the settings of hyperparameters follow the original works in [5] and [7]. For the feature exploitation module, the size of the experience replay is 10^6, the batch size is 4096, and the learning rate α is 0.01. The step size of each action is set according to the output of the baselines, which is 0.005 for HandFoldingNet and 1.5 for JGR-P2O.

3D mean error, success rate, and the running time on each frame (fps) are introduced to evaluate the performance of the framework. 3D mean error is used to compute the overall accuracy of the framework. Success rate is used to further evaluate the distribution of joint errors, which denotes the percentage of success frames where the worst joint error is below a threshold. The running time on each frame (fps) is introduced to evaluate the real-time performance of the framework.

Table 1. Test Result of HandFoldingNet on ICVL Dataset

Method	Mean error (mm)	# Params	Time (ms)
Backbone	6.34	0.78M	4.06
Backbone+Original	5.95	1.28M	4.82
Backbone+S-DQN	**5.95**	**0.78M**	**4.43**

Table 2. Test Result of JGR-P2O on NYU Dataset

Method	Mean error (mm)	# Params	Time (ms)
Backbone	8.63	0.72M	4.79
Backbone+Original	8.29	1.37M	6.04
Backbone+S-DQN	**8.34**	**0.72M**	**5.16**

3.2 Ablation Studies

Ablation studies are first performed to show the effectiveness of the proposed strategy. To achieve fair comparison, experiments are performed according to previous work on the baselines. For HandFoldingNet, ICVL is used, and NYU is used for JGR-P2O.

The results of the experiments are shown in Table 1 and Table 2. In fact, the feature extraction module can output coarse joint locations. Thus, the first row of the tables shows the accuracy of the feature extraction module, the next two rows show the accuracy based on the original feature exploitation module and the proposed S-DQN, respectively. From the tables above, improvement on accuracy based on S-DQN is 0.95 mm and 0.59 mm on each baseline, respectively. The overall accuracy of the proposed framework equals the original work on HandFoldingNet, and shows a 0.05 mm difference on JGR-P2O. Meanwhile, Fig. 4a and 4c show the 3D joint errors in detail, where the largest decreases of the joint error are 0.56 mm and 0.51 mm. The accuracy on some joint is even better than the originals, such as the Ring R joint in Fig. 4a and the Mid R joint in Fig. 4c. In addition, from Fig. 4b and 4d, the increases of the success rate under 5 mm are and 3.95% and 3.29%. Thus, the proposed strategy based on CNNs and S-DQN shows good effectiveness on maintaining the overall accuracy.

Moreover, the running time of the framework shows an obvious decrease comparing to the original baselines. From Table 1 and Table 2, the running time of S-DQN is 0.37 ms, and the running time of the original exploitation modules is 0.76 ms and 1.25 ms, according to the results on our own platform. Thus, the proposed strategy shows decreases of 0.39 ms and 0.88 ms on running time. The reason for reducing the running time is that the shallow representation network used in S-DQN contains only 170 parameters, which is a significant decrease comparing to previous works. Meanwhile, the accuracy of S-DQN is

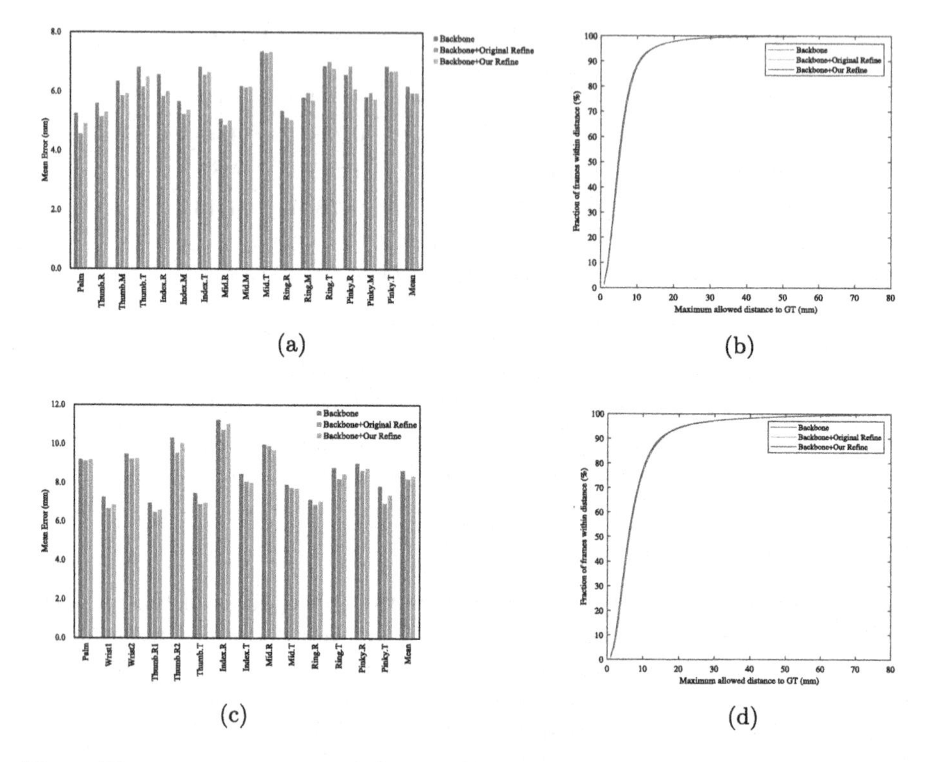

Fig. 4. The results of ablation studies, (a) and (b) shows the joint accuracy and success rate using HandFoldingNet to extract features, (c) and (d) shows the joint accuracy and success rate using JGR-P2O to extract features.

not sacrificed using the parameter-parsimonious network, as shown in Table 1 and Table 2.

In summary, the ablation studies show effectiveness of S-DQN-based feature exploitation module, and the strategy successfully improves real-time performance without sacrificing accuracy.

3.3 Comparison with State-of-the-Art

Following most previous works, the state-of-the-art methods on NYU and ICVL are listed in Table 3 to evaluate the performance of our strategy, where the running speeds are listed in fps to show real-time performance for fair comparisons. From Table 3, the accuracy of our strategy ranks 3rd and 4th on both baselines on the corresponding datasets. Meanwhile, the number of parameters of our framework ranks the smallest two, and the running speed of our strategy reaches the best and second in 2D-based and 3D-based methods, respectively. VirtualView [4], AWR [10] and the original HandFoldingNet [5] show better accuracy, while our strategy achieves better real-time performance. Note that the speed of the

Table 3. Comparisons on state-of-the-arts on NYU and ICVL datasets, where mean errors (in mm), number of parameters and the speed (in fps) is listed.

Method	Mean error (mm)		#Params	Speed (fps)	Input Type
	NYU	ICVL			
DeepModel [24]	17.04	11.56	–	–	2D
REN-4×6×6 [19]	13.39	7.63	–	–	2D
REN-9×6×6 [19]	12.69	7.31	–	–	2D
DeepPrior++ [15]	12.24	8.10	–	–	2D
Pose-REN [1]	11.81	6.79	–	–	2D
DenseReg [18]	10.2	7.30	5.8M	27.8	2D
CrossInfoNet [6]	10.08	6.73	23.8M	124.5	2D
A2J [21]	8.61	6.46	44.7M	105.1	2D
JGR-P2O [7]	8.29	6.02	1.4M	111.2	2D
SRNet [23]	9.17	6.15	3.3M	–	2D
VirtualView [4]	6.40	4.76	–	26.71	2D
AWR [10]	7.18	5.98	460M	–	2D
JGR-P2O+Ours	**8.35**	–	**0.72M**	**139.7**	**2D**
3DCNN [9]	14.1	–	–	215	3D
V2V-PoseNet [14]	8.42	6.28	457.5M	3.5	3D
SHPR-Net [2]	10.78	7.22	–	–	3D
HandPointNet [8]	10.54	6.94	2.58M	48.0	3D
HandFoldingNet [5]	8.58	5.95	1.28M	84.0	3D
HandFoldingNet+Ours	–	**5.95**	**0.78M**	**79.2**	**3D**

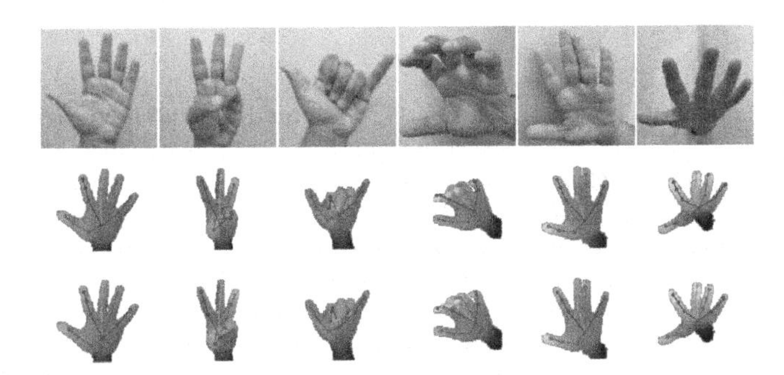

Fig. 5. Visualization of 3D hand pose estimation results of ICVL dataset, where the first row is a realistic view of the hand pose in the dataset, the second row shows the predicted locations on the corresponding depth maps, and the third row shows the groundtruth.

listed frameworks are referenced from their original works. In fact, the speeds on the original baselines are 76.8 fps and 124.4 fps on our GPU. Thus, the actual improvements on running speed on both baselines are 2.4 fps and 15.3 fps, with the preprocessing time of 8.2 ms and 2.0 ms, respectively. The visualization of some representative hand poses based on our strategy is also shown in Fig. 5, which further shows the accuracy of the proposed strategy.

Therefore, the accuracy the proposed strategy reaches state-of-the-art, and the running speed reaches the best, which shows effectiveness of the proposed strategy.

4 Conclusions

In this paper, the process of 3D hand pose estimation is first decomposed into feature extraction and feature exploitation, a novel cascaded framework is then proposed to improve real-time performance without sacrificing accuracy. In the framework, feature extraction is seen as a dimension reduction process, which is achieved using CNNs. Feature exploitation is considered as a path optimization process, which is performed using S-DQN. Ablation studies and extensive experiments are carried out to evaluate the effectiveness of the strategy on two representative baselines. The results show that the proposed framework saves 0.39 ms and 0.88 ms testing time per sample, and the speed is increased by 2.4 fps and 15.3 fps, respectively. Meanwhile, the accuracy remains state-of-the-art based on S-DQN, which shows the effectiveness of the proposed strategy.

Efficient 3D hand pose estimation is achieved using the proposed parameter-parsimonious strategy, while the estimation accuracy can be further improved. Future work includes introducing multi-modal feature extraction and novel exploitation methods to obtain better performance.

References

1. Chen, X., Wang, G., Guo, H., Zhang, C.: Pose guided structured region ensemble network for cascaded hand pose estimation. Neurocomputing **395**, 138–149 (2020)
2. Chen, X., Wang, G., Zhang, C., Kim, T.K., Ji, X.: SHPR-Net: deep semantic hand pose regression from point clouds. IEEE Access **6**, 43425–43439 (2018)
3. Chen, Z., Wang, X., Zhou, Y., Zou, L., Jiang, J.: Content-aware cubemap projection for panoramic image via deep Q-learning. In: Proceedings of International Conference on Multimedia Modeling, pp. 304–315 (2020)
4. Cheng, J., et al.: Efficient virtual view selection for 3D hand pose estimation. arXiv preprint arXiv:2203.15458 (2022)
5. Cheng, W., Park, J.H., Ko, J.H.: HandFoldingNet: A 3D hand pose estimation network using multiscale-feature guided folding of a 2D hand skeleton. In: Proceedings of the IEEE/CVF International Conference on Computer Vision, pp. 11260–11269 (2021)
6. Du, K., Lin, X., Sun, Y., Ma, X.: CrossInfoNet: Multi-task information sharing based hand pose estimation. In: Proceedings of the IEEE/CVF Conference on Computer Vision and Pattern Recognition, pp. 9896–9905 (2019)
7. Fang, L., Liu, X., Liu, L., Xu, H., Kang, W.: JGR-P2O: joint graph reasoning based pixel-to-offset prediction network for 3D hand pose estimation from a single depth image. In: Proceedings of the European Conference on Computer Vision, pp. 120–137 (2020)

8. Ge, L., Cai, Y., Weng, J., Yuan, J.: Hand PointNet: 3D hand pose estimation using point sets. In: Proceedings of the IEEE Conference on Computer Vision and Pattern Recognition, pp. 8417–8426 (2018)
9. Ge, L., Liang, H., Yuan, J., Thalmann, D.: Robust 3D hand pose estimation from single depth images using multi-view CNNs. IEEE Trans. Image Process. **27**(9), 4422–4436 (2018)
10. Huang, W., Ren, P., Wang, J., Qi, Q., Sun, H.: AWR: adaptive weighting regression for 3D hand pose estimation. In: Proceedings of AAAI Conference on Artificial Intelligence, vol. 34, pp. 11061–11068 (2020)
11. Li, H., Chen, J., Hu, R., Yu, M., Chen, H., Xu, Z.: Action recognition using visual attention with reinforcement learning. In: Proceedings of International Conference on Multimedia Modeling, pp. 365–376 (2019)
12. Li, Z., Zhang, X.: Deep reinforcement learning for automatic thumbnail generation. In: Proceedings of International Conference on Multimedia Modeling, pp. 41–53 (2019)
13. Mnih, V., et al.: Human-level control through deep reinforcement learning. Nature **518**(7540), 529–533 (2015)
14. Moon, G., Chang, J.Y., Lee, K.M.: V2V-PoseNet: voxel-to-voxel prediction network for accurate 3D hand and human pose estimation from a single depth map. In: Proceedings of the IEEE Conference on Computer Vision and Pattern Recognition, pp. 5079–5088 (2018)
15. Oberweger, M., Lepetit, V.: DeepPrior++: Improving fast and accurate 3D hand pose estimation. In: Proceedings of the IEEE International Conference on Computer Vision Workshops, pp. 585–594 (2017)
16. Tang, D., Jin Chang, H., Tejani, A., Kim, T.K.: Latent regression forest: Structured estimation of 3D articulated hand posture. In: Proceedings of IEEE/CVF Conference on Computer Vision and Pattern Recognition, pp. 3786–3793 (2014)
17. Tompson, J., Stein, M., Lecun, Y., Perlin, K.: Real-time continuous pose recovery of human hands using convolutional networks. ACM Trans. Graphics (ToG) **33**(5), 1–10 (2014)
18. Wan, C., Probst, T., Van Gool, L., Yao, A.: Dense 3D regression for hand pose estimation. In: Proceedings of the IEEE Conference on Computer Vision and Pattern Recognition, pp. 5147–5156 (2018)
19. Wang, G., Chen, X., Guo, H., Zhang, C.: Region ensemble network: towards good practices for deep 3D hand pose estimation. J. Vis. Commun. Image Represent. **55**, 404–414 (2018)
20. Wang, Y., Zhang, L., Wang, L., Wang, Z.: Multitask learning for object localization with deep reinforcement learning. IEEE Trans. Cogn. Dev. Syst. **11**(4), 573–580 (2019)
21. Xiong, F., et al.: A2J: Anchor-to-joint regression network for 3D articulated pose estimation from a single depth image. In: Proceedings of IEEE/CVF Conference Computer Vision and Pattern Recognition, pp. 793–802 (2019)
22. Zeng, C., et al.: Learning compliant grasping and manipulation by teleoperation with adaptive force control. In: 2021 IEEE/RSJ International Conference on Intelligent Robots and Systems (IROS), pp. 717–724 (2021)
23. Zhang, X., Zhang, F.: Differentiable spatial regression: a novel method for 3D hand pose estimation. IEEE Trans. Multimedia **24**, 166–176 (2022)
24. Zhou, X., Wan, Q., Zhang, W., Xue, X., Wei, Y.: Model-based deep hand pose estimation. In: Proceedings of the Twenty-Fifth International Joint Conference on Artificial Intelligence, pp. 2421–2427 (2016)

ICDAR: Intelligent Cross-Data Analysis and Retrieval

EvIs-Kitchen: Egocentric Human Activities Recognition with Video and Inertial Sensor Data

Yuzhe Hao[1(✉)], Kuniaki Uto[1], Asako Kanezaki[1], Ikuro Sato[1,2], Rei Kawakami[1], and Koichi Shinoda[1]

[1] Tokyo Institute of Technology, Meguro, Tokyo, Japan
yuzhe@ks.c.titech.ac.jp, {uto,kanezaki,isato,shinoda}@c.titech.ac.jp, reikawa@sc.e.titech.ac.jp
[2] Denso IT Laboratory, Inc., Shibuya, Tokyo, Japan

Abstract. Egocentric Human Activity Recognition (ego-HAR) has received attention in fields where human intentions in a video must be estimated. The performance of existing methods, however, are limited due to insufficient information about the subject's motion in egocentric videos. We consider that a dataset of egocentric videos along with two inertial sensors attached to both wrists of the subject to obtain more information about the subject's motion will be useful to study the problem in depth. Therefore, this paper provides a publicly available dataset, EvIs-Kitchen, which contains well-synchronized egocentric videos and two-hand inertial sensor data, as well as interaction-highlighted annotations. We also present a baseline multimodal activity recognition method with two-stream architecture and score fusion to validate that such multimodal learning on egocentric videos and intertial sensor data is more effective to tackle the problem. Experiments show that our multimodal method outperforms other single-modal methods on EvIs-Kitchen.

Keywords: Egocentric video · Inertial sensor · Multimodal dataset

1 Introduction

Methods of understanding human behaviors from sensors have been attracting much attentions recently. Human Activity Recognition (HAR) is a task to infer the type of action that a human subject conducts from a video. This is one of the basic computer-vision tasks used in real applications. More recently, with the development of wearable devices, it has become easier to obtain first-person-view videos, also known as egocentric videos. These videos contain subject-centric views, from which the subject motion from one scene to the other is directly observable and the subject's intent can be often inferred. Benefited from this, HAR task can be extended to interaction-highlighted actions, where more interactions between subjects and objects are involved ("cut carrot", "grab apple", etc.), rather than just the traditional subject-only actions ("sitting down", "stand up", etc.). HAR using egocentric video (ego-HAR) fits the needs of many applications in contemporary society, such as monitoring the behavior of subjects using

D.-T. Dang-Nguyen et al. (Eds.): MMM 2023, LNCS 13833, pp. 373–384, 2023.
https://doi.org/10.1007/978-3-031-27077-2_29

virtual reality (VR) or augmented reality (AR) devices, automotive applications, and health monitoring in medical applications.

On the other hand, egocentric video has several shortcomings in HAR. First, the entire body pose of action subjects cannot be generally observed from the video [16,23]. This likely deteriorates recognition performance for certain types of action categories due to the relatively insufficient visual information. Second, video sequences often involve severe and frequent camera motion which causes difficulties to extract consistent features from videos.

To alleviate the first problem, Singh *et al.* [23] specifically focused to capture hand motion, which highly correlates with many action types, by attention mechanisms. But, it suffers from obvious limitation when hands do not appear in the video. The utilization of optical flow in the two-stream deep neural networks (DNNs) [22] suffer from the second problem. In principle, optical flow can provide detailed motion information, but in practice, severe and frequent camera motion can generate a large amount of noisy flow, which degrades motion features. These problems seem to remain inherently as far as one relies only on egocentric videos.

Another approach is to utilize multi-modal settings, in which camera and other sensor(s) simultaneously capture the scene. This approach has been shown to be effective in different domains, such as video-audio [1,20], RGB-depth [5], and egocentric videos and inertial sensor [25]. The inertial sensor provides a mode about basic motion parameters, such as linear / angular velocity and acceleration. It is lightweight and relates only to motion, uncontaminated by visual noise. Hence, for ego-HAR, the IMU sensor data (S-mode) and the egocentric video (V-mode) can be complementary to each other. There have been a few studies along this direction, such as third-person-view HAR with inertial sensor data [2], and ego-HAR with head sensor data [24,25].

There exists at least two challenges to implement S-mode into an ego-HAR system effectively: (1) S-mode sensors should be placed in appropriate positions of body for the purpose of action recognition. If they are placed in inappropriate positions, the information would be too irrelevant to action categories or too redundant to V-mode. (2) V-mode and S-mode should be well synchronized. Human action naturally involves correlated motions of body parts; therefore, synchronization among sensors is important to capture these correlations.

To attack these challenges, we make the following contributions:

- We release a new multi-modal ego-HAR dataset, "EvIs-Kitchen" dataset, in which input data are captured by a head-mounted egocentric camera, an inertial sensor attached to the left-hand wrist, and the same inertial sensor attached to the right-hand wrist. The wrist sensors provide two-hand motion, from which hand-object or hand-hand interactions can be observed together with egocentric camera. All inputs are frame-level synchronized. The dataset contains sequences of typical activities in kitchen and action labels with detailed annotations about interactions.
- We propose a V-S multi-modal baseline model with score fusion for our EvIs-Kitchen dataset. We evaluated our 3D-ResNet-based model against LSTM-based model [24] and transformer-based model [21] that are extended from

the original, and confirmed that ours yielded best accuracy for the kitchen action recognition task.

2 Related Work

2.1 Egocentric Human Activities Recognition (Ego-HAR)

Deep learning approaches have shown their powerful feature extraction ability and have achieved great success in third-person-view HAR [22,28], while early studies of ego-HAR mostly used the conventional image processing methods with manually-crafted image features [14]. Especially the two-stream network [22] has been applied by many ego-HAR studies. In this method, two branches extract features from RGB frames and optical flows independently, then those two features are concatenated and fed to a final FC layer to obtain the prediction results. Kwon *et al.* [13] add a temporal long pooling layer at the end of two-stream CNN to extract temporal features. Lu *et al.* [17] use 3DCNN as the feature extractor for each branch to extract spatio-temporal features.

To be robust against the egocentric motion of the camera and the insufficient information about the subject's body gestures in egocentric videos, many studies use an attention mechanism that focuses on hands of the subject, as hands play an important role in many actions. Singh *et al.* [23] include the pre-annotated mask of hands in the network inputs in order to learn interactions between hands, objects, and eyes. Ma *et al.* [18] add a hand segmentation module and a localization module at the RGB branch of the two-stream network [22], in order to predict the object label, which is referred to as the *noun* label, and use the optical flow branch to predict the action label, which is referred to as the *verb* label. Kapidis *et al.* [11] use a hand track module to obtain the attention information. However, these methods require the hands to be visible in the videos.

Many studies have started to use other attention mechanisms to enhance the two-stream network's ability to extract effective features. Sudhakaran *et al.* [26] use an LSTM module to provide long short-term attention. Lu *et al.* [16] add a spatial attention network using gaze features [6] to highlight the location where the subject is looking at, and use a Bi-LSTM layer in each branch to enhance their extraction ability of temporal features. Wang *et al.* [30] use a more general object recognition module to locate the attention on what the subject is manipulating.

The above video-only methods tend to suffer from the degraded quality of the estimated optical flow due to the unstable egocentric motion of the camera. The heavy computational cost of calculating the flow is also an issue. Relatively insufficient visual information in egocentric video also limits their performance.

2.2 Video-sensor Multi-modal Methods

Multi-modal methods which combine several different modes have been emerged and shown to be effective [1,5,12,20,27]. For HAR and ego-HAR, several studies use the light-weight inertial sensor data as a new mode.

Chen *et al.* [2] use the data from a single inertial sensor located on a subject's right wrist (or right thigh) and a third-person-view depth map as inputs, and they extract conventional hand-crafted features for HAR. Song *et al.* [25] utilize videos and inertial sensor data for ego-HAR. Their inertial sensor is attached on a subject's head, and fisher linear discriminator is applied to classify the samples. Song *et al.* [24] use an LSTM network to extract features from inertial sensor data. A two-stream CNN is used to extract features from egocentric video, and features are fused at the decision level.

In previous studies, there is only a single inertial sensor installed on the subject. The features from a single sensor may not represent the whole action well. Also, the coarse precision of synchronization between two modalities can result in the correlated information between modalities being ignored. As the number of sensors increases and the precision of synchronization improves, multimodal methods will perform better in ego HAR.

2.3 Ego-HAR Related Datasets

There are several datasets for ego-HAR. EPIC-Kitchen dataset [3,4] is about kitchen activity and is the largest ego-HAR dataset, however, no inertial sensor data is provided in this dataset. GTEA Gaze [6] and GTEA Gaze+ [15] datasets are also kitchen-activity datasets. They were collected with eye-tracking glasses which provide gaze information. Gaze information is still based on the visual field; thus, it is not sufficient to infer the motion of the subject. EPIC-Tent dataset [10] is an ego-HAR dataset that takes camping as the activity theme, which involves full-body actions. Due to the device limitation, they collect the inertial sensor data of the subject's head, which is hardly involved in most actions. Also, their features are largely overlapped with those obtained by the egocentric video camera at the same position. Egocentric Multimodal Activity (EMA) dataset [24,25] provides egocentric video and inertial sensor data also collected from the subject's head as EPIC-Tent. Its action classes are all subject-highlighted actions such as "sit down" and "stand up", which have clear action characteristics and hardly involve the interaction between subjects and objects in the scene.

These datasets do not fully utilize the interaction correlation and the complementarity between egocentric video and inertial sensor data. Thus, a dataset with more improvements on these limitations is needed for multi-modal ego-HAR.

3 EvIs-Kitchen Dataset

3.1 Overview

We built **E**gocentric **V**ideo and **I**nertial **S**ensor data Kitchen activity dataset (EvIs-Kitchen dataset). Different from the EMA [24,25] and EPIC-Tent [10] which merely use an inertial sensor attached on the subject's head, we use

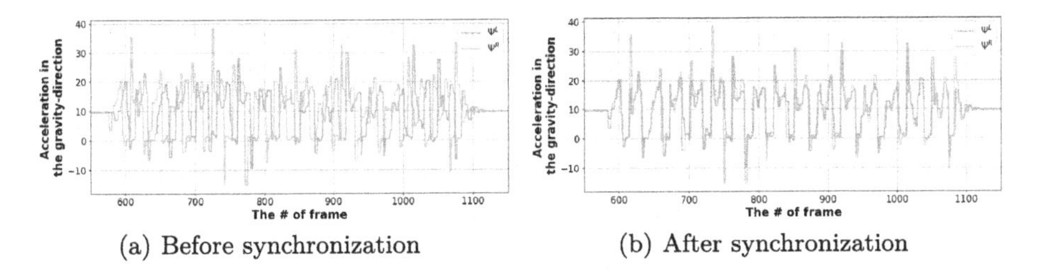

(a) Before synchronization (b) After synchronization

Fig. 1. Visualization of the L and R jump-patterns of S-mode before/after shifting the signals by $argmax\ C_{\text{shift}}(z)$. The blue one represents the jump-pattern of the left wrist, the orange one represents that for the right wrist.

two inertial sensors attached on the left and right wrists. In contrast to third-person-view HAR V-S multimodal dataset [2], which includes inertial sensor data from only right hand (or right thigh), our inertial sensor data are from both wrists. Many human actions involve two-hand motions in action-specific manners. Kitchen activities would be the typical examples and our dataset can provide potential action characteristics involving interactions between two hands.

Our dataset consists of sequences of common kitchen activities as it involves rich interactions among the subject's body, object, and environment. The data are collected from 12 subjects (8 males and 4 females), each of which cooks 7 different recipes. The details of actions vary from subject to subject since they have different cultural backgrounds and are allowed to cook in their own ways. This makes the contents of our dataset not only more diverse but also more realistic. We show some of our classes in Fig. 3.

We used a GoPro 5 camera with a head-band to collect egocentric video data, and used two Fitbit Ionic watches, one was placed at the left wrist and the other was at the right wrist. The watch contains 3 kinds of sensors: (1) An accelerometer provides 3-axis linear acceleration (a_x, a_y, a_z). (2) A gyroscope provides 3-axis angular velocity $(\omega_x, \omega_y, \omega_z)$. (3) An orientation sensor provides orientation vector in 4-digit quaternion form (a, b, c, d), which overcomes the inconsistency problem at the position $0/2\pi$ in Euler angle form's orientation vector. All of the collected data streams were in 30 FPS.

The samples in our dataset are action segments manually cut from a long video clip. Their temporal lengths vary from 0.5 to 30 s. Each sample is annotated with a verb class representing the subject's action, and a noun class representing the object that the subject is manipulating. The verb classes are mainly related to motion information, while the noun classes are mainly related to visual information. There are 4,527 action samples with 56 noun classes and 35 verb classes in total. The number of samples of each class also varies according to their frequency in the kitchen activities.

3.2 Synchronization

Synchronization among different modes is an essential step in the multi-modal task. The different modes share some common characteristics since they represent the same action. Synchronization is to ensure and further enhance this commonality. It may not be important in score fusion method, but it is crucial to understanding interactions among different modes, and it is necessary for future advanced fusion method.

The synchronization is conventionally done with timestamps. However, it is hard to obtain the timestamps of each frame of video with GoPro 5. In addition, the timestamps of both V-mode and S-mode are not accurate due to the inevitable transferring and hardware delay.

We utilize a novel method to synchronize the different modes by synchronizing their first series of frames, which contain a common action with clear characteristics. At the beginning of recording, we ask the subject to jump vertically multiple times while keeping a rigid upright posture for a fixed period of time (contains T_j frames), which causes a common and characteristic jump-pattern Ψ among all modes ($\Psi(t)$ equals to zero at outside of $[0, T_j]$). Then we synchronize different modes by matching their Ψ.

For synchronizing S-modes of left and right wrists, we first utilize their accelerometer data in the gravity-direction as their jump-patterns Ψ^{L} and Ψ^{R}, respectively, which contain clear oscillation waveforms. Then we calculate the shift-correlated coefficient C_{shift} of them to find out at which position they are best-matched:

$$C_{\text{shift}}(z) = \sum_{t=-T_j}^{T_j} \Psi^{\mathrm{L}}(t-z)\, \Psi^{\mathrm{R}}(t). \tag{1}$$

where z is the shifted distance. At $z \in [-T_j, T_j]$ with maximum $C_{\text{shift}}(z)$ the two jump-patterns of two S-modes of left and right wrists are synchronized. We use Ψ^{S} to represent the synchronized jump-pattern of S-mode, which is averaged from synchronized Ψ^{L} and Ψ^{R}. Figure 1 shows the performance of our synchronization.

For synchronizing V-mode with S-mode, we need to extract the jump-pattern Ψ^{V} in V-mode, which is caused by the vertical camera motion. We utilize the Scale-Invariant Feature Transform (SIFT) features [19] to for temporal calibration. First, we find the set of all SIFT-matched point pairs for two adjacent frames t and $t+1$:

$$P_t = \{((x_t^{(i)}, y_t^{(i)}), (x_{t+1}^{(i)}, y_{t+1}^{(i)})) \mid i = 0,1,...,n_t\}. \tag{2}$$

where $(x_t^{(i)}, y_t^{(i)})$ is the coordinate of i-th matched pairs in frame t (x denotes the horizontal direction and y denotes the vertical direction). n_t is the number of SIFT-matched point pairs between t and t+1 frames..

Then we calculate the average SIFT vertical displacement d_t between frames t and $t+1$:

$$d_t = \frac{1}{n_t} \sum_{i=1}^{n_t} (y_{t+1}^{(i)} - y_t^{(i)}). \tag{3}$$

Finally, we obtain the entire jump-pattern of V-mode Ψ^{V} by calculating the d_t for every two adjacent frames in V-mode:

$$\Psi^{\mathrm{V}}(t) = d_t,\ t = 0,1,...,T_j. \tag{4}$$

With the extracted Ψ^{V}, we synchronize V-mode and S-mode by calculating the C_{shift} of Ψ^{V} and Ψ^{S}, which is the same way as synchronizing Ψ^{R} and Ψ^{L}. In our EvIs-Kitchen dataset, all sample segments are cut from the synchronized long sequence obtained by this method.

4 V-S Multi-modal Ego-HAR

We provide a classic two-stream score-fusion architecture as the baseline for the V-S multi-modal ego-HAR task. It contains a video branch to process the egocentric video data, and a sensor branch to process the inertial sensor data of left and right wrists. Finally, we apply a learnable weights score fusion module to fuse the results from each branch.

4.1 Video (V) Branch

We used a R(2+1)D-18 model [7,29] as the Video Branch network to extract features from egocentric video, which uses ResNet-18 [8] architecture with (2+1)D convolution as the basic convolution block. It is the current state-of-the-art method for video-only ego-HAR task. Considering the efficiency and information redundancy in the V-mode, we uniformly sample 32 frames from each raw V-mode sample as the input for the Video Branch.

The output is a classification vector $\boldsymbol{f}^{\mathrm{V}}$:

$$\boldsymbol{f}^{\mathrm{V}} = [f_1^{\mathrm{V}}, f_2^{\mathrm{V}}, \cdots, f_i^{\mathrm{V}}, \cdots, f_C^{\mathrm{V}}]^\top, \tag{5}$$

where C is the number of classes, and element f_i^{V} represents the classification score of i-th class.

4.2 Sensor (S) Branch

The architecture of the sensor branch is divided into two stages: (1) Front feature processing stage and (2) Main feature extraction stage. The entire front feature processing stage are shown in Fig. 2. During this stage, different kinds of sensors are encoded separately, then fused into comprehensive sensor features. Comparing to simultaneously linear projecting all kinds of inertial sensor data into sensor features in the previous works [9,24,25], our sensor-wise decomposition and fusion processes (Step (1) and (2) in Fig. 2) enlarge the input S-mode to S-features with larger channel dimension, which includes the prior knowledge of each kind of sensor. It is easier to search for more meaningful and comprehensive information with such S-features. The output S-features of this stage will be used as the input to the Main feature extraction stage.

We use two sub-branches to process the inertial sensor data from left and right hand separately, then applying score fusion at the end of these two sub-branches for getting the final output of S-branch.

During the main feature extraction stage, we proposed a ResNet-Sensor-3D (Rsen3D) network to do this task. Rsen3D is based on the 3D-ResNet-18 architecture. Because S-features do not contain any spatial structure, the spatial kernel size of all the convolution blocks are 1×1. Similar to V-branch, the final output of S-branch is also a classification vector $\boldsymbol{f}^{\mathrm{S}} = [f_1^{\mathrm{S}}, f_2^{\mathrm{S}}, \cdots, f_i^{\mathrm{S}}, \cdots, f_C^{\mathrm{S}}]^\top$.

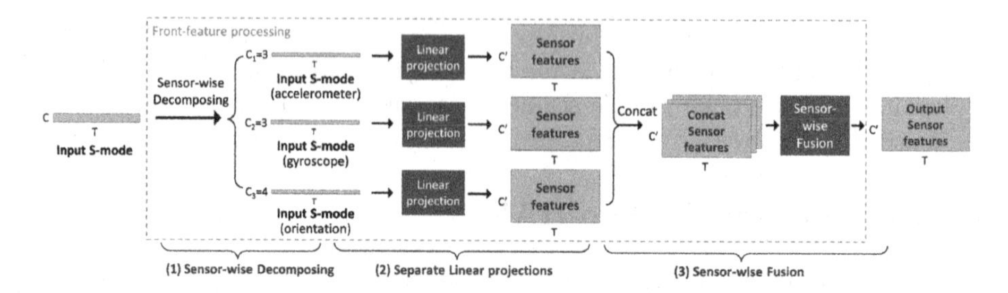

Fig. 2. The front feature processing stage. $C = 10$ means 10 channels in each frame's S-mode; Among them, C_1, C_2, C_3 represent accelerometer-related channels (a_x, a_y, a_z), gyroscope-related channels (g_x, g_y, g_z) and orientation-related channels (o_a, o_b, o_c, o_d); C' represents the channel dimension of linear projected features.

4.3 V-S Score Fusion

For fusing the classification vector of V-branch and S-branch, we apply an adaptive pooling method. The score of each class in fused classification vector $\boldsymbol{f}^{\mathrm{F}} = [f_1^{\mathrm{F}}, f_2^{\mathrm{F}}, \cdots, f_i^{\mathrm{F}}, \cdots, f_C^{\mathrm{F}}]^\top$ is the weighted sum of corresponding elements in $\boldsymbol{f}^{\mathrm{V}}$ and $\boldsymbol{f}^{\mathrm{S}}$, and each class has its own specified weights set:

$$f_i^{\mathrm{F}} = w_i^{\mathrm{V}} f_i^{\mathrm{V}} + w_i^{\mathrm{S}} f_i^{\mathrm{S}}, \quad i = 1, 2, 3, ..., C \tag{6}$$

where w_i^{V} and w_i^{S} are the weights for V- and S-classification vectors for the i-th class, respectively. All these weights are learnable during the training. The fused classification vector $\boldsymbol{f}^{\mathrm{F}}$ is our final multi-modal classification result.

5 Experiments and Analysis

We separate ego-HAR into 2 sub-tasks, *verb* recognition and *noun* recognition. This separation helps us to analyze how different sensors plays important roles depending on tasks. For these two tasks, the network structures are the same, and only the numbers of classes are different.

First, we compare 3 different inertial sensor networks: (1) Our proposed Rsen3D-18 network, (2) sensor LSTM network (SenLSTM) [24] with 2 layers and 128 feature channels, and (3) transformer for sensor (T4sen) [21]. Then we compare the multi-modal method with video-only and sensor-only methods.

All the models were trained and tested on our EvIs-Kitchen dataset, where we chose 1/4 of all samples (samples of Subject 4, 10, 11) as the test set which contains 1,167 samples, and keep the rest as the training set which contains 3,360 samples. Both sets have similar class frequency distribution and the same gender ratio (Male : Female = 2 : 1). All inertial sensor data were zero-padded to the same temporal length 400. We trained them with batch size 16 for 50 epochs in each experiment. We used Adam optimizer and the learning rate started at 4e–4 and reduced to a tenth every 15 epochs.

Table 1. Comparison among sensor-only models.

Sensor model	Noun acc@1(%)	Verb acc@1(%)
Rsen3D-18 (ours)	**22.88**	**52.61**
SenLSTM [24]	21.50	49.95
T4sen [21]	20.85	40.87

5.1 Comparison Between Inertial Sensor Processing Methods

The existing methods to process inertial sensor data vary according to the input data. To compare such different sensor-only models reasonably, we used the same front feature processing method introduced in Sect. 4.2 for all models. For the SenLSTM model [24], we used 2 LSTM layers and set the feature dimension to 64. For T4sen model [21], we used 6 layers and 8 heads, and set the feature dimension also to 64.

The overall top 1 accuracy comparisons on *noun* task and *verb* task of this set of experiments are shown in Table 1. Our proposed Rsen3D-18 model achieves the best performances among three candidates. Since the features extracted from different representation levels are well-inherited and enhanced through residual structure of Rsen3D, more semantic information encoded in its final output. We believe this is the reason it outperforms the other methods.

5.2 Comparison Between Multi-modal and Single-modal Methods

During this set of experiments, we trained the single-modal methods end-to-end. For video-only method (R(2+1)D-18 model), we loaded the pre-trained weights provided by torch-vision package. It is pre-trained on Kinetics-400 dataset (a massive third-person-view HAR dataset). For the multi-modal method, we first loaded the trained parameters from single-modal experiments for V- and S-branches, then froze these two branches and just trained and fine-tuned the last fusion classifier layer. The experimental results are shown in Table 2. Our proposed multi-modal method improves the overall-average top-1 accuracy by 2.57 points on *noun* task, and by 2.66 points on *verb* task.

The detailed class-wise comparisons are shown in Fig. 3. The multi-modal method is not always better than video-only method for every class. The improvement on *verb* task is significant (Fig. 3(b)). Most of verb classes achieved better recognition results since inertial sensor data mainly bring motion information. For example, "cut" involves continuous upside-down movements of hands; "stir", "transfer" and "pour" also involve clear and distinct short-time motion patterns; The most significant improvement is on "throw-around". This action always appears with a noun class "patty". It means throwing a patty from one hand to the other to get rid of the air inside of patty. The characteristic of this action is very clear and unique. It also involves two hands' interaction, which is well-captured by EvIs-Kitchen dataset's inertial sensor data.

However, some classes suffer from degradation. Such classes are, for example, "pluck" (remove the green leaf on strawberry or tomato), "clean" (clean up the white stuffs on the surface of orange), and "spread" (break the orange from the

Table 2. Comparison among multi-modal and single-modal methods

Method	Model	Noun acc@1(%)	Verb acc@1(%)
Sensor-only	Rsen3D-18	22.88	52.61
Video-only	R(2+1)D-18 [29]	77.72	77.72
Multi-modal	Two-stream Score-fusion	**80.29**	**80.38**

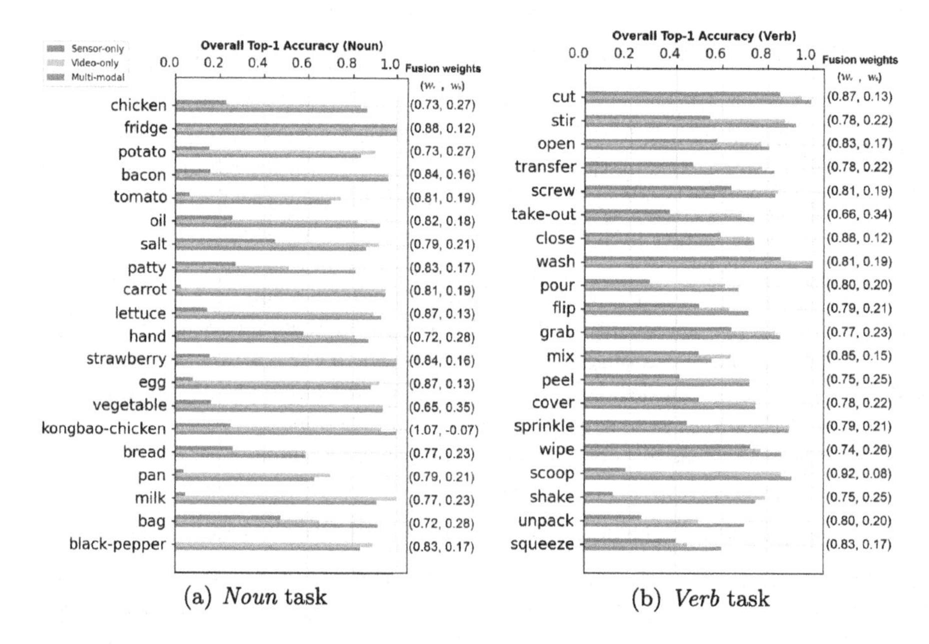

Fig. 3. Top 1 accuracy comparison among experiments in Table 2. The left column is the learned weights of each classes for fusion. We picked up the top-20 most frequent classes among all classes to show in this figure. (Color figure online)

ball-shape to several pieces). These actions cannot be recognized well because they involve finger-level motions, which are not captured by our wrist-level inertial sensors. Another reason for the degradation is the action characteristics that may vary significantly even inside of the same class. For example, "screw" (screw the lid from a bottle or a jar) can be different regarding the spinning direction, as some subjects tend to lay the jar down at first then screw its lid. The direction and patterns of "mix" also can vary according to subjects' habits. These difference can cause confusion to S-branch's recognition, eventually degrades the overall performance.

For the *noun* task (Fig. 3(a)), the improvement is not as significant as the *verb* task, because the *noun* task is highly relied on visual information, which is not provided by the inertial sensor data. But still, performance improvements are observed for those noun classes involving distinctive motion patterns. For example, the classes "oil", "patty", "kongbao-chicken" and "bag" always appear with a certain verb class as a combination such as "pour oil", "throw-around patty", "stir kongbao-chicken" and "tear bag". Because inertial sensor data bring additional motion information, the recognition on these classes are enhanced. Degradation on the *noun* task is observed to be somewhat severer than the *verb* task. Inertial sensor signals can be often irrelevant to certain noun classes, and this may cause contamination of visual information after the V-S feature fusion.

6 Conclusion

In this work, we introduce our self-built well-synchronized V-S multi-modal dataset, EvIs-Kitchen. It contains inertial sensor data of both wrists, providing richer information about the subject motion including hand-hand interaction, and the detailed class labels about objects and actions. we provide allow further research possibility in this field. We also propose a baseline multi-modal method that yields decent performance of ego-HAR with the help of inertial sensor data. The experiments show our proposed multi-modal method outperforms the video-only method by 2.57 percents on the *noun* task and by 2.66 percents on the *verb* task. However, the detailed class-wise analysis also shows some accuracy drops on certain classes with our proposed multi-modal method due to the confusion from the motion diversity. To further improve the performance of multi-modal method, we will explore improving the performance of S-branch, for decreasing the confusion it brings, and adding more information communication between V-branch and S-branch, for achieving better fusion performance.

Acknowledgement. This work is an outcome of a research project, Development of Quality Foundation for Machine-Learning Applications, supported by DENSO IT LAB Recognition and Learning Algorithm Collaborative Research Chair (Tokyo Tech.). It was also supported by JST CREST JPMJCR1687.

References

1. Afouras, T., Chung, J.S., Senior, A., Vinyals, O., Zisserman, A.: Deep audio-visual speech recognition. IEEE Trans. Pattern Anal. Mach. Intell. **44**(12), 8717–8727 (2018)
2. Chen, C., Jafari, R., Kehtarnavaz, N.: Utd-mhad: A multimodal dataset for human action recognition utilizing a depth camera and a wearable inertial sensor. In: ICIP, pp. 168–172 (2015)
3. Damen, D., et al.: Scaling egocentric vision: The epic-kitchens dataset. In: ECCV, pp. 720–736 (2018). https://doi.org/10.1007/978-3-030-01225-0_44
4. Damen, D., et al.: Rescaling egocentric vision: collection, pipeline and challenges for epic-kitchens-100. IJCV **130**(1), 33–55 (2022)
5. Eitel, A., Springenberg, J.T., Spinello, L., Riedmiller, M., Burgard, W.: Multimodal deep learning for robust rgb-d object recognition. In: IROS, pp. 681–687 (2015)
6. Fathi, A., Li, Y., Rehg, J.M.: Learning to recognize daily actions using gaze. In: ECCV, pp. 314–327 (2012). https://doi.org/10.1007/978-3-642-33718-5_23
7. Ghadiyaram, D., Tran, D., Mahajan, D.: Large-scale weakly-supervised pre-training for video action recognition. In: CVPR, pp. 12046–12055 (2019)
8. He, K., Zhang, X., Ren, S., Sun, J.: Deep residual learning for image recognition. In: CVPR, pp. 770–778 (2016)

9. Imran, J., Raman, B.: Multimodal egocentric activity recognition using multi-stream cnn. In: ICVGIP, pp. 1–8 (2018)
10. Jang, Y., Sullivan, B., Ludwig, C., Gilchrist, I., Damen, D., Mayol-Cuevas, W.: Epic-tent: An egocentric video dataset for camping tent assembly. In: ICCV Workshops, pp. 0–0 (2019)
11. Kapidis, G., Poppe, R., Van Dam, E., Noldus, L., Veltkamp, R.: Egocentric hand track and object-based human action recognition. In: SmartWorld/SCALCOM/UIC/ATC/CBDCom/IOP/SCI, pp. 922–929 (2019)
12. Kazakos, E., Nagrani, A., Zisserman, A., Damen, D.: Epic-fusion: Audio-visual temporal binding for egocentric action recognition. In: CVPR, pp. 5492–5501 (2019)
13. Kwon, H., Kim, Y., Lee, J.S., Cho, M.: First person action recognition via two-stream convnet with long-term fusion pooling. Pattern Recogn. Lett. **112**, 161–167 (2018)
14. Laptev, I., Marszalek, M., Schmid, C., Rozenfeld, B.: Learning realistic human actions from movies. In: CVPR, pp. 1–8. IEEE (2008)
15. Li, Y., Ye, Z., Rehg, J.M.: Delving into egocentric actions. In: CVPR, pp. 287–295 (2015)
16. Lu, M., Li, Z.N., Wang, Y., Pan, G.: Deep attention network for egocentric action recognition. IEEE Trans. Image Process. **28**(8), 3703–3713 (2019)
17. Lu, M., Liao, D., Li, Z.N.: Learning spatiotemporal attention for egocentric action recognition. In: ICCV Workshops, pp. 4425–4434 (2019)
18. Ma, M., Fan, H., Kitani, K.M.: Going deeper into first-person activity recognition. In: CVPR, pp. 1894–1903 (2016)
19. Ng, P.C., Henikoff, S.: Sift: predicting amino acid changes that affect protein function. Nucleic Acids Res. **31**(13), 3812–3814 (2003)
20. Owens, A., Efros, A.A.: Audio-visual scene analysis with self-supervised multisensory features. In: ECCV, pp. 631–648 (2018). https://doi.org/10.1007/978-3-030-01231-1_39
21. Shavit, Y., Klein, I.: Boosting inertial-based human activity recognition with transformers. IEEE Access **9**, 53540–53547 (2021)
22. Simonyan, K., Zisserman, A.: Two-stream convolutional networks for action recognition in videos. arXiv preprint arXiv:1406.2199 (2014)
23. Singh, S., Arora, C., Jawahar, C.: First person action recognition using deep learned descriptors. In: CVPR, pp. 2620–2628 (2016)
24. Song, S., et al.: Multimodal multi-stream deep learning for egocentric activity recognition. In: CVPR Workshops, pp. 24–31 (2016)
25. Song, S., Cheung, N.M., Chandrasekhar, V., Mandal, B., Liri, J.: Egocentric activity recognition with multimodal fisher vector. In: ICASSP, pp. 2717–2721 (2016)
26. Sudhakaran, S., Escalera, S., Lanz, O.: Lsta: Long short-term attention for egocentric action recognition. In: CVPR, pp. 9954–9963 (2019)
27. Tang, Y., Wang, Z., Lu, J., Feng, J., Zhou, J.: Multi-stream deep neural networks for rgb-d egocentric action recognition. IEEE Trans. Circuits Syst. Video Technol. **29**(10), 3001–3015 (2018)
28. Tran, D., Bourdev, L., Fergus, R., Torresani, L., Paluri, M.: Learning spatiotemporal features with 3d convolutional networks. In: ICCV, pp. 4489–4497 (2015)
29. Tran, D., Wang, H., Torresani, L., Ray, J., LeCun, Y., Paluri, M.: A closer look at spatiotemporal convolutions for action recognition. In: CVPR, pp. 6450–6459 (2018)
30. Wang, X., Wu, Y., Zhu, L., Yang, Y.: Symbiotic attention with privileged information for egocentric action recognition. In: AAAI, vol. 34, pp. 12249–12256 (2020)

COMIM-GAN: Improved Text-to-Image Generation via Condition Optimization and Mutual Information Maximization

Longlong Zhou, Xiao-Jun Wu(✉), and Tianyang Xu

The School of Artificial Intelligence and Computer Science, Jiangnan University, Wuxi 214122, People's Republic of China
{wu_xiaojun,tianyang.xu}@jiangnan.edu.cn

Abstract. Language-based image generation is a challenging task. Current studies normally employ conditional generative adversarial network (cGAN) as the model framework and have achieved significant progress. Nonetheless, a close examination of their methods reveals two fundamental issues. First, the discrete linguistic conditions make the training of cGAN extremely difficult and impair the generalization performance of cGAN. Second, the conditional discriminator cannot extract semantically consistent features based on linguistic conditions, which is not conducive to conditional discrimination. To address these issues, we propose a condition optimization and mutual information maximization GAN (COMIM-GAN). To be specific, we design (1) a text condition construction module, which can construct a compact linguistic condition space, and (2) a mutual information loss between images and linguistic conditions to motivate the discriminator to extract more features associated with the linguistic conditions. Extensive experiments on CUB-200 and MS-COCO datasets demonstrate that our method is superior to the existing methods.

Keywords: cGAN · Condition optimization · Mutual information maximization

1 Introduction

In recent years, deep learning has greatly accelerated the development of image fusion [3], target tracking [23–26], action recognition [15], text-to-image generation and other research fields. Generative Adversarial Networks (GANs) [4] have been widely used in these fields. Text-to-Image generation (T2I) works are mostly built on conditional generation adversarial networks [11], aiming to build a text-to-image mapping. Most previous methods [13,18,22,27,29] are generally employ muti-stage generative adversarial network. They first utilize sentence embedding and random noise to generate a blurred low-resolution image, and then refine the fine-gained features with word embedding progressively. In this framework, multiple generators are entangled and difficult to train stably. Moreover, the process of refinement involves a word attention mechanism whose

D.-T. Dang-Nguyen et al. (Eds.): MMM 2023, LNCS 13833, pp. 385–396, 2023.
https://doi.org/10.1007/978-3-031-27077-2_30

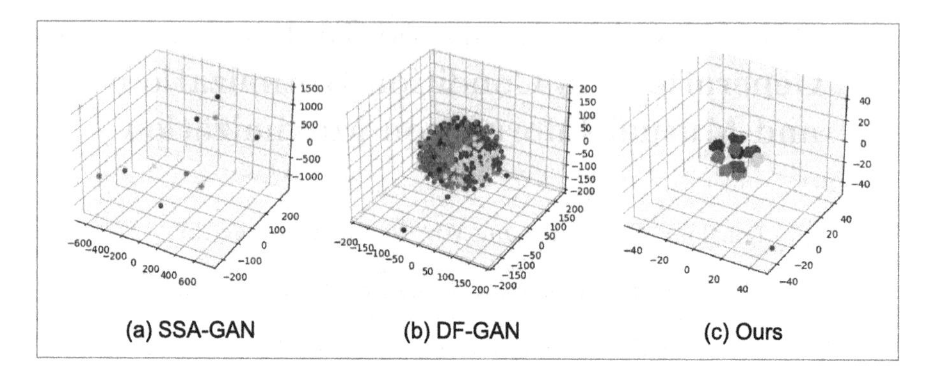

Fig. 1. Visualization results of the text conditions on CUB dataset. We randomly sample the text condition of an image 1000 times during training and apply t-SNE [10] to map them to a three-dimensional space.

computational consumption increases exponentially with the increase of image resolution. To overcome these flaws, we adopt a single-stage generative network without the word attention mechanism, which has shown tremendous advantages in recent works [8,16,19].

Although the current works using single-stage generative network have obtained impressive results, there still remain two fundamental problems. First, the discrete text conditions make the training of models extremely difficult and impair the generalization performance. As shown in Fig. 1, SSA-GAN [8] lazily use manually annotated text descriptions as conditions (see Fig. 1(a)), but DF-GAN [19] thoughtfully concatenates the text embedding with random noise (see Fig. 1(b)). Although splicing random noise can construct continuous text condition space, it will cause semantic entanglement. We believe that it is necessary to maintain independence between multiple texts of an image. Second, if the features extracted from real image and generated image in discriminator are both irrelevant to the given text condition, the conditional discriminate network will be useless. Conversely, if the discriminator can extract more image features related to the given text condition, the conditional discriminate network will be more effective. For this purpose, improving the guidance of text condition in feature extraction stage of discriminator is extremely important. However, existing T2I studies neglect this issue.

To deal with the aforementioned issues, we put forward a condition optimization and mutual information maximization approach dubbed as COMIM-GAN (see Fig. 2). For the first issue, we design a text condition construction module (TCCM), which uses $Top - K$ most similar sentence embeddings to construct an interpolation fusion condition, and concatenates the fused condition with the given sentence embedding as a joint text condition. The joint text condition can be considered as a random sampling in the neighborhood of the given sentence embedding (see Fig. 1(c)), which expands the text condition space without distorting the original semantics. For the second issue, we propose a mutual information loss between the image representation and the joint text condition based on contrast learning [6,7,20]. Maximizing the mutual information between real image and the joint text condition in discriminator helps the discriminator to capture more features associated with the text condition. Moreover, maximizing the mutual information between generated image and the joint text condition in generator helps the generator to generate semantically consistent images.

In summary, this work has the following contributions:

- We propose a novel model COMIM-GAN with conditional optimization and mutual information maximization. To our best knowledge, we are the first to introduce mutual information maximization in T2I generation task.
- We design a plug and play text condition construction module to optimize the text condition space, which optimizes the training process and improves the generalization performance of our model.
- A novel mutual information loss stimulates discriminator to extract more image features associated with the given text condition and urges the generator generate semantically consistent images.
- Extensive experiments confirm that our COMIM-GAN outperforms most advanced methods.

2 Related Work

2.1 Text-to-Image Generation

T2I generation task is firstly proposed in GAN-INT-CLS [12], which uses one generator and one discriminator to synthesize text-dependent images. To generate high-resolution images, StackGAN [27] divides the generation of high-resolution images from text descriptions into two stages: first, synthesize a simple low-resolution image, and then gradually improve the resolution and details. To further enrich the details of generated images, AttnGAN [22] utilizes a word attention mechanism, which calculates a word-context condition vector for every pixel position in feature map to modify the image features. However, AttnGAN neglects that each word has a different importance in the textual description. Therefore, DM-GAN [29] designs a memory writing gate to redefine the word content, which can highlight the important word information.

In order to enrich the details of the generated image, it is necessary to introduce fine-grained word information, but this also brings enormous computation. For simplicity and efficiency, Souza et al. [16] proposes an efficient neural architecture based on BigGAN-Deep. To fully utilize the sentence information, DF-GAN [19] discards the batch normalization layer, and stacks the affine transformation layer more times in generator, in addition, DF-GAN proposes a Matching Aware zero-centered Gradient Penalty (MA-GP) to help the discriminator converge. SSA-GAN [8] proposes a Semantic-Spatial Aware Module to highlight the spatial position of the object in image, so as to achieve accurate control of text-to-image information fusion.

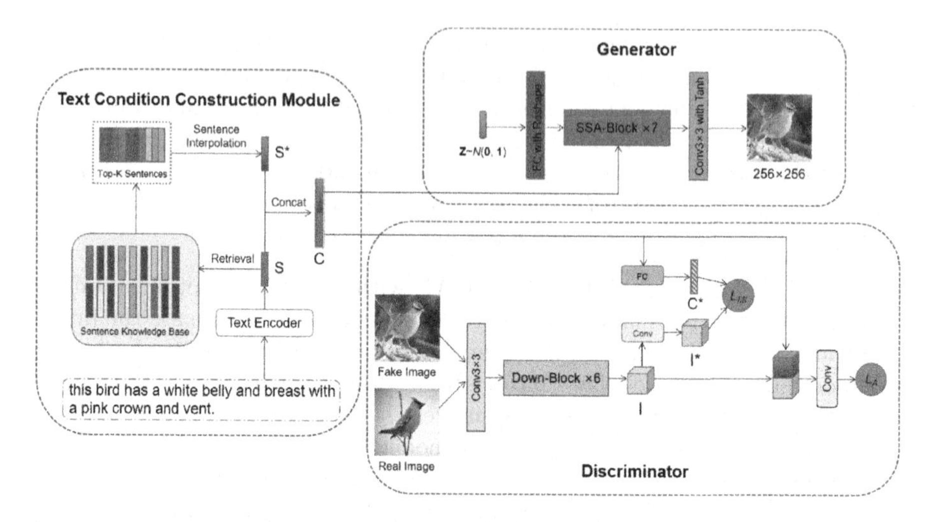

Fig. 2. Our COMIM-GAN has a Text Condition Construction Module (TCCM), a generator and a discriminator. The TCCM constructs a joint text condition c for generator and discriminator. There is a mutual information loss L_{MI} between the image representation and the corresponding joint text condition, which helps the conditional discriminator work better.

2.2 Mutual Information

Mutual information is widely concerned in representation learning [1], aiming at minimizing information loss in data dimensionality reduction and feature extraction. InfoGAN [2] introduces the concept of mutual information into Generative Adversarial Networks for the first time, which optimizes the performance of GANs by maximizing the mutual information between observations and a small subset of latant variables. It is shown in [20] the mutual information between local and global features is the lower bound of the mutual information between real image and global features. Therefore, InfoMax-GAN [7] maximizes the mutual information between local and global features in discriminator, which helps the discriminator extract more features. Inspired by InfoMax-GAN, we maximize the mutual information between images and the corresponding text conditions in this work, which can enhance the identification ability of the conditional discriminator.

3 COMIM-GAN

We provide the network structure of our COMIM-GAN in Fig. 2. Our COMIM-GAN follows the structure of SSA-GAN [8], it has 7 SSA-Blocks in generator and 6 Down-Blocks in discriminator. Different from other previous works [8,19], we design a Text Condition Construction Module (TCCM) to optimize the text condition space. Furthermore, we propose a mutual information loss to urge the discriminator to extract more image features associated with the given text condition.

3.1 Text Condition Construction Module

Figure 2 illustrates the Text Condition Construction Module (TCCM). First, we utilize a pre-trained text encoder [22,29] to process all sentences in the dataset into sentence embeddings, then use these embeddings to construct a sentence knowledge base for sentence retrieval. Next, we process the input sentence into a sentence embedding, and retrieve the most similar K sentence embeddings $\{s_1, ..., s_K\}$ in the sentence knowledge base through cosine similarity, then we construct a fused sentence embedding s^* by interpolation fusion strategy. The fused sentence embedding s^* can be calculated as follows [16]:

$$s^* = \sum_{i=1}^{K} \left[s_i \times \frac{e^{m_i}}{\sum_{j=1}^{K} e^{m_j}} \right], \tag{1}$$

where m_i is a scalar randomly sampled from gaussian distribution. Finally, we concatenate the fused sentence embedding s^* and the input sentence embedding s as a joint text condition c to control image generation.

Our TCCM is a plug and play module, it expands the text condition space (see Fig. 1(c)), which facilitates the training of conditional GAN and improve the generalization performance of T2I generation models. Moreover, the joint text conditions constructed by TCCM based on one same text description are semantically similar to this text description, which helps the generator to synthesize semantically accurate images. On the other hand, the joint text conditions constructed by TCCM based on different text descriptions of an image still retain the differences in semantics, which ensures the diversity of generated images.

3.2 Mutual Information Maximization

In this section, we put forward a mutual information loss [7] between the image representation and the text condition based on contrast learning. First, We force the discriminator to extract more image features associated with the text condition by maximizing the mutual information between real image representation and the corresponding text condition, which helps the discriminant network efficiently distinguish the real or fake images, and promotes the discriminator to converge better. The mutual information loss between the real image representation and the corresponding text condition in discriminator is defined as:

$$L_{RMI} = -[\sum_{i=1}^{N} \sum_{k=1}^{M} log \frac{exp(R(x_i^k, s_i))}{\sum_{j=1}^{N} exp(R(x_j^k, s_i))}] \tag{2}$$

where $\{x_i, s_i\}$ is the positive sample and $\{x_j, s_i\}$ is the negative sample, x_i^k represents the kth position vector of the real image x_i, $M = H * W$, $R(\cdot)$ denotes the matching score between text and image.

In addition, based on the inherent advantages of contrast learning, maximizing the mutual information between fake image presentation and the corresponding text condition in generator will help the generator synthesize semantically consistent images. The mutual information loss between the fake image representation and the corresponding text condition in generator is defined as:

$$L_{FMI} = -[\sum_{i=1}^{N}\sum_{k=1}^{M} log\frac{exp(R(\hat{x}_i^k, s_i))}{\sum_{j=1}^{N} exp(R(\hat{x}_j^k, s_i))}] \tag{3}$$

3.3 Objective Function

Adversarial Loss. For conditional adversarial learning, we utilise the hing loss [8,16,19] to precisely identify the semantic consistency between the image and text. The adversarial loss for the discriminator is defined as:

$$\begin{aligned} L_{adv}^{D} =&\mathbb{E}_{x\sim p_{data}}\left[max(0, 1 - D(x, s))\right] \\ &+ \frac{1}{2}\mathbb{E}_{x\sim p_G}\left[max(0, 1 + D(\hat{x}, s))\right] \\ &+ \frac{1}{2}\mathbb{E}_{x\sim p_{data}}\left[max(0, 1 + D(x, \hat{s}))\right] \end{aligned} \tag{4}$$

where x and $\hat{x}$ represent the real image and fake image respectively, s and $\hat{s}$ represent the matching text condition and mismatching text condition respectively. The corresponding adversarial loss for the generator is defined as:

$$L_{adv}^{G} = -\mathbb{E}_{x\sim p_G}\left[D(\hat{x}, s)\right] \tag{5}$$

DAMSM Loss. Deep Attentional Multimodal Similarity Model (DAMSM) Loss was first proposed in AttnGAN [22], it calculates the similarity between image and text at sentence and word levels, which can improves the semantic consistency of the generated images. The DAMSM loss is defined as:

$$\begin{aligned} L_{DAMSM} = -[&\sum_{i=1}^{N} log\frac{exp(\gamma R(x_i, s_i))}{\sum_{j=1}^{N} exp(\gamma R(x_i, s_j))} \\ &+\sum_{i=1}^{N} log\frac{exp(\gamma R(x_i, s_i))}{\sum_{j=1}^{N} exp(\gamma R(x_j, s_i))}] \end{aligned} \tag{6}$$

where $\{x_i, s_i\}_{i=1}^{N}$ are the image-text pairs, $R(\cdot)$ represents the matching score between images and texts, γ is a smoothing factor.

MA-GP Loss. Match-Aware Zero-Centered Gradient Penalty (MA-GP) Loss [19] can smoothen the convergence surface of the discriminator, which alleviates the problem that the discriminator oscillates near the optimal solution and cannot converge in the later period of training. The MA-GP loss is defined as:

$$L_{MA-GP} = \mathbb{E}_{x\sim p_{data}}\left[(||\nabla_x D(x, s)||_2 + ||\nabla_s D(x, s)||_2)^p\right] \tag{7}$$

Total Generator Loss. Based on adversarial loss L^G_{adv}, we add the DAMSM loss L_{DAMSM} and the mutual information loss between fake images and texts L_{FMI}, so the whole generator loss is defined as:

$$L_G = L^G_{adv} + \lambda_1 L_{DAMSM} + \lambda_2 L_{FMI} \tag{8}$$

Total Discriminator Loss. Based on adversarial loss L^D_{adv}, we add the MA-GP loss L_{MA-GP} and the mutual information loss between real images and texts L_{RMI}, so the whole discriminator loss is defined as:

$$L_D = L^D_{adv} + \lambda_3 L_{MA-GP} + \lambda_4 L_{RMI} \tag{9}$$

4 Experiments

In this section, The CUB [21] and COCO [9] benchmark datasets and the evaluation metrics are first introduced. Then, we quantitatively and qualitatively evaluate our COMIM-GAN model. Finally, we present the ablation studies.

4.1 Datasets

Following many T2I generation works, our COMIM-GAN is evaluated on CUB [21] and COCO [9] datasets. 8855 training images and 2933 test images make up the CUB dataset. 82783 training images and 40504 test images make up the COCO dataset. Each image in CUB dataset is accompanied by ten text descriptions and is clipped and preprocessed to guarantee that the object image size ratio for the bird bounding box is greater than 0.75. Each image in COCO dataset is accompanied by five text descriptions.

4.2 Evaluation Metrics

We use the Inception Score (IS) [14] and Fréchet Inception Distance (FID) [5] as our evaluation metrics in this work. The IS computes KL-divergence between the conditional class distribution and the marginal class distribution. It can be formulated as:

$$IS = exp(\mathbb{E}_{x \sim p_g} D_{KL}(p(y|x) \parallel p(y))) \tag{10}$$

where x is the synthetic images and y is the class label predicted by a pre-trained Inception-v3 network [17]. The FID computes the Fréchet distance between the distribution of the synthetic images and the ground truth images. The FID can be formulated as:

$$FID(r, g) = ||\mu_r - \mu_g||_2^2 + trace(\Sigma_r + \Sigma_g - 2(\Sigma_r \Sigma_g)^{1/2}) \tag{11}$$

where r and g represent the representation vector of real images and synthetic images, μ_r, μ_g, Σ_r and Σ_g represent the mean and covariance of real image distribution and synthetic image distribution, respectively.

Fig. 3. Examples (CUB: 1st - 4th columns), COCO: (5th - 8th columns) are generated by DF-GAN [19], SSA-GAN [8] and our proposed COMIM-GAN.

To compute IS and FID, we randomly select text descriptions from the test dataset and generate 30000 images (256 × 256 resolution) for all methods. Previous studies [19,28] indicate that the IS metric entirely fails in evaluating the synthesized images for the COCO dataset. So, we skip comparing the IS on the COCO dataset.

4.3 Quantitative Results

As shown in Table 1, our COMIM-GAN achieves the state-of-the-art results on two common datasets. Compared with our baseline SSA-GAN [8], our COMIM-GAN improves the IS from 5.17 to 5.32 and decreases the FID from 16.58 to 13.58 on CUB test dataset. In addition, our COMIM-GAN miraculously decreases the FID from 19.37 to 14.07 on COCO test dataset, achieving 27.36% improvement. As is known to all, higher IS means higher quality and lower FID means higher diversity of synthetic images. The Quantitative results indicate that our COMIM-GAN model has significant advantages over other advanced methods in generating high-quality images and improving the diversity of images.

4.4 Qualitative Results

We compare the images synthesized by our COMIM-GAN with those by DF-GAN [19] and SSA-GAN [8] from two aspects: text-to-image semantic consistency and image quality.

As shown in Fig. 3, our method can better comprehend the semantics of text descriptions and then synthesize images with consistent content, exhibiting more similar to ground truth images according to the semantic information. For instance, in the 2nd column, "white bill" is given in the text, however, the images synthesized by DF-GAN and SSA-GAN contain a black bill while the image synthesized by our COMIM-GAN correctly contains a white bill. More examples can be found here, in the 6th column, our method can generate some bananas with clear texture but the images generated by DF-GAN and SSA-GAN are deformed and unreal, in the 8th column, the boat generated by our method is more real than DF-GAN and SSA-GAN. The results of qualitative analysis demonstrate that our model can synthesize ground-reality images and maintain sufficient semantic consistency between text and image.

Table 1. Comparing the results of our COMIM-GAN with other advanced methods based on IS and FID metrics. The bold is best.

Methods	CUB		COCO
	IS ↑	FID ↓	FID ↓
StackGAN [27]	3.70 ± 0.04	51.89	74.05
AttnGAN [22]	4.36 ± 0.03	23.98	35.49
SEGAN [18]	4.67 ± 0.04	18.17	32.28
DM-GAN [29]	4.75 ± 0.07	16.09	32.64
DAE-GAN [13]	4.42 ± 0.04	15.12	28.12
DF-GAN [19]	5.10 ± 0.04	14.81	19.32
SSA-GAN [8]	5.07 ± 0.04	15.61	–
SSA-GAN(fine-tune) [8]	5.17 ± 0.08	16.58	19.37
COMIM-GAN	**5.32 ± 0.05**	**13.58**	**14.07**

Table 2. The performance of the Text Condition Construction Module (TCCM) and Mutual Information Maximization (MIM) on CUB dataset.

ID	Components		IS↑	FID↓
	TCCM	MIM		
0	–	–	5.07 ± 0.04	15.61
1	✓	–	5.21 ± 0.07	14.26
2	–	✓	5.15 ± 0.06	14.38
3	✓	✓	**5.32 ± 0.05**	**13.58**

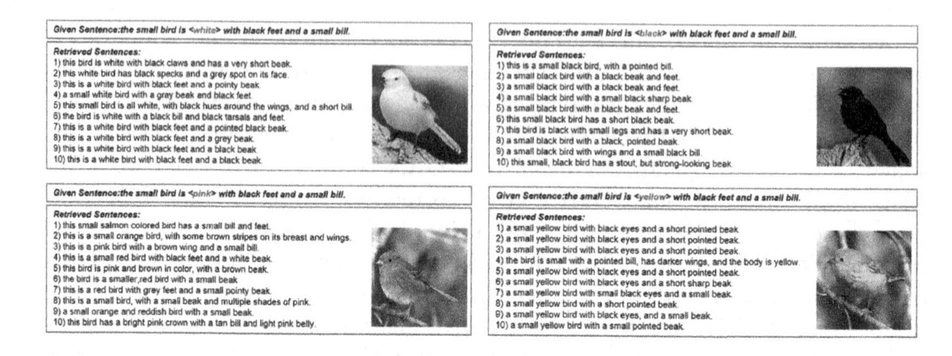

Fig. 4. Examples generated by our COMIM-GAN on CUB test dataset. When we change the color description of the bird in one sentence, we can get different retrieval results and generate images with corresponding colors.

4.5 Ablation Studies

We first quantitatively or qualitatively analyze the advantages of the Text Condition Construction Module (TCCM) and the Mutual Information Maximization (MIM), then we analyze the generalization performance of our COMIM-GAN.

TCCM and MIM. As shown in Table 2, when we use the TCCM in our baseline SSA-GAN [8], we can increase the IS from 5.07 to 5.21, and decrease the FID from 15.61 to 14.26. Moreover, when we use the MIM in our baseline, we can increase the IS from 5.07 to 5.15, and decrease the FID from 15.61 to 14.38. Quantitative results indicate that the TCCM and MIM can enhance the quality and diversity of fake images. Obviously from Table 2, we get the best results when the TCCM and MIM are used together in this work.

Generalisation Ability. In this part, we evaluate the sensitivity of our method to the text descriptions by tuning the key attributes. As shown in Fig. 4, randomly select a sentence from the test dataset and change the color attribute, we can get the fake images with the corresponding colors. This proves that our method is sensitive to text input and has a strong generalisation ability. However, we find that the sentences retrieved by our method have a small amount of repetition. For instance, the 9th and 10th sentences in the upper left corner of Fig. 4. After careful analysis, we find that two images with similar content may have the same text description in datasets, but the probability of repetition is negligible, so it has no impact on our method.

5 Conclusion

In this work, we propose a neoteric T2I generation model COMIM-GAN. We design a plug and play Text Condition Construction Module (TCCM) to optimize the text condition space. In TCCM, we build a joint text condition from a sentence knowledge base, the joint text condition can stabilize the training of COMIM-GAN and improve the diversity of generated images. Furthermore, we introduce two mutual information losses to maximize the mutual information between the joint text condition and the image representation, which can enhance the generalization ability of generator and the convergence performance of discriminator. Extensive experiments on two common datasets confirm that COMIM-GAN significantly outperforms other advanced methods.

References

1. Bengio, Y., Courville, A., Vincent, P.: Representation learning: a review and new perspectives. IEEE Trans. Patt. Anal. Mach. Intell. **35**(8), 1798–1828 (2013)
2. Chen, X., Duan, Y., Houthooft, R., Schulman, J., Sutskever, I., Abbeel, P.: Infogan: Interpretable representation learning by information maximizing generative adversarial nets. In: Advances in Neural Information Processing Systems, vol. 29 (2016)
3. Cheng, C., Wu, X.J., Xu, T., Chen, G.: Unifusion: a lightweight unified image fusion network. IEEE Trans. Instrument. Measure. **70**, 1–14 (2021)
4. Goodfellow, I., et al.: Generative adversarial nets. In: Advances in Neural Information Processing Systems, vol. 27 (2014)
5. Heusel, M., Ramsauer, H., Unterthiner, T., Nessler, B., Hochreiter, S.: Gans trained by a two time-scale update rule converge to a local nash equilibrium. In: Advances in Neural Information Processing Systems, vol. 30 (2017)
6. Hjelm, R.D., et al.: Learning deep representations by mutual information estimation and maximization. arXiv preprint arXiv:1808.06670 (2018)
7. Lee, K.S., Tran, N.T., Cheung, N.M.: Infomax-gan: Improved adversarial image generation via information maximization and contrastive learning. In: Proceedings of the IEEE/CVF Winter Conference on Applications of Computer Vision, pp. 3942–3952 (2021)
8. Liao, W., Hu, K., Yang, M.Y., Rosenhahn, B.: Text to image generation with semantic-spatial aware gan. In: Proceedings of the IEEE/CVF Conference on Computer Vision and Pattern Recognition, pp. 18187–18196 (2022)
9. Lin, T.-Y., et al.: Microsoft COCO: common objects in context. In: Fleet, D., Pajdla, T., Schiele, B., Tuytelaars, T. (eds.) ECCV 2014. LNCS, vol. 8693, pp. 740–755. Springer, Cham (2014). https://doi.org/10.1007/978-3-319-10602-1_48
10. Van der Maaten, L., Hinton, G.: Visualizing data using t-sne. J. Mach. Learn. Res. **9**(11), 2579–2605 (2008)
11. Mirza, M., Osindero, S.: Conditional generative adversarial nets. Computer Science. pp. 2672–2680 (2014)
12. Reed, S., Akata, Z., Yan, X., Logeswaran, L., Schiele, B., Lee, H.: Generative adversarial text to image synthesis. In: International Conference on Machine Learning, pp. 1060–1069. PMLR (2016)
13. Ruan, S., et al.: Dae-gan: Dynamic aspect-aware gan for text-to-image synthesis. In: Proceedings of the IEEE/CVF International Conference on Computer Vision, pp. 13960–13969 (2021)
14. Salimans, T., Goodfellow, I., Zaremba, W., Cheung, V., Radford, A., Chen, X.: Improved techniques for training gans. In: Advances in Neural Information Processing Systems, vol. 29, pp. 2234–2242 (2016)

15. Shen, Z., Wu, X.J., Xu, T.: Fexnet: foreground extraction network for human action recognition. IEEE Trans. Circ. Syst. Video Technol. **32**(5), 3141–3151 (2021)
16. Souza, D.M., Wehrmann, J., Ruiz, D.D.: Efficient neural architecture for text-to-image synthesis. In: 2020 International Joint Conference on Neural Networks (IJCNN)., pp. 1–8. IEEE (2020)
17. Szegedy, C., Vanhoucke, V., Ioffe, S., Shlens, J., Wojna, Z.: Rethinking the inception architecture for computer vision. In: Proceedings of the IEEE Conference on Computer Vision and Pattern Recognition, pp. 2818–2826 (2016)
18. Tan, H., Liu, X., Li, X., Zhang, Y., Yin, B.: Semantics-enhanced adversarial nets for text-to-image synthesis. In: Proceedings of the IEEE/CVF International Conference on Computer Vision, pp. 10501–10510 (2019)
19. Tao, M., Tang, H., Wu, F., Jing, X.Y., Bao, B.K., Xu, C.: Df-gan: A simple and effective baseline for text-to-image synthesis. In: Proceedings of the IEEE/CVF Conference on Computer Vision and Pattern Recognition, pp. 16515–16525 (2022)
20. Tschannen, M., Djolonga, J., Rubenstein, P.K., Gelly, S., Lucic, M.: On mutual information maximization for representation learning. arXiv preprint arXiv:1907.13625 (2019)
21. Wah, C., Branson, S., Welinder, P., Perona, P., Belongie, S.: The caltech-ucsd birds-200-2011 dataset (2011)
22. Xu, T., et al.: Attngan: Fine-grained text to image generation with attentional generative adversarial networks. In: Proceedings of the IEEE Conference on Computer Vision and Pattern Recognition, pp. 1316–1324 (2018)
23. Xu, T., Feng, Z.H., Wu, X.J., Kittler, J.: Learning adaptive discriminative correlation filters via temporal consistency preserving spatial feature selection for robust visual object tracking. IEEE Trans. Image Process. **28**(11), 5596–5609 (2019)
24. Xu, T., Feng, Z.H., Wu, X.J., Kittler, J.: An accelerated correlation filter tracker. Pattern Recogn. **102**, 107172 (2020)
25. Xu, T., Feng, Z., Wu, X.J., Kittler, J.: Adaptive channel selection for robust visual object tracking with discriminative correlation filters. Int. J. Comput. Vision **129**(5), 1359–1375 (2021)
26. Xu, T., Wu, X.J., Kittler, J.: Non-negative subspace representation learning scheme for correlation filter based tracking. In: 2018 24th International Conference on Pattern Recognition (ICPR), pp. 1888–1893. IEEE (2018)
27. Zhang, H., Xu, T., Li, H., Zhang, S., Wang, X., Huang, X., Metaxas, D.N.: Stackgan: Text to photo-realistic image synthesis with stacked generative adversarial networks. In: Proceedings of the IEEE International Conference on Computer Vision, pp. 5907–5915 (2017)
28. Zhang, Z., Schomaker, L.: Dtgan: Dual attention generative adversarial networks for text-to-image generation. In: 2021 International Joint Conference on Neural Networks (IJCNN), pp. 1–8. IEEE (2021)
29. Zhu, M., Pan, P., Chen, W., Yang, Y.: Dm-gan: Dynamic memory generative adversarial networks for text-to-image synthesis. In: Proceedings of the IEEE/CVF Conference on Computer Vision and Pattern Recognition, pp. 5802–5810 (2019)

A Study of a Cross-modal Interactive Search Tool Using CLIP and Temporal Fusion

Jakub Lokoč and Ladislav Peška(✉)

Faculty of Mathematics and Physics, Charles University, Prague, Czechia
{jakub.lokoc,ladislav.peska}@matfyz.cuni.cz

Abstract. Recently, the CLIP model demonstrated impressive performance in text-image search and zero classification tasks. Hence, CLIP was used as a primary model in many cross-modal search tools at evaluation campaigns. In this paper, we show a study performed with the model integrated to a successful video search tool at the respected Video Browser Showdown competition. The tool allows more complex querying actions on top of the primary model. Specifically, temporal querying and Bayesian like relevance feedback were tested as well as their natural combination – temporal relevance feedback. In a thorough analysis of the tool's performance, we show current limits of cross-modal searching with CLIP and also the impact of more advanced query formulation strategies. We conclude that current cross-modal search models enable users to solve some types of tasks trivially with a single query, however, for more challenging tasks it is necessary to rely also on interactive search strategies.

Keywords: Multimedia retrieval · User study · Cross-modal search

1 Introduction

Thanks to web-scale training datasets, modern deep architectures, and powerful computing clusters, it becomes easier to face cross-modal search problems [16]. As a recent example, the CLIP model [11] clearly demonstrated the true potential of huge train datasets[1]. The model combines image and language processing trainable branches, that can be efficiently trained with image-text pairs and contrastive divergence. As a result, the model shows impressive performance in cross-modal image retrieval tasks. Furthermore, the reported evidence is not limited to a wide spectrum of narrow benchmarks. The top performing systems at the interactive search challenges like Video Browser Showdown (VBS) [12] or Lifelog Search Challenge (LSC) [4] utilized CLIP based models for ranking. We remind that the competitions employ larger multimedia datasets with arbitrary "everyday-life"contents. Hence, general purpose models (like CLIP) are required as any type of concept can be the target of searching.

Although an effective primary retrieval model is a key factor to solve search tasks, there is still a chance that a currently issued query is not suitable for a

[1] Hundreds of millions of image-text pairs were used.

D.-T. Dang-Nguyen et al. (Eds.): MMM 2023, LNCS 13833, pp. 397–408, 2023.
https://doi.org/10.1007/978-3-031-27077-2_31

task. This leads to a natural need for users in the search loop, where inspection of visualized result sets and query reformulation represent basic highly appreciated interactions. Indeed, the top performing systems at VBS 2022 [1,5,7] utilized not only CLIP based ranking, but also additional interactive search features.

In this paper, we present a detailed analysis of the performance of CVHunter, which placed among the top teams. The list of basic features of CVHunter is recapitulated in Sect. 2, detailing both ranking and browsing approaches. In Sect. 3, we provide a thorough analysis of results available thanks to in-build logging mechanisms. The analysis reveals statistics about types of queries used, their effectiveness, and also frequent interaction patterns.

2 CVHunter

CVHunter is an interactive search tool first developed for LSC 2021, but relying on an already existing feature extraction pipeline. Although it is a newly developed tool, it builds on reports about previously successful systems SOMHunter [6,15] and VIRET [9,10]. The project uses .NET framework's WPF for rapid development and also simple project architecture which was shown to be sufficient (by SOMHunter). Hence, it is a matter of a few weeks (for experienced VBS participants) to implement a tool prototype for a competition and test various types of hypotheses. As mentioned before, the tool relies on an existing feature extraction pipeline that extracts a list of selected frames from a video collection and a CLIP feature vector for each selected frame.

Note that while the tool was evaluated on content-based video retrieval tasks, its possible usage is broader as it can handle any kind of temporally ordered images. One of the additional use-cases is to search within a stream of artificial visualizations corresponding to the quantized time series, such as electroencephalograms [3]. We believe that for this kind of data, cross-modal search and human-in-the-loop paradigms gain on importance considerably.

2.1 Input Data and Basic Ranking Approaches

The input for the system are three files produced by an external extraction pipeline (python scripts). The first (largest) file contains extracted video frame thumbnails with a given sampling rate (mostly 4fps), allowing smooth playback of actions. This image file is indexed by a second meta-data file, containing a table with attributes of each extracted frame. Specifically, each row contains video ID, shot ID, frame number, pointer to the image file, and an indicator whether the frame is selected/representative. The third file contains CLIP vectors for all representative frames. CVHunter loads the table and all feature vectors, while image thumbnails in the largest file are accessed on demand. We note that for the current VBS collection [14] comprising 2300 h hours of video, the 16 GB RAM memory approaches the limits for all the loaded meta-data.

Based on the aforementioned data structure, the cross-modal search is possible only for the subset of extracted frames with available CLIP features (more

Algorithm 1. $TemporalQuery(textQuery_1, textQuery_2, Dataset, c)$

Require: $f_{CLIP} : Img \rightarrow R^n$, $f_{CLIP\_WS} : Text \rightarrow R^n$
  $q_1 = f_{CLIP\_WS}(textQuery_1)$
  $q_2 = f_{CLIP\_WS}(textQuery_2)$

  for each $frame \in Dataset$ **do**
    $frame.score_1 \leftarrow \delta_{cos}(q_1, f_{CLIP}(frame.Image))$
    $frame.score_2 \leftarrow \delta_{cos}(q_2, f_{CLIP}(frame.Image))$
  end for

  for each $frame \in Dataset$ **do**
    $maxScore_2 = MaxScore_2FromCNextFrames(frame, c)$
    $frame.overallScore = -frame.score1 * maxScore_2$
  end for

  $SortByOverallScore(Dataset)$ ▷ Shows items with the highest overall score

than 2.5 M frames). In order to obtain the feature vector for a text query, an external web service ($f_{CLIP\_WS} : Text \rightarrow R^n$) is called. With the query vector, the basic ranking of all database objects is computed using the cosine distance. The tool supports also temporal [8,9] and context aware queries, allowing description of temporally close shots. The temporal query is reminded in Algorithm 1.

In addition, a Bayesian relevance feedback approach [2] was integrated as well. This approach allows relevance feedback sessions, where operators select positive examples reflecting their search needs and then the system updates maintained relevance scores for all indexed frames. Each session starts with a text query and then can continue with relevance feedback iterations. Please note that frame CLIP vectors are pre-computed (using $f_{CLIP} : Img \rightarrow R^n$) and stored in the CLIP vector file. Since the aforementioned search features were available already in the predecessors of CVHunter, in the next section we focus on details of a newly tested search feature – temporal relevance feedback.

2.2 Temporal Relevance Feedback

In order to express search needs for a sequence of video shots, temporal feedback allows to specify positive examples for both components of an initial temporal query. For simplicity, we assume a video frame object containing fields $Image$, $score_1$, $score_2$, and $overallScore$.

Algorithm 2 shows steps used to compute new scores of frame objects based on the currently available scores and two (temporally ordered) example images. Let's assume that $score_1$ and $score_2$ store the actual relevance scores of a frame f_i with respect to a temporal query. For example, for a temporal query"gray cat → green tree", $score_1$ contains relevance score with respect to query "gray cat", while $score_2$ contains relevance score with respect to query "green tree".

Algorithm 2. $TemporalRelevanceFeedback(example_1, example_2, Dataset, \alpha, c)$

Require: $f_{CLIP} : Img \rightarrow R^n$
$NegativeExamples \leftarrow Dataset.CurrentDisplay.Sample(k)$

for each $frame \in Dataset$ **do**
  $n_{Sum} \leftarrow 0$
  for each $negative \in NegativeExamples$ **do**
    $n_{Sum} += e^{-\delta_{cos}(f_{CLIP}(negative.Image), f_{CLIP}(frame.Image))/\alpha}$
  end for

  if $example_1 != null$ **then**
    $e_1 \leftarrow e^{-\delta_{cos}(f_{CLIP}(example_1), f_{CLIP}(frame.Image))/\alpha}$
    $frame.score_1 \leftarrow frame.score_1 * e_1/(e_1 + n_{Sum})$
  end if

  if $example_2 != null$ **then**
    $e_2 \leftarrow e^{-\delta_{cos}(f_{CLIP}(example_2), f_{CLIP}(frame.Image))/\alpha}$
    $frame.score_2 \leftarrow frame.score_2 * e_2/(e_2 + n_{Sum})$
  end if
end for

$NormalizeScores(Dataset)$ ▷ Max $score_1 \rightarrow 1$, Max $score_2 \rightarrow 1$

for each $frame \in Dataset$ **do**
  $maxScore_2 = MaxScore_2FromCNextFrames(frame, c)$
  $frame.overallScore = frame.score1 * maxScore_2$
end for

$SortByOverallScore(Dataset)$ ▷ Shows items with the highest overall score

The standard temporal query ends with a temporal fusion considering selected representative frames following f_i in the video. However, for temporal relevance feedback, these scores are further updated with respect to the provided example images. For example, let's assume that out of the available options (e.g., current display), the operator can select the best matching examples reflecting his/her mental image of a grey cat for the first part of the query and analogically selects the best representative of a green tree for the second part.

For simplicity, Algorithm 2 considers that only one example is given for each temporal query part (CVHunter supports multiple examples). In addition to positive examples, the algorithm uses also implicit negative examples sampled from the current display. Using a Bayesian like update formula, both $score_i$ fields are updated with corresponding values $e_i/(e_i + n_{Sum})$ and normalized, while the temporal fusion is computed afterwards. Please note that the temporal fusion works also for cases when an example image is provided only for one part of the temporal query. In such case, scores corresponding to the other query part are left intact (except normalization). According to our preliminary experiments, we set the parameter $\alpha = 1$. Furthermore, in order to prioritize high scores obtained by the original text search, the score was powered by 4 in the beginning of each relevance feedback session. Based on our observations, α and power values are dependent and the employed combination of values was competitive. In the future, we plan more thorough experimental evaluations on this topic.

2.3 Visualization and Browsing Options

As demonstrated at interactive search challenges, every successful tool requires also visualization and browsing capabilities. Standard approaches integrated to CVHunter comprise grid visualization of thumbnails, scrolling in the result set, inspection of representative video frames for a selected thumbnail, row-based visualization of context frames with possibility of row-based scrolling, frame close up display with video preview, kNN display showing the nearest frames (in the CLIP space) to a selected thumbnail, and similarity search in a single video. For row-based visualization showing context frames, the display component filters results that were already displayed as context frames of a frame with a better rank. In addition, the tool supports a filter showing just top-k items from a video, where k corresponds to a number 1–9.

3 Analysis of the Results

The Video Browser Showdown competition runs at the DRES server [13], which logs all meta-data for performed known-item search tasks. Hence, the timestamps of task start, task end, and ground truth (target video segment info) are available for after-competition log analysis. CVHunter uses synchronized[2] timestamps to store top-k video frames for each query ($k = 10^4$ in most cases) as well as selected browsing actions (video inspection, close-up detail). Hence, it is possible to identify task related actions, reconstruct basic search sequences and identify the best achieved ranks of correct video frames in logged result sets. By "a correct frame“, we mean that the frame is from a searched video segment (KIS task ground truth). In the analysis, we report the best rank of a correct frame as a basic measure of query effectiveness. If a correct frame was not found in the log, graphs show the corresponding point as $> 10^4$ or at coordinate 20000.

3.1 Text Query Variance Among Operators

With cross-modal search allowing convenient free-form text querying, there is no guarantee to obtain similar queries for the same target by different operators. Upon a closer inspection of the logs corresponding to individual operators, we can observe several notable differences. In general, Operator 1 wrote longer textual

[2] After logging to the DRES server, the server timestamp and local timestamp were stored in CVHunter logs. Since CVHunter used timestamps based on seconds and there might be a short communication delay, the presented times of actions are close approximations of the exact moments.

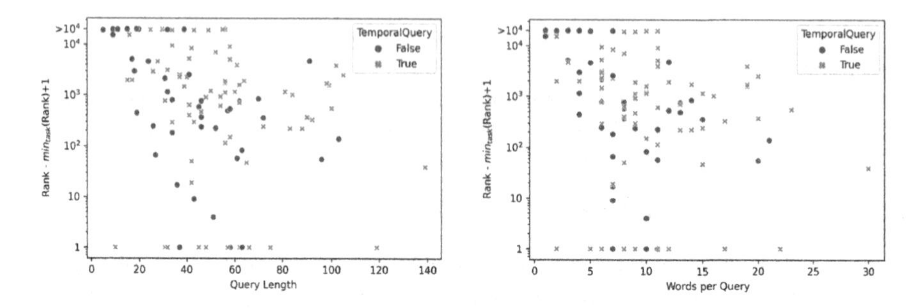

Fig. 1. Dependence of the correct item's rank on the query length (left) and the number of words per query (right). Note that in order to adjust for varying task difficulty, the minimal rating per task was subtracted.

queries (both w.r.t. number of chars and number of words) and he/she also utilized temporal queries more often (in 67% of cases as compared to 48% for Operator 2). In turn his/her query results more often contained searched items within top-k (see Table 1) and also the median of searched item's rank was lower (730 vs 1154).

These results illustrate that there can be substantial differences among capabilities of individual users to operate tools such as CVHunter, which should be taken into account during the evaluation process. Furthermore, it may seem that query's relevance (i.e., the rank of searched items) can correlate with some generic query features such as its length. Nonetheless, evaluating such dependencies is extremely tricky due to the varying complexity of individual tasks. For instance, if the searched item represents some very rare concept within the dataset, even a simple (short) query can return it among top few results. To account for the task variability, but still evaluate the link between queries length and relevance, we proceeded as follows. First, for the sake of results stability, we only kept the tasks for which at least 3 text queries were generated. Than, in order to normalize results w.r.t. task difficulty, we subtracted the minimal per-task rank from all corresponding results and depict them as scatter plots on Fig. 1. Indeed, we can observe a moderate negative correlation between query ranks and both words count (Pearson correlation: −0.42) and characters count (Pearson correlation: −0.39). Although this cannot be a universal recommendation as the exceptions are plenty, it seems that writing a thorough textual description of the searched item often pays off.

Table 1. Text query statistics per individual operators. "Words" denote the average volume of words per query, "Length" denote the average length (volume of chars) of the query and "top-x" denote the volume of queries, where the searched object had rank <= top-x.

	Total	Words	Length	top-10	top-20	top-50	top-100	top-200	top-500	$> 10^4$
Op. 1	69	10.97	57.90	9	10	11	14	15	28	8
Op. 2	60	6.97	38.26	5	8	9	11	11	17	11

3.2 Search Process

Figure 3 and Fig. 4 visualize search processes of two operators (the faster one solves a task for the whole team) in all VBS 2022 visual and textual KIS tasks. Each point in the graph corresponds either to a text query (including temporal variants), relevance feedback action (including temporal variant), kNN search (only a few points), or selected browsing actions (video inspection and close-up detail inspection). We emphasize that frequent browsing actions like scrolling or video preview are not part of the visualization. All query points are placed on x-axis based on the elapsed time, while y-axis corresponds to the rank of the first correct item in the logged result set (20000 for unknown ranks). Browsing actions use the same y coordinate as the previous query and their duration is displayed using a solid horizontal line. The dotted line is just a linear interpolation, grouping actions of the same user. Please note that both figures show only the rank of correct frames based on a ground truth video segment, while there appear also other "incorrect" frames from the correct video. Based on these related video frames, users can still recognize the similar content, focus on the video, and find a correct frame using video browsing actions. Therefore, existence of such best video frame rank (identified in task logs of both operators) is illustrated with a dotdash line for challenging tasks. This line does not show when the item appeared. We also note that the submission times are taken from server logs.

In both figures, it is apparent that some known-item search tasks can be effectively solved with a cross-modal search approach (e.g., t07, t10, v03). Please note that the query can be effective even without a temporal query composition (e.g., t01, t03, t09, v10, v12). Although the first (text) query action requires some time (thinking, translation), in the mentioned cases the searched target frame appeared with a good rank and after that there was a correct submission (rectangle symbol). Nevertheless, browsing was used also for relatively easy tasks.

Contrary to the easy known-item search tasks, both figures reveal also more complicated search processes where the target items did not appear at good ranks or where a correct item was overlooked (e.g., in task v06). These cases illustrate the need for interactive query reformulation even for state-of-the-art cross-modal approaches like CLIP. It is also important to focus on effective visualization of the results. Indeed, in some cases it is not clear how the scene looks like (e.g., in task t08 a correct item was among top-10), or a correct known frame can be overlooked due to attention focus. Anyway, the challenging tasks are exactly those, where interactive approaches may help. We may observe that in visual KIS tasks the relevance feedback approach may significantly help with the rank of searched items. Operator 2 used the temporal relevance feedback to solve visual KIS tasks 1 and 8. In both cases, the previous text search only provided a starting point for selection of promising example images. We also admit that in

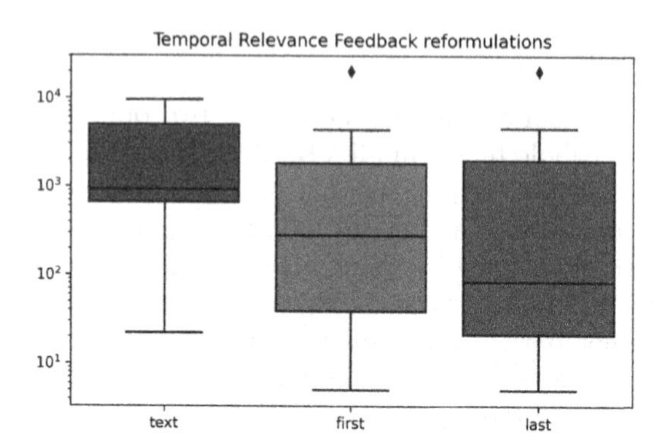

Fig. 2. Boxplots of rankings before and after the temporal relevance feedback reformulations. First denotes ranks after the first iteration of relevance feedback, last denote ranks after the last iteration of relevance feedback.

Table 2. All temporal feedback search sessions from VBS 2022 logs, each starting with a temporal text query (TQ) and each consisting of 1–4 example image feedback iterations. Numbers in brackets show the numbers of example images for the first or for the second part of the temporal query. Red color indicate cases, where the method did not improve the rank of a searched target.

	Rank TQ	Rank I1		Rank I2		Rank I3		Rank I4	
1	549	1261	(1/0)	4532	(0/1)				
2	4855	3436	(2/3)	1177	(3/2)				
3	909	176	(1/0)	37	(2/0)				
4	1018	122	(1/1)						
5	843	384	(2/0)						
6	74	22	(1/1)						
7	7576	4323	(1/1)						
8	704	44	(1/1)						
9	5214	955	(1/0)	194	(1/0)	43	(1/0)	19	(1/0)
10	9437	$> 10^4$	(1/1)						
11	928	9	(1/1)						
12	22	5	(1/1)						

some cases (e.g., in task t06 or v08) relevance feedback can be contra-productive due to unknown visual appearance of target items or similarity model. Yet, Fig. 3 also shows that relevance feedback helped to solve task t11.

3.3 Temporal Feedback Query Statistics

In the last part of our analysis, we focus on the performance of temporal relevance feedback actions, detailed in Algorithm 2. Figure 2 shows three box plots. The first box plot illustrates distribution of ranks of target frames right after a temporal query. The high average rank indicates that these are usually the situations, where users are not satisfied with the results and think about a more complex search strategy. The second box plot shows distribution of ranks after the first relevance feedback iteration, while the last box plot shows the distribution of obtained ranks after the last feedback iteration.

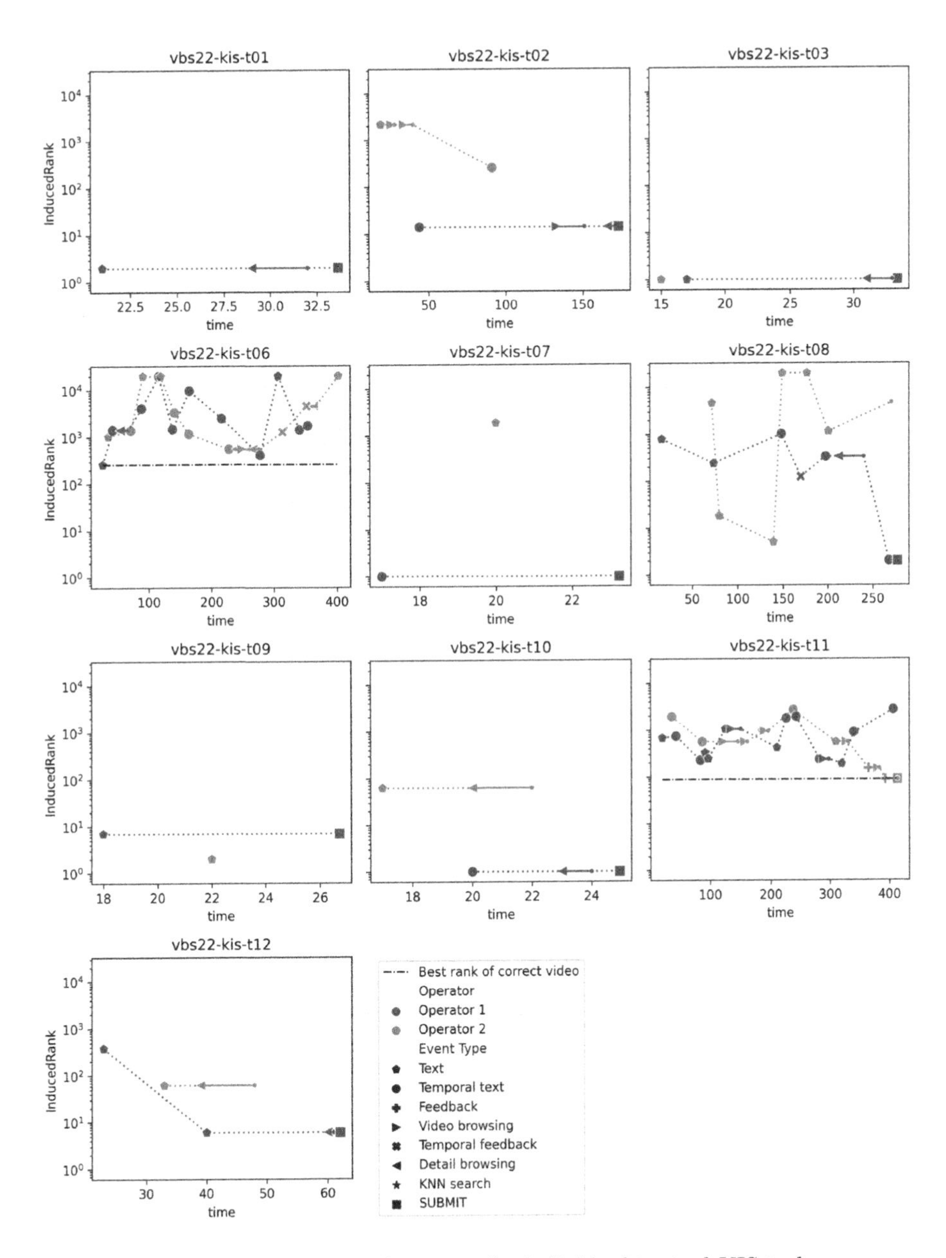

Fig. 3. Interactive search process for individual textual KIS tasks.

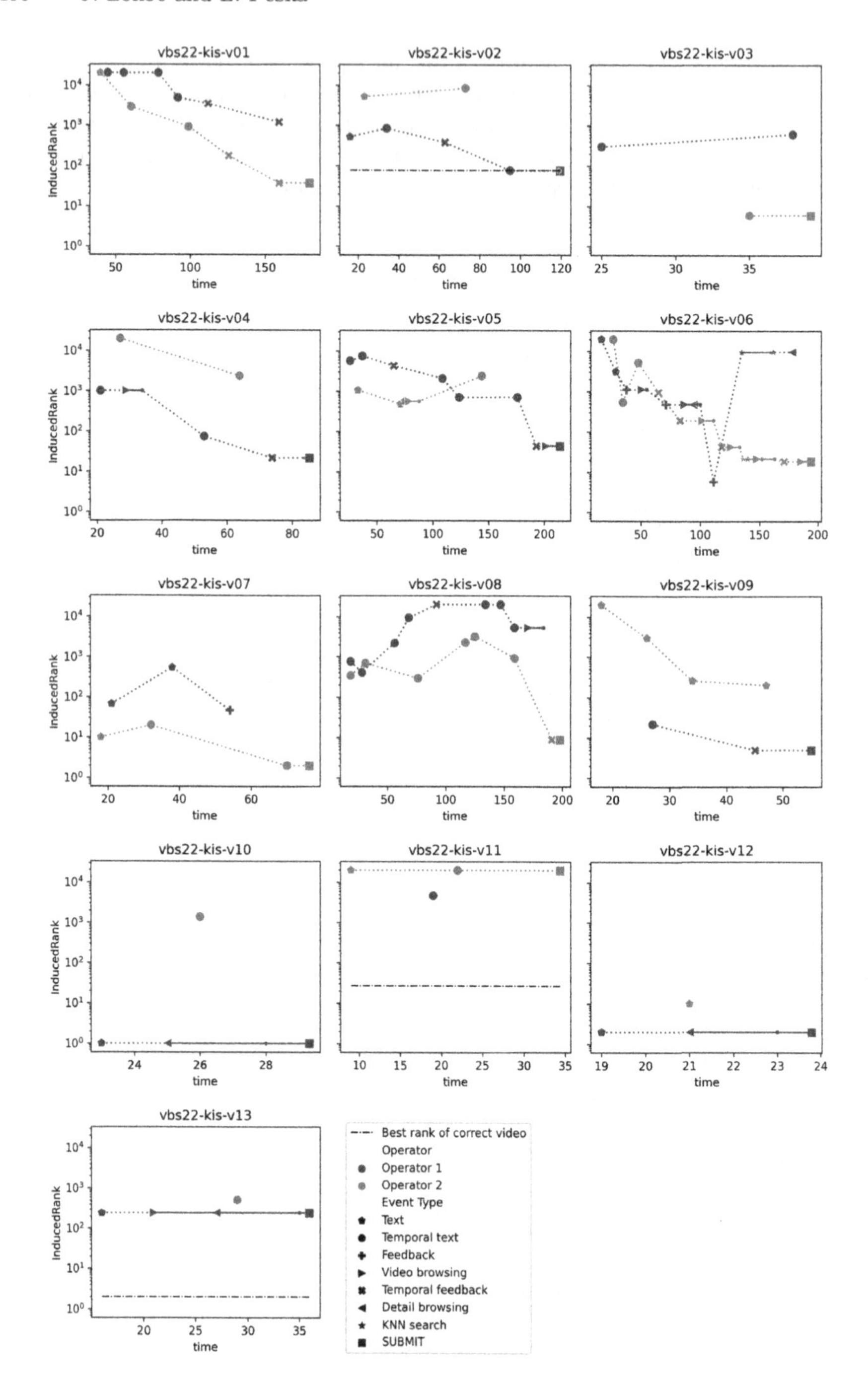

Fig. 4. Interactive search process for individual visual KIS tasks.

Although the number of queries is not high enough for statistical significance, we observe promising improvements (please note log scale on y axis). Table 2 shows the effect of each feedback action individually. Each row shows one search session starting with a temporal text query, followed by temporal relevance feedback action(s). We may observe that only in three cases the feedback action was contra-productive. This indicates that operators were able to select suitable example images and the feedback model benefited from the examples. Lines 3, 4, 8, 9, 11 show that the feedback process can converge to a correct submission rapidly.

4 Conclusions

This paper presents a cross-modal search study performed with an interactive video search tool utilizing the state-of-the-art CLIP model. The study was performed at the Video Browser Showdown 2022 where 23 known-item search tasks were presented. The results show that a single query search strategy helps only in specific cases, while advanced query composition strategies are important for more challenging known-item search tasks. Variance among users is discussed as well, pointing on different query statistics and their effect. The study also reveals promising performance of temporal relevance feedback approach, which substantially helped to solve several visual known-item search tasks.

In the future, we plan to organize studies to thoroughly investigate effectiveness of cross-modal approaches combined with temporal relevance feedback queries. We plan to compare temporal and non-temporal variants of queries as well as to better understand optimal parameter settings for model variants.

Acknowledgments. This paper has been supported by Czech Science Foundation (GAČR) project 22-21696S. Source codes and raw data are available from https://github.com/lpeska/MMM2023_CVHunter_analysis.

References

1. Amato, G., et al.: VISIONE at video browser showdown 2021. In: Lokoč, J., et al. (eds.) MMM 2021. LNCS, vol. 12573, pp. 473–478. Springer, Cham (2021). https://doi.org/10.1007/978-3-030-67835-7_47
2. Cox, I., Miller, M., Omohundro, S., Yianilos, P.: Pichunter: Bayesian relevance feedback for image retrieval. In: International Conference on Pattern Recognition. vol. 3, pp. 361–369. IEEE (1996), https://doi.org/10.1109/ICPR.1996.546971
3. Gao, Y., Gao, B., Chen, Q., Liu, J., Zhang, Y.: Deep convolutional neural network-based epileptic electroencephalogram (eeg) signal classification. Front. Neurol. **11** (2020). https://doi.org/10.3389/fneur.2020.00375
4. Gurrin, C., et al.: Introduction to the third annual lifelog search challenge (lsc'20). In: International Conference on Multimedia Retrieval, pp. 584–585. ACM (2020). https://doi.org/10.1145/3372278.3388043

5. Hezel, N., Schall, K., Jung, K., Barthel, K.U.: Efficient Search and Browsing of Large-Scale Video Collections with Vibro. In: Þór Jónsson, B., Gurrin, C., Tran, M.-T., Dang-Nguyen, D.-T., Hu, A.M.-C., Huynh Thi Thanh, B., Huet, B. (eds.) MMM 2022. LNCS, vol. 13142, pp. 487–492. Springer, Cham (2022). https://doi.org/10.1007/978-3-030-98355-0_43
6. Kratochvíl, M., Veselý, P., Mejzlík, F., Lokoč, J.: SOM-Hunter: video browsing with relevance-to-som feedback loop. In: Ro, Y.M., et al. (eds.) MMM 2020. LNCS, vol. 11962, pp. 790–795. Springer, Cham (2020). https://doi.org/10.1007/978-3-030-37734-2_71
7. Lokoč, J., Mejzlík, F., Souček, T., Dokoupil, P., Peška, L.: Video search with context-aware ranker and relevance feedback. In: Þór Jónsson, B., Gurrin, C., Tran, M.-T., Dang-Nguyen, D.-T., Hu, A.M.-C., Huynh Thi Thanh, B., Huet, B. (eds.) MMM 2022. LNCS, vol. 13142, pp. 505–510. Springer, Cham (2022). https://doi.org/10.1007/978-3-030-98355-0_46
8. Lokoč, J., et al.: A W2VV++ case study with automated and interactive text-to-video retrieval. In: International Conference on Multimedia. ACM (2020). https://doi.org/10.1145/3394171.3414002
9. Lokoč, J., Kovalčík, G., Souček, T., Moravec, J., Čech, P.: A framework for effective known-item search in video. In: International Conference on Multimedia, pp. 1777–1785. ACM (2019). https://doi.org/10.1145/3343031.3351046
10. Peška, L., Kovalčík, G., Souček, T., Škrhák, V., Lokoč, J.: W2VV++ BERT model at VBS 2021. In: Lokoč, J., Skopal, T., Schoeffmann, K., Mezaris, V., Li, X., Vrochidis, S., Patras, I. (eds.) MMM 2021. LNCS, vol. 12573, pp. 467–472. Springer, Cham (2021). https://doi.org/10.1007/978-3-030-67835-7_46
11. Radford, A., et al.: Learning transferable visual models from natural language supervision. CoRR abs/2103.00020 (2021). https://arxiv.org/abs/2103.00020
12. Rossetto, L., et al.: Interactive video retrieval in the age of deep learning-detailed evaluation of VBS 2019. IEEE Trans. Multimedia **23**, 243–256 (2020). https://doi.org/10.1109/TMM.2020.2980944
13. Rossetto, L., Gasser, R., Sauter, L., Bernstein, A., Schuldt, H.: A system for interactive multimedia retrieval evaluations. In: Lokoč, J., et al. (eds.) MMM 2021. LNCS, vol. 12573, pp. 385–390. Springer, Cham (2021). https://doi.org/10.1007/978-3-030-67835-7_33
14. Rossetto, L., Schuldt, H., Awad, G., Butt, A.A.: V3C – a research video collection. In: Kompatsiaris, I., Huet, B., Mezaris, V., Gurrin, C., Cheng, W.-H., Vrochidis, S. (eds.) MMM 2019. LNCS, vol. 11295, pp. 349–360. Springer, Cham (2019). https://doi.org/10.1007/978-3-030-05710-7_29
15. Veselý, P., Mejzlík, F., Lokoč, J.: SOMHunter V2 at video browser showdown 2021. In: Lokoč, J., et al. (eds.) MMM 2021. LNCS, vol. 12573, pp. 461–466. Springer, Cham (2021). https://doi.org/10.1007/978-3-030-67835-7_45
16. Wang, K., Yin, Q., Wang, W., Wu, S., Wang, L.: A comprehensive survey on cross-modal retrieval. CoRR abs/1607.06215 (2016). http://arxiv.org/abs/1607.06215

A Cross-modal Attention Model for Fine-Grained Incident Retrieval from Dashcam Videos

Dinh-Duy Pham[1], Minh-Son Dao[2](✉), and Thanh-Binh Nguyen[1]

[1] AISIA Research Lab-University of Science-Vietnam National University, Ho Chi Minh City, Vietnam
ngtbinh@hcmus.edu.vn

[2] National Institute of Information and Communications Technology, Tokyo, Japan
dao@nict.go.jp

Abstract. Dashcam video has become popular recently due to the safety of both individuals and communities. While individuals can have undeniable evidence for legal and insurance, communities can benefit from sharing these dashcam videos for further traffic education and criminal investigation. Moreover, relying on recent computer vision and AI development, a few companies have launched the so-called AI dashcam that can alert drivers to near-risk accidents (e.g., following distance detection, forward collision warning) to improve driver safety. Unfortunately, even though dashcam videos create a driver's travel log (i.e., a traveling diary), little research focuses on creating a valuable and friendly tool to find any incident or event with few described sketches by users. Inspired by these observations, we introduce an interactive incident detection and retrieval system for first-view travel-log data that can retrieve fine-grained incidents for both defined and undefined incidents. Moreover, the system gives promising results when evaluated on several public datasets and popular text-image retrieval methods. The source code is published at https://github.com/PDD0911-HCMUS/Cross_Model_Attention

Keywords: Dashcam video · Incident detection and retrieval · Travel log · Interactive searching · Intelligent cross-data retrieval · Text-image matching

1 Introduction

Intelligent transportation systems have become essential topics to improve the quality of the current transportation and reduce street accidents in each country. Moreover, with the development of telecommunication and information technology during the last few years, people have had more and more chances to integrate such systems into their vehicles and receive the associated support any time they drive. To build an intelligent transportation system, different companies utilize the recently advanced techniques in computer vision, signal processing, and

D.-T. Dang-Nguyen et al. (Eds.): MMM 2023, LNCS 13833, pp. 409–420, 2023.
https://doi.org/10.1007/978-3-031-27077-2_32

natural language processing along with sensors to detect real-time behaviors and interactions of people and other transportation systems on the street. And then, the systems can recommend quick actions to avoid incidents [1].

Using dash cams installed in vehicles is increasingly ubiquitous in different countries. Therefore, people can easily record all things happening during their road trips and provide potential digital evidence for any traffic accidents related to or a relevant crime scene like a moving surveillance camera [2]. In addition, it is worth noting that these dash cams could monitor in different directions: in front of the vehicles, behind them, or even inside the cars. Kim and colleagues studied the difference in dashcam video sharing's motives and privacy concerns among multiple countries [3]. Based on these advantages, people started utilizing dashcams as an economical and common resource to enhance traveling safety. Evans et al. [4] presented a brief review of the urban road incident detection research and highlighted potential enhancements for road incident detection algorithms (IDAs). Adamova showed the importance of using dashcams in real-time activities for different vehicle views to enhance drivers' safety levels when moving on a street [5]. Bazilinskyy et al. [6] investigated the importance of better understanding road traffic safety using dashcam videos collected from four locations worldwide.

Connecting dashcams mounted on vehicles provides a large IoT that requires Big Data techniques to handle and extract useful information. Mohandu et al. [7] provided a detailed survey of the current big data techniques applied in different intelligent transportation systems and listed several open problems related to big data analytics in this domain.

Considering emerging trends in incident detection using dashcam videos mentioned above, we observe there is an utmost requirement for a quick query tool that can return a relevant incident image with the free-style textual query from users. Beyond the methods of detecting incidents that produce a database of short videos indexed by incident types, users need a tool that not only searches by incident types but also by user-defined types. In other words, users can imagine what the accident scene looks like and assume that a system can return what they want exactly. Nevertheless, while regular users can generate less detailed queries, experts (e.g., police officers, insurance staff, detectives) can create more detailed ones. Hence, the requirement of multi-level fine-grained textual incident retrieval is authentic and in very high demand.

One of the solutions for the challenge is the image-text retrieval approach [8] and adversarial text-to-image synthesis [9], where people can find or generate the relevant image by describing their textual query. These approaches relied heavily on a vast training database where perfect pairs (image, text) are created carefully. Applications that want to deal with text-image retrieval either reutilize a big pre-trained model and downstream to smaller datasets or create a new model trained on a particular dataset. Both mentioned approaches do not suit text-image incident retrieval. The key is that popular relevant text-image pre-trained models do not cover the incident case where how an incident happens and what an accident scene looks like are searching content with high frequency.

In light of these observations, we propose a novel interactive incident retrieval system for first-view travel-log data that can overcome the mentioned challenges. Furthermore, it is worth noting that the proposed approach can efficiently adapt to new datasets, models, and domains for an interactive incident retrieval system. We measure the performance of our framework and its components on different datasets, users, and models. The experimental results show the promising performance of the proposed method and provide further applications in exploiting and exploring first-view travel-log data and contributing to smart mobility, where safety is the priority.

The paper is organized as follows: Sect. 1 introduced the research motivation and related works. Section 2 defines the problem statements and proposed solution. Section 3 explains the methodology. Section 4 discusses the experimental results. Section 5 concludes the work and points out future works.

2 Problem Statements

In this section, we introduce the problem we want to tackle. The problem plays at the core of our system that aims to fine-grained retrieve incidents from user's textural queries with additional constraint supports.

Problem Statement: Given an image dataset DI and the textual query Q whose content normally is a various-granularity cognitive-based expression of a want-to-be-found incident. Let $I_j \in DI$ be the jth image. The problem here is to construct the function $F(Q, DI) \rightarrow \{I_k\}$ to find a set of best-matched pair $\{(Q, I_k)\}$. Q is the set of captions, descriptions, or query sentences of the image.

Proposed Solution: Let EI_{ij} and MEI_{ij} be the ith image embedding extracted from I_j and the model used to extract this object. Here, the image embedding terminology expresses features (both high semantic and low levels) extracted from I_j. Let ET_j and MET_j be the text embedding of the annotation or caption or query of the query Q_j and the model used to build up to ET_j. Let JS be the joint representation space where the query Q, $\{I_j\}$ are projected. Inside JS, we will find the best match embedding vectors of Q and $\{I_j\}$ interactively with the judgment made by users. In other words, we try to construct a joint representation space where a textual query and its relevant images have the minimum similarity distance. In this space, information from both data modalities is exchangeable. Moreover, with the support of users, we can tune the similarity measure to decrease the intra-class errors and increase the inter-class distance.

3 Methodology

This section describes the methodology used to solve the mentioned problem. We start from the method overview, continued by system components individually or reciprocally to help readers have inter-and intra-component visions.

3.1 A Cross-Modal Attention Model: An Overview

Figure 1 illustrates the method's overview. We leverage the text-image cross-modal attention to create the joint representation space. We build only the encoder to generate the joint representation space and use the information retrieval technique to carry out the query progress. In general, an in-context pair of image and text is proceeded to generate associative embedding vectors. These vectors are pushed through the multi-head attention module to encode. The encoded vectors go through several full connections to form a final joint representation space that will further be utilized for querying.

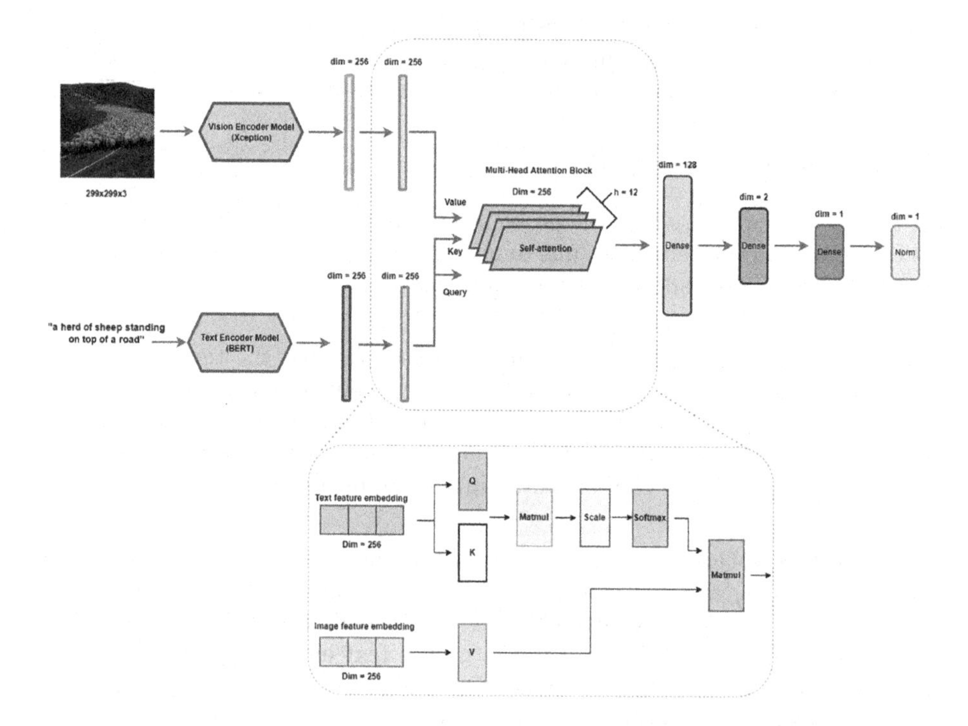

Fig. 1. A cross-modal attention model: an overview

3.2 Vision Encoder

We utilize the Xception [10] model that pre-trained on the ImageNet dataset to build our embedding vector. We retrain the model with all frozen layers on our dataset. We also add two full connection layers as the end of the backbone with the output is a 256-dimension vector. We normalize the input image size 299×299. The top-left part of Fig. 1 illustrates this encoder.

3.3 Text Encoder

We use the smaller BERT models referenced in the published article [11] as the core of our text encoder. It is well known that the standard BERT recipe (including model architecture and training objective) is effective on a wide range of model sizes beyond BERT-Base and BERT-Large. The smaller BERT [12] models are considered for limited computational resources, and they can be fine-tuned in the same manner as the original BERT models. However, they are most effective in knowledge distillation, where a more extensive and accurate teacher produces the fine-tuning labels. Our goal is to enable research in institutions with fewer computational resources and encourage the community to seek directions of innovation alternatives to increasing model capacity. We also add two full connection layers as the end of the backbone with the output for embedding text to a 256-dimensions vector. The top-left (second line) part of Fig. 1 illustrates this encoder.

3.4 Multi-Head Attention Block

To build our attention component, we inherit the spirit from the attention article [13]. We reuse this attention structure with a modification of (V (Vector), Q (Query), K(Key)) settings. In our research, we set K and Q for image embedding vectors and V for text embedding vectors. The bottom part of Fig. 1 illustrates this block structure.

3.5 Multi-Head Cross-Modal Attention

We design our multi-head cross-modal attention block by consuming the outputs of the vision and text encoders as Value and Query-Key, respectively. We have run a survey to collect all possible accident scene descriptions from volunteers and police traffic accident records. The authors discuss people's subitizing in [14], which usually is less than four objects at once that people can pay attention to. Considering three areas of an accident scene image (i.e., left, right, and center), we come up with 12 objects that may be concerned. Hence, we set the multi-head as 12 to emphasize a maximum of 12 suspect objects/zones/phrases primarily associated with incident scene descriptions. At the end of this component, we add three full connection layers to decrease the dimension and increase the feature's salience.

Loss F unction: We compute the pairwise dot-product similarity between each $caption_i$ and $image_i$ as the predictions to calculate the loss. The target similarity between $caption_i$ and $image_i$ is computed as the average of the (dot-product similarity between $caption_i$ and $image_i$) and (the dot-product similarity between $image_i$ and $image_j$). Then we utilize cross-entropy to compute the loss between the targets and the predictions.

3.6 Fine-Grained Incident Retrieval

The Fine-Grained Incident Retrieval aims to enhance the searching accuracy against the travel-log dataset (e.g., dashcam, lifelog camera), incredibly satisfying the free-style textual queries that primarily explain the accident scene in the user's cognition. In this case, searching by only an incident class could not meet users' requirements.

For example, the class-based incident query returns all incident moments that met searching criteria such as "*find all car-crash incidents that happened last week.*" with "car-crash" being an accident class/type. Nevertheless, when users (e.g., police officers, lawyers, and insurance companies) want to find more detailed incident scenes, we have to run fine-grained searching by shrinking the search space accordingly. For example, a fine-grained query could be "*find the incident where a white car hit a blue truck from behind and happened near an intersection.*"

Assuming we have the joint representation space created by training the multi-head cross-modal attention model on a dataset. Then we want to make a search engine that can consume a text query and utilize the joint representation space to find the set of images with maximum similarity measure. We apply the approximate similarity matching approach [15] that utilizes frameworks like ScaNN [16], Annoy, or Faiss [17] to work with a large number of images in a real-time mode.

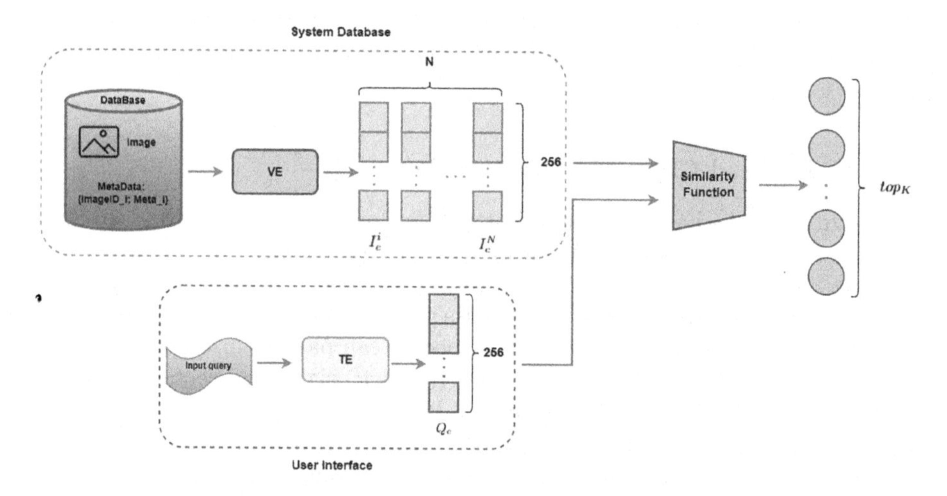

Fig. 2. Overview system image retrieval

We define the similarity function as follows:

$$Dot_{similarity} = Q_e \cdot I_{es}$$

where Q_e is Query embedding that user input and I_{es} is all image embedding vectors using Vision Encoder to predict after training. $Q_e = TE(Q)$ and $I_{es} = VE(I_s)$ the TE and QE is a model encoder after complete training progress, Q is a user's querying input, and I_s is a list of all images that we have in our dataset. After encoding, the Q_e and I_{es} become Q_{1*256} and I_{N*256}, respectively, N is the number of images that we have in the dataset. Hence, the $Dot_{similarity}$ has size $1 * N$. Finally, the search engine returns values and indices of the k largest entries. Figure 2 illustrates the search engine.

The System Database is a services backend independent of the User Interface part. The System Database part will create a set of embedding vectors and store it in the system as the database when the user inputs a query. The User Interface part will create a Query embedding vector, and the system will use the similarity function to get top-K results.

4 Experimental Results

This section describes, explains, and discusses datasets, working environments, evaluation, and comparisons with other methods. It should be noted that this project is ongoing, and not all system components have gained the best results.

4.1 Dataset

We use various datasets gathered from public sources and created by ourselves. The former included 8-classes of incidents dataset [24], BDD1000K [23], and RetroTruck [22] datasets and the latter contains I4W [25] datasets. The 8-classes dataset contains around 12K positive samples for eight incident classes whose volumes are illustrated in Fig. 3. The RetroTruck dataset has 254 videos (25 fps) of normal and 56 abnormal driving scenes. The I4W contains 600 videos (15 min/video) recorded repeatedly in four courses in Tokyo center, Japan.

Fig. 3. 8-classes dataset [24]: A sample

While the former already has incident labels, the latter needs manual labeling for incidents. First, we asked volunteers to label incidents recorded in the latter manually. Second, we merge all datasets to form an extensive dataset for our evaluation. Finally, we asked volunteers to label data and created a structured data set of the form $D_k = \{(I_k; [T_k^i])\}$ where I is an image, and $[T_k^i]$ is a list of captions similar with I's content, emphasize on the incident aspect. We organize the dataset as $\{key; value\}$ where *key* represents an image's ID, and *value* is for the image's description (i.e., image label/annotation/caption), and store them (i.e., key, value) in JSON format. The significant criterion that differed from our dataset is that we asked the volunteers to describe an image of an accident scene in detail as if they reported the accident they witnessed to police staff.

We also use CIFAR and MS-COCO datasets with five captions for each image to compare our method. In this case, we do not emphasize the particular incident images. In contrast, we want to see how well our method can work with other datasets of broad domains.

4.2 Experimental Setup

We set the working environment as follows: Keras and Tensorflow have the same version 2.8.0, with CUDA 10.1 running on Python 3.8, using GPU NVIDIA RTX 3060 12 GB. We simulate online mode by considering that one video is continuously sent to the system from a dashcam. The hyperparameters of our model are denoted in Table 1.

Table 1. Model parameters

Parameter	Value
Pre-trained model	Xception and BERT
Trainable params	frozen all layers
Scale normalized image size	299 × 299
Batch size	32
Number epoch	21
Optimizer	Adam
Learning rate	0.0001
Reduce learning rate schedule	Reduce learning rate on plateau
Reduce learning rate factor	0.2
Training batch	70% dataset
Validation batch	20% dataset
Testing batch	10% dataset
Training time	around 3.5 h
Query Time	around 1.2 s
Vision encoder parameters	21,452,328 (Trainable: 590K, Non-trainable: 20.8M)
Text encoder parameters	28,961,281 (Trainable: 197K, Non-trainable: 28.7M)
Total parameters of model	53,560,107 (Trainable: 3.9M, Non-trainable: 49.6M)

4.3 Results

We implement different fine-grained text-image incident retrieval methods to compare with our model. We utilize Knowledge Graph (KG) to analyze text query features to form a $\{source - edge - target\}$ tuple. Then, we use FAISS [17] to calculate the correlation between the text query and images stored in our indexed databases. We also develop a Dual Encoder (DE) method with two Encode blocks (i.e., one for image and one for texts). Then we perform the math of the correlation between a text query and images with the query relationship. We merge these two methods by bagging to generate the third one. This model outputs a sub-dataset from the original dataset that correlates with the query. All these three methods run on both MS-COCO and our dataset for comparison.

We also reproduce three well-known methods DSRAN [18], VSRN [19], and SCAN [20], for comparison. The first two methods run on the MS-COCO dataset, and the last runs on our dataset.

Table 2. Evaluation on MS-COCO dataset

Method	p@10	p@50	p@100	p@200
KG	3	12	23	52
DE	5	32	60	136
DE + KG	2	30	55	77
SCAN [20]	7	40	67	143
DSRAN [18]	**8**	45	**77**	163
VSRN [19]	6	37	63	80
Cross-Modal Attention	7	**47**	70	**170**

Table 3. Evaluation on our dataset

Method	p@10	p@50	p@100	p@200
KG	6	31	54	69
DE	8	42	87	159
DE + KG	5	35	68	83
SCAN [20]	7	45	93	172
Cross-Modal Attention	**8**	**47**	**98**	**196**

Since the freestyle query primarily represents a subjective perspective of users when imagining an accident scene, it is hard to have an automatic judgment mechanism. Hence, we asked volunteers to help us evaluate each method's results mentioned above. We ask one person who does not familiar with the topic as a naive user and another with experience as an expert user.

Volunteers handle our system to make 50 different queries. The volunteers are free to use the interactive GUI, introduced in our previous work [21], to add more attributes, topologies, and classes under their cognition of what they imagine their queries. The volunteer gives P@10 score (i.e., precision at top 10 results) for each query at the first round, then records the number of querying loops when the system returns no more new results (i.e., reaches the maximum R-precision). Then, we report each method's P@10, P@50, P@100, and P@200 to form the Table 2 and 3. Our model gives a better result than others, both in the particular incident domain and in broader domains, except DSRAN at P@10 and P@100.

Table 4 denotes ten random queries and their results with the confirmation of volunteers about right or wrong. Images with their red-box boundaries indicate wrong answers. We could see that the portion of correct answers in top K-results is relatively high.

Table 4. Multimodal fine-grained text-image cross-modal attention model: querying samples. Images with red-box boundaries are wrong results

Query	Results
a car with fire smoke pouring out of it	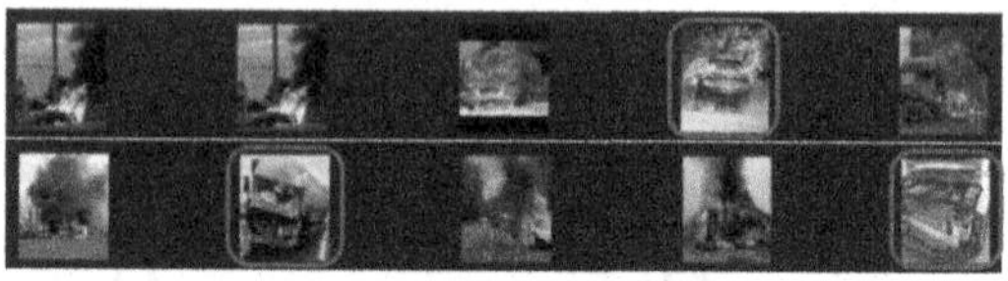
a dump truck is parked on the side of the road	
a dump truck that is on the ground	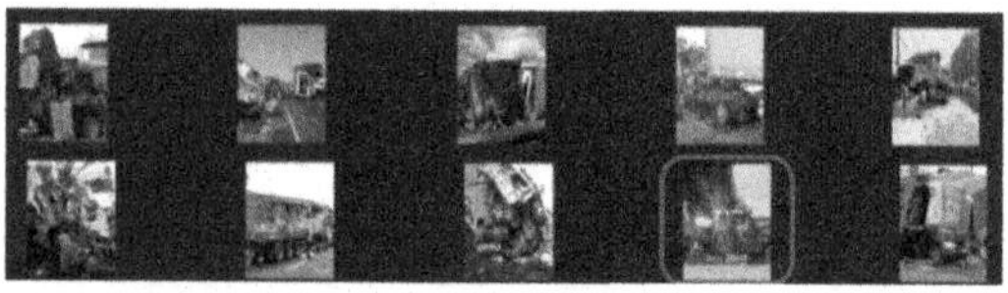
man standing on top of a pile of rocks	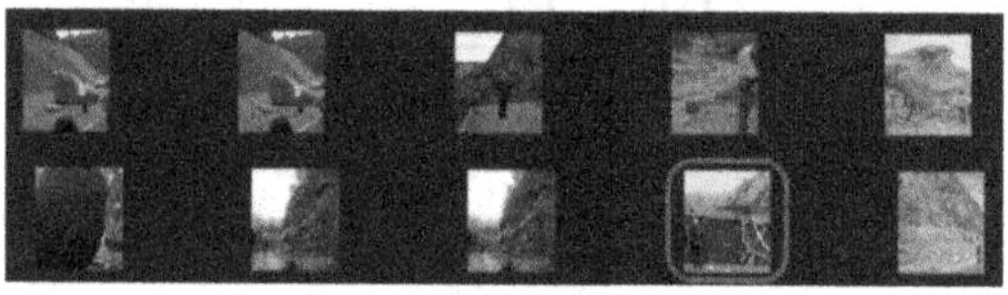
a road that has some trees on it	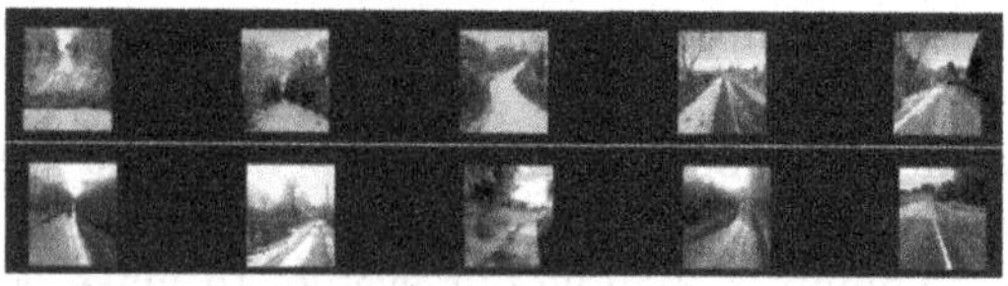
brok en down tree on the side of a road	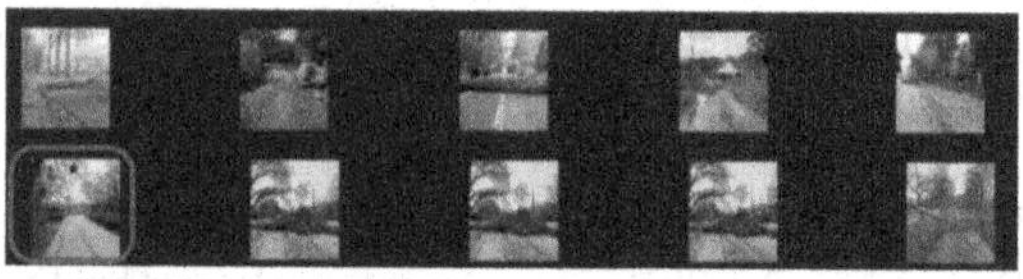
car is driving down a snowy road	

5 Conclusions

This paper introduces a fine-grained text-image incident retrieval using multi-modal cross-modal attention with the FAISS index. The former aim to create a joint representation space where textual and visual features can compensate and assist each other to build the link of salient zones together. We emphasize the specific domain: incident, where the description of the accident scene requires more detail and the interaction among traffic objects (e.g., car, truck, tree, pedestrian, flood) and object's attributes (e.g., color, size, position) are the utmost attention. To our best knowledge, these criteria have not been mentioned in any work. Besides, the FAISS index can support running on a massive database in real-time mode. We evaluate our model on different datasets and models with the judgment of both naive and expert users. The experimental results show our advantages and open a new approach for exploiting and exploring first-view travel-log data and contributing to smart mobility, where safety is the priority.

We will consider the position of textual and visual embedding vectors in the future since, as we mentioned, the traffic object's interaction is one of the incident domain's significant criteria. We also will ask more volunteers to enrich the labels of each incident image to have better cross-link between textual and visual components. We will investigate the decoder with the transformer to get direct relevant images with the query without utilizing the FAISS index. We also want to benefit from pre-train vast models (e.g., Visual Genome) of objects and their topology in an image and downstream them into our dataset, which could bring more accuracy and alleviate the burden of manually labeling.

Acknowledgement. The results of this study are based on collaborative research on "Research and Development of Interactive Visual Lifelog Retrieval Method for Multimedia Sensing" between National Institute of Information and Communications Technology, Japan and University of Science, Vietnam National University - Ho Chi Minh City, Vietnam from April 2020 to March 2022.

References

1. Xu, Y., Liang, X., Dong, X.. Chen, W.: Intelligent Transportation System and Future of Road Safety. In: 2019 IEEE International Conference On Smart Cloud (SmartCloud), pp. 209–214 (2019)
2. Lee, K., Choi, J., Park, J., Lee., S.:D Metadata-driven Dashcam Analysis System. DFRWS APAC, Your Car Is Recording (2021)
3. Kim, J., Park, S., Lee, U.: Dashcam Witness: video sharing motives and privacy concerns across different nations. IEEE Access. **8**, 110425–110437 (2020)
4. Evans, J., Waterson, B., Hamilton, A.: Evolution and Future of Urban Road Incident Detection Algorithms. J. Transport. Eng. Part A: Syst. **146** (2020)
5. Adamová, V.: Dashcam as a device to increase the road safety level. In: International Conference on Innovations in Science and Education (CBU), pp. 1–5 (2020)
6. Bazilinskyy, P., Eisma, Y., Dodou, D., Winter, J.: Risk perception: a study using dashcam videos and participants from different world regions. Traffic Inj. Prev. **21**, 347–353 (2020)

7. Mohandu, A., Kubendiran, M.: Survey on Big Data Techniques in Intelligent Transportation System (ITS). Materials Today: Proceedings (2021)
8. Cao, M., Li, S., Li. T., Nie, L., Zhang, M.: Image-text Retrieval: a survey on recent research and development, https://doi.org/10.48550/arXiv.2203.14713
9. Frolov, S., Hinz, T., Raue, T., Hees, J., Dengel, A.: Adversarial Text-to-Image Synthesis: A Review, https://doi.org/10.48550/arXiv.2101.09983
10. Chollet, F. Xception: Deep Learning with Depthwise Separable Convolutions. CoRR. abs/1610.02357 (2016). http://arxiv.org/abs/1610.02357
11. Turc, I., Chang, M., Lee, K., Toutanova, K.: Well-Read Students Learn Better: The Impact of Student Initialization on Knowledge Distillation. CoRR. abs/1908.08962 (2019). http://arxiv.org/abs/1908.08962
12. Turc, I., Chang, M., Lee, K., Toutanova, K.: Well-Read Students Learn Better: On the Importance of Pre-training Compact Models. ArXiv Preprint ArXiv:1908.08962v2 (2019)
13. Vaswani, A., et al.: Attention Is All You Need. CoRR. abs/1706.03762 (2017). http://arxiv.org/abs/1706.03762
14. Tian, Y., Chen, L.: Cross-modal attention modulates tactile subitizing but not tactile numerosity estimation. Attention, Perception, Psychophys. **80**(5), 1229–1239 (2018). https://doi.org/10.3758/s13414-018-1507-x
15. Wang, M., Xu, X., Yue, Q., Wang, Y.A.: Comprehensive Survey and Experimental Comparison of Graph-Based Approximate Nearest Neighbor Search. CoRR. abs/2101.12631 (2021). https://arxiv.org/abs/2101.12631
16. Guo, R., et al.: Accelerating Large-Scale Inference with Anisotropic Vector Quantization. International Conference On Machine Learning. (2020). https://arxiv.org/abs/1908.10396
17. Johnson, J., Douze, M., Jégou, H.: Billion-scale similarity search with GPUs. IEEE Trans. Big Data **7**, 535–547 (2019)
18. Wen, K., Gu, X., Cheng, Q.: Learning Dual Semantic Relations with Graph Attention for Image-Text Matching. CoRR. abs/2010.11550 (2020). https://arxiv.org/abs/2010.11550
19. Li, K., Zhang, Y., Li, K., Li, Y., Fu, Y.: Visual Semantic Reasoning for Image-Text Matching. CoRR. abs/1909.02701 (2019). http://arxiv.org/abs/1909.02701
20. Lee, K., Chen, X., Hua, G., Hu, H., He, X.: Stacked Cross Attention for Image-Text Matching. CoRR. abs/1803.08024 (2018). http://arxiv.org/abs/1803.08024
21. Dao, M., Pham, D., Nguyen, M., Nguyen, T., Zettu, K.: MM-trafficEvent: An Interactive Incident Retrieval System for First-view Travel-log Data. In: 2021 IEEE International Conference On Big Data (Big Data), pp. 4842–4851 (2021)
22. Haresh, S., Kumar, S., Zia, M.Z., Tran, Q.H.: Towards anomaly detection in dashcam videos. In: IEEE Intelligent Vehicles Symposium (IV), 2020, pp. 1407–1414
23. Yu, F., et al.: Bdd100k: A diverse driving dataset for heterogeneous multitask learning (2020)
24. Levering, A., Tomko, M., Tuia, D., Khoshelham, K.: Detecting unsigned physical road incidents from driver-view images. IEEE Trans. Intell. Veh. **6**(1), 24–33 (2021)
25. Zhao, P., Dao, M.-S., Nguyen, N.-T., Nguyen, T.-B., Dang-Nguyen, D.-T., Gurrin, C.: Overview of mediaeval 2020 insights for wellbeing: Multimodal personal health lifelog data analysis. In: MediaEval (2020)

Textual Concept Expansion with Commonsense Knowledge to Improve Dual-Stream Image-Text Matching

Mingliang Liang(✉), Zhuoran Liu, and Martha Larson(✉)

Radboud University, Nijmegen, Netherlands
{m.liang,z.liu,m.larson}@cs.ru.nl

Abstract. We propose a Textual Concept Expansion (TCE) approach for creating joint textual-visual embeddings. TCE uses a multi-label classifier that takes a caption as input and produces as output a set of concepts that are used to expand, i.e., enrich the caption. TCE addresses the challenge of the limited number of concepts common between an image and its caption by leveraging general knowledge about the world, i.e., commonsense knowledge. Following a recent trend, the commonsense knowledge is acquired by creative use of the training data. We test TCE within a popular dual-stream approach, Consensus-aware Visual-Semantic Embedding (CVSE). This popular approach leverages a graph that encodes the co-occurrence of concepts, which it takes to represent a consensus between the textual and visual modality that captures commonsense knowledge. Experimental results demonstrate an improvement of image-text matching when TCE is used for the expansion of the background collection and the query. Query expansion, not possible in the original CVSE, is particularly helpful. TCE can be extended in the future to make use of data that is similar to the target domain, but is drawn from an additional, external data set.

1 Introduction

Learning joint representations for images and text is the backbone of image-text matching. An important research direction focuses on dual-stream models that map the representations of images and texts into a joint space with the goal of aligning global-level representations [6,7] or local-level representations [9,12,25] of images and text. Recently, a growing amount of research has attempted to exploit commonsense knowledge to enrich visual-textual embeddings [20,24]. A promising approach is to consider commonsense knowledge in the form of concept co-occurrences, i.e., knowledge about the concepts that occur together in the real world. This knowledge is reflected in the co-occurrence of concept-words in text data and objects in image data. The same co-occurrences appear across different modalities, allowing vision and language to inform each other.

D.-T. Dang-Nguyen et al. (Eds.): MMM 2023, LNCS 13833, pp. 421–433, 2023.
https://doi.org/10.1007/978-3-031-27077-2_33

In this paper, we propose Textual Concept Expansion (TCE), an approach for creating joint textual-visual embeddings that automatically expands captions with a set of textual concepts. We build on the Consensus-Aware Visual-Semantic Embedding (CVSE) approach, which is a dual-stream model that exploits commonsense concept co-occurrence knowledge learned from the textual modality to complement textual captions [25]. Our novelty and main contribution are to use a transformer as a multi-label classifier to expand the textual captions in the background collection (T-BC) and textual query set (T-Q). In contrast, the original CVSE is not able to expand textual queries to make them richer and more effective. Also, to our knowledge, previous work on image-text matching has not used a text classifier for this purpose.

Image	Captions	Concepts	Concept Expansion
	a cat is sitting on some jeans and a purse with a dog on the floor in the background	cat, jeans, purse, dog	black, white, pillow, woman, person, sofa
	a black cat laying down on an old purse	black, cat, purse	dog, white, pillow, jeans, woman, person, sofa
	a black and white cat is sitting by a purse	black, white, cat, purse	dog, white, pillow, jeans, woman, sofa
	a cat using a woman's purse as a pillow	cat, woman, purse, pillow	dog, black, white, jeans, woman, sofa
	a cat lounges on the legs of a person laying on a sofa	cat, person, sofa	dog, black, white, pillow, woman, jeans

Fig. 1. An example of information missing in captions associated with an image. The concepts occurring in captions are shown in the "Concepts" column. Observe that no single individual caption contains all possible concepts. Our Textual Concept Expansion (TCE) can expand captions with predicted concepts, illustrated in the "Concept Expansion" column, to enrich the textual representation of an image.

The problem that we address and the solution that we propose are illustrated in Fig. 1. To understand the problem, consider the image and the captions in the "Captions" column and the concepts occurring in these captions in the "Concepts" column. These concepts vary from image to image. For example, observe that there is a dog in the background of the image, but not all captions include the concept "dog". This missing information in the textual modality of the image-text pair may hinder the training of the model and it may also compromise the test performance.

Our solution is built on the insight that joint representations and image-text matching can be improved if we can predict the missing concepts of captions in order to enrich textual representations along each stage of the image-text matching pipeline. Making more effective use of the training data will help us to acquire the necessary information to do so. In Fig. 1, we see that the five captions associated with the image in the training set, which were written by five different people, contain complementary concepts. In our approach, we will merge the five captions associated with each training image for the calculation of concept co-occurrence and we will merge the concepts in the captions to use as ground truth to train a multi-label classifier. In this way, we introduce more commonsense knowledge into the creation of the joint textual-visual embeddings. The classifier is able to take the captions as input and produce a concept expansion which resembles the one shown in Fig. 1.

Currently, research focuses both on *acquiring* and *exploiting* commonsense knowledge to create joint vision-language embeddings that improve image-text matching. In terms of acquisition, commonsense knowledge has often been acquired from outside of the current task data set as external knowledge from a similar domain [14,20]. However, recent work has shown that improvement does not necessarily require external knowledge. Instead, useful concept co-occurrences can be extracted directly from the task data [20,25]. Our work follows the path of using existing training resources more effectively. These approaches can later be extended to incorporate external data, should such data become available. Our work is set apart by careful attention to the status of the background collection. A cross-modal retrieval system should not make use of the connection between images and ground truth captions in the background collection.

In terms of the exploitation of commonsense knowledge, recent work has focused mainly on image-based approaches to enrich image representations and use external resources. Key examples include [8,14,20]. In contrast, our focus is on expanding textual captions and on making more effective use of training data that is already available.

We choose to build on CVSE because it is the leading approach that has demonstrated the usefulness of concept co-occurrence information that is extracted from the textual modality. CVSE is a dual-stream model, but we note that the idea of TCE could be extended to single-stream models in future work. We observe that CVSE suffers from two issues, *incomplete graph* and *doubtful use of background collection captions.* We first propose an adapted version of CVSE, CVSE-a, that addresses the incomplete graph issue. Then, we introduce TCE, which addresses the issue of the limited common concepts between an image and individual captions illustrated in Fig. 1 and also strictly avoids doubtful use of the background collection. Finally, we carry out an extensive experimental analysis of TCE within CVSE-a (i.e., TCE-CVSE-a), also in comparison with other state-of-the-art approaches. In sum, this paper makes the following contributions:

- We use a carefully designed oracle experiment to demonstrate the potential of textual concept expansion to substantially improve the performance of image-text matching.
- We introduce Textual Concept Expansion (TCE) and show that it improves the performance of bi-directional cross-modal retrieval with respect to the original CVSE and other state-of the-art approaches.[1]

[1] https://github.com/Anastasiais-ml/TCE-CVSE.

- We demonstrate the importance of textual query expansion and also of using expansion at all states of the image-text matching pipeline, without resorting to doubtful use of the background collection.

The remainder of the paper is structured as follows. In Sect. 2, we cover the relevant literature. Then, in Sect. 3, we explain CVSE and our adapted model CVSE-a, before presenting our Textual Concept Expansion (TCE) approach. Sect. 4 presents results and Sect. 5 presents a conclusion and an outlook.

2 Related Work

2.1 Image-text Matching

Dual-stream models for image-text matching learn representations of images and text in a joint representation space using two encoders. Frome *et al.* proposed DeViSE, the first dual-stream model that maps image embeddings to pre-trained text embeddings by a CNN model with a projection layer [7]. Following this paradigm, subsequent research has improved the learning of joint textual-visual representations by using different strategies, including emphasizing hard negative samples [6], employing multiple different representations of both visual and textual modalities [21], or extracting image regions to align words for stacked cross model attention [12]. Later, pre-trained single-stream models with cross-attention (i.e., cross-attention models) were developed [11,13,14,17], where the key component is a transformer that allows multi-modal data interactions directly [11,13,14]. Compared to dual-stream models, the cross-attention model significantly improves the performance of the visual-language representation model. However, the working mechanism of the transformer requires each sample pair to be calculated through a feed-forward pass during test, which adds a considerable computation overhead [11,13,14,16]. Additionally, the heavy computation takes extra time, which compromises the matching performance in a real-world deployment. Our paper only considers the dual-stream model for the sake of practicality, with a focus on improving the representation learning.

2.2 Dual-stream Models with Commonsense Knowledge

Commonsense knowledge is known to benefit dual-stream models [20,25]. In some cases, this knowledge is extracted from an external dataset [24] or is provided by an external pre-trained model [4,14,20]. As an alternative to external knowledge, the training data itself can also be effectively used to provide commonsense knowledge, e.g., concept vocabulary or concept-co-occurrence matrix, that assists image-text matching. Graph conventional networks (GCNs) are often used to model concept-co-occurrence and have proven to be a good way to represent relationships between concepts [3,20,25]. ML-GCN [3] learns object label embeddings with a GCN that can model concept-co-occurrence to improve multi-label classification. CVSE [25] aims to improve the performance of image-text matching by transferring the module of label embedding in ML-GCN to the image-text matching task in order to learn from the cross-modal relationships within data, which is referred to as consensus knowledge. Different from previous research, in this paper, we predict concept labels beyond the concepts that are literally included in the caption. Specifically, we train a text-based multi-label classifier on the training set. Our concept expansion method is trained by making more effective use of training data, in which each caption is associated with concept labels extracted from all relevant captions of an image. During the test, we use the trained classifier to expand captions in the background collection (T-BC) and query set (T-Q). The trained multi-label classifier can take captions as input and predict concepts not initially included in the input captions.

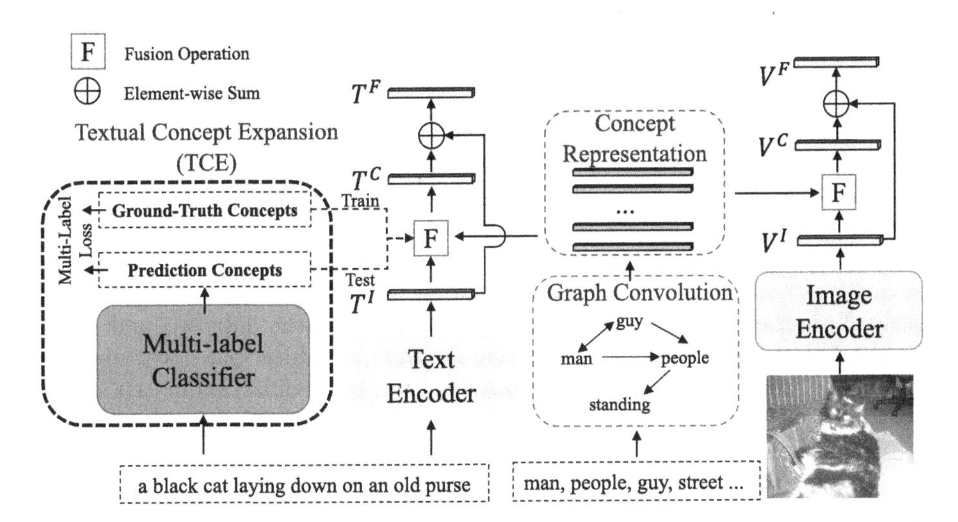

Fig. 2. Textual Concept Expansion CVSE (TCE-CVSE) framework for image-text matching. The multi-label classifier is trained using the captions in the training set. The ground truth concept labels are derived from the sister captions associated with the same image, which promotes the alignment of text and image representations. During test, the TCE module is used to expand the textual queries and the textual captions in the background collection. It predicts concepts that may co-occur in the image. The image-text matching score is the cosine distance between the text representation (T^F) and the image representation (V^F).

3 Method

In this section, we first present the technical details of CVSE and discuss two issues with CVSE. Then we go on to present our own Textual Concept Expansion (TCE) approach. Figure 2 provides the picture of the combined pipeline. On the right is the original CVSE architecture and on the left, enclosed in the dashed line, is our TCE approach, which extends this architecture by adding a means to expand training captions, background collection, and textual queries while avoiding doubtful use of background collection captions.

3.1 Consensus-Aware Visual-Semantic Embedding (CVSE)

Consensus-Aware Visual-Semantic Embedding (CVSE) [25] consists of three modules, which we describe here in turn:

Concept Instantiation: CVSE first selects the top q most frequent concepts of three types (i.e., Object, Motion and Property) from the whole data set as concepts vocabulary. It restricts the ratio of different types of concepts to Object:Motion:Property = 7:2:1 [25]. In [25], the frequency of the concepts is calculated on the whole dataset. This is unreasonable because the test set is unknown when we train our model. To avoid concepts of the test set leaking to the training stage, we select the concepts only on the training set in all our experiments.

Concept Correlation Graph Building: A correlation graph over the selected concepts is built on the basis of the conditional probability matrix $P(C_i|C_j)$ where the probability is the co-occurrence probability of concept C_i and concept C_j appearing in the same caption,

$$P_{i,j} = M_i/N_j \tag{1}$$

where N_j is the number of times concept C_j occurs in the data set, and M_i is the number of times concept Ci co-occurs with C_j. To filter out noisy edges, a threshold τ is used to binarize the matrix $p_{i,j}$ to obtain binarized matrix A.

$$A_{i,j} = \begin{cases} 0 & \text{if } P_{i,j} < \tau \\ 1 & \text{if } P_{i,j} \geq \tau \end{cases} \tag{2}$$

To alleviate the problem of over-smoothing in the graph, w is used to re-weight the weights a concept itself and other correlated concepts in the matrix A'.

$$A'_{i,j} = \begin{cases} (w/\sum_{i=1,i\neq j}^{C} A_{i,j}) & \text{if } i \neq j \\ 1 - w & \text{if } i = j \end{cases} \tag{3}$$

More details on the correlational matrix can be found in [3,25].

Incomplete Graph: Note that CVSE builds the correlation graph by only exploiting the concepts that co-occur within one caption, which contradicts the CVSE textual instantiation that leverage all five captions. Especially, since the MS COCO dataset [15] has short captions, most of the concepts in a caption only co-occur with one or two other concepts. Consequently, the concept correlation matrix is sparse after filtering noisy edges. To address this issue, we propose to use all relevant captions associated with one image to calculate the correlation graph, which includes more related concepts. We also show that a correlation graph that is built on more captions can improve the image-text matching performance (see Sect. 4). We name the adjusted version of CVSE that leverages all relevant captions associated with one image as the *CVSE-adapted* (*CVSE-a*).

Consensus-Aware Representation Learning: The pipeline of CVSE is shown in the right of Fig. 2. First, taking the concept correlation matrix $A'_{i,j}$ as input, the GCN outputs a concept representation $Z \in \mathbb{R}^{q \times d}$, where d is the dimension of concept embedding Z_i. For the image stream, a pre-trained model Faster-RCNN [1,18] is used to extract representations of image regions. For the text stream, a bi-directional GRU [19] be used to encode sentences to embedding. CVSE incorporates the instance-level image and text representation (T^I, V^I) by a self-attention layer [23,25]. Then, the instance-level representations query the concept representations to obtain consensus-level representation (T^C, V^C):

$$T^C = \sum_{i}^{q} (\alpha L_i^t + (1-\alpha) S_i^T) \cdot Z_i \tag{4}$$

$$V^C = \sum_{i}^{q} S_i^V \cdot Z_i \tag{5}$$

where $S_i^T \in \mathbb{R}^{q \times 1}$ and $S_i^V \in \mathbb{R}^{q \times 1}$ are correlation scores between, respectively, V^I and T^I and the concept representation Z_i. The concept label $L^t \in \mathbb{R}^{q \times 1}$ is the intersection between the words in the caption and the words in the concept vocabulary. If a word occurring in the caption appears in the concept vocabulary, the concept word will be labeled 1 in the concept label vector, and otherwise it will be labeled 0. α is used to control the proportional contribution of the correlation scores (S^I) and concept labels (L^t) to generate the textual consensus-level representation T^C.

The consensus-level representations (T^C, V^C) are then fused into (T^I, V^I) to obtain the final fused representation (T^F, V^F):

$$T^F = \beta T^I + (1-\beta) T^C \tag{6}$$

$$V^F = \beta V^I + (1-\beta) V^C \tag{7}$$

where β is the weight between the instance-level representations, T^I and V^I, and the consensus-level representations, T^C and V^C. As shown in Eq. 4, the concept labels help the model re-weight the textual representation after querying the concept embedding in the text stream. A Kullback-Leibler (KL) divergence D_{KL} between the correlation scores is also incorporated to adjust the correlation scores. Taken together, a triplet loss [7] is formulated as:

$$L = \lambda_1 L_{rank}(V^F, T^F) + \lambda_2 L_{rank}(V^I, T^I) + \lambda_1 L_{rank}(V^C, T^C) + \lambda_4 D_{KL} \tag{8}$$

During test, we calculate the similarity score between the text representation (T^F) and image representation (V^F) with the cosine distance.

Doubtful Use of Background Collection Captions: During the training of CVSE, concept labels are extracted from all captions associated with the image. During test, CVSE uses text-to-text similarity calculated on background collection captions to expand both background collection and textual queries. We call this use of the background collection "doubtful" for the following reason.

Following the standard cross-modal setup [6,21], the calculation of such text-to-text similarities should be excluded because if the text ground truth (captions) of the background collection were available at test time during text-to-image retrieval, the retrieval would not be truly crossing modalities.

3.2 Textual Concept Expansion (TCE)

We propose TCE to tackle CVSE's doubtful use of background collection issue and provide effective textual conception expansion for both background collection (T-BC) and query (T-Q). TCE is a multi-label classifier that takes a caption as input and produces a set of relevant concepts. Different from [20], the training of TCE does not need external knowledge but rather incorporates commonsense knowledge by making more effective use of the training data.

TCE predicts a set of concept labels for an input caption. TCE consists of a pre-trained BERT [5] model and a multi-label classification layer, and a multi-label classification loss is adopted:

$$L_{ML} = \sum_{c=1}^{C} y^c log(\sigma(\hat{y}^c)) + (1 - y^c) log(1 - \sigma(\hat{y}^c)) \tag{9}$$

where $\sigma(\cdot)$ is the sigmoid function, $\hat{y}^c$ and y^c are the predicted score and ground truth label. The training data of TCE are captions labeled $\{y_1, y_2, \ldots, y_i\}$ with multiple concept labels $y_i = \{c_1, c_2, \ldots, c_i\}$, where c_i is the concept appearing in the caption. Comparing with CVSE, we select the top q most frequent concepts of the three concept types without restricting the ratio to constitute concept vocabulary. Specifically, each image in MS COCO is associated with five captions, and then each caption can be labeled with concepts from all five associated captions. During test, TCE, which has already been trained, predicts the concept labels of an input caption to carry out concept expansion. The predicted concept labels (L^t) are used to enrich the text representation for image-text matching. In the next section, we evaluate the performance of TCE building on CVSE-a.

4 Experiments

In this section, we introduce the experimental dataset, the implementation details of TCE-CVSE-a, and a description of the selection of the vocabulary of concepts. Then, we present oracle experiments that provide evidence of the potential of TCE to improve image-text matching, especially text-to-image retrieval. The next experiment provides a comparison of TCE-CVSE-a with the original CVSE and other state-of-the-art approaches. Finally, we show that applying TCE in different parts of the image-text matching pipeline can influence performance differently, and TCE works the best when applied in all parts.

4.1 Datasets and Implementation Details

Datasets and Evaluation Metrics. We evaluate our approach on the MS COCO [15] dataset. Following [10], the dataset is split into 113287 training images, 1000 validation images, and 5000 test images, and each image is annotated with 5 captions. The validation set is used for model selection during training. All 5000 test images are split into 5 folds of 1000 images, and we report the average performance of the 5 folds. For image-to-text matching, for each fold, we use every image as a query in turn, and query the background collection of the captions associated with the 1000 images in that fold. For text-to-matching, for each fold, we use all five captions associated with each image as a query in turn, and query the background collection of the 1000 images. For both image retrieval and text retrieval, we measure the performance by Recall at K (R@K) where K = 1, 5, and 10 and their average value (mR). To implement TCE-CVSE-a, we follow the setup of the original CVSE as described in the [25] to set $\alpha = 0.35$ in Eq. (4) and λ_1, λ_2, λ_3, $\lambda_4 = 3, 5, 1, 2$ in Eq. (8), but we remove Confidence Scaling (CS) function and set $\tau = 0.5$ in Eq. (2), $\beta = 0.9$ in Eqs. (7) and (6) and follow ML-GCN [3] to set $w = 0.2$ in Eq. (3) . More details on the setup are available in the original CVSE paper [25].

Implementation Details of TCE. For the multi-label classifier of TCE, we build the training set on the MS COCO training split, where each caption is labeled with the union of the set of concept labels that are extracted from all five relevant captions. Improving upon [25], the concept vocabulary is the top 300 concepts of types Object, Motion, and Property in the MS COCO training split. To train the multi-label classifier, an AdamW optimizer with a learning rate of 2e−5 and 5% of training steps for warm-up. We set the confidence score to 0.2 based on the validation performance.

Table 1. Oracle experimental results on MS COCO 1K test. We used the ground truth (GT) concept labels to expand the captions to test the performance of the model.

Approach	CE			Text Retrieval			Image Retrieval			mR
	Train	T-BC	T-Q	R@1	R@5	R@10	R@1	R@5	R@10	
CVSE-a (Oracle)	GT	GT	GT	77.4	94.5	97.5	63.0	91.0	96.2	86.6
	✗	GT	GT	73.9	94.0	97.5	56.8	86.7	93.8	83.8
	GT	GT	✗	77.4	94.5	97.5	56.6	86.1	93.3	84.2

Table 2. Evaluation of image-text matching approaches on MS COCO 1K test. In the concept expansion (CE) column, Train is training stage, T-BC is the textual background collection, and T-Q is textual query. TCE is used to expand concepts both T-BC and T-Q. We compare TCE with CVSE's use of KNN (including the background collection) for textual concept expansion.

Approach	CE			Text Retrieval			Image Retrieval			mR
	Train	T-BC	T-Q	R@1	R@5	R@10	R@1	R@5	R@10	
VSE++ [6]	✗	✗	✗	64.7	–	95.9	52.0	–	92.0	–
PVSE [21]	✗	✗	✗	69.2	91.6	96.6	55.2	86.5	93.7	82.1
SCAN [12]	✗	✗	✗	72.7	94.8	98.4	58.8	88.4	94.8	83.6
CVSE [25]	GT	KNN	✗	74.1	94.1	97.4	55.6	86.1	93.4	83.5
TCE-CVSE-a (Ours)	GT	TCE	TCE	74.6	94.4	98.1	57.6	87.4	94.0	84.4

Concept Vocabulary Selection. We used NLTK tool [2] and Stanford POS tagger [22] to tag words. To build the concept vocabulary, we select the top-q frequency words appearing in the vocabulary among three types of words (Object, Motion and Property) in the training dataset rather than restrict the ratio of different types of concepts. All our experiments use this strategy to select concepts. We also reproduce the CVSE's strategy of selecting concepts in the MS COCO [15] dataset. Note that our concept vocabulary has only a 74.6% overlapping with the original CVSE paper [25].

4.2 Experimental Results

Textual Concept Expansion by Ground Truth. We first show the potential of textual concept expansion through two oracle experiments, in which we deliberately pass the ground truth concept labels to the background collection and the query. The ground truth concept labels for each image-text pair are extracted from the five associated captions and then used for training and testing. Comparing row 1 and 2 in Table 1, we observe that if we do not expand with concepts during the training stage, expanding the background collection and the query is ineffective. The mR increases substantially (from 83.8% to 86.6%) when we expand the training data in addition to the background collection and the query. Comparing row 1 and 3 in Table 1, we observe the importance of query expansion. The performance of image retrieval substantially decreases without query expansion, specifically, R@1 decreases from 63.0% to 56.6%. Next, we go on to examine how the potential of concept expansion just demonstrated translates into improvement in a standard experimental setting.

Table 3. Evaluation of concept expansion in different parts of CVSE on MS COCO 1K test.

Approach	CE			Text Retrieval			Image Retrieval			mR
	Train	T-BC	T-Q	R@1	R@5	R@10	R@1	R@5	R@10	
CVSE	✗	✗	✗	73.9	94.1	97.5	56.8	86.8	93.8	83.8
CVSE	✗	KNN	✗	73.9	94.0	97.5	56.8	86.8	93.8	83.8
CVSE	✗	✗	KNN	73.9	94.1	97.5	56.8	86.8	93.8	83.8
CVSE-a	GT	✗	✗	72.5	94.1	97.4	56.6	86.1	93.3	83.3
CVSE-a	GT	KNN	✗	74.5	94.2	97.5	56.6	86.1	93.3	83.7
CVSE-a*	GT	KNN	KNN	74.5	94.2	97.5	57.3	86.8	93.7	84.0
TCE-CVSE-a (Ours)	GT	TCE	TCE	74.6	94.4	98.1	57.6	87.4	94.0	84.4

Textual Concept Expansion. We show that TCE is a more effective approach than CVSE, due to its expansion of the background collection and textual queries. As shown in the last row of Table 2, TCE-CVSE-a outperforms CVSE and other approaches for both image and text retrieval. Results for VSE++, PVSE, and SCAN are from the literature. The added possibility of text query expansion, makes TCE-CVSE-a more effective than CVSE, especially for image retrieval. Informal inspection of results has revealed that in some cases TCE-CVSE-a helps increase the ranking of a relevant image that CVSE neglects.

Concept Expansion in Different Parts. To further investigate the contributions of concept expansion in different parts of the image-text matching pipeline, we investigate different combinations of expansion for training data, background collection, and queries. The experimental results in Table 3 demonstrate that TCE is effective when concepts expansion is applied training data, background collection and queries (row 7). In rows 1-3, we see the performance of the original CVSE with KNN expansion (using the background collection) without expanding the training data. We conjecture that the poor performance is due to training and testing inconsistency. In rows 4-6, we see the performance when training data expansion is added. Note that the best performance is achieved in row 6 by an illicit oracle version (CVSE-a*), where the ground truth information is used for query expansion. In row 7, we see that TCE-CVSE-a can fully exploit the benefits of textual concept expansion, improving over CVSE with the original doubtful use of the background collection and even with illicit use of query expansion. Finally, we note that by, comparing row 4 in Table 2 and row 5 in Table 3, we can observe a slight improvement in mR by building the correlation graph using concepts occurring in all five captions of an image.

5 Conclusion and Outlook

In this paper, we propose a textual concept expansion (TCE) approach that uses a multi-label classifier to improve the performance of image-text matching. The multi-label classifier expands captions by predicting commonly co-occurring concepts with the goal of improving representation learning. We demonstrate that concepts that are common between multiple captions associated with an image can be exploited to improve textual representations in the joint representation space. Experimental results demonstrate that our proposed TCE can effectively improve the performance of image-text matching. Currently, TCE only leverages the training data in the image-text matching task. In the future, TCE can be further improved by leveraging an additional external data set. A pre-trained TCE can be used as external knowledge for image-matching tasks in a similar domain. Additionally, future work on integrating TCE into single-stream cross-attention models is promising.

References

1. Anderson, P., et al.: Bottom-up and top-down attention for image captioning and visual question answering. In: IEEE Conference on Computer Vision and Pattern Recognition, pp. 6077–6086 (2018)
2. Bird, S., Klein, E., Loper, E.: Natural language processing with Python: analyzing text with the natural language toolkit. " O'Reilly Media, Inc". (2009)
3. Chen, Z., Wei, X., Wang, P., Guo, Y.: Multi-label image recognition with graph convolutional networks. In: IEEE Conference on Computer Vision and Pattern Recognition, pp. 5177–5186 (2019)
4. Cheng, M., et al.: Vista: vision and scene text aggregation for cross-modal retrieval. In: IEEE Conference on Computer Vision and Pattern Recognition, pp. 5184–5193 (2022)
5. Devlin, J., Chang, M.W., Lee, K., Toutanova, K.: BERT: Pre-training of deep bidirectional transformers for language understanding. In: Human Language Technology Conference of the North American Chapter of the Association for Computational Linguistics, pp. 4171–4186 (2019)
6. Faghri, F., Fleet, D., Kiros, J., Fidler, S.: VSE++: improving visual-semantic embeddings with hard negatives. In: British Machine Vision Conference, pp. 1–12 (2018)
7. Frome, A., et al.: Devise: a deep visual-semantic embedding model. In: Advances in Neural Information Processing Systems, pp. 2121–2129 (2013)
8. Huang, Y., Wu, Q., Song, C., Wang, L.: Learning semantic concepts and order for image and sentence matching. In: IEEE Conference on Computer Vision and Pattern Recognition, pp. 6163–6171 (2018)
9. Karpathy, A., Joulin, A., Li, F.: Deep fragment embeddings for bidirectional image sentence mapping. In: Advances in Neural Information Processing Systems, pp. 1889–1897 (2014)
10. Karpathy, A., Li, F.: Deep visual-semantic alignments for generating image descriptions. In: IEEE Conference on Computer Vision and Pattern Recognition, pp. 3128–3137 (2015)

11. Kim, W., Son, B., Kim, I.: ViLT: vision-and-language transformer without convolution or region supervision. In: International Conference on Machine Learning, pp. 5583–5594 (2021)
12. Lee, K.-H., Chen, X., Hua, G., Hu, H., He, X.: Stacked cross attention for image-text matching. In: Ferrari, V., Hebert, M., Sminchisescu, C., Weiss, Y. (eds.) ECCV 2018. LNCS, vol. 11208, pp. 212–228. Springer, Cham (2018). https://doi.org/10.1007/978-3-030-01225-0_13
13. Li, L.H., Yatskar, M., Yin, D., Hsieh, C.J., Chang, K.W.: VisualBERT: a simple and performant baseline for vision and language. arXiv:1908.03557 (2019)
14. Li, X., et al.: OSCAR: object-semantics aligned pre-training for vision-language tasks. In: Vedaldi, A., Bischof, H., Brox, T., Frahm, J.-M. (eds.) ECCV 2020. LNCS, vol. 12375, pp. 121–137. Springer, Cham (2020). https://doi.org/10.1007/978-3-030-58577-8_8
15. Lin, T.-Y., et al.: Microsoft COCO: common objects in context. In: Fleet, D., Pajdla, T., Schiele, B., Tuytelaars, T. (eds.) ECCV 2014. LNCS, vol. 8693, pp. 740–755. Springer, Cham (2014). https://doi.org/10.1007/978-3-319-10602-1_48
16. Lu, J., Batra, D., Parikh, D., Lee, S.: VilBERT: pretraining task-agnostic visiolinguistic representations for vision-and-language tasks. In: Advances in Neural Information Processing Systems, pp. 13–23 (2019)
17. Radford, A., et al.: Learning transferable visual models from natural language supervision. In: International Conference on Machine Learning, pp. 8748–8763 (2021)
18. Ren, S., He, K., Girshick, R.B., Sun, J.: Faster R-CNN: towards real-time object detection with region proposal networks. In: Advances in Neural Information Processing Systems, pp. 91–99 (2015)
19. Schuster, M., Paliwal, K.K.: Bidirectional recurrent neural networks. IEEE Trans. Signal Process. **45**(11), 2673–2681 (1997)
20. Shi, B., Ji, L., Lu, P., Niu, Z., Duan, N.: Knowledge aware semantic concept expansion for image-text matching. In: International Joint Conference on Artificial Intelligence, pp. 5182–5189 (2019)
21. Song, Y., Soleymani, M.: Polysemous visual-semantic embedding for cross-modal retrieval. In: IEEE Conference on Computer Vision and Pattern Recognition, pp. 1979–1988 (2019)
22. Toutanova, K., Klein, D., Manning, C.D., Singer, Y.: Feature-rich part-of-speech tagging with a cyclic dependency network. In: Human Language Technology Conference of the North American Chapter of the Association for Computational Linguistics, pp. 252–259 (2003)
23. Vaswani, A., et al.: Attention is all you need. In: Advances in Neural Information Processing Systems, pp. 5998–6008 (2017)
24. Vo, D.M., Chen, H., Sugimoto, A., Nakayama, H.: NOC-REK: novel object captioning with retrieved vocabulary from external knowledge. In: IEEE Conference on Computer Vision and Pattern Recognition, pp. 18000–18008 (2022)
25. Wang, H., Zhang, Y., Ji, Z., Pang, Y., Ma, L.: Consensus-aware visual-semantic embedding for image-text matching. In: Vedaldi, A., Bischof, H., Brox, T., Frahm, J.-M. (eds.) ECCV 2020. LNCS, vol. 12369, pp. 18–34. Springer, Cham (2020). https://doi.org/10.1007/978-3-030-58586-0_2

Generation of Synthetic Tabular Healthcare Data Using Generative Adversarial Networks

Alireza Hossein Zadeh Nik[1,2], Michael A. Riegler[1,3(✉)], Pål Halvorsen[1,4], and Andrea M. Storås[1,4]

[1] SimulaMet, Oslo, Norway
michael@simula.no
[2] University of Stavanger, Stavanger, Norway
[3] University of Tromsø, Tromsø, Norway
[4] OsloMet, Oslo, Norway

Abstract. High-quality tabular data is a crucial requirement for developing data-driven applications, especially healthcare-related ones, because most of the data nowadays collected in this context is in tabular form. However, strict data protection laws complicates the access to medical datasets. Thus, synthetic data has become an ideal alternative for data scientists and healthcare professionals to circumvent such hurdles. Although many healthcare institutions still use the classical de-identification and anonymization techniques for generating synthetic data, deep learning-based generative models such as generative adversarial networks (GANs) have shown a remarkable performance in generating tabular datasets with complex structures. This paper examines the GANs' potential and applicability within the healthcare industry, which often faces serious challenges with insufficient training data and patient records sensitivity. We investigate several state-of-the-art GAN-based models proposed for tabular synthetic data generation. Healthcare datasets with different sizes, numbers of variables, column data types, feature distributions, and inter-variable correlations are examined. Moreover, a comprehensive evaluation framework is defined to evaluate the quality of the synthetic records and the viability of each model in preserving the patients' privacy. The results indicate that the proposed models can generate synthetic datasets that maintain the statistical characteristics, model compatibility and privacy of the original data. Moreover, synthetic tabular healthcare datasets can be a viable option in many data-driven applications. However, there is still room for further improvements in designing a perfect architecture for generating synthetic tabular data.

Keywords: Synthetic data generation · Deep learning · Medical data

D.-T. Dang-Nguyen et al. (Eds.): MMM 2023, LNCS 13833, pp. 434–446, 2023.
https://doi.org/10.1007/978-3-031-27077-2_34

1 Introduction

The use of machine learning (ML) in medicine has shown promising results to solve different tasks such as automatic detection of gastrointestinal diseases [1], radiology applications [2] and mental health [3]. However, most ML models are known to be 'data-hungry', meaning that they should be developed on a large amount of data in order to perform well. In the healthcare domain, access to large datasets can be challenging because of privacy protection regulations and lack of data, e.g., due to rare or neglected diseases [4,5]. Consequently, synthetic data generation is considered an attractive alternative in order to create vast amounts of data [6,7]. Synthetic data generation aims to synthesize new data through automated processes that preserve the underlying structure and statistical properties of the original sensitive data to prevent people's privacy from being compromised. In the medical field, synthetic health data can help medical practitioners to share data without any privacy violations and use the synthetic data in addition to the original health data itself [8]. Most of the time, synthetic data generation is focused on imaging data. Nevertheless, much of the healthcare data, including electronic healthcare record (EHR)s, are collected in tabular form or as time-series data (e.g., from biomedical sensors).

While several data generation methods have been proposed, e.g., variational auto-encoders [9] and probabilistic Bayesian networks [10], generative adversarial network (GAN)s have shown excellent results. A GAN consists of two competing models referred to as the generator and the discriminator [11]. The generator tries to learn how to generate synthetic data that resembles the original data. The discriminator, on the other hand, tries to distinguish synthetic samples generated by the generator from original samples. During training of the GAN, the generator and the discriminator compete against each other, which ideally leads to improved performance for both models [11]. After training, the generator can be applied to generate synthetic data. Although synthetic data generation has become very popular, the real effect of the generated data is not well understood and researched yet.

Consequently, this paper focuses on generation and evaluation of synthetic tabular healthcare data. We train different types of GANs developed for tabular data generation on healthcare datasets of different sizes and with different numbers of variables, column data types, feature distributions and inter-variable correlations. After training, the resulting GANs are used to generate synthetic tabular data. Moreover, a comprehensive evaluation framework is defined to evaluate the quality of the synthetic records and the viability of each model in preserving the patients' privacy. We evaluate the strengths and weaknesses of each model based on statistical similarity metrics, ML-based evaluation scores, and distance-based privacy metrics.

The rest of this paper is organized as follows: Sect. 2 presents related work. The method and data is outlined in Sect. 3, and the results are included in Sect. 4. Section 5 provides a discussion of the findings, and conclusions are drawn in Sect. 6.

2 Background and Related Work

The development of deep learning-based tabular synthetic data generative models has been an active research area for the scientific community in recent years. Specifically, a plethora of publications has proposed GAN-based generative models for synthesizing tabular data. MedGAN, developed by Choi et al. [12], is one of the first architectures designed to generate discrete aggregated healthcare patient records. The authors proposed using a pre-trained auto-encoder to circumvent the problems of generating discrete values in GANs. While the generator learns the continuous latent codes in the training process, the pre-trained decoder, placed between the generator and discriminator, translates the generator's output to the original data format and passes it to the discriminator. TableGAN is one of the first GAN-based models developed to simultaneously generate tabular datasets containing both numerical and categorical columns [13]. The generator and discriminator in this tabular synthesizer are adopted based on deep convolutional neural networks to capture inter-variable dependencies between columns. Moreover, an auxiliary classifier is incorporated into the training process to increase the semantic integrity of the generated samples. Lei Xu et al. introduced tabular GAN (TGAN) [14] and conditional tabular GAN (CTGAN) [15] to create high-quality tabular datasets of different data types. In TGAN the generator is a recurrent neural network with long-short-term memory (LSTM), while the discriminator is a multilayer perceptron (MLP). On the other hand, CTGAN uses a conditional generator and a training-by-sampling technique to tackle the challenge of generating imbalanced data. Both CTGAN and TGAN models use a mode-specific normalization technique to deal with the complexity of generating multi-modal numerical columns. However, in contrast to the TGAN architecture, the generator in CTGAN is a fully connected neural network. Zhao et al. [16] adopted the core features of CTGAN and TableGAN models to handle the highly imbalanced categorical features and to improve generating skewed multi-modal and long-tailed continuous columns. They proposed conditional TableGAN (CTABGAN) to model tabular datasets of mixed types, including categorical and continuous features. Not only does CTABGAN use a novel conditional generator, but it also uses classification, information and generator losses in the training process.

Recently, the research community has focused on protecting the generative models against malicious attacks compromising the privacy and integrity of sensitive medical information. Several research papers, such as [17–20], proposed using differential privacy in the generation process to circumvent this. However, in complex use cases, it has been demonstrated that the quality of the synthetic records would decrease significantly when the noise is added in the generation process to ensure differential privacy constraints. For elaboration on GANs for tabular healthcare data generation, the interested reader is referred to [21].

Although many tabular synthesizers are proposed for healthcare generation tasks, most are designed for specific medical applications. For instance, many papers investigating the tabular data generation in the healthcare domain use the MIMIC III clinical database [22] to exclusively synthesize patients' ICD-9 codes (diagnostic codes). However, this paper intends to study the strengths and weaknesses of the synthetic generative models in the healthcare domain that are not application-specific. In other words, we investigate the GAN-based models' capabilities of generating tabular healthcare datasets that are both representative to most medical applications and contain various data types.

3 Data and Method

In order to explore how different properties of tabular data are captured by the generative models, the models are trained on four different tabular data sets from the healthcare domain. The datasets are of various sizes and include different data types and distributions. The four datasets are the Epileptic Seizure, Thyroid and Diabetes datasets, as well as selected tables from the MIMIC III data. An overview of the datasets is provided in Table 1.

Table 1. The datasets applied in this work for synthetic tabular data generation. All target columns are binary categorical.

Dataset	#Train	#Test	Target Name	Explanation of Target Name (yes/no)
Epileptic Seizure	9,000	2,500	Y	Epileptic seizure
Thyroid	7,100	2,000	BinaryClass	Thyroid disease
Diabetes	70,000	19,000	Readmitted	Hospital readmission
MIMIC III	31,900	9,000	Hospital_Expire_Flag	Survived

The Epileptic Seizure dataset, originally from [23], includes electroencephalogram (EEG) recordings from 500 patients. A preprocessed version of the dataset available at Kaggle is applied [24]. Converting the EEG signals from time series to a tabular format results in a dataset consisting of $11,500$ rows and 178 columns. The column values are discrete and range from -1850 to 1750.

The Thyroid dataset from the UCI Machine Learning Repository [25] includes $9,172$ rows representing patient records, and 20 columns with information about the patients, including whether they are diagnosed with thyroid disease or not. The column indicating whether the patient is on antithyroid therapy or not is highly imbalanced, with the majority of patients not being on antithyroid therapy. Missing values are imputed using the SimpleImputer from the scikit-learn library [26].

The Diabetes dataset [27] includes medical characteristics about diabetic patients that were admitted to hospitals in the United States. The columns representing diagnostic information (diag_1 - 3) exhibits multiple modes. The dataset also contains columns with discrete values with small and large value-ranges. Following preprocessing, the dataset has $89,053$ rows (inpatient encounters) and 29 columns.

The MIMIC III dataset [22] contains 26 tables including information such as diagnoses, prescribed medications and laboratory measurements for patients admitted to critical care units. For this work, we join seven of these tables in order to create a dataset containing $40,895$ rows and 14 columns. Each row represents a patient and includes demographic and medical details during a stay at the intensive care unit at the hospital. The column representing the length of patient's stay is highly skewed to the right with some outliers that stayed in the hospital for a very long time. Patients above 90 years old did not have their true age registered. For these patients, we randomly assigned them ages between 90 and 100 years, with decreasing probability with increasing age.

Four different GAN architectures are trained for generation of tabular healthcare data: TGAN [14], CTGAN [15], CTABGAN [16] and Wasserstein GAN with gradient penalty (WGAN-GP) [28]. The WGAN-GP is used as a baseline model because this is a general GAN that has proven robust to the mode collapse problem [28]. Consequently, we want to explore whether the GAN architectures developed specifically for tabular data generation outperform the more general WGAN-GP. The code used to run all experiments including details regarding the model architectures and hyperparameters is available on GitHub[1].

All the experiments are conducted using Python 3.8 as the primary programming language. The proposed tabular data generation models are implemented using Tensorflow [29], except the CTABGAN model, which is built with Pytorch [30]. Furthermore, the experiments are conducted on the University of Stavanger's GPU cluster (Gorina6) on an Nvidia Tesla V100 machine equipped with 32 GB of memory. However, the evaluations and comparisons are conducted on a Desktop PC with specifications of AMD Ryzen 5 5600G with 8 GB of memory.

The models are evaluated based on their ability to generate realistic synthetic data and protect the privacy of the individuals. Two different methods are applied to evaluate the abilities of the GANs to generate realistic samples: Statistical resemblance and a ML model-based approach. For statistical resemblance, we include the marginal column distributions for the original and synthetic data. Ideally, the distributions should be similar. Regarding the ML model-based approach, ML models are trained to classify samples as real or synthetic. If the quality of the synthetic data is high, the classifiers will not be able to distinguish between real and synthetic data. Predictive ML models are also trained on either real or synthetic data and then the predictions on the same test set are compared. For realistic generated data, the predictive performance should not be too different between the models trained on real and fake data, respectively. To evaluate privacy protection, the Euclidean distances are calculated between the synthetic samples and their original counterparts. Consequently, the distribution of pairwise distances between each synthetic record and its nearest original neighbor is achieved. Ideally, the resulting distribution has a large mean and small standard deviation when evaluating the models through the lens of privacy. However, a large mean also indicates poor quality of the synthetic data, meaning that the distance-based privacy metric is inversely proportional to the ML-based evaluation scores. We therefore evaluate them simultaneously to find the overall best-performing model in terms of generating realistic samples while still preserving privacy.

[1] https://github.com/ds-anik/Synthetic_Tabular_Healthcare_Data_Generation.

4 Results

This section presents the experimental results, starting with the column distributions for the generated datasets in Sect. 4.1, then providing the ML-based evaluation scores in Sect. 4.2 and finally presenting the metrics for privacy preservation in Sect. 4.3.

4.1 Column Distributions

Figure 1 compares the marginal distributions (top row) and the cumulative distributions (bottom row) of the diag_2 numerical column in the Diabetes dataset. It is clear that the marginal distribution of the original data has a dominant peak at 450 and multiple lower peaks around it. The WGAN-GP implementation lacks any specific normalization to detect various modes in the numerical features and generates a simple normal distribution around the dominant peak, i.e. it faces mode-collapse. Although the Wasserstein GAN loss function was introduced to circumvent the mode-collapse issue, we observe that it is not applicable to detect complex multi-modal distributions as in our case. The other three models clearly excel WGAN-GP in capturing the modes of the diag_2 column, probably because they apply a mode-specific normalization technique for dealing with multi-modal continuous columns. Comparing the cumulative distributions of all the models demonstrate that the CTABGAN architecture outperforms the other architectures in generating skewed multi-modal numerical columns.

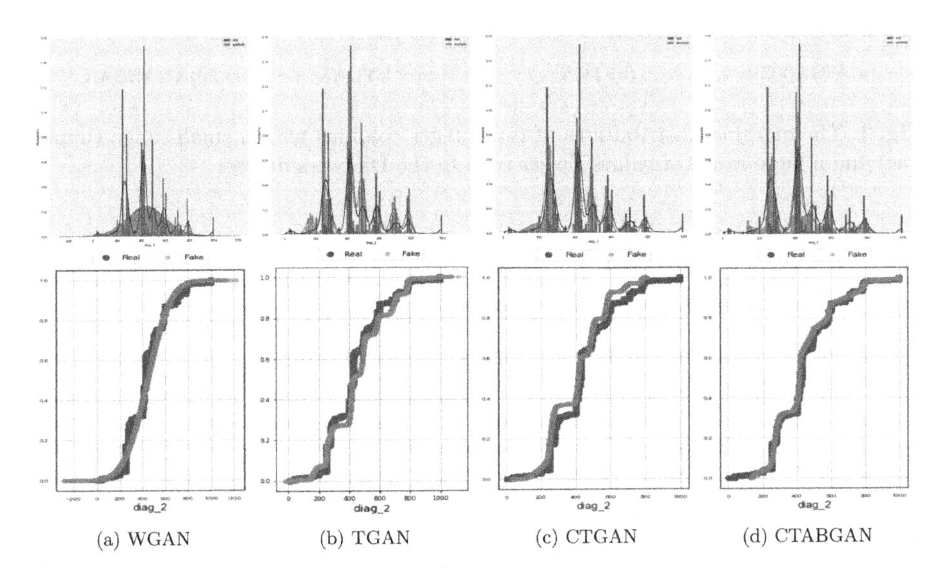

(a) WGAN (b) TGAN (c) CTGAN (d) CTABGAN

Fig. 1. The marginal and cumulative probability distributions for the multi-modal *diag_2* column in the Diabetes dataset.

Figure 2 shows the marginal probability distributions of two discrete numerical columns in the Diabetes dataset. One column has a small range of integer values and the other has a wide range. We observe that for small-range discrete numerical columns, the original and synthetic distribution tend to resemble perfectly. However, the models' performances significantly drop when generating integer columns with a wide range of values. While the marginal distribution of the original data resembles a simple Gaussian distribution, the probability distributions resulting from the TGAN, CTGAN, and CTABGAN consist of several modes. WGAN-GP, on the other hand is better at generating wide-ranging integer variables.

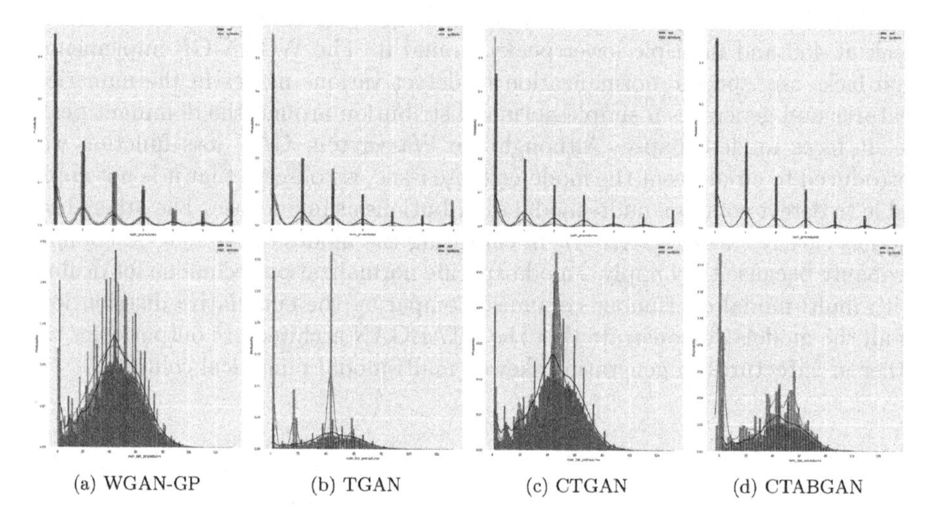

(a) WGAN-GP (b) TGAN (c) CTGAN (d) CTABGAN

Fig. 2. The marginal distributions of two integer columns with a small range (upper row) and a large range of values (lower row) in the Diabetes dataset.

4.2 Machine Learning-based Evaluation

To compare the inferential ability of the original and synthetic datasets, we train a set of predictive models on both the real and fake datasets and compare their predictive capability using the real data. Since all the chosen datasets include a categorical target column, we use the Macro-F1 score to evaluate the predictive capabilities of the models. The F1 score is the harmonic mean of the sensitivity and precision, taking both of them into account. The value ranges from 0 to 1, where 1 is best. Macro-F1 calculates the F1 score for each category and compute unweighted mean of the F1 scores. Macro-F1 is used instead of accuracy due to the imbalanced nature of many categorical features across the investigated datasets. The goal is to verify if the same insights are derived from real and fake datasets when trained on an equally tuned ML model, not picking the best classifier. Thus, we exclude hyper-parameter tuning for each predictive model and compare the GANs based on the average Macro-F1 scores of the classifiers. In addition to comparing the inferential abilities on original and synthetic data, we train logistic regression (LR) and support vector machine (SVM) classifiers on the labelled original and fake datasets to evaluate whether the models are able to distinguish original and fake samples. The normalized area under the receiver operating characteristic (AUROC) score is used for model evaluation. If the synthetic data is inseparable from the original one, the unnormalized AUROC score would be 0.5, indicating that the classifier is guessing randomly and unable to distinguish the real and fake classes. However, since most of the evaluation metrics in our setting are in the range of 0 to 1, we normalize the classification result to 1 minus the average AUROC score. A normalized AUROC score of 1 is best and means that the real and synthetic data are inseparable. A synthetic dataset with low Macro-F1 difference and high normalized AUROC is considered ideal. Tables 2 and 3 show the absolute difference in Macro-F1 scores of the decision tree (DT), random forest (RF), LR and MLP classifiers trained on the original and synthetic datasets, respectively. WG, TG, CT and CTAB stand for WGAN-GP, TGAN, CTGAN and CTABGAN respectively. The normalized AUROC scores for the LR and SVM classifiers are reported in the last two rows of both tables.

Table 2. The difference of the Macro-F1 classification scores and the ML detection scores in the Diabetes and MIMIC III datasets. The best results for each dataset are highlighted in bold. Abbreviations: ΔF1 = difference in Macro-F1 classification score.

	Diabetes				MIMIC III			
	WG	TG	CT	CTAB	WG	TG	CT	CTAB
ΔF1-DT	0.083	0.056	0.030	**0.016**	0.189	0.068	0.052	**0.031**
ΔF1-RF	0.122	0.082	0.044	**0.023**	0.120	0.030	0.020	**0.010**
ΔF1-LR	0.129	0.050	0.046	**0.009**	0.094	**0.001**	0.015	0.007
ΔF1-MLP	0.143	0.087	0.044	**0.013**	**0.015**	0.035	0.039	0.028
ΔF1-average	0.119	0.068	0.041	**0.015**	0.104	0.033	0.031	**0.019**
AUROC-LR	0.330	0.520	0.540	**0.700**	0.220	0.620	0.730	**0.790**
AUROC-SVM	0.110	0.290	0.330	**0.560**	0.110	0.390	0.480	**0.540**

If we average the Macro-F1 differences across all four classifiers in the Diabetes dataset, we find that the CTABGAN model has the lowest average score of 0.015, followed by CTGAN and TGAN models with average scores of 0.041 and 0.068. This pattern is repeated in the MIMIC III and Thyroid datasets, with the CTABGAN model outperforming others in terms of the difference in Macro-F1 classification scores followed by the CTGAN, TGAN, and WGAN-GP. The reason why CTABGAN is the best performing model in these datasets is due to the modified conditional GAN architecture and an additional information loss term in the optimization process. Although the Diabetes, MIMIC III, and Thyroid datasets follow a similar pattern regarding the average Macro-F1 scores, there is a larger gap between the CTABGAN and CTGAN in the Diabetes dataset compared to the ones in the other two datasets. This can be related to the multi-modal nature of the numerical columns in the Diabetes dataset and how the CTABGAN model successfully generates this type of numerical distribution, benefiting from an extended condition vector in its architecture.

However, in the Epileptic dataset, the WGAN-GP outperforms other models regarding the average Macro-F1 differences. WGAN-GP is the best performing model with an average score of 0.085, followed by CTABGAN, TGAN, and CTGAN, with average scores of 0.18, 0.21, and 0.23, respectively. The Epilpetic dataset includes 178 integer columns with wide ranges. Similar to our interpretation in Sect. 4.1, we find that although the mode-specific normalization approach in the CTGAN, TGAN, and CTABGAN is well-suited for numerical columns with complex distributions, it may prevent the model from reaching an ideal optimum in the smaller datasets with the discrete numerical variables (integers).

Table 3. The difference of the Macro-F1 classification scores and the ML detection scores in the Thyroid and Epileptic datasets. The best results for each dataset are highlighted in bold. Abbreviations: ΔF1 = difference in Macro-F1 classification score.

	Thyroid				Epileptic			
	WG	TG	CT	CTAB	WG	TG	CT	CTAB
ΔF1-DT	0.300	0.240	0.200	**0.100**	**0.070**	0.210	0.210	0.180
ΔF1-RF	0.260	0.220	0.190	**0.100**	**0.050**	0.280	0.250	0.230
ΔF1-LR	0.140	0.090	**0.050**	**0.050**	**0.010**	**0.010**	0.080	0.030
ΔF1-MLP	0.240	0.170	0.120	**0.080**	**0.210**	0.350	0.390	0.280
ΔF1-average	0.230	0.180	0.140	**0.080**	**0.085**	0.210	0.230	0.180
AUROC-LR	0.530	0.700	0.700	**0.780**	**0.770**	0.330	0.450	0.730
AUROC-SVM	0.380	0.600	0.530	**0.620**	**0.620**	0.250	0.380	0.590

The LR and SVM classifiers' normalized AUROC scores follow the same pattern as the average Macro-F1 differences. Due to the auxiliary classifier in the CTABGAN's architecture and the classification loss term in its optimization process, it is much harder to distinguish the synthetic data generated from the CTABGAN model from the original data. Consequently, the normalized AUROC scores of the CTABGAN model outperform the scores from the other models for the Diabetes, MIMIC III, and Thyroid datasets. Again, WGAN-GP's score exceeds the scores of the other models for the Epileptic dataset, followed by CTABGAN, CTGAN, and TGAN.

Overall, the CTABGAN model achieves the best ML cross-testing scores for the Diabetes, MIMIC III, and Thyroid datasets. This is because of its modified conditional GAN architecture and an additional information loss term in its optimization process. However, WGAN-GP outperforms the other architectures for the Epileptic dataset as the normalization techniques in TGAN, CTGAN, and CTABGAN are less suitable for the wide-ranging integer columns in this dataset. The same pattern is observed for the normalized AUROC scores: The CTABGAN model is the best-performing model at generating synthetic records that are indistinguishable from the original ones, except from the Epileptic dataset where WGAN-GP is ranked highest.

4.3 Preserved Privacy

We also evaluate the GAN's potential to preserve the privacy of sensitive data. This evaluation category is especially important in the healthcare domain, where patients share sensitive and private information. Suppose a patient's confidential data is to be re-identified by accessing the synthetic data. In that case, the patient's sensitive information is undoubtedly leaked in the synthetic dataset, and the GAN simply replicates the original records when generating new ones.

Table 4. The distribution of Euclidean distances between synthetic records and their closest original counterparts. The format is *mean* ± *std*.

Model	Diabetes	MIMIC III	Thyroid	Epileptic
WGAN-GP	3.10 ± 0.46	1.37 ± 1.14	1.88 ± 0.97	7.55 ± 8.45
TGAN	2.69 ± 0.61	0.93 ± 0.80	1.39 ± 1.05	7.81 ± 9.26
CTGAN	2.71 ± 0.64	0.97 ± 0.90	1.34 ± 1.08	9.10 ± 9.18
CTABGAN	3.02 ± 0.46	1.11 ± 0.84	1.76 ± 1.06	8.08 ± 9.27

Table 4 shows the mean and standard deviation of the distance to closest record distributions. In the Diabetes, MIMIC III, and Thyroid datasets, the WGAN-GP model maintains the largest distance between the original and synthetic data (lowest privacy risk). This verifies the results in Sect. 4.2, as the WGAN-GP model was the worst-performing regarding ML utilities. Interestingly, in contrast to the results for ML utilities, we observe that the CTABGAN is the second best-performing model in all the datasets. Although the CTABGAN model outperforms the CTGAN and TGAN models in the Diabetes, MIMIC III, and Thyroid datasets, it preserves the privacy more when generating synthetic records.

In the Epileptic dataset, the CTGAN and WGAN-GP models are the best and worst models regarding privacy preservation with means of 9.10 and 7.55, respectively. This verifies the results for ML utilities as the WGAN-GP is the best-performing and CTGAN is the worst-performing one in the Epileptic dataset.

5 Discussion

From the results in Sect. 4.1, we observe that CTABGAN outperforms the other GANs when generating data that are distributed in a similar way as the original data. This is also true regarding skewed multi-modal numerical columns. Although both CTGAN and CTABGAN use a conditional generator and the training-by-sampling technique to handle generating imbalanced categorical features, it is observed that the CTGAN model slightly outperforms the CTABGAN architecture when dealing with highly imbalanced categorical variables. This is because of the modification of the CTABGAN's conditional generator to capture skewed multi-modal distributions more effectively compared to the CTGAN model. Besides the one-hot encoded representations for categorical columns, the extended conditional vector in CTABGAN includes the mode of the numerical columns, increasing its performance in capturing modes with less weight. In Fig. 1, it is clear that WGAN-GP faces mode collapse, generating a simple normal distribution. Although the Wasserstein GAN loss function was introduced to circumvent the mode-collapse issue, we observe that it is not applicable to detect complex multi-modal distributions, as in our case.

One of the few areas where there is room for further improvement is the generation of discrete numerical columns, as none of the proposed models makes any distinction between the continuous and discrete numerical features. This is confirmed in Fig. 2 for the column with wide-ranging values in the Diabetes dataset and is also observed in Sect. 4.2, where the models struggle with the wide-ranging, discrete numerical columns in the Epileptic dataset. TGAN, CTGAN, and CTABGAN produces multi-modal distributions even though the original data shows a single mode. This occurs due to the mode-specific normalization implemented in the mentioned GANs. Although this technique is well-suited for continuous columns with complex distributions, it may prevent the model from reaching an ideal optimum for the discrete numerical variables. The exception is the WGAN-GP architecture, which might explain why the WGAN-GP model nicely captures the wide-ranging discrete column in the Diabetes dataset and is ranked highest for the ML-based metrics for the Epileptic dataset. However, WGAN-GP shows poor performance in capturing multimodal distributions as well as for the ML-based metrics for the other datasets.

When considering the comparison of column distributions, ML-based evaluation scores and privacy preserving metrics simultaneously, the CTABGAN model outperforms the other models in generating realistic samples while also preserving the privacy of the original samples in all datasets.

6 Conclusion

In conclusion, the obtained visual and quantitative results demonstrate that synthetic healthcare data can be a reliable substitute for original data. Of the tested architectures, CTABGAN seems to be most promising for generating realistic synthetic tabular healthcare data without leaking individuals' sensitive information. However, for datasets containing columns of wide-ranging integer values, a vanilla WGAN-GP might be more appropriate. Generating data using GANs eliminates the need for traditional anonymization and obfuscation techniques which are too risky and negatively impact the data utilities. For future work, further developments in the models' architectures can potentially improve the performance of synthetic data generation. Moreover, there is much room for the proposed data generating models to improve training convergence in small-sized datasets and generate discrete numerical columns. Development of generative models that also handle other types of data such as free text, ordinal values and time series should be explored. Finally, we only investigate the strengths and weaknesses of the GAN-based models in generating healthcare tabular datasets. However, it would be interesting to test other generative models like variational autoencoders, Gaussian copula and Bayesian networks in the medical use cases.

References

1. Tavanapong, W., Oh, J., Riegler, M., Khaleel, M.I., Mitta, B., De Groen, P.C.: Artificial intelligence for colonoscopy: past, present, and future, IEEE Journal of Biomedical and Health Informatics
2. Choy, G.: Current applications and future impact of machine learning in radiology. Radiology **288**(2), 318 (2018)
3. Shatte, A.B., Hutchinson, D.M., Teague, S.J.: Machine learning in mental health: a scoping review of methods and applications. Psychol. Med. **49**(9), 1426–1448 (2019)
4. van de Sande, D., et al.: Developing, implementing and governing artificial intelligence in medicine: a step-by-step approach to prevent an artificial intelligence winter, BMJ Health & Care Informatics 29 (1)
5. Rajkomar, A., Dean, J., Kohane, I.: Machine learning in medicine. N. Engl. J. Med. **380**(14), 1347–1358 (2019)
6. Thambawita, V., et al.: DeepSynthBody: the beginning of the end for data deficiency in medicine. In: 2021 International Conference on Applied Artificial Intelligence (ICAPAI), pp. 1–8. IEEE (2021)
7. Goncalves, A., Ray, P., Soper, B., Stevens, J., Coyle, L., Sales, A.P.: Generation and evaluation of synthetic patient data. BMC Med. Res. Methodol. **20**(1), 1–40 (2020)
8. Rashidian, S., et al.: SMOOTH-GAN: towards sharp and smooth synthetic EHR data generation. In: Michalowski, M., Moskovitch, R. (eds.) AIME 2020. LNCS (LNAI), vol. 12299, pp. 37–48. Springer, Cham (2020). https://doi.org/10.1007/978-3-030-59137-3_4
9. Kingma, D.P., Welling, M.: Auto-encoding variational bayes (2013)
10. Gogoshin, G., Branciamore, S., Rodin, A.S.: Synthetic data generation with probabilistic Bayesian networks. Math. Biosci. Eng. MBE **18**(6), 8603 (2021)
11. Goodfellow, I., et al.: Generative adversarial nets. In: Ghahramani, Z., Welling, M., Cortes, C., Lawrence, N., Weinberger, K., (Eds.), Advances in Neural Information Processing Systems, vol. 27, Curran Associates Inc., (2014)
12. Choi, E., Biswal, S., Malin, B., Duke, J., Stewart, W.F., Sun, J.: Generating multi-label discrete patient records using generative adversarial networks. In: Doshi-Velez, F., Fackler, J., Kale, D., Ranganath, R., Wallace, B., Wiens, J., (Eds.), Proceedings of the 2nd Machine Learning for Healthcare Conference, vol. 68 of Proceedings of Machine Learning Research, pp. 286–305. PMLR (2017)

13. Park, N., Mohammadi, M., Gorde, K., Jajodia, S., Park, H., Kim, Y.: Data synthesis based on generative adversarial networks. Proceedings of the VLDB Endowment **11**(10), 1071–1083 (2018)
14. Xu, L., Veeramachaneni, K.: Synthesizing tabular data using generative adversarial networks (2018)
15. Xu, L., Skoularidou, M., Cuesta-Infante, A., Veeramachaneni, K.: Modeling tabular data using conditional GAN. In: Wallach, H., Larochelle, H., Beygelzimer, A., Alché-Buc, F.D., Fox, E., Garnett, R., (Eds.), Advances in Neural Information Processing Systems, vol. 32, Curran Associates Inc., (2019)
16. Zhao, Z., Kunar, A., Birke, R., Chen, L.Y.: CTAB-GAN: effective table data synthesizing. In: Balasubramanian, V.N., Tsang, I., (Eds.), Proceedings of The 13th Asian Conference on Machine Learning, vol. 157 of Proceedings of Machine Learning Research, pp. 97–112. PMLR (2021)
17. Xie, L., Lin, K., Wang, S., Wang, F., Zhou, J.: Differentially private generative adversarial network. arXiv preprint arXiv:1802.06739
18. Torkzadehmahani, R., Kairouz, P., Paten, B.: DP-CGAN: differentially private synthetic data and label generation. In: Proceedings of the IEEE/CVF Conference on Computer Vision and Pattern Recognition Workshops (2019)
19. Torfi, A., Fox, E.A., Reddy, C.K.: Differentially private synthetic medical data generation using convolutional GANs. Inf. Sci. **586**, 485–500 (2022)
20. Jordon, J., Yoon, J., Van Der Schaar, M.: PATE-GAN: generating synthetic data with differential privacy guarantees. In: International Conference on Learning Representations (2018)
21. Coutinho-Almeida, J., Rodrigues, P.P., Cruz-Correia, R.J.: GANs for tabular healthcare data generation: a review on utility and privacy. In: Soares, C., Torgo, L. (eds.) DS 2021. LNCS (LNAI), vol. 12986, pp. 282–291. Springer, Cham (2021). https://doi.org/10.1007/978-3-030-88942-5_22
22. Johnson, A.E., et al.: MIMIC-III, a freely accessible critical care database. Scientific Data 3 (160035)
23. Andrzejak, R.G., Lehnertz, K., Mormann, F., Rieke, C., David, P., Elger, C.E.: Indications of nonlinear deterministic and finite-dimensional structures in time series of brain electrical activity: dependence on recording region and brain state. Phys. Rev. E 64 (061907)
24. Harun-Ur-Rashid, Supriya, Epileptic seizure recognition (2018)
25. Dua, D., Graff, C.: UCI machine learning repository (2017)
26. Pedregosa, F., et al.: Scikit-learn: machine learning in Python. J. Mach. Learn. Res. **12**, 2825–2830 (2011)
27. Strack, B., et al.: Impact of HbA1c measurement on hospital readmission rates: analysis of 70,000 clinical database patient records, BioMed Research International (2014)
28. Gulrajani, I., Ahmed, F., Arjovsky, M., Dumoulin, V., Courville, A.C., Improved training of wasserstein GANs. In: Guyon, I., et al. (Eds.), Advances in Neural Information Processing Systems, vol. 30, Curran Associates Inc., (2017)
29. Abadi, M.: TensorFlow: large-scale machine learning on heterogeneous systems. Software available from tensorflow.org (2015)
30. Paszke, A., et al.: PyTorch: an imperative style, high-performance deep learning library. In: Wallach, H., Larochelle, H., Beygelzimer, A., d' Alché-Buc, F., Fox, E., Garnett, R., (Eds.), Advances in Neural Information Processing Systems 32, Curran Associates Inc., pp. 8024–8035 (2019)

FL-Former: Flood Level Estimation with Vision Transformer for Images from Cameras in Urban Areas

Quoc-Cuong Le[3], Minh-Quan Le[1,2], Mai-Khiem Tran[1,2], Ngoc-Quyen Le[4], and Minh-Triet Tran[1,2](✉)

[1] University of Science, VNU-HCM, Ho Chi Minh City, Vietnam
{lmquan,tmkhiem}@selab.hcmus.edu.vn
[2] Viet Nam National University, Ho Chi Minh City, Vietnam
tmtriet@hcmus.edu.vn
[3] Board of Management of Saigon High-Tech-Park, Ho Chi Minh City, Vietnam
lequoccuong@tphcm.gov.vn
[4] Southern Regional Hydrometeorological Center, Ho Chi Minh City, Vietnam

Abstract. Flooding in urban areas is one of the serious problems and needs special attention in urban development and improving people's living quality. Flood detection to promptly provide data for hydrometeorological forecasting systems will help make timely forecasts for life. In addition, providing information about rain and flooding in many locations in the city will help people make appropriate decisions about traffic. Therefore, in this paper, we present our FL-Former solution for detecting and classifying rain and inundation levels in urban locations, specifically in Ho Chi Minh City, based on images recorded from cameras using Vision Transformer. We also build the HCMC-URF dataset with more than 10 K images of various rainy and flooding conditions in Ho Chi Minh City to serve the community's research. Finally, we propose the software architecture and construction of an online API system to provide timely information about rain and flooding at several locations in the city as extra input for hydrometeorological analysis and prediction systems, as well as provide information to citizens via mobile or web applications.

Keywords: Flood classification · Vision Transformer · Urban Flood Dataset

1 Introduction

The development of smart cities is one of the popular trends in the world, helping to create a better living environment for people. One of the essential issues to building a smart city is the collection and timely provision of highly accurate data for management and forecasting systems to assist appropriate decisions or plans.

Q.-C. Le and M.-Q. Le—Equal contribution.

D.-T. Dang-Nguyen et al. (Eds.): MMM 2023, LNCS 13833, pp. 447–459, 2023.
https://doi.org/10.1007/978-3-031-27077-2_35

Along with the increasing speed of construction and urbanization, the design to control and limit flooding in urban areas is a challenging problem and needs special attention. In addition to building professional and specialized observation and data collection stations to record environmental conditions as well as hydrometeorological parameters, taking advantage of the traffic camera system to provide more input information for hydrometeorological forecasting systems is a potential and effective solution. Therefore, in this paper, we propose a simple yet effective solution to estimate with high accuracy the level of flooding in urban areas using visual information recorded from the existing traffic camera system.

We propose FL-Former, a Vision Transformer-based solution, to classify four levels of raining and flooding in urban areas: no rain, light rain, heavy rain, and flood. We also collect and label 10K actual images recorded from traffic cameras in Ho Chi Minh City in 2022 for research purposes, named HCMC-Urban Rain and Flood Dataset (HCMC-URF). The experimental results show that the proposed baseline method achieves high accuracy (over 80%) in detecting and classifying the degree of flooding in urban areas.

Besides, to serve real applications, we have proposed a software architecture and realized an online API system to provide data on inundation as well as rainfall to add more. Input data for hydrometeorological analysis and forecasting systems at Southern Regional Hydrometeorological Center, Ho Chi Minh City, Vietnam. Our API system also provides information visualized on the map, helping people have more information about the rain and flooding situation in Ho Chi Minh City.

The structure of this paper is as follows. In Sect. 2, we briefly review recent work on flood detection or flood level classification, together with some vision techniques for scene analysis and understanding that we utilize in this paper for flood classification. Then we propose in Sect. 3 our method using Vision Transformer, namely FL-Former, to classify different levels of flooding in urban areas from images captured by cameras. We present our process to collect and label data for rain and flood in Ho Chi Minh city, and the experimental results in this dataset using our proposed method in Sect. 4. Our proposed architecture and implementation for online APIS to provide information on flood from cameras are presented in Sect. 5. Finally, the conclusion and future work are discussed in Sect. 6.

2 Related Work

2.1 Flood Estimation

Based only on free satellite images and open-source software, Notti et. al. propose a semi-automatic method for flood mapping [12]. Another method to detect automatically near real-time flood using Suomi-NPP/VIIRS data is proposed by Li et. al. [7]. Detecting people and vehicles in danger is also considered in the warning system framework proposed by Giannakeris et. al. [4]. Oddo et. al. integrate a socioeconomic damage assessment model with a near real-time flood remote sensing and decision support tool (NASA's Project Mekong) [13]. Shen et. al. review theories and algorithms of flood inundation mapping using SAR data, together with a discussion of their strengths and limitations, focusing on the level of automation, robustness, and accuracy [17]. Li et. al. assess the roles of SAR intensity and interferometric coherence in urban flood detection using multi-temporal TerraSAR-X data [8]. Shahabi et. al. propose a new flood susceptibility mapping technique [16]. Anusha et. al. discuss the thresholding and unsupervised classification methodologies, to find the inundated areas due to incessant rains and the rise of water level in Rapti and Ghaghara Rivers [1]. To assist efforts to operationalize deep learning algorithms for flood mapping at a global scale. Bonafilia et. al. introduce Sen1Floods11, a surface water data set including raw Sentinel-1 imagery and classified permanent water and flood water [2]. Using UAV-based aerial imagery, Munawar et. al. propose a CNN-based method for flood detection from the images of the disaster zone [11].

2.2 Contrastive Language-Vision Pretraining

Visual representation learning under text supervision has received a lot of interest as a result of the success of Contrastive Language-Image Pre-Training (CLIP [14]) derivatives in several fields [5,6,10,15]. A potential approach that draws on a considerably larger supply of supervision is learning about pictures straight from raw text. On a dataset of 400 million (image, text) pairings gathered from the internet, the authors show that the straightforward pre-training task of guessing which caption goes with which picture provides an effective and scalable technique to learn SOTA image representations from scratch. Natural language is employed to refer to previously taught visual ideas (or describe brand-new ones) after pre-training, enabling zero-shot transfer of the model to subsequent tasks. CLIP offers a powerful English text-image representation. There are works on teaching bilingual text-image representation and multilingual text-image representation to broaden the language of CLIP models.

2.3 Vision Transformer

For image classification, the earliest ViT study used a transformer encoder on coarse, non-overlapping picture patches, and thus needs a lot of training data upfront. Vision Transformer (ViT [3]), which requires far fewer CPU resources to train than state-of-the-art convolutional networks, achieves great results when applied to several mid-sized or tiny picture recognition benchmarks after pre-training on vast volumes of data. Since its debut, Vision Transformers have been enthusiastically adopted and used for various visual identification tasks employing priors such as local structure modeling and multi-scale feature hierarchies. ViT's suggestion to apply bidirectional self-attention on picture patches rather than pixels immediately pushed Transformer's image classification performance closer to SOTA. More recent approaches begin to model ViT architectures after CNNs and modify them to resemble CNNs [9]. Tokens are gradually combined to downsample the feature resolutions with lower computation costs and larger embedding sizes.

3 Methodology

3.1 Data Collection

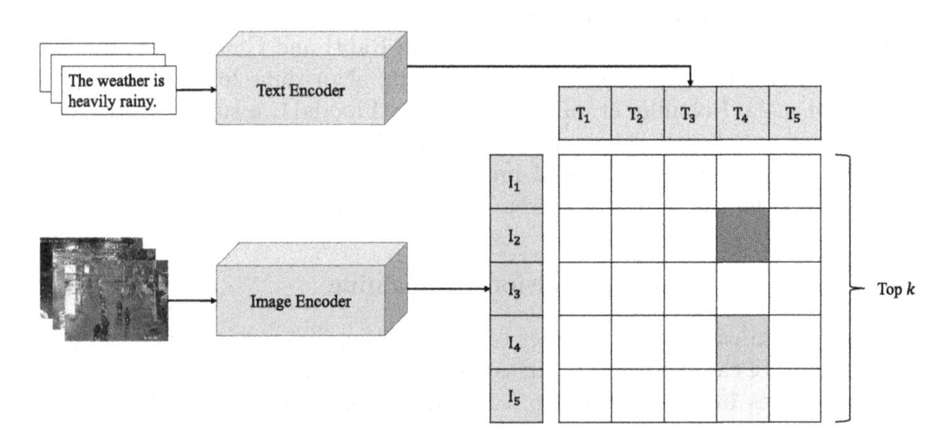

Fig. 1. Pseudo labeling training strategy with contrastive language-image pretraining.

From the camera system of Ho Chi Minh City, we crawl traffic videos over four months and obtained 25,075 videos, which takes 22.571 terabytes. From the collected videos, we extract frames to create an image dataset of the traffic system in Ho Chi Minh City for the later flood evaluation task. We assess the flood level of the city via the image classification problem. In particular, given an image of a traffic video, we classify it into 4 groups: normal case (no rain), light rain, heavy rain, and flood. With respect to training an image classifier, we need a dataset comprising images and their corresponding category. To construct a dataset without time-consuming and labor-saving, we leverage a pseudo-labeling training strategy with contrastive language-image pretraining (CLIP). To be specific, we compute the feature vectors of every image in the dataset using a pre-trained image encoder. Likewise, we encode pre-defined text queries via a pre-trained text encoder to extract the representation of these queries. Later on, we leverage extracted text features to retrieve the top k similar images with respect to certain text queries, see Fig. 1. For instance, we query a sentence "The weather is heavily rainy." Top k images whose encoded vectors have the most cosine similarity with the query representation are grouped into the category heavy rain. In this way, we not only leverage abundant unlabeled data but also save plenty of labor forces and time. Totally, we obtain a training dataset of 10000 images with 4 classes. Figure 3 illustrates our pseudo-labeling dataset for flood level estimation via contrastive language-image pretraining.

3.2 Flood Evaluation Meets Image Classification

In this part, we present a simple yet effective method that is able to evaluate the flood level in Ho Chi Minh city. Specifically, we formulate the problem of flood evaluation as the image classification task. Given an image of a traffic scene, we feed it into a deep neural network to predict the category it belongs to, including normal case, light rain, heavy rain, and flood. We adapt the idea of Vision Transformer, which divides an image into fix-sized patches, linearly embeds each of them, adds positional encoding data, and feeds a sequence of tokens into a Transformer encoder.

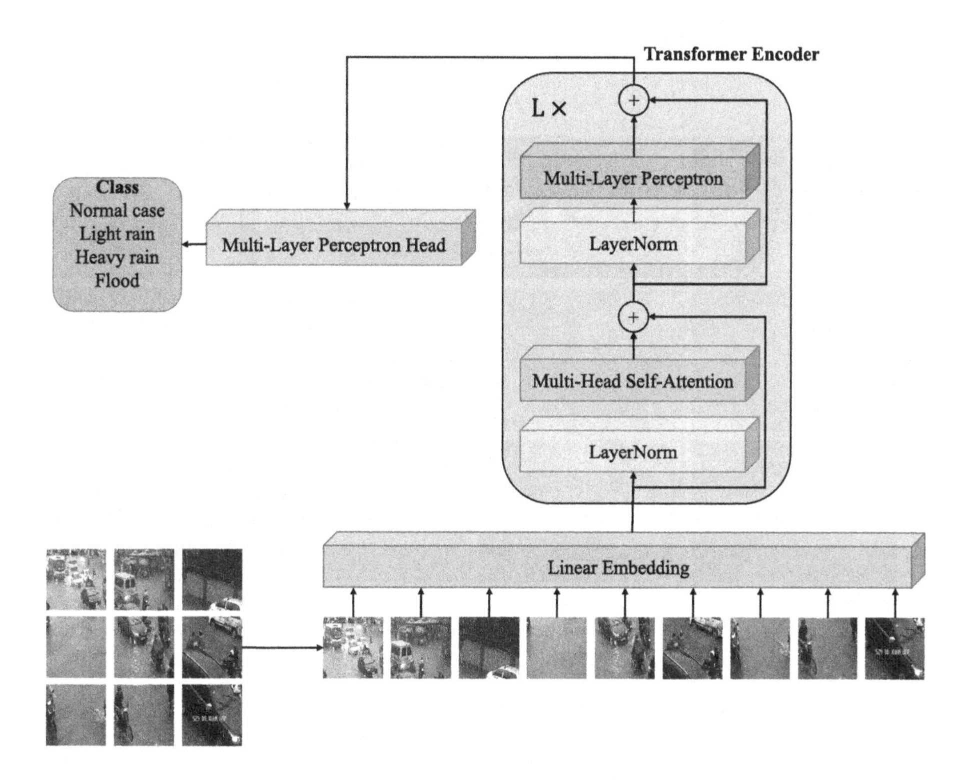

Fig. 2. The architecture of Vision Transformer.

Figure 2 illustrates the architecture of the Vision Transformer. We split a 2D image $\mathbf{x} \in \mathbb{R}^{H \times W \times C}$ into a sequence of N patches where $N = HW/P^2$ and (P, P) is the size of each patch. A sequence of patches is denoted as $\mathbf{x}_p \in \mathbb{R}^{N \times P^2 \times C}$. Later on, we feed these patches into linear embedding, multi-head self-attention [18] and multi-layer perceptron layers as follows:

$$
\begin{aligned}
\mathbf{z}_0 &= \left[\mathbf{x}_{\text{class}}\,; \mathbf{x}_p^1\mathbf{W}; \mathbf{x}_p^2\mathbf{W}; \cdots; \mathbf{x}_p^N\mathbf{W}\right] + \mathbf{W}_{pos} \\
\mathbf{z}'_\ell &= \text{MultiHead} - \text{SelfAttention}\left(\text{LayerNorm}\left(\mathbf{z}_{\ell-1}\right)\right) + \mathbf{z}_{\ell-1} \\
\mathbf{z}_\ell &= \text{MultiLayer} - \text{Perceptron}\left(\text{LayerNorm}\left(\mathbf{z}'_\ell\right)\right) + \mathbf{z}'_\ell, \\
\mathbf{y} &= \text{LayerNorm}\left(\mathbf{z}_L^0\right)
\end{aligned}
\tag{1}
$$

where $\mathbf{W} \in \mathbb{R}^{(P^2 \cdot C) \times D}$, $\mathbf{W}_{pos} \in \mathbb{R}^{(N+1) \times D}$, D is a latent vector size.

4 Experiments

4.1 Experimental Settings

Fig. 3. Examples of the pseudo-labeling classification dataset for flood level estimation via contrastive language-image pretraining.

Along with 10000 images collected via pseudo-labeling in the training dataset, we collect and manually label 2000 images to construct a testing dataset for evaluation. In the training stage, we leverage the stratified K-fold training strategy to improve the robustness of the model as well as leverage the total amount of training data. Specifically, the training set is split into 5 folds with the same proportion of categories. Then, we iteratively utilize 4 folds for training and the other for validation. After training k models ($k = 5$), we perform inference on the testing dataset to generate softmax vectors and compute the mean over k different vectors from k models. Consequently, the final result is decided by the average of k models, which consolidates the model's effectiveness and avoids overfitting.

Table 1. We train the ViT model on our dataset with 5 folds training strategy. Then, we compute the mean of softmax vectors to ensemble these 5 models to increase the effectiveness and immunity to overfitting. Moreover, data augmentation methods help improve the performance of the model.

Model	Evaluation	Accuracy	
		w/o aug.	w/ aug.
Fold 0	validation	80.97%	82.63%
	test	80.21%	82.59%
Fold 1	validation	81.76%	83.12%
	test	80.45%	82.70%
Fold 2	validation	80.36%	81.82%
	test	81.25%	83.92%
Fold 3	validation	82.65%	83.11%
	test	81.58%	82.94%
Fold 4	validation	83.44%	84.51%
	test	82.67%	84.32%
Average Softmax	test	**85.33%**	**87.18%**

4.2 Implementation Details

We leverage Vision Transformer (ViT-Base16) as the image classifier. The framework was implemented on PyTorch based on the source code of the original paper. We also applied simple augmentation methods: random resizing, random cropping, translation, rotation, and flipping. The ViT classifier was trained for 100 epochs with a batch size of 16, a base learning rate of 0.001, a warmup learning with step sizes of 500, and a cosine learning rate scheduler. The model was trained on a single RTX 3090 24 GB.

4.3 Experimental Results

In this part, we evaluate the performance of our proposed method on the collected testing dataset. We demonstrate the performance of each k-fold model independently and also show the performance of the average results. Moreover, we include the testing results of ViT model with and without data augmentation methods for the ablation study.

Table 1 illustrates the performance of the model when training on a specific split. The average of k models achieves the best results with a significant gain of 2.66% from the fold 4 model. Furthermore, a dramatic increase of approximately 2% is witnessed in the model's accuracy when applying data augmentation strategies.

5 Systems Architecture

5.1 Data Acquisition System

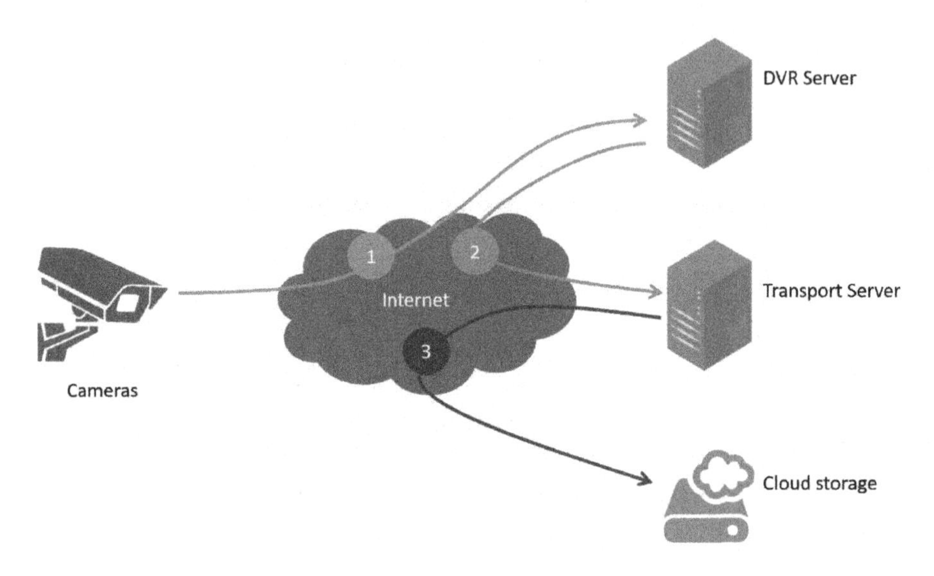

Fig. 4. Data acquisition system components

Data acquisition system connects, acquires and performs a number of validation on video data. Ideally, it fetches data from any surveillance systems, as long as there are APIs to do so, and upload the data onto a pre-configured cloud storage service. Solution-wise, the data acquisition system consists of the following components: (i) Cameras, (ii) DVR server, (iii) Transport server, and (iv) Cloud storage. The function and responsibilities of each component are as follows:

Cameras: The hot spots are established from site surveying and public flooding feedback channel. At these hot spots, the cameras are deployed. Due to the nature of the deployment requirements, which specifies camera locations to spread across the area, some very far from the central server, the cameras are inherently IP cameras. They are the source of data to the system, which carries information about flooding status. The quantity and quality of data generated from the cameras are subject to device specifications, including sensor type, sensitive wavelengths, viewing angle, encoding capabilities, and connectivity capabilities.

DVR Server: Digital Video Recording Servers are responsible for collecting imagery data, provisioning the cameras, and providing temporary storage for new data. In our installment, the server was a desktop running iVMS-4200 software from HikVision. In DVR servers, the specifications that impact solution operations are video encoding-decoding capabilities (e.g., speed and format, such as H.264 or H.265) for storage, local storage capacity, connectivity, network ingress bandwidth, and compatible standards that cameras can inter-operate.

Transport Server: The transport server essentially checks and forwards video and imagery data from DVR Server to cloud storage for permanent large-scale storage, in which aspects such as access bandwidth or storage capacity can be dynamically scaled according to the operator's budget. In our deployment, we make use of 8 virtual machine instances on a cloud infrastructure provider to download from 2 DVR servers over ISAPI, an open, REST-like HTTP API that was offered by HikVision.

Cloud Storage: Cloud storage services provide data owners abilities to store, organize, manage, share, and backup the data that is stored externally, accessible on the Internet, and maintained by third-party cloud storage providers, such as DropBox, Google, and Backblaze to name a few. The services are accessible on the Internet, hence the contents can be accessed on demand regardless of origin.

The system comprises of three independently executing processes:

Data Acquisition Process: The cameras generate data while working. Data generated from cameras may be encoded and sent to a pre-configured DVR server over the Internet for collection and processing.

Transport Process: Data generated from cameras after being collected and stored locally on DVR server will be downloaded to transport servers. In our installment, a total of 8 transport servers were instantiated to download videos from 32 cameras from DVR server. Downloaded data is checked for duplication and corruption before being uploaded to cloud storage. In our deployment, we noticed that the HikVision DVR server sometimes combines and split videos according to our query, which has resulted in duplicate content. As a workaround, we query the video list by the hour (thus, fewer splitting and combining actions happen) and check the resultant list against duplication with file hashing. Any videos that duplicate what we have known will be discarded. We also noticed that the connection initiation between transport servers and DVR servers often fails, which we suspect is due to the use of dynamic DNS of the latter system and the relatively long distance between the two servers (approximately over 1000 km geographically). If the transport server is executed as a role on the same machine that runs the DVR server, the downloading process is effectively nullified, leaving only the validation work.

Upload Process: After a video passed the validation steps, the transport server will try to upload the data to the server. In our deployment, we have never encountered over 10 retries across all 8 transport servers. However, in case the upload fails, the transport server will employ a back-off strategy so as not to overload the local cloud storage ingress gateway, which can lead to warnings and service disruptions.

Interoperability System: The interoperability system is created to consume, analyze and offer insights on video data stored on a pre-configured cloud storage service. In our deployment, the data resides on cloud storage from data acquisition systems. An interoperability system is created with similar characteristics to that of a data acquisition system, which is loose-coupled, contained, and modularized. This allows for easy replacement of one functionality without affecting the rest of the system.

The execution flow of the system consists of two independent flows. First, the cloud storage service is watched for new data. When new data is available, it will be downloaded from cloud storage onto storage that is local to the analysis service. Then, the analysis service analyzes and indexes the new data with its result. Second, the results are offered over HTTP REST APIs from web service (Figs. 4 and 5).

The system consists of three services being run on a processing server. The services are:

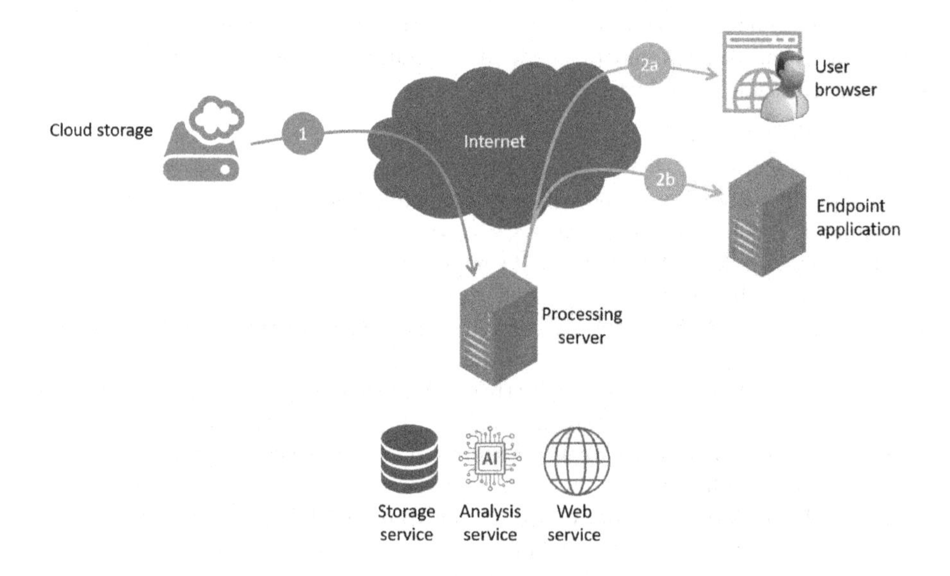

Fig. 5. Interoperability system components

Analysis Service: The analysis service consists of the model described in Sect. 3. When sufficiently trained, the model can analyze input data and give results. This service is the main and most important service overall. Also included in this service are specific weights and possibly model structures for each camera. This way, analysis can be automatically carried out on all data, and data from each camera goes through a fine-tuned analysis process that produces the highest quality results.

Storage Service: The storage service formats, labels, store, and provide lookup capabilities on results analyzed from input data. This works by creating an abstraction layer to separate the concern of the logical place to store the result, whether the back-end is filed on a file system or entities in a database. Multiple storage places can be specified to achieve last-level redundancy, similar to RAID. Furthermore, the lookup functionality provides the ability to retrieve past results given a filter.

Web Service: The web service provides the information under data transmitted over an HTTP connection. Functionality-wise, the web service simply transforms the data into a standard representation that downstream consumers (e.g., user browser or endpoint application) can receive. In our deployment, we wanted to achieve the best compatibility. Thus, we opted for REST API over HTTP protocol. This provides purely information with minimal representation. The output is in JSON format and can be easily consumed by downstream applications. We designed the following API endpoints for consumers to get the information:

- /api/rain and /api/flood provide information about the raining intensity and flooding intensity, respectively, at a given location, optionally at a specific time. A specific location is required and passed as "location_id" parameter. A specific time input is optional, passed as a "timestamp" parameter.
- Similarly, /api/level provides information about the water level

6 Conclusion

In this paper, we introduce the HCMC-RFL dataset, collected over 4 months in the raining season in 2022 from traffic cameras in Ho Chi Minh City. This dataset consists of 10K images labeled according to 4 levels of urban rain and flooding. We also propose FL-Former algorithm using Vision Transformer to estimate with high accuracy the degree of rain and flooding in urban areas from traffic camera images. We also develop an online API system that allows us to provide additional data for hydrometeorological analysis and forecasting systems.

Currently, we continue to collect more data by enriching our dataset HCMC-URFD, and from creating two specialized datasets focusing on raining and flooding in urban areas with images captured from the traffic cameras in cities. In the newly constructed dataset for flooding, we consider four levels of inundation: no flooding, mild flooding, flooding, and severe flooding. Similarly, we also consider four levels of rain for the rain dataset: no rain, small rain, regular rain, and heavy rain. Besides, we also improve the accuracy and processing speed of the urban flooding estimation algorithm.

Acknowledgements. This work is supported by the Department of Science and Technology, Ho Chi Minh City, in the project 97/2020/HD-QPTKHCN.

References

1. Anusha, N., Bharathi, B.: Flood detection and flood mapping using multi-temporal synthetic aperture radar and optical data. Egypt. J. Remote Sens. Space Sci. **23**(2), 207–219 (2020)
2. Bonafilia, D., Tellman, B., Anderson, T., Issenberg, E.: Sen1floods11: a georeferenced dataset to train and test deep learning flood algorithms for sentinel-1. In: 2020 IEEE/CVF Conference on Computer Vision and Pattern Recognition Workshops (CVPRW), pp. 835–845 (2020)
3. Dosovitskiy, A., et al.: An image is worth 16x16 words: transformers for image recognition at scale. In: 9th International Conference on Learning Representations, ICLR 2021, Virtual Event, Austria, 3–7 May 2021. OpenReview.net (2021)
4. Giannakeris, P., Avgerinakis, K., Karakostas, A., Vrochidis, S., Kompatsiaris, I.: People and vehicles in danger - a fire and flood detection system in social media. In: 2018 IEEE 13th Image, Video, and Multidimensional Signal Processing Workshop (IVMSP), pp. 1–5 (2018)
5. Kim, G., Kwon, T., Ye, J.C.: DiffusionCLIP: text-guided diffusion models for robust image manipulation. In: Proceedings of the IEEE/CVF Conference on Computer Vision and Pattern Recognition (CVPR), pp. 2426–2435 (2022)
6. Li, M., et al.: Clip-event: Connecting text and images with event structures. In: Proceedings of the IEEE/CVF Conference on Computer Vision and Pattern Recognition (CVPR), pp. 16420–16429 (June 2022)
7. Li, S., et al.: Automatic near real-time flood detection using Suomi-NPP/VIIRS data. Remote Sens. Environ. **204**, 672–689 (2018)
8. Li, Y., Martinis, S., Wieland, M.: Urban flood mapping with an active self-learning convolutional neural network based on TerraSAR-X intensity and interferometric coherence. ISPRS J. Photogramm. Remote. Sens. **152**, 178–191 (2019)
9. Mangalam, K., et al.: Reversible vision transformers. In: 2022 IEEE/CVF Conference on Computer Vision and Pattern Recognition (CVPR), pp. 10820–10830 (2022)
10. Materzyńska, J., Torralba, A., Bau, D.: Disentangling visual and written concepts in clip. In: Proceedings of the IEEE/CVF Conference on Computer Vision and Pattern Recognition (CVPR), pp. 16410–16419 (2022)
11. Munawar, H.S., Ullah, F., Qayyum, S., Khan, S.I., Mojtahedi, M.: UAVs in disaster management: application of integrated aerial imagery and convolutional neural network for flood detection. Sustainability **13**(14), 7547 (2021)
12. Notti, D., Giordan, D., Caló, F., Pepe, A., Zucca, F., Galve, J.P.: Potential and limitations of open satellite data for flood mapping. Remote Sens. **10**(11), 1673 (2018)
13. Oddo, P.C., Ahamed, A., Bolten, J.D.: Socioeconomic impact evaluation for near real-time flood detection in the lower Mekong river basin. Hydrology **5**(2), 23 (2018)
14. Radford, A., et al.: Learning transferable visual models from natural language supervision. In: Meila, M., Zhang, T. (eds.) Proceedings of the 38th International Conference on Machine Learning, ICML 2021, 18–24 July 2021, Virtual Event. Proceedings of Machine Learning Research, vol. 139, pp. 8748–8763. PMLR (2021)
15. Sanghi, A., et al.: Clip-forge: towards zero-shot text-to-shape generation. In: Proceedings of the IEEE/CVF Conference on Computer Vision and Pattern Recognition (CVPR), pp. 18603–18613 (2022)

16. Shahabi, H., et al.: Flood detection and susceptibility mapping using sentinel-1 remote sensing data and a machine learning approach: Hybrid intelligence of bagging ensemble based on k-nearest neighbor classifier. Remote Sens. **12**(2), 266 (2020)
17. Shen, X., Wang, D., Mao, K., Anagnostou, E., Hong, Y.: Inundation extent mapping by synthetic aperture radar: a review. Remote Sens. **11**(7), 879 (2019)
18. Vaswani, A., et al.: Attention is all you need. In: Advances in Neural Information Processing Systems, pp. 5998–6008 (2017)

The NCKU-VTF Dataset and a Multi-scale Thermal-to-Visible Face Synthesis System

Tsung-Han Ho, Chen-Yin Yu, Tsai-Yen Ko, and Wei-Ta Chu(✉)

National Cheng Kung University, Tainan, Taiwan
tsaiyenk@andrew.cmu.edu, wtchu@gs.ncku.edu.tw

Abstract. We propose a multi-scale thermal-to-visible face synthesis system to achieve thermal face recognition. A generative adversarial network is constructed by one generator that transforms a given thermal face into a face in the visible spectrum, and three discriminators that consider multi-scale feature matching and high-frequency components, respectively. In addition, we provide a new paired thermal-visible face dataset called VTF that mainly contains Asian subjects captured in various visual conditions. This new dataset not only poses technical challenges to thermal face recognition, but also enables us to point out the race bias issue in current thermal face recognition methods. Overall, the proposed system achieves the state-of-the-art performance in both the EURECOM and NCKU-VTF datasets.

1 Introduction

With the rage of COVID-19, infrared face images have attracted much attention because they are formed based on heat signals from the skin tissues and are directly used for monitoring temperature. They also have been widely used in many domains like night-time surveillance and access control. Compared to face images in the visible spectrum, infrared face images are less affected by visual appearance changes caused by lighting/illumination variations.

Thermal face recognition aims to identify a face captured in the thermal spectrum (mid-wave infrared and long-wave infrared) by finding the most similar face captured in the visible spectrum. It is thus a cross-spectrum matching task. Two types of methods have been proposed in past studies. The first type focuses on extracting features from thermal faces and visual faces, and then embeds features into a common feature space so that faces of the same identity can be mapped into a locality [5,15]. The second type focuses on transforming thermal faces into their visible counterparts and adopting face recognition methods widely developed [2,4,12]. In this work, we focus on the second type.

Because of the modality gap, the relationship between thermal faces and visible faces is complicated and may be non-linear. We thus need a powerful mapping from the thermal spectrum to the visible spectrum while preserving the identity information. Thanks to recent development of generative adversarial

D.-T. Dang-Nguyen et al. (Eds.): MMM 2023, LNCS 13833, pp. 463–475, 2023.
https://doi.org/10.1007/978-3-031-27077-2_36

networks (GANs), we can synthesize visible faces from given thermal faces, and the synthesized faces can be matched against a gallery of visible faces [4].

Currently large-scale thermal face benchmarks with high variations are not widely available. The UND Collection X1 [3] consists only of paired thermal and visible faces in frontal view. The Nagoya dataset [11] is a small-scaled dataset containing paired thermal and visible faces in low resolution. The EURECOM dataset [13] contains faces in different poses, expressions, occlusions, and illuminations, and is more challenging. However, subjects in it are mainly Caucasian. We are wondering if there is a data bias issue and are curious how current methods work for other races. In this work we collect a large-scale high-variation dataset in high resolution, which mainly contains subjects from Asia. This dataset is named as National Cheng Kung University Visible-Thermal Face (NCKU-VTF) dataset.

Overall contributions of this work are summarized as follows.

- We present a GAN-based thermal to visible face synthesis system with the ideas of multi-scale synthesis.
- We collect a new large-scale benchmark consisting of high-variations paired thermal and visible faces where subjects mainly come from Asia.
- We evaluate on the EURECOM dataset and the NCKU-VTF dataset, and show state-of-the-arts thermal face recognition performance.

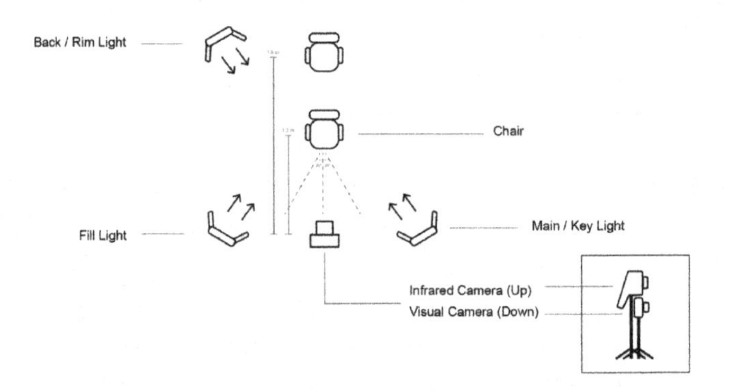

Fig. 1. Illustration of the setting to collect the dataset.

2 NCKU-VTF Dataset

2.1 Environment Setup

Figure 1 illustrates the room setting to collect the dataset. Temperature of this room is constantly set to 24 °C. During the shooting process, external light was blocked by curtains. We used three identical 2900 lux stand lamps for lighting, with the color temperature set to 3,300 K. The intensity of the lighting is set to 100% for the key light and the backlight, while set to 50% for the fill light.

We used a FLIR E75 Advanced Thermal Camera to capture thermal images, and a Sony α7III digital single-lens reflex (DSLR) camera with a lens of a 28–70mm focal length and a 3.5-5.6 aperture size to capture visible images. While shooting, we used default setting of the thermal camera, and the DSLR camera was set to 1/30 shutter speed, F5.6, and ISO 640.

2.2 Settings of Image Collection

The participants were mainly recruited from the school's public social media platform. After obtaining consent from the participants, they were requested to rest in the room for five to ten minutes to eliminate possible external influences on body temperature. During the shooting process, the staff gave the participants instructions to follow and provided them with the objects required for the occlusion category, like a cap. Each participant was requested to bring his/her own mask to create another type of occlusion. The visible and thermal images were taken simultaneously in the order detailed below.

This visible and thermal paired face database is with ground truths of identity and facial landmarks. The range of the participants' ages is between 19 to 57, the gender ratio of males to females is 1 to 3, and the ethnic distribution includes Indonesian, Colombian, Iranian, and Taiwanese. The collection settings mainly refer to the EURECOM dataset [13] with some extensions.

Each subject was captured two times separated by at least 7 days but not exceeding two weeks. We say images are captured in "two runs" in the following. In each run 30 pairs of visual and thermal face images were captured per person. The total number of face images in the NCKU-VTF dataset is 6,000. The variations are described as follows:

- Standard: The standard appearance of each subject is a face with no hair occlusion and in neutral emotion. The standard illumination is set to the key light and the fill light.
- Distance: Four pairs of face images captured with standard appearance, standard illumination, neutral expression in two different distances. As shown in Fig. 1, the chair can be moved to two positions, which are far from the camera by 1.2 m and 1.8 m, respectively. For one distance, the captured faces were captured in focus; while for the other distance, the faces were intentionally captured out of focus.
- Head pose: Eight pairs of face images were captured with standard appearance, standard illumination, neutral expression in eight different head poses: up, down, right at 30°, right up at 30°, right down at 30°, left at 30°, left up at 30°, and left down at 30°.
- Illumination: Five pairs of face images were captured with standard appearance, frontal view, neutral expression in five different illuminations: key light, fill light, rim light, all lights on, and all lights off.

- Expression: Six pairs of face images were captured with standard appearance, standard illumination, frontal view with six different facial expressions: neutral, happy, angry, sad, close eyes, and surprised.
- Occlusion: Seven pairs of face images were captured with standard illumination, frontal view, neutral expression with different types of occlusions: face mask, eyeglasses, sunglasses, cap, right eye occluded by the right hand, left eye occluded by the left hand, and face shape occluded by hair or bangs.

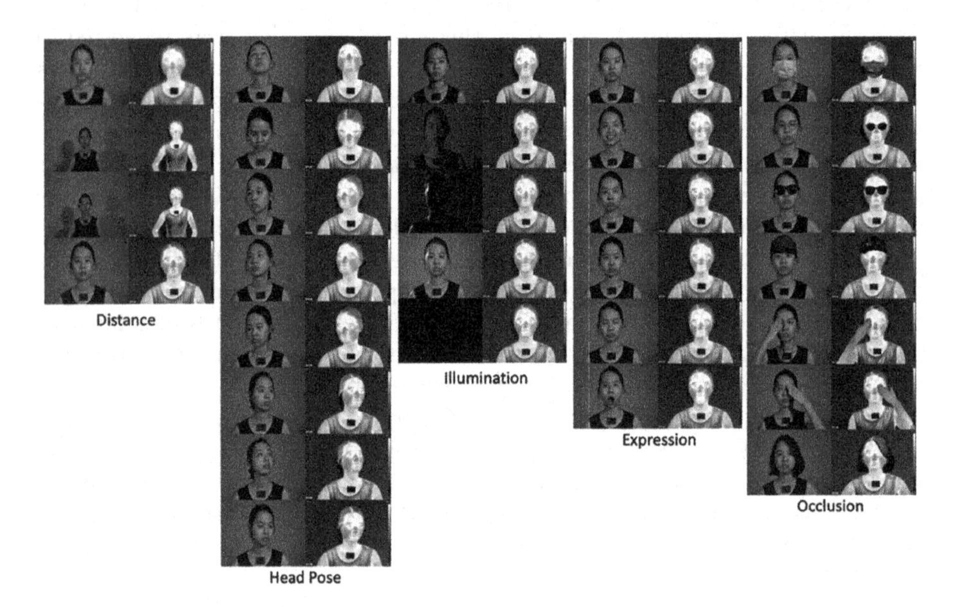

Fig. 2. Sample face pairs of the NCKU-VTF dataset.

Note that the face pair in the standard setting is the same as the first in-focus face pair. Therefore, we have totally 30 pairs of face images captured from each subject in each run. There are thus 60 pairs of face images per person when two runs of capturing were finished. Figure 2 shows sample face pairs of a subject in the NCKU-VTF dataset.

In addition to identity of each subject, we also provide accurate facial landmark labels for both thermal faces and visible faces. In order to ease the burden of landmark labeling, we first adopted HRNet [17] to detect 68 facial landmarks on visible faces. Positions of detected landmarks were then mapped to the corresponding thermal faces. Notice that the thermal face and visible face captured simultaneously are not perfectly aligned, and calibration is needed before mapping landmark coordinates. As shown in Fig. 2, each subject is attached with a cold black object on his/her neck, which four corners can be clearly seen in both thermal images and visible images. Based on coordinates of these four corners, we can find the affine transformation matrix to represent transformation between both images. With this matrix, coordinates of thermal facial landmarks are mapped to visible faces. Also notice that positions of detected landmarks are not perfect. Therefore, we need to manually fix landmark coordinates before mapping. Figure 3 shows samples of facial landmarks on thermal-visible face pairs.

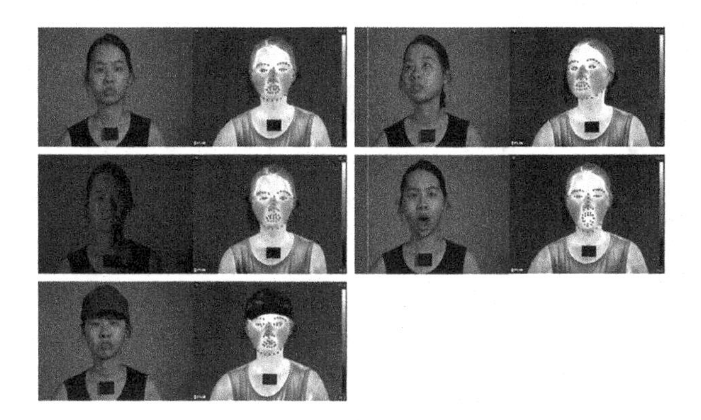

Fig. 3. Sample facial landmarks in the NCKU-VTF dataset.

3 The Proposed Method

Given a thermal face $\boldsymbol{T}$, we need to find a function f that synthesizes a face in the visible spectrum, i.e., $f(\boldsymbol{T}) = \boldsymbol{Q}$. Assuming that we have a set of visible faces as the gallery set $\mathcal{G}$, and the synthesized face image is the probe $\boldsymbol{Q}$, thermal face recognition is done by matching the probe with images in $\mathcal{G}$. The probe is recognized as the individual who is in $\mathcal{G}$ and has the face most similar to $\boldsymbol{Q}$.

3.1 Multi-scale Synthesis Framework

We propose a multi-scale synthesis model to do image transformation. This model consists of a generator $\boldsymbol{G}$ and two discriminators $\boldsymbol{D}_1$ and $\boldsymbol{D}_2$. Figure 4 illustrates the proposed framework. The generator is built based on the architecture proposed in [10], which consists of a convolutional layer, a series of residual blocks and a transposed convolutional layer. A thermal face of 512×512 pixels (after resizing) is passed through these components, and a synthesized visible face of 512×512 pixels is output.

We develop multi-scale discriminators $\boldsymbol{D}_1$ and $\boldsymbol{D}_2$, which have identical architecture, to examine images at two different scales. Specifically, we downsample the synthesized image and the corresponding visible image from 512×512 to 256×256. The 512×512 synthesized image and visible image are fed to $\boldsymbol{D}_1$, while the 256×256 versions are fed to $\boldsymbol{D}_2$.

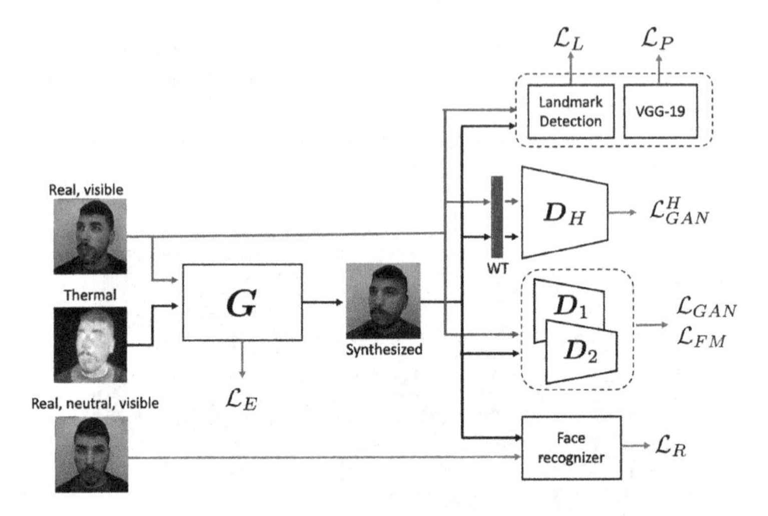

Fig. 4. The proposed multi-scale synthesis framework, associated with auxiliary models like facial landmark detection, perceptual feature extraction, high-frequency discriminator, and face recognizer.

3.2 Loss Functions

Assume that the training dataset consists of thermal-visible face pairs $\{(t_i, x_i)\}$, where t_i is a thermal face image and x_i is the corresponding visible face image. We describe details of loss functions in the following.

Adversarial Loss $\mathcal{L}_{GAN}$: With multi-scale discriminators $\boldsymbol{D}_1$ and $\boldsymbol{D}_2$, the adversarial loss $\mathcal{L}_{GAN}$ is defined as

$$\begin{aligned}\mathcal{L}_{GAN}(\boldsymbol{G}, \boldsymbol{D}_1, \boldsymbol{D}_2) = \mathbb{E}_x[\log \boldsymbol{D}_1(x) + \log \boldsymbol{D}_2(x)] \\ + \mathbb{E}_t[\log(1 - \boldsymbol{D}_1(\boldsymbol{G}(t)) + \log(1 - \boldsymbol{D}_2(\boldsymbol{G}(t))].\end{aligned} \tag{1}$$

Feature Matching Loss $\mathcal{L}_{FM}$: Features of visible faces are extracted by the discriminator. Conceptually we hope features of a synthesized face should be similar to its visible counterpart. The feature matching loss $\mathcal{L}_{FM}$ is defined as:

$$\mathcal{L}_{FM}(\boldsymbol{G}, \boldsymbol{D}_k) = \mathbb{E}_{t,x}\Big[\frac{1}{2}\sum_{k=1}^{2}\sum_{i=1}^{M}\frac{1}{N_i}[\|\boldsymbol{D}_k^{(i)}(x) - \boldsymbol{D}_k^{(i)}(\boldsymbol{G}(t))\|_1]\Big], \tag{2}$$

where $\boldsymbol{D}_k^{(i)}$ represents the output feature maps of the ith layer of $\boldsymbol{D}_k$, M is the total number of layers, and N_i is the number of elements in feature maps. The notation $\|\cdot\|_1$ denotes the L_1 norm.

Perceptual Loss $\mathcal{L}_P$: This loss is designed to measure the perceptual difference between a synthesized image and its visible counterpart.

$$\mathcal{L}_P(\boldsymbol{G}) = \mathbb{E}_{t,x}||\Phi_P(\boldsymbol{G}(t)) - \Phi_P(x)||_1, \tag{3}$$

where Φ_P denotes the perceptual features extractor, which actually is the VGG-19 model [16] pre-trained on the ImageNet dataset [14]. The features are extracted at multiple layers, and are then concatenated.

Feature Embedding Loss $\mathcal{L}_E$: In the generation process, we hope features extracted from the thermal face can be embedded into the ones similar to the features extracted from the corresponding visible face.

$$\mathcal{L}_E(\boldsymbol{G}) = \mathbb{E}_{t,x}\|\Phi_E(t) - \Phi_E(x)\|_1, \tag{4}$$

where $\Phi_E(t)$ and $\Phi_E(x)$ denote the feature maps output by the third residual block of the generator, respectively.

Facial Landmark Loss $\mathcal{L}_L$: To improve synthesis details on eyes, nose, and mouth, we consider positions of facial landmarks. We use a facial landmark detector Φ_L proposed in [1] to detect 68 facial landmarks, and calculate the normalized mean error.

$$NME(\boldsymbol{l}_1, \boldsymbol{l}_2) = \sum \frac{\|\boldsymbol{l}_1 - \boldsymbol{l}_2\|_2}{Kd}, \tag{5}$$

where $\boldsymbol{l}_1$ and $\boldsymbol{l}_2$ are the coordinates of landmarks on two faces. The value d is the interpupillary distance between the centers of two eyes, and the value K is the number of facial landmarks. The landmark loss $\mathcal{L}_L$ is defined as:

$$\mathcal{L}_L(\boldsymbol{G}) = \mathbb{E}_{t,x}NME(\Phi_L(\boldsymbol{G}(t)), \Phi_L(x)), \tag{6}$$

where Φ_L conceptually denotes the landmark detection module that outputs coordinates of landmarks $\boldsymbol{l}$'s.

Face Feature Loss $\mathcal{L}_R$: Preserving identity information is certainly needed to achieve accurate face recognition. We calculate the face feature loss $\mathcal{L}_R$ by measuring the distance between identity-specific features.

$$\mathcal{L}_R(\boldsymbol{G}) = \mathbb{E}_{t,\bar{x}}MSE(\Phi_R(\boldsymbol{G}(t)), \Phi_R(\bar{x})), \tag{7}$$

where $\Phi_R(\cdot)$ is the function representing a ResNet-50-based face recognizer that extracts identity-specific features, and $\bar{x}$ is the neutral face of an individual. Specifically, the output feature maps of the last residual block are averagely pooled to be the identity-specific features. The function $MSE(\cdot,\cdot)$ is the mean square error between two features.

Overall, combining all aforementioned loss functions, the loss of the proposed model is formulated as:

$$\mathcal{L}_{Mul} = \mathcal{L}_{GAN} + \lambda_{FM}\mathcal{L}_{FM} + \lambda_P\mathcal{L}_P + \lambda_E\mathcal{L}_E + \lambda_L\mathcal{L}_L + \lambda_R\mathcal{L}_R. \tag{8}$$

The weights to combine different losses are empirically set as: $\lambda_{FM} = 10$, $\lambda_P = 10$, $\lambda_E = 1$, $\lambda_L = 20$, and $\lambda_R = 1$.

3.3 High-Frequency Information

When working on face editing, Gao et al. [7] directly added high-frequency components to the generation process to make results more promising. In this work,

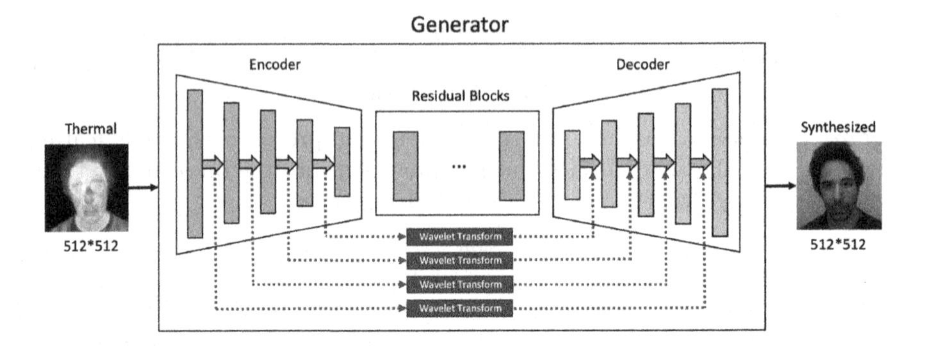

Fig. 5. Illustration of the structure of the thermal-to-visible generator.

we also investigate whether adding high-frequency components into the thermal-to-visible generator provides performance gain. Figure 5 illustrates structure of the generator, which consists of five convolutional layers (denoted as E_1 to E_5), followed by nine residual blocks, and then five de-convolutional layers (denoted as D_1 to D_5). The convolutional layers act as an encoder to extract features from the thermal face, the residual blocks implement the transformation function, and the de-convolutional layers act as a decoder to make detailed synthesis.

We especially apply the Haar wavelet transform to feature maps output by each convolutional layer, and then add the high-high (HH) part, high-low part (HL), and low-high (LH) part together to represent the high-frequency components. Let $\mathsf{W}(f(E_i))$ denote the high-frequency components computed from the output $f(E_i)$ of E_i. We argue that a good synthesized face should have high-frequency components similar to the corresponding visible counterpart. To achieve this, we extract their high-frequency components and feed them into a discriminator $\boldsymbol{D}_H$. The adversarial loss considering the high-frequency components is formulated as:

$$\mathcal{L}_{GAN}^{H}(\boldsymbol{G}, \boldsymbol{D}_H) = \mathbb{E}_x[\log \boldsymbol{D}_H(\mathsf{W}(x))] + \mathbb{E}_t[\log(1 - \boldsymbol{D}_H(\mathsf{W}(\boldsymbol{G}(t)))]. \tag{9}$$

Combining Eq. (9) with Eq. (8), the whole loss function for the proposed method is

$$\mathcal{L}_{All} = \mathcal{L}_{Mul} + \mathcal{L}_{GAN}^{H}. \tag{10}$$

The proposed model is trained by solving the minimax game:

$$\min_{\boldsymbol{G}} \max_{\boldsymbol{D}_1, \boldsymbol{D}_2, \boldsymbol{D}_H} \mathcal{L}_{All}(\boldsymbol{G}, \boldsymbol{D}_1, \boldsymbol{D}_2, \boldsymbol{D}_H). \tag{11}$$

The proposed method is implemented by PyTorch. Training thermal-visible face pairs are randomly flipped during the training phase to augment variations. The Adam optimizer is used to find optimal network parameters. The number of training epochs is 120. The learning rate is set as 0.0002 in the first 80 epochs, and is reduced by a factor of $(1 - 0.025 * n)$ after n subsequent epochs [18]. The batch size is 1.

4 Experimental Results

4.1 Experimental Settings

We evaluate the proposed approach based on the collected NCKU-VTF dataset and the EURECOM VIS-TH dataset [13]. The EURECOM dataset were collected from 50 subjects of different ages, genders, and ethnicities. Visible faces and thermal faces were captured in pairs for each subject. There are 21 different capturing settings, and thus this dataset contains $50 \times 2 \times 21 = 2100$ images in total. To recognize the synthesized face, we follow the settings in [12] where the LightCNN [19] model pretrained on the CASIA-Web Face [20] and MS-Celeb-1M [8] datasets is used as the face recognizer. The probe image's features extracted by LightCNN are compared with features of gallery images.

Table 1. Recognition accuracies of two methods on the NCKU-VTF dataset.

	Vis.	Thm.	Multi-Syn [4]	SAGAN [6]	Our
Distance (4)	81.00	6.50	12.50	14.75	22.25
Expression (6)	91.50	7.83	23.00	38.00	36.33
Head Pose (8)	66.60	11.20	13.75	20.00	17.13
Occlusion (7)	45.29	4.43	12.29	16.14	15.57
Illumination (5)	70.00	11.75	23.20	42.60	41.60
Average	70.88	8.34	16.95	26.30	26.58

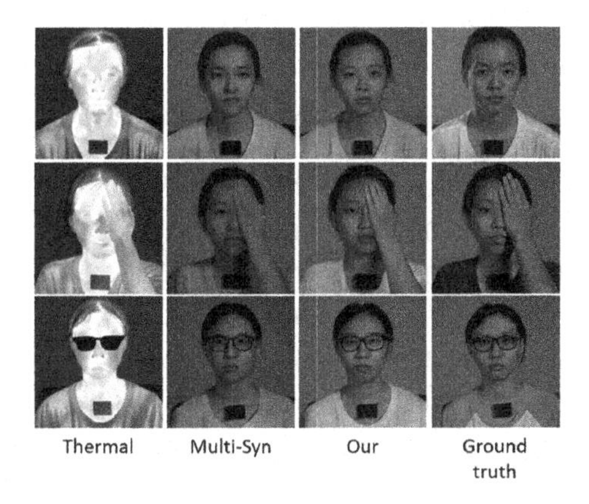

Fig. 6. Sample synthesis results obtained by two methods on the NCKU-VTF dataset.

4.2 Performance on the NCKU-VTF Dataset

Table 1 shows recognition accuracies of three methods on the NCKU-VTF dataset. As can be seen, our work consistently outperforms [4], which shows the effectiveness of adding high-frequency components. The method based on a self-attention generative adversarial network (SAGAN) [6] is also compared. Our method outperforms SAGAN, but the margin is relatively small. This shows the effectiveness of the self-attention module used in [6], and gives us hints to improve our generator network in the future.

Figure 6 shows sample synthesis results obtained by [4] and our method on the NCKU-VTF dataset. Comparing with ground truths, generally the proposed method synthesizes more facial details than [4].

4.3 Performance on the EURECOM Dataset

Table 2 shows recognition accuracies of various methods on the EURECOM dataset. The numbers shown in the parentheses of the leading column is the number of faces for each corresponding individual. For Pix2Pix, TV-GAN, and CRN, we directly quote the values provided in [12]. As can be seen, the proposed method outperforms the state of the arts.

Figure 7 shows sample synthesis results obtained by different methods. As can be seen, the CRN method synthesizes more informative details compared to Pix2Pix and TV-GAN, but some facial parts remain blurry. Our method gives much clearer synthesized results on important facial parts like eyes, nose, and mouth. The cases of occlusion and pose variations generally give rise to more challenges, and the synthesis results are generally worse.

Table 2. Recognition accuracies of various methods on the EURECOM dataset.

	Visible	Thermal	Pix2Pix [9]	TV-GAN [21]	CRN [12]	Our
Neutral (1)	100.00	32.00	48.00	54.00	82.00	90.00
Expression (6)	99.66	23.00	37.33	38.33	67.66	63.00
Head pose (4)	80.50	12.50	14.50	15.50	30.00	29.00
Occlusion (5)	98.80	14.40	16.40	25.00	44.80	39.20
Illumination (5)	87.20	15.60	29.60	35.20	63.60	76.00
Average	95.23	19.50	29.17	33.61	57.61	59.44

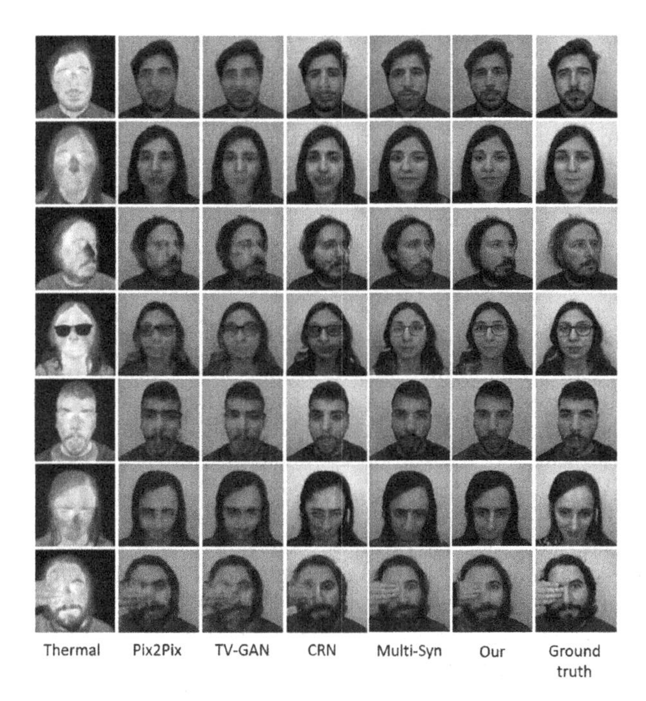

Fig. 7. Sample transformation results on the EURECOM dataset.

5 Conclusion

A multi-scale face synthesis system is presented to transform thermal faces into visible faces. This synthesis system is built based on the concept of adversarial learning, where one generator is constructed to do transformation, and three discriminators are constructed to examine features and high-frequency components in synthesized results and real faces. We provide a new paired thermal-visible face dataset that mainly contains Asians with many visual variations. With this new dataset, we evaluate performance of the proposed method.

Acknowledgement. This work was funded in part by Qualcomm through a Taiwan University Research Collaboration Project and in part by the National Science and Technology Council, Taiwan, under grants 111-3114-8-006-002, 110-2221-E-006-127-MY3, 108-2221-E-006-227-MY3, 107-2923-E-006-009-MY3, and 110-2634-F-006-022.

References

1. Bulat, A., Tzimiropoulos, G.: How far are we from solving the 2D & 3D face alignment problem? In: Proceedings of International Conference on Computer Vision, pp. 1021–1030 (2017)
2. Chatterjee, S., Chu, W.T.: Thermal face recognition based on transformation by residual U-net and pixel shuffle upsampling. In: Proceedings of International Conference on Multimedia Modelling (2020)

3. Chen, X., Flynn, P.J., Bowyer, K.W.: IR and visible light face recognition. Comput. Vis. Image Underst. **99**(3), 332–358 (2005)
4. Chu, W.T., Huang, P.S.: Thermal face recognition based on multi-scale image synthesis. In: Proceedings of International Conference on Multimedia Modelling (2021)
5. Chu, W.T., Wu, J.N.: A parametric study of deep perceptual model on visible to thermal face recognition. In: Proceedings of IEEE International Conference on Visual Communications and Image Processing (2018)
6. Di, X., Riggan, B.S., Hu, S., Short, N.J., Patel, V.M.: Polarimetric thermal to visible face verification via self-attention guided synthesis. In: Proceedings of IAPR International Conference on Biometrics (2019)
7. Gao, Y., et al.: High-fidelity and arbitrary face editing. In: Proceedings of IEEE/CVF Conference on Computer Vision and Pattern Recognition (2021)
8. Guo, Y., Zhang, L., Hu, Y., He, X., Gao, J.: MS-Celeb-1M: a dataset and benchmark for large-scale face recognition. In: Leibe, B., Matas, J., Sebe, N., Welling, M. (eds.) ECCV 2016. LNCS, vol. 9907, pp. 87–102. Springer, Cham (2016). https://doi.org/10.1007/978-3-319-46487-9_6
9. Isola, P., Zhu, J., Zhou, T., Efros, A.: Image-to-image translation with conditional adversarial networks. In: Proceedings of IEEE/CVF Conference on Computer Vision and Pattern Recognition, pp. 5967–5976 (2017)
10. Johnson, J., Alahi, A., Fei-Fei, L.: Perceptual losses for real-time style transfer and super-resolution. In: Leibe, B., Matas, J., Sebe, N., Welling, M. (eds.) ECCV 2016. LNCS, vol. 9906, pp. 694–711. Springer, Cham (2016). https://doi.org/10.1007/978-3-319-46475-6_43
11. Kresnaraman, B., Deguchi, D., Takahashi, T., Mekada, Y., Ide, I., Murase, H.: Reconstructing face image from the thermal infrared spectrum to the visible spectrum. Sensors **16**(4), 568 (2016)
12. Mallat, K., Damer, N., Boutros, F., Kuijper, A., Dugelay, J.: Cross-spectrum thermal to visible face recognition based on cascaded image synthesis. In: Proceedings of International Conference on Biometrics (2019)
13. Mallat, K., Dugelay, J.: A benchmark database of visible and thermal paired face images across multiple variations. In: Proceedings of International Conference of the Biometrics Special Interest Group (2018)
14. Russakovsky, O., et al.: ImageNet large scale visual recognition challenge. Int. J. Comput. Vision **115**(3), 211–252 (2015). https://doi.org/10.1007/s11263-015-0816-y
15. Sarfraz, M.S., Stiefelhagen, R.: Deep perceptual mapping for thermal to visible face recognition. In: Proceedings of British Machine Vision Conference (2015)
16. Simonyan, K., Zisserman, A.: Very deep convolutional networks for large-scale image recognition. In: Proceedings of International Conference on Learning Representations (2015)
17. Wang, J., et al.: Deep high-resolution representation learning for visual recognition. IEEE Transactions on Pattern Analysis and Machine Intelligence (2020)
18. Wang, T., Liu, M., Zhu, J., Tao, A., Kautz, J., Catanzaro, B.: High-resolution image synthesis and semantic manipulation with conditional GANs. In: Proceedings of IEEE/CVF Conference on Computer Vision and Pattern Recognition, pp. 8798–8807 (2018)
19. Wu, X., He, R., Sun, Z., Tan, T.: A light CNN for deep face representation with noisy labels. IEEE Trans. Inf. Forensics Secur. **13**(11), 2884–2896 (2018)

20. Yi, D., Lei, Z., Liao, S., Li, S.: Learning face representation from scratch (2014). arXiv:1411.7923
21. Zhang, T., Wiliem, A., Yang, S., Lovell, B.: TV-GAN: generative adversarial network based thermal to visible face recognition. In: Proceedings of International Conference on Biometrics (2018)

Link-Rot in Web-Sourced Multimedia Datasets

Viktor Lakic, Luca Rossetto(✉), and Abraham Bernstein

Department of Informatics, University of Zurich, Zurich, Switzerland
rossetto@ifi.uzh.ch

Abstract. The Web is increasingly used as a source for content of datasets of various types, especially multimedia content. These datasets are then often distributed as a collection of URLs, pointing to the original sources of the elements. As these sources go offline over time, the datasets experience decay in the form of link-rot. In this paper, we analyze 24 Web-sourced datasets with a combined total of over 270 million URLs and find that over 20% of the content is no longer available. We discuss the adverse effects of this decay on the reproducibility of work based on such data and make some recommendations on how they could be mediated in the future.

Keywords: Link rot · Dataset rot · Online datasets · Reproducibility

1 Introduction

Multimedia datasets enjoy increasing popularity, largely driven by the ever increasing need for training data of large machine learning models. The Web offers a convenient source from which to compile such datasets. Such collections of Web content are then often distributed as a list of URLs, directly pointing to the original source of the individual contained content elements. While this distribution method is convenient, it comes with a substantial drawback. Since commonly, the authors of a collection are not the authors of its content, they have no control over the availability of the contained elements. When individual elements become unavailable, the links contained within the collection break and the collection degrades, calling into question the reproducibility of any results obtained by using the dataset.

In this paper, we analyze the link-rot of Web-sourced datasets of various types. To do this, we query the original sources of 24 different datasets published between 2009 and 2022 and observe the overall success rate as well as the different error responses. We find that of the roughly 270 million URLs queried, over 20% do no longer return the expected content.

The remainder of this paper is structured as follows: after providing a brief overview of related work in Sect. 2, Sect. 3 introduces the 24 different datasets that were used in this study. Section 4 outlines the methodology used for the analysis of the datasets and their online availability before Sect. 5 details the results. We offer some concluding remarks in Sect. 6.

D.-T. Dang-Nguyen et al. (Eds.): MMM 2023, LNCS 13833, pp. 476–488, 2023.
https://doi.org/10.1007/978-3-031-27077-2_37

2 Related Work

Hyperlinks that do not or no longer point to their intended resource are a well-known occurrence in the Web. While at the very least a source of annoyance to the everyday Web-user, the impact of such 'broken' links can be more far-reaching. The Hiberlink project [22] studied the extent of the preservation of linked online resources in scholarly publications. Zhou et al. [32] and Klein et al. [13] found that already in 2014, over 20% of surveyed scholarly articles included Web links and that this rate appeared to be increasing. Experiments presented in [31] estimate that roughly 36% of all URLs contained in published scientific papers no longer point to their intended resource.

Also, outside of scholarly publications, the decay of online references has undesirable consequences. Dividino et al. [7] observe increasing error responses across a large number of linked open data endpoints over time and Zittrain et al. [34] find that 50% of URLs in US Supreme Court opinions no longer point to their original content. As several of these studies only consider `400` and `500` HTTP responses as rotten links, it is safe to assume that the actual number is probably even higher as many sites use 'soft-404s', which respond with a `200 OK` code but display a more user-friendly `404` error message [16].

While several suggestions and recommendations have been made to address these issues over the years [3,21], the problem still persists.

With the increased activities in the area of machine learning research in recent years, the need for ever more training data has become apparent and the Web offers a convenient source of such data in various modalities. Early Web-sourced multimedia datasets such as ImageNet [6], have been dwarfed by more recent collections such as LAION-400M [24] or LAION-5B.[1] The latter are exclusively distributed as a collection of URLs to the individual image sources, since both the size of the collection as well as the uncertain licensing of the individual contained items makes an other form of distribution inconvenient.

There has so far been no comprehensive analysis on the rate of (link-)rot experienced by such collections and its consequence for reproducibility of work based on them.

3 Datasets

For our analysis, we use 24 different Web-sourced datasets originally released for different purposes and containing different kinds of documents. The complete list of all considered Datasets is shown in Table 1.

The age range of the considered datasets spans roughly 13 years, the oldest one being from 2009 while the most recent one was published in 2022. The number of URLs per dataset ranges from 10^4 to 10^9, with the consequence that the two largest datasets together account for 81% of all URLs. Since our focus is on multimedia datasets, 15 of them contain references to videos, 6 contain images, with one of them containing both. The remaining 4 datasets are comprised of

[1] https://laion.ai/blog/laion-5b/.

Table 1. Datasets Overview sorted by number of URLs

Dataset	#URLs	Hosts	Content	Release
YouTube Speakers [23]	1'111	YouTube	Video	2013
Evoked Expressions in Video [28]	5'155	YouTube	Video	02.2021
RealEstate10K [33]	7'255	YouTube	Video	05.2018
Columbia Consumer Video [11]	9'317	YouTube	Video	04.2011
ActivityNet [8]	19'994	YouTube	Video	06.2015
V3C [20]	28'450	Vimeo	Video	10.2018
IACC [17]	39'937	Archive.org	Video	2010
What's Cookin' [15]	47'935	YouTube	Video	03.2015
PS-Battles [10]	102'028	Mixed	Image	04.2018
Document Similarity Triplets [5]	120'515	Wikipedia arxiv.org	Other	2014
NUS-WIDE [4]	257'789	Flickr	Image	07.2009
YouTube-BoundingBoxes [18]	285'410	YouTube	Video	02.2017
Kinetics 700–2020 [27]	643'459	YouTube	Video	10.2020
CORD-19 [30]	1'122'751	Mixed	Other	03.2022
Sports-1M [12]	1'133'158	YouTube	Video	04.2014
AudioSet [9]	2'084'320	YouTube	Video	06.2017
Conceptual Captions [25]	3'333'652	Mixed	Image	07.2018
Google Metadata for Datasets [2]	3'602'027	Mixed	Other	05.2020
YouTube-8M [1]	5'410'112	YouTube	Video	09.2016
Open Images Dataset V6 [14]	9'178'275	Flickr	Image	02.2020
Wikilinks [26]	10'888'549	Mixed	Other	10.2012
ImageNet [6]	14'197'119	Mixed	Image	2011
YFCC100M [29]	100'000'000	Flickr	Video Image	02.2016
Web Video in Numbers [19]	121'781'244	YouTube Vimeo	Video	07.2017
Total	274'299'562			
Total unique	270'828'632			

URLs referencing other resource types and are included for comparison. Out of the 15 datasets containing references to Web video, 11 exclusively use YouTube as a host. An additional two exclusively contain references to Vimeo[2] and the Internet Archive[3] respectively. The remaining one contains references to both YouTube and Vimeo. This uniformity in hosts is an almost exclusive property of the video datasets. Of the collections of URLs linking to images, only one of them uses one host exclusively; Flickr.[4] All the other image datasets as well as all datasets referencing other content have a diverse set of hosts from all over the Web.

[2] https://vimeo.com.
[3] https://archive.org.
[4] https://flickr.com.

Across all the analyzed datasets, there are a total of 270‘828‘632 unique URLs, which means that out of the 274‘299‘562 ones that form the naive total, 3‘470‘930 are duplicates. For the analyses presented in Sect. 5, we treat all datasets independently and can therefore ignore this overlap.

4 Methodology

To query the URLs, we used a custom distributed setup with a central coordination node and multiple worker instances hosted on various cloud providers. Workers would request batches of URLs from the coordinator via a simple REST API, query all URLs in the batch and report their findings back. The URLs were queried between April and June 2022 with some additional queries in August.

For Vimeo and YouTube, the workers could make use of APIs provided by the platforms in order to efficiently query the status of videos. This enabled the worker to request the status of multiple videos within one request, thereby reducing the required number of requests. Using the APIs of these platforms was reasonable since the set of possible errors reasonably to be expected from one of these platforms was limited to a well-formed response about a video no longer being available. For this analysis, we did not distinguish between semantic reasons for the unavailability of a video, such as a video being deleted by the user, regionally restricted, set to private, or removed by the platform for copyright violations.

For all other platforms, a worker would query an URL directly via a HTTP request. In case the worker could not establish a HTTP connection, it would retry at a later point in time for a maximum of three tries. If the 3rd try failed to establish any connection, it would record the type of network error (i.e., name resolution failure, connection refusal, connection timeout, etc.) as a result for that URL. In cases where a HTTP connection could be established, the worker would first check the status code. If a well-formed HTTP error is reported, the worker accepts it as a response for that URL. If a `200 OK` status is returned, the worker validates the actually returned content. It checks if the content type corresponds to the expected data (e.g., if a request to an URL ending in ‘.jpg’ has the content type ‘image/jpg’) and if the content itself is valid (e.g., if the returned image can be decoded). This is done in order to identify both inherently broken documents as well as ‘soft-404s’ which are human-readable error responses delivered with a valid status code, rendering them no longer machine readable.

Once a worker has queried all URLs of its current batch, it sends the obtained results back to the coordinator and requests a next batch. The coordinator stores all received results persistently for subsequent analysis.

5 Results

This section discusses the results obtained from the analysis of the datasets introduced in Sect. 3.

Table 2. Total number of available and unavailable elements per dataset

Dataset	Total	Available	Unavailable	Availability
YouTube Speakers	1‘111	1‘111	0	100.0%
PS-Battles	102‘028	101‘676	352	99.7%
Document Similarity Triplets	120‘515	119‘347	1‘168	99.0%
IACC	39‘937	38‘132	1‘805	95.5%
RealEstate10K	7‘255	6‘871	384	94.7%
Kinetics 700–2020	643‘459	600‘265	43‘194	93.3%
EEV	5‘155	4‘743	412	92.0%
YouTube-BoundingBoxes	285‘410	261‘539	23‘871	91.6%
Open Images Dataset V6	9‘178‘275	8‘237‘848	940‘427	89.8%
AudioSet	2‘084‘320	1‘804‘406	27‘9914	86.6%
YFCC100M	100‘000‘000	86‘388‘275	13‘611‘725	86.4%
What’s Cookin’	47‘935	40‘905	7‘030	85.3%
Conceptual Captions	3‘333‘652	2‘845‘237	48‘8415	85.3%
ActivityNet	19‘994	16‘846	3‘148	84.3%
YouTube-8M	5‘410‘112	4‘557‘852	852‘260	84.2%
Sports-1M	1‘133‘158	923‘190	209‘968	81.5%
V3C	28‘450	22‘729	5‘721	79.9%
NUS-WIDE	257‘789	205‘471	52‘318	79.7%
Google Metadata for Datasets	3‘602‘027	2‘784‘720	817‘307	77.3%
CORD-19	1‘122‘751	863‘362	259‘389	76.9%
Web Video in Numbers	121‘781‘244	90‘691‘509	31‘089‘735	74.5%
Columbia Consumer Video	9‘317	6‘893	2‘424	74.0%
Wikilinks	10‘888‘549	6‘495‘786	4‘392‘763	59.7%
ImageNet	14‘197‘119	7‘144‘800	7‘052‘319	50.3%
Total	274‘299‘562	214‘164‘681	60‘134‘881	78.1%

Table 2 shows the number of available and unavailable elements for each dataset as well as the fraction of availability. The mean-average availability across all datasets is 78.1%, meaning 21.9% of all queried URLs did not return the content referenced when the dataset was compiled. Out of the 24 tested datasets, only one is still completely available from the original sources. The most ‘link-rotten’ dataset is ImageNet, which is also among the oldest among the tested ones. In the roughly 11 years between its publication and our analysis, almost half of its original sources became unavailable.

While the age of a dataset is certainly a contributing factor to the amount of Link-Rot it shows, it lacks clear explanatory power. This can be seen in Fig. 1 which shows the amount of rot per dataset on the vertical and the time of its release on the horizontal axis. In this figure, the three datasets that are either composed of one type of content from two hosts or two content types from one host (i.e., Document Similarity Triples, YFCC100M, and Web Video in Numbers) are split in two to consider each host or type separately. The correlation between the age of an observed dataset and the amount of rot it experiences is 0.19 ($p = 0.348$). When only considering the video datasets sourced

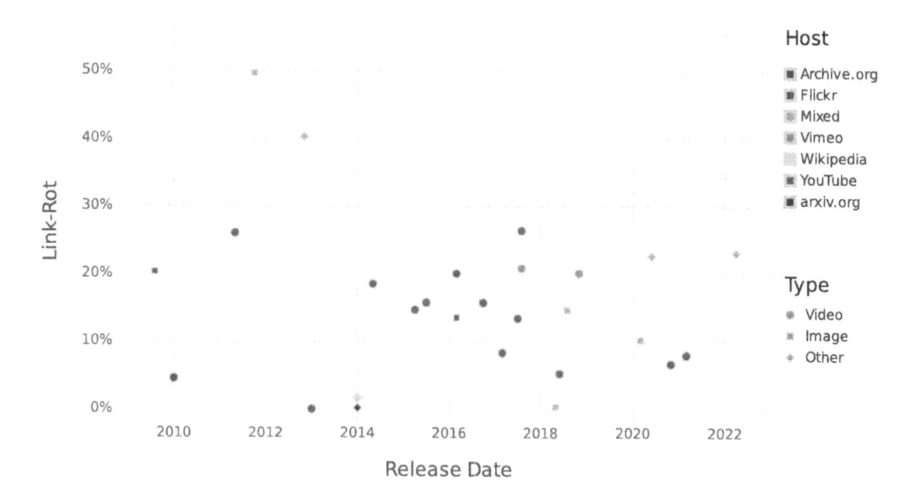

Fig. 1. Amount of Link-Rot relative to Dataset Release Date by Type and Host

from YouTube, which from the largest subset, the correlation grows to 0.35 ($p = 0.257$). The two datasets with the worst availability are both sourced from the Web at large and are both roughly a decade old at the time of sampling. Their availability is much lower when compared to other datasets of similar age that were sourced from one single hosting platform. It therefore stands to reason that the general decay rate in the Web might be higher compared to that of a dedicated content hosting platform (correlation: 0.73, $p = 0.064$). This makes intuitive sense, since a dedicated platform might have a stronger interest to guard against unintended content loss.[5] In contrast to more broadly sourced collections, datasets sourced from only one source do however have a single point of failure, as they would become completely unavailable if the one host they are sourced from would become permanently unavailable (however unlikely this might be for these particular platforms).

In contrast to the video hosting platforms, the sources used for the document and image datasets can all be delivered with a simple HTTP requests and are hence easier to obtain. Since communication is happening directly via HTTP rather than some platform-specific mechanism, a different set of failure states needs to be handled in case the request does not return the expected content. Table 3 shows the different errors that the workers encountered when querying the content of the various datasets.

The first failure case listed in the table consists of a request returning a well-formed HTTP response containing *invalid content*. This predominantly occurs in the form of so-called 'soft-404s' which is a human-readable error page delivered with a `200 OK` status code. This category also include requests that return

[5] This is to be seen as a general trend rather than a definite insight, as due to lack of release dates for individual URLs and the resulting crudity of the analysis, none of the correlations pass any reasonable threshold for statistical significance.

incomplete or otherwise unreadable files. These kinds of invalid responses can often be observed in *Conceptual Captions* and *ImageNet* which are both composed of images sourced from all over the Web.

The next category of problems encompasses everything happening one or more layers below HTTP, resulting in no valid HTTP connection being established. This includes connection refusals, connection timeouts, name resolution errors, etc. These errors predominantly occur when querying datasets from a large range of hosts.

The remaining categories consist of various well-formed HTTP errors. These can primarily be divided into two groups: client-side errors (`HTTP 4XX`) and server-side errors (`HTTP 5XX`). All error codes outside of these ranges are grouped into the *Other* column in this table. Those are largely composed of status codes that are not officially defined in the HTTP specification.[6]

Among the client side errors, the `404 Not Found` is the most common one, which is to be expected, as it is the appropriate reply for a request for a resource that does not exist. This is in contrast to `410 Gone`, which describes a resource that was valid once but no longer exists. Other common responses in this category include `401 Unauthorized` and `403 Forbidden`, both dealing with requests for which additional permissions would be required, as well as `400 Bad Request` that here appears to be used as a more generic catch-all error response. Several other error codes in the 400-range were observed as well, many of them lacking an official definition.[7]

In the remaining category of server-side errors, the most commonly observed is the generic `500 Internal Server Error`. Other commonly observed errors from this category include `502 Bad Gateway`, `503 Service Unavailable`, and `504 Gateway Timeout`. The less commonly observed status codes include a wide range, including many for which no official definition exists.[8]

An obvious solution to the decay of a dataset though link-rot is to produce an archived copy of all of its content, before it has a chance to decay. To do this, it is however necessary that all content elements of a dataset are licensed in such a way that allows their redistribution. Table 4 shows that out of the analyzed datasets, only 5 exclusively contain content that comes with such permissive licenses and for only 8 of the datasets, we were able to confirm that archival copies independent from the contents original source are available.

[6] Observed status codes below 400 and above 600: 101, 300, 301, 302, 303, 304, 307, *600*, *617*, *651*, *670*, *724*, *750*, *903*, *999*. Italicized codes have no generally accepted definition.

[7] Observed error codes between 400 and 500: 400, 401, 402, 403, 404, 405, 406, 407, 408, 409, 410, 412, 414, 415, 416, 417, 418, *419*, 420, 421, 422, 423, 424, 426, 429, 444, *445*, *447*, 449, *451*, *456*, *463*, *465*, *470*, *471*, *473*, *477*, *478*, *479*, *490*, *493*, *498*. Italicized codes have no generally accepted definition.

[8] Observed error codes between 500 and 600: 500, 501, 502, 503, 504, 505, 506, 507, 508, 509, 511, *512*, *520*, *521*, *522*, *523*, *524*, *525*, *526*, *529*, *530*, *533*, *534*, *535*, *543*, *555*, *556*, *567*, *591*. Italicized codes have no generally accepted definition.

Table 3. Distribution of reason for missing elements from Image and Document datasets. Most relevant HTTP error status codes are listed explicitly, others are shown in aggregate. *Invalid Content* counts instances of valid HTTP responses that delivered either invalid or incomplete documents or documents of a different datatype. *No HTTP Connection* counts instances where no HTTP connection could be established to the remote server. Relevant error codes include: `400 Bad Request`, `401 Unauthorized`, `403 Forbidden`, `404 Not Found`, `410 Gone`, `500 Internal Server Error`, `502 Bad Gateway`, `503 Service Unavailable`, and `504 Gateway Timeout`. *Other* codes include both officially defined HTTP status codes as well as such for which no official definition and agreed-upon semantics exists.

Dataset	Unavailable	Invalid Content	No HTTP Connection	HTTP 4XX						HTTP 5XX					Other
				400	401	403	404	410	Other	500	502	503	504	Other	
PS-Battles	352	30	36	2	1	50	197	2	12	8	2	7	0	5	0
Document Similarity Triplets	1‘168	128	0	0	0	0	1‘040	0	0	0	0	0	0	0	0
IACC	1‘805	1	86	0	0	91	101	0	0	1‘526	0	0	0	0	0
NUS-WIDE	52‘318	0	0	0	0	0	35‘141	17‘177	0	0	0	0	0	0	0
Conceptual Captions	488‘415	66‘228	113‘993	13‘933	3‘666	63‘818	187‘610	3175	23‘549	5‘467	1420	4‘669	25	759	9
CORD-19	259‘389	0	10‘967	0	7	176‘583	5‘618	3‘452	2‘142	124	8	60‘488	0	0	0
Google Metadata for Datasets	817‘307	0	49‘611	254	1	540‘485	169‘119	5‘613	49‘189	886	51	2‘098	0	0	0
Open Images Dataset V6	940‘427	1	13	0	0	0	734‘184	206‘171	0	19	39	0	0	0	0
Wikilinks	4‘392‘763	0	1‘486‘965	12‘289	7‘351	679‘336	1‘605‘625	132‘304	356‘554	50‘951	3‘763	35‘848	902	17‘841	1‘659
ImageNet	7‘052‘319	720‘890	1‘785‘527	27‘642	4‘635	462‘426	3‘377‘523	406‘918	181‘403	29‘716	5‘784	4‘0977	2‘291	4‘874	347
YFCC100M	13‘611‘725	3	2‘586	0	0	1	9‘860‘215	3‘746‘659	0	294	1‘966	0	1	0	0

Table 4. Availability of dataset content under a permissive license and availability of a complete copy of the linked content. *Google Metadata for Datasets*, *Web Video in Numbers*, and *Wikilinks* are omitted from this table, as these datasets are only concerned with the links themselves and not the actual content they point to.

Dataset	Permissive licence	Copy available
ActivityNet		✓
AudioSet		
Columbia Consumer Video		
Conceptual Captions		
CORD-19		✓
Document Similarity Triplets	✓	
Evoked Expressions in Video		
IACC	✓	✓
ImageNet		✓
Kinetics 700–2020		✓
NUS-WIDE		
Open Images Dataset V6	✓	✓
PS-Battles		
RealEstate10K		
Sports-1M		
V3C	✓	✓
What's Cookin'		
YFCC100M	✓	✓
YouTube Speakers		
YouTube-8M		
YouTube-BoundingBoxes		

6 Conclusion

We presented an analysis of 24 Web-sourced datasets and found that of the combined 270 million URLs, over 20% no longer point to the intended content. Only one of the 24 datasets was unaffected by this link-rot while the most affected has decayed by almost half. We found that datasets sourced from the Web at large experience the largest amount of link-rot and the greatest diversity in observed errors during querying. While the amount of link-rot of a dataset can only increase over time, age is only one factor among several. Only for a small subset of the studied datasets were we able to confirm that a complete archival copy exists and is available. Even fewer datasets are exclusively composed of content that would easily allow for such re-distribution.

Since it is not foreseeable that the need for diverse datasets of various modalities will decrease, it is to be expected that the observed link-rot will have increasingly adverse effects of the reproducibility of scientific work that makes use of such data. Authors of such datasets, especially when sourcing the content from the Web, should therefore take care that the content of their collections is licensed in such a way that enables re-distribution. Further, they should endeavor to keep a complete reference copy of all the content of their dataset and make at least those parts of it available upon request that can no longer be obtained from the original sources. In cases where the collections become too large to be easily handled, it might become necessary to employ distributed storage solutions so that the ones interested in using a dataset can contribute to its hosting, thereby ensuring future availability while sharing the incurred resource requirements.

Acknowledgements. This work has been partially supported by the Swiss National Science Foundation, Project "MediaGraph" (Grant Number 202125).

References

1. Abu-El-Haija, S., et al.: Youtube-8m: a large-scale video classification benchmark. CoRR abs/1609.08675 (2016), http://arxiv.org/abs/1609.08675
2. Brickley, D., Burgess, M., Noy, N.: Google dataset search: building a search engine for datasets in an open web ecosystem. In: The World Wide Web Conference, pp. 1365–1375. WWW 2019, Association for Computing Machinery, New York, NY, USA (2019). https://doi.org/10.1145/3308558.3313685
3. Burnhill, P., Mewissen, M., Wincewicz, R.: Reference rot in scholarly statement: threat and remedy. Insights 28(2) (2015). https://doi.org/10.1629/uksg.237
4. Chua, T.S., Tang, J., Hong, R., Li, H., Luo, Z., Zheng, Y.T.: NUS-WIDE: a real-world web image database from national university of Singapore. In: Proceedings of the ACM International Conference on Image and Video Retrieval. CIVR 2009, Association for Computing Machinery, New York, NY, USA (July 8–10, 2009). https://doi.org/10.1145/1646396.1646452
5. Dai, A.M., Olah, C., Le, Q.V.: Document embedding with paragraph vectors. In: NIPS Deep Learning Workshop (2014)
6. Deng, J., Dong, W., Socher, R., Li, L.J., Li, K., Fei-Fei, L.: ImageNet: a large-scale hierarchical image database. In: 2009 IEEE Conference on Computer Vision and Pattern Recognition, pp. 248–255. IEEE (2009). https://doi.org/10.1109/CVPR.2009.5206848
7. Dividino, R., Kramer, A., Gottron, T.: An investigation of HTTP header information for detecting changes of linked open data sources. In: Presutti, V., Blomqvist, E., Troncy, R., Sack, H., Papadakis, I., Tordai, A. (eds.) ESWC 2014. LNCS, vol. 8798, pp. 199–203. Springer, Cham (2014). https://doi.org/10.1007/978-3-319-11955-7_18
8. Caba Heilbron, F., Victor Escorcia, B.G., Niebles, J.C.: ActivityNet: a large-scale video benchmark for human activity understanding. In: Proceedings of the IEEE Conference on Computer Vision and Pattern Recognition, pp. 961–970 (June 2015). https://doi.org/10.1109/CVPR.2015.7298698
9. Gemmeke, J.F., et al.: Audio set: an ontology and human-labeled dataset for audio events. In: 2017 IEEE International Conference on Acoustics, Speech and Signal Processing (ICASSP), pp. 776–780. IEEE, New Orleans, LA (2017). https://doi.org/10.1109/ICASSP.2017.7952261

10. Heller, S., Rossetto, L., Schuldt, H.: The PS-battles dataset - an image collection for image manipulation detection. CoRR abs/1804.04866, arXiv:1804.04866 (Apr 2018). https://doi.org/10.48550/ARXIV.1804.04866
11. Jiang, Y.G., Ye, G., Chang, S.F., Ellis, D., Loui, A.C.: Consumer video understanding: a benchmark database and an evaluation of human and machine performance. In: Proceedings of ACM International Conference on Multimedia Retrieval (ICMR), oral session, pp. 1–8. ICMR 2011, Association for Computing Machinery, New York, NY, USA (2011). https://doi.org/10.1145/1991996.1992025
12. Karpathy, A., Toderici, G., Shetty, S., Leung, T., Sukthankar, R., Fei-Fei, L.: Large-scale video classification with convolutional neural networks. In: 2014 IEEE Conference on Computer Vision and Pattern Recognition, CVPR 2014, Columbus, OH, USA, June 23–28, 2014, pp. 1725–1732. IEEE Computer Society (2014). https://doi.org/10.1109/CVPR.2014.223
13. Klein, M., et al.: Scholarly context not found: one in five articles suffers from reference rot. PLOS ONE 9(12), 1–39 (12 2014). https://doi.org/10.1371/journal.pone.0115253
14. Krasin, I., et al.: OpenImages: a public dataset for large-scale multi-label and multi-class image classification. (2017)
15. Malmaud, J., Huang, J., Rathod, V., Johnston, N., Rabinovich, A., Murphy, K.: What's cookin'? Interpreting cooking videos using text, speech and vision (2015). https://doi.org/10.48550/ARXIV.1503.01558
16. Meneses, L., Furuta, R., Shipman, F.: Identifying "Soft 404" error pages: analyzing the lexical signatures of documents in distributed collections. In: Zaphiris, P., Buchanan, G., Rasmussen, E., Loizides, F. (eds.) TPDL 2012. LNCS, vol. 7489, pp. 197–208. Springer, Heidelberg (2012). https://doi.org/10.1007/978-3-642-33290-6_22
17. Over, P., Awad, G., Smeaton, A.F., Foley, C., Lanagan, J.: Creating a web-scale video collection for research. In: Proceedings of the 1st Workshop on Web-scale multimedia corpus, pp. 25–32. WSMC 2009, Association for Computing Machinery, New York, NY, USA (2009). https://doi.org/10.1145/1631135.1631141
18. Real, E., Shlens, J., Mazzocchi, S., Vanhoucke, V., Pan, X.: Youtube-boundingboxes: a large high-precision human-annotated dataset for object detection in video. In: Proceedings of the IEEE Conference on Computer Vision and Pattern Recognition (CVPR), pp. 7464–7473 (July 2017). https://doi.org/10.48550/ARXIV.1702.00824
19. Rossetto, L., Schuldt, H.: Web video in numbers - an analysis of web-video metadata (2017). https://doi.org/10.48550/ARXIV.1707.01340
20. Rossetto, L., Schuldt, H., Awad, G., Butt, A.A.: V3C – a research video collection. In: Kompatsiaris, I., Huet, B., Mezaris, V., Gurrin, C., Cheng, W.-H., Vrochidis, S. (eds.) MMM 2019. LNCS, vol. 11295, pp. 349–360. Springer, Cham (2019). https://doi.org/10.1007/978-3-030-05710-7_29
21. Sanderson, R., Phillips, M., de Sompel, H.V.: Analyzing the persistence of referenced web resources with memento. CoRR abs/1105.3459 (2011). http://arxiv.org/abs/1105.3459
22. Sanderson, R., Van de Sompel, H., Burnhill, P., Grover, C.: Hiberlink: towards time travel for the scholarly web. In: Proceedings of the 1st International Workshop on Digital Preservation of Research Methods and Artefacts, p. 21. Association for Computing Machinery, New York, NY, USA (2013). https://doi.org/10.1145/2499583.2500370

23. Schmidt, L., Sharifi, M., Lopez-Moreno, I.: Large-scale speaker identification. In: 2014 IEEE International Conference on Acoustics, Speech and Signal Processing (ICASSP), pp. 1650–1654. IEEE (2014). https://doi.org/10.1109/ICASSP.2014.6853878
24. Schuhmann, C., et al.: LAION-400M: open dataset of clip-filtered 400 million image-text pairs. CoRR abs/2111.02114 (2021). https://arxiv.org/abs/2111.02114
25. Sharma, P., Ding, N., Goodman, S., Soricut, R.: Conceptual captions: a cleaned, hypernymed, image alt-text dataset for automatic image captioning. In: Proceedings of the 56th Annual Meeting of the Association for Computational Linguistics (Volume 1: Long Papers), pp. 2556–2565. Association for Computational Linguistics, Melbourne, Australia (Jul 2018). https://doi.org/10.18653/v1/P18-1238, https://aclanthology.org/P18-1238
26. Singh, S., Subramanya, A., Pereira, F., McCallum, A.: Wikilinks: a large-scale cross-document coreference corpus labeled via links to Wikipedia. Tech. Rep. UM-CS-2012-015, University of Massachusetts, Amherst (2012)
27. Smaira, L., Carreira, J., Noland, E., Clancy, E., Wu, A., Zisserman, A.: A short note on the kinetics-700-2020 human action dataset (2020). https://doi.org/10.48550/ARXIV.2010.10864
28. Sun, J.J., Liu, T., Cowen, A.S., Schroff, F., Adam, H., Prasad, G.: EEV: a large-scale dataset for studying evoked expressions from video. arXiv preprint arXiv:2001.05488 abs/2001.05488 (2021). https://doi.org/10.48550/ARXIV.2001.05488
29. Thomee, B., et al.: YFCC100M: the new data in multimedia research. Communications of the ACM 59(2), 64–73 (Jan 2016). https://doi.org/10.1145/2812802
30. Wang, L.L., et al.: CORD-19: the covid-19 open research dataset. In: Proceedings of the 1st Workshop on NLP for COVID-19 at ACL 2020. Association for Computational Linguistics (Jul 2020). https://www.aclweb.org/anthology/2020.nlpcovid19-acl.1

31. Zhou, K., Grover, C., Klein, M., Tobin, R.: No more 404s: predicting referenced link rot in scholarly articles for pro-active archiving. In: Proceedings of the 15th ACM/IEEE-CS Joint Conference on Digital Libraries, pp. 233–236. JCDL 2015, Association for Computing Machinery, New York, NY, USA (2015). https://doi.org/10.1145/2756406.2756940
32. Zhou, K., Tobin, R., Grover, C.: Extraction and analysis of referenced web links in large-scale scholarly articles. In: IEEE/ACM Joint Conference on Digital Libraries, JCDL 2014, London, United Kingdom, September 8–12, 2014, pp. 451–452. IEEE Computer Society, New York, NY, USA (2014). https://doi.org/10.1109/JCDL.2014.6970220
33. Zhou, T., Tucker, R., Flynn, J., Fyffe, G., Snavely, N.: Stereo magnification: learning view synthesis using multiplane images. ACM Trans. Graph. (Proc. SIGGRAPH) 37(4), 1–12 (07 2018). https://doi.org/10.1145/3197517.3201323
34. Zittrain, J., Albert, K., Lessig, L.: Perma: scoping and addressing the problem of link and reference rot in legal citations. Legal Inf. Manag. **14**(2), 88–99 (2014). https://doi.org/10.1017/S1472669614000255

People@Places and ToDY: Two Datasets for Scene Classification in Media Production and Archiving

Werner Bailer(✉) and Hannes Fassold

Joanneum Research, Graz, Austria
{werner.bailer,hannes.fassold}@joanneum.at

Abstract. In order to support common annotation tasks in visual media production and archiving, we propose two datasets which cover the annotation of the bustle of a scene (i.e., populated to unpopulated), the cinematographic type of a shot as well as the time of day and season of a shot. The dataset for bustle and shot type, called People@Places, adds annotations to the Places365 dataset, and the ToDY (time of day/year) dataset adds annotations to the SkyFinder dataset. For both datasets, we provide a toolchain to create automatic annotations, which have been manually verified and corrected for parts of the two datasets. We provide baseline results for these tasks using the EfficientNet-B3 model, pretrained on the Places365 dataset.

Keywords: Datasets · Neural networks · Cinematography · Video retrieval

1 Introduction

While research has moved from image classification to object detection, segmentation and other more advanced topics, performing classifications of images or entire shots of videos is still a practically relevant task in describing visual content in order to make it findable. This task occurs when describing newly arriving content for production purposes (e.g., news) or annotating large amounts of otherwise sparsely documented content in media archives. Locations are among the three most frequently used search facets in video archive search [13]. For many purposes in visual content creation, place categories (i.e., street, shopping mall) are needed rather than named locations. Automatically labeling images or video shots with such location categories is a typical classification problem, and the Places365 dataset [31] is a very well known resource for this task. However, in a practical setting, there are other key properties of the scene, that are relevant to judge whether a shot is usable or not.

First, it is important to know whether the scene is "empty", or there are people or vehicles visible. We call this property "bustle", i.e., whether there are traces of people being active in that scene or not. While it has always been an

D.-T. Dang-Nguyen et al. (Eds.): MMM 2023, LNCS 13833, pp. 489–501, 2023.
https://doi.org/10.1007/978-3-031-27077-2_38

important query criteria to explicitly look for a quiet or busy view of the scene, the recent COVID-19 pandemic has made that a much requested feature, as depending on the level of restrictions valid at that time, news reports require either empty or populated street scenes.

Second, the shot type (sometimes called shot size) is a key cinematographic property, which determines the importance of a subject, and the context in which a particular shot can be used. The shot type is typically defined by the height ratio of the depicted persons in relation to the view.

Third, for outdoor shots the time of day and the season are important properties. A news editor searching outdoor shots of a building (e.g., house of parliament) wants to find shots that match the season of the story, as well as day or nighttime. For more scenic views, a sunrise or sunset shot is often requested.

Although these are not uncommon properties of content, there are hardly any datasets covering these properties – in particular, datasets with sizes useful for applying deep learning. We propose an automatic workflow to add relevant annotations to these datasets, performing manual annotations where required. In particular, the contributions of this paper are:

- We propose the People@Places dataset, based on Places365, adding bustle (6 classes) and shot type (9 classes) annotations.
- We propose the ToDY (time of day/year) dataset, based on Skyfinder [17], adding time of day (5 classes) and season (4 classes) annotations.
- We provide a baseline for the classification tasks on these datasets, using an efficient state of the art approach.
- We provide the toolchains that were used to create the two datasets, which can be used to replicate this approach for other datasets.

The rest of this paper is organized as follows. After discussing related work in Sect. 2, Sect. 3 describes the dataset creation process for People@Places, and Sect. 4 describes the creation process for ToDY. We present experimental results using the baseline in Sect. 5 and Sect. 6 concludes the paper.

2 Related Work

We review related work on location, shot type and time of day/season classification, and for the detectors used for automatic dataset annotation. To the best of our knowledge, there is no existing work on bustle classification. The closest tasks seem to be people counting or crowd estimation, but those differ as we consider both persons and vehicles, while we are not interested in the exact numbers.

For *location type classification*, many traditional classification architectures, such as the VGG or ResNet families have been applied. Global covariance pooling is proposed in [26] to capture richer features and improve generalization. One variant of this approach, iterative matrix square root normalized covariance pooling network (iSQRT-COV-Net) used to be the best performing method on Places365, while RS-VGG16 [18] is a recent method proposing a compact model derived from

VGG16. In the last few months, vision transformer models such as ViT have taken the lead [28]. A recent extension using large transformer models (86-632M parameters), and self-supervision using masked autoencoders (MAE) is to the best of our knowledge currently the best performing model for classification on Places365.

Like other computer vision tasks, *shot type* (sometimes referred to as *shot size*) *classification* is primarily addressed with deep learning approaches, either approached using CNNs directly for classification [20], using general semantic segmentation [4] or focusing on separating the subject from the background and feeding the regions into a two-stream network [19]. One issue with shot type classification is that the datasets used in many works are not accessible, as they rely on materials from motion picture films that cannot be distributed due to copyright restrictions.

The *classification of time of day* and *season* is a topic that seems to be somewhat neglected. An early work, [7] proposes a system for season classification, but relies on color histograms and the amount of exposed skin of the depicted persons rather than on training samples. The TRECVID semantic indexing task [3] included daytime/nighttime as concepts, and the task was addressed both with traditional machine learning as well as early deep learning methods. However, except for the limitation to only two classes, the resolution and quality of this dataset is quite limited. The Youtube-8M dataset [1] covers some of the relevant classes (sunset, sunrise, night, autumn and winter), while the rest of the times of day and seasons are missing. Some vocabularies from the broadcast domain cover time of day (e.g., EBU LocationTime[1]) or season (e.g., TV-Anytime Weather [9]), but no annotated images are provided in this context.

For annotating the dataset for bustle and shot type with vehicles, persons and size of the (partial) persons in the image, we employ object detection, face detection and human pose detection. We employ YoloV4-CSP [25], which combines the CSP-Net proposed in YoloV4 [5] with an efficient model scaling strategy [25], a combination which provides us a highly accurate detector with a low inference time. RetinaFace [8] was chosen as one of the top performing methods on the challenging WIDER Face [29] hard split. For human pose detection, we employ the ROMP algorithm [22]. We chose this method because it is one of the top performing methods on a very realistic (and consequently difficult) dataset named 3D Poses in the Wild [16]. Furthermore, in contrast to other methods (like [14]) which performs also quite well on this dataset, it is a computationally efficient single-stage method which does the pose detection for all persons occurring in the image simultaneously.

One could argue that recent advances in foundation models including both vision and language such as Florence [30] or Flava [21] will sooner or later eliminate the need of training classifiers for particular task, and instead allow adaptation of models using few or no samples. This may be true, however, we strongly believe that this does not eliminate the need for specific datasets that allow the evaluation of such generic models on these tasks.

[1] https://www.ebu.ch/metadata/ontologies/ebucore/ebucore_LocationTimeType.html.

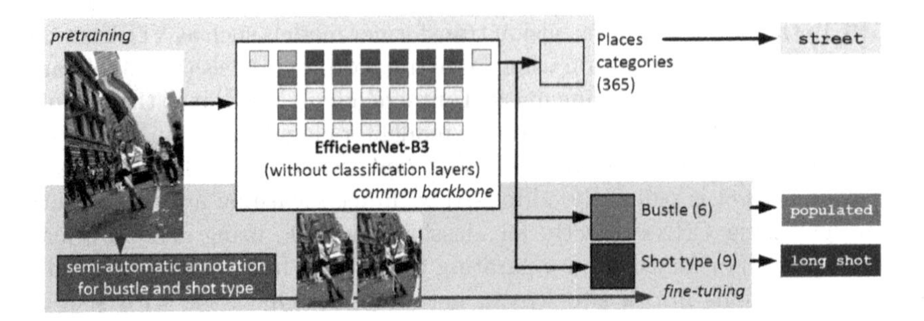

Fig. 1. Pretrained on fine-grained places categories, the backbone of the network is used to train classification heads for supercategories, bustle and shot type. Bustle/shot type annotations are created automatically (manually corrected for the validation set).

3 People@Places: Dataset for Bustle and Shot Type Classification

We amend the Places365-Standard dataset (high resolution images) with per image annotations for bustle and shot type. For bustle, we define six classes from entirely unpopulated to populated, resulting from discussions with domain experts from media production and archiving. The classification treats few large persons or vehicles separately, in order to address cases where those are in the focus of the image. Otherwise the classes use a combination of the number and size of objects, expressed by the image area covered together by these objects (see Table 1).

For shot types, there are a number of taxonomies that differ in the level of detail. All of them use the size of the main person depicted in the shot as reference. We use the IPTC NewsCodes scene types[2], and the lists proposed by Arijon [2], Galvane [11] and Rao et al. [19] as sources, but decided to go for a finer classification (see Table 1). As the annotations of the Places365 test split are not provided (as part of a benchmark) we work with the training and validation splits in this paper, to which we have full access.

The dataset creation process is semi-automatic, where automatic annotation is performed for the entire dataset, and manual verification is performed for the validation split. The process for creating the annotations is shown in Fig. 2. The bustle classes depend on the presence of persons and vehicles, thus object detections for these classes are used. While person detections give a coarse indication about the size of depicted person, it is not clear which part of the person is visible. Human pose estimation and face detection are used to complement this information. In detail the process consists of the following steps.

Object Detection. We run YOLOv4 CSP [25] (trained on MS COCO) over all the images, considering all detections with a score ≤ 0.1 as no occurrence. From the

[2] https://cv.iptc.org/newscodes/scene/.

Table 1. Definition of bustle and shot type classes.

Class	Definition
Bustle	
unpopulated	no persons or vehicles
few people	< 3 persons, no vehicles, area < 10%
few vehicles	< 3 vehicles, no persons, area < 20%
few large	< 3 people/vehicles, any area
medium	< 11 people/vehicles, area < 30%
populated	more people/vehicles or covering larger area
Shot type	
extreme close-up	detail of face
close-up	head
medium close-up	cut under chest
tight medium shot	cut under waist
medium shot	cut under crotch
medium full shot	cut under knee
full shot	person fully visible
long shot	person 1/3 of frame height
extreme long shot	person <1/3 of frame height

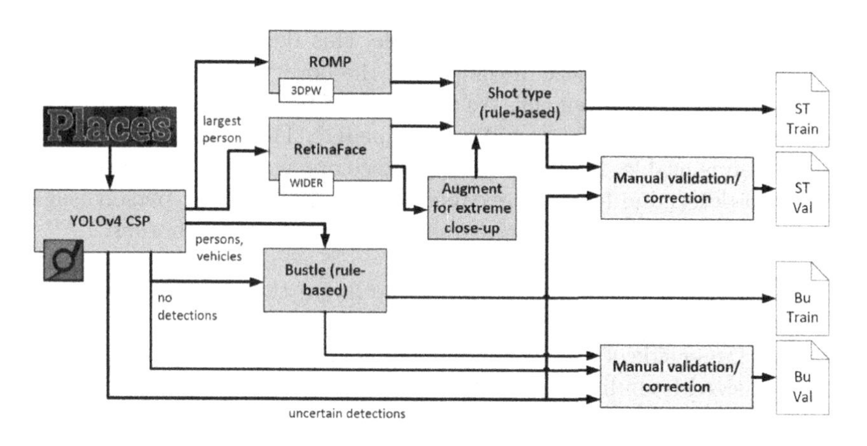

Fig. 2. Dataset creation process for bustle and shot type annotations.

remaining detections, those with a score ≥ 0.5 are kept as reliable. Detections between these thresholds are considered uncertain, and the images are excluded. From the detections, persons and vehicles (i.e., the classes bicycle, car, motorcycle, airplane, bus, train, truck, boat) are kept. Based on the criteria defined in Table 1, the bustle annotation is created. In addition, the tallest person is selected and output as annotation.

Face Detection. Face detection is performed using RetinaFace [8], with a model trained on WIDER Face [29], on all images that contain person detections. Multiple faces may overlap the tallest person, and it is not always straight forward to identify the correct one. We keep the face region with the largest size of the intersecting area, weighted by the detection confidence, i.e. $sc_f = (F \cap P)c_f$, where F is the face region, P is the person region and c_f is the confidence reported by the face detector.

Human Pose Detection. We use the ROMP [22] human pose detector (trained on 3DPW [16]), applied to a cropped out image of the tallest detected person (resp. the visible part of it). We obtain a 2D skeleton (SMPL [15] with 54 points), of which we use 10 (pelvis, left/right foot, head, left/right hip, thorax, left/right knee, spine).

Person Size Estimation. In order to filter unreliable detections, we filter pose and face detections for which $\max(w_D, h_D) \geq \tau \min(w_P, h_P)$, where w and h denote width and height, D denotes the pose/face detection bounding box and P denotes the person detection bounding box. τ is set to 0.1 for faces, and 0.6 for poses. If a reliable pose is found, we use it for person size estimation. We use the legs only if they appear to be stretched, i.e. head and at least one foot are on different sides of a horizontal line through the pelvis point, and the hip to feet distance is larger than the thorax to pelvis distance. If the legs are used, we check if feet and hip are on different sides of the knee (at least for one leg), otherwise we ignore the feet. If head to feet is visible, this determines the person size, otherwise we estimate the size of the part of the body not considered reliable to get the overall size measurement. We use ratios of body proportions from [10], a compact visualisation can be found on Wikipedia[3]. This is also done if only the face detection is usable. If neither pose nor face are available, we use the person detection to determine long and extreme long shots from the person height, if the person bounding box does not extend to the lower image border.

Augmentation for Extreme Close-Up. As we found that extreme close-ups are rare in the dataset, we augment it by sampling cropped images from all close-up shots. If the larger side of a face bounding box is at least $s_{\min}$ pixels, we determine a randomly sized bounding box with $w \in [s_{\min}, 0.75w_D]$ and $h \in [s_{\min}, 0.75h_D]$, with $s_{\min}$=175.

Verification (Validation Split Only.) For verification, we import the set of images into the CVAT annotation tool[4]. Each image's bustle and shot type annotation is initialized from the automatic annotation. A single annotator reviewed and corrected around 1,300 images. The accuracy of the automatically created annotations against the manually checked ones is provided in Table 4.

Data Sampling. From the training set we randomly sample 100K images per class, for validation we sample 100 images per class from the manually corrected set (images used for the bustle and shot type tasks may partly overlap which is not an issue since they are treated as independent classification problems).

[3] https://en.wikipedia.org/wiki/Drawing.
[4] https://github.com/openvinotoolkit/cvat.

The annotations for bustle and shot type as well as the code of the toolchain used to create it are provided at https://github.com/wbailer/PeopleAtPlaces.

4 ToDY: Dataset for Time of Day and Season

In order to build a dataset, we need a large scale outdoor dataset. We amend the Skyfinder dataset, which is a subset of the Archive of Many Outdoor Scenes (AMOS) dataset [17] dataset, consisting of about 1,500 weather webcam images per camera from 53 webcams, each covering one or multiple years. The images come with location (see Fig. 3 left for a plot), date and time metadata, image timestamps (in UTC), basic weather conditions and a number of derived attributes. We aim to label each of the images with time of day and season based on the available metadata. The time of day classes and their definitions are listed in Table 2, the season classes are the meteorological seasons [24], i.e., spring, summer, fall and winter.

As the location of the webcams from which the images were collected are known, as well as the dates and times when the images were taken, we can derive the season from the date and the hemisphere, and we can determine the time of the day based on the sun's position. We calculate the sun's elevation over/under the horizon at the location and time of the image, using the PyEphem[5] library. Note that this calculation will assume a horizon in a flat landscape, not considering any mountains or buildings. We are aware of this limitation, but still assume that the calculated position will be a useful approximation of the real situation.

There are multiple definitions of dusk and dawn, and we use the one for civil dusk/dawn [6], which defines begin of dusk/end of dawn when the sun is 6° below the horizon. While the begin of sunrise/end of sunset is clearly defined with the upper tip of the sun disk being just/still visible, there is not such a clear definition of the end of sunrise/begin of sunset. As the visual effect of sunrise/sunset extends beyond the point where the sun is fully visible, we chose to set this mark at the sun being 3° above the horizon. A visualization of those definitions is shown in Fig. 3 (right). In addition, it needs to be considered whether a location is sufficiently far north/south, so that polar night or day occur, and thus no sunset/sunrise happens.

Based on this information, we derive season and time of day images for each image in the dataset. However, we observe three main issues with the data: (i) noisy images, in particular during nighttime, (ii) incomplete images (due to data loss when transmitting the image from the camera) and (iii) inaccurate timestamps. In order to estimate the noise level, we use the mask for the sky region provided for the Skyfinder dataset, as the sky region does hardly contain structures with strong gradients. We split the image into 8×8 patches, and we calculate the standard deviation of all patches containing at least 80% sky, and determine the noise level as the median of the standard deviations in these patches. In order to handle incomplete images, we calculate a RGB histogram of the image, and remove all images where one value covers more than 50% of the pixels of the image.

The time provided in the metadata should match the time stamp of the downloaded image file, when corrected by the UTC offset. However, even with

[5] https://rhodesmill.org/pyephem/.

a tolerance of 15 min, this does not hold for about 2/3 of the images. This is in particular a problem for classifying twilight, sunset and sunrise, as this inaccuracy may change the correct class. As we cannot tell which of the two times is correct, we decided to manually check the images. We import the set of images into the CVAT annotation tool[6], and initialize the time of day with the automatically determined value. About 10K images have been manually checked and the annotations have been corrected when necessary.

The toolchain also supports augmentation of the data by cropping versions of the images with a smaller portion of sky region. From the sky annotations of the dataset, a horizon line is determined as the 0.9 quantile of lowest sky pixels in each column. Then images with the same aspect ratio as the original image but different fractions of the height above this horizon line are sampled. As the annotations are global, they are still valid for the modified images.

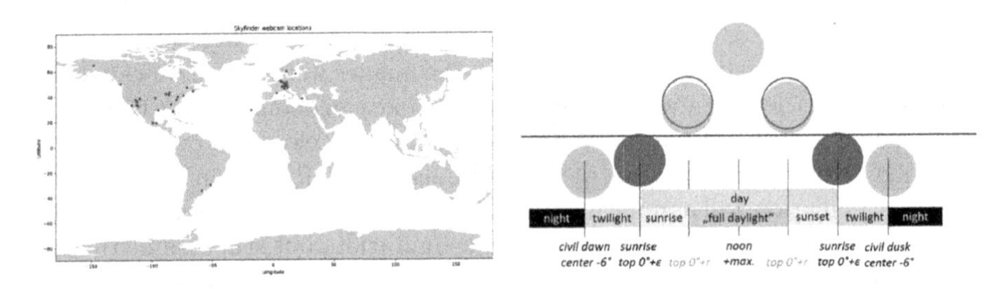

Fig. 3. Location of webcams in the Skyfinder dataset (left), visualization of the times of day used (right).

Table 2. Definition of time of day classes.

Class	Definition
Night	night time
Twilight	before sunrise/after sunset, using the definition of civil twilight
Sunrise	sun above horizon, until fully above horizon
sunset	sun above horizon, after being fully above horizon
Fulldaylight	sun completely above horizon
Day	day time, i.e. fulldaylight, sunrise or sunset (not used as a separate class, can be derived from the other classes)

We split the resulting season and time of day annotations into balanced training and test sets. This results in 2,790 training files and 311 validation files per class for season, and 986 training files and 110 validation files per class for time of day.

The annotations for time of day and season as well as the code of the toolchain used to create it are provided at https://github.com/wbailer/ToDY.

[6] https://github.com/openvinotoolkit/cvat.

5 Experiments

5.1 Baseline

We use EfficientNet-B3 [23] as the baseline model for location type classification and as a common backbone for all tasks. EfficientNet is a family of DNNs that differ in terms of number of parameters and performance. According to [23], the B3 variant provides a good tradeoff, and variants with better performance will have a significantly higher number of parameters. We train the model using the Pytorch Image Models framework (TIMM) [27], with a learning rate of 0.016 for 75 epochs.

To put the results of the model in relation to the state of the art, we compare the performance of the model on the validation set of the Places365 dataset against MAE [12], iSQRT-COV-Net [26] and RS-VGG16 [18]. However, all these methods have a significantly higher number of parameters as EfficientNet-B3. Still, its performance is slightly better than that of RS-VGG16. The results are summarized in Table 3. Throughout the paper, we use accuracy at rank 1 (acc@1) as the main metric.

Table 3. Comparison on Places365 validation (365 classes).

Method	no. params	acc@1	acc@5
MAE (ViT-H) [12]	632M	60.3	–
iSQRT-COV-Net [26]	>26M	56.320	86.270
RS-VGG16 [18]	19M	51.680	82.040
EfficientNet-B3	12M	51.874	82.825

5.2 People@Places

Table 4. Performance for bustle and shot type. Toolchain refers to the toolchain in Sect. 3, E2E refers to an end-to-end trained classifier.

Method	bustle	bustle0	bustle1	shot type	
	acc@1	acc@1	acc@1	acc@1	acc±1@1
Toolchain	81.020	95.892	95.538	56.726	70.604
E2E	66.337	84.158	81.683	50.715	67.437

The results for bustle and shot type classification are provided in Table 4. We compare the results of the computationally quite demanding annotation toolchain as described in Sect. 3 with the classifier trained on the datasets. The models are trained for 25 epochs (50 for shot type) with a learning rate of 0.016. For bustle classification, we observe that the results obtained from the classifier are significantly worse than that obtained with the detectors in the annotation process. To investigate this further, we introduce two binary variants of the problem: *bustle0* classifies class *unpopulated* against all others, and *bustle1* classifies {*unpopulated, few people, few vehicles*} against all others. It turns out that in these cases the performance of the classifier is closer to that of the detector toolchain. Our interpretation is that the network can well discriminate the presence of people or vehicles, but responds similarly for images with different count or size of objects, which makes it more difficult to discriminate the intermediate classes. This means that if a binary bustle classification is needed, this can be done efficiently with the classifier, while for the multi-class problem, the (computationally more expensive) detector-based approach used for dataset annotation provides better results.

For shot type classification, we observe that the results come closer to that of the annotation toolchain, but still stay below. We observe that many of the wrongly classified shots are those in nearby classes (e.g., medium shot vs. medium full shot). We thus add an evaluation metric for measuring classification into the correct or an adjacent class, which we call $acc\pm1@1$. We can observe that the performance of the annotation toolchain is in this case significantly higher, and additionally the gap between the performance of the classifier and the toolchain is reduced. For practical cases in editing, shots with similar types (which may be border cases) might already be a useful result.

5.3 ToDY

Table 5. Top-1 accuracy for time of day and season classification using EfficientNet-B3. The pretraining column specifies the base model being used, ToD+ refers to the time of day annotations after manual revision.

Pretraining	ToD acc@1	ToD+ acc@1	Season acc@1
None	63.918	20.000	28.310
ImageNet	52.577	66.182	84.225
Places365	54.639	69.818	86.197

The results for bustle and shot type classification are provided in Table 5. The models are trained for 450 epochs for season and 1,000 epochs for time of day (stopping early if a performance ceiling is reached) with a learning rate of 0.016. We compare the performance when training EfficientNet-B3 from scratch and from models pretrained on ImageNet and Places365. We provide two results for time of day: ToD refers to the automatically generated annotations, and ToD+ to the annotations after manual corrections. Overall, the performance starting from a pretrained model is better than starting from scratch, and pretraining on Places365 provides slightly better results than pretraining on ImageNet. We assume this is due to the fact that Skyfinder images are more similar to images in many categories in Places365 than to those in ImageNet. The results of 86%

accuracy for season and almost 70% accuracy for time of day show that the resulting classifiers are practically usable.

There is one anomaly, which is the high score for training time of day with automatically generated annotations from scratch. However, this seems to be due to a particular initialization, which already yields this score in the initial iteration, from which it does not change significantly during training. When applying the manual corrections, the performance falls to random when training from scratch.

6 Conclusion

We have proposed two datasets to address relevant classification tasks in visual media production and archiving: one addressed bustle and shot type classification, the other season and time of day classification. We provide toolchains for generating the additional annotations, as well as the datasets, which include manually verified and corrected subsets. The datasets are useful for classifying these properties in images, and the toolchains enable adding these annotations to other similar datasets with limited manual effort. As a baseline, we provide experimental results using EfficientNet-B3 for the four tasks on the two datasets.

Acknowledgments. The research leading to these results has been funded partially by the program "ICT of the Future" by the Austrian Federal Ministry of Climate Action, Environment, Energy, Mobility, Innovation and Technology (BMK) in the project "TailoredMedia" and from the European Union's Horizon 2020 research and innovation programme, under grant agreement n° 951911 AI4Media (https://ai4media.eu). The authors would like to thank Martin Winter, Hermann Fürntratt and Stefanie Onsori-Wechtitsch for support with the face detector and annotation tool setup, and Levi Herrich for checking and correcting the time of day annotations.

References

1. Abu-El-Haija, S., et al.: Youtube-8m: A large-scale video classification benchmark. arXiv preprint arXiv:1609.08675 (2016)
2. Arijon, D.: Grammar of the film language. Silman-James Press (1991)
3. Awad, G., Snoek, C.G., Smeaton, A.F., Quénot, G.: Trecvid semantic indexing of video: a 6-year retrospective. ITE Trans. Media Technol. Appl. **4**(3), 187–208 (2016)
4. Bak, H.Y., Park, S.B.: Comparative study of movie shot classification based on semantic segmentation. Appl. Sci. **10**(10), 3390 (2020)
5. Bochkovskiy, A., Wang, C.Y., Liao, H.Y.M.: Yolov4: Optimal speed and accuracy of object detection. ArXiv abs/2004.10934 (2020)
6. Boggs, S.: Seasonal variations in daylight, twilight, and darkness. Geogr. Rev. **21**(4), 656–659 (1931)
7. Cheng, P., Zhou, J.: Automatic season classification of outdoor photos. In: 2011 Third International Conference on Intelligent Human-Machine Systems and Cybernetics. vol. 1, pp. 46–49. IEEE (2011)

8. Deng, J., Guo, J., Zhou, Y., Yu, J., Kotsia, I., Zafeiriou, S.: Retinaface: Single-stage dense face localisation in the wild. arXiv preprint arXiv:1905.00641 (2019)
9. ETSI: Ts 102 822-3-1 v1.9.1 - broadcast and on-line services: Search, select, and rightful use of content on personal storage systems (tv-anytime); part 3: Metadata; sub-part 1: Phase 1 - metadata schemas. Tech. rep. (2015)
10. Fairbanks, A.T., Fairbanks, E.F.: Human proportions for artists. Fairbanks Art and Books (2005)
11. Galvane, Q.: Automatic Cinematography and Editing in Virtual Environments. Ph.D. thesis, Université Grenoble Alpes (ComUE) (2015)
12. He, K., Chen, X., Xie, S., Li, Y., Dollár, P., Girshick, R.: Masked autoencoders are scalable vision learners. arXiv preprint arXiv:2111.06377 (2021)
13. Huurnink, B., Hollink, L., Van Den Heuvel, W., De Rijke, M.: Search behavior of media professionals at an audiovisual archive: a transaction log analysis. J. Am. Soc. Inform. Sci. Technol. **61**(6), 1180–1197 (2010)
14. Kissos, I., Fritz, L., Goldman, M., Meir, O., Oks, E., Kliger, M.: Beyond Weak Perspective for Monocular 3D Human Pose Estimation. In: Bartoli, A., Fusiello, A. (eds.) ECCV 2020. LNCS, vol. 12536, pp. 541–554. Springer, Cham (2020). https://doi.org/10.1007/978-3-030-66096-3_37
15. Loper, M., Mahmood, N., Romero, J., Pons-Moll, G., Black, M.J.: SMPL: a skinned multi-person linear model. ACM Trans. Graphics (Proc. SIGGRAPH Asia) **34**(6), 248:1–248:16 (2015)
16. von Marcard, T., Henschel, R., Black, M.J., Rosenhahn, B., Pons-Moll, G.: Recovering accurate 3D human pose in the wild using IMUs and a moving camera. In: Ferrari, V., Hebert, M., Sminchisescu, C., Weiss, Y. (eds.) ECCV 2018. LNCS, vol. 11214, pp. 614–631. Springer, Cham (2018). https://doi.org/10.1007/978-3-030-01249-6_37
17. Mihail, R.P., Workman, S., Bessinger, Z., Jacobs, N.: Sky segmentation in the wild: An empirical study. In: IEEE Winter Conference on Applications of Computer Vision (WACV),D pp. 1–6 (2016). https://doi.org/10.1109/WACV.2016.7477637,acceptance rate: 42.3%
18. Qassim, H., Verma, A., Feinzimer, D.: Compressed residual-vgg16 cnn model for big data places image recognition. In: 2018 IEEE 8th Annual Computing and Communication Workshop and Conference (CCWC), pp. 169–175. IEEE (2018)
19. Rao, A., et al.: A unified framework for shot type classification based on subject centric lens. In: Vedaldi, A., Bischof, H., Brox, T., Frahm, J.-M. (eds.) ECCV 2020. LNCS, vol. 12356, pp. 17–34. Springer, Cham (2020). https://doi.org/10.1007/978-3-030-58621-8_2
20. Savardi, M., Signoroni, A., Migliorati, P., Benini, S.: Shot scale analysis in movies by convolutional neural networks. In: 2018 25th IEEE International Conference on Image Processing (ICIP), pp. 2620–2624. IEEE (2018)
21. Singh, A., et al.: Flava: A foundational language and vision alignment model. In: Proceedings of the IEEE/CVF Conference on Computer Vision and Pattern Recognition, pp. 15638–15650 (2022)
22. Sun, Y., et al.: Monocular, one-stage, regression of multiple 3d people. In: Proceedings of the IEEE/CVF International Conference on Computer Vision, pp. 11179–11188 (2021)
23. Tan, M., Le, Q.: Efficientnet: Rethinking model scaling for convolutional neural networks. In: International Conference on Machine Learning, pp. 6105–6114. PMLR (2019)
24. Trenberth, K.E.: What are the seasons? Bull. Am. Meteor. Soc. **64**(11), 1276–1282 (1983)

25. Wang, C.Y., Bochkovskiy, A., Liao, H.Y.M.: Scaled-yolov4: Scaling cross stage partial network. In: Proceedings of the IEEE/CVF Conference on Computer Vision and Pattern Recognition, pp. 13029–13038 (2021)
26. Wang, Q., Xie, J., Zuo, W., Zhang, L., Li, P.: Deep cnns meet global covariance pooling: Better representation and generalization. IEEE Trans. Pattern Anal. Mach. Intell. **43**(8), 2582–2597 (2020)
27. Wightman, R.: Pytorch image models. https://github.com/rwightman/pytorch-image-models (2019). https://doi.org/10.5281/zenodo.4414861
28. Xiao, T., Dollar, P., Singh, M., Mintun, E., Darrell, T., Girshick, R.: Early convolutions help transformers see better. In: Advances in Neural Information Processing Systems, vol. 34 (2021)
29. Yang, S., Luo, P., Loy, C.C., Tang, X.: Wider face: A face detection benchmark. In: Proceedings of the IEEE Conference on Computer Vision and Pattern Recognition, pp. 5525–5533 (2016)
30. Yuan, L., et al.: Florence: A new foundation model for computer vision. arXiv preprint arXiv:2111.11432 (2021)
31. Zhou, B., Lapedriza, A., Khosla, A., Oliva, A., Torralba, A.: Places: a 10 million image database for scene recognition. IEEE Trans. Pattern Anal. Mach. Intell. **40**(6), 1452–1464 (2017)

ScopeSense: An 8.5-Month Sport, Nutrition, and Lifestyle Lifelogging Dataset

Michael A. Riegler[1,3], Vajira Thambawita[1], Ayan Chatterjee[1], Thu Nguyen[1], Steven A. Hicks[1], Vibeke Telle-Hansen[2], Svein Arne Pettersen[3], Dag Johansen[3], Ramesh Jain[1,4], and Pål Halvorsen[1,2](✉)

[1] SimulaMet, Oslo, Norway
paalh@simula.no
[2] Oslo Metropolitan University, Oslo, Norway
[3] UIT The Artic University of Norway, Tromsø, Norway
[4] University of California Irvine, CA, USA

Abstract. Nowadays, most people have a smartphone that can track their everyday activities. Furthermore, a significant number of people wear advanced smartwatches to track several vital biomarkers in addition to activity data. However, it is still unclear how these data can actually be used to improve certain aspects of people's lives. One of the key challenges is that the collected data is often massive and unstructured. Therefore, a link to other important information (e.g., when, what, and how much food was consumed) is required. It is widely believed that such detailed and structured longitudinal data about a person is essential to model and provide personalized and precise guidance. Despite the strong belief of researchers about the power of such a data-driven approach, respective datasets have been difficult to collect. In this study, we present a unique dataset from two individuals performing a structured data collection over eight and a half months. In addition to the sensor data, we collected their nutrition, training, and well-being data. The availability of nutrition data with many other important objectives and subjective longitudinal data streams may facilitate research related to food for a healthy lifestyle. Thus, we present a sport, nutrition, and lifestyle logging dataset called *ScopeSense* from two individuals and discuss its potential use. The dataset is fully open for researchers, and we consider this study as a potential starting point for developing methods to collect and create knowledge for a larger cohort of people.

1 Introduction

Data is the new gold, is a phrase commonly touted. In healthcare, data is critical as their proper collection and analysis can improve the quality of life and even save lives. Nowadays, people record their daily activities digitally through a significant number of devices and sensors. Such digital recording of a person's activity and lifestyle is referred to as *lifelogging* [15]. Recording and analyzing such lifelog data provide a great opportunity for studying an individual's life experience.

D.-T. Dang-Nguyen et al. (Eds.): MMM 2023, LNCS 13833, pp. 502–514, 2023.
https://doi.org/10.1007/978-3-031-27077-2_39

It can help monitor a person's activity to improve health and well-being [24], help recover memories of past events [31], or analyze social behaviour [6,21]. Lifelogs are also sources of wast rich data for interesting research. For instance, Chokr et al. [4] described a machine learning approach to predict the number of calories from food images, and De Choudhury et al. [5] described the impact of interaction on social media to influence mental health. Therefore, lifelogs containing various types of information from a person's daily lifestyle are valuable in numerous contexts.

Although lifelogs might contain data highly valuable for research, they are often not generally available to researchers. A lifelog is typically not stored centrally by a single service that can be easily accessed, but rather exists as the union of data stored in a large number of online and offline data silos [19]. Still, some datasets exist, and existing lifelogging datasets [18] usually contain a person's daily life activities automatically captured and recorded using smartphone applications, wearable devices, and other sensors. One example is the NTCIR Lifelog test collection [14] consisting of lifelogging datasets for the NTCIR-12/13/14 lifelog tasks, which was first released at the NTCIR-12 conference [13]. The images in this dataset are captured by wearable cameras carried by two different lifeloggers. Some work has been done with similar datasets, for example, retrieving moment of interest [8,21]. However, a key challenge in lifelogging research is the poor availability of test collections [7]. Hence, there is a need for more available lifelog datasets, especially ones collected over longer time periods and with multiple modalities and purposes.

Capturing daily life events is also something many sports professionals do. Athletes have kept written training diaries for a long time using both pen-and-paper and, more recently, digital logging systems. Now, the use of wearables to measure activity and its intensity in both top sport and among the regular physically active population help to improve performance, recovery, and other aspects of health [9]. A challenge is to make sense of the data, and often, the captured data is limited to self-reports since activity logs from smartwatches and phones are hard to understand. Thus, there are still steps needed for integration of data [11] and to find standardized ways to analyze, evaluate, and present data [10]. Another problem in the area of sport is that professional athletes do not control the captured data by themselves, and they need the assistance of coaches, physicians, or technical support staff [19]. This process adds the burden of informed consent, authorization, and privacy. Furthermore, a trainer or team doctor does not have time to properly evaluate the myriads of sensor data from the athletes to possibly detect relevant performance data that could be used to improve training.

Moreover, dietary intake is an important factor affecting metabolic regulation, thereby affecting human health and fitness. However, individual response to dietary components has been known for a long time [22,23,29] where evidence has shown that people eating identical meals show high variability in metabolic response, such as blood glucose and lipids [2,12,16,25,27,33]. Hence, one set of dietary advice may not be equally efficient for all individuals. To improve healthy lifestyle and reduce the risk of diseases, we need to investigate individual variation using multi-dimensional and high-resolution continuous time-series of markers of individual health and fitness. Furthermore, we need to investigate how various factors correlate. With new technology and analytical tools, we may be able to develop more personalized approaches, but to achieve this, access to relevant data is crucial.

To aid these efforts, automatic methods to analyze sensor data and the quantification of self-reports may play a crucial role in retrieving the information that people may need. To be able to perform these analyses with the increasing volume of data produced by different devices, new methods and tools are needed. *ScopeSense* is made available in an effort to enable the development of such support systems. We provide a starting point by combining the idea of lifelogging data collection with activity logging. Multiple activity-specific analyses can be performed on such data as predicting sports performance, weight loss, or gain, but there is a lack of available datasets. We have therefore logged several objectives and subjective parameters of a person's daily life together with all food and drink intake. Specifically, we have used the following systems to capture the data:

- The Apple Watch series 6 smartwatch to track 24/7 activity and heart rate, and training sessions, electrocardiogram (ECG), sleep, blood oxygen, etc.
- The PMSys sports logging app to track subjective wellness parameters such as sleep, mood, stress, fatigue, muscle soreness; training load with session rating of perceived exertion (sRPE) and length; and injuries with body location and severity.
- The Lifesum food tracking app where all intake of food and drinks are self-reported and logged, together with a persons weight, and the content of nutrients are calculated.

ScopeSense contains logging data between 8. February 2021 and 20. October 2021 (255 days) from two individuals. To the best of our knowledge, *ScopeSense* is the only available dataset to combine both subjective and objective parameters combining daily lifestyle and activities over a long period of time. The dataset is openly available for research.

In the rest of the paper, we describe the data collection procedure in Sect. 2, and in Sect. 3, we describe the dataset in detail. Section 4 presents some initial analysis of the data. Furthermore, we provide examples of possible use-cases and applications of the dataset in Sect. 5, before we conclude the paper in Sect. 6.

2 Data Collection

Based on experiences from previous datasets [30], we conjecture a need for even more details in a lifelog covering activity, wellness, and nutrition data, collected over longer time periods. In this respect, *ScopeSense* is a 255 days dataset containing the lifelogs of two individuals with objective biometrics and activity data, food and nutrition, and subjective wellness and exercise load. During the data collection phase, there was a goal of regular running and strength workout sessions. The participants gave written consent to publish and distribute the data. Moreover, the data collection has been reported to the Norwegian centre

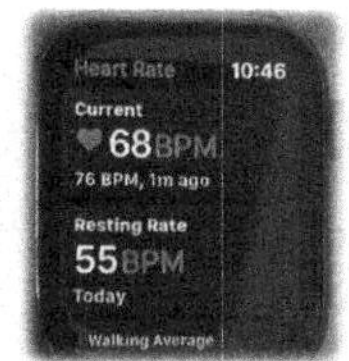

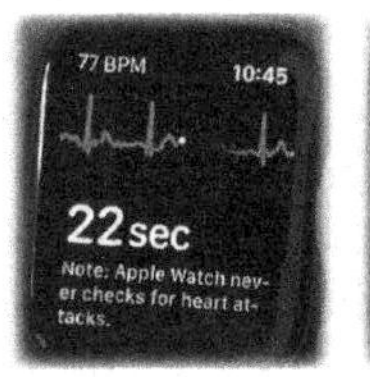

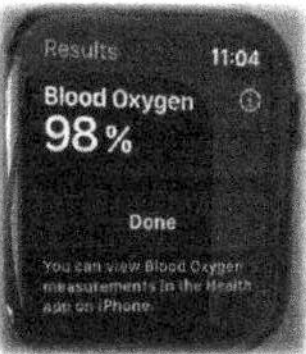

Fig. 1. Examples of logging using Apple Watch version 6.

for research data (NSD), and been assessed and approved (reference #294827). Additionally, our application to the regional committees for medical and health-care research ethics (REK) concluded that no approval from was needed for this data collection (reference #506192).

2.1 Objective Biometrics and Activity Data

We used an Apple Watch (series 6, hardware version 6.2, software version 7.6.1), in combination with an iPhone, to collect objective biometric and activity data. This watch was chosen because it was equipped with the most available sensors in one device at that time. Most of the biometric data were recorded automatically by wearing the smartwatch 24/7; however, for some measurements, we manually started a recording using the various apps on the smartwatch (Fig. 1):

- All exercise and training data were logged using the Apple workout app. It mainly starts the type of exercise one specifies and ends the session when it is finished.
- An ECG recording was made once every day to measure the heart's rhythm and electrical activity. This is a 30 s measurement holding in the clock button.
- Blood oxygen is measured by starting the app and allowing the sensors to work for 15 s.
- All other parameters are recorded automatically, like heart rate, sleep, number of hours where the person has been in a vertical position, flights (floors) climbed, counted steps walked/ran, energy used, etc., but one may use the apps to monitor your values.

All the data collected by the smartwatch is stored and extracted from the Apple Health application which works like a central data hub for the Apple devices.

2.2 Food and Nutrition

To collect dietary intake and nutrition data, we used the Lifesum[1] mobile app on iPhones with a premium subscription. Basically, Fig. 2 shows the procedure used for collecting dietary intake data. The participants selected meal or snack

[1] https://lifesum.com.

(Fig. 2a), and then, they inserted the portion of each components (Fig. 2b), either searching in the existing database or scanning the barcode printed on the packaging. Thereafter, they added the estimated amount (e.g., weight in grams). When known, the exact type of a component was inserted, like the type of bread, to get the exact types of nutrients. However, the exact type is sometimes unknown, e.g., when eating out. To quantify the amounts, we used the information provided on the packaging, e.g., while consuming a box of yogurt, we measured the amount of liquid a cup or a glass can contain. Moreover, we weighted the amounts included in a portion of a particular meal, e.g., the weight of a slice of cheese or a spoon of jam. A challenge was when eating out where it is practically infeasible to estimate accurately the individual ingredients used in a dish and the total volume, and if so, an approximate number is reported. To ease the logging, the app supports making favorites and bookmarks of the ingredients, and also to make predefined meals and dishes. Here, we also logged the weight of the participants every day. Moreover, the app sent push-messages to remind participants to log both dietary intake and weight. Data from Lifesum is included in the Apple Health data, but also extracted the data in own comma separated value (CSV) files directly from the Lifesum system.

(a) Select meal.

(b) Type & amount.

Fig. 2. Logging food using Lifesum.

2.3 Subjective Wellness, Training Load, and Injuries

We used the PMSys system[2] to collect subjective data regarding wellness, training load, and injuries. Figure 3 depicts an example of normal reporting sequences. The PMSys system is an online sports logging system where athletes can monitor for example individual training load, daily subjective wellness parameters, and injuries [26,32]. Wellness has typically been reported once a day through a sequence of questionnaires, as shown in Fig. 3. The wellness data was reported in the morning.

Training load or Session Rating of Perceived Exertion (sRPE) is a metric calculated from the product of the session length and the reported Rating of Perceived Exertion (RPE), i.e., reported similarly as shown in Fig. 3. The perceived training load is reported after every training session. Finally, the injuries questionnaire is recommended completed once a week, regardless of having an injury or not, but here the participants mainly reported the injury when one occurred, where the participants press on a body part to indicate a minor or

[2] https://forzasys.com/pmSys.html.

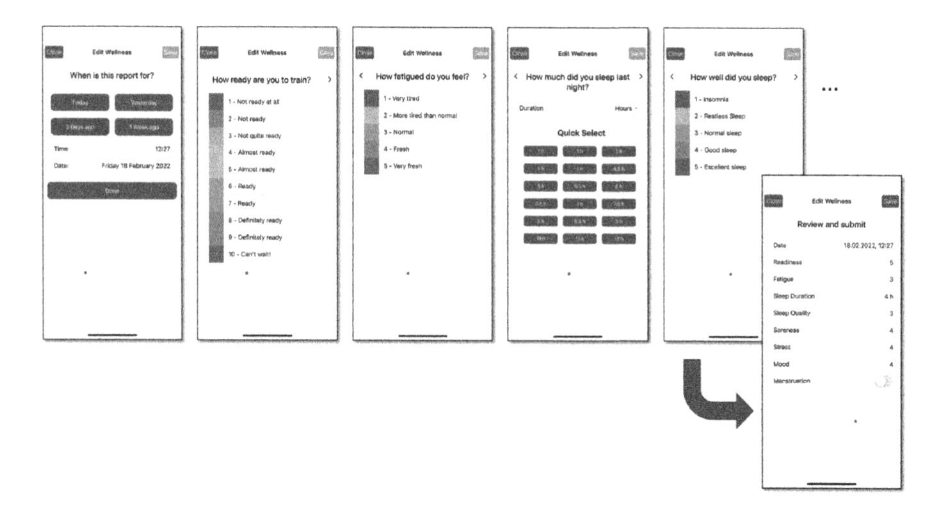

Fig. 3. Subjective parameters logged using PMSys, exemplified by wellness.

major injury or pain. To ensure timely reporting, PMSys sends scheduled push notifications directly to the participants' smartphones. All data was extracted at the end of the logging period from the system into CSV files.

2.4 Data Anonymization

To prevent any identity disclosure of the participants, the data has been processed to remove all ids. All occurrences of names in devices, like GPS coordinates, have been changed to a random value or removed completely to make it impossible to re-identify the participants. Exact GPS position data is also not important for the sport-related analysis besides features that can be extracted from the data, such as speed, distance, and elevation level which are part of the dataset.

3 Dataset Overview

ScopeSense contains logging data from two male persons, in the age range of 40–50 years, between 8. February 2021 and 20. October 2021 (255 days) in Norway. The participants have followed no particular food regime and have been exercising regularly during the period. The dataset is organized according to the way it is collected, even though one could alternatively organize the data according to the type of data. We divided and organized data into three folders for each participant as shown in Fig. 4.

The dataset is fully available and open for free use for researchers at the well-known Open Science Framework (OSF)[3] and at the Simula dataset site[4].

[3] https://osf.io/v5acr/.

[4] https://datasets.simula.no/scopesense.

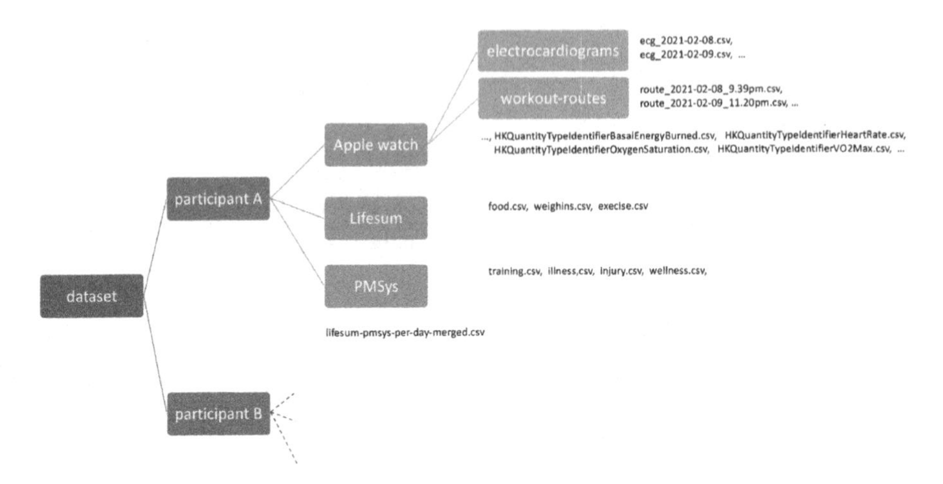

Fig. 4. Dataset organization. The structure for participants A and B is the same. Blue boxes represent the folder structure of the dataset. (Color figure online)

The dataset is free to use for research and teaching purposes under the license Attribution-NonCommercial 4.0 International (CC BY-NC 4.0)[5].

3.1 Apple Watch Data

We have exported all data from the Apple health system. The raw data comes in XML-format[6], but we have extracted the various parameters as CSV-files. There is one record for each of the measurements, where each record contains the parameter type, the source (which is anonymized), the type of device (hardware and software), the unit, dates, and the value(s). The Apple health system stores a large number of health parameters, including step count, heart rate, resting heart rate, and blood oxygen saturation. Other parameters include sleep, energy burned (basal/active), time standing, distance (walked/ran), and walking speed. The ECG data-series captured once a day is stored in its own sub-folder.

Each training session contains a training record in Apple health. Specifically, combining information from various files, one can extract information like type of training, duration, start and end time, total active energy (kcal), total distance (km), average heart rate and heart rate during training, step count, and speed. Each run session is also collected as a series of points along the route as shown in Fig. 5 giving, for example, time, speed, and elevation, but where the actual GPS coordinates are removed for privacy issues. All these routes are stored in a separate sub-folder.

3.2 Lifesum Data

The Lifesum system collects as much as possible of consumed dietary (eating and drinking) information. Every report in the system is contained in one row of

[5] https://creativecommons.org/licenses/by-nc/4.0/.
[6] https://developer.apple.com/documentation/healthkit.

```
time,lon,lat,elevation,speed,course,hAcc,vAcc
...
2021-02-19T17:15:15Z,43.448116,94.243819,224.995163,1.859252,231.84137,0.851156,0.798664
2021-02-19T17:15:16Z,43.448125,94.243802,225.067017,1.99339,230.884521,0.860994,0.803635
2021-02-19T17:15:17Z,43.448133,94.243784,225.131119,2.111444,232.588455,0.874346,0.810753
2021-02-19T17:15:18Z,43.44814,94.243765,225.185623,2.213889,232.957123,0.891013,0.81974
...
```

Fig. 5. Example records a run route session.

the table with the date of the report, which meal it is connected to (or a snack), the "name" of the food (including a title, potential manufacturer, and content), and the amount (both for the used metric and in grams). Subsequently, as can be viewed in the last part of each row, the reported intake is used to calculate the number of calories, various types of carbohydrates, various types of fat, protein, sodium, potassium, and cholesterol.

In addition, the Lifesum system has logged the participant's daily weight. The participants' weight is given in each row, and also the calculated body mass index (BMI) based on the participants' height.

Finally, we have also merged Lifesum and PMSys data (see below) in a day-by-day manner. These records are stored in the `lifesum-pmsys-per-day-merged.csv` file which contains one row per day.

3.3 PMSys Data

In terms of subjective PMSys reporting, the raw data is contained in the CSV files:

- Wellness includes parameters like time and date, fatigue, mood, readiness, sleep duration (number of hours), sleep quality, soreness (and soreness area), and stress. Fatigue, sleep quality, soreness, stress, and mood all have a 1–5 scale. Score 3 is normal, scores 1–2 are below normal, and 4–5 are scores above normal. Sleep length is just a measure of how long the sleep was in hours, and readiness (scale 1–10) is an overall subjective measure of how ready you are to exercise, i.e., 1 means not ready at all, and 10 indicates that you are exceptionally ready. In elite sport, readiness is used to tune if the athlete should push or pull the training load. In the CSV file, there are columns for each wellness parameter, which contain columns for the date and for the wellness parameter value for each of the participants.
- A training load report contains a training session's time, type of activity, perceived exertion (RPE), and duration in the number of minutes. This is, for example, used to calculate the session's training load or sRPE (RPE × duration). The data is stored as one tab in the spreadsheet for each of the participants. There is one line for each session with the date, daily summarized load, the calculated session RPE (sRPE), the experienced RPE, the session length, and then various calculated metrics like weekly load, monotony, strain, acute chronic workload ration, and chronic training load (over 28 and 42 d). If more than one session per day has been logged, the additional lines will have just the sRPE, RPE, and length of the session, which are then used to calculate the total load parameters in the first line.

– Injury and illness are reported in separate files having one line per incident with date and symptom/place of pain.

Table 1. Important features for the different self reported values in PMSys.

Readiness	Mood	Fatigue	Stress	Sleep Quality	Sleep Duration	Soreness	Daily Load
calories burned, fatigue, soreness, mood	fat, stress, sleep duration, readiness	potassium, soreness, stress, sleep quality, readiness	carbs, fatigue, mood, sleep quality	bmi, proteins, fatigue, mood, stress, sleep duration	soreness, mood, stress, sleep quality	weight, fatigue, readiness	weight, bmi, calories burned, carbs (fiber), fat unsaturated, proteins, fatigue, soreness, readiness

4 Initial Experiments

We provided some simple baseline experiments to provide an initial idea about how the dataset can be used and test its usefulness. For all experiments, we used 60% of the data for training and validation, and the remaining 40% as a test dataset. We prepossessed the data into a-value-per-day records (also included in the dataset) and used a subset of the data (weight_kg, bmi, calories_burned per day, caloeries per day, carbs per day, carbs_fiber per day, carbs_sugar per day, fat per day, fat_saturated per day, fat_unsaturated per day, potassium per day, protein per day, sodium per day, daily load, fatigue, soreness, mood, stress, sleep quality, sleep duration and readiness) where one instance represents one day of the collection period. The split between the train and test datasets was performed randomly and an equal number of instances per participant was included in the splits.

For the first experiment, we explored which features are important to predict different self-reported values. Specifically, we explored the important features for Readiness, Mood, Fatigue, Stress, Sleep Quality, Sleep Duration, Soreness, and Daily Load. We used Correlation-based Feature Subset Selection with Best First Ranker to select and rank the features. Table 1 shows the results where we observe that different features affect different aspects of ones well-being. For example, we can see that potassium is an important feature to predict fatigue, which is reasonable as a symptom is to feel tired/fatigue if a person has a too low level of this chemical element [28]. This shows that the dataset holds the potential to discover different aspects between self-reported and measured values.

For the second experiment, we trained a simple machine learning model to predict readiness (similar to [20,32]) which is seen as one of the most important factors in a sport context. As a baseline, we used ZeroR which predicts the average of the data. In addition, we trained two regressors using RandomForest and the SMOReg support vector machine. For ZeroR, the mean absolute error and root mean squared error were 1.29 and 1.6103, respectively, compared to

0.9944 and 1.3599 for SMOReg, and 0.9897 and 1.298 for RandomForest. From this initial simple analysis, we observe that some of the subjective measurements can be predicted using just a subset of features.

5 Example Applications of the Dataset

Healthcare systems are undergoing a major transformation. It used to be reactive and symptom-based, where a doctor primarily was consulted when being ill. Doctors diagnosed the sickness followed by proper medical treatment. Recently, system biologists and healthcare researchers started adopting a different perspective. This perspective was introduced as a P4 approach to medicine [1,17]. The P4 approach is based on a predictive, preventive, personalized, and participatory approach that uses data about a person as the main driver in devising an approach that is not reactive but proactive. This approach requires that each person is considered a unique system and is modeled using longitudinal data. This approach was not practical until smartphones, wearable devices, and associated advances in machine learning and cloud computing came along. Therefore, ubiquitous health monitoring is required for seamless data collection and subsequent applications of artificial intelligence and advanced healthcare technologies.

In medicine, an emerging concept is N-of-1 approach, where longitudinal data about a person is used to model the person rather than collecting population data and considering a person a sample of this population. P4 approaches for all aspects of health and wellness require collecting objective as well as subjective data. As discussed in this paper, some of the data is relatively easy to collect, while others require careful planning, collection, and structured organization. Such an approach is inherently multimodal and requires applications of traditional multimedia systems competence in addition to emerging and novel machine learning approaches.

Valid, privacy-preserved, relevant, and accurate data is fundamental to this approach. In this paper, we presented our initial experience with relatively invasive data collection with data collected over a long period of time. This paper presents this data for researchers with different interests to explore approaches that may be suitable. We hope that this data is the beginning of a collection of such data and sharing in the community to enable exploration of approaches related to health. We want to emphasize the role of food data as a very effective data stream. Food containing essential nutrients has always been considered vital for a healthy human being, and the relation between food and exercise has been investigated for a long time showing the importance of eating correctly to perform best [3]. Yet, approaches for understanding the long-time effect of what, when, where, and how much food affects different aspects of physical and psychological health are only partly explored due to the unavailability of data. Just exploring if and how food intake affects physical activity or exercise patterns can give us simple, useful, and interesting information, especially on an individual basis. Moreover, subjective readiness and other wellness parameters can be validated for "real" coinciding changes against objective heart rate variability data from the Apple watch. We hope that the dataset presented in this paper will jump-start this process, and possibly enable insights into how food, readiness for exercise, execution of exercise, and general wellness are related. Moreover, since the data is collected over a long period of time, it's worth analyzing it locally in shorter time intervals to consider if changes in some features, such as dining preferences and activity, may change between seasons with a potential causal effect on wellness.

6 Conclusion

We have presented the *ScopeSense* dataset, containing both objective and subjective longitudinal parameters from activity, wellness, food, and biometrics, potentially enabling the development of several interesting analysis applications. Our initial experiments show that such analyses are possible, but we conjecture the dataset has greater potential beyond what we have demonstrated in this paper. It is for our peer colleagues to use to expand and support this line of technology support for proactive medicine.

We are using our initial experiences with this dataset collection to currently expand further work to include a larger and more diverse cohort with even more parameters monitored. Also, we are developing a wide collection of machine learning applications that can sift through and analyze the data while reporting to enable next-generation proactive medicine to handle massive amounts of heterogeneous data in close to real-time. The goal is to have a personalized digital health screening service analyzing data and, e.g., detecting anomalies and concerning deviations providing potential for rapid response and intervention.

References

1. Auffray, C., Charron, D., Hood, L.: Predictive, preventive, personalized and participatory medicine: back to the future. Genome Med. **2**(8), article 57 (2010)
2. Berry, S.E., Valdes, A.M., et al.: Human postprandial responses to food and potential for precision nutrition. Nature Med. **26**(6), 964–973 (2020)
3. Burke, L.M., Cox, G.R.: The complete guide to food for sports performance: Peak nutrition for your sport. Allen & Unwin (2010)
4. Chokr, M., Elbassuoni, S.: Calories prediction from food images. In: Proceedings of the 29th Innovative Applications of Artificial Intelligence (IAAI) Conference (2017)
5. De Choudhury, M., Kiciman, E., et al.: Discovering shifts to suicidal ideation from mental health content in social media. In: Proceedings of the Conference on Human Factors in Computing Systems (CHI), pp. 2098–2110 (2016)
6. Dinh, T.D., Nguyen, D.H., Tran, M.T.: Social relation trait discovery from visual lifelog data with facial multi-attribute framework. In: Proceedings of the International Conference on Pattern Recognition Applications and Methods (ICPRAM), pp. 665–674 (2018)

7. Duane, A., Gurrin, C.: User interaction for visual lifelog retrieval in a virtual environment. In: Kompatsiaris, I., Huet, B., Mezaris, V., Gurrin, C., Cheng, W.-H., Vrochidis, S. (eds.) MMM 2019. LNCS, vol. 11295, pp. 239–250. Springer, Cham (2019). https://doi.org/10.1007/978-3-030-05710-7_20
8. Duane, A., Gurrin, C.: Baseline analysis of a conventional and virtual reality lifelog retrieval system. In: Ro, Y.M., et al. (eds.) MMM 2020. LNCS, vol. 11962, pp. 412–423. Springer, Cham (2020). https://doi.org/10.1007/978-3-030-37734-2_34
9. Düking, P., Achtzehn, S., et al.: Integrated framework of load monitoring by a combination of smartphone applications, wearables and point-of-care testing provides feedback that allows individual responsive adjustments to activities of daily living. Sensors **18**(5), 1632 (2018)
10. Düking, P., Fuss, F.K., et al.: Recommendations for assessment of the reliability, sensitivity, and validity of data provided by wearable sensors designed for monitoring physical activity. JMIR mHealth and uHealth **6**(4), e102 (2018)
11. Düking, P., Stammel, C., et al.: Necessary steps to accelerate the integration of wearable sensors into recreation and competitive sports. Current Sports Medicine Reports **17**(6), 178–182 (2018)
12. Gaundal, L., Myhrstad, M.C.W., et al.: Beneficial effect on serum cholesterol levels, but not glycaemic regulation, after replacing sfa with pufa for 3 d: a randomised crossover trial. British J. Nutrition **125**(8), 915–925 (2021)
13. Gurrin, C., Joho, H., et al.: NTCIR Lifelog: The first test collection for lifelog research. In: Proceedings of the International ACM SIGIR Conference on Research and Development in Information Retrieval, pp. 705–708 (2016)
14. Gurrin, C., Joho, H., et al.: Overview of NTCIR-13 Lifelog-2 task. In: Proceedings of the 13th NTCIR Conference on Evaluation of Information Access Technologies. NTCIR (2017)
15. Gurrin, C., Smeaton, A.F., Doherty, A.R.: Lifelogging: personal big data. Found. Trends Inform. Retrieval **8**(1), 1–125 (2014)
16. Hansson, P., Holven, K.B., et al.: Meals with similar fat content from different dairy products induce different postprandial triglyceride responses in healthy adults: A randomized controlled cross-over trial. J. Nutrition **149**(3), 422–431 (2019)
17. Hood, L., Friend, S.H.: Predictive, personalized, preventive, participatory (p4) cancer medicine. Nature Rev. Clin. Oncol. **8**(3), 184–7 (2011)
18. Ionescu, B.: ImageCLEF 2019: Multimedia Retrieval in Medicine, Lifelogging, Security and Nature. In: Crestani, F., et al. (eds.) CLEF 2019. LNCS, vol. 11696, pp. 358–386. Springer, Cham (2019). https://doi.org/10.1007/978-3-030-28577-7_28
19. Johansen, H., Gurrin, C., Johansen, D.: Towards consent-based lifelogging in sport analytic. In: He, X., Luo, S., Tao, D., Xu, C., Yang, J., Hasan, M.A. (eds.) MMM 2015. LNCS, vol. 8936, pp. 335–344. Springer, Cham (2015). https://doi.org/10.1007/978-3-319-14442-9_40
20. Kulakou, S., Ragab, N., et al.: Exploration of different time series models for soccer athlete performance prediction. Eng. Proc. **18**(1), 37 (2022)
21. Le, N.K., Nguyen, D.H., et al.: HCMUS at the NTCIR-14 Lifelog-3 Task. In: Proceedings of the NTCIR Conference on Evaluation of Information Access Technologies, pp. 48–60 (2019)
22. Miller, J.Z., Weinberger, M.H., et al.: Blood pressure response to dietary sodium restriction in healthy normotensive children. Am. J. Clin. Nutrition **47**(1), 113–119 (1988)
23. Morris, C., O'Grada, C., et al.: Identification of differential responses to an oral glucose tolerance test in healthy adults. PLoS ONE **8**(8), e72890 (2013)

24. Nag, N., Jain, R.: A navigational approach to health: actionable guidance for improved quality of life. Computer **52**(4), 12–20 (2019)
25. Ordovas, J.M., Ferguson, L.R., et al.: Personalised nutrition and health. BMJ 361 (2018)
26. Pettersen, S.A., Johansen, H.D., et al.: Quantified soccer using positional data: A case study. Front. Physiol. **9**, 866 (2018)
27. Retterstøl, K., Svendsen, M., et al.: Effect of low carbohydrate high fat diet on ldl cholesterol and gene expression in normal-weight, young adults: a randomized controlled study. Aherosclerosis **279**, 52–61 (2018)
28. Sjøgaard, G.: Potassium and fatigue: the pros and cons. Acta Physiologica Scandinavica **156**(3), 257–264 (1996)
29. Sweeney, J.S.: Dietary factors that influence the dextrose tolerance test: a preliminary study. Archives Internal Med. **40**(6), 818–830 (1927)
30. Thambawita, V., Hicks, S.A., et al.: Pmdata: A sports logging dataset. In: Proceedings of the ACM Multimedia Systems Conference (MMSys), pp. 231–236 (2020). https://doi.org/10.1145/3339825.3394926
31. Truong, T.D., Dinh-Duy, T., et al.: Lifelogging retrieval based on semantic concepts fusion. In: Proceedings of the ACM Workshop on The Lifelog Search Challenge (LSC), pp. 24–29 (2018)
32. Wiik, T., Johansen, H.D., et al.: Predicting peek readiness-to-train of soccer players using long short-term memory recurrent neural networks. In: Proc. of the IEEE International Conference on Content-Based Multimedia Indexing (CBMI), pp. 1–6 (2019)
33. Zeevi, D., Korem, T., et al.: Personalized nutrition by prediction of glycemic responses. Cell **13**(5), 1079–1094 (2015)

Fast Accurate Fish Recognition with Deep Learning Based on a Domain-Specific Large-Scale Fish Dataset

Yuan Lin[1], Zhaoqi Chu[2], Jari Korhonen[3], Jiayi Xu[4], Xiangrong Liu[4], Juan Liu[2(✉)], Min Liu[5], Lvping Fang[5], Weidi Yang[5], Debasish Ghose[1], and Junyong You[6]

[1] School of Economics, Innovation, and Technology, Kristiania University College, Oslo, Norway
{yuan.lin,debasish.ghose}@kristiania.no

[2] School of Aerospace Engineering, Xiamen University, Xiamen, China
23020220157220@stu.xmu.edu.cn, cecyliu@xmu.edu.cn

[3] School of Natural and Computing Sciences, University of Aberdeen, Aberdeen, UK
jari.korhonen@abdn.ac.uk

[4] School of Information Science and Technology, Xiamen University, Xiamen, China
xrliu@xmu.edu.cn

[5] School of Ocean and Earth, Xiamen University, Xiamen, China
{minliuxm,lpfang,wdwang}@xmu.edu.cn

[6] Norwegian Research Centre (NORCE), Bergen, Norway
juyo@norceresearch.no

Abstract. Fish species recognition is an integral part of sustainable marine biodiversity and aquaculture. The rapid emergence of deep learning methods has shown great potential on classification and recognition tasks when trained on a large scale dataset. Nevertheless, some practical challenges remain for automating the task, e.g., the lack of appropriate methods applied to a complicated fish habitat. In addition, most publicly accessible fish datasets have small-scale and low resolution, imbalanced data distributions, or limited labels and annotations, etc. In this work, we aim to overcome the aforementioned challenges. First, we construct the OceanFish database with higher image quality and resolution that covers a large scale and diversity of marine-domain fish species in East China sea. The current version covers 63,622 pictures of 136 fine-grained fish species. Accompanying the dataset, we propose a fish recognition testbed by incorporating two widely applied deep neural network based object detection models to exploit the facility of the enlarged dataset, which achieves a convincing performance in detection precision and speed. The scale and hierarchy of OceanFish can be further enlarged by enrolling new fish species and annotations. Interested readers may ask for access and re-use this benchmark datasets for their own classification tasks upon inquiries. We hope that the OceanFish database and the fish recognition testbed can serve as a generalized benchmark that motivates further development in related research communities.

D.-T. Dang-Nguyen et al. (Eds.): MMM 2023, LNCS 13833, pp. 515–526, 2023.
https://doi.org/10.1007/978-3-031-27077-2_40

Keywords: Fish recognition · Deep learning · Data augmentation · Convolutional neural networks

1 Introduction

The rapid emergence of computer vision and machine learning technologies is expected to offer new tools for exploring the ocean. Observation of the population and distribution of marine species are among the important tasks which give valuable insights to address marine biodiversity in terms of the variety, distinctiveness, and complexity of marine life [1]. Reliable algorithms that assist and automatize the recognition of marine species are in great demand. Fish species classification and recognition is a compelling research field with potential applications that cover a broad range of industries, including but not limited to fishery.

Traditional methods of identifying fish species are in general using shape and texture feature extraction [2–4]. However, the main drawbacks of feature based approach comes from its sensitivity to background noise, lack of generalization and difficulties of finding discriminating features, especially when the task deals with recognizing sub-ordinate object classes. In recent years, deep convolutional neural networks (CNN) have shown impressive results and large potential for identifying fish species. A deep-learning architecture that is composed of two principal component analysis (PCA) based convolutional layers, spatial pyramid pooling, and a linear SVM classifier was proposed in [5] to recognize fish from ocean surveillance videos, and it achieved 98.64% accuracy. In [6], AlexNet [7] was trained via transfer learning for automatic fish species classification from underwater video source, achieving an accuracy of 99.45%. Deep learning methods have been applied to fish recognition competition on the Kaggle challenge named "The Nature Conservancy Fisheries Monitoring" [8], whereas the deployed fish benchmarks are, however, small-scale and of poor image quality. There are some practical challenges remaining for identifying fish species. First, fish recognition is hindered by the poor image quality, uncontrolled objects and unconstrained natural habitat especially in wild circumstances. Second, the number of fish species in marine ecology is very large, and the distribution of fish species is highly characterized by regions or territories. Third, in real-world applications, data assessment, e.g. labeling and annotation, is expensive as it involves time-consuming and labour-intensive process.

The success of deep learning in classification and recognition tasks lies in the enormous explosion of data, as proven by the significant achievements leveraged by large-scale open datasets, such as ImageNet [9], COCO [10], and VOC [11] before being fine-tuned to new applications. However, most large-scale visual databases are not dedicated to fish species. For example, ImageNet provides a hierarchical framework that contains over 14 million hand-annotated images and 20 thousand ambiguous categories. However, it contains little fishery data, and the hierarchical architecture and label annotations do not align with our study. Similar limitation was addressed by Microsoft COCO, which shows the limitations of dataset in terms of volume and category. Other widely used datasets, e.g. TinyImage [12], ESPdataset [13] either only provide low-quality fish subsets, or most of the dataset is not publicly accessible.

To the best of our knowledge, the scale, accuracy and diversity of most open fish image datasets is insufficient. One popular fish database is Fish4Knowledge [14], where the fish data is acquired from a live video dataset resulting in 27,370 verified fish images covering 23 representative species. The dataset has been used by the 2015 Sea Clef contest [15] for fish species classification [24]. However, the video quality (640×480 pixels) and the number of species (mainly coral reef fishes in Southern China Sea) remain quite limited. QUT fish dataset, mostly collected in a constrained and plain environment [16], consists of only 3,960 labelled images of 468 species. DeepFish dataset, proposed in [17], contains approximately 40,000 underwater images from 20 habitats in the marine-environments of tropical Australia. However, the main purpose of the dataset is to assess fish-habitat associations in challenging, even inaccessible environments, thus with limited generalization. Another open fishery images database lately proposed by [18] contains 86,029 images of 34 object classes, making it the largest and most diverse public dataset of fisheries EM imagery to-date, as claimed by the author. However, the resolution restriction of EM images can inhibit further comprehensive progress. Therefore, there is a need of constructing a reliable fish dataset of large scale, accuracy and diversity of species.

In this work, we introduce a novel domain-specific dataset, named OceanFish, with 63,622 high resolution (1088×816 pixels) PASCAL VOC formatted images [11] covering 136 categories mainly inhabiting in East China Sea. Interested readers may ask for access and re-use this benchmark datasets for their own classification tasks upon inquiries. We hope that the scale, accuracy and diversity of the fish dataset can offer unparalleled opportunities to researchers in a variety of research communities and beyond. Accompanying the dataset, we also implement a fine-tuned fish recognition model by incorporating two of the most widely used object recognition models, Faster R-CNN [21] and Single Shot MultiBox Detector (SSD) [23], and investigate their performance on OceanFish. As a result, we obtain a comprehensive testbed, which can be further tailored to include new algorithms for specialized problem setups.

In summary, the main contribution of this work lie in two folds: a large scale, high quality dataset OceanFish, and a deep learning based framework to recognize fish species. Specifically, the contribution of OceanFish is highlighted in the following aspects: the dataset is domain specific, large scale, the image quality is enhanced, and the application aspects clarified.

The remainder of the paper is structured as follows: We present OceanFish in Sect. 2, where data augmentation mechanisms are elaborated as well. In Sect. 3, we present our generalized fish recognition framework, and explain the experimental results. Finally, we conclude in Sect. 4.

2 OceanFish

2.1 Constructing the Dataset

One ambition of this work is to construct a high quality benchmark dataset that covers a large scale and diversity of marine-domain fish species, and its scale and hierarchy can be further enlarged by enrolling new fish species and annotations. We believe that the dataset will facilitate developing advanced, large-scale deep learning algorithms. We build our own customerized dataset OceanFish by shooting video clips for collected fish samples from different angels to create variation, with implementation details described as below:

1. Specify the photographing conditions and device for image acquisition. A light background which is contrast to the fish color is required to ensure that fish's outline is clearly visible. We use D65 light source with a color temperature of 6500K.
2. Collect fish characteristics by shooting from various angles. The video clip is taken from carefully chosen angles i.e., 30 or 45°C, with the camera rotating for about 60 s. Subsequent acquisition of images are extracted from the video frames.
3. Categorize and annotate images according to the chosen naming convention.
4. Split the dataset into training, test, and validation parts. 60% of the entire dataset is for training, 20% for testing, and 20% for validation.

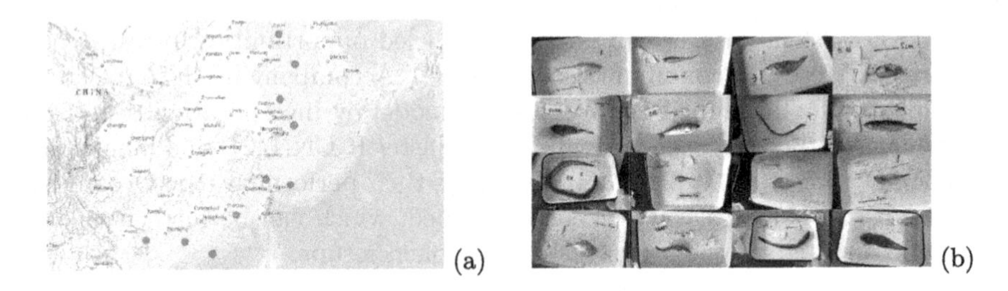

Fig. 1. (a)Locations and territory where OceanFish categories are acquired (b)16 representative fish categories in OceanFish

Images for OceanFish dataset were collected for 136 fish species, mainly inhabiting in the East China Sea, and the total volume is 63,622 pictures of 1088 × 816 pixels in PASCAL format. The locations from where the fish were acquired and 16 different fish species as examples are shown in Fig. 1, and the distribution of OceanFish images among categories is shown in Fig. 2.

A comparison between different fish datasets is shown in Table 1. Note that as well as Fish4Knowledge and Rockfish, the images of OceanFish are taken in a controlled environment which eliminate background noise, allowing a reliable and comprehensive analysis. Overall, the OceanFish dataset achieves an overall upgrade in comparison to the previous fish datasets in terms of scale, accuracy, and diversity of species.

PASCAL VOC2007 was applied to create the dataset, and LabelImg software was used to manually label and annotate images, and to create XML files automatically. XML files contain recorded information about the fish sample locations and classes.

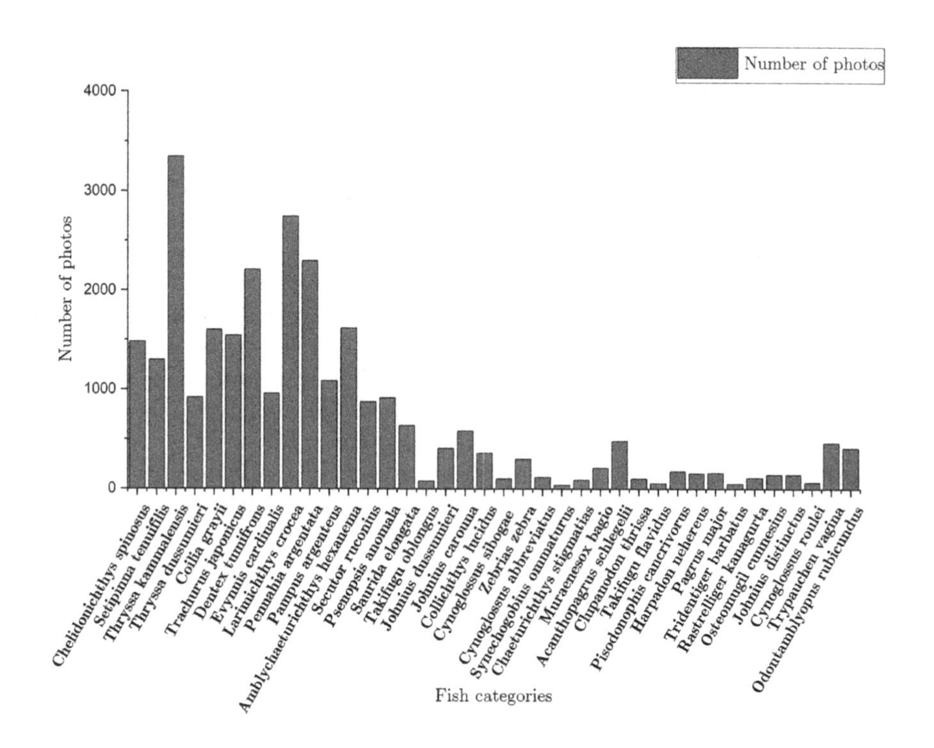

Fig. 2. OceanFish image distribution among different categories

2.2 Data Augmentation

Data augmentation, used to increase the number of training samples by applying a set of transformation to the images, has proven to improve the performance of the neural network models, especially when learning from small scale datasets. It can also be beneficial in yielding faster convergence, decreasing the risk of overfitting, and enhancing generalization [26]. In this work, most images are shot in the same setup environment. The similarity of image layout, background noise, spatial properties, texture, or overall shapes can lead to a classification model with a compromised generalization capability. To reduce the risk of overfitting, we applied label-preserving transformations to artificially enlarge the dataset.

Table 1. Comparison between fish dataset statistics

Dataset	No. of Images	Tasks	Resolution	Environment Type	Has background
OceanFish (ours)	63,622	Detection	1088×816	Controlled	No
QUT	3,960	Classification	480×360	Controlled	No
Fish4Knowledge	1.6B	Detection	352×240	On site	No
RockFish	4,307	Detection	1280×720	On site	Yes

In this work, we employed three data augmentation methods: color jitter, random cropping, and affine transformation. Color jitter randomly changes the brightness, contrast, and saturation of an image instead of the RGB channels. We adopt the color jitter mechanism employed by Facebook AI Research for the reimplementation of ResNet-101 [27]. Random cropping prevents a CNN from overfitting to specific features by changing the apparent features in an image. In this work, we applied Scale Jittering that is used by VGG and ResNet-101 networks [29]. Affine transformations generate duplicate images that are shifted, zoomed in/out, rotated, flipped, and distorted can be generated for image augmentation. We have applied a set of affine transformations for image augmentation, and both the original images and the duplicates were used as input to the neural networks.

Most existing learning algorithms produce inductive learning bias towards the frequent (majority) classes if training data are not balanced, resulting in poor minority class recognition performance. Data augmentation also aims at alleviating this problem and improving data generalizability. In this work, we chose a data equalization mechanism inspired by the implementation of AlexNet [7]. The procedure is described as follows:

1. Get a list of the original images, sorted by label.
2. Calculate the number of samples in each category and save the largest number as M.
3. Generated a new list of size M for each category, and fill in the new list by looping the original content repeatedly, until the number of samples in the new list reaches M.
4. Random shuffle the new list with M samples.
5. To avoid overfitting, the aforementioned color jitter, scale jittering, and affine transformations are applied for image augmentation.

3 Methods and Experiments

3.1 CNN Based Object Detection

Since the 2010 s, there has been a trend towards utilizing and continuously improving CNNs in object detection tasks. CNN based object detection algorithms can be divided into two major categories: region proposal based approaches such as R-CNN [19], Fast R-CNN [20], and Faster R-CNN [21], which yield high detection accuracy in terms of mean Average Precision (mAP) in different object detection scenarios; and regression based end-to-end approaches such as You Only Look Once (YOLO) [22] and Single Shot MultiBox Detector (SSD) [23], which predict bounding box coordinates and class probabilities straight from image pixels, thus achieving fast detection at the cost of decreased accuracy. Faster R-CNN and SSD are applied in this work, and their concepts are depicted in Fig. 3. Note that we have focused on the original meta-architecture for clarity and simplicity, although many improvements have been introduced to the original architectures since their release.

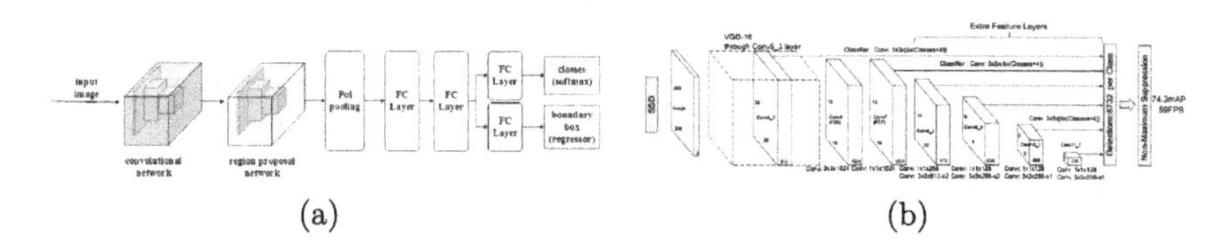

Fig. 3. Two representative CNN based object detection framework (a) Faster R-CNN (b) Single Shot MultiBox Detector (SSD)

3.2 Experiments and Performance Evaluation

In this section, we carry out a series of experiments to evaluate the performance of Faster R-CNN and SSD on OceanFish. Both models contain a deep learning backbone network as feature extraction module. We perform fish recognition on OceanFish by fine-tuning the pre-trained model. TensorFlow is used for all the experiments. Models are trained on NVIDIA GeForce GTX 1080Ti 11G GPUs with a total batch size of four. Dataset is divided into training, validation, and test sets in a proportion $3:1:1$. The learning rate is initialized to $5 \cdot 10^{-4}$. Figure 4 shows a successful detection of Decapterus maruadsi, with both classification and localization information.

First, we choose ResNet-101, ResNet-50 [27], and VGG-16 [29] as backbone feature extractors for Faster R-CNN and evaluate their performance on a subset of OceanFish with 20 fish species and 400 images in each category. The experimental results show that ResNet-101 achieves the highest accuracy with the mAP of 0.994, compared to ResNet-50 with the mAP of 0.988, and VGG-16 with the mAP of 0.986. Therefore, we select ResNet-101 backbone as the feature extractor for Faster R-CNN and it applies to the following experiments.

In Fig. 5(a), we depict the total loss in respect with training iterations for Faster R-CNN. The total loss is gradually decreasing and converging as the number of iterations increases. The Ap value distribution of a subset of fish species is shown in Fig. 5(b), where the average mAP for the complete dataset is approximately 0.9712.

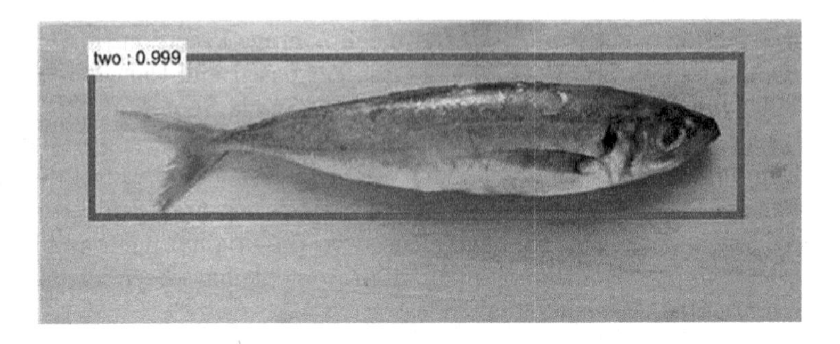

Fig. 4. A successful detection of fish specie Decapterus maruadsi

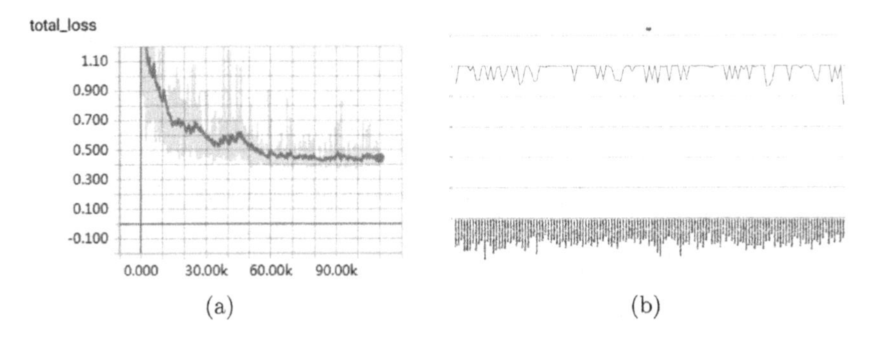

(a) (b)

Fig. 5. Faster R-CNN: (a) Total loss v.s. Number of training iterations, (b) Ap value distribution of a subset of fish species

The results in Table 2 show that Faster R-CNN has relatively low miss rate for all categories, which means that the framework in general detects and localizes the target properly. We apply both false positive rate (FPR) and true positive rate (TPR) in the confusion matrix to evaluate the classification performance. Relatively large false positive error rate is observed for several fish species, indicating that the proportion of misclassified items in the whole set cannot be ignored. For example, among the 669 images of Thryssa kammalensis in OceanFish, 38 images are falsely detected as Thryssa dussumieri, five images as Setipinna tenuifilis, and two images as Pampus argenteus.

In the second part of the experiments, we apply VGG-16 as SSD backbone network. Figure 6(a) shows that the total loss is gradually decreasing and converging as the number of iterations increases. After 70,000 iterations, the SSD model yields the recognition accuracy with mAP of 0.8156 over the complete OceanFish dataset, which indicates a slight performance degradation in comparison with Faster R-CNN. Moreover, the Ap value of fish species demonstrates more variance among categories, as shown in Fig. 6(b).

As shown in Table 3, SSD and Faster R-CNN achieve satisfactory localization accuracy, since both detection algorithms have utilized bounding box regression for fine-tuning of object locations. It has also been observed that SSD and Faster

Table 2. Fish species with top 10 highest error rate classified by Faster R-CNN

Category	Name	No. of Images	Miss Rate	False Positive
19	Johnius carouna	313	0	16.6%
3	Thryssa kammalensis	669	0.89%	7.3%
120	Johnius trewavasae	269	0.37%	15.2%
10	Pennahia argentatus	460	1.3%	8.9%
9	Larimichthys crocea	549	1.4%	4.9%
61	Saurida tumbil	116	0%	18.1%
20	Collichthys lucidus	181	0.55%	7.7%
16	Saurida elongata	223	0.9%	4.9%
24	Cynoglossus abbreviatus	238	1.3%	4.6%
22	Cynoglossus sibogae	148	1.4%	7.4%

Table 3. Fish species with top 10 highest error rate classified by SSD

Category	Name	Total Images	Miss Rate	False Positive
19	Thryssa kammalensis	669	0.9%	33%
3	Trachinocephalus myops	389	0.77%	49.9%
120	Pennahia argentatus	460	1.3%	39.6%
10	Odontamblyopus lacepedii	433	0.46%	31.2%
9	Saurida elongata	223	2.2%	130%
61	Johnius carouna	313	0%	58.2%
20	Lagocephalus wheeleri	406	0.49%	28.8%
16	Chelidonichthys kumu	297	%	37.4%
24	Upeneus bensasi	431	0.46%	23.2%
22	Decapterus maruadsi	257	0%	36.6%

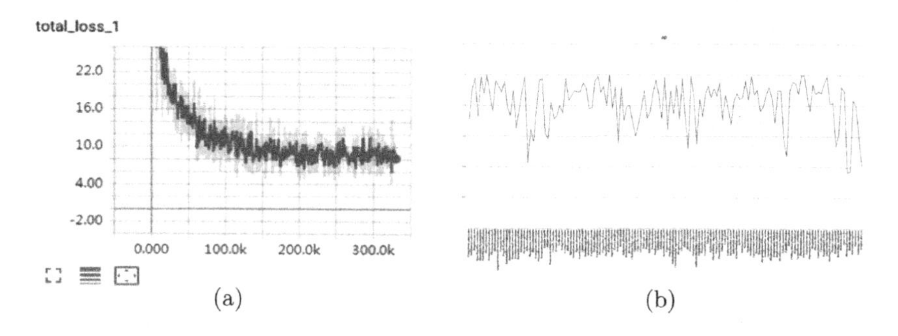

Fig. 6. SSD: (a) Total loss v.s. Number of training iterations (b)Ap value distribution among selected fish species

R-CNN present the accuracy/speed tradeoff due to the essential architecture difference. Although SSD shows faster speed in comparison with Faster R-CNN, it introduces a certain performance degradation. For example, among the total 669 images of Thryssa kammalensis in OceanFish, 221 images were falsely detected as other fish species, yielding false positive error rate exceeding 30%.

Table 4 shows the performance of our testbed in terms of mAP value. Note that Faster R-CNN and SSD has achieved mAP of 97.12% and 81.56% respectively when applied to OceanFish dataset. It can be seen that the overall detection performance improves significantly, noting that previous models, i.e., AlexNet and CIFAR-10 [28] only accomplish mAP of 60.90% and 71.10% for classifying the fish subset of ImageNet. The result shows that our proposed testbed together with fine-grained dataset yield satisfactory results, matching our observed trends that large-scale fine-grained datasets play a vital role in deep learning tasks.

Table 4. Performance of our testbed in terms of mAP value based on OceanFish

Model	Dataset	Pre-processing	Backbone network	Precision rate
Faster R-CNN (ours)	63622 images 136 categories	Data augmentation	ResNet-101	97.12%
SSD (ours)	63622 images 136 categories	Data augmentation	VGG-16	81.56%

4 Conclusion

In this work, we introduced a novel domain-specific dataset named OceanFish and tailored to marine domain fish classification and recognition tasks. The dataset consists of 63,622 high resolution images covering 136 clusters, and it is therefore much larger than the earlier publicly accessible fish datasets. The dataset can be further enlarged by enrolling new fish species and annotations. We hope that the OceanFish database can serve as a benchmark that motivates further development in related research communities.

We implemented a deep learning based fish recognition testbed by incorporating Faster R-CNN and SSD, and investigated their performance on OceanFish dataset comprehensively. We demonstrated through extensive experiments that the model achieves satisfactory classification accuracy and speed. It can be further tailored and used to create specialized algorithms in different problem setups, and thus it has provided us with potential of applying the generalized approach together with domain specific large scale dataset to smart farming and aquaculture production in terms of productivity, food security, and sustainability.

Experimental results show that our model may introduce false detections in some fish categories, and the Ap values demonstrate certain variance among categories. The reason behind might lie in the model architecture and inadequate examples of some categories. To improve the performance further, we may improve model architecture, or refine training procedure, including changes in loss functions, data preprocessing, and optimization methods etc., in our future work. Last while not least, we encourage researchers to enlarge OceanFish dataset, in both scale and hierarchy, to provide upgraded capability.

References

1. Goulletque, P., et al.: The importance of marine biodiversity. Biodiversity in the Marine Environment, pp 1–13 (2014)
2. Yi-Haur, S., et al.: Fish observation, detection, recognition and verfication in the real world. In: Proceedings of the International Conference on Image Processing, Computer Vision, and Pattern Recognition(IPCV), p. 1, (2012)
3. Katy, B., et al.: Fish species recognition from video using SVM classifier. In: Proceedings of the 3rd ACM International Workshop on Multimedia Analysis for Ecological Data, pp. 1–6 (2014)
4. Mehdi, R., et al.: Automated fish detection in underwater images using shape based level sets. Photogram. Record. **30**(149), 46–62 (2015)
5. Qin, H.W., et al.: DeepFish: accurate underwater live fish recognition with a deep architecture. Neurocomputing **187**, 49–58 (2016)
6. Tamou, A.B., et al.: Underwater live fish recognition by deep learning. In: International Conference on Image and Signal Processing, pp. 275–283 (2018)
7. Krizhevsky, A., et al.: ImageNet classification with deep convolutional neural networks. In: Proceedings of the 25th International Conference on Neural Information Processing Systems, vol. 1, pp. 1097–1105 (2012)
8. https://www.kaggle.com/c/the-nature-conservancy-fisheries-monitoring . Kaggle Competition. The Nature Conservancy Fisheries Monitoring (2017)
9. Deng, J., et al.: ImageNet: a large-scale hierarchical image database. In: IEEE Conference on Computer Vision and Pattern Recognition, 2009. CVPR (2009)
10. Lin, T.-Y., et al.: Microsoft COCO: common objects in context. In: Fleet, D., Pajdla, T., Schiele, B., Tuytelaars, T. (eds.) ECCV 2014. LNCS, vol. 8693, pp. 740–755. Springer, Cham (2014). https://doi.org/10.1007/978-3-319-10602-1_48
11. Everingham, M., et al.: The PASCAL visual object classes (VOC) challenge. Int. J. Comput. Vis. **88**(2), 303–338 (2010). https://doi.org/10.1007/s11263-009-0275-4
12. Torralba, A., et al.: 80 million tiny images: a large dataset for nonparametric object and scene recognition. IEEE Trans. Pattern Anal. Mach. Intell. **30**(11), 1958–1970 (2008)
13. Ahn, L.V., et al.: Labeling images with a computer game. In: CHI04 Proceedings of the SIGCHI Conference on Human Factors in Computing Systems, pp. 319–326 (2004)
14. Fisher, R. et al.: Overview of the Fish4Knowledge project. In: Fish4Knowledge: Collecting and Analyzing Massive Coral Reef Fish Video Data, pp. 1–17 (2016)
15. Alexis, J., et al.: Life CLEF 2015: multimedia life species identification challenges. In: Experimental IR Meets Multilinguality, Multimodality, and Interaction. Springer International Publishing, pp. 462–483 (2015)
16. Anantharajah, K., et al.: Local inter-session variability modelling for object classification. In: IEEE Winter Conference on Applications of Computer Vision, pp 309–316 (2014)
17. Saleh, A., et al.: A realistic fish-habitat dataset to evaluate algorithms for underwater visual analysis. Sci. Rep. **10**, 14671 (2020)

18. J. Key, et al.: The fishnet open images database: a dataset for fish detection and fine-grained categorization in fisheries. In: 8th Workshop on Fine-Grained Visual Categorization at CVPR (2021)
19. Girshick, R., et al.: Feature hierarchies for accurate object detection and semantic segmentation. In: Proceedings of the IEEE Conference on Computer Vision and Pattern Recognition, vol. 2014, pp. 580–587 (2014)
20. Girshick, R.: Fast R-CNN. In: Proceedings of the IEEE International Conference on Computer Vision, vol. 2015, pp. 1440–1448 (2015)
21. Ren, S.Q., et al.: Faster R-CNN: towards real-time object detection with region proposal networks. In: Advances in Neural Information Processing Systems, vol. 2015, pp. 91–99 (2015)
22. Redmon, J., et al.: You only look once: unified, real-time object detection. In: Proceedings of the IEEE Conference on Computer Vision and Pattern Recognition, pp. 779–788 (2016)
23. Liu, W., et al.: SSD: single shot multibox detector. In: Leibe, B., Matas, J., Sebe, N., Welling, M. (eds.) ECCV 2016. LNCS, vol. 9905, pp. 21–37. Springer, Cham (2016). https://doi.org/10.1007/978-3-319-46448-0_2
24. Villon, S., et al.: Coral reef fish detection and recognition in underwater videos by supervised machine learning: comparison between deep learning and HOG + SVM methods. In: International Conference on Advanced Concepts for Intelligent Vision Systems, ACIVS, pp. 160–171 (2016)
25. Deng, J., et al.: http://www.image-net.org/challenges/LSVRC/2012/. In: ILSVRC-2012 (2012)
26. Takahashi, R., Matsubara, T.: Data augmentation using random image cropping and patching for deep CNNs. In: arXiv (2018)
27. He, K., et al.: Deep residual learning for image recognition. In: Proceedings of the IEEE Computer Society Conference on Computer Vision and Pattern Recognition (CVPR2016), pp. 770–778 (2016)
28. Krizhevsky, A.: Learning multiple layers of features from tiny images. Technical report, University of Toronto, pp. 1–60 (2009)
29. Simonyan, K., et al.: Very deep convolutional networks for large-scale image recognition. In: arXiv:1409.1556 (2014)

GIGO, Garbage In, Garbage Out: An Urban Garbage Classification Dataset

Maarten Sukel(✉), Stevan Rudinac, and Marcel Worring

University of Amsterdam, Amsterdam, The Netherlands
m.m.sukel@uva.nl

Abstract. This paper presents a real-world domain-specific dataset, which facilitates algorithm development and benchmarking on the challenging problem of multimodal classification of urban waste in street-level imagery. The dataset, which we have named "GIGO: Garbage in, Garbage out," consists of 25k images collected from a large geographic area of Amsterdam. It is created with the aim of helping cities to collect different types of garbage from the streets in a more sustainable fashion. The collected data differs from existing benchmarking datasets, introducing unique scientific challenges. In this fine-grained classification dataset, the garbage categories are visually heterogeneous with different sizes, origins, materials, and visual appearance of the objects of interest. In addition, we provide various open data statistics about the geographic area in which the images were collected. Examples are information about demographics, different neighborhood statistics, and information about buildings in the vicinity. This allows for experimentation with multimodal approaches. Finally, we provide several state-of-the-art baselines utilizing the different modalities of the dataset.

1 Introduction

The number of people living in cities is increasing steadily, which also increases the pressure on local governments to maintain the public spaces in a more efficient and sustainable manner. Therefore cities are actively researching new methods of resolving urban issues, one of the most significant being the accumulation of waste. Multimodal approaches have been proposed to help route citizen reports about such issues effectively [7,23]. However, recent advances in multimedia and computer vision make it possible to collect and process information about public spaces using real-time street-level imagery. Such techniques can help to quickly identify and address urban issues. For example, the City of Amsterdam recently conducted experiments in which street-level imagery was streamed in real-time to collect information about urban issues like garbage on the streets [24]. This in turn can be an additional source of actionable intelligence for local governments to increase the effectiveness of the garbage collection process.

To allow for experimentation with multimedia techniques in urban waste management, we publish a dataset with this work. We name the dataset after

D.-T. Dang-Nguyen et al. (Eds.): MMM 2023, LNCS 13833, pp. 527–538, 2023.
https://doi.org/10.1007/978-3-031-27077-2_41

a) Collected dataset of street-level imagery with a focus on garbage

b) Annotated with domain experts which allows for multimodal experiments

Fig. 1. To create the "GIGO" dataset images a) collected over several months are b) annotated with domain experts and further curated into the eventual dataset of 25k images with classes *garbage, garbage bag, litter, bulky waste, and cardboard.*

the well-known machine learning principle "Garbage in, Garbage out," which we shorten as "GIGO". The main task is to classify images based on the presence of several types of garbage illustrated in Fig. 1, such as *garbage, garbage bags, cardboard, bulky waste*, and *litter* using visual content and associated geospatial metadata. The dataset, which contains 25k annotated images, is collected as part of a system to process and analyze street-level imagery in real-time [24]. The images are taken by driving through the city using a service vehicle with two smartphones attached to the windshield, both capturing images of the road, sidewalks, and surrounding urban spaces. Images in our GIGO dataset were annotated in collaboration with domain experts knowledgeable about the process of identifying the location and, afterward, collecting different types of garbage on the city streets.

In the past, well-known benchmarking initiatives such as TREC [10], MediaEval [15] and TRECVID [22] were also motivated by real-world problems, allowing for alignment between companies, government agencies, and the research community. Here we create an additional possibility to research such a real-world problem with high societal impact while providing a test bed that can push multimedia research forward.

Existing fine-grained visual classification benchmarks are often broadly defined and have categories that are not urban domain-specific. Datasets like Stanford cars [13], FGVC Aircraft [19] and CUB-200-2011 [29], contain cars, aircraft, and birds but little to no additional multimedia channels. We expect that an effective combination of different modalities featured in our GIGO dataset

(a) The area of the image used to classify litter.

(b) The area of the image used to classify the bulky waste.

Fig. 2. Images generated using Gradcam [21].

is likely to improve performance further. The task we focus on in this paper is fine-grained multimodal garbage classification in street-level imagery. This is a particularly challenging task for several reasons. The regions of interest are difficult to identify due to the dynamic and noisy background and the varying lighting conditions of the images. As shown in Fig. 1, the classes are visually heterogeneous with differences in sizes and materials. In addition, while images intentionally captured by humans often follow established photographic rules about e.g., scene composition, images captured from a fixed position on a vehicle have a much larger variety of object locations. To illustrate the effect of varying object sizes and positioning on the image, we have used Gradcam [21] to create a heatmap of the relevant image regions when performing classification with ResNet. Litter, in particular, can be present in a tiny fraction of the image only, as seen in Fig. 2a. However, depending on the perspective and distance, bulky waste with a higher physical volume can cover only a small portion of the image, as seen in Fig. 2b. The dynamic urban environment in which the data was collected presents a challenging test-bed for developing and evaluating novel multimodal approaches.

Following the first law of geography [28], images captured close to each other are generally more likely to have related semantics, thus, potentially object classes. Also, scenes recorded in the same neighborhood might have similar visual backgrounds creating noise in the classification, which could be resolved by incorporating information about the image context. Finally, the geospatial data collected with the images and the open data about the surrounding geographic area also allows for experimentation with the predictive potential of the geospatial patterns that occur within the dataset.

Since garbage on the streets of a city is always produced by the activity of people in the area there are many regularities in how the garbage is produced throughout a geographic area. And thus, we see a large potential in using open data provided by the local government, such as neighborhood statistics and information about the function of buildings. Shops tend to produce a different type of garbage than residential areas, so that information is likely valuable for

 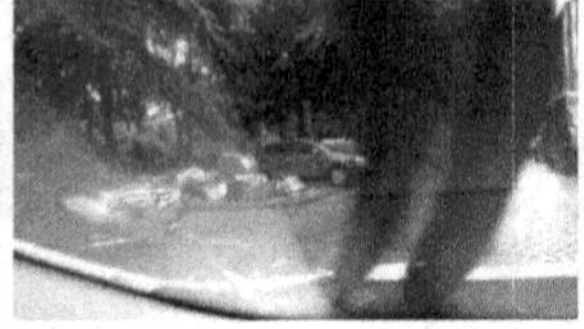 

Fig. 3. Images containing the cardboard class, which can have many materials, shapes, and colors. (Color figure online)

increasing the performance of garbage classification. By releasing the dataset and discussing possible improvements using multimodal techniques, we aim to open up the research in this direction for the community (Figs. 3, 4 and 5).

For this work, we have experimented with several state-of-the-art classification approaches, both visual and multimodal, to create an idea of how the data can be used and to establish a baseline. The main contributions of this paper are as follows:

- We publish a large collection of annotated street-level imagery focusing on the fine-grained multimodal classification of urban garbage.
- We perform experimentation with state-of-the-art computer vision approaches to multi-label classification of street-level images.
- We experiment with multimodal approaches to the problem and discuss possible improvements.

2 Related Work

2.1 Related Datasets

Common datasets and benchmarks have been a cornerstone of development in multimedia as they allow for comparing methods and validating them in different domains. Good examples of this are tasks such as video content search [18] and lifelog retrieval systems [8]. The GIGO dataset released with this paper will add a new type of data to benchmark and additional challenges to solve.

A classic example of a field driven by benchmarking datasets is computer vision. One of the most used datasets is ImageNet [14], which facilitates research into image classification. Another well-known dataset is MS-COCO [16] which is used for benchmarking object detection approaches. These datasets consist of web-scraped images annotated through public crowdsourcing. Due to this method of collecting images, the images are frequently taken with an object of interest in focus that also has a certain position according to well-established principles of image aesthetics. The reason for this is that when a person takes a photo, they often follow these practices consciously or unconsciously. In addition, the variety between the images taken in different "generic" contexts is much different from those taken from a mounted perspective in an urban setting.

For collecting the GIGO dataset it was not possible to use web scraping on social media and other sources, as is common with other datasets. The dataset needed to be collected on the city's streets in a dynamic urban environment.

In an urban context, a well-known dataset with a similar angle and perspective is the Cityscapes [5] dataset. The classes in this dataset like *building, road, vegetation, street light,* and *car* are more abundant and frequent in an urban environment than the classes presented in the GIGO dataset. Since the classes of the GIGO dataset are more imbalanced and occur less often, it creates different challenges for obtaining a good performance. The same goes for datasets created for the task of autonomous driving like the nuScenes [3] dataset. Within this task, there have been significant advancements in using the cameras and other sensors for understanding the environment by doing semantic segmentation, vehicle detection, and semantic reasoning [27]. However, the task differs from the one presented in this paper since the temporal and geographical frequency of such images is much higher.

The classes in the GIGO dataset are visually heterogeneous and annotated by domain experts. There are several other well-known fine-grained datasets, like Stanford cars [13], CUB-200-2011 [29], and FGVC Aircraft [19]. These collections are composed and annotated by experts since the determination of the classes requires domain-specific knowledge. In addition, such image collections are more balanced than what would occur in a real-world situation, where the frequency of encountered classes varies significantly. Thus we have decided it is best for the released dataset to have a realistic imbalance. Apart from having visually heterogeneous classes, the GIGO dataset also has noisy backgrounds creating an extra dimension to the task.

2.2 Classification in Urban Context

There are plentiful state-of-the-art techniques that perform well on the image classification task. Models like ResNet [11] have proven capable of good performance. On the other hand, more recent networks like EfficientNet [26] achieve similar or better results on benchmarks, while having a faster inference speed making them suitable for large-scale processing of real-time street-level imagery. These techniques are however often evaluated on images without a geospatial context. The difference with the task presented in this work is that the classification of garbage classes in an urban context allows for new ways of processing information that can potentially improve the performance of this task.

Similar but different techniques like 3D-CNN [20] are also used on sequences of images instead of a single frame since pictures that are temporally close often contain similar information. However, the images of the GIGO dataset have a less direct temporal connection. The images are thus unlikely to be similar to the previous and next images as in videos. There is however a strong spatiotemporal link that can be made between the frames.

More recently transformers [6,17] have shown even better results for image classification. This is promising since they also have the potential to harness the context of the images, which makes them a likely candidate to expand into

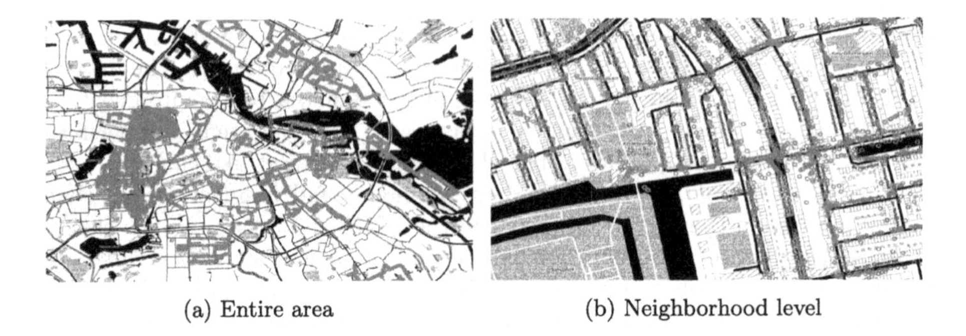

(a) Entire area (b) Neighborhood level

Fig. 4. The geographical distribution of images

a more context-aware type of classification. Recently research also focused on using temporal relationships in combination with transformers [33]. In addition, a recent survey [31] shows transformers are capable of various multimodal tasks.

The use of graphs has been proven useful in an urban context because of the strong spatial and temporal dynamics between flows in a city. Graphs and graph neural networks have been successfully used to predict traffic flow in cities [2,32, 34], which are also heavily influenced by human behavior. Graphs have also been used in combination with visual data. For example, finding patterns on satellite imagery in an urban setting [1,12]. Similar approaches are also interesting to experiment with on the GIGO dataset. In addition, since the images are collected in the same geographical area, there are likely performance gains to be made by combining this information.

3 The GIGO Dataset

The GIGO dataset and additional contextual data sources mentioned in the paper can be found here: https://doi.org/10.21942/uva.20750044.

Data Collection. The data used for the experiments is collected during the period between 2020-02-18 and 2021-06-09 by frequently driving with a vehicle that had two cameras attached (as seen in Fig. 1) to the windshield. While driving these devices were capturing images of the right and left sides of the street. The images have a resolution of 1280×780. Smartphone cameras are used due to their connectivity and ubiquity. A total of 25k images have been collected and annotated.

Privacy. In order to ensure there is no privacy-sensitive information within the released dataset, several measures have been taken. Both license plates and people have been automatically removed from the images using object detection. Since these techniques are likely to miss some objects of interest, afterward the images were manually inspected to see if any information was missed to improve the filtering of privacy-sensitive information further.

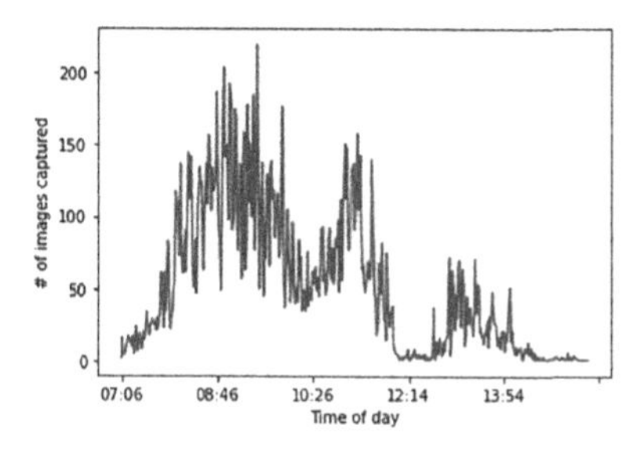

Fig. 5. The distribution of time of the day the images were collected

Table 1. Class descriptions and train and test set distribution

Class name	Description	Train N	Test N
Garbage	Any garbage, or 0 when clean	7768	1584
Garbage bags	Garbage bags	1716	242
Cardboard	Cardboard waste	3851	541
Bulky waste	Bulky waste, for example, a sofa or a fridge	4312	754
Litter	The image contains litter, like cans or bottles	3832	1032

Data Annotation. In collaboration with domain experts specializing in garbage collection in urban environments, classes were determined, and an annotation guideline was created. These guidelines were then followed and all images were manually annotated. The annotation process was started by annotating the *garbage* class for all images. This is because the other classes can only occur when there is any type of garbage shown in the image. The other classes used are as follows: *garbage bags, cardboard, litter, and bulky waste.* These all have different visual representations and different spatial-temporal characteristics. The definition of the classes can be found in Table 1.

Train Test Split. The data is split into 19.827 train images and 5.172 test images. This is done based on the days they were captured to ensure that the scenes found in the photos do not overlap. This also allows for evaluation similar to a real-life setup of such a system, where the situation on the street and conditions are changing daily due to the dynamic nature of cities. Sixty days of recording are thus used for training, and seven days of recording are used for testing.

Additional Contextual Data. The area in which images are collected has additional data available, as illustrated in Fig. 6a and 6b, which can be used for experiments. Information like the function of buildings in the area could be valuable since a residential building will likely have different dynamics regarding garbage production in the area than an office building. Similarly, the location of

(a) Building functions like housing, working, or shopping.

(b) Different types of garbage collection points.

Fig. 6. Examples of contextual data that is available

garbage containers is likely to affect the positioning of the different classes. In addition, open data statistics about e.g. demographics, economy, and the land area used in different neighborhoods could help describe patterns that occur with the positioning of garbage.

4 Baseline Experiments

4.1 Evaluation

To allow for comparison with other multi-label classification approaches, we compute the commonly used mAP metric. In addition to that, for every class, the F1 score is calculated. A reason for this is that both false positives and false negatives are undesired within the domain of the task. Another reason for evaluating using the F1 score is the class imbalance, which needs to be taken into account.

4.2 Experiments

Several state-of-the-art image classification models have been adopted and evaluated to create a baseline. In particular, we experiment with ResNet [11], EfficientNet [25], VisionTransformers [6] and SwinTransformers [17]. The network's last layer is converted into a sequential layer consisting of a linear layer and a sigmoid layer to make the network's output multi-class. The models are pretrained on ImageNet and afterward finetuned on the GIGO dataset.

To assess the value of the multimodal information that can be found in the GIGO dataset, a GNN was trained on a geospatial graph representation of the images (Fig. 7). The nodes of the graph were created using an image representation from the penultimate layer of a ResNet50 network. The edges between these nodes were created using the spatiotemporal proximity between the nodes, taking into account the distance and the time between the capturing

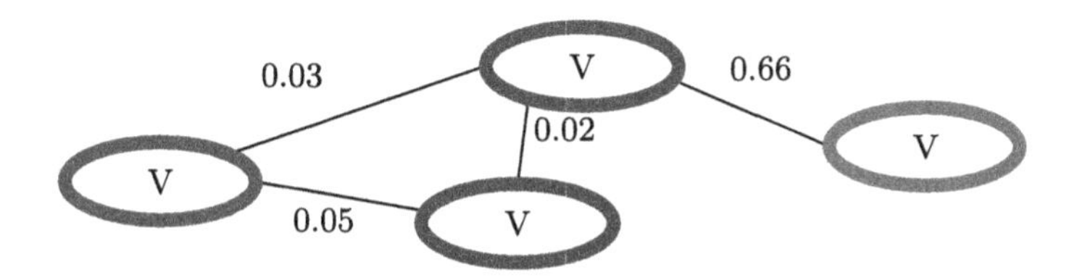

Fig. 7. A schematic portion of the multimodal graph, where the nodes (V) are a visual feature vector and the edges features are the normalized geographic distance between them.

of the images. Whenever the distance between nodes was lower than 250 m and the time of recording was lower than a minute, they were connected with an edge-weighted by the normalized Haversine distance between the coordinates of the image. The GNN used consisted of a SAGEConv [9] layer, followed by a ReLU and a dropout layer. Afterward, this is passed into another SAGEConv layer, followed by the Jumping Knowledge architecture introduced in [30]. The model is finalized by a linear layer and a sigmoid layer outputting the five classes.

4.3 Results

As illustrated in Table 2, overall the best performing architecture is the Vision-Transformer [6]. It performs best for all individual classes when comparing the F1 score and has the best MAP overall. ResNet and EfficientNet have a similar MAP, but the classes have different top performers when using these architectures.

Multimodal GNN. The second best-performing model is the multimodal GNN. While the multimodal baseline uses simple multimodal features only, it outperforms the most common image classification techniques, which indicates that it is indeed possible to use geospatial data to further improve performance.

The multimodal context of the GIGO dataset allows for further experimentation due to the variety of new approaches suitable for this task and the availability of a multitude of potentially interesting auxiliary data sources. The most

Table 2. The MAP for the best performing models and the F1- score calculated per class.

model	MAP	Garbage	Bulky waste	Garbage bag	Cardboard	Litter
VisionTransformer	0.47	0.80	0.83	0.74	0.82	0.73
Multimodal GNN	0.46	0.80	0.81	0.72	0.82	0.72
ResNet152	0.45	0.80	0.80	0.69	0.80	0.73
EfficientNet_b5	0.45	0.77	0.82	0.73	0.81	0.74
SwinTransformer	0.40	0.77	0.79	0.66	0.79	0.66

considerable potential improvement in this task is incorporating information about the surroundings of the scene. This could be demographic information or a result of combining of several images taken on the same route. There are several approaches suitable for harnessing such information about the image context. This can for example be done with graph neural networks. Another interesting development is the use of transformers for similar tasks.

Transformers. The use of metadata of the images, like location and time of day, could be integrated into transformers using a similar approach to [4], where the metadata of skin disease images is used to improve the visual classification. As mentioned in the related work, transformers can efficiently handle several modalities, and thus, it seems likely that using this technique will improve performance on the GIGO dataset.

5 Conclusion

This paper presents the GIGO dataset. The dataset is composed of 25k images, annotated in collaboration with experts specializing in the process of garbage collection in urban environments. The task presented is the multi-label classification of garbage in street-level imagery. The task is particularly challenging due to the classes being visually heterogeneous in shape, size, and location in the image. Similar to the classes of interest, the backgrounds are also noisy since the images are captured in real-world conditions by a vehicle moving through a dynamic urban environment. In addition to the task and dataset, we created several baselines using visual and multimodal information and discuss possible methods for building on such approaches. The task of garbage classification on street-level imagery has a strong geo-temporal context, allowing for the benchmarking of a wide variety of multimodal approaches. We believe that opening up GIGO dataset to the multimedia community can facilitate research into this topic, which can help make urban life more sustainable and increase the livability of cities.

References

1. Albert, A., Kaur, J., Gonzalez, M.C.: Using convolutional networks and satellite imagery to identify patterns in urban environments at a large scale. In: Proceedings of the 23rd ACM SIGKDD International Conference on Knowledge Discovery and Data Mining, pp. 1357–1366 (2017)
2. An, J., et al.: IGAGCN: information geometry and attention-based spatiotemporal graph convolutional networks for traffic flow prediction. Neural Netw. **143**, 355–367 (2021)
3. Caesar, H., et al.: nuScenes: a multimodal dataset for autonomous driving. In: Proceedings of the IEEE/CVF Conference on Computer Vision and Pattern Recognition, pp. 11621–11631 (2020)

4. Cai, G., Zhu, Y., Wu, Y., Jiang, X., Ye, J., Yang, D.: A multimodal transformer to fuse images and metadata for skin disease classification. The Visual Computer, pp. 1–13 (2022)
5. Cordts, M., et al.: The cityscapes dataset for semantic urban scene understanding. In: Proceedings of the IEEE Conference on Computer Vision and Pattern Recognition, pp. 3213–3223 (2016)
6. Dosovitskiy, A., et al.: An image is worth 16x16 words: transformers for image recognition at scale. arXiv preprint arXiv:2010.11929 (2020)
7. Moya, M.G., Phan, T.T., Gatica-Perez, D.: Zurich like new: analyzing open urban multimodal data. In: Proceedings of the 1st International Workshop on Multimedia Computing for Urban Data, pp. 1–8 (2021)
8. Gurrin, C., et al.: [invited papers] Comparing approaches to interactive lifelog search at the lifelog search challenge (LSC2018). ITE Trans. Media Technol. Appl. **7**(2), 46–59 (2019)
9. Hamilton, W.L., Ying, R., Leskovec, J.: Inductive representation learning on large graphs. In: Proceedings of the 31st International Conference on Neural Information Processing Systems, pp. 1025–1035 (2017)
10. Harman, D.: Overview of the first TREC conference. In: Proceedings of the 16th Annual International ACM SIGIR Conference on Research and Development in Information Retrieval, pp. 36–47. SIGIR 1993, Association for Computing Machinery, New York, NY, USA (1993). https://doi.org/10.1145/160688.160692. https://doi.org/10.1145/160688.160692
11. He, K., Zhang, X., Ren, S., Sun, J.: Deep residual learning for image recognition. In: Proceedings of the IEEE Conference on Computer Vision and Pattern Recognition, pp. 770–778 (2016)
12. Hong, D., Gao, L., Yao, J., Zhang, B., Plaza, A., Chanussot, J.: Graph convolutional networks for hyperspectral image classification. IEEE Trans. Geosci. Remote Sens. **59**(7), 5966–5978 (2021)
13. Krause, J., Stark, M., Deng, J., Fei-Fei, L.: 3D object representations for fine-grained categorization. In: 4th International IEEE Workshop on 3D Representation and Recognition (3dRR-13). Sydney, Australia (2013)
14. Krizhevsky, A., Sutskever, I., Hinton, G.E.: ImageNet classification with deep convolutional neural networks. In: Advances in Neural Information Processing Systems 25 (2012)
15. Larson, M., et al.: Automatic tagging and geotagging in video collections and communities. In: Proceedings of the 1st ACM International Conference on Multimedia Retrieval. ICMR 2011, Association for Computing Machinery, New York, NY, USA (2011). https://doi.org/10.1145/1991996.1992047. https://doi.org/10.1145/1991996.1992047
16. Lin, T.-Y., et al.: Microsoft COCO: common objects in context. In: Fleet, D., Pajdla, T., Schiele, B., Tuytelaars, T. (eds.) ECCV 2014. LNCS, vol. 8693, pp. 740–755. Springer, Cham (2014). https://doi.org/10.1007/978-3-319-10602-1_48
17. Liu, Z., et al.: Swin transformer: hierarchical vision transformer using shifted windows. In: Proceedings of the IEEE/CVF International Conference on Computer Vision, pp. 10012–10022 (2021)
18. Lokoč, J., et al.: Is the reign of interactive search eternal? findings from the video browser showdown 2020. ACM Trans. Multimedia Comput. Commun. Appl. (TOMM) **17**(3), 1–26 (2021)
19. Maji, S., Rahtu, E., Kannala, J., Blaschko, M., Vedaldi, A.: Fine-grained visual classification of aircraft. arXiv preprint arXiv:1306.5151 (2013)

20. Qiu, Z., Yao, T., Mei, T.: Learning spatio-temporal representation with pseudo-3D residual networks. In: proceedings of the IEEE International Conference on Computer Vision, pp. 5533–5541 (2017)
21. Selvaraju, R.R., Cogswell, M., Das, A., Vedantam, R., Parikh, D., Batra, D.: Grad-CAM: visual explanations from deep networks via gradient-based localization. In: Proceedings of the IEEE International Conference on Computer Vision, pp. 618–626 (2017)
22. Smeaton, A.F., Over, P., Kraaij, W.: Evaluation campaigns and TRECVid. In: Proceedings of the 8th ACM International Workshop on Multimedia Information Retrieval, pp. 321–330. MIR 2006, Association for Computing Machinery, New York, NY, USA (2006). https://doi.org/10.1145/1178677.1178722. https://doi.org/10.1145/1178677.1178722
23. Sukel, M., Rudinac, S., Worring, M.: Multimodal classification of urban micro-events. In: Proceedings of the 27th ACM International Conference on Multimedia, pp. 1455–1463 (2019)
24. Sukel, M., Rudinac, S., Worring, M.: Urban object detection kit: a system for collection and analysis of street-level imagery. In: Proceedings of the 2020 International Conference on Multimedia Retrieval, pp. 509–516 (2020)
25. Tan, M., Le, Q.: EfficientNet: rethinking model scaling for convolutional neural networks. In: International Conference on Machine Learning, pp. 6105–6114. PMLR (2019)
26. Tan, M., Le, Q.: EfficientNetV2: smaller models and faster training. In: International Conference on Machine Learning, pp. 10096–10106. PMLR (2021)
27. Teichmann, M., Weber, M., Zoellner, M., Cipolla, R., Urtasun, R.: MultiNet: real-time joint semantic reasoning for autonomous driving. In: 2018 IEEE Intelligent Vehicles Symposium (IV), pp. 1013–1020. IEEE (2018)
28. Tobler, W.R.: A computer movie simulating urban growth in the detroit region. Econ. Geograp. **46**, 234–240 (1970). http://www.jstor.org/stable/143141
29. Wah, C., Branson, S., Welinder, P., Perona, P., Belongie, S.: The Caltech-UCSD Birds-200-2011 Dataset. Tech. Rep. CNS-TR-2011-001, California Institute of Technology (2011)
30. Xu, K., Li, C., Tian, Y., Sonobe, T., Kawarabayashi, K.i., Jegelka, S.: Representation learning on graphs with jumping knowledge networks. In: International Conference on Machine Learning, pp. 5453–5462. PMLR (2018)
31. Xu, P., Zhu, X., Clifton, D.A.: Multimodal learning with transformers: a survey. arXiv preprint arXiv:2206.06488 (2022)
32. Yu, B., Yin, H., Zhu, Z.: Spatio-temporal graph convolutional networks: a deep learning framework for traffic forecasting. arXiv preprint arXiv:1709.04875 (2017)
33. Zhang, H., Hao, Y., Ngo, C.W.: Token shift transformer for video classification. In: Proceedings of the 29th ACM International Conference on Multimedia, pp. 917–925 (2021)
34. Zhao, L., et al.: T-GCN: a temporal graph convolutional network for traffic prediction. IEEE Trans. Intell. Transp. Syst. **21**(9), 3848–3858 (2019)

Marine Video Kit: A New Marine Video Dataset for Content-Based Analysis and Retrieval

Quang-Trung Truong[1(✉)], Tuan-Anh Vu[1], Tan-Sang Ha[1], Jakub Lokoč[2], Yue-Him Wong[3], Ajay Joneja[1], and Sai-Kit Yeung[1]

[1] Hong Kong University of Science and Technology, Clear Water Bay, Hong Kong
qttruong@connect.ust.hk
[2] FMP, Charles University, Prague, Czech Republic
[3] Shenzhen University, Shenzhen, China

Abstract. Effective analysis of unusual domain specific video collections represents an important practical problem, where state-of-the-art general purpose models still face limitations. Hence, it is desirable to design benchmark datasets that challenge novel powerful models for specific domains with additional constraints. It is important to remember that domain specific data may be noisier (e.g., endoscopic or underwater videos) and often require more experienced users for effective search. In this paper, we focus on single-shot videos taken from moving cameras in underwater environments, which constitute a nontrivial challenge for research purposes. The first shard of a new Marine Video Kit dataset is presented to serve for video retrieval and other computer vision challenges. Our dataset is used in a special session during Video Browser Showdown 2023. In addition to basic meta-data statistics, we present several insights based on low-level features as well as semantic annotations of selected keyframes. The analysis also contains experiments showing limitations of respected general purpose models for retrieval. Our dataset and code are publicly available at https://hkust-vgd.github.io/marinevideokit.

Keywords: Known-item search · Underwater video · Dataset statistics

1 Introduction

In order to facilitate the development of multimedia retrieval and analysis models, the research community establishes and uses various benchmark multimedia datasets [2,9,9,14,21,21–23,23]. The datasets usually provide so-called ground-truth annotations and allow repeatable experimental comparison with state-of-the-art methods.

The source of large benchmark collections is often a video sharing platform (e.g., Youtube, or Vimeo) with specific licensing of the content. For example, the

D.-T. Dang-Nguyen et al. (Eds.): MMM 2023, LNCS 13833, pp. 539–550, 2023.
https://doi.org/10.1007/978-3-031-27077-2_42

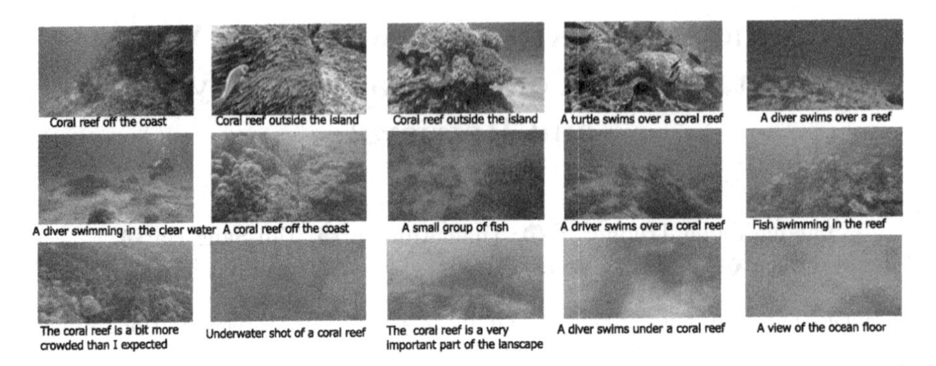

Fig. 1. Several examples of dataset video frames and their ClipCap descriptions.

respected Vimeo Creative Commons collection dataset (V3C) [18] contains several thousand hours of videos downloaded from the Vimeo platform. Although only videos with the creative commons license were used for the dataset, it is available only with an agreement form indicating possible changes in the future. Hence, it is beneficial to design also new video datasets with a limited number of copyright owners, limiting potential future changes to the dataset (experiment repeatability). Designing datasets with highly challenging content is also necessary, even for a limited number of data items. This aspect is especially important for interactive search evaluation campaigns [7,8] addressing a broad community of researchers from different multimedia retrieval areas. Indeed, for many research teams (especially smaller ones), it might be more feasible to participate in a difficult challenge over 10–100 hours of videos, rather than a challenge over 10.000 h and more.

In general everyday videos, there appear many common classes of objects, and thus even a larger collection can be effectively filtered with text queries. On the other hand, a domain specific collection (i.e., one cluster with a lower variance of common keywords) might already be challenging for lower sizes of collections. Therefore, we selected underwater marine videos with the seafloor, coral reefs, and various biodiversity where potentially effective keywords are unknown to ordinary users. Furthermore, unlike common "everyday" videos, underwater videos pose additional obstacles for multimedia analysis and retrieval models. These challenges include low visibility, blurry shots, varying sizes and poses of objects, a crowded background, light attenuation, and scattering, among others [10]. Therefore, not only general purpose models but also domain specific classifiers require novel ideas and breakthroughs to reach human-level accuracy.

This paper presents a first fragment of a new "Marine Video Kit" dataset, currently intended mainly for content-based retrieval challenges. It is composed of more than 1300 underwater videos from 36 locations worldwide and at different times across the year. Whereas the long-term ambition for this dataset might be even a video-sharing platform with controllable extensions, for now, we present a manually organized set of directories comprising videos, selected frames, and various forms of meta-data. So far, the meta-data comprises available video attributes (location, time) and pre-computed captions as well as embeddings for the set of selected video frames. In the future, we plan to provide also more annotations for object detection, semantic segmentation, object tracking, etc.

The remainder of this paper is structured as follows: Sect. 2 gives an overview of related works, Sect. 3 introduces the details of the Marine Video Kit dataset, and finally, Sect. 4 concludes the paper.

2 Related Work

Numerous datasets were created for the object detection and segmentation task in order to better comprehend marine life and ecosystems. In this section, we briefly review some recent works for Marine-related datasets.

The Brackish [16] is an open-access underwater dataset containing annotated image sequences of starfish, crabs, and fish captured in brackish water with varying degrees of visibility. The videos were divided into categories according to the primary activity depicted in each one. A bounding box annotation tool was then used to manually annotate each category's 14,518 frames, producing 25,613 annotations in total.

MOUSS dataset [4] is gathered by a horizontally-mounted, grayscale camera that is placed between 1 and 2 m above the sea floor and is illuminated solely by natural light. In most cases, the camera remains stationary for 15 min at a time in each position. There are two sequences in the MOUSS datasets: MOUSS seq0 and MOUSS seq1. The MOUSS seq0 includes 194 images, all of which belong to the Carcharhini-formes category, and each image has a resolution of 968 by 728 pixels. There is only one category in the MOUSS seq1, which is called Perciformes. Each image has a resolution of 720 by 480 pixels. A human expert was responsible for assigning each of the species labels.

WildFish [24] is a large-scale benchmark for fish recognition in the wild. It consists of 1,000 fish categories and 54,459 unconstrained images. This benchmark was developed for the field of the classification task. In the field of image enhancement, the database known as Underwater ImageNet [5] is made up of subsets of ImageNet [3] that contain photographs taken underwater. As a result, the distorted and undistorted sets of underwater images were allowed to have 6,143 and 1,817 pictures, respectively. Fish4Knowledge [6] gave an analysis of fish video data.

OceanDark [20] is a novel low-lighting underwater image dataset that was created to quantitatively and qualitatively evaluate the proposed framework. This dataset was developed in the field of image enhancement and consisted of images that were captured using artificial lighting sources.

The Holistic Marine Video Dataset [11], also known as HMV, is a long video that simulates marine videos in real time and annotates the frames of HMV with scenes, organisms, and actions. The goal of this dataset is to provide a large-scale video benchmark with multiple semantic aspect annotations. On the other hand, they also provide baseline experiments for reference on HMV for three tasks: the detection of marine organisms, the recognition of marine scenes, and the recognition of marine organism actions.

3 Marine Video Kit Dataset

In this section, we will show details about our Marine Video Kit Dataset, which provides a dense, balanced set of videos focusing on marine environment, hence enriching the pool of existing marine dataset collections.

Many types of cameras were used to build the presented first shard of the Marine Video Kit dataset, such as Canon PowerShot G1 X, Sony NEX-7, OLYMPUS PEN E-PL, Panasonic Lumix DMC-TS3, GoPro cameras, and consumer cellphones cameras. The dataset consists of a larger number of single-shot videos without post-processing. Unlike common video collections taken by crawling search engines, the presented dataset focuses solely on marine organisms captured during diving periods. The typical duration of each video is about 30 s.

To illustrate the specifics of the underwater environment, we present 3D color histograms showing differences for different parts of the dataset. For a larger set of selected frames (extracted with 1fps), we also analyze semantic descriptions automatically extracted by the ClipCap model [15]. For a randomly selected subset of 100 frames, the automatically generated descriptions are compared with manually created annotations in a known-item search experiment.

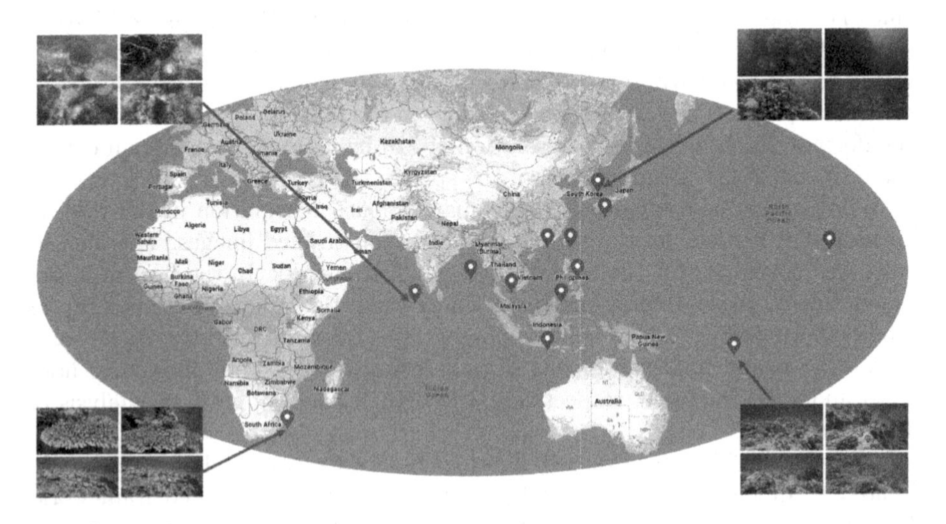

Fig. 2. The world map illustrates countries/regions where we capture data around the world (map source: Google Maps).

3.1 Acquiring the Dataset

Different ocean areas have their own marine biodiversities, such as various animals, plants, and microorganisms. Therefore, to create a diverse dataset for the research community, we have visited and captured data from 11 different regions and countries during daytime and nighttime (unless bad weather or other circumstances happen). We present a world map illustrating 11 countries/regions where the videos were captured Fig. 2.

We categorize the captured videos in terms of their location and time, then utilize OpenCV library [1] to process all data into one unifying format (JPG for images, MP4 for videos). Due to the variety of capturing devices, our raw videos have different resolutions, from the HD (720p) resolution to Ultra HD (4K), with a frame rate of 30 fps. Therefore, we also utilize FFmpeg library [19] to convert all data to low and high resolution for different research purposes, such as video retrieval, super-resolution, object detection, segmentation, etc.

3.2 Dataset Structure

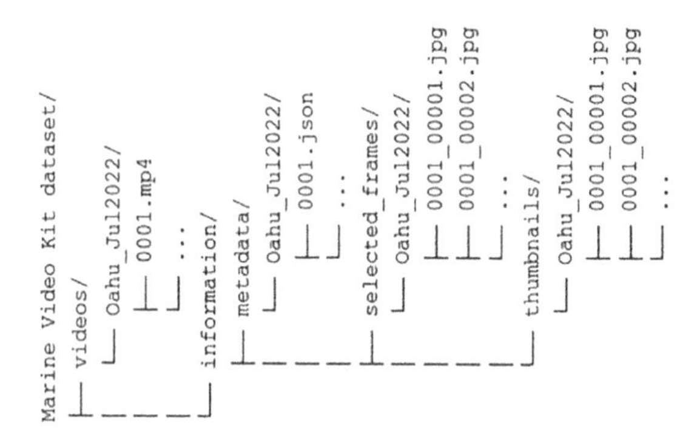

Fig. 3. Directory structure of the Marine Video Kit dataset.

In this section, we show our data's directory structure, which organizes different aspects of the data. As shown in Fig. 3, there are two sub-directories for video and its supplementary information.

For each video directory, we format their name as *location_time* pattern to explicitly represent the time and location that they were captured. For example, "Oahu_Jul2022' was captured at Oahu - the third-largest of the Hawaiian Islands, in July 2022.

For each information directory, the *selected_frames* directory stores all frames that are evenly selected one frame per second and are kept in the original resolution, while the *thumbnails* directory stores the same frame but in down-scaled resolution. Finally, the *metadata* directory contains the associated meta information of each video in JSON format for easier sharing and parsing information.

This metadata file contains semantic and statistical information, such as video name, duration, height, width, camera device, directory, license, and reference information, such as ClipCap captions.

3.3 Dataset Statistics

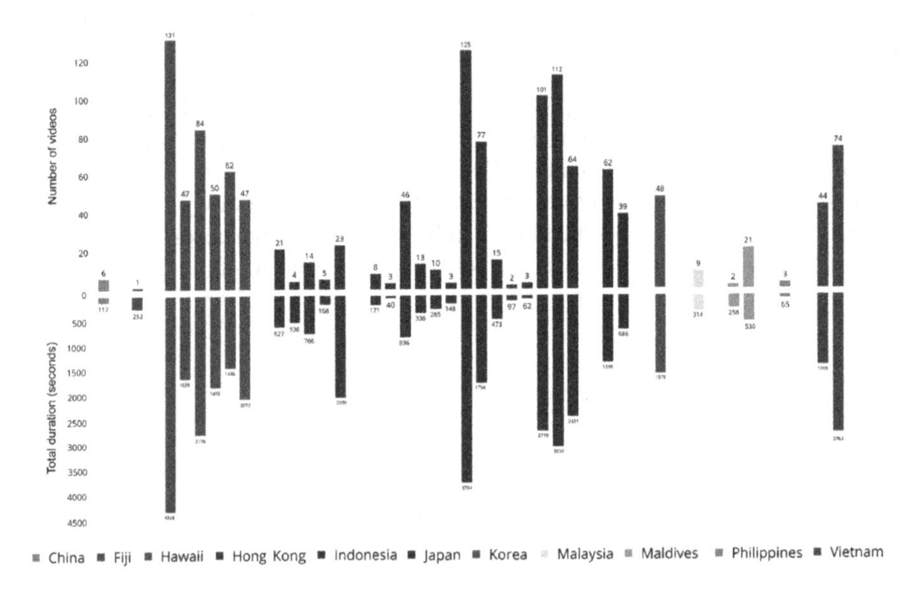

Fig. 4. The figure shows the number of videos and overall time duration for each region.

We capture data from 11 different regions and countries during the time from 2011 to 2022. There is a total of 1379 videos with a length from 2 s to 4.95 min, with the mean and median duration of each video is 29.9 s, and 25.4 s, respectively. The total duration is slightly above 12 h, however, the diving time is significantly larger, up to a thousand hours. Figure 4 shows the number of videos, and the total length of videos varies by category.

To represent marine videos from directories in color space, 3D color histograms are used to aggregate pixels to a $256 \times 256 \times 256$ array for 3D representation. The 3D visualization of each directory is presented in Fig. 5, 6 with 3D points colored by RGB histogram values based on corresponding places in 3D spaces, and the sizes are proportional to the color densities. We used color quantization and normalization for a more informative illustration of 3D color histogram densities.

A big challenge when using Marine Video Kit dataset for retrieval tasks is that the marine environment has changed considerably, and the captured data is heavily affected by lighting variations or caustics, thus causing low visibility. Figure 5 illustrates regions under low lighting, and Fig. 6 mentions regions under good lighting.

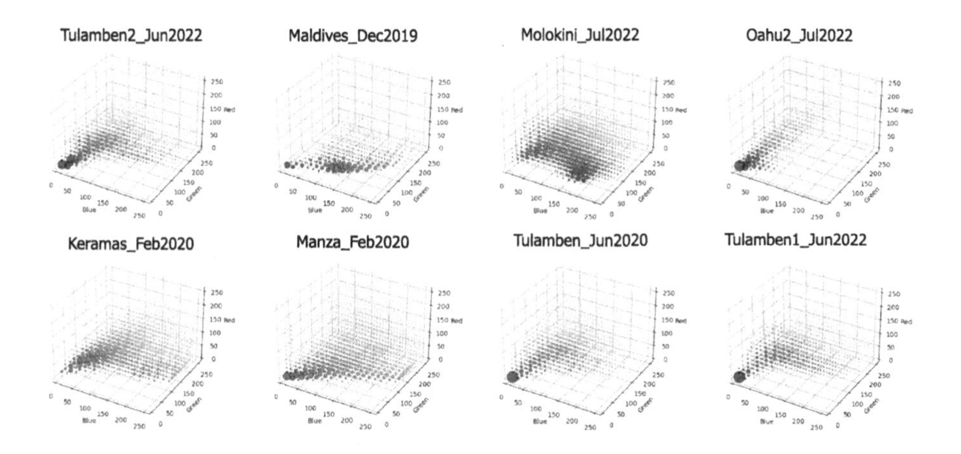

Fig. 5. Illustration of video frame colors present in regions under low illumination using 3D color histograms.

3.4 ClipCap Captions

To provide semantic information for the set of uniformly selected frames, ClipCap [15] architecture was employed. ClipCap elegantly combines rich features of the CLIP model [17] with the powerful GPT-2 language model [12]. Specifically, a prefix is computed for the image feature vector, while the language model continues to generate the caption text based on the prefix. Additionally, ClipCap is efficient in diverse datasets, which motivates us to caption selected frames in the marine dataset. Captions given by ClipCap are utilized as an automatically generated caption attribute in video metadata files.

We adopt ClipCap for selected frames in Fig. 7 that are mentioned in metadata files. The process to generate captions consists of two steps to ensure content relevance. The captions of selected frames are automatically outputted by the ClipCap model; after that, we regularly inspect the captions to remove unrelated descriptions. Figure 1 shows selected frame captions of ClipCap that are utilized to describe semantic information of the marine dataset, and Fig. 8 presents frequencies for individual words in frame captions.

3.5 Known-Item Search Experiment

In this section, we show that the new dataset represents a challenge for an information retrieval task. Specifically, a known-item search experiment is performed using a subset of all selected video frames O_i (extracted with 1fps, about 40K frames), and their CLIP representations [17]. The experiment assumes a set of pairs $[q_i, t_i]$, where t_i is a CLIP embedding vector representing a target (searched) video frame O_i, and q_i is a CLIP embedding vector representing a query description of the target image O_i. Using one pair, it is possible to rank all selected video frames based on their cosine distance to the query vector q_i.

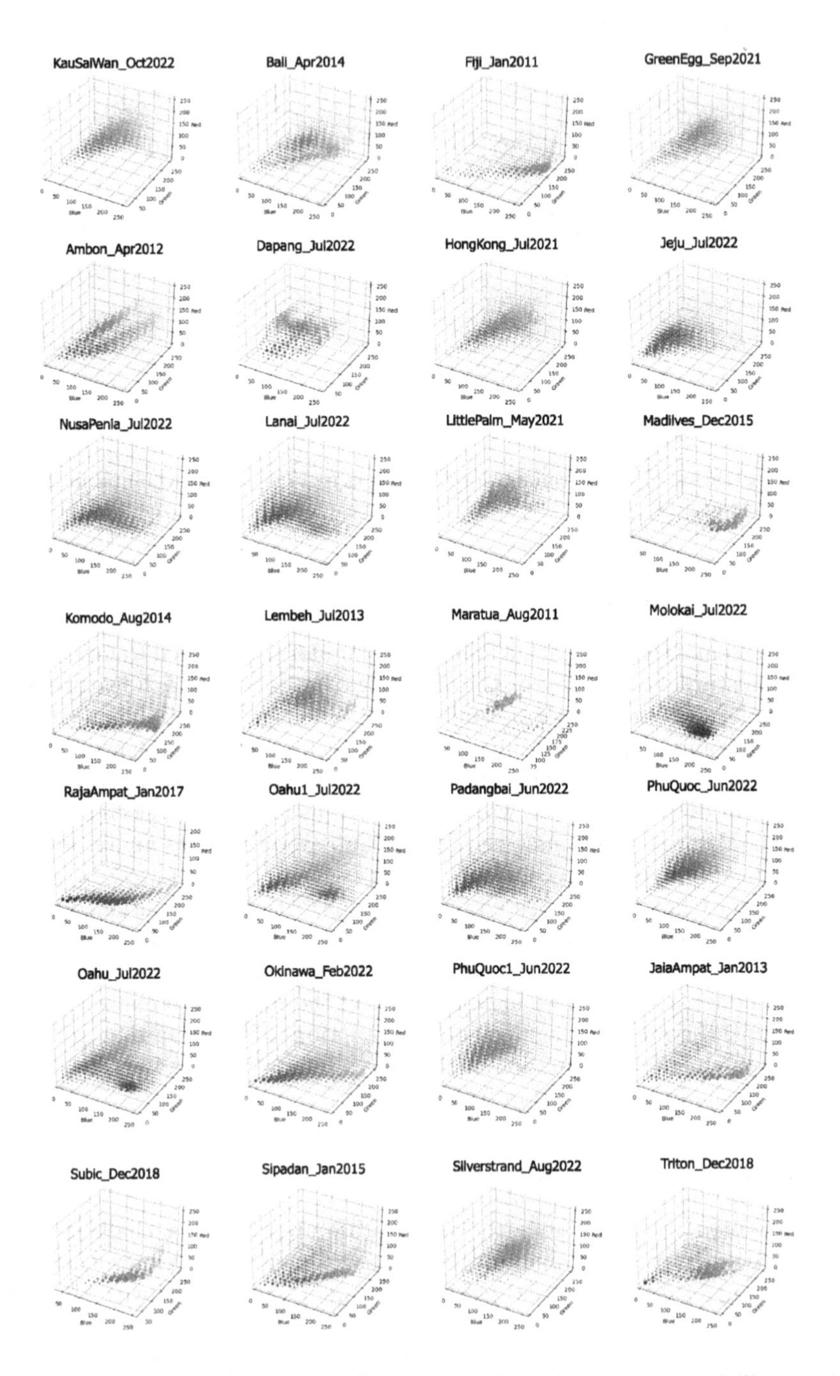

Fig. 6. Illustration of video frame colors present in regions under good illumination using 3D color histograms.

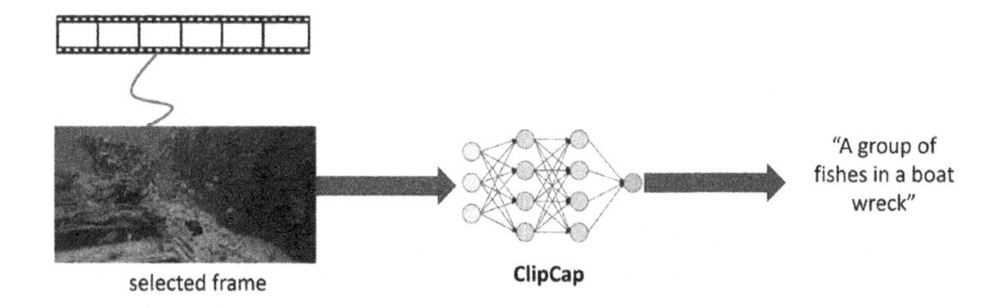

Fig. 7. The captioning architecture. A selected frame is extracted from marine video, then fed into the ClipCap model to output the caption of the selected frame.

From this ranking, the rank of the target image O_i can be identified and stored. Repeating this experiment for all available pairs $[q_i, t_i]$, a distribution of ranks of searched items can be analyzed.

Figure 9 shows the result of our preliminary KIS experiment, where 100 pairs $[q_i^{novice}, t_i]$ and 100 pairs $[q_i^{expert}, t_i]$ were created manually. Novice and VBS Expert annotators described randomly selected target images, where the VBS Expert is not an expert in the marine domain but has experience with query formulation from the Video Browser Showdown. ClipCap annotations were computed for the same target images as well. Using CLIP embeddings, ranks of target images were computed for the novice, VBS Expert, and ClipCap queries. Figure 9 shows that the overall distributions of ranks are similar for ClipCap and novice users, while the VBS expert was able to reach a better average and median rank. In the future, we plan more thorough experiments. We also note that employed random selection of target images often does not lead to unique items, and users did not see the search results for the provided 100 text descriptions/queries. Both notes affect the overall distribution of ranks.

Since ClipCap descriptions are available for all database frames, we also present another KIS experiment for ClipCap based queries with similar average target rank as novice users in Fig. 9. Figure 10 was evaluated for 4000 pairs $[q_i, t_i]$, where the target image descriptions were obtained using the ClipCap approach. We may observe that only about 30% of all queries are allowed to find the tar-

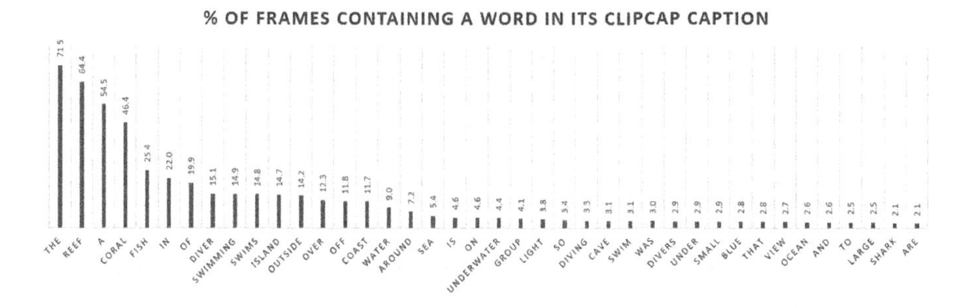

Fig. 8. Occurrence of words in frame captions. Computed for a subset of the dataset.

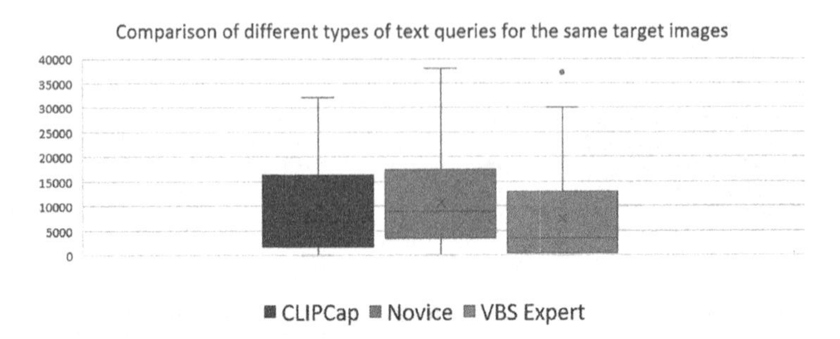

Fig. 9. Ranks for ClipCap, Novice, and VBS Expert text queries for 100 target images.

get in the top-ranked 2000 dataset items. Even for the small dataset. Although the CLIP retrieval model helps, it is indeed a way more difficult challenge than searching common videos with a large number of different concepts (e.g., compared to the results of a similar study here [13]). Furthermore, searching top 2000 items is not trivial. Looking at the top 500 close-ups of the histogram, only 6.6% of target items can be found in the top 100. Hence, we conclude that limited vocabulary (i.e., ClipCap like queries) and similar content make the known-item search challenge difficult for the proposed dataset even with the respected CLIP model.

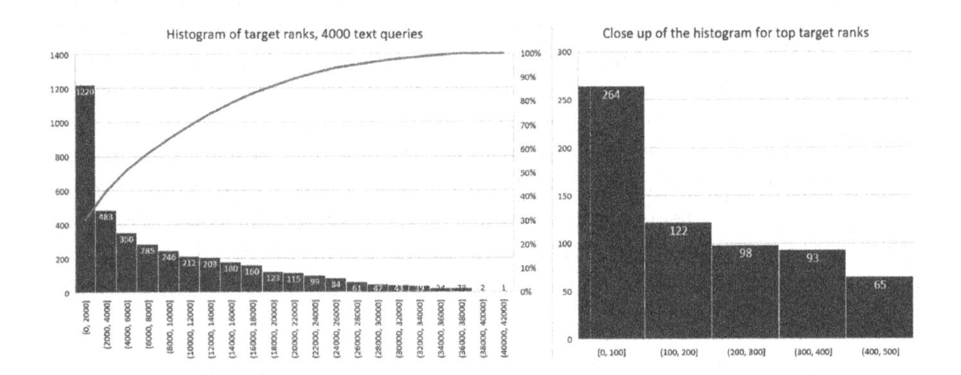

Fig. 10. 4000 KIS queries using ClipCap captions, histogram bins aggregate ranks of target images for the corresponding queries.

4 Conclusion

We provide the Marine Video Kit dataset, single-shot videos challenging for content-based analysis and retrieval. To provide a first insight of the new marine dataset, basic statistics based on meta-data, low-level color descriptors, and ClipCap semantic annotations are presented. We present a baseline retrieval study for the Marine Video Kit dataset to emphasize the domain's specificity. Our experiments show that the similarity of content in the dataset causes difficulties for a respected cross-modal based know-item search approach. We hope that our dataset with the baseline for content-based retrieval will accelerate considerable progress in the marine video retrieval area.

We plan to extend the dataset in the future with new annotations and videos from new environments. Besides, additional computer vision tasks over the dataset, such as semantic segmentation or object detection, could be prepared for the research community. Furthermore, motion analysis, fish counting, or detection tasks are also meaningful information for retrieval applications.

Acknowledgements. This research project is partially supported by an internal grant from HKUST (R9429), the Innovation and Technology Support Programme of the Innovation and Technology Fund (Ref: ITS/200/20FP), the Marine Conservation Enhancement Fund (MCEF20107), Charles University grant (SVV-260588), and the Innovation Team Project of Universities in Guangdong Province (No. 2020KCXTD023).

Disclaimer. Any opinions, findings, conclusions, or recommendations expressed in this material do not necessarily reflect the views of HKLTL, CAPCO, HK Electric, and the Marine Conservation Enhancement Fund.

References

1. Bradski, G.: The OpenCV Library. Dr. Dobb's Journal of Software Tools (2000)
2. Chen, J., Chen, X., Ma, L., Jie, Z., Chua, T.S.: Temporally grounding natural sentence in video. In: Proceedings of the 2018 Conference on Empirical Methods in Natural Language Processing (2018)
3. Deng, J., Dong, W., Socher, R., Li, L.J., Li, K., Fei-Fei, L.: Imagenet: a large-scale hierarchical image database. In: 2009 IEEE Conference on Computer Vision and Pattern Recognition (2009)
4. Derya, A., Anthony, H., Suchendra, B.: Mouss dataset (2018)
5. Fabbri, C., Islam, M.J., Sattar, J.: Enhancing underwater imagery using generative adversarial networks. In: 2018 IEEE International Conference on Robotics and Automation (ICRA) (2018)
6. Fisher, R.B., Chen-Burger, Y.H., Giordano, D., Hardman, L., Lin, F.P., et al.: Fish4Knowledge: collecting and analyzing massive coral reef fish video data, vol. 104. Springer (2016)
7. Gurrin, C., et al.: Introduction to the fifth annual lifelog search challenge. In: International Conference on Multimedia Retrieval (2022)
8. Heller, S., et al.: Interactive video retrieval evaluation at a distance: comparing sixteen interactive video search systems in a remote setting at the 10th video browser showdown. Int. J. Multimed. Inf. Retr. **11**(1), 1–18 (2022)
9. Krishna, R., Hata, K., Ren, F., Fei-Fei, L., Niebles, J.C.: Dense-captioning events in videos. In: International Conference on Computer Vision (ICCV) (2017)
10. Levy, D., Levy, D., Belfer, Y., Osherov, E., Bigal, E., Scheinin, A.P., Nativ, H., Tchernov, D., Treibitz, T.: Automated analysis of marine video with limited data. In: 2018 IEEE/CVF Conference on Computer Vision and Pattern Recognition Workshops (CVPRW) (2018)

11. Li, Q., Li, J., Shi, Z., Gu, Z., Zheng, H., Zheng, B., Li, J.: A holistic marine video dataset. In: OCEANS 2021: San Diego - Porto (2021)
12. Li, X.L., Liang, P.: Prefix-tuning: Optimizing continuous prompts for generation. arXiv preprint arXiv:2101.00190 (2021)
13. Lokoč, J., Souček, T.: How many neighbours for known-item search? In: Similarity Search and Applications - 14th International Conference, SISAP 2021 Proceedings (2021)
14. Mithun, N.C., Li, J., Metze, F., Roy-Chowdhury, A.K.: Learning joint embedding with multimodal cues for cross-modal video-text retrieval. In: Proceeding of International Conference on Multimedia Retrieval (ICMR). ACM (2018)
15. Mokady, R., Hertz, A., Bermano, A.H.: Clipcap: Clip prefix for image captioning. arXiv preprint arXiv:2111.09734 (2021)
16. Pedersen, M., Haurum, J.B., Gade, R., Moeslund, T.B., Madsen, N.: Detection of marine animals in a new underwater dataset with varying visibility. In: The IEEE Conference on Computer Vision and Pattern Recognition (CVPR) Workshops (2019)
17. Radford, A., et al.: Learning transferable visual models from natural language supervision. In: International Conference on Machine Learning, pp. 8748–8763. PMLR (2021)
18. Rossetto, L., Schuldt, H., Awad, G., Butt, A.A.: V3c-a research video collection. In: International Conference on Multimedia Modeling (2019)
19. Tomar, S.: Converting video formats with ffmpeg. Linux Journal (2006)
20. Tunai, P.M., Alexandra, B.A., Maia, H.: A contrast-guided approach for the enhancement of low-lighting underwater images. J. Imaging **5**(10), 79 (2019)
21. Xu, J., Mei, T., Yao, T., Rui, Y.: Msr-vtt: A large video description dataset for bridging video and language. In: 2016 IEEE Conference on Computer Vision and Pattern Recognition (CVPR) (2016)
22. Youngjae, Y., Jongseok, K., Gunhee, K.: A joint sequence fusion model for video question answering and retrieval. In: Proceeding of European Conference on Computer Vision (ECCV) (2018)
23. Zhou, L., Xu, C., Corso, J.J.: Towards automatic learning of procedures from web instructional videos. In: AAAI Conference on Artificial Intelligence (2018)
24. Zhuang, P., Wang, Y., Qiao, Y.: Wildfish: a large benchmark for fish recognition in the wild. In: Proceeding of ACM Multimedia Conference on Multimedia Conference (2018)

SNL: Sport and Nutrition Lifelogging

Arctic HARE: A Machine Learning-Based System for Performance Analysis of Cross-Country Skiers

Tor-Arne S. Nordmo[1,3](✉), Michael A. Riegler[1,2,3], Håvard D. Johansen[1,3], and Dag Johansen[1,3]

[1] UiT: The Arctic University of Norway, Tromsø, Norway
tor-arne.s.nordmo@uit.no
[2] SimulaMet, Oslo, Norway
[3] University of Oslo, Oslo, Norway

Abstract. Advances in sensor technology and big data processing enable new and improved performance analysis of sport athletes. With the increase in data variety and volume, both from on-body sensors and cameras, it has become possible to quantify the specific movement patterns that make a good athlete.

This paper describes *Arctic Human Activity Recognition on the Edge* (Arctic HARE): a skiing-technique training system that captures movement of skiers to match those against optimal patterns in well-known cross-country techniques. Arctic HARE uses on-body sensors in combination with stationary cameras to capture movement of the skier, and provides classification of the perceived technique. We explore and compare two approaches for classifying data, and determine optimal representations that embody the movement of the skier. We achieve higher than 96% accuracy for real-time classification of cross-country techniques.

Keywords: Machine learning · Activity recognition · Distributed systems · Embedded systems · Ski technique classification

1 Introduction

Analyzing the performance of athletes is becoming practically feasible with the growth of technology. Performance analysis of athletes is the act of quantifying sports performance in order to develop an understanding that can inform the conscious or unconscious choices done by the athlete in order to enhance their performance [14]. Multiple products allow coaches and athletes to review their performance manually [18,22]. Physiological data is also often obtained through on-body sensors [4,7].

Endurance and technique are related aspects that are significant factors for quantifying the performance of an athlete. It is important to make systems that can be used outside of the laboratory environment in order to realistically capture these aspects. Real-time analysis can provide immediate feedback to an athlete while training. This allows them to quickly adjust their movement to

D.-T. Dang-Nguyen et al. (Eds.): MMM 2023, LNCS 13833, pp. 553–564, 2023.
https://doi.org/10.1007/978-3-031-27077-2_43

improve their performance. Sports produce very large amounts of data that can be analyzed, however it can be difficult to make general analysis software due to the differences between the sports and differences between data sources.

We have developed a distributed system that can perform real-time performance analysis on the edge using on-body sensors. We have also explored a video-based approach to compare with the sensor-based approach. We apply this system to the field of cross-country skiing. This system allows ski athletes to receive real-time feedback while they are training.

The remainder of this paper is structured as follows: In Sect. 2 we give a description of *Arctic HARE* and its subcomponents. In Sect. 3 we briefly explain the acquisition and preprocessing of the data used. Then, Sect. 4 provides an experimental evaluation of the system with a brief discussion of the results. Then, we give an overview of related work in Sect. 5. Finally, we conclude the paper in Sect. 6.

2 Related Work

Øyvind Gløersen and Gilgien [5] automatically detected cycle length (how far the skier moves during a cycle), cycle duration and sub-techniques (Offset/V1 skate (V1), V2/One skate (1SK), and V2 Alt/Two skate (2SK)) using Differential Global Navigation Satellite System (dGNSS) measurements of the head of the skier. They based the cycle on the lateral velocity of the skier. They achieved an accuracy between 98% and 100%, depending on how many skiers the model was trained. There are some drawbacks with the dGNSS approach. The need for stationary base stations that need time to calibrate, and need to be placed so all of them can communicate with each other [13]. Obstacles such as trees and buildings, and differences in elevation make this placement non-trivial.

Rindal et al. [16] use two Integrated Measurement Unit (IMU)s on the skier's arm and chest and a Multi-layer Perceptron (MLP) to do classification on the sensor data. The dataset they use consists of 10 skiers performing 6 techniques. A cycle detection method is used to split the sensor data based on the cycles. The split data is then interpolated or decimated, to ensure equal length on all splits. They achieved good results with ≈93% accuracy on a relatively large dataset (over 8000 cycles/feature vectors). The Arctic HARE system utilizes the same technique for cycle detection, but uses different machine learning methods, and explores different sensor distributions on the body of the skier to maximize accuracy while minimizing the number of sensors.

Rassem et al. [15] use deep-learning algorithms on 3D accelerometer data from cross-country skiing. They tested Convolutional Neural Network (CNN)s, different versions of Long Short-Term Memory (LSTM)s, and a MLP for classifying data from 1SK and 2SK skating. They segmented the accelerometer data by using a window over 1 s with 50% overlap. The Arctic HARE system also utilizes deep-learning models for classifying the different sub-techniques, but explores different types of data from more classes and employs different preprocessing methods before training.

There is an issue with the papers by Øyvind Gløersen and Gilgien [5] and by Rindal et al. [16] due to either not describing their data fully or not having a uniform dataset. Having a non-uniform dataset and evaluating a model based on its accuracy can be deceptive due to the accuracy paradox [11]. We circumvent this because of our balanced datasets and we use f1-score as a metric as well.

3 Arctic HARE

Arctic HARE is a prototype we have devised that aims to provide real-time performance analysis for cross-country skiers. It consists of multiple components for gathering and processing activity data. Multiple on-body Inertial Measurement Unit (IMU) sensors and a camera is used for measuring body movement. The sensor data is processed on a mobile computation device, and the video data is processed on a cloud node. Either data source can be used for classifying the sub-technique performed by the skier. We have applied machine learning methods to achieve this. These models can be trained in a cloud system. An overview of the architecture can be seen in Fig. 1.

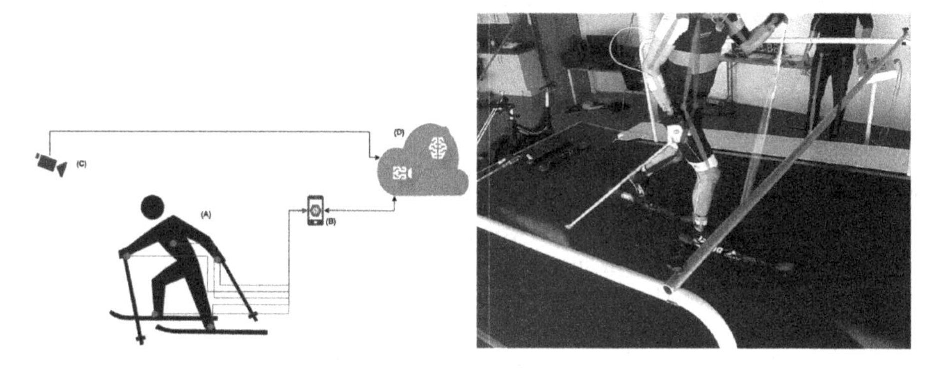

Fig. 1. Architectural overview of the *Arctic HARE* system. (A) illustrates the distribution of the sensors on the body of the user. They are connected to the mobile device (B) which, along with the static camera (C), communicates with the cloud (D). Also shows output of OpenPose result overlayed over video. Note that the head is outside of the view of the camera and the lack of points on the treadmill controller. This is discussed further in Sect. 6.2.

3.1 IMU Sensor Suit

A system was built to gather sensor data from the user's limbs. Five IMU sensors were distributed onto forearms, calves and chest of the user which were connected to a Raspberry Pi. The sensors themselves measure acceleration, magnetic field and orientation in three orthogonal directions, and are located on the limbs of the user. These sensors' values across the x, y and z directions comprise the total 46 features that the *Arctic HARE* system uses in its feature vectors, including a timestamp.

The sensor locations were chosen to be on the limbs in order to properly quantify the movement of the user. The sensor on the chest was added because it is commonly used [3,16] and it captures the average overall movement of the user. It can also be used as a reference point for the other sensors to see how much they move with regards to the torso of the user.

3.2 Video System

A camera was mounted on the wall next to the user at a single stationary angle, which was used for the video data. The camera was connected to a black box embedded system for storage that would automatically save video data. The video data is at 42 FPS with a resolution of 1440×1080.

The video approach was chosen due to the recent increase in interest and success of computer vision both in industry and academia [9]. Exploring multiple methods of recording human movement can also have advantages for different kinds of movement and in multiple scenarios. Thus comparing or combining the video approach with the IMU-based approach seemed like an interesting area to explore.

4 Data Acquisition and Preprocessing

During data acquisition a professional ski athlete wears the sensor suit and is filmed with a stationary camera while skiing with roller skis on a large treadmill. The data was recorded from seven young national-class male elite skiers skiing at their respective marathon speeds. The skiers' chosen speeds and inclines were relatively similar for the respective sub-techniques. The treadmill speed and incline was adjusted during transitions between sub-techniques. The incline also did not change dynamically as they would in the field. This causes a problem regarding the data width, i.e. the data cannot represent the full range of realistic speeds and inclines. However, due to the nature of the system, which allows for edge computing, it can easily be tested in the field and can acquire data from more realistic conditions.

The subset of classical and skating sub-techniques that we consider are:

Diagonal Stride (DIA) is a classical technique where the skier moves their arms and legs in opposition, similar to how you walk. It is used mainly on uphills [12]. It is the only one of these techniques where the arms move asymmetrically.

Double Poling (DP) is another classical technique used while going slightly downhill or at high speeds. It is done by only pushing against the snow with the poles at the same time, with very little movement of the legs [12].

Double Poling with Kick (DPK) is a classical technique similar to DP, but it also involves a kick. The kick alternates between the left and right foot. This technique is used for traveling across rolling terrain for long distances when conditions are too fast for DIA, but too slow for DP [12].

V1 skate (V1) is a skating technique, though it is quite different from other skating techniques. It is regarded as the best way to go uphill. It is done in sequence by first pushing the poles down, then planting one ski before planting the other ski [12].

One skate (1SK) is another skating technique. It is called "one skate" because there is one poling action for every leg push. This technique is often used on gentle terrain. It is also known as gear 2 [12].

Two skate (2SK) is a skating technique, named similarly to the one above because the is one poling action for every other leg push. This is a high speed technique. It is also known as gear 3 [12].

4.1 Preprocessing

Sensor Data. The IMU sensor data is used to determine cycle length by detecting peaks in the data. The method used is similar to what is described in [16]. The z-axis data of the gyroscopic sensor on the right arm is filtered using a gaussian low-pass filter over 15 samples to remove high-frequency noise. The peaks of the signal are then detected using a first-order difference approximation of the derivative to find where the slope is zero. The indices of these peaks are then saved in a file to be used for splitting both the IMU and video data into sequences. The length of the longest cycle is stored in the configuration file described in Sect. 5 in order for new data sequences to be padded to the appropriate length.

After this, multiple data sets were generated, one for each of the sensor distributions. Each distribution was identified by a 5 bit code, where a 1 or a 0 indicated whether IMU sensor data from that specific sensor was present in the respective data set. These datasets were used for the comparison in Sect. 6.1.

Cycle length changes based on which sub-technique is used. Measuring the sub-techniques for five minutes each is not a guarantee for a completely balanced dataset, however the dataset was approximately uniform. The total amount of feature vectors in the training dataset was approximately 8, 000.

Video Data. The raw video files are stored as multiple MPEG transport stream (`.ts`) files that are then concatenated together to one MPEG-4 (`.mp4`) file for each skier. It was then split up into the respective classes based on the IMU sensor data splits. The video sequences are then split into individual frames which are resized to 256 by 192 pixels with 3 channels (R, G, B) and concatenated as sequences of $256 \times 192 \times 3$-tensors where the sequences correspond to cycles similar to what is done with the IMU data. The frames are resized to fit into the Inception-v3 model. This dependency on the IMU sensor data to create the video cycles can be circumvented by for example using OpenPose for detecting cycles instead, or using a set interval duration as a sliding window. Before training of the neural networks the tensor sequences are padded with zero-tensors (which corresponds to completely black images) so that all sequences are of the same length. The splitting procedure is very parallelizable, therefore each step was

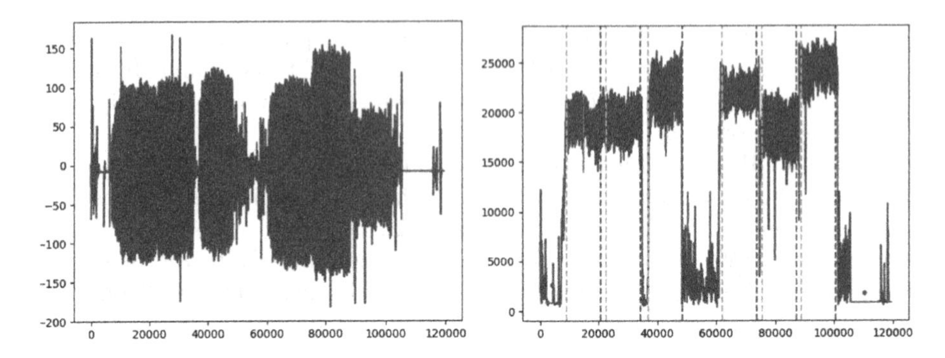

Fig. 2. The raw output of the z-axis of the gyroscope on the right arm, followed by the same signal after it has been through the annotation process. The red dots represent the cluster means and the green and red dashed lines represent the beginning and end of each class respectively. (Color figure online)

parallelized over the video sequences by using the `multiprocessing` module of `Python`.

The video data was considerably smaller with only 4 sub-techniques performed by 2 skiers. This reduced the amount of possible training data by a considerable amount to approximately 2,000 feature vectors. It was therefore important to explore methods for reducing the dimensionality of the data, as was discussed above.

4.2 Data Annotation

Annotation of the IMU and video data is a crucial part in creating the training dataset. It is possible to do this by visual inspection, as can be seen in Fig. 2, where there are six segments with different mean amplitudes. However, this process can be slow and the classes can be difficult to discern from each other, so alternative methods were explored. The goal was to split and annotate the data automatically.

The method we ended up using was the unsupervised clustering method K-means on the sensor data after convolving the absolute value of it with a 1-vector with 100 entries. This effectively scales the amplitudes of the signal outputs during technique performances, but not between them. We tried to locate 9 clusters; the six sub-techniques and three pauses located at the beginning, between the classical and skating parts, and at the end. The interval of a certain sub-technique is determined to be at the mean of each cluster: $\text{Interval}_k = \mu_k \pm \frac{L}{2}$, where L is the total duration of a sub-technique performance in number of measurements ($\approx$12,000). If the pauses are too long it can interfere with the clustering, but these can be trimmed or avoided during data acquisition by starting and stopping the program at times closer to the technique exercises. The final result can be seen in the second image in Fig. 2. The reason this works is because we know the order and duration of the techniques that are performed. We can therefore visually inspect whether the clusters make sense.

5 Machine Learning Methods

All of the models for the IMU sensor data and video data were created using *Keras*[1] [23], a neural network framework running on top of *TensorFlow*[2] [1]. *Keras* provides an API for constructing neural networks with a layer abstraction. Dropout was used to prevent overfitting. The sequences were padded to the same length so that they could be batched uniformly.

5.1 On IMU Sensor Data

The choice of using an LSTM-based network for classification of the IMU sensor data is based on good results on time series data [6]. The IMU data was concatenated based on a cycle detection method described in Sect. 4.1 and padded before being classified by an LSTM network. Datasets for each sensor configuration was created and fed through the network to determine the best configuration (see Sect. 6.1). The model we used has two LSTM layers followed by a fully-connected layer and then a Dropout layer. The number of neurons of the LSTM layers was set to 50 while testing the different sensor configurations as a initial experimental setup.

After a suitable sensor configuration was found, grid search was used to determine optimal hyperparameters while trying to make the model as small as possible, which was a requirement in order to make training and classification more efficient. The number of units in the LSTM layers were varied between 10 and 128 units and the dropout layer's rate varied between 0.1 and 0.7.

5.2 On Video Data

Convolutional Neural Networks (CNN) have achieved state-of-the-art results on image data [19] and are widely used in industry today [10]. Combining this with RNNs was therefore expected to yield good results.

Two different CNN-based methods were used for feature extraction; one based on extracting general features of the frames and one that uses pose estimation from the frames to determine the locations of the body parts of the skier.

The first method is done by sending the processed video data sequences through Inception-v3 [19], which is a CNN trained on the ImageNet dataset [17]. We use all the layers except the final layers which are fully-connected and used for classification. The frames reduced from $256 \times 192 \times 3$ tensors to 2048-dimensional vectors. The vectors are combined into sequences similar to what is done with

[1] https://keras.io/.
[2] https://www.tensorflow.org/.

IMU data. Then these output sequences are run through an LSTM network which classifies the sequences into the different sub-technique classes.

Openpose [8], which is another CNN-based application, was the second method used for feature extraction. Openpose extracts the positions of the limbs of the skier in the video frames. These positions, which can be seen in the second image in Fig. 1, are used as features in a 36-dimensional vector which are combined into sequences, similar to what is done above.

The LSTM architectures were also tuned by a grid-search, but were not constrained by trying to minimize the network size as what was done in Sect. 5.1. Due to the fact that the features of the training data are different and the feature vectors have a different dimensionality, one needs to determine which architecture works for the given data.

6 Evaluation

To collect data and perform experiments we used a Raspberry Pi 3 model B connected to five ADXL345 IMUs [21] as our mobile computation device. The Raspberry Pi ran Raspbian 4.4 and the server ran Ubuntu 16.04.4 LTS. A Raspberry Pi was chosen due to having similar hardware limitations, size and architecture as modern smartphones, but with a simple interface for connecting the wired IMU sensors. Two different cloud instances were used, one rack server with 16 GB of memory and two quad-core Intel Xeon X5355 CPUs running at 2.66 GHz and a 1 Gbit/s ethernet network interface for training the sensor-based model. A tower computer with quad-core Xeon E5-1620 CPU running at 3.70 GHz and a Titan Xp GPU with 12 GB RAM and 3584 CUDA cores running at a base rate of 1417 MHz MHz was used for training and classification with the video-based model.

We apply repeated stratified k-fold cross-validation in order to evaluate the predictive ability of the models, with $k = 7$, giving a training/test split of approximately 85%/15%. The repetition is done the get more stable results and to avoid bias of the estimator [2].

6.1 Comparing Sensor Distributions

We wanted to investigate an optimal distribution of sensors on the body of a ski athlete that minimizes the number of sensors. This is because we avoid the problem with the *curse of dimensionality* [20], and we wanted to make the sensor suit easier to apply and less interfering for the skier.

First, we performed an exhaustive search over the entire grouped feature space. The space is grouped because subsets of the features belong to different IMU sensors. As was defined in Sect. 4.1, the different sensor distributions are represented by a 5-bit code.

The $(50, 50)$ layer configuration used for the results above contains a total of $34,306$ parameters that need to be trained. The top 5 sensor configurations that

Table 1. Table of the top 5 sensor configurations with the best accuracy.

Sensor conf.	Mean acc.	Std. dev.
00110	96.94%	2.71%
11000	93.71%	5.18%
01100	93.28%	4.94%
10010	91.94%	6.20%
01001	91.20%	6.87%

Table 2. Table of the top hyperparameter choices with means and standard deviations of accuracy and f1-score.

Hyperparameters	Accuracy	F1-Score
20,10,0.1	79.25%, 13.95%	70.92%, 27.34%
32,64,0.2	82.45%, 10.99%	78.52%, 29.17%
50,50,0.1	96.11%, 3.98%	96.33%, 4.29%
64,64,0.2	95.87%, 5.48%	92.49%, 3.31%
128,64,0.4	96.77%, 2.26%	94.74%, 4.02%

contain less than 4 IMU sensors, with corresponding accuracies can be seen in Table 1.

The best sensor configuration seems to be 00110, which corresponds to the IMU sensors on the left leg and on the right arm. As a result, this configuration will be explored further. The grid search space is between 10 and 128 for the two LSTM layers, and between 0.1 and 0.7 for the dropout rate.

It is important to note how the hyperparameters scale the number of weights that need to be trained and used for calculating predictions; a $(20, 10)$ network consists of $4,426$ trainable parameters, a $(50, 50)$ network consists of $34,306$ trainable parameters, and a $128, 128$ network consists of $207,622$ trainable parameters. We want to reduce the size of the model by minimizing the hyperparameters as well. Based on the results in Table 2, we want to keep the $50, 50$ configuration due to the high accuracy and f1-score and the fact that it is a relatively small network. Results from 5 times repeated 7-fold cross-validation.

Finally, we will test the classification rate using the final sensor distribution. This efficacy test will be done to determine if the IMU-based approach can handle real-time feedback. The live classification application fills a buffer corresponding to approximately 5.3 s before detecting the peaks and concatenating the feature vectors. The time from getting the first feature vector in the buffer to classification was determined to be $5.32\,\text{s} \pm 0.11\,\text{s}$. Thus the time it took to process the feature vectors is inconsequential to the rate of classification and the true bottleneck lies in the fact that classification cannot occur before the buffer is full.

The right leg appeared in multiple configurations that got the best accuracies. The reason for this might be that the skiers most likely are all right-handed, and therefore probably favor the right leg as well. This can be significant, particularly for less symmetric movements such as V1. The best configurations include IMU sensors on a leg and an arm. Combining the IMU sensors on a leg and an arm makes sense considering most movements are highly symmetrical and those limbs encompass the entire movement well.

6.2 Comparison of the Video-Based Methods

Both of the methods tested on the video data are examples of transfer learning. We wanted to determine which method would produce the best features to use in order to classify the video data correctly. It is important to extract the relevant information from the video so that the models are trained correctly. This is related to the curse of dimensionality. We do not want the models to learn irrelevant particularities about the video data. Therefore we conjectured that OpenPose will give a better result, given that Inception-v3 gives a more general representation of the entire frame.

First we determined the best LSTM layer configuration for the respective feature extraction methods. This was done via a grid search as explained in Sect. 5.2. The number of the units in the LSTM layers ranged between 32 and 256, and the Dropout rate between 0.1 and 0.7.

Based on the results above we chose $(256, 32)$ and $(64, 128)$ as the number of units respectively for the LSTM layers after Inception-v3 and OpenPose feature extraction. The accuracy and F1-scores using both feature extraction methods is given in Table 3. We also evaluate efficacy of the two methods.

Table 3. Accuracy and F1-scores with corresponding standard deviations using the different feature extraction methods. Average time for processing a 2 s video file is also shown.

Feature extraction	Accuracy	F1-score	Processing time per file
Inception-v3	94.33%, 4.90%	93.32%, 7.18%	20.55 s ± 0.08 s
OpenPose	86.69%, 13.13%	87.94%, 17.23%	10.44 s ± 0.12 s

7 Conclusion

We have designed, implemented and experimentally evaluated *Arctic HARE*, a machine learning-based system for performance analysis of cross-country skiers. In this work we wanted to explore two different approaches to automatic performance analysis of cross-country skiers. We wanted to minimize the number of IMU sensors while achieving acceptable accuracy and see whether the video-based approach is viable by exploring feature extraction methods.

The system was built with this mind by creating the sensor suit with five IMU sensors distributed across the body, and using multiple feature extraction methods on the video data. We also implemented several applications for data acquisition and live classification in order to test the system. The novelty of our work is two-fold. Firstly, the IMU-based part of the system is located on the body of the skier and gives feedback in real-time which allows for in-field usage. Secondly, our exploration of video-based activity recognition with possible applications.

We determined from the results of the experimental evaluation of the system that, for the task of performance analysis of cross-country skiers, it is possible to achieve high classification accuracy using only two IMU sensors. Also, based on the processing time needed for the video data and the fact that the IMU sensors can easily be used in multiple environments, the IMU-based approach is more fitting for the task compared to video. However, the video-based approach can be used for specific areas, such as post-race analysis of the finish line.

Despite cross-country skiing being our domain of research in this text, numerous other sports share similar periodic patterns. Therefore, we can apply *Arctic HARE* to these as well, and we are currently exploring other fields such as football and running.

Acknowledgements. This work is partially funded by the Research Council of Norway project number 274451 and Lab Nord-Norge ("Samfunnsløftet").

References

1. Abadi, M., et al.: Tensorflow: a system for large-scale machine learning. In: OSDI, vol. 16, pp. 265–283 (2016)
2. Bengio, Y., Grandvalet, Y.: No unbiased estimator of the variance of k-fold cross-validation. J. Mach. Learn. Res. **5**, 1089–1105 (2004)
3. Casale, P., Pujol, O., Radeva, P.: Activity Recognition from Single Chest-Mounted Accelerometer, March 2014. https://www.researchgate.net/publication/260425987_Activity_Recognition_from_Single_Chest-Mounted_Accelerometer
4. FitnessKeeper, I.: Runkeeper (2018). https://runkeeper.com/. Accessed 21 May 2018
5. Øyvind Gløersen, Gilgien, M.: Classification of Ski Skating Techniques using the Head's Trajectory for use in GNSS Field Applications (2016). https://www.researchgate.net/publication/311735263_Classification_of_Ski_Skating_Techniques_using_the_Head's_Trajectory_for_use_in_GNSS_Field_Applications
6. Graves, A.: Supervised sequence labelling. In: Supervised Sequence Labelling with Recurrent Neural Networks, pp. 5–13. Springer, Heidelberg (2012)
7. Halvorsen, P., et al.: Bagadus: an integrated system for arena sports analytics: a soccer case study. In: Proceedings of the 4th ACM Multimedia Systems Conference, pp. 48–59. ACM (2013)
8. Hidalgo, G.: OpenPose: Real-time multi-person keypoint detection library for body, face, and hands estimation. https://github.com/CMU-Perceptual-Computing-Lab/openpose (2017)
9. Kishor, N.: Top 8 Technology Trends for 2018 You Must Know About (2018). http://houseofbots.com/news-detail/2653-4-top-8-technology-trends-for-2018-you-must-know-about. Accessed 15 May 2018
10. Lambert, F.: Tesla's new head of AI and Autopilot Vision comments on his new role (2017). https://electrek.co/2017/06/21/tesla-ai-autopilot-vision/. Accessed 25 May 2018
11. Lao, R.: Machine Learning — Accuracy Paradox (2017). https://www.linkedin.com/pulse/machine-learning-accuracy-paradox-randy-lao. Accessed 16 May 2018
12. McKenney, K.: Cross-country ski techniques with video examples (2014). http://crosscountryskitechnique.com/

13. Navipedia: Differential gnss (2014). https://gssc.esa.int/navipedia/index.php?title=Differential_GNSS&oldid=13309. Accessed 3 May 2018
14. O'Donoghue, P.: Research Methods for Sports Performance Analysis (2010)
15. Rassem, A., El-Beltagy, M., Saleh, M.: Cross-country skiing gears classification using deep learning. arXiv preprint arXiv:1706.08924 (2017)
16. Rindal, O.M.H., Seeberg, T.M., Tjønnås, J., Haugnes, P., Sandbakk, Ø.: Automatic classification of sub-techniques in classical cross-country skiing using a machine learning algorithm on micro-sensor data. Sensors **18**(1), 75 (2017)
17. Russakovsky, O., et al.: Imagenet large scale visual recognition challenge. Int. J. Comput. Vision **115**(3), 211–252 (2015)
18. Software, Q.: Quintic software: Biomechanical 2D Video Analysis (2018). https://www.quinticsports.com/software/. Accessed 21 May 2018
19. Szegedy, C., Vanhoucke, V., Ioffe, S., Shlens, J., Wojna, Z.: Rethinking the inception architecture for computer vision. In: Proceedings of the IEEE Conference on Computer Vision and Pattern Recognition, pp. 2818–2826 (2016)
20. Theodoridis, S., Koutroumbas, K.: Pattern Recognition, 4 edn. Academic Press (2009)
21. www.analog.com: Adxl345 (2013). http://www.analog.com/en/products/mems/accelerometers/adxl345.html
22. www.hudl.com: Hudle: One platform to help the whole team improve (2018). https://www.hudl.com/products/hudl. Accessed 21 May 2018
23. www.keras.io: Keras: The python deep learning library (2018). https://keras.io/. Accessed 10 Mar 2018

Soccer Athlete Data Visualization and Analysis with an Interactive Dashboard

Matthias Boeker[1,2(✉)] and Cise Midoglu[1,3]

[1] Simula Metropolitan Center for Digital Engineering (SimulaMet), Oslo, Norway
matthias@simula.no
[2] Oslo Metropolitan University (OsloMet), Oslomet, Norway
[3] Forzasys, Fornebu, Norway

Abstract. Soccer is among the most popular and followed sports in the world. As its popularity increases, it becomes highly professionalized. Even though research on soccer makes up for a big part in classic sports science, there is a greater potential for applied research in digitalization and data science. In this work we present `SoccerDashboard`, a user-friendly, interactive, modularly designed and extendable dashboard for the analysis of health and performance data from soccer athletes, which is open-source and publicly accessible over the Internet for coaches, players and researchers from fields such as sports science and medicine. We demonstrate a number of the applications of this dashboard on the recently released SoccerMon dataset from Norwegian elite female soccer players. `SoccerDashboard` can simplify the analysis of soccer datasets with complex data structures, and serve as a reference implementation for multidisciplinary studies spanning various fields, as well as increase the level of scientific dialogue between professional soccer institutions and researchers.

Keywords: Dashboard · Association football · Soccer · Wellness · Athlete performance

1 Introduction

Soccer is a tremendously popular sport, both in professional and amateur capacity, and is consequently a part of the daily lives of millions of people worldwide. As a frequently practiced human bodily activity, soccer is also subject to much interest from medical professionals as well as researchers within a healthcare context. Professional soccer teams have a large stake in monitoring the wellness and performance of athletes, in order to avoid injuries and protect the health of the athletes, as well as improve their game performance in professional competition. However, until recent years, information about soccer players' physical state and wellness has been collected only by pen and paper. Crunching numbers into formulas by hand or using third party spreadsheet software to compute reports still require a lot of time. In addition, teams who have followed the trend of digitalization have largely relied on in-house or closed-source commercial tools

D.-T. Dang-Nguyen et al. (Eds.): MMM 2023, LNCS 13833, pp. 565–576, 2023.
https://doi.org/10.1007/978-3-031-27077-2_44

for data visualization and analysis. This has made it challenging for researchers to work on real data from athletes, for more advanced scientific purposes, and in an anonymized but open-access and reproducible manner.

In this work, we focus on the fundamental requirements for conducting reproducible research on soccer athlete wellness and performance data, including the digitalization, cloudification, accessibility, and user-friendly representation of the data. We design and implement a dashboard called `SoccerDashboard`, which allows for data aggregation and statistics generation, as well as more advanced investigations in the context of subjective wellness, injury, and training load for professional athletes, including time series forecasting and correlation analysis. As an open-source, modular, and easy-to-extend platform, `SoccerDashboard` can be used by the scientific community for a plethora of soccer-related analyses. Overall, the contributions of this paper are as follows: (1) We provide `SoccerDashboard` as a publicly accessible web-based dashboard, which can be used by researchers, sports and healthcare professionals to analyze soccer athlete datasets in depth, from multiple perspectives, and in a reproducible fashion. (2) We provide a preliminary analysis and overview of the recently released large-scale multivariate soccer athlete health and performance monitoring dataset "SoccerMon" using our dashboard. (3) We provide an open source software repository which can be used as an inspiration for similar data structuring and analysis efforts on various athlete datasets.

The rest of this paper is organized as follows. In Sect. 2, we provide background information and a brief overview of related work, as well as describe the relevant parts of the SoccerMon dataset in the context of lifelogging, health and performance monitoring for professional athletes. In Sect. 3, we elaborate on our implementation of a modular, easy-to-extend, user-friendly, and interactive dashboard for the analysis of athlete data, called `SoccerDashboard`. In Sect. 4, we demonstrate a number of analyses currently supported by `SoccerDashboard` on the SoccerMon dataset. In Sect. 5, we discuss our research, provide notes on potential future work, and conclude the paper.

2 Background and Related Work

Digital Health Monitoring. The future of healthcare is influenced significantly by digital transformation, which allows for radically interoperable data and artificial intelligence (AI) applications. Open, secure platforms are expected to drive much of this change, allowing easy access to relevant data, improving quality of care and delivering value to patients, healthcare practitioners, hospitals, and governments. Digital health monitoring solutions facilitate care by relying on information and communication technology (ICT), ubiquitous devices such as wearables, and participatory sensing.

Mobile Health (mHealth) is a practice of medicine and public health supported by mobile devices. mHealth encourages reuse of code to create an open platform for collecting health data, but also following a standard for data capture. The objective of mHealth is to make it easier to collect data and exchange the data between different systems and platforms, but also so that the patient can collect and share their own personal health data as they wish. *Open mHealth*[1] is a registered non-profit organization which revolves around building an architecture with shared data standards. It has created a set of standardized frameworks with optimized data schemes for clinical usage. Instead of collecting data only from closed systems, the frameworks enable the possibility of retrieving all types of data from third party systems, such as wearables. *Ohmage* [17] is an open-source project which has become a platform for collecting health data. It is created to collect data from users by either manual registration through a mobile application or by letting the application collect data automatically (continuous data streams). All captured data is timestamped, geocoded and uploaded to the Ohmage server for analysis. Ohmage is a product of many participatory sensing systems combined to provide a generic platform with the possibility to customize for different scenarios. *District Health Information Software (DHIS)*[2], is an open-source project and tool for collecting data and validation of data initiated and created at University of Oslo and is widely used in Africa and Asia. The system collects health data in order to group and discover outbreaks of lethal viruses such as Malaria and Ebola. DHIS2 was created as a Health Information System (HIS), but in the later years incorporated the participatory sensing approach for data collection. *MilanLab*[3] is one of the most well-known laboratories working on athlete health, which started operating in March 2002. The goal was to provide the best possible analysis and management of individual athletes in AC Milan and the Italian national team, prevent injuries and thereby increase the average age of a professional athlete in AC Milan. The lab itself had the responsibility of assessing athletes in all types of areas (neurology, biochemistry, psychology etc.) in all stages of an athlete's stay in the club, whether it is in the pre-signing stage or during a season. *PmSys Athlete Monitoring System*[4] is developed by Forzasys in collaboration with a number of research organizations within the context of the Female Football Research Centre (FFRC), and employs participatory sensing principles, where elite female soccer players are asked to provide daily subjective self-reports via iOS or Android based mobile applications, including various training load, wellness, injury and illness metrics.

Athlete Health and Performance. Monitoring athlete health and performance can be undertaken in various contexts. *Training load:* One commonly used self-reported measure in sports science is the rating of perceived exertion (RPE). The RPE allows athletes to rate their perceived exertion on a scale of $[0 - 10]$ ranging from *very weak* to *very strong* and is an indicator for physical training load [1,11,15]. The athletes basically ask themselves after the training session *How was my workout?* [9]. A reason for collecting subjective RPE data over third-party measurement devices is the fact the objective measurement devices, such as heart rate monitors, cannot tell how an athlete actually

[1] https://www.openmhealth.org/.
[2] https://dhis2.org/.
[3] https://www.acmilan.com/en/club/venues/vismara/milan-lab/.
[4] https://forzasys.com/pmSys.html.

felt about the intensity of the session. Hence, research differentiates between internal and external training load. Internal training load is measured by metrics like RPE. External training load comprises objective measured performance indicators such as the number of sprints and total distance covered [5]. Various metrics can be derived from the RPE, like session RPE. Session RPE is the product of session duration and RPE. The accumulation of the session RPE per day is called the daily training load, to give two important examples. We describe more metrics and their formulas in Table 1. The session RPE and daily training load are most frequently used to monitor athletes' training cycles. Long-term observation of the training load allows trainers to monitor (daily, weekly) mesocycles and (monthly, yearly) macrocycles [9]. Monitoring allows an appropriate periodization of training load and an identification of training regimes [9,14]. Derived metrics such as Monotony, Strain, etc., which are introduced in Table 1 can further improve such decision-making. Metrics about training load are also consulted to prevent the occurrence of injuries and to avoid training overload. It is well documented in research literature that training load is related to the number of injury occurrences [2,5,18]. The acute chronic workload (also defined in Table 1) stands out in its relation to the likelihood of injury occurrence. A survey by Cross et al. [4] concludes that either too high or too low acute chronic work load increases the likelihood of injury. Finding the right amount of training load is hence crucial to protect the athletes from injuries [5,6,10]. However, an athlete's physiological state is influenced by more factors than only the training load. Therefore, further analysis on self-reported measures of well-being or wellness are needed.

Wellness: Self reported questionnaires on the wellness of athletes can reflect their adaptiveness to specific training regimes, loads or competition [7]. The term wellness includes emotional states like mood or stress, fatigue and readiness. The analysis by Gastin et al. [7] shows the practical relevance of self reported metrics on the psychometric state of athletes and how trainers need to consider them to tailor training plans to an individual level. Govus et al. [8] suggest that monitoring subjective wellness may give insights into an athlete's capacity to perform within a training session. It is important to lay out the relationship between an athlete's state of wellness and training load. An athlete's state of wellness can affect her self-reported training load. E.g., an accumulated level of fatigue and soreness can potentially change the self rating of perceived exertion or vice versa. A central metric for the wellness of athletes is defined as the readiness to train. The question about the readiness to train covers a broad range of emotions and is in itself reflected by exhaustion and fatigue. Monitoring and analyzing this metric can give trainers crucial insights into the general wellness of an athlete. Wiik et al. [19] and Kulakou et al. [12] applied machine learning techniques to forecast readiness up to one week ahead. The ability to predict readiness in the near future allows trainers to respond early to potential dips in athlete wellness and determine optimal training loads.

Illness and Injury: Another aspect of athlete healthcare which can benefit from a lifelogging context is the tracking of injuries and illnesses. In [3] by Oslo Sports

Trauma Research Centre, a new approach to prospective monitoring of illness and injury in elite athletes is proposed. The authors introduced a method for monitoring athletes based on weekly questionnaires. The surveys were sent to athletes over email weekly. After an athlete has finished the survey, a project coordinator collects them and assembles a report based on the responses. Then, coordinator sends the report to medical staff, which follows up with the athlete. The weekly questions are to capture health problems such as injury and illness for a athlete. The survey starts with four key questions. If the athlete choose the minimal value for each questions (no problems), the questionnaire is finished for the week. However, if the athlete reported anything else than the minimum value, the questionnaire continues by asking them whether their problem is an illness or an injury. If the athlete reported an injury problem, the following question is where the injury is located, and if it was an illness problem, the athlete is asked for symptoms.

Table 1. For each training session, athletes report session duration and the overall Rating of Perceived Exertion (RPE). These subjective parameters are then used to calculate a variety of different measures of training load as shown in this table.

Metric	Description	Formula
Session RPE (sRPE)	The workload of a single session depending on the duration and the reported RPE values	$RPE \cdot duration$
Training Load	The sum of sRPE during a day	$\sum sRPE$ per day
Weekly Load (WL)	The sum of sRPE over the last 7 days	$\sum sRPE$ per week
Acute Training Load (ATL)	The current level of fatigue (average sRPE over the last 7 days)	$7^{-1} \sum_{n=i}^{i+7} DL_i$
Chronic Training Load (CTL)	The cumulative training dose that builds up over a longer period of time (average sRPE over the last 28 or 42 days)	$x^{-1} \sum_{n=i}^{i+x} DL_i$, x = 28 or 42
Acute Chronic Work Load (ACWR)	An indication of whether an athlete is in a well-prepared state, or at an increased risk of getting injured (ATL divided by CTL)	$ATL \cdot CTL^{-1}$
Monotony	Reflection of training variation across the last 7 days (mean sRPE divided by the standard deviation (SD) = ATL / SD)	$ATL \cdot SD^{-1}$
Strain	Reflection of the overall training stress from the last 7 days (total weekly sRPE multiplied with Monotony)	$WL \cdot Monotony$

SoccerMon Dataset. Athlete data has been collected by the PmSys system from 6 teams over more than 3 years, and a part of this data has been recently made available as the SoccerMon dataset in [16]. We use the subjective reports available from this dataset as a starting point in building `SoccerDashboard`.

Wellness metrics in the SoccerMon dataset include fatigue, mood, sleep duration, sleep quality, stress, and readiness to play, which have been reported daily by each athlete in each team. The latter metric is of special significance, as it has been shown that there is a correlation between an athlete's perceived readiness to play and their game performance, as well as their risk of injury during gameplay. In Sect. 4, we introduce an ARIMA-based time series prediction application implemented in `SoccerDashboard`, which operates on individual athlete time series of reported readiness, with configurable prediction window size, as well as an athlete wellness overview chart which presents the wellness metrics for a given athlete, aggregated over a selected period of time. *Injury information* is included in the SoccerMon subjective daily reports and use an enumerated list of potential injury areas, including Head/Neck, Stomach, Back, Groin/Hip, and Leg, Thigh, Knee, and Foot with right/left indicators. There are 2 levels of severity: *minor* and *major*. In Sect. 4, we present an injury overview visualization for teams which aggregates the number of injuries per location and intensity level across the complete dataset for a given team. *Training load metrics* included in the SoccerMon dataset are based on overall Rating of Perceived Exertion (RPE) values reported by each athlete for each training session, which are then used to calculate a variety of other training load metrics as described in Table 1. In Sect. 4, we present a percentile-based representation of these training metrics along with athlete wellness metrics, aggregated over a desired period of time across all athletes of a team, for a given team.

3 Dashboard Implementation

`SoccerDashboard` is implemented in Python as a modular and interactive web-based dashboard for the analysis of sports data, in particular the SoccerMon dataset. Streamlit is used for agile cloud deployment, enabling dynamic updates and unrestricted public access. **Data management:** The dashboard is based on the SoccerMon dataset. To allow easy access to the data and data processing, we structure and store the data in a class representation. Each player is an instance of a Python dataclass *SoccerPlayer* which is described by various attributes. Each attribute contains information extracted from the SoccerMon dataset, like reported injuries, wellness parameters or training loads. The *SoccerPlayer* class instances are organised themselves in a dataclass *Team*. A team instance includes its respective player instances and the game performances for several matches. This hierarchical structure of the players simplified the analysis of team metrics but still allows easy retrieval of information about individual players. **Code:** The dashboard as well as the data processing and storage are handled in Python in a functional manner. The project is hosted by GitHub under https://github.com/simula/soccermon and version control is undertaken via GitHub. We use Poetry for dependency management as it ensures repeatable installs and usage. **Visualization and deployment:** Streamlit is an open-source app framework for Machine Learning and Data Science teams, which allows for deploying web-based applications in the cloud. It supports continuous integration and agile redeployment through GitHub, and a variety of visualization options with Python-based declarations. We use Streamlit to establish `SoccerDashboard` as a publicly accessible web app under https://soccer-dashboard.simula.no.

4 Analysis

`SoccerDashboard` currently supports 7 different types of analyses, grouped into 2 categories according to the display of information for an individual player vs. an entire team. Each category is presented as a dashboard page, selectable from the control panel on the left, and each page contains multiple tabs corresponding to individual analyses. Figure 1 presents an overview of the dashboard controls which can be used to navigate across different pages and tabs.

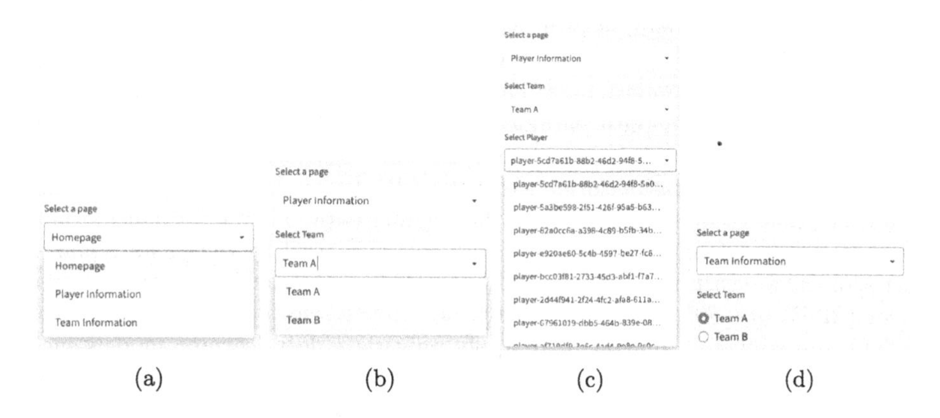

Fig. 1. `SoccerDashboard` controls: (a) Page selection, (b) Player information: team selection, (c) Player information: player selection, (d) Team information: team selection.

4.1 Player Information Page

The player information page contains analyses and visualizations related to individual players, which are represented with anonymized UUIDs in the dataset. The information presented in this page allows trainers to conduct in-depth analysis and tailored training plans.

Readiness Forecast. Forecasting an athlete's readiness to train can help trainers to plan future training session on even personal level or team lineups. Forecasts allow trainers to identify problematic developments of athletes' wellness before they might occur. However, forecasting the future is never an easy task especially when the underlying data is a self reported measure. Therefore, predictions too far into the future can be difficult. Our goal is to forecast the readiness to train for the next seven days, three days and the next day. Next to the their readiness to train, each athlete recorded various wellness features. All of these features contain valuable information to forecast the readiness into the future. We use an Autoregressive Moving Average (ARIMA) model with exogenous regression for the forecasting of readiness, also called regression model with ARIMA errors. The ARIMA model is transparent and explainable. Based on the best fitting parameter combination, information can be derived on how the values of the past days will affect future readiness. This makes the model interpretable, which is particularly useful in multivariate models, as in this case. Moreover, ARIMA models copes with seasonality and cyclic effects within the time series. We fit a model for each player. Iterative imputation was applied to address the missing values in the data [13]. Imputation was performed only for the forecast model, but omitted for other purposes.

The evaluation of the model is done in two parts. First, we evaluate the goodness-of-fit of various ARIMA models. We evaluated seven different parameter variations. The goodness-of-fit was measured by the corrected Akaike information criterion (AICc) and the Bayesian information criterion (BIC). Moreover, we tested if the underlying time series are stationary, which is assumed by the ARIMA model. The best parameter combination for the Readiness ARIMA model is a first order autoregressive process and a second order moving average process. The forecasting performance of the ARIMA model was evaluated using a test set that accounted for 20% of the original data. The performance evaluated at different forecasting window sizes and measured by the Mean Squared Error of the forecast and true data. The ARIMA models achieve a mean square error (MSE) of 0.895, 0.915, and 0.964 for window sizes of 1 day, 3, days, and 7 days, respectively. Kulakou et al. apply univariate deep learning models such as long short-term memory and gated recurrent units to forecast readiness. The authors trained the models on the whole team and achieved the best MSE of 0.68. For models trained for a single player, the best result of the deep learning models is an MSE of 1.25 for a window size of 1 day. Therefore, the multivariate ARIMA performs better for a single player. Figure 2a presents a sample forecast for a random player from Team B using a prediction window of 7 days.

Wellness Overview. We present the wellness parameters for each player in a spider or radar chart. Such charts provide a convenient visualisation of multi-dimensional and ordinal data. We aggregate the wellness parameters over time and provide selectable time ranges like last week, last fortnight, last month and last year. Selectable time ranges allow trainers differentiate between most recent wellness and a overall state of wellness. It is important to recall the subjectivity of the metrics visualised. The analysis of a certain metric at a given time point might not give much information. Rather the increase or decline of a metrics over time give indications on an athletes state of wellness. Due to the multi-dimensionality of the spider chart, trainers can analysis potential dependencies between wellness factors. E.g. the stress level and the soreness might increase and decrease simultaneously, hence trainers gain a better understanding on how resting periods could decrease mental agitation. Figure 2b presents the wellness overview chart of a random player from Team A, aggregated over the last year.

Training Load Overview. The training load overview module visualised the ACWR over a certain time range. We used a bar plot to visualise the ACWR as it gives an intuitive understanding and an easier visual comparison of different loads. Furthermore, we analysed the distribution of the ACWR and identify high training load days and low training load days. The identification of high training load days is based on the upper 75% quartile of the athletes past data. The identification of low training load days is based on the lower 25% quartile. The identified days are colored red and blue respectively. In Sect. 2 we discussed the importance of the ACWR score for injury prevention and for determining optimal training loads. The visualisation allows the analysis of a full yearly scale up to more zoomed in weekly visual analysis of ACWR. Thus, it accounts for analysis regarding macro- and mesocycles.

4.2 Team Information Page

The team information page contains analyses and visualizations related to entire teams, which are represented with anonymized letters (Team A, Team B) in the dataset.

Aggregated Metrics. This module presents a quartile-based representation of wellness and training metrics, aggregated over a desired period of time across all players of a team, for a given team. Figure 2d presents a sample visualization of the Strain metric aggregated over the period 01.06.2020 - 31.12.2021 for Team A. The purpose of the module is to provide trainers with an adequate overview of the state of the entire team. The quartile-based representation shows the range distribution for the team. Wider ranges indicate that the team perceive different wellness states or exertion rates. Trainers can reflect on their training regimes, if these effects are intended e.g. when the training regimes are highly individualised or not. This allows trainers to better assess the mood or stress within the team. Questions arise such as: How do game days, injuries or new training modules affect the mood or stress within the team?

Injury Overview. In this module, we provide a summary of injuries that have occurred in a given team since the beginning of the dataset (2020 for Soccer-Mon). Figure 2e presents a sample for Team A. The visualization distinguishes between the location of the injury and the severity of the injury (minor and severe injuries).

Training Load Overview. The training load overview for the entire team is presented as a table with the mean values and standard deviation for each training load metric per player. Figure 2f presents the overview for Team A. Both tables can be downloaded as *.csv* files. The functionality is provided to allow users their own analysis of their team.

Correlation Analysis. This module presents the correlation matrix of the wellness metrics and the training load metrics. The correlation matrix shows linear relationships between certain metrics. The relationship among wellness and the relationship among training load and wellness parameters are of special interest. In Table 1 we describe the derivation of the training load metrics

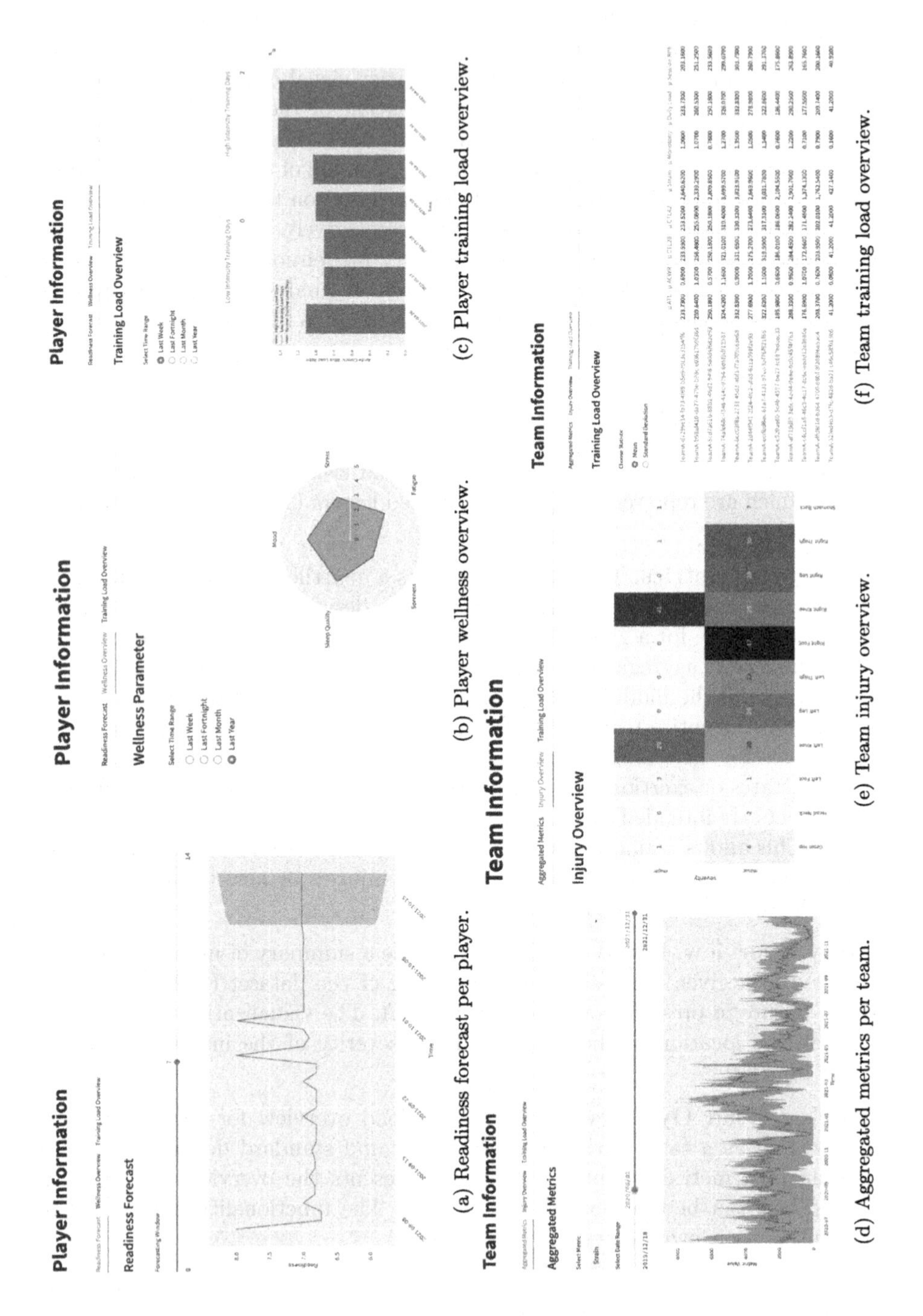

Fig. 2. `SoccerDashboard` player and team information pages, sample visualizations.

from the RPE. Naturally, this results in high correlation among these metrics. In Sect. 2 we discussed potential interaction between the wellness of a athlete and training loads. The correlation analysis underlines these hypotheses.

5 Conclusion and Future Work

In this work, we present an interactive and modular dashboard for the visual analysis of soccer athlete data including wellness, injury, and training load metrics, called `SoccerDashboard`. Such a dashboard for visualizing aggregated statistics and presenting various data analyses is of interest to a number of stakeholders including trainers, team managers, athletes, healthcare professionals, and researchers. Possible applications based on the use cases described in Sect. 4 range from informed training session planning to athlete injury monitoring and prevention, and the integration of AI-based modeling. `SoccerDashboard` is a proof-of-concept implementation which is intended to provide the scientific community with a means of conducting repeatable studies on open soccer datasets. It is a user-friendly and extendable open-source framework which allows for easy access and customization for researchers, as opposed to commercial and in-house solutions.

As `SoccerDashboard` is currently tailored towards the SoccerMon dataset, our next goal is to use further abstractions in the code for data handling so that it can dynamically adapt to different inputs and be used as a cross-dataset compatible visualization platform. We are also planning to add support for integrating different types of metrics such as advanced performance tracking parameters derived from GPS data, and additional modules for machine learning applications such as intra-team player clustering, neural network based time series forecasts for wellness and injury risk prediction, overall team comparisons, and player performance benchmarking. We believe that `SoccerDashboard` can serve as a reference implementation for mutidisciplinary studies spanning various fields, and increase the level of scientific dialogue between professional soccer institutions and researchers.

References

1. Borg, G.A.: Psychophysical bases of perceived exertion. Medicine & science in sports & exercise (1982)
2. Carey, D.L., Crossley, K.M., Whiteley, R., Mosler, A., Ong, K.L., Crow, J., Morris, M.E.: Modeling training loads and injuries: the dangers of discretization. Med. Sci. Sports Exercise **50**(11), 2267–2276 (2018)
3. Clarsen, B., Rønsen, O., Myklebust, G., Flørenes, T.W., Bahr, R.: The Oslo sports trauma research center questionnaire on health problems: a new approach to prospective monitoring of illness and injury in elite athletes. Br. J. Sports Med. **48**(9), 754–760 (2014)
4. Cross, M.J., Williams, S., Trewartha, G., Kemp, S.P., Stokes, K.A.: The influence of in-season training loads on injury risk in professional rugby union. Int. J. Sports Physiol. Perform. **11**(3), 350–355 (2016)

5. Drew, M.K., Finch, C.F.: The relationship between training load and injury, illness and soreness: a systematic and literature review. Sports Med. **46**(6), 861–883 (2016)
6. Gabbett, T.J., Domrow, N.: Relationships between training load, injury, and fitness in sub-elite collision sport athletes. J. Sports Sci. **25**(13), 1507–1519 (2007)
7. Gastin, P.B., Meyer, D., Robinson, D.: Perceptions of wellness to monitor adaptive responses to training and competition in elite Australian football. J. Strength Conditioning Res. **27**(9), 2518–2526 (2013)
8. Govus, A.D., Coutts, A., Duffield, R., Murray, A., Fullagar, H.: Relationship between pretraining subjective wellness measures, player load, and rating-of-perceived-exertion training load in American college football. Int. J. Sports Physiol. Perform. **13**(1), 95–101 (2018)
9. Haddad, M., Stylianides, G., Djaoui, L., Dellal, A., Chamari, K.: Session-RPE method for training load monitoring: validity, ecological usefulness, and influencing factors. Front. Neurosci. **11**, 612–613 (2017)
10. Hulin, B.T., Gabbett, T.J., Blanch, P., Chapman, P., Bailey, D., Orchard, J.W.: Spikes in acute workload are associated with increased injury risk in elite cricket fast bowlers. Br. J. Sports Med. **48**(8), 708–712 (2014)
11. Impellizzeri, F.M., Rampinini, E., Coutts, A.J., Sassi, A., Marcora, S.M., et al.: Use of RPE-based training load in soccer. Med. Sci. Sports Exerc. **36**(6), 1042–1047 (2004)
12. Kulakou, S., Ragab, N., Midoglu, C., Boeker, M., Johansen, D., Riegler, M.A., Halvorsen, P.: Exploration of different time series models for soccer athlete performance prediction. Eng. Proc. **18**(1), 37 (2022)
13. Little, R.J., Rubin, D.B.: Statistical analysis with missing data, vol. 793. John Wiley & Sons (2019)
14. Loturco, I., Nakamura, F., Kobal, R., Gil, S., Pivetti, B., Pereira, L., Roschel, H.: Traditional periodization versus optimum training load applied to soccer players: effects on neuromuscular abilities. Int. J. Sports Med. **37**(13), 1051–1059 (2016)
15. Marynowicz, J., Kikut, K., Lango, M., Horna, D., Andrzejewski, M.: Relationship between the session-RPE and external measures of training load in youth soccer training. J. Strength Conditioning Res. **34**(10), 2800–2804 (2020)
16. Midoglu, C., Boeker, M., Winther, A.K., Pettersen, S.A., Johansen, D., Riegler, M., Halvorsen, P., Hicks, S.: SoccerMon (2022). https://doi.org/10.17605/OSF.IO/URYZ9
17. Ramanathan, N., et al.: ohmage: an open mobile system for activity and experience sampling. In: 2012 6th International Conference on Pervasive Computing Technologies for Healthcare (PervasiveHealth) and Workshops, pp. 203–204 (2012). doi: 10.4108/icst.pervasivehealth.2012.248705
18. Rogalski, B., Dawson, B., Heasman, J., Gabbett, T.J.: Training and game loads and injury risk in elite Australian footballers. J. Sci. Med. Sport **16**(6), 499–503 (2013)
19. Wiik, T., Johansen, H.D., Pettersen, S.A., Baptista, I., Kupka, T., Johansen, D., Riegler, M., Halvorsen, P.: Predicting peek readiness-to-train of soccer players using long short-term memory recurrent neural networks. In: 2019 International Conference on Content-Based Multimedia Indexing (CBMI), pp. 1–6. IEEE (2019)

Sport and Nutrition Digital Analysis: A Legal Assessment

Bjørn Aslak Juliussen[1](✉), Jon Petter Rui[2,3], and Dag Johansen[1]

[1] Department of Computer Science, UiT The Arctic University of Norway, Tromsø, Norway
bjorn.a.juliussen@uit.no
[2] Faculty of Law, University of Bergen, Bergen, Norway
[3] Faculty of Law, UiT The Arctic University of Norway, Tromsø, Norway

Abstract. This paper presents and evaluates legal aspects related to digital technologies applied in the elite soccer domain. Data Protection regulations in Europe clearly indicate that compliance-by-design is needed when developing and deploying such technologies. This is particularly true when health data is involved, but a complicating factor is that the distinction between what is health data or not is unclear. Add to the fact that modern analysis algorithms might deduce personal medical-related data when correlating and sifting through what might seem more harmless data in isolation. We conclude with a set of recommendations rooted in current legal frameworks for developers of sports and wellness systems where privacy and data protection can be at risk.

Keywords: Personal data protection · Sport and nutrition analysis · General data protection regulation · Machine learning analysis · Sensitive data · Health data

1 Introduction

Performance development in international elite soccer is depending on evidence-based insights and best practices. Rapidly-increasing use of massive amounts of heterogeneous digital data and associated analyzes related to individual athletes, their teams, and opponents are fuelling this pervasive trend. Both qualitative and quantitative data are monitored, collected, stored, analyzed, and used. In a top-flight international soccer club, there are often more specialized staff and coaches than there are players. This includes a significant number of data engineers, statisticians, data analysts, nutritionists, and computer scientists.

The digital transformation in elite soccer is not a goal in itself, but an important toolkit for improved player development, player recruitment, team composition, and competition strategy. To illustrate with a common example, a head coach has as a rule a specific style-of-play for the match team. This includes how to organize the fundamental structure of the 11 players on the pitch, specific coordination patterns among sub-groups of players, what to do in specific attack

D.-T. Dang-Nguyen et al. (Eds.): MMM 2023, LNCS 13833, pp. 577–588, 2023.
https://doi.org/10.1007/978-3-031-27077-2_45

situations, the role of each in a defensive structure and the like. Individual athletes are prepared for or even selected for specific roles in the team adhering to this style. The cohort of players will be developed in such an explicit and quantifiable framework, where specific collaboration models and related player position profiles are made. As such, a specific player position might have required properties such as extreme top speed, endurance, and specific technical skills, and analysts typically use massive amounts of relevant data to find the most proper candidate. This candidate can be in the internal player cohort, but also external as potential recruitment when possible. This selected candidate can vary throughout the competition season depending on, for instance, the opponent team or relative development compared with club colleagues competing for the same position, so this analysis is a dynamic and ongoing process.

This digital transformation in elite soccer can be a double-edged sword. On the one hand, are the obvious benefits a coach or individual athlete gains from this type of evidence-based insight. This again can result in, for instance, specific training interventions, adjustments in concrete behaviour outside the training arena such as sleep pattern, and even change in dietary habits. On the other hand, digitizing the personal aspects of an athlete throughout not only the time spent on training facilities, but even 24/7, might violate privacy and even security aspects. Anecdotal experiences from a decade of collaboration with elite soccer teams indicate that this problem does not always get proper attention.

This paper is addressing this contrasting problem as follows. First, we give a brief overview of the practical potential of state-of-art technology in the elite soccer domain. Next, we will discuss relevant legal frameworks related to this domain and its use of digital technologies. We then recommend specific actions to mitigate the legal risk with digital technology applied in the sports domain. Lastly, we conclude.

2 Athlete Monitoring Technologies

The soccer domain is infused with digital technologies, and their applicability has become mainstream for clubs well below the elite athlete level. Particularly elite clubs have embraced this new development, but notice that such clubs guard their technology and practical use with secrecy. Considered a competitive advantage and potential tipping point for the club, obtaining concrete insights from those external to the club might be a challenge. This includes researchers who want to investigate and publish openly on such closed cohorts. Fortunately, we have been on the inside of elite soccer teams and national teams for over a decade carrying out fundamental research in close collaboration with these use partners [3,10–12,15,18]. The rest of this section will briefly outline our experience with digital systems and their practical use in this context.

2.1 Digital Video

Video footage made its foray into the elite soccer domain decades ago and is considered very important, if not the most, for performance development. The video comes from many sources and captures previous games, opponent teams, public best-practices and drills, internal team training sessions, and even individual athletes carrying out specific drills. The list is longer.

Soccer clubs can obtain videos from past games from numerous sources. One such is a commercial broadcast channel, even if the specific video has been made for a public audience. This might be a problem for soccer analysis since replays and detailed close-up views might lose some aspects of importance for an expert. Coach staff analyzing the match might rather prefer video footage from the entire soccer field while the game unfolds, not just where the ball currently is located and covered by the public broadcaster.

A decade ago, we were challenged by our collaborators in the soccer domain to research and develop a video system covering the entire pitch. It was not an option to use a single-camera due to low resolution, but we managed to build a series of novel Bagadus systems [18] stitching together video streams from multiple cameras covering the entire field. This stitching is a very computationally demanding process, but using specialized GPU hardware and fine-tuned software modules we managed to produce such a panoramic stitch from 5 different 4K cameras in real-time. The analyst could now zoom into specific details of the pitch, and even replay this content in the locker room during the break in the middle of the game. We are pleased to observe that similar systems have later been implemented by industry and deployed at many soccer stadiums throughout the world. This includes high-end solutions at Allianz Arena in Munich from SAP and Panasonic, and numerous low-end systems with a single camera setup.

Body-carried cameras can also be used by coaching staff moving about on the larger pitch during a training session. We have built such analysis systems using, for instance, GoPro cameras, Panasonic Lumix on hand-carried gimbals, chest body cameras, and even Google glasses worn by coaches. The applicability of such videos was primarily for player feedback during or following a specific drill sequence. Such drills can be on the pitch up-close, for instance, capturing a goalkeeper in action to determine bodily balance details, a specific sprint acceleration session, a technical drill, or how a player is acting in a man-to-man situation on a corner.

The analysis of video data is still primarily manual where soccer experts attempt to find specific events in the video. This can be to detect minor details or what exactly went wrong in a situation, or the opposite, to find positive sequences to, for instance, reinforce learning positively for the players involved. Some commercial actors also tout support from artificial intelligence (AI) analysis solutions, but we consider these at a relatively early functional stage.

Video can also be used to extract quantified data, either semi-automatically or fully automated. Algorithmic parsing of video by feature extraction or feature tracking software is commonly used. This type of extraction can be used to gain insight into how many meters a specific player or entire team covers during a match or the number of successful passes. This feature extraction process is often semi-automated as humans in the loop evaluate and adjust the algorithmic output; examples, where feature extraction software has problems, are during player occlusion and for accurate ball detection. Enterprise companies have emerged that produce semi-automated solutions in this domain, but their result is often produced with a delay of a day or so after the sports event.

2.2 Positional Data

Video state-of-art analysis software is not accurate and timely enough, and it has primarily been a toolkit for resource-rich clubs. Nevertheless, public broadcasters have increased the use of video-extracting software for their audience. An example of applicability is to illustrate and visualize quantified data with colourful heat maps of player positions at half-time or in the minutes following a game. For sports scientists and coaching staff in elite clubs, this type of video feature extraction is seldom accurate enough in a performance development context. They have found alternatives in the use of Internet of Things (IoT) based solutions.

Alternatives to capture positional data have emerged in the last decade and are now commonly used in many mid- to top-flight teams. This works as follows. A body-worn sensor captures the positions of the athlete wearing it on the back with a frequency 10 Hz or more, and these IoT devices are either radio- or GPS-based. A radio-based system has antennas installed around the soccer pitch and receives signals from the sensors worn by the athletes. A back-end server computes the position of the players in soft real-time with relatively high accuracy. A GPS-based system is satellite-based and has less accuracy, but is more portable as no stationary antennas are needed.

We have been using both radio-based and GPS-based systems with our player cohorts, the former type for a decade. Such a system provides high precision down to 10 cm error deviation and is an important toolkit for sports science research [2,14,21]. Also, this technology provides ground truth data that coaches can use to determine the specific performance parameters and overall load of the individual athletes. This is frequently used when planning and executing practices during a micro-cycle (week), and real-time feedback from the system enables immediate and customized interventions for specific athletes. In particular, a common performance indicator in elite soccer is the number of high-intensity runs and distance, and physical coaches even intervene during a training session and inform the athletes that they need more (or less). If the athlete does not take this feedback into account while practising with the cohort, extra sprint meters can be added afterwards to meet the desired training goal. In some cases, specific players have even been pulled out of practice before it ended due to overload worries.

2.3 Physiological Data

Positional data systems provide massive amounts of physiological data. Additionally, these IoT devices might have other sensors included like a heart rate monitor. This is clearly a medical sensor device, and similar sensor data can be obtained from, for instance, a smartwatch. Other types of biosensors can be packaged, for instance, as patches glued on the body.

The coaching staff also have other sources of data to profile their athletes. An example is a regular body weight, and another is as simple as a device for measurement of heights (to take into account load for younger players when in growth periods). Despite simple devices, this data is important and is logged in the digital profiles of the players. A more digital toolkit is a force platform that measures, for instance, the jumping, landing, and isometric movements of an athlete. Coaching staff or the players themselves also input data manually into digital logs from, for instance, a weight- and strength session.

A medical device like a DEXA scanner is not uncommon in elite clubs, a scanner that determines body composition by passing low-radiation x-ray beams through tissues. This is used to determine tissue composition in the body, which includes the relative difference of muscle mass in the left and right foot or a change from the previous scan (particularly when recovering from an injury).

2.4 Medical Data

A DEXA scanner is not necessarily operated by a medical doctor. In one of the clubs we have collaborated the most with, the physical coach and head of sports science are in charge. Other medical data is also involved, particularly when a player is injured. If the injury does not need medical attention, non-medical staff can log this data. This is entering a grey zone with regard to special data confidentiality procedures that medical staff must follow with regard to what can be shared and by whom. It is no longer in the grey zone when the athlete is considered a patient undergoing treatment by the medical staff in the club.

We have frequently observed that blood samples are taken from athletes during intense training periods, but not necessarily by medical staff. Lactate testing is actually testing the blood of an athlete to calculate their lactate threshold, or the intensity of experience before the lactate increases exponentially.

2.5 Qualitative Data

Proper recovery relaxation and nutrition habits matter very much for the healthy development of an athlete. An elite player will not be able to reach their full potential without these two aspects seriously addressed. Numerous devices exist for monitoring recovery like sleep and physical activity external to the training facilities and diet. We have worked with clubs using sensor armbands, and some have used even more invasive monitoring technology in beds and bedrooms. This has been considered very invasive and was only used for a few weeks. With diet, pictures have been taken of meals and uploaded to a club-operated server, often supplemented with textual input of nutrition details. This has proven to be a tedious task for most athletes with frequent dropouts of such logging regimes.

Subjective data from athletes themselves, however, can provide important insights that digital technologies cannot properly (yet). One example is subjective measurements of the important recovery part of an athlete's daily life which includes, for instance, sleep quality and time slept. This type of data can be manually entered into a digital log by the players themselves, or as observed in one U.S. Major League Soccer elite club, reported orally and logged by coaching staff as part of the morning assessment of each player and their readiness to train that day. In one Premier League Club in England visited half a decade ago, an iPad was even mounted on the wall in the locker room, with players lining up to report their session Rating of Perceived Exertion (sRPE). So much for privacy.

We have developed a more private system, PMSys [11], that is app-based and reports and quantifies aspects including an athlete's mood, sleep quality, readiness to train, any illness symptoms, and the like. More recently, this system has also become a toolkit for many elite female soccer teams in Norway. Notably, this system was totally re-designed and re-implemented before the GDPR came effective in May 2018. The next section will return to such regulatory aspects.

3 Legal Aspects of Sport and Nutritional Digital Analysis

3.1 Legal Bases for Processing Personal Data in Sports Analysis

The General Data Protection Regulation (GDPR) regulates the processing of personal data in the European Union and the European Economic Area [16]. Processing is defined broadly in the regulation compared to traditional use in computer science. GDPR Art 4(2) defines that "processing means any operation or set of operations which is performed on personal data or on sets of personal data, whether or not by automated means (...)".

Personal data is defined in Article 4(1) and Recital 26 of the regulation as any information relating to an identified or identifiable natural person. The identification of a natural person can be both direct and indirect by reference to an identifier of a natural person or to one or more factors specific to the physical, physiological, genetic, mental, economic, cultural, or social identity of a person.

In the context of sport and nutritional analysis, any information directly or indirectly linked to a natural person processed in an electronic system would be under the scope of the regulation [9]. This includes, for example, performance monitoring or nutritional logging of athletes in a sports club.

Personal data belongs to the natural person it identifies. The GDPR builds on the notion of informational self-determination.[1] A sports club processing personal data, therefore, need to rely on one of the legal bases in Article 6 of the GDPR for the processing to be lawful. With regard to the processing of sports and nutritional data carried out by a sports club, the legal bases a) Consent, b) Performance of a contract or the legal basis in f) Legitimate interests after a proportionality assessment are the potential relevant legal bases.

3.2 Consent as the Legal Basis for Processing Personal Data

Article 7(1)(a) requires that the entity processing the personal data relying on consent shall be able to demonstrate and prove that the data subject has consented.

[1] See, Articles 13,15,17 GDPR.

The consent should be intelligible and in an easily accessible form with clear and plain language. The consent should be informed and given by a clear affirmative act establishing a freely given, specific, and unambiguous indication of the data subject's agreement, according to Recital 32 of the GDPR.

Recitals 42 and 43 of the GDPR lays down the requirements for a given consent: *"Consent should not be regarded as freely given if the data subject has no genuine or free choice or is unable to refuse or withdraw consent without detriment"*. Furthermore, according to Recital 43, consent should *"not provide a valid legal ground for the processing of personal data in a specific case where there is a clear imbalance between the data subject and the controller* (...)".

The European Data Protection Board (EDPB), tasked under Article 68 of the GDPR to ensure the consistent interpretation of the GDPR in different member states across the EU and EEA, has discussed the use of consent as a legal basis in digital sports analysis in sports clubs. The EDPB has stated:

"Athletes may request monitoring during individual exercises in order to analyse their techniques and performance. On the other hand, where a sports club takes the initiative to monitor a whole team for the same purpose, consent will often not be valid, as the individual athletes may feel pressured into giving consent so that their refusal of consent does not adversely affect teammates" [8].

Both the recitals of the GDPR and the statement from the EDBP denote problematic aspects of relying on consent as the legal basis for processing personal data in sports analysis. Both the pressure to consent and the power imbalance between the athlete and the club indicate that consent may not be a valid basis for processing personal data in such a situation.

3.3 Performance of a Contract as the Legal Basis

Professional athletes are often employed by a team. The performance of a contract might therefore be a potential basis for processing personal data. According to Article 6(1)(b) of the GDPR, processing shall be lawful if the processing is *necessary* for the performance of a contract to which the data subject is party. The assessment of the necessity of collecting and analysing personal data combines a fact-based assessment of the processing *"for the objective pursued and of whether it is less intrusive compared to other options for achieving the same goal"* [7]. The performance of a contract could only represent a valid legal basis for processing if less intrusive means of processing could not achieve the same goal. When assessing whether Article 6(1)(b) is a valid basis for processing personal data in sports and nutritional analysis in a team or a sports club, it is vital to assess whether the same goal as the goal achieved by the analysis could be achieved by less intrusive means than the athletes logging their performance and food intake. There is, therefore, a need to assess the specifics of each case under question in order to conclude that Article 6(1)(b) could be a valid basis for sports and nutritional digital analysis.

3.4 Legitimate Interests as the Legal Basis

GDPR Article 6(1)(f) could be a relevant legal basis for processing personal data in sports analysis. Under Article 6(1)(f), the processing must first be necessary for the purposes of "the legitimate interests" pursued by the sports club. The interests pursued through sports analysis would, most likely, be to improve the performance of the athletes or the total performance of the team. Such a system also has the potential to be used to control the athletes. Such a control regime comes with the potential to give athletes that have not complied with the exercise or food regime a sanction. The predecessor of the EDPB has in their opinion on the concept of *legitimate interests* held that measures of controls are harder to justify as legitimate interest [22]. Lastly, the legitimate interests established under Article 6(1)(f) must override the data subjects' interests in protecting their personal data and privacy.

Systems for sports and nutritional analysis often process data regarded as sensitive personal data under Article 9(1), such as health data. According to Article 9(2), Article 6(1)(f) is not a valid legal basis for processing sensitive personal data. Article 6(1)(f) is therefore not a valid legal basis for systems both processing personal and sensitive personal data.

Recently, professional athletes have launched legal initiatives related to protecting the privacy of performance data in various jurisdictions [1,17,20]. Under the GDPR, controllers or processors relying on an invalid legal basis or breaching other data protection regulations, risk receiving administrative fines of up to 20 000 000 EURO or 6 % of global annual turnover, according to GDPR Article 83(5). The various lawsuits from athletes illustrate the importance of thoroughly assessing the legal basis of processing personal data in sports analysis.

3.5 Processing of Sensitive Personal Data

The GDPR differentiates between personal data under Article 4(1) and sensitive personal data in Article 9. The most relevant category of sensitive personal data in sports analysis is data concerning health. Health data under Article 9(1) is defined as *"all data pertaining to the health status of a data subject which reveal information relating to the past or the current physical or mental health status of a data subject"*, according to Recital 35 of the GDPR. The Court of Justice of the European Union (CJEU) has interpreted the definition of health data "in a wide interpretation" and included information regarding all aspects of the health of an individual [4].

In a sport and nutritional analysis context, information on, for instance, known diseases and allergies would be included in the definition of health data, but also information on sleep quality, menstrual cycle, or information on respiratory rate, blood oxygen, and heart rate.

As a general rule, the processing of sensitive personal data is prohibited under Article 9(1). When processing personal data in a digital sports analysis context, it is, therefore, very significant to categorise whether the personal data is also regarded as sensitive personal data under Article 9(1). A recent judgement from the CJEU [5], analyses the understanding of when personal data under Article 4 also could be regarded as sensitive personal data under Article 9.

After analysing the legal sources, the CJEU concluded in paragraph 127 of the Judgement that: "(...) Personal data that are liable indirectly to reveal sensitive information concerning a natural person" should not be excluded from the strengthed protection regime of Article 9 GDPR.

The potential effect the judgement from the CJEU would have on sport and nutrition analysis could be illustrated by an example: Suppose that athletes in a female sports team log information on food and nutrition. Some of the athletes change their diet slightly in one week of the month in a repetitive pattern. Under the new understanding of sensitive personal data from the CJEU, the logging of food intake would, potentially, be regarded as health data under Article 9(1) if it is possible to deduce health information from the personal data.

Processing of sensitive personal data under Article 9(1) of the GDPR requires that both the conditions in Article 6(1)(a-f) of the GDPR and one of the exemptions from the prohibition in Article 9(1) are fulfilled for the processing to be lawful.

Article 9(2) opens for processing sensitive personal data under ten different exemptions. The most relevant legal basis for processing sensitive health data in a cohort of athletes under Article 9(2) GDPR is explicit consent under Article 9(2)(a). Such consent must list the purposes of the processing and should also list the nature of the sensitive personal data. Explicit consent is not that much different from consent under Article 6(1)(b). The key difference is that consent could be expressed through behaviour, while explicit consent must be oral or written and the entity processing the data must keep a record of the explicit consent.

3.6 Compliance by Design and by Default

According to Articles 5(1)(c) and (f), the processor or controller processing personal data should limit the amount of personal data to what is necessary for the purpose of the processing and process the data in a manner that ensures the confidentiality and security of the personal data. A sports analysis system both processing personal and sensitive personal data needs to have some limitations on access to sensitive personal data, in order to respect the principles of purpose limitation and the confidentiality requirements expressed in Article 5.

Article 25 of the GDPR lays down requirements for data protection by design and by default. In order to fully comply with these requirements, it is important to assess data protection aspects in the process of designing the software and continuously throughout the lifetime of the system.

3.7 Third-Party Applications and Third Countries

Digital analysis for sports purposes in sports clubs frequently applies third-party and off-the-shelf applications for logging exercise and food intake.

The GDPR regulates the processing of personal data in the EU. In order to safeguard the rights relating to personal data for data subjects from this area, the regulation has a general prohibition of transfers of personal data from the EU to countries outside of the union, known as third countries.

This general prohibition comes with some exemptions, among others countries pre-approved by the European Commission (EC). In July 2020 the CJEU invalidated such a pre-approval regarding the U.S. due to indiscriminate U.S. surveillance programs [6]. After the invalidation, transfers to the U.S. have to rely on, for instance, the EC Standard Contractual Clause decision. As part of this legal basis, exporters of personal data are required to conduct an impact assessment, assessing whether the data protection laws and practices of the receiving third country are essentially equivalent EU data protection law.

Third-country transfers in digital sports analysis could be illustrated with an example: A Norwegian football club mandates their players to log their exercise in Strava and their food intake in Lifesum. Strava's servers are located in San Francisco and they transfer personal data from the users in the EEA to the U.S. servers [19]. Lifesum also transfer data to the U.S. and other third countries to be processed by third-party providers that process data on behalf of Lifesum [13]. In the example, the football club determines the purposes and means of the outputs of the logged data in the applications and has the role of the controller under Article 4(7) of the GDPR. The football club might be held responsible for the transfer of personal data to third countries on a somewhat inadequate legal basis. In such a situation the more opportune approach would be to use applications that do not transfer personal data to third countries.

4 Mitigating Legal Risks: Recommendations

Based on the legal analysis above, the following recommendations for compliance with EU data protection law in digital sport and nutrition analysis should be carefully considered:

1. If a sport club processes personal data that is regarded as sensitive personal data under Article 9(1), *explicit* consent is the potential valid legal basis under both Articles 6(1)(b) and 9(2) of the GDPR.
2. The sports club should introduce measures to ensure that such a consent is freely given, that the data subjects are informed of the purposes of the processing and the type of data processed, that the consent is without pressure from neither the club nor teammates, and that there is no negative consequence for not consenting to the processing of personal data.
3. If off-the-shelf applications for sport and nutritional logging are used in a sports club, it is safer to choose a provider that does not transfer data to third countries. The use of providers that transfer data to third countries requires an examination of whether the receiving jurisdiction offers essentially equivalent personal data protection.
4. Private data obtained from elite athletes might be very relevant and interesting for a broader external audience. Proper security platforms and procedures must ensure data confidentiality and integrity of such valuable data.

5. A role-specific security approach where personal data is further classified based on sensitivity is recommended and particularly required when health data is involved. A strict access regime must be in place, even internally, to ensure that, health (related) data is not shared with non-medical staff. A particular concern is when data is moved outside the administrative domain of the enterprise (club), and this off-loading of data must be carefully validated in a compliance context.
6. A compliance-by-design software development process is highly recommended, where compliance is a first-order non-functional property from the specification of the software to be built and throughout the life-cycle of the software.

5 Concluding Remarks

A digital transformation has happened in elite sports over the last decade. Athletes are monitored while practising or during gameplay, and this monitoring extends to cover life outside the sports facilities. Personal data is not only monitored, but stored, federated and combined, analysed, shared at least internally among club staff, and further used. This can be problematic in a legal context, and we have briefly evaluated the applicability of the technology in a legal framework in Europe. The developers of systems built for the sports domain must be aware of and take into account these laws and regulations. We, therefore, advise a compliance-by-design approach when such software is developed and deployed.

Notice that we have used elite soccer as a use case to enable proper and concrete evaluation of more general legal frameworks. Our findings and recommendations though extend beyond a specific sport like soccer and should be generally applicable to any sport using similar digital solutions.

References

1. ABC News, Data collection in Australian sport leading to privacy issues for athletes (2022). https://www.abc.net.au/news/2022-04-15/data-collection-in-australian-sport-leading-to-privacy-issues/100993950
2. Baptista, I., Johansen, D., Figueiredo, P., Rebelo, A., Pettersen, S.A: Positional Differences in Peak- and Accumulated Training Load Relative to Match Load in Elite Football. 2019. Sports 8. https://doi.org/10.3390/sports8010001
3. Baptista, I., Johansen, D., Seabra, A., Pettersen, S. A.: Position specific player load during match- play in a professional football club. PloS one 13. (2014). https://doi.org/10.1371/journal.pone.0198115
4. Case C-101/01 Lindqvist (2003) EU:C:2003:596
5. Case C-184/20 OT Vyriausioji tarnybinės etikos komisija (2022) EU:C:2022:601
6. Case C-311/18 Data Protection Commissioner v Facebook Ireland Limited and Maximillian Schrems (2020) EU:C:2020:559
7. EDPB, Guidelines 2/2019. Version 2.0. Adopted on 8 October 2019
8. EDPB, Guidelines 3/2019. Version 2.0. Adopted on 29 January 2020

9. Flanagan, C.A: Stats entertainment: the legal and regulatory issues arising from the data analytics movment in association football. Entertainment Sports Law J. **19**(1). (2022). https://doi.org/10.16997/eslj.1082
10. Johansen, D., et al.: Search-based composition, streaming and playback of video archive content. Multimed. Tools Appl. **61**, 419–445 (2012). https://doi.org/10.1007/s11042-011-0847-5
11. Johansen, H.D., Johansen, D., Kupka, T., Riegler, M.A., Halvorsen, P: Scalable infrastructures for efficient real-time sports analytics. In: ICMI'20 Companion: Companion Publication of the 2020 International Conference on Multimodal Interaction, pp. 230–234. (2020). https://doi.org/10.1145/3395035.3425300
12. Johansen, D., Stenhaug, M., Hansen, R.B.A., Christensen, A., Høgmo, P.M: Muithu: smaller footprint, potentially larger imprint. In: Seventh International Conference on Digitial Information Mangement (ICDIM), pp. 205–214. (2012). https://doi.org/10.1109/ICDIM.2012.6360105
13. Lifesum, A.B.: Privacy Policy (2022). https://lifesum.com/policy/. Accessed 17 Sept 2022
14. Pedersen, S., Welde, B., Sagelv, E.H., Heitmann, K.A., Randers, M.B., Johansen, D., Pettersen, S.A.: Associations between maximal strength, sprint, and jump height and match physical performance in high-level female football players. Scand. J. Med. Sci. Sports **32**, 54–61 (2022). https://doi.org/10.1111/sms.14009
15. Pettersen, S. A., et al.: Soccer video and player position dataset. In: MMSys'14: Proceedings of the 5th ACM Multimedia Systems Conference, pp. 18–23 (2014). https://doi.org/10.1145/2557642.2563677
16. Regulation (EU) 2016/679 of the European Parliament and of the Council of Europe of 27 April 2016 on the protection of natural persons with regard to the processing of personal data and on the free movement of such data, and repealing Directive 95/47/EC (General Data Protection Regulation) OJ L 119
17. SportTechie, Sportradar CEO Carsten Koerl Calls Wearables Data 'A Huge Market' (2022). https://www.sporttechie.com/sportradar-ceo-monitors-collective-bargaining-with-eye-on-whether-sportsbooks-can-receive-a-players-wearable-data
18. Stensland, H.K., et al.: Bagadus: an integrated real-time system for soccer analytics. ACM Trans. Multimed. Comput. Communc. Appl. **10**(1). (2014). https://doi.org/10.1145/2541011
19. Strava, Privacy Policy (2021). https://www.strava.com/legal/privacy. Accessed 17 Sept 2022
20. The Football, What is 'Project Red Card' (2022). https://www.foottheball.com/explainer/what-is-project-red-card-russell-slade-gdpr-player-data-lawsuit-illegal-collection-processing-implications/
21. Winther, A.K., Baptista, I.A.M-D.V., Pedersen, S., Thomsen, M.B.R., Krustrup, P., Johansen, D., Pettersen, S.A.: Position specific physical performance and running intensity fluctuations in elite women's football. Scandinavian J. Med. Sci. Sports. **2022**(32), 105–114. https://doi.org/10.1111/sms.14161
22. WP29, Opinion 06/2014. Adopted on 9 April 2014

Towards Deep Personal Lifestyle Models Using Multimodal N-of-1 Data

Nitish Nagesh(✉), Iman Azimi, Tom Andriola, Amir M. Rahmani, and Ramesh Jain

University of California Irvine, Irvine, CA, USA
{nnagesh1,azimii,tom.andriola,a.rahmani}@uci.edu, jain@ics.uci.edu

Abstract. The rise of wearable technology has enabled users to collect data about food, exercise, sleep, bio-markers, and other lifestyle parameters continuously and almost unobtrusively. However, there is untapped potential in developing personal models due to challenges in collecting longitudinal data. Therefore, we collect N-of-1 dense multimodal data for an individual over three years that encompasses their food intake, physical activity, sleep, and other physiological parameters. We formulate hypotheses to examine relationships between these parameters and test their validity through a combination of correlation, network mapping, and causality techniques. While we use correlation analysis and GIMME (Group Iterative Multiple Model Estimation) network plots to investigate the association between parameters, we use causal inference to estimate causal effects and check the robustness of causal estimates by performing refutation analysis. Through our experiments, we achieve statistical significance for the causal estimate thereby validating our hypotheses. We hope to motivate individuals to collect and share their long-term multimodal data for building personal models thereby revolutionizing future health approaches.

Keywords: Lifelogging · Multimodal longitudinal data · Food and activity dataset · Causal inference · Personal models

1 Introduction

Society exists only as a mental concept; in the real world there are only individuals.
– Oscar Wilde

Most research in the healthcare domain is based on collecting data for a group of individuals and building disease models based on population-level data. However, each individual is uniquely shaped by their genetics, environment, and lifestyle. Until recently it was not feasible to collect a large volume of longitudinal data for a person and hence it was common to consider that an individual is a sample of the population. With technological advancements, we can now collect a large volume of data for every individual and develop N-of-1 approaches to build deep personal models [9,19].

D.-T. Dang-Nguyen et al. (Eds.): MMM 2023, LNCS 13833, pp. 589–600, 2023.
https://doi.org/10.1007/978-3-031-27077-2_46

The longitudinal data for an individual should be collected using a multitude of sensors that observe different physical properties as dense signals collected at different time intervals. This multimodal data is essential for collecting detailed personal information that can be analyzed, managed, and used for building lifestyle models [1,16]. Long-term health problems are increasing in the United States and worldwide due to sedentary behavior and poor diet choices. According to the World Health Organization, obesity has tripled since 1975 and is a major risk factor for noncommunicable diseases such as diabetes, cancers, and cardiovascular diseases [22]. Obesity can be kept under check through a nutritious diet accompanied by physical activity [8,24]. Food and physical activity affect the health and well-being of individuals [20,42].

These studies are conventionally conducted on groups of people, who share a characteristic, such as ethnicity, age, and health condition. The correlation between ultra-processed foods and higher body mass index was investigated in children and adults separately [6]. A study on the elderly with higher coffee consumption showed a higher rate of developing mild cognitive impairment [34]. The association between the consumption of sugar-sweetened beverages and the risk of obesity and cardio-metabolic diseases was examined in adults [28]. Physical activity has been conventionally assessed using self-reported questionnaires or surveys, where the users were asked to provide information about their daily activities and physical health over months. For example, the obesity and physical activity of 374 participants aged 7 to 18 years were investigated in two years [12]. In another study, the relationship between body-mass index and physical activity of over 2000 women was assessed annually in a longitudinal study [17]. Moreover, wearable devices have been recently proposed to track physical activity ubiquitously. To this end, smartwatches and smartphones have been employed to monitor different populations such as pregnant women, the elderly, and children suffering from Parkinson's disease and cancer [5,7,37].

These studies extract the relationship in the data by investigating the patterns obtained from the populations over time. Such evaluations can be beneficial to understand the overall behavior of entities. However, they are not sufficient to understand an individual's health condition, create models, and perform decision-making. Population-based models fail to capture a person's genetics, lifestyle, physiological characteristics, and environmental conditions in their analysis. On the other hand, personal models – representations of the person – are required to provide tailored healthcare services based on the person's characteristics. For example, a longitudinal study on pregnant women showed lower sleep quality throughout pregnancy [2]. However, the values might be in an acceptable range based on the population data. To address this problem, N-of-1 or single-subject clinical trials have been proposed to include a single patient in a study assessing solely her health condition [19]. However, the N-of-1 method necessitates long-term data collection to acquire physiological and lifestyle data from the user ubiquitously through smart devices.

In this paper, we make the following contributions:

1. N-of-1 trial for collecting food consumption, physical activity, sleep, and physiological parameter data using a smartphone, smart ring, and smart scale.
2. Formulating hypothesis to examine the relationship between parameters in collected longitudinal data.
3. Testing and validating hypothesis through correlation, GIMME (Group Iterative Multiple Model Estimation), and causal inference techniques.

2 Data Collection, Dataset, and Personal Model

We conduct an N-of-1 long-term longitudinal study to collect multidimensional food, physical activity, and health data. The study aims to monitor the user's health condition and its relation with nutrient intake and lifestyle data. We collect the data using smart devices such as Oura Ring, Arboleaf Weighing Scale, and a Smartphone with varying levels of granularity over more than three years.

- The participant was asked to log their food during each meal using the Cronometer application (https://cronometer.com). The application tracks calories and parses macro-nutrient and micro-nutrient information based on meal inputs. Food and nutrition logs are available from July 1, 2019, to June 14, 2022.
- The participant was also instructed to wear an Oura ring both during the day and at the night, over twenty hours. Oura ring (https://ouraring.com) is a smart ring that captures information about a user's sleep, heart rate, heart rate variability, physical activity, body temperature, and a slew of other parameters. Oura ring data was captured between April 14, 2020, to June 11, 2022.
- The subject was asked to periodically monitor their body vitals by using the arboleaf smart weighing scale (https://www.arboleaf.com). The device captures body fat percentage, body mass index, visceral fat, muscle mass, and other biomarkers. Smart scale readings are available between Jan 05, 2020, to June 08, 2022.
- The participant was asked to record their height, weight, systolic blood pressure, and diastolic blood pressure. This data along with integrated Oura ring and Arboleaf scale data was stored in the Apple Health Kit from October 15, 2018, to June 14, 2022.

While efforts were made to collect all the parameters daily, data from the smart weighing scale is sparser with an average collection frequency of three times per week, due to the lower variation in parameters such as body weight and body fat. Data in such a high-dimensional space require analysis and modeling techniques to assess the relationships between the parameters (e.g., food energy and stress level). Our longitudinal data – collected from a single person – enable us to find these relationships for the person, extract individualized rules and thresholds, create personalized models, and perform hypothesis testing based on the user's conditions.

3 Data Analysis Methods

In this section, we outline the correlation analysis, Group Iterative Multiple Model Estimation (GIMME) method, and causal inference techniques used to analyze longitudinal data collected in our study.

3.1 Correlation Analysis

We first investigate the correlation of the paired parameters in the dataset. To this end, the Spearman correlation technique [21] is exploited to measure the monotonic association between two parameters. We analyze the correlation between the individual data categories viz. food and nutrition data from the Cronometer app, filtered Oura ring data about physical activity, and body mass-related parameters from the Arboleaf weighing scale.

3.2 GIMME Method

We leverage the Group Iterative Multiple Model Estimation (GIMME) technique to investigate the positive and negative cause and effect between variables [3,10]. In contrast to the correlation method, GIMME assess the relationship between individual parameters and account for the effect of time on the variables. The method builds a model for the person, showing the associations over time and how the parameters are related. GIMME is utilized for longitudinal multivariate data, where the analysis of lead-lag relationships between the parameters is needed. For example, the method can investigate if there is a relationship between the consumption of certain food and the physical activity level on the same day and the next day. GIMME is scalable and can be used for processing multiple data collections. The confidence of effects detection is determined by the number of samples but not the number of users [10].

3.3 Causal Inference

Causal inference [13,23] is a technique used to determine cause and effect relationships between parameters. While correlation analysis describes associations between variables, causal paths provide scope for explaining, for example, if a certain treatment has an impact on an outcome. To understand the statistical significance and causality patterns of food and activity parameters, we use the DoWhy causal inference library [32] developed by Microsoft Research. Traditionally, this inference engine has been used for economics and business purposes. We adopt the causal inference flowchart as shown in Fig. 1 for our food and activity dataset.

The most important component of causal inference modeling is generating the causal input graph based on domain knowledge. Causality occurs because of the assumptions leading to certain identifications, and we perform statistical analysis based on the data input. We use the backdoor identification criteria to check if the treatments cause the outcome i.e. check if modifying the treatment causes changes in the outcome while all other factors are kept constant. Next, we model the causal effect based on the proposed causal graph parameters and inputs. We use linear regression models to estimate the outcome of the causal effect model. Using statistical techniques, we estimate the causal expressions. Finally, we check the validity of the estimate through different refutation techniques. The most important aspect is to check the robustness of the estimate to validate our initial hypothesis. The results of the refutation tests do not necessarily prove the correctness of the estimate but only reiterate the confidence in the estimate.

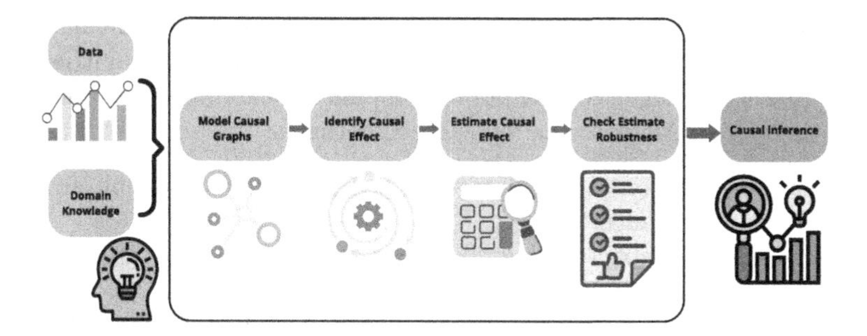

Fig. 1. Causal inference flowchart. Based on [32]

We use the following refutation analysis methods:

- Random Common Cause: It is a method to check the effect of adding random covariates on the causal estimate. If there is a limited or negligible change in the effect, it implies that the original assumption is correct.
- Placebo Treatment Refuter: It is a method of adding random covariate as a treatment or cause. After data analysis, if the new estimate tends to zero, it means that the assumptions are correct.
- Data Subset Refuter: It measures the causal variation effect across different subsets of the data. Correct initial assumptions are validated by limited variation compared to the original causal effect.

4 Experiments, Results and Discussion

In this section, we first formulate four hypotheses and test their validity through correlation and causation techniques using our dataset for building robust personal models.

4.1 Hypothesis Formation and Testing

Caffeine intake affects heart variability (HRV) [18,35]. The precise causal effect depends on the amount of caffeine, time of consumption, and underlying health conditions. Caffeine consumption leads to increased physical activity performance [26,39]. Energy balance is the difference between energy intake and energy expenditure [40]. A positive energy balance leads to obesity. Caffeine intake leads to a decrease in energy balance thereby contributing to weight loss [11,41] Reduction of body fat and weight can be achieved by incorporating physical activity into an individual's routine [14]. Physical activity and negative energy balance lead to improvement in heart rate variability [15,33]. Energy balance is affected by alcohol consumption [43], sugar intake, [36] and macro-nutrient intake such as carbohydrates, fibers, and protein [31]. Furthermore, macro-nutrients [29], alcohol [25,27] and sugar consumption [4,30] affect heart rate variability.

We, therefore, formulate the following hypotheses:

1. Caffeine consumption causes changes in heart rate variability, physical activity, and energy balance.
2. Physical activity impacts energy balance and average heart rate variability.
3. Energy balance affects heart rate variability.
4. Alcohol, sugar, carbohydrates, protein, and fiber are factors that affect energy balance and heart rate variability.

Through different correlation and causation techniques, we test the validity of the hypotheses using our N-of-1 longitudinal data.

4.2 Correlation Analysis

We observed a high correlation between energy intake and macro-nutrients, such as carbohydrates, fats, and protein. We also found a high correlation between caffeine with rest time and activity time (low, medium, and high). From the oura ring correlation analysis, we observed a high correlation between average HRV and the number of calories burned through the user's physical activity, which in turn is positively related to the number of steps and equivalent walking distance. The highest contributor to the number of steps is medium activity time when compared with the low and high activity time. Moreover, our findings show a high correlation between all parameters measured through the arboleaf weighing scale including body weight, BMI, body fat percentage, visceral fat, and muscle mass. We select certain metrics that have significant individual correlation values to plot a cross-correlation map as shown in Fig. 2. We use the derived correlations to investigate causal relationships between the hypothesis variables and contributing factors.

4.3 GIMME Analysis

We examine the inter-dependencies between energy intake through food and body physiology parameters such as body mass index (BMI) and basal metabolic

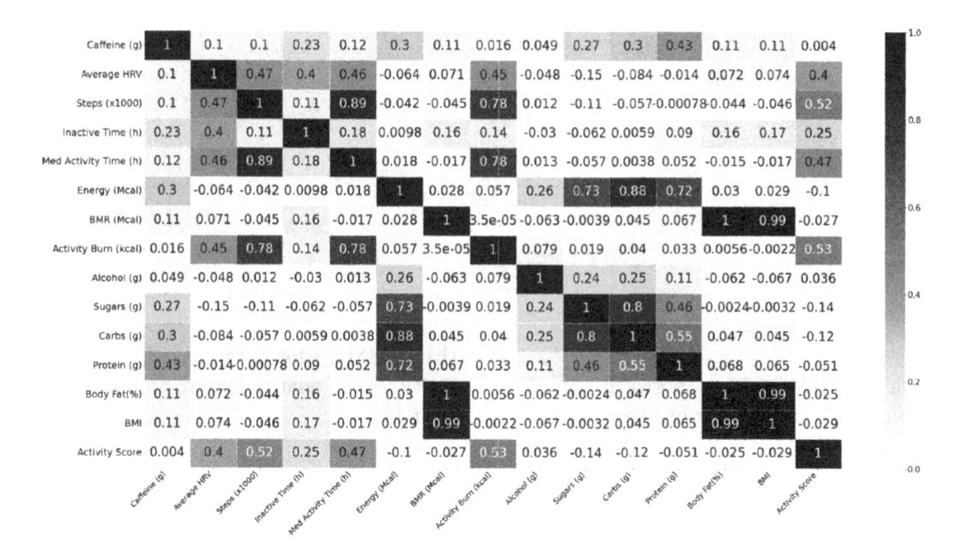

Fig. 2. Correlation heatmap between a subset of parameters affecting food intake, physical activity, sleep, and body physiology

rate (BMR) using GIMME [3] and plot the network map in Fig. 3a. The solid red lines show a positive contemporaneous effect, a solid blue line indicates a negative contemporaneous effect, a dashed red line shows a positive lag effect, and a dashed blue line implies a negative lag effect. There is a strong positive contemporaneous effect between BMI and BMR implying the concurrence of both these parameters. A higher BMR in turn leads to higher energy intake with a lag indicating a higher tendency to consume more calories after prolonged periods of resting. Higher carbs and fat intake contribute to higher energy consumption, but there is a weak lag effect due to energy intake on BMI. The positive and negative lead-lag effect between macro-nutrients, such as carbs, fiber, and protein, suggests that there is a bi-directional relationship between these parameters.

Additionally, we model the effect on caffeine, energy burned through activity, activity time, HRV, and energy intake through the GIMME network. Although there is a high correlation between steps and medium activity time, the GIMME network path and strength enable us to model causal relationships. Another interesting insight generated from the network is that inactive time in a day leads to taking more steps the following day. However, increasing physical activity through walking leads to less inactive time on the same day. Further, higher calorie intake is concurrent with burning more calories through physical activity and a greater number of steps the previous day leading to higher food intake the next day. Caffeine has a bidirectional negative lead and lag effect on energy burned through activity.

Despite its versatility, GIMME has certain limitations. The maximum number of recommended variables for network analysis is 20. As the number of variables increases, the network becomes bigger and more complicated. Further,

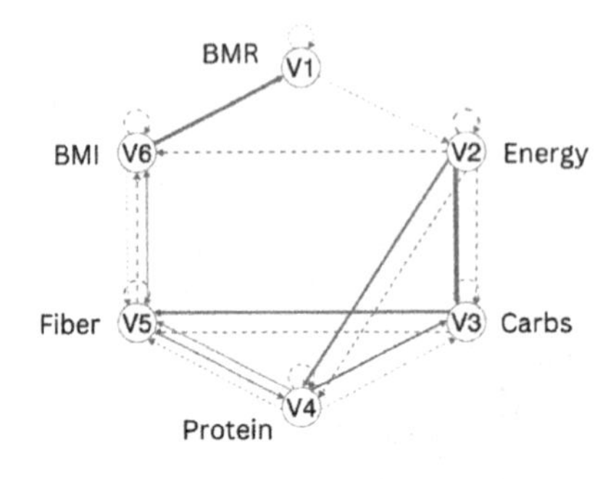

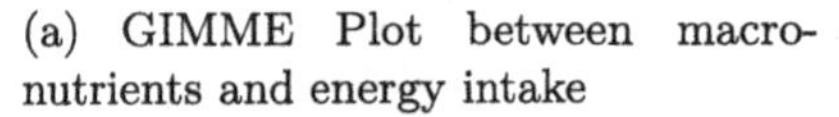
(a) GIMME Plot between macronutrients and energy intake

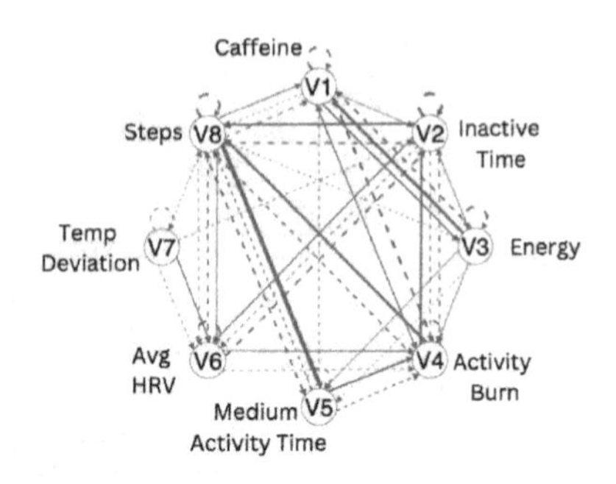

(b) GIMME Plot between caffeine, activity levels, energy intake, and energy burned

Fig. 3. GIMME network analysis plots

for the analysis to work without errors, there should be sufficient variation between the row values in a specific column to obtain statistically significant results. Although GIMME is designed for group analysis, the variables should be consistent across all individuals, which presents a data collection and processing challenge.

4.4 Causal Inference

We model the hypotheses as a causal graph after reviewing the literature and consulting expert physicians as shown in Fig. 4. Using the DoWhy library for causal inference [32], we identify the target estimand. We then estimate the causal effect of the identified estimand and determine its statistical significance. We tabulate the results for causal estimates and the p-value in Table 1. For the hypothesized causal graph, the p-value is 0.0332 which is less than 0.05 thereby implying the statistical significance of the hypotheses.

We perform refutation analysis to check the robustness of our causal estimate. As shown in Table 2, we observe minimal change in the new estimated causal

Table 1. Causal inference estimates and statistical significance

Parameters	Values
Exposure variable	Caffeine (g)
Outcome variable	Average HRV
Estimand	$AverageHRV \sim Caffeine(g) + Caffeine(g) * Fat(g) + Caffeine(g) * Protein(g) + Caffeine(g) * Alcohol(g) + Caffeine(g) * Carbs(g) + Caffeine(g) * Sugarsg)$
Causal Effect Estimate	7.52979
Causal Estimate p-value	0.0332

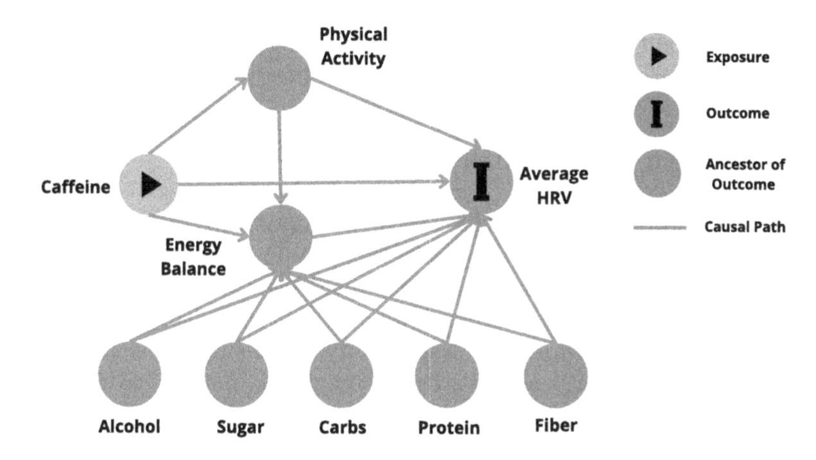

Fig. 4. Causal Graph Input. The green button with the enclosed arrow represents the treatment variable. The blue button with an enclosed ' (Color figure online)I' is the outcome variable. The blue circle represents the ancestor to the outcome variable. The green arrow paths show causality. Notations adopted from DAGitty [38], a tool for drawing directed acyclic graphs (DAG).

Table 2. Refutation analysis to validate robustness of causal estimates

Refutation test	New estimated effect	Estimated p-value
Random cause	7.52306	0.46
Placebo treatment	0.02564	0.4941
Subset of data	7.64812	0.4372

effect in comparison to the original causal effect estimate. While performing refutation tests while adding an independent random variable as a common cause and while using only a subset of the data, the p-value greater than 0.05 implies that the estimate is not incorrect.

The p-value from the placebo test checks if the new estimated effect is notably different from zero. In our hypotheses, since the p-value is greater than 0.05, adding a random covariate as a placebo treatment does not invalidate the new estimated effect. Although a limited variation of the causal effect after refutation analysis does not prove the correctness of the estimate, it improves confidence in the estimate.

We check the effect of causal variation on different subsets of the data. As evidenced by the results, there is no variation between the new estimated effect and the initial estimated value. Limited variation of the causal effect after refutation analysis does not prove the correctness of the estimate but does improve the confidence in the estimate. Therefore, by using correlation and causality analysis we show the value of applying the population-level hypothesis to N-of-1 data. This paves the way for investigating the relationships between the different features in the food and activity dataset and testing the validity of hypotheses present in literature for building better personal lifestyle models.

5 Conclusion and Future Work

Health science professionals have indicated the importance of continuous longitudinal data to model the importance of lifestyle on physiology. Technological advancements have enabled the building of evidence-based models at an individual level rather than a population level. We collect multimodal N-of-1 data encompassing food, nutrition, physical activity, and body physiology parameters for more than three years. We believe that this dataset is unique in its duration as well as the heterogeneity of lifestyle and physiological data. Using this data, we presented our early experiments in understanding the relationships between lifestyle parameters and physiology. We initiate the process to transition away from population-level studies to individual data collection due to the higher scope for developing personalized lifestyle models. By building deep personal lifestyle models, doctors and individual users can work in tandem to reduce chronic diseases and help patients achieve their health goals. Using a combination of exploration and exploitation techniques, we validate our hypotheses and achieve statistical significance thereby solidifying the soundness of our approach. We have end-to-end control of the dataset collection, processing, interpretation, and analysis framework thereby allowing us to champion the cause of individual data collection and personalized recommendations. We plan to make this dataset available for future use to the research community enabling deeper exploration.

References

1. Acosta, J.N., Falcone, G.J., Rajpurkar, P., Topol, E.J.: Multimodal biomedical ai. Nature Medicine pp. 1–12 (2022)
2. Azimi, I., et al.: Personalized maternal sleep quality assessment: an objective IoT-based longitudinal study. IEEE Access **7**, 93433–93447 (2019)
3. Beltz, A.M., Gates, K.M.: Network mapping with gimme. Multivar. Behav. Res. **52**(6), 789–804 (2017)
4. Benichou, T., et al.: Heart rate variability in type 2 diabetes mellitus: a systematic review and meta-analysis. PLoS ONE **13**(4), e0195166 (2018)
5. Cai, G., Huang, Y., Luo, S., Lin, Z., Dai, H., Ye, Q.: Continuous quantitative monitoring of physical activity in Parkinson's disease patients by using wearable devices: a case-control study. Neurol. Sci. **38**(9), 1657–1663 (2017)
6. Costa, C., Rauber, F., Leffa, P.S., Sangalli, C., Campagnolo, P., Vitolo, M.R.: Ultra-processed food consumption and its effects on anthropometric and glucose profile: a longitudinal study during childhood. Nutr. Metab. Cardiovasc. Dis. **29**(2), 177–184 (2019)
7. Coughlin, S.S., Caplan, L.S., Stone, R.: Use of consumer wearable devices to promote physical activity among breast, prostate, and colorectal cancer survivors: a review of health intervention studies. J. Cancer Surviv. **14**(3), 386–392 (2020). https://doi.org/10.1007/s11764-020-00855-1

8. Fock, K.M., Khoo, J.: Diet and exercise in management of obesity and overweight. J. Gastroenterol. Hepatol. **28**, 59–63 (2013)
9. Gabler, N.B., Duan, N., Vohra, S., Kravitz, R.L.: N-of-1 trials in the medical literature: a systematic review. Medical care, pp. 761–768 (2011)
10. Gates, K.M., Molenaar, P.C.: Group search algorithm recovers effective connectivity maps for individuals in homogeneous and heterogeneous samples. Neuroimage **63**(1), 310–319 (2012)
11. Harpaz, E., Tamir, S., Weinstein, A., Weinstein, Y.: The effect of caffeine on energy balance. J. Basic Clin. Physiol. Pharmacol. **28**(1), 1–10 (2017)
12. Herman, K.M., Craig, C.L., Gauvin, L., Katzmarzyk, P.T.: Tracking of obesity and physical activity from childhood to adulthood: the physical activity longitudinal study. Int. J. Pediatr. Obes. **4**(4), 281–288 (2009)
13. Hernán, M.A., Robins, J.M.: Causal inference (2010)
14. Hill, J.O., Commerford, R.: Physical activity, fat balance, and energy balance. Int. J. Sport Nutr. Exerc. Metab. **6**(2), 80–92 (1996)
15. Hottenrott, K., Hoos, O., Esperer, H.D.: Heart rate variability and physical exercise. current status. Herz **31**(6), 544–552 (2006)
16. Jain, R.: Lifeblood of health is data. IEEE Multimed. **29**(1), 128–135 (2022)
17. Kimm, S.Y., Glynn, N.W., Obarzanek, E., Kriska, A.M., Daniels, S.R., Barton, B.A., Liu, K.: Relation between the changes in physical activity and body-mass index during adolescence: a multicentre longitudinal study. The Lancet **366**(9482), 301–307 (2005)
18. Koenig, J., et al.: Impact of caffeine on heart rate variability: a systematic review. J. Caffeine Res. **3**(1), 22–37 (2013)
19. Lillie, E.O., Patay, B., Diamant, J., Issell, B., Topol, E.J., Schork, N.J.: The n-of-1 clinical trial: the ultimate strategy for individualizing medicine? Pers. Med. **8**(2), 161–173 (2011)
20. Melzer, K., Kayser, B., Saris, W.H., Pichard, C.: Effects of physical activity on food intake. Clin. Nutr. **24**(6), 885–895 (2005)
21. Myers, L., Sirois, M.J.: Spearman correlation coefficients, differences between. Encyclopedia of statistical sciences 12 (2004)
22. Ortega, F.B., Lavie, C.J., Blair, S.N.: Obesity and cardiovascular disease. Circ. Res. **118**(11), 1752–1770 (2016)
23. Pearl, J.: Causal inference. Causality: objectives and assessment, pp. 39–58 (2010)
24. Philippou, C., Andreou, E., Menelaou, N., Hajigeorgiou, P., Papandreou, D.: Effects of diet and exercise in 337 overweight/obese adults. Hippokratia **16**(1), 46 (2012)
25. Ralevski, E., Petrakis, I., Altemus, M.: Heart rate variability in alcohol use: a review. Pharmacol. Biochem. Behav. **176**, 83–92 (2019)
26. Ruxton, C.: The impact of caffeine on mood, cognitive function, performance and hydration: a review of benefits and risks. Nutr. Bull. **33**(1), 15–25 (2008)
27. Ryan, J., Howes, L.: Relations between alcohol consumption, heart rate, and heart rate variability in men. Heart **88**(6), 641–642 (2002)
28. Santos, L.P., Gigante, D.P., Delpino, F.M., Maciel, A.P., Bielemann, R.M.: Sugar sweetened beverages intake and risk of obesity and cardiometabolic diseases in longitudinal studies: a systematic review and meta-analysis with 1.5 million individuals. Clinical Nutrition ESPEN (2022)
29. Sauder, K.A., Johnston, E.R., Skulas-Ray, A.C., Campbell, T.S., West, S.G.: Effect of meal content on heart rate variability and cardiovascular reactivity to mental stress. Psychophysiology **49**(4), 470–477 (2012)

30. Schroeder, E.B., et al.: Diabetes, glucose, insulin, and heart rate variability: the atherosclerosis risk in communities (aric) study. Diabetes Care **28**(3), 668–674 (2005)
31. Schutz, Y.: Macronutrients and energy balance in obesity. Metabolism **44**, 7–11 (1995)
32. Sharma, A., Kiciman, E.: Dowhy: an end-to-end library for causal inference. arXiv preprint arXiv:2011.04216 (2020)
33. Soares-Miranda, L., Sattelmair, J., Chaves, P., Duncan, G.E., Siscovick, D.S., Stein, P.K., Mozaffarian, D.: Physical activity and heart rate variability in older adults: the cardiovascular health study. Circulation **129**(21), 2100–2110 (2014)
34. Solfrizzi, V., et al.: Coffee consumption habits and the risk of mild cognitive impairment: the Italian longitudinal study on aging. J. Alzheimers Dis. **47**(4), 889–899 (2015)
35. Sondermeijer, H.P., van Marle, A.G., Kamen, P., Krum, H.: Acute effects of caffeine on heart rate variability. Am. J. Cardiol. **90**(8), 906–907 (2002)
36. Swithers, S.E., Martin, A.A., Davidson, T.L.: High-intensity sweeteners and energy balance. Physiol. Behav. **100**(1), 55–62 (2010)
37. Teixeira, E., et al.: Wearable devices for physical activity and healthcare monitoring in elderly people: a critical review. Geriatrics **6**(2), 38 (2021)
38. Textor, J., Hardt, J., Knüppel, S.: Dagitty: a graphical tool for analyzing causal diagrams. Epidemiology **22**(5), 745 (2011)
39. Tripette, J., et al.: Caffeine consumption is associated with higher level of physical activity in Japanese women. Int. J. Sport Nutr. Exerc. Metab. **28**(5), 474–479 (2018)
40. Webber, J.: Energy balance in obesity. Proc. Nutrition Soc. **62**(2), 539–543 (2003)
41. Westerterp-Plantenga, M., Diepvens, K., Joosen, A.M., Bérubé-Parent, S., Tremblay, A.: Metabolic effects of spices, teas, and caffeine. Physiol. Behav. **89**(1), 85–91 (2006)
42. Wetter, A.C., et al.: How and why do individuals make food and physical activity choices? Nutr. Rev. **59**(3), S11–S20 (2001)
43. Yeomans, M.R.: Alcohol, appetite and energy balance: is alcohol intake a risk factor for obesity? Physiol. Behav. **100**(1), 82–89 (2010)

Capturing Nutrition Data for Sports: Challenges and Ethical Issues

Aakash Sharma[1(✉)], Katja Pauline Czerwinska[2,3], Dag Johansen[1], and Håvard Dagenborg[1]

[1] UiT The Arctic University of Norway, Tromsø, Norway
aakash.sharma@uit.no
[2] Volda University College, Volda, Norway
[3] RheinMain University of Applied Sciences, Wiesbaden, Germany

Abstract. Nutrition plays a key role in an athlete's performance, health, and mental well-being. Capturing nutrition data is crucial for analyzing those relations and performing necessary interventions. Using traditional methods to capture long-term nutritional data requires intensive labor, and is prone to errors and biases. Artificial Intelligence (AI) methods can be used to remedy such problems by using Image-Based Dietary Assessment (IBDA) methods where athletes can take pictures of their food before consuming it. However, the current state of IBDA is not perfect. In this paper, we discuss the challenges faced in employing such methods to capture nutrition data. We also discuss ethical and legal issues that must be addressed before using these methods on a large scale.

Keywords: Nutrition · Food images · Privacy · AI · Dietary assessment

1 Introduction

Sports science, just like many other fields, has benefited from data-based research and advanced computing applications. Athletes are interesting subjects to study for nutritionists, sports scientists, doctors, and researchers from many more fields. Nutrition plays an important role in an athlete's performance and health [8]. Some nutrients, such as amino acid-electrolyte, can affect performance in endurance athletes [38]. Such effects of nutrition on performance can be applicable to the general population as well. In some cases, however, researchers discover placebo effects, such as that of electrolytes [23]; or refute old notions, for example that athletic performance requires a diet high in proteins.

Research in sports sciences has focused on improving athletic performance, game strategies, and optimizing training processes. Body-worn sensors, such as ZXY sport tracking [28], capture an athlete's movements during a game. Other body-worn sensors, such as smart watches, can capture a multitude of individual performance indicators, on and off the field. However, smartwatches are targeted more toward individuals than teams. A combination of sensors with advanced

D.-T. Dang-Nguyen et al. (Eds.): MMM 2023, LNCS 13833, pp. 601–612, 2023.
https://doi.org/10.1007/978-3-031-27077-2_47

video-based systems [15] can digitize gameplays and offer stakeholders analytical tools to improve game strategies. Looking at individual athletes, nutrition plays an important role in their performance. However, researchers have also discovered effects of an athlete's nutrition on other aspects beyond their athletic performance. Diets can be linked to an athlete's mood [36], behavior [19], and even mental well-being [22,30].

The robustness of nutrition research relies on the validity and reliability of the chosen dietary assessment method [20]. Capturing an individual's diet is challenging [7,20]. Most studies have relied on self-report methods to capture dietary intakes [20]. Self-report methods either rely on real-time recording or methods of recall [26]. Real-time recording methods, for example, food diaries require an individual to record every food or beverage they consume in real-time. Such methods are considered to be labor-intensive. The methods of recall (e.g. 24-h dietary recall) rely on an individual's self-assessment of their diet. In 24-h dietary recall, an individual is asked to describe the food they consumed over one day. Self-report methods (e.g., food records) are frequently utilized despite the evident inaccuracy of these methods in assessing energy and nutrient intake [12,37]. Memory-based methods are flawed with inaccuracies and depend heavily on the subjects' ability to recall their diet from their memory. Subar et al. [37] argue that despite the inaccuracies, self-reported data can be used to inform dietary guidance and public health policy. However, in sports sciences, as the focus is on personalized dietary interventions for athletes, the accuracy of dietary intake assessment methods is crucial.

IBDA methods that assess food intake via images of foods have overcome many of the limitations of traditional self-report. Taking images using a smartphone camera is less labor intensive than noting down food in a food diary. Deep learning methods and Artificial Intelligence (AI) can identify food from images taken with a smartphone camera [18,21]. In a lab, such as a cafeteria setting, digital photography has proven to be an unobtrusive and accurate method for assessing food provision, plate waste, and food intake. Many feasibility studies [4,16,25,29] have shown promise in using Image-Based Dietary Assessment (IBDA) methods for capturing nutrition data. However, state-of-the-art IBDA methods are not precise enough to capture dietary information accurately in every possible scenario. In their review, Höchsmann and Martin [16] highlight the benefits of taking food images over traditional dietary assessment methods, such as food records. They report that the technology to recognize food and portion size correctly is still in its infancy but promises real-time feedback to the user, reduces biases, and increases the efficacy of dietary interventions. Inferences from one's dietary data can lead to discoveries that may benefit both the athlete and the team. Long-running epidemiological studies can also incorporate these methods to accurately capture, analyze, and validate the effects of dietary interventions in a population.

Individuals' perceptions of the sensitivity of their data and the necessary privacy protections can change over time [2,32]. Perceived infringements to privacy can result in a complete rejection of new technology even if they are not based on evidence [2]. It is therefore necessary to address and mitigate such risks by providing accountability and transparency in a new system. Wachter and Mittelstadt [40] argue that individuals have little control and oversight over how their personal data is used to draw inferences about them. And the inferences derived from one's data might be considered sensitive by the individual while the legal frameworks might not provide the necessary protections. Recent laws, such as the General Data Protection Regulation (GDPR), have been hailed as promising as they provide an individual (or a data subject) with rights to know about (Article 13–15), rectify (Article 16), delete (Article 17), object to (Article 21), or port (Article 20) personal data. However, these rights are significantly curtailed for inferences. The GDPR also provides insufficient protection against sensitive inferences (Article 9) or remedies to challenge inferences or important decisions based on them (Article 22(3)).

In the following paragraphs, we will introduce nutrition data, its relevance for dietary interventions, and methods for dietary assessments. We will present our findings on privacy perceptions of food images followed by inferences and sensitivity as reported by the participants. Later, we discuss our findings, their relevance to sports sciences, and our conclusion.

2 Background

Nutrition data has its use in many research fields. Large cohort population studies are a frequently used tool to investigate epidemics such as obesity, heart, and coronary disease, diabetes, and cancer. The most frequently used tools are based on self-reporting either through interviews, questionnaires, or combinations [7,20,37]. The goal is to quantify the health status of each individual as well as of the population as a whole. More precisely that means describing relevant health parameters in objective and quantifiable metrics including anthropometrics (height, weight, etc.), biochemistry, and clinical issues. In a hospital setting, data sources may include blood samples, DNA, x-ray images, EMG, ECG, blood oxygen (VO_2) uptake, body composition scans, and many more. Applying these methods to large population studies presents many challenges. It can be too expensive and too invasive to apply such methods preemptively to a population as a whole. Also, since most interventions include dietary adjustments, it is necessary to collect data on what people are eating and drinking. In epidemiology, this is usually called *dietary assessment*. Epidemiologists investigate patterns and causes of diseases and other outcomes that affect people. A dietary assessment provides key data for epidemiologists studying all topics connected to the food intake of both populations as a whole, and on the individual level. The same methods to collect data are used by researchers, who devise and suggest food and dietary policies for the public, sports nutritionists, who manage the diet of athletes, physicians helping the chronically ill manage their diet-related diseases, and individuals doing self-interventions for their own health development.

The classical methods used to perform dietary assessment are few but clearly defined, and have stayed largely unchanged since their inception [14]. Most existing methods for dietary assessment are based on self-reporting. A recognized shortcoming of self-reporting methods for dietary intake is that they tend to have a bias that weakens the quality of the data [31], and several studies have questioned the validity of data from such methods [42]. Types of bias range from errors in how well subjects remember what they ate, misreporting due to practicalities and required effort to report, to the unwillingness to share certain details of their diet. The resulting discrepancies may be very large [43]. In fact, data collected by the U.S. National Health and Nutrition Examination Survey (NHANES) in the period 1971–2010 are not physiologically plausible when compared to estimated total energy expenditure [3]. Many computer scientists have seen this as an opportunity to contribute to the development of technology-based scientific methods for dietary assessment. While the methods used in dietary science can be described as various forms of subjectively validated self-reporting, methods developed in computer science are typically based on objective measurements from validated devices and sensors. A wide array of methods and technologies have been developed, but none of these seem to have gained enough traction to replace the standard methods of self-reporting in epidemiology.

Work done in this field is published in a wide array of publications, including dietary science, healthcare technology, bio-engineering, sensor, computer vision, and multimedia journals. Two previous surveys focused on studies on just one method, image-based dietary intake reporting [6,13]. These surveys are published in dietary science journals, and both compared studies gathered solely from medical and dietary publications. They also both compared effort and precision in these studies, with traditional methods as a reference. A third previous survey did a systematic review across multiple technologies, made a taxonomy to classify and describe the different technologies, but focused on readiness to deploy in low-income countries and published in a biology journal [9]. A fourth, earlier survey covered a broad set of technologies from an epidemiological perspective [17], but did not go into detail on each method, nor did it develop a taxonomy to describe them. Its focus was on how new technologies could be used in epidemiological studies. Finally, Public Health Nutrition [1] collected eight articles on automated dietary assessment as proof of the interest in methods and approaches that can be used for the automated collection and processing of data on dietary intake. The stated intent by its editor was to illustrate the width and the scope of current research efforts, but only the editorial mentioned all the specialized hardware solutions that are under development [24].

Dietary assessment is a complicated task that aims to quantify one's diet [7]. Surveys of developments in dietary assessments [6,10,16] highlight the importance of issues and the promise to mitigate some of them. Nutritionists and epidemiologists still use methods with a lot of participant and researcher burden [10]. Newer technological advancements are often ignored or disqualified due to a lack of acceptance in the community. Measurement errors in self-report dietary assessment methods [12,20] in epidemiological studies can be ignored due to large-scale data collection [37] as researchers are still able to use self-report methods for statistical analysis among a large population. However, in sports, personalized interventions are necessary. The same methods cannot be applied to the relatively small number of participants in athletic teams. Thus, to improve data collection and analysis, newer computer-assisted methods are being designed to improve the ease and accuracy of nutrition data collection and analysis. IBDA is one such promising method.

Computer-assisted methods are designed to reduce the labor required by participants and researchers. Some of these methods are very intuitive and embrace ubiquitous tools such as mobile phones. Once trained, these methods allow participants to conduct these assessments for a long time without needing supervision. Intuitive and easy-to-use tools are required in order to capture more nutrition data among participants. Such tools become necessary as more data is needed for personalized dietary interventions, which are more effective than general population recommendations [27]. Some of these newer tools may be perceived as intrusive by the participants. Participation in research studies relies heavily on trust, and any damage to their reputation can have severe consequences for organizations that obtain data based on informed consent. Thus, it is crucial to understand challenges and ethical issues before implementing and using new tools for capturing data. In the next Section, we describe a study in which we investigated some of these challenges.

3 Our Experiences with Privacy Perceptions of Food Images

We investigated common perceptions of computer systems using food images for dietary assessment. The study also delves into perceived risks and data-sharing behaviors. In the study [32], we investigated the privacy perceptions of 105 individuals by using a web-based survey. We analyzed these perceptions along with perceived risks in sharing dietary information with third parties.

We conducted the study by using a web-based questionnaire. The questionnaire was developed using close-ended questions for their statistical analyses. Some questions are repeated in the questionnaire to reduce biased context. Responses were collected between February and June 2019. The survey was designed to record participants' perceptions in the following scenarios: (1) Scenario A, the participant having to record and share dietary data as an athlete; (2) Scenario B, the possibility and severity of privacy leak from one's dietary data; and (3) Scenario C, sharing dietary data and reports among different social groups. We analyzed the data and performed P-value checks for hypotheses testing.

We now present a brief summary of the results. One of the goals of computer-assisted methods is to reduce labor. In terms of effort, approximately 80.9% (85/105) of the participants agreed that capturing diet records by using a phone camera is easier than writing down their dietary intake. Similar to writing down, individuals can also record audio to describe their diets accurately to record them. However, more than four-fifths of the participants (86/105, 81.9%) preferred capturing photos over recording their diets by voice. Only some participants were undecided about preferring image capture over writing down or recording audio (7/105, 6.7% and 8/105, 7.6%; respectively). Overall, the IBDA method seems to be the preferred self-report method for dietary assessment by the participants.

People take pictures and share them with their social groups through different channels. We investigate the sharing behavior in the context of food images. Taking a food picture and sharing it is considered as a familiar practice among the general population. About half of the participants (52/105, 49.6%) had previously posted food images on social networks. Even those participants, who lacked experience (53/105, 50.4%) in posting food images, showed similar attitudes toward the ease of recording their diet intakes by using photography. For a successful IBDA-based study, a participant must record every dietary intake whether a meal or a beverage. Compared to the irregular posting of images on social networks, adherence to recording diets requires extra effort from a participant. Some may perceive it as intrusive. When we asked about the intrusiveness of such a requirement, about two-thirds of the participants (69/105, 65.7%) indicated that it would be intrusive.

In terms of sharing behavior, participants reported varied preferences depending on the context of the metadata. When done in real-time, recording a diet can generate metadata such as time and location from a smartphone. We asked participants to report their preferences on sharing such metadata with their different social groups. We used *Family*, *Friends*, *Doctor*, *Team*, and *Fans* as the target social groups, as they would be relevant for an athlete. In comparison to the social group family(79/105, 75.2%), the participants were more willing to share food images with their doctors (94/105, 89.5%). Only a quarter (26/105, 24.8%) of the participants showed a willingness to share food pictures with their sports team while nearly half (47/105, 44.8%) showed a willingness to share food pictures with friends. In terms of the metadata associated with dietary data, the willingness to share drops further. Only 68.6% (72/105) of the participants agreed to share the time of the meal with their families in comparison to 82.9% (87/105) sharing the time of the meal with their doctors. The time of a meal is in fact an important consideration for elite athletes and their coaches for restitution and training planning. The full analysis can be referred to in our paper [32].

4 Inferences and Sensitivity

The state of AI-based inferences from food images is still in its infancy [16]. Researchers are exploring novel methods to extract, analyze, and infer useful information from food images [6,10]. There might be some information about the participant that can be inferred from their food images, which the subjects did not consent to. Such inferences can be perceived as privacy infringements by the participants and affect the perception towards the use of IBDA methods. As noted earlier, perceived infringements on privacy can affect how users may adopt new technology [2]. Based on the data collected in our study [32], we present what participants perceived as possible to infer from their food image dataset, which has been anonymized. Additionally, they also indicate how sensitive they think

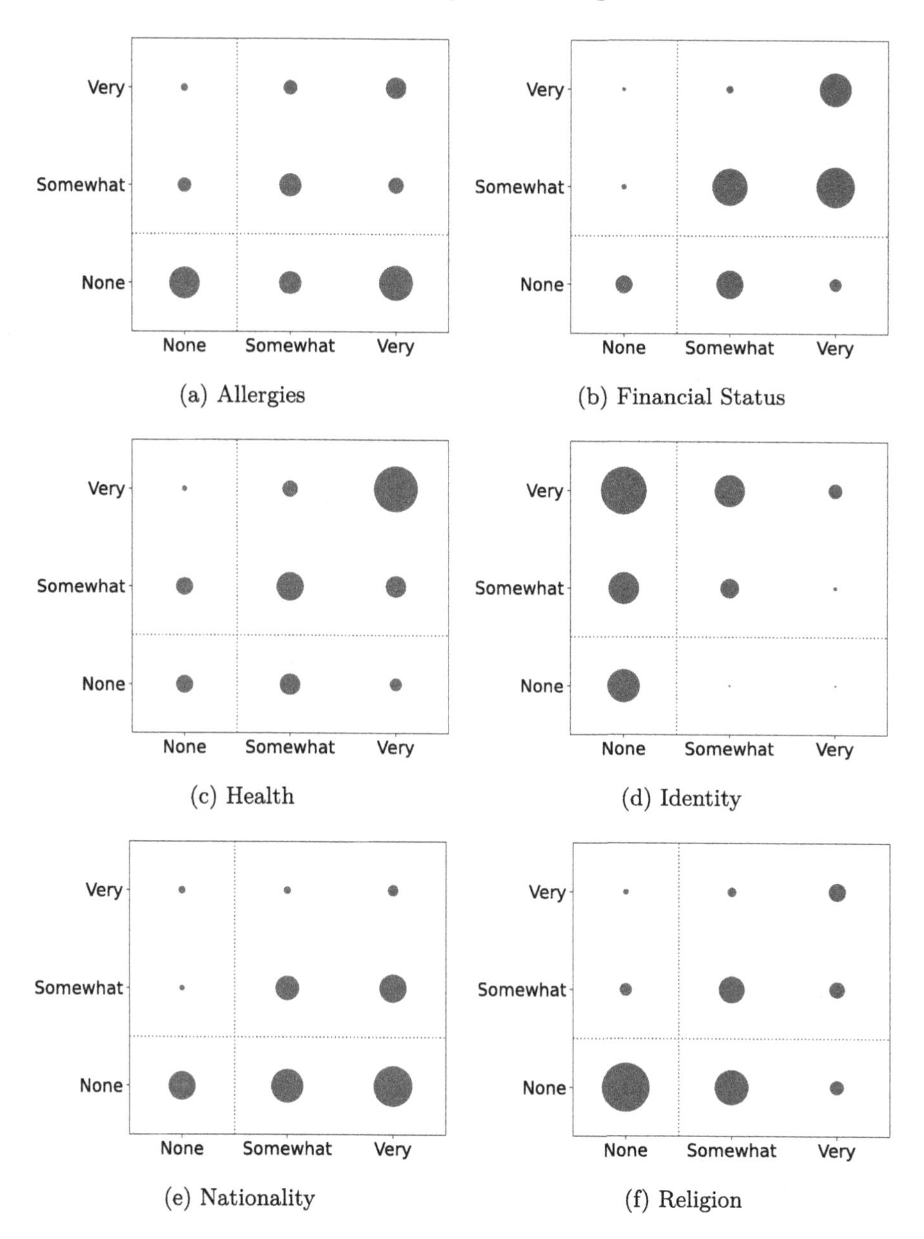

(a) Allergies (b) Financial Status (c) Health (d) Identity (e) Nationality (f) Religion

Fig. 1. Perceived *Likelihood* (x-axis) vs *Level of Concern* (y-axis) for different inferences from an individual's food image dataset as reported by participants (n = 105).

such information is to them. The participants indicate their responses as *none*, *somewhat*, or *very* for both the likelihood and their level of concern. We plot the graphs based on responses from the participants (n = 105). The attributes that are concerning to the participants will be indicated by the size of bubbles above the dotted blue line. Similarly, the attributes, which the participants think can be derived from their food images, will be indicated by the size of bubbles on the right side of the blue-dotted line. The participants were not given any scientific input on the current state of the research. The results are purely based on their knowledge and thinking.

Figure 1 shows the *Likelihood* and the *Level of Concern* if information about an individual can be inferred from their food image dataset. Note that these perceptions about the likelihood of inference might not be based on reality but perceived by the participants. However, we consider them as these perceptions may affect one's willingness to use IBDA methods to record their nutrition data. In Fig. 1a, the participants report that it might be possible to infer their allergies from food images. However, they do not seem to be very concerned about such information. It may be that they perceive allergy information as useful for dietary interventions for recommending a better diet. The possibility of deriving one's financial status (Fig. 1b) from their food image seems concerning to the participants. And they report that it is likely to be derived from the images.

Additionally, health information (Fig. 1c) is another area where participants report a higher level of concern if inferred from their food dataset. The perceived likelihood of such inference is also high among the participants. The participants do not believe that one's identity (Fig. 1d) can be inferred from their food image dataset. However, the level of concern about one's identity is high, which is in line with earlier research [5,41]. Inferring one's nationality (Fig. 1e) or religion (Fig. 1f) is not as concerning as one might believe. However, in cases of discrimination, these might be protected under the law. Participants reported that it is very likely that an individual's nationality can be inferred from their food image dataset. They also reported that it is not very concerning to derive one's religion from their food dataset. However, only 16% (17/105) of the participants reported following a religious diet.

5 Discussion

Nutrition data allow researchers to understand novel relations between our food and its effect on our bodies. The research's impact is beyond the field of sports sciences. Traditional methods suffer from limitations that restrict nutrition data collection on a larger scale, outside a lab, and in longer temporal periods for longitudinal studies. Novel computer-assisted methods to capture nutrition data provide a good opportunity to increase the scale of the research. Combined with deep learning and AI, and the increasing scale of nutrition research, personalized dietary interventions can benefit from evidence-based insights impacting healthy developments. The fields of sports sciences, epidemiology, and nutrition will benefit from precise, non-biased, large-scale nutrition data collection. As it is with other fields, the challenges and ethical issues need to be addressed as well.

IBDA-based methods provide a scalable solution to capture nutrition data. As reported in our study [32], taking photos of food as a method to self-report diet is preferred over other traditional methods. It still requires some labor on the participant's part to take the picture. Passive video surveillance methods to capture nutrition using video cameras in a canteen (or a lab) can reduce the participant's labor further. However, the privacy implications and the required GDPR consent can become a challenge [11] as individuals need to give their explicit consent. The collected nutrition data can be useful for nutritionists, doctors, and researchers. Sharing of such data requires consent from the participants as they might be continuously adding more data for personalized dietary interventions. In our study [32], individuals reported varying behaviors for sharing data, as knowing what the data is being used for increases the chance of it being shared. No single policy can be applied to indicate their sharing preferences for all users as the perception of their privacy and their sharing behavior can be highly variable.

People's perception of their privacy evolves over time [39]. And so does their perception of what is possible to infer from their data. Their perceptions may change over time due to knowledge or growing coverage of privacy issues in the media. As the sensitivity of the inference changes, a participant's consent to sharing may change as well. GDPR and other modern laws provide subjects/participants with extended rights to their data. Existing role-based access control systems for sharing the data may result in a one-time-only check for data access. Furthermore, the sensitivity of one's information is also not static. System support for managing sensitivity and evolving perceptions requires dynamic policies that are attached to the data and not the system. Such a system should also be able to handle accountability to provide subjects with information about the data being used. For sports science, a large amount of athlete data collected on and off the field presents a big challenge of data ownership and privacy concerns [35]. Such a system can be crucial in sports where athletes' participation is crucial and relies on their trust in the team and institution.

We have built *Lohpi* [33,34], a distributed system to support dynamic policies. We argue for a decentralized approach, where different institutions can process data on their infrastructures and maintain control of data assets, rather than a central one-fits-all service. The policy enforcement is done at the nodes hosting data assets and updated through a resilient metadata distribution substrate. The metadata distribution substrate is implemented using gossip-based data exchanges, which is a probabilistic data dissemination scheme. We have primarily used it in the medical domain, but also for sharing soccer data from female top athletes in Norway. Privacy and security have been first-order design principles, and the system can easily be adapted to support nutrition data (both media files and extracted information). Notably, this system enables sensitive data to be shared between collaborators in a controlled manner. We argue that a decentralized service that maintains metadata, a global view of all data usage, and active policies combined with local monitoring and security enforcement can provide automated compliance checking.

6 Conclusion

Taking food images is an interesting method for capturing nutrition data. It potentially requires much less labor when compared to traditional dietary assessment methods, and almost no training as individuals are already familiar with capturing images with their smartphones. The captured data can be fed to AI-based systems that can provide real-time dietary interventions toward a specified goal. Such systems present interesting opportunities for sports where stakeholders (coaches, managers, and owners) are interested in improving athletes' performance, maintaining good health, and their mental well-being.

We presented our findings on perceptions towards capturing nutrition data using food pictures. These perceptions play a vital role in the acceptance of new technology. Any perceived risk or sensitive information must be handled carefully with accountability and transparency in the system. New laws and regulations can also introduce changes to an information's sensitivity. A system must support these changing perceptions and information sensitivity.

Acknowledgement. The authors would like to thank Lars Brenna and the anonymous reviewers for their feedback. This work was funded in part by the Research Council of Norway project numbers 263248 and 274451.

References

1. Public Health Nutrition. Public Health Nutrition 22(7) (2019)
2. Adams, A.: The implications of users' multimedia privacy perceptions on communication and information privacy policies. In: Proceedings of Telecommunications Policy Research Conference, pp. 1–23 (1999)
3. Archer, E., Hand, G.A., Blair, S.N.: Validity of US nutritional surveillance: national health and nutrition examination survey caloric energy intake data, 1971–2010. PLoS ONE **8**(10), e76632 (2013)
4. Ashman, A.M., Collins, C.E., Brown, L.J., Rae, K.M., Rollo, M.E.: Validation of a smartphone image-based dietary assessment method for pregnant women. Nutrients **9**(1), 73 (2017)
5. Berendt, B., Günther, O., Spiekermann, S.: Privacy in e-commerce: stated preferences vs. actual behavior. Commun. ACM **48**(4), 101–106 (2005)
6. Boushey, C., Spoden, M., Zhu, F., Delp, E., Kerr, D.: New mobile methods for dietary assessment: review of image-assisted and image-based dietary assessment methods. Proc. Nutrition Soc., 1–12 (2016)
7. Brenna, L., Johansen, H.D., Johansen, D.: A survey of automatic methods for nutritional assessment. arXiv preprint arXiv:1907.07245 (2019)
8. Buskirk, E.R.: Diet and athletic performance. Postgrad. Med. **61**(1), 229–236 (1977)
9. Coates, J., Bell, W.F., Colaiezzi, B., Cisse, C.: Scaling up dietary data for decision-making in Africa and Asia: new technological frontiers. FASEB J. **30**(1 Supplement), 669–715 (2016)
10. Dao, M.C., Subar, A.F., Warthon-Medina, M., Cade, J.E., Burrows, T., Golley, R.K., Forouhi, N.G., Pearce, M., Holmes, B.A.: Dietary assessment toolkits: an overview. Public Health Nutr. **22**(3), 404–418 (2019)

11. Etteldorf, C.: EDPB publishes guidelines on data processing through video devices. Eur. Data Prot. L. Rev. **6**, 102 (2020)
12. Frongillo, E.A., Baranowski, T., Subar, A.F., Tooze, J.A., Kirkpatrick, S.I.: Establishing validity and cross-context equivalence of measures and indicators. J. Acad. Nutr. Diet. **119**(11), 1817–1830 (2019)
13. Gemming, L., Utter, J., Mhurchu, C.N.: Image-assisted dietary assessment: a systematic review of the evidence. J. Acad. Nutr. Diet. **115**(1), 64–77 (2015)
14. Gibson, R.S.: Principles of Nutritional Assessment. Oxford University Press, USA (2005)
15. Halvorsen, P., et al.: Bagadus: an integrated system for arena sports analytics: a soccer case study. In: Proceedings of the 4th ACM Multimedia Systems Conference, pp. 48–59 (2013)
16. Höchsmann, C., Martin, C.K.: Review of the validity and feasibility of image-assisted methods for dietary assessment. Int. J. Obes. **44**(12), 2358–2371 (2020)
17. Illner, A., Freisling, H., Boeing, H., Huybrechts, I., Crispim, S., Slimani, N.: Review and evaluation of innovative technologies for measuring diet in nutritional epidemiology. Int. J. Epidemiol. **41**(4), 1187–1203 (2012)
18. Kawano, Y., Yanai, K.: Real-time mobile food recognition system. In: Proceedings of the IEEE Conference on Computer Vision and Pattern Recognition Workshops, pp. 1–7 (2013)
19. Killer, S.C., Svendsen, I.S., Jeukendrup, A., Gleeson, M.: Evidence of disturbed sleep and mood state in well-trained athletes during short-term intensified training with and without a high carbohydrate nutritional intervention. J. Sports Sci. **35**(14), 1402–1410 (2017)
20. Kirkpatrick, S.I., Baranowski, T., Subar, A.F., Tooze, J.A., Frongillo, E.A.: Best practices for conducting and interpreting studies to validate self-report dietary assessment methods. J. Acad. Nutr. Diet. **119**(11), 1801–1816 (2019)
21. Liu, C., Cao, Yu., Luo, Y., Chen, G., Vokkarane, V., Ma, Y.: DeepFood: deep learning-based food image recognition for computer-aided dietary assessment. In: Chang, C.K., Chiari, L., Cao, Yu., Jin, H., Mokhtari, M., Aloulou, H. (eds.) ICOST 2016. LNCS, vol. 9677, pp. 37–48. Springer, Cham (2016). https://doi.org/10.1007/978-3-319-39601-9_4
22. Martins, L.B., Braga Tibães, J.R., Sanches, M., Jacka, F., Berk, M., Teixeira, A.L.: Nutrition-based interventions for mood disorders. Expert Rev. Neurother. **21**(3), 303–315 (2021)
23. McClung, M., Collins, D.: "Because I know it will!": placebo effects of an ergogenic aid on athletic performance. J. Sport Exercise Psychol. **29**(3), 382–394 (2007)
24. Mikkelsen, B.E.: Man or machine? will the digital transition be able to automatize dietary intake data collection? Public Health Nutr. **22**(7), 1149–1152 (2019)
25. Naaman, R., et al.: Assessment of dietary intake using food photography and video recording in free-living young adults: a comparative study. J. Acad. Nutr. Diet. **121**(4), 749–761 (2021)
26. Naska, A., Lagiou, A., Lagiou, P.: Dietary assessment methods in epidemiological research: current state of the art and future prospects. F1000Research 6 (2017)
27. Ordovas, J.M., Ferguson, L.R., Tai, E.S., Mathers, J.C.: Personalised nutrition and health. Bmj 361 (2018)
28. Pettersen, S.A., et al.: Soccer video and player position dataset. In: Proceedings of the 5th ACM Multimedia Systems Conference, pp. 18–23 (2014)
29. Prinz, N., Bohn, B., Kern, A., Püngel, D., Pollatos, O., Holl, R.W.: Feasibility and relative validity of a digital photo-based dietary assessment: results from the Nutris-Phone study. Public Health Nutr. **22**(7), 1160–1167 (2019)

30. Scarmeas, N., Anastasiou, C.A., Yannakoulia, M.: Nutrition and prevention of cognitive impairment. Lancet Neurol. **17**(11), 1006–1015 (2018)
31. Schoeller, D.A.: Limitations in the assessment of dietary energy intake by self-report. Metabolism **44**, 18–22 (1995)
32. Sharma, A., Czerwinska, K.P., Brenna, L., Johansen, D., Johansen, H.D., et al.: Privacy perceptions and concerns in image-based dietary assessment systems: questionnaire-based study. JMIR Hum. Factors **7**(4), e19085 (2020)
33. Sharma, A., Nilsen, T.B., Brenna, L., Johansen, D., Johansen, H.D.: Accountable human subject research data processing using lohpi. In: Proceedings of the ICTeSSH 2021 Conference, July 2021. https://doi.org/10.21428/7a45813f.80ebd922
34. Sharma, A., Nilsen, T.B., Czerwinska, K.P., Onitiu, D., Brenna, L., Johansen, D., Johansen, H.D.: Up-to-the-minute Privacy Policies via gossips in Participatory Epidemiological Studies. Frontiers in Big Data p. 14 (2021)
35. Socolow, B., Jolly, I.: Game-changing wearable devices that collect athlete data raise data ownership issues. World Sports Advocate **15**(7), 15–17 (2017)
36. Soh, N.L., Walter, G., Baur, L., Collins, C.: Nutrition, mood and behaviour: a review. Acta Neuropsychiatrica **21**(5), 214–227 (2009)
37. Subar, A.F., et al.: Addressing current criticism regarding the value of self-report dietary data. J. Nutr. **145**(12), 2639–2645 (2015)
38. Tai, C.Y., et al.: An amino acid-electrolyte beverage may increase cellular rehydration relative to carbohydrate-electrolyte and flavored water beverages. Nutr. J. **13**(1), 1–7 (2014)
39. Tsay-Vogel, M., Shanahan, J., Signorielli, N.: Social media cultivating perceptions of privacy: A 5-year analysis of privacy attitudes and self-disclosure behaviors among Facebook users. New Media Soc. **20**(1), 141–161 (2018)
40. Wachter, S., Mittelstadt, B.: A right to reasonable inferences: re-thinking data protection law in the age of big data and ai. Colum. Bus. L. Rev., p. 494 (2019)
41. Wang, Y., Norice, G., Cranor, L.F.: Who is concerned about what? a study of American, Chinese and Indian users' privacy concerns on social network sites. In: McCune, J.M., Balacheff, B., Perrig, A., Sadeghi, A.-R., Sasse, A., Beres, Y. (eds.) Trust 2011. LNCS, vol. 6740, pp. 146–153. Springer, Heidelberg (2011). https://doi.org/10.1007/978-3-642-21599-5_11
42. Westerterp, K.R., Goris, A.H.: Validity of the assessment of dietary intake: problems of misreporting. Current Opinion Clin. Nutrition Metabolic Care **5**(5), 489–493 (2002)
43. Witschi, J.C.: Short-term dietary recall and recording methods. Nutritional Epidemiology **4**, 52–68 (1990)

VBS: Video Browser Showdown

VISIONE at Video Browser Showdown 2023

Giuseppe Amato, Paolo Bolettieri, Fabio Carrara, Fabrizio Falchi, Claudio Gennaro, Nicola Messina, Lucia Vadicamo(✉), and Claudio Vairo

ISTI-CNR, Via G. Moruzzi 1, 56124 Pisa, Italy
{giuseppe.amato,paolo.bolettieri,fabio.carrara,fabrizio.falchi, claudio.gennaro,nicola.messina,lucia.vadicamo,claudio.vairo}@isti.cnr.it

Abstract. In this paper, we present the fourth release of VISIONE, a tool for fast and effective video search on a large-scale dataset. It includes several search functionalities like text search, object and color-based search, semantic and visual similarity search, and temporal search. VISIONE uses ad-hoc textual encoding for indexing and searching video content, and it exploits a full-text search engine as search backend. In this new version of the system, we introduced some changes both to the current search techniques and to the user interface.

Keywords: Content-based video retrieval · Video search · Information search and retrieval · Surrogate Text Representation · Multi-modal retrieval

1 Introduction

Video Browser Showdown (VBS) [11,14,19] is an annually-held international competition for video search on a large-scale dataset (V3C1 + V3C2) composed of 17,235 videos for a total duration of 2300 h hours [20]. It consists of three different tasks: visual and textual known-item search (KIS) and ad-hoc video search (AVS). In the 2022 competition, our system VISIONE [3] ranked first in the KIS visual task, and third in the entire competition, behind Vibro [12] and CVHunter [13] ranked, respectively, first and second. In the 2023 edition of the competition, some important changes will be introduced. The duration of the target scene to be found in the KIS tasks will be reduced from 20 s to only 3 s. Moreover, there will be a dedicated session issuing tasks in the Marine Video Kit dataset [21], which is composed of highly redundant videos taken from moving cameras in underwater environments.

In this paper we present the fourth version of VISIONE [1–4,7] which includes important changes compared to the previous version. We implemented a new model for our text-to-image retrieval tool, called ALADIN [16], which replaces

D.-T. Dang-Nguyen et al. (Eds.): MMM 2023, LNCS 13833, pp. 615–621, 2023.
https://doi.org/10.1007/978-3-031-27077-2_48

the TERN feature [15] used last year. In this version, we also included a text-to-video retrieval tool exploiting the state-of-the-art CLIP2Video [9] network. We improved our web interface to speed up the performance in AVS tasks by adding a mechanism for quickly selecting the most similar frames for each video. Finally, we implemented a clustering method to reduce the number of frames shown in the result set so that it is easier for the user to find the target video. All these changes will be described in more detail in Sect. 3.

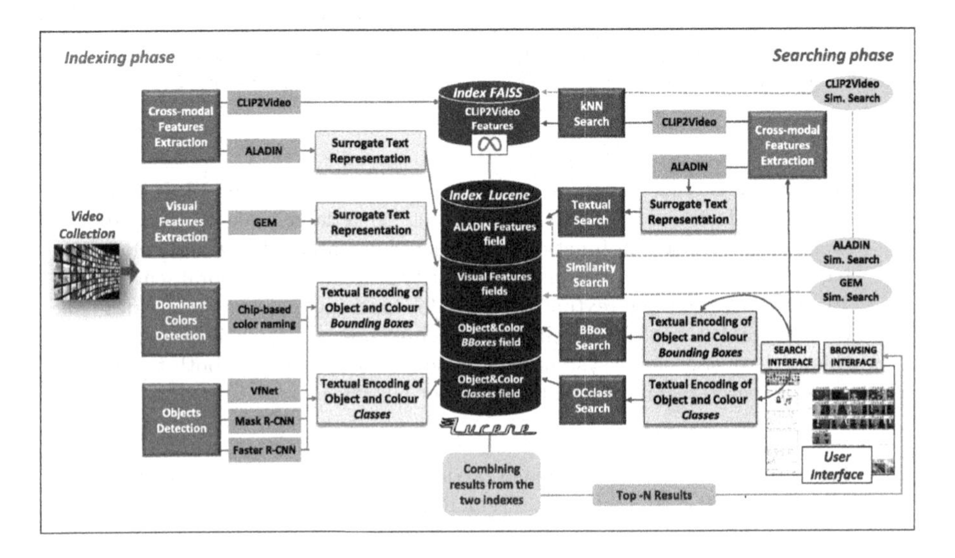

Fig. 1. VISIONE system architecture

2 System Overview

VISIONE integrates several search functionalities that allow a user to search for a target video segment by formulating textual and visual queries, which can be also combined with a temporal search. In particular it supports *free text search*, *spatial color and object search*, *visual similarity search*, and *semantic similarity search*. The system architecture is summarized in Fig. 1, while a screenshot of the user interface is shown in Fig. 2.

To support the free text search and the semantic similarity search, we employ two cross-modal feature extractors based on, respectively, CLIP2Video [9] and ALADIN [16] pre-trained models. For the object detection, we use three models: VfNet [23] (trained on COCO dataset), Mask R-CNN [10] (trained on LVIS dataset), and a Faster R-CNN+Inception ResNet V2[1] (trained on the Open Images V4). The color annotation process relies on two chip-based color naming

[1] http://tfhub.dev/google/faster_rcnn/openimages_v4/inception_resnet_v2/1.

techniques [6,22]. Finally, the visual similarity search is based on comparing GEM [18] features. We employ two indexes: the first to store the CLIP2Video features (searched using the Facebook FAISS library[2]), and the second to store all the other descriptors (searched using Apache Lucene[3]). Note that to index the extracted descriptors with Lucene, we designed special text encodings, based on the Surrogate Text Representations (STRs) approach [2,5,8].

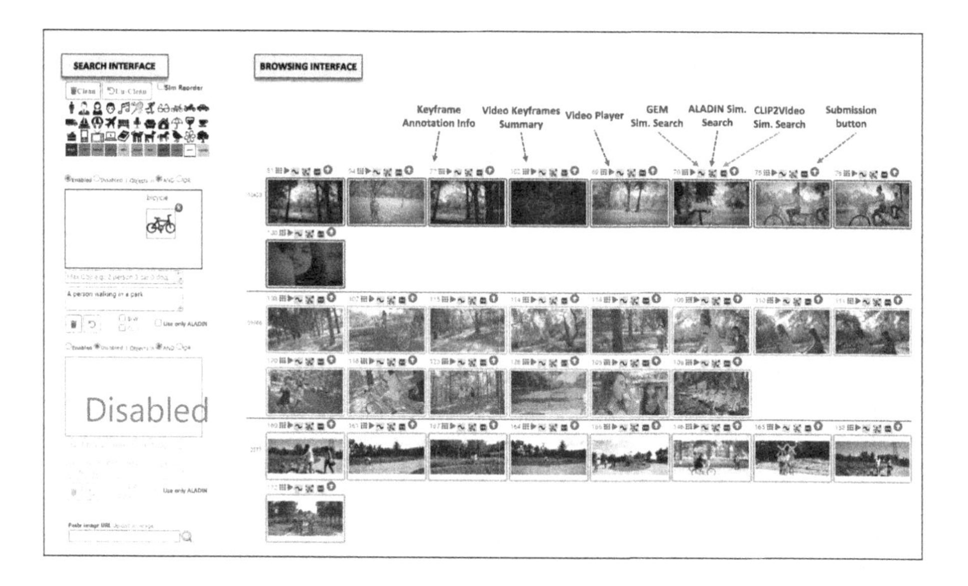

Fig. 2. User interface

3 Recent Changes to the VISIONE System

Compared to last year's system description [3], we modified some features used for object, similarity, and text search, as described below[4].

Using ALADIN for Text-to-Image Retrieval. We developed a new cross-modal retrieval deep neural network, called ALADIN (ALign And DIstill Network) [16]. ALADIN first produces high-effective scores by aligning at fine-grained level images and texts. Then, it learns a shared embedding space – where an efficient kNN search can be performed – by distilling the relevance scores obtained from the fine-grained alignments. We empirically found that this network is able to

[2] https://github.com/facebookresearch/faiss.
[3] https://lucene.apache.org/.
[4] Please note that some of these changes were already integrated in VISIONE some weeks before the last VBS competition.

compete with state-of-the-art vision-language Transformers while being almost 90 times faster at inference time.

ALADIN visual features can also be used to perform an image-to-image similarity search, which showed remarkable performance in semantic image retrieval.

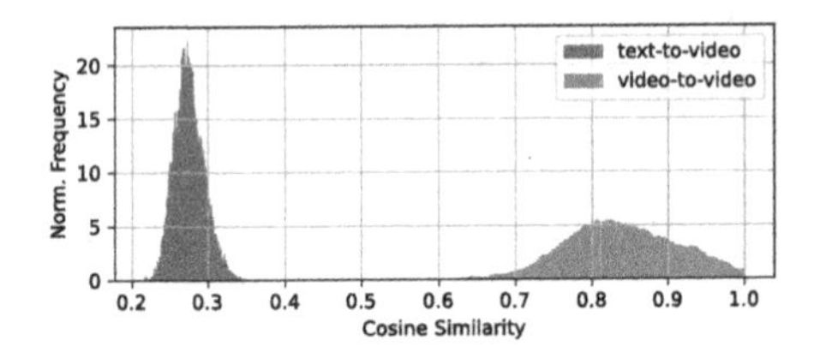

Fig. 3. Distribution of cosine similarities between text-video and video-video CLIP2Video features.

Using CLIP2Video for Text-to-Video Retrieval. In order to deeply understand videos, in particular temporal correlations and actions among multiple frames of a shot, we use CLIP2Video [9], which is one of the state-of-the-art networks for text-to-video retrieval. We re-engineered the code for easily extracting fixed-sized descriptors for texts and images that can be compared with cosine similarity. However, we found some problems in post-processing these features using our STR representation for textual-based indexing [5]. In particular, looking at Fig. 3, we noticed that the distribution of the cosine similarities of the CLIP2Video features has a very low mean value in the text-to-image cross-modal setup. This may happen if element-wise products underlying the dot-product computation have a negative sign, which in turn implies that there could be a lot of mixed-sign factors. This is a bad scenario for the STR representation, given that the CReLU operation at the core of the STR method zeroes out the contribution from mixed-sign factors. Therefore, for the CLIP2Video features, the approximated cosine similarity computed in the STR representation badly approximates the original one. For these reasons, we indexed and searched these cross-modal features with FAISS, using an exact search and an 8-bit scalar quantization for reducing the index size in memory.

The visual features extracted using CLIP2Video are also employed for a semantic reverse video search, where a video segment displayed in the results can be used as a query to search other video clips semantically similar to it.

Improvements in the Browsing Interface. To improve the user's browsing experience, we included the possibility of displaying a short preview of a video clip by right-clicking on one of the results displayed in the user interface. We also introduced multiple selections of frames to submit during AVS tasks and the ability to submit a given instant of a video directly from the videoplayer.

Object Search. We used three object detectors trained on three different datasets (COCO, LVIS, and Open Images V4), which have different classes. We built a mapping of these classes using a semi-automatic procedure in order to have a unique final list of 1,460 classes[5]. We also generated a hierarchy for each class, using wordnet[6], which is used for query expansion both at index time and at runtime.

Planned Changes. The current display of results is based on a grouping of images by video, and for each video, only the 20 keyframes with higher scores are displayed to the user. One of the main drawbacks of such visualization is the presence of many near duplicate keyframes, which burdens the visualization without adding distinctive information. We, therefore, plan to use hierarchical clustering techniques to improve the result visualization.

Moreover, we plan to exploit the Whisper model [17] to add a speech-to-text functionality that would facilitate issuing a textual query by dictating it to the system instead of typing it.

4 Conclusions and Future Work

This paper presents the novelties introduced in the VISIONE system for participating to the next edition of the Video Browser Showdown. To further improve the system, in the future we would like to exploit relevant feedback techniques, which may speed up and increase the number of correct results during AVS tasks. Moreover, we plan to investigate novel STR encoding techniques to effectively transform CLIP2Video features into textual documents. This would allow us to get rid of the FAISS index and be able to use a single scalable full-text search engine for indexing all the different descriptors.

Acknowledgements. This work was partially funded by AI4Media - A European Excellence Centre for Media, Society and Democracy (EC, H2020 n. 951911) and INAROS, CNR4C program (Tuscany POR FSE CUP B53D21008060008).

References

1. Amato, G., et al.: VISIONE at VBS2019. In: Kompatsiaris, I., Huet, B., Mezaris, V., Gurrin, C., Cheng, W.-H., Vrochidis, S. (eds.) MMM 2019. LNCS, vol. 11296, pp. 591–596. Springer, Cham (2019). https://doi.org/10.1007/978-3-030-05716-9_51
2. Amato, G., et al.: The VISIONE video search system: exploiting off-the-shelf text search engines for large-scale video retrieval. J. Imaging **7**(5), 76 (2021)
3. Amato, G., et al.: VISIONE at video browser showdown 2022. In: Þór Jónsson, B., et al. (eds.) MMM 2022. LNCS, vol. 13142, pp. 543–548. Springer, Cham (2022). https://doi.org/10.1007/978-3-030-98355-0_52

[5] https://doi.org/10.5281/zenodo.7194300.
[6] https://wordnet.princeton.edu/.

4. Amato, G., et al.: VISIONE at video browser showdown 2021. In: Lokoč, J., et al. (eds.) MMM 2021. LNCS, vol. 12573, pp. 473–478. Springer, Cham (2021). https://doi.org/10.1007/978-3-030-67835-7_47
5. Amato, G., Carrara, F., Falchi, F., Gennaro, C., Vadicamo, L.: Large-scale instance-level image retrieval. Inf. Process. Manage. **57**, 102100 (2019)
6. Benavente, R., Vanrell, M., Baldrich, R.: Parametric fuzzy sets for automatic color naming. JOSA A **25**(10), 2582–2593 (2008)
7. Bolettieri, P., et al.: An image retrieval system for video. In: Amato, G., Gennaro, C., Oria, V., Radovanović, M. (eds.) SISAP 2019. LNCS, vol. 11807, pp. 332–339. Springer, Cham (2019). https://doi.org/10.1007/978-3-030-32047-8_29
8. Carrara, F., Vadicamo, L., Gennaro, C., Amato, G.: Approximate nearest neighbor search on standard search engines. In: Skopal, T., Falchi, F., Lokoč, J., Sapino, M.L., Bartolini, I., Patella, M. (eds.) SISAP 2022. LNCS, vol. 13590, pp. 214–221. Springer, Cham (2022). https://doi.org/10.1007/978-3-031-17849-8_17
9. Fang, H., Xiong, P., Xu, L., Chen, Y.: Clip2video: mastering video-text retrieval via image clip. arXiv preprint arXiv:2106.11097 (2021)
10. He, K., Gkioxari, G., Dollár, P., Girshick, R.: Mask R-CNN. In: Proceedings of the IEEE International Conference on Computer Vision, pp. 2961–2969 (2017)
11. Heller, S., et al.: Interactive video retrieval evaluation at a distance: comparing sixteen interactive video search systems in a remote setting at the 10th video browser showdown. Int. J. Multimed. Inf. Retrieval **11**(1), 1–18 (2022)
12. Hezel, N., Schall, K., Jung, K., Barthel, K.U.: Efficient search and browsing of large-scale video collections with vibro. In: Þór Jónsson, B., et al. (eds.) MMM 2022. LNCS, vol. 13142, pp. 487–492. Springer, Cham (2022). https://doi.org/10.1007/978-3-030-98355-0_43
13. Lokoč, J., Mejzlík, F., Souček, T., Dokoupil, P., Peška, L.: Video search with context-aware ranker and relevance feedback. In: Þór Jónsson, B., et al. (eds.) MMM 2022. LNCS, vol. 13142, pp. 505–510. Springer, Cham (2022). https://doi.org/10.1007/978-3-030-98355-0_46
14. Lokoč, J.: Is the reign of interactive search eternal? findings from the video browser showdown 2020. ACM Trans. Multimed. Comput. Commun. Appl. **17**(3), 1–26 (2021)
15. Messina, N., Falchi, F., Esuli, A., Amato, G.: Transformer reasoning network for image-text matching and retrieval. In: 2020 25th International Conference on Pattern Recognition (ICPR), pp. 5222–5229. IEEE (2021)
16. Messina, N., et al.: Aladin: distilling fine-grained alignment scores for efficient image-text matching and retrieval. arXiv preprint arXiv:2207.14757 (2022)
17. Radford, A., Kim, J.W., Xu, T., Brockman, G., McLeavey, C., Sutskever, I.: Robust speech recognition via large-scale weak supervision. Technical report, OpenAI (2022)
18. Revaud, J., Almazan, J., Rezende, R., de Souza, C.: Learning with average precision: Training image retrieval with a listwise loss. In: International Conference on Computer Vision, pp. 5106–5115. IEEE (2019)
19. Rossetto, L., et al.: Interactive video retrieval in the age of deep learning - detailed evaluation of VBS 2019. IEEE Trans. Multimed., 1 (2020)
20. Rossetto, L., Schuldt, H., Awad, G., Butt, A.A.: V3C – a research video collection. In: Kompatsiaris, I., Huet, B., Mezaris, V., Gurrin, C., Cheng, W.-H., Vrochidis, S. (eds.) MMM 2019. LNCS, vol. 11295, pp. 349–360. Springer, Cham (2019). https://doi.org/10.1007/978-3-030-05710-7_29

21. Truong, Q.T., et al.: Marine video kit: a new marine video dataset for content-based analysis and retrieval. In: MultiMedia Modeling - 29th International Conference, MMM 2023, Bergen, Norway, January 9–12, 2023. Springer (2023)
22. Van De Weijer, J., Schmid, C., Verbeek, J., Larlus, D.: Learning color names for real-world applications. IEEE Trans. Image Process. **18**(7), 1512–1523 (2009)
23. Zhang, H., Wang, Y., Dayoub, F., Sunderhauf, N.: VarifocalNet: an IoU-aware dense object detector. In: 2021 IEEE/CVF Conference on Computer Vision and Pattern Recognition (CVPR). IEEE, June 2021

Traceable Asynchronous Workflows in Video Retrieval with vitrivr-VR

Florian Spiess[1], Silvan Heller[1], Luca Rossetto[2], Loris Sauter[1], Philipp Weber[1(✉)], and Heiko Schuldt[1]

[1] Department of Mathematics and Computer Science, University of Basel, Basel, Switzerland
{florian.spiess,silvan.heller,loris.sauter,heiko.schuldt}@unibas.ch, phil.weber@stud.unibas.ch

[2] Department of Informatics, University of Zurich, Zurich, Switzerland
rossetto@ifi.uzh.ch

Abstract. Virtual reality (VR) interfaces allow for entirely new modes of user interaction with systems and interfaces. Much like in physical workspaces, documents, tools, and interfaces can be used, put aside, and used again later. Such asynchronous workflows are a great advantage of virtual environments, as they enable users to perform multiple tasks in an interleaved manner. However, VR interfaces also face new challenges, such as text input without physical keyboards, and the analysis of such asynchronous workflows. In this paper we present the version of vitrivr-VR participating in the Video Browser Showdown (VBS) 2023. We describe the current state of our system, with a focus on improvements in text input methods and logging of asynchronous workflows.

Keywords: Video browser showdown · Virtual reality · Interactive video retrieval · Content-based retrieval

1 Introduction

Multimedia retrieval can be viewed as a mostly linear workload: a user formulates their information need as a query and receives results accordingly. However, in many cases, particularly when the information need is fuzzy and unrefined, users will iterate on their original query, by refining it with insights gained from previous results. It may even occur, that newer results alter the user's understanding of previous results, prompting them to return to a previous query to reevaluate its results. Such asynchronous workflows are particularly useful for analytical workloads, where results from different queries may need to be compared and contrasted. While they provide a great advantage for many use cases, these convoluted workflows are much more challenging to analyze for usage patterns.

Text input is a requirement of many applications, but with the increasing popularity of cross-modal retrieval, it is of particular importance in multimedia retrieval. Despite several existing approaches, fast and reliable text input is still a challenge in virtual reality (VR).

D.-T. Dang-Nguyen et al. (Eds.): MMM 2023, LNCS 13833, pp. 622–627, 2023.
https://doi.org/10.1007/978-3-031-27077-2_49

vitrivr-VR is a VR multimedia retrieval and analytics research system leveraging VR to support efficient text input and enabling asynchronous analytical workflows. To evaluate the effectiveness of its approach, vitrivr-VR participates in multimedia retrieval campaigns, such as the Video Browser Showdown (VBS) [4], in which it has participated multiple times in the past [9,10].

In this paper, we describe the version of vitrivr-VR participating in the VBS 2023, with a focus on asynchronous workflows and our improved methods for text input. Section 2 describes the system and its components, Sect. 3 introduces our novel text input method, Sect. 4 relates our efforts to record and analyze asynchronous interactions, and Sect. 5 concludes.

2 vitrivr-VR

vitrivr-VR is a an experimental VR multimedia retrieval and analytics system. The vitrivr-VR stack consists of three components: the *Cottontail DB* column store [3], the *Cineast* retrieval engine [8], and the *vitrivr-VR*[1] virtual reality user interface. The user interface of vitrivr-VR is developed in Unity with C# and is capable of running on any XR platform supporting OpenXR, but has been designed for and tested on the HTC Vive Pro and Valve Index. To enable multimedia retrieval, vitrivr-VR communicates with Cineast through a RESTful API provided with OpenAPI specifications. Cineast enables a variety of retrieval features, such as OCR and ASR search, as well as color and deep-learning based features. Cottontail DB provides Boolean, text and kNN vector space retrieval through a gRPC connection.

The main focus of vitrivr-VR is to explore and exploit the affordances of VR user interfaces. As such, vitrivr-VR provides a sophisticated yet simple interaction system, on which the three main phases of multimedia analytical workloads are built: *query formulation*, *result set exploration*, and *media item inspection*.

The interaction system of vitrivr-VR aims to be as expressive as possible, while maintaining simplicity and intuitiveness by resembling the interaction with real-world objects. To this end, the system supports two main kinds of interaction: *grabbing* objects and UI elements by utilizing the grab gesture on Index controllers or the grip button on other controllers, and *triggering* objects and UI elements by pressing the controller trigger while hovering over an object or UI element. Through only these two interactions, interface elements can be designed to allow complex interactions. In addition to these two primary interactions, the interaction system also supports pointer-style interactions with canvas-based UI elements. Building on this interaction system, vitrivr-VR provides interfaces for a multitude of input modalities such as text, concept, and pose based query terms. Previous iterations of the VBS have shown that being able to include temporal order in a query is an advantage. For this reason vitrivr-VR allows users to specify the temporal order of query terms by grabbing and dragging small representations of them into the desired order.

[1] https://github.com/vitrivr/vitrivr-vr.

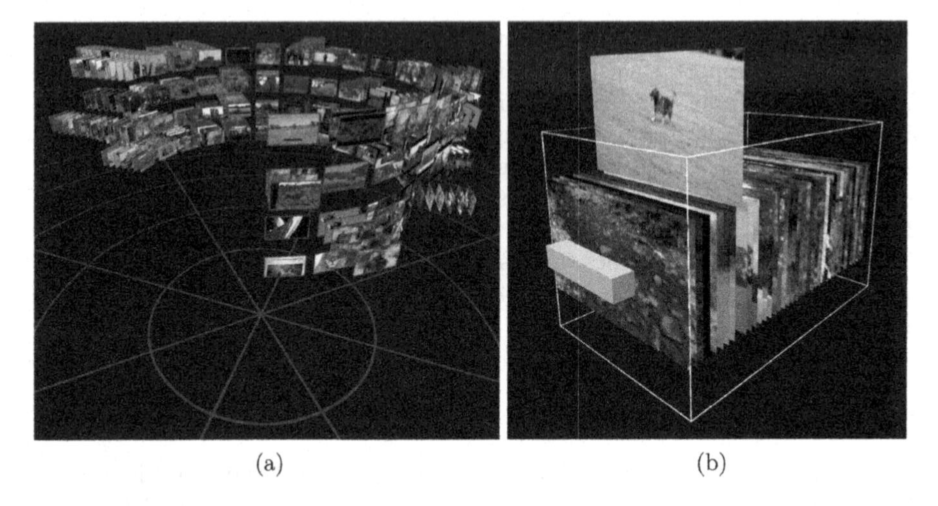

(a) (b)

Fig. 1. Screenshots of the vitrivr-VR user interface; temporal results view (1a) and multimedia drawer segment view (1b).

Once a query has been formulated and results have been returned, they are displayed through a cylindrical results display. Results displays for temporal queries show each result sequence as a stack that can be grabbed and pulled towards the user for easier viewing, as seen in Fig. 1a. The cylindrical results displays can be rotated to reveal further results. Once a video segment of interest has been identified, it can be popped out of the results display and positioned freely and independently in virtual space. This video detail view allows the video associated with the selected clip to be played and explored. To gain an even better overview of the selected video, the detail view allows a multimedia drawer view to be created. This multimedia drawer, which can be seen in Fig. 1b, displays the most representative frame of every video segment of a video in a temporally ordered horizontal stack, similar to files in a drawer. A handle at the front of the drawer allows it to be extended, increasing the gap between frames and allowing them to be viewed more clearly. By moving the controller through the drawer and hovering over a frame, it is moved above the drawer, allowing it to be viewed individually. By selecting a hovered frame, the associated video detail view is skipped to the respective video segment.

3 Text Input in Virtual Reality

With the current trend of multimedia retrieval towards machine learning enabled text-based query methods such as CLIP [7] and our own visual-text co-embedding [9], effective text input methods are becoming a necessity for multimedia retrieval systems. While this does not pose much of a problem for conventional desktop-based multimedia retrieval systems, fast and reliable text input in VR is still a difficult challenge. Many different text input methods have already

Fig. 2. The word-gesture keyboard used in vitrivr-VR, as a word is being input. The red sphere indicates the controller, the red line is the graph drawn by the user and the small display above the keyboard shows the current prediction.

been developed for VR [1,2,11], with none being able to match the speed and precision of physical keyboards [5]. These different methods vary greatly in text input speed and flexibility of their input vocabulary, and as such are differently well suited for different tasks and use-cases. To take advantage of the strengths and overcome the weaknesses of these methods, vitrivr-VR uses a combination.

As a primary text input method, vitrivr-VR uses an on-device speech-to-text approach based on Mozilla DeepSpeech[2] and publicly available as a Unity Package Manager (UPM) package[3]. While speech-to-text methods are unable to match physical keyboards in precision and input flexibility, usually being limited to a predetermined set of words or phonemes, they allow fast, hands-free text input. The two main limitations of speech-to-text methods are the used vocabulary and misinterpretations of speech. Both of these limitations lead to incorrect or missing words in the transcribed text and require manual correction, often on the level of individual characters. Such corrections are usually difficult if not impossible using speech-to-text. For this purpose, vitrivr-VR offers a VR word-gesture keyboard as a secondary text-input method.

Previously, vitrivr-VR used a very simple virtual touch keyboard as secondary text input method. Although this method functioned well to type individual words or make small corrections in VR, the simplicity of the approach resulted in some limitations. The touch activation of the keyboard, while allowing intuitive pressing of the virtual keys without the use of controller buttons, resulted in occasional accidental inputs due to the user moving a controller through it outside of their field of view. This touch activation also caused occasional double inputs, due to the controller passing fully through a key and reactivating it when moving back. Another limitation of this simple approach is the

[2] https://github.com/mozilla/DeepSpeech.

[3] https://github.com/Spiess/deep-speech-upm.

very limited typing speed when writing entire words. To address these limitations we have replaced the simple touch keyboard with our VR word-gesture keyboard implementation shown in Fig. 2. To avoid accidental input, our word-gesture keyboard requires the press of the controller trigger to register a button press. Our implementation is based on SHARK[2] [6] and allows the fast input of individual words through gestures on the keyboard. To input a word, its graph must be traced on the keyboard, starting from the key of the first character and passing through each character of the word in order using straight line paths between characters. Our word-gesture keyboard is available as an open-source UPM package on GitHub[4].

4 Interaction Logging for Asynchronous Workflows

vitrivr-VR enables users to jump back to results displays of previous queries and to keep selected video detail views in the virtual space as potential result candidates, or for future reference. These features, combined with the ability to place results in a corner of virtual space and only come back to them after doing something unrelated, enable highly asynchronous workflows and facilitate analytical usage through the comparison of query result sets and items.

While these asynchronous workflows are in many regards a strength of vitrivr-VR, they make analysis of interaction data much more difficult. The VBS has often encouraged, and sometimes required, that in addition to task solution submissions participating systems log interaction data and raw query results. While this is enough data to allow submitted results to be traced back to their origin query for most conventional systems that only support linear workflows, doing the same in vitrivr-VR is often very difficult, and in certain cases impossible. By exploring beyond the exact returned set of results through interaction methods like the multimedia drawer, and keeping them in the virtual space over the course of several queries, it quickly becomes ambiguous which query and interactions led to any specific submission.

To capture the flow of data, we improved the interaction logging of vitrivr-VR with identifying information allowing the path of interactions resulting in a submission to be uniquely identified. Through the improved data logging, we are not only able to determine in greater detail which query and interactions led to a submission, but also allows us to analyze the different paths that can lead to a submission.

5 Conclusion

In this paper we present the vitrivr-VR system, in the state in which we intend to participate in the VBS 2023. We describe the system with a focus on changes to text input and logging of asynchronous interactive workflows.

[4] https://github.com/Philipp1202/WGKeyboard.

Acknowledgements. This work was partly supported by the Swiss National Science Foundation through projects "Participatory Knowledge Practices in Analog and Digital Image Archives" (contract no. 193788) and "MediaGraph" (contract no. 202125).

References

1. Boletsis, C., Kongsvik, S.: Controller-based text-input techniques for virtual reality: an empirical comparison. Int. J. Virtual Reality **19**(3), 2–15 (2019). https://doi.org/10.20870/IJVR.2019.19.3.2917
2. Chen, S., Wang, J., Guerra, S., Mittal, N., Prakkamakul, S.: Exploring word-gesture text entry techniques in virtual reality. In: CHI Conference on Human Factors in Computing Systems. ACM (2019). https://doi.org/10.1145/3290607.3312762
3. Gasser, R., Rossetto, L., Heller, S., Schuldt, H.: Cottontail DB: an open source database system for multimedia retrieval and analysis. In: International Conference on Multimedia, pp. 4465–4468. Association for Computing Machinery, New York, NY, USA (2020). https://doi.org/10.1145/3394171.3414538
4. Heller, S., et al.: Interactive video retrieval evaluation at a distance: Comparing sixteen interactive video search systems in a remote setting at the 10th video browser showdown. Int. J. Multimedia Inf. Retrieval **11**(1), 1–18 (2022). https://doi.org/10.1007/s13735-021-00225-2
5. Kim, J.R., Tan, H.Z.: A study of touch typing performance with keyclick feedback. In: IEEE Haptics Symposium, pp. 227–233 (2014). https://doi.org/10.1109/HAPTICS.2014.6775459
6. Kristensson, P.O., Zhai, S.: SHARK^2: a large vocabulary shorthand writing system for pen-based computers. In: Symposium on User Interface Software and Technology, pp. 43–52. ACM (2004). https://doi.org/10.1145/1029632.1029640
7. Radford, A., et al.: Learning transferable visual models from natural language supervision. arXiv (2021). https://doi.org/10.48550/ARXIV.2103.00020
8. Rossetto, L.: Multi-modal video retrieval, Ph. D. thesis, University of Basel (2018). https://doi.org/10.5451/unibas-006859522
9. Spiess, F., et al.: Multi-modal video retrieval in virtual reality with vitrivr-VR. In: Þór Jónsson, B., et al. (eds.) MMM 2022. LNCS, vol. 13142, pp. 499–504. Springer, Cham (2022). https://doi.org/10.1007/978-3-030-98355-0_45
10. Spiess, F., Gasser, R., Heller, S., Rossetto, L., Sauter, L., Schuldt, H.: Competitive interactive video retrieval in virtual reality with vitrivr-VR. In: Lokoč, J., et al. (eds.) MMM 2021. LNCS, vol. 12573, pp. 441–447. Springer, Cham (2021). https://doi.org/10.1007/978-3-030-67835-7_42
11. Yu, C., Gu, Y., Yang, Z., Yi, X., Luo, H., Shi, Y.: Tap, dwell or gesture? Exploring head-based text entry techniques for HMDs. In: CHI Conference on Human Factors in Computing Systems, pp. 4479–4488. ACM (2017). https://doi.org/10.1145/3025453.3025964

Video Search with CLIP and Interactive Text Query Reformulation

Jakub Lokoč, Zuzana Vopálková, Patrik Dokoupil, and Ladislav Peška(✉)

SIRET Research Group, Department of Software Engineering, Faculty of Mathematics and Physics, Charles University, Prague, Czech Republic
{jakub.lokoc,ladislav.peska}@matfyz.cuni.cz

Abstract. Nowadays, deep learning based models like CLIP allow simple design of cross-modal video search systems that are able to solve many tasks considered as highly challenging several years ago. In this paper, we analyze a CLIP based search approach that focuses on situations, where users cannot find proper text queries to describe searched video segments. The approach relies on suggestions of classes for displayed intermediate result sets and thus allows users to realize missing words and ideas to describe video frames. This approach is supported with a preliminary study showing potential of the method. Based on the results, we extend a respected known-item search system for the Video Browser Showdown, where more challenging visual known-item search tasks are planned.

Keywords: Interactive video retrieval · Deep features · Text-image retrieval

1 Introduction

The Video Browser Showdown competition (VBS) [6,9] represents a respected international event, where different teams try to interactively solve challenging tasks over a non-trivial video collection (currently V3C [11] and DRES server [10] are used). The competition focuses on known-item search and ad-hoc search tasks [4,5] that cannot be automatically solved with a given single text query (at least yet). The last several years have shown that the key to success is to design interactive search interfaces on top of the most recent deep learning models. For example, the top three known-item search systems at VBS 2022 [1–3] used CLIP [8] based models allowing effective free-form text search and a result set inspection component providing additional querying options.

The VBS 2023 event introduces several changes. First, visual known item search target segments can be shorter and thus the teams will have less options to select highly specific objects for querying. Indeed, with CLIP and temporal query fusion some teams reached impressive times to solve visual KIS tasks at VBS 2022. Second, a domain specific collection will be used for search tasks within a homogeneous cluster. Specifically, a marine videos collection [12] will

D.-T. Dang-Nguyen et al. (Eds.): MMM 2023, LNCS 13833, pp. 628–633, 2023.
https://doi.org/10.1007/978-3-031-27077-2_50

be used where generally known keywords like fish, coral reef, or swimming cannot sufficiently filter the result set.

In this paper, we present an approach tackling with the newly introduced challenges, yet still keeping the system simple for users. Let's emphasize that shorter visual KIS target segments and domain specific videos increase demands on text query specification. Hence, we focus on interactive query reformulation with a system providing suggestions of class labels. This approach already proved to be promising [7] with "classical" deep network based classification approaches like GoogleNet. For VBS 2023, we plan to test this approach using a CLIP model that represents state-of-the-art in joint embedding as well as zero shot classification. Our assumption is that classes proposed by CLIP zero-shot classification could help to enhance text queries for CLIP.

2 Interactive Text Query Reformulation with CLIP

Text query reformulation is a standard search strategy, where users change text query if the result set does not show promising items. The users rely on intuition, concept selection as well as language skills to select a new item or describe the same item in different words. However, there might be situations where users cannot find proper words. Hence, computer aided suggestions may represent an important feedback for users trying to describe searched items. At the same time, users can better understand (and learn) how the network links visual data with text. The basic idea is to show enhanced results of the initial text query, where for each displayed image the system provides also keywords assigned to the image. The system can show also an aggregation of the most frequent keywords assigned to top-k images in the result set. Please note that this approach is an alternative to kNN based browsing. Instead of using image's embedding, only the selected entity label(s) are taken from the example image.

To provide preliminary quantitative performance estimates, we performed a study (with four users) that revealed potential of this approach. First, in order to obtain keywords for each database image (i.e., selected video frame), the CLIP model was used as a zero-shot classifier for a list of preselected classes (different sets for V3C and marine datasets). Based on the classification results, top classes were stored for each image. Then, the study was performed with a simple web-based tool, where users see a randomly selected target known image (i.e., search objective, information need instance) and can write a text query to find it. The result set of each query is enhanced with suggested keywords (classes) obtained in the offline zero-shot classification step. This enhancement is illustrated in Fig. 1 for the marine video dataset. Users can click on the suggested keywords and thus interactively update the text query (manual corrections are allowed as well, see Table 1). Once the query is updated, a new result set can be obtained. The system tracks the position of the searched target image and logs its position/rank for each query. For the purposes of the study, only one iteration of reformulations was allowed and the set of selected target known images was limited to images with fish or coral reefs. The study used image dataset with 22K extracted frames.

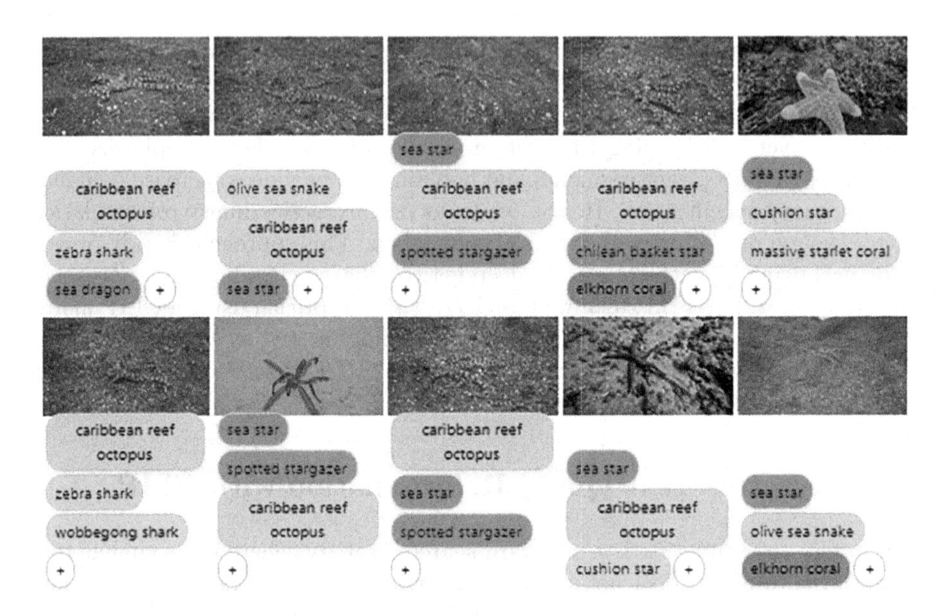

Fig. 1. Illustration of an enhanced result set with classes from zero-shot CLIP classification. Colors indicate frequency of classes (in top-10 classes for each image), where the green color indicates that the class is rare. (Color figure online)

Table 1. Two examples, where interactive text query reformulation significantly helped to improve the rank of the searched target image. We may observe a shift from general queries to expert-like descriptions.

Text query before and after reformulation	Rank
Corals and small black fish	1371
Smooth mounding coral, threespot damselfish, nassau grouper	18
Three rays, sand, coral	3965
Three rays, sand, coral, blue spotted ribbontail ray, porcupine ray	4

Based on the logged position/rank of the target frame in each iteration, Fig. 2 shows the effect of text search and query reformulation for more challenging tasks. We may observe that some targets were easy to find already with the first query, while on average there is a clear improvement in the first reformulation step in searches with two iterations. In some cases the reformulation can be contra-productive (see some negative values in boxplot RI1–RI2). Nevertheless, suggestions often help users to improve the text query for the same target and also gradually "teach" users the important domain knowledge (i.e., correspondence between visual patterns and class labels). Please note that it is not guaranteed that CLIP assigns correct class labels, especially in domain specific collections. However, as long as CLIP assigns wrong labels (e.g., fish species) consistently to visual features, their usage for searching in the join space might still be effective.

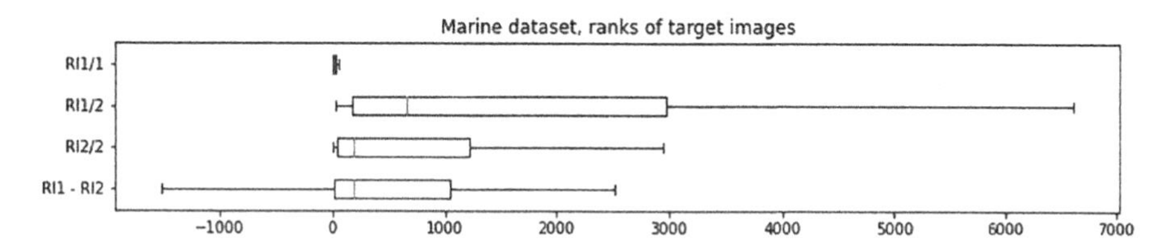

Fig. 2. Ranks of searched target images for 66 searches with just one iteration RI1/1 and 122 searches with two iterations RI1/2, RI2/2. The difference between ranks of the first and second iteration (in searches with two iterations) is displayed as RI1–RI2.

3 VBS Participation

The aforementioned method is general enough to be integrated to any known-item search interactive video search system supporting CLIP based search. This year, we plan to continue with CVHunter [3] that implements standard useful browsing features (result set scrolling, video inspection, video preview, etc.) on top of several CLIP-based ranking approaches. Since the presented scenes from V3C will be shorter and marine video dataset contains single-shot videos, we do not plan to integrate the general context-aware ranker[1] (to keep minimalist design). However, a special case of context-aware ranking, temporal queries, could still be useful in cases the short segment contains two distinct scenes. In other words, CVHunter will still allow to consider a context of searched frames. The tool will also keep Bayesian relevance feedback, including its temporal variant.

The new class suggestion feature will be integrated into the system in two ways. First, a summary of suggested class labels from top results of a query can be presented in an available text panel (e.g., application form caption). Second, top class labels of each image can be presented below the image in an info panel, allowing simple text utilization using mouse click. Furthermore, all available class labels for an image can be shown when a detail window is opened.

[1] According to our VBS 2022 log analysis, the general context-aware ranker was almost not used at all anyway, users preferred temporal queries.

4 Conclusions

We analyzed an interactive search approach for VBS allowing users to select promising class labels from provided suggestions. According to the performed study, the approach provides a promising performance in situations where users find it more difficult to find proper text queries. This approach is integrated to the system designed for the Video Browser Showdown competition, providing another interesting interactive search option.

Acknowledgment. This work has been supported by Charles University grant SVV-260588.

References

1. Amato, G., et al.: VISIONE at video browser showdown 2022. In: Þór Jónsson, B., et al. (eds.) MMM 2022. LNCS, vol. 13142, pp. 543–548. Springer, Cham (2022). https://doi.org/10.1007/978-3-030-98355-0_52
2. Hezel, N., Schall, K., Jung, K., Barthel, K.U.: Efficient search and browsing of large-scale video collections with Vibro. In: Þór Jónsson, B., et al. (eds.) MMM 2022. LNCS, vol. 13142, pp. 487–492. Springer, Cham (2022). https://doi.org/10.1007/978-3-030-98355-0_43
3. Lokoč, J., Mejzlík, F., Souček, T., Dokoupil, P., Peška, L.: Video search with context-aware ranker and relevance feedback. In: Þór Jónsson, B., et al. (eds.) MMM 2022. LNCS, vol. 13142, pp. 505–510. Springer, Cham (2022). https://doi.org/10.1007/978-3-030-98355-0_46
4. Lokoč, J., et al.: A task category space for user-centric comparative multimedia search evaluations. In: Þór Jónsson, B., et al. (eds.) MMM 2022. LNCS, vol. 13141, pp. 193–204. Springer, Cham (2022). https://doi.org/10.1007/978-3-030-98358-1_16
5. Lokoč, J., Bailer, W., Schoeffmann, K., Münzer, B., Awad, G.: On influential trends in interactive video retrieval: video browser showdown 2015–2017. IEEE Trans. Multimedia **20**(12), 3361–3376 (2018). https://doi.org/10.1109/TMM.2018.2830110
6. Lokoč, J., et al.: Is the reign of interactive search eternal? findings from the video browser showdown 2020. ACM Trans. Multimedia Comput. Commun. Appl. (TOMM) **17**(3) (2021). https://doi.org/10.1145/3445031
7. Peska, L., Mejzlík, F., Soucek, T., Lokoc, J.: Towards evaluating and simulating keyword queries for development of interactive known-item search systems. In: Gurrin, C., Jónsson, B.Þ., Kando, N., Schöffmann, K., Chen, Y.P., O'Connor, N.E. (eds.) Proceedings of the 2020 on International Conference on Multimedia Retrieval, ICMR 2020, Dublin, Ireland, 8–11 June 2020, pp. 281–285. ACM (2020). https://doi.org/10.1145/3372278.3390726
8. Radford, A., et al.: Learning transferable visual models from natural language supervision. CoRR abs/2103.00020 (2021). https://arxiv.org/abs/2103.00020
9. Rossetto, L., et al.: Interactive video retrieval in the age of deep learning-detailed evaluation of VBS 2019. IEEE Trans. Multimedia **23**, 243–256 (2020). https://doi.org/10.1109/TMM.2020.2980944

10. Rossetto, L., Gasser, R., Sauter, L., Bernstein, A., Schuldt, H.: A system for interactive multimedia retrieval evaluations. In: Lokoč, J., et al. (eds.) MMM 2021. LNCS, vol. 12573, pp. 385–390. Springer, Cham (2021). https://doi.org/10.1007/978-3-030-67835-7_33
11. Rossetto, L., Schuldt, H., Awad, G., Butt, A.A.: V3C – a research video collection. In: Kompatsiaris, I., Huet, B., Mezaris, V., Gurrin, C., Cheng, W.-H., Vrochidis, S. (eds.) MMM 2019. LNCS, vol. 11295, pp. 349–360. Springer, Cham (2019). https://doi.org/10.1007/978-3-030-05710-7_29
12. Truong, Q.T., et al.: Marine video kit: a new marine video dataset for content-based analysis and retrieval. In: Dang-Nguyen, D., et al. (eds.) MMM 2023. LNCS, vol. 13833, pp. xx–yy. Springer, Cham (2023)

Perfect Match in Video Retrieval

Sebastian Lubos[1(✉)], Massimiliano Rubino[2], Christian Tautschnig[2], Markus Tautschnig[2], Boda Wen[2], Klaus Schoeffmann[3], and Alexander Felfernig[1]

[1] Graz University of Technology, Graz, Austria
slubos@ist.tugraz.at
[2] Streamdiver GmbH, Klagenfurt am Wörthersee, Austria
[3] Klagenfurt University, Klagenfurt am Wörthersee, Austria

Abstract. This paper presents the first version of our video search system *Perfect Match* for the Video Browser Showdown 2023 competition. The system indexes videos from the large V3C video dataset and derives visual content descriptors automatically. Furthermore, it provides an interactive video search *user interface (UI)*, which implements approaches from the domain of *critiquing-based* recommendation, to enable the user to find the desired video segment as fast as possible.

Keywords: Video retrieval · Interactive video search · Critiquing

1 Introduction

Determining which video fits the requirements is becoming an increasingly complex task with the growing amount of available videos. Hence, efficient retrieval of videos from large catalog is an important challenge.

The *Video Browser Showdown (VBS)* is an annual competition that encourages researchers to develop innovative approaches to tackle this challenge. During the event, two types of tasks are defined [9]. Firstly, during the *Ad-hoc Video search (AVS)* task, a textual description of video content is given, with the aim to find as many scenes as possible, in a limited time interval, that match the description. Secondly, in the *Known-item search (KIS)* task, the requested video segment is known and participants need to find it as fast as possible.

For the VBS 2023, two datasets, *V3C1* and *V3C2*, from the *Vimeo Creative Commons Collection* are used [11]. They comprise 17235 videos with a total duration of around 2300h. Additionally, a special marine video dataset with underwater and scuba diving videos will be used in an additional challenge [13].

In order to enable searching inside this large video collection, we developed a system that automatically indexes the videos from the dataset and infers visual content descriptors from the video key frames. Our *Perfect Match* video search application receives user input in an interactive fashion via the UI, and uses it to compose queries for the indexed videos, including the inferred metadata, in order to find the perfect match, i.e., the video segment that solves the task. In

D.-T. Dang-Nguyen et al. (Eds.): MMM 2023, LNCS 13833, pp. 634–639, 2023.
https://doi.org/10.1007/978-3-031-27077-2_51

our application, requirements are collected in an interactive manner by adopting *critiquing-based* approaches, which form a special type of *knowledge-based recommender systems* [5], rather than posting a single search query.

The remainder of this paper is organized as follows. First we give an introduction to critiquing-based recommendation in Sect. 2. Thereafter, the system is presented in detail in Sect. 3, followed by the conclusion in Sect. 4.

2 Critiquing

Knowledge-Based Recommender Systems form a special type of recommenders which suggest items based on explicit knowledge about the item catalog and user preferences. In contrast to other recommender systems based on content or past user interactions, preferences are not acquired beforehand but rather determined in a conversational recommendation process [5].

Critiquing-Based Recommendations [3,4] form a subtype of knowledge-based recommender systems, that use different similarity measures to retrieve items which are most similar to given user requirements. A characteristic of those systems is that, rather than specifying all requirements beforehand in a query-like way, requirements are specified in a navigation-based fashion. The navigation (also browsing-based approach) supports situations in which users might not know beforehand exactly what they are searching for. They are guided through the process in order to improve the process of finding the item they need.

Typically, a user specifies some raw initial requirements and receives first suggestions of items that fulfill the requirements. The user reviews the suggested item and specifies an additional requirement that is not fulfilled yet, i.e., critique. In the following, the system suggests a new alternative the user should review. These steps are repeated until the user is satisfied with the recommendation, or no further alternatives are available.

The integration of those critiquing-based aspects for the purpose of video search is explained in the following.

3 Perfect Match - System Overview

Our *Perfect Match* system consists of two processing stages with different purposes. Firstly, an offline phase is used to index the video segments of the V3C dataset and automatically extract the video content metadata. Secondly, in the online phase, user search queries are composed and submitted to the API, which ranks the frames based on the query and indexed video metadata.

Separating the processing in two phases was a necessary design decision to reach decent performance when using the UI for searching in the online phase. Video indexing and metadata extraction is a very hardware-intensive and time-consuming task. Performing it on-demand in the online phase would not be feasible. Hence, the online phase solely exploits the results of this processes by querying the persisted metadata, which is fast enough for a good user experience. In the following sections, the details of the phases are further explained.

3.1 Video Indexing

The first step in the video indexing part is to create a database reference for each video in the V3C dataset. A unique id is assigned to each video. Then, *FFmpeg*[1] is used to extract and index selected frames of the videos with a unique id.

Due to performance and redundancy reasons, it is not practicable to extract the visual content descriptors for every frame in the videos. Therefore, we decided to sample frames from each video in a regular interval, and use those for analysis. Each of the frames is analyzed for visual content, and printed texts. The results of this analysis are stored in the database. For each frame, we store the information which entities have been recognized, how often they have been recognized, and which confidence for recognition was provided by the models.

For concept detection *AlexNet* [7] is used, which is a pretrained *Convolutional-Neural-Network (CNN)*, to identify the content of frames. It is trained on the *ILSVRC2010* dataset [12]. This model classifies images with 1000 different classes describing their content. Furthermore, we apply the object detection algorithm *YOLOv5* [6] to identify objects inside the frames. It is trained to classify 80 different objects of the *COCO* dataset [8].

Additionally, *CLIP* [10], a neural network trained to learn visual concepts with natural language supervision is used. It is used to predict the likeliest caption of a fiven input image. We combine it with the *Food101* [2], and *SUN* [14] datasets, which enable classification of recognized food, and categorization of scenes respectively.

Finally, *optical character recognition (OCR)* is applied for recognition of printed textual characters in images.

The combination of those visual content descriptors is then used to query the database in order to find the desired video frame.

3.2 Search Queries and Ranking

The retrieval of matching frames is split into a query and ranking phase. In the first phase, candidate frames that match the requirements specified by the user are identified. In the second phase, those candidates are ranked by a score that determines how good they match the requirements.

Our application offers the possibility to filter the video catalog by specifying concepts, objects, and OCR values that should be included in the desired video segment. For those selected values, the user can further refine them by expressing critiques concerning the importance of single values on a 3-point-scale with the values "less important", "neutral", or "very important". This is, for example, useful if a specific object is predominantly in the searched video segment. Those weightings will influence the score in the rating phase.

Additionally, in the interactive process, users can criticize the currently shown proposal with three different options. Firstly, the user can state that the current frame does not belong to the searched video segment, which will exclude it

[1] https://ffmpeg.org.

from the candidates. Secondly, the user can mention that the video the current frame is derived from does not contain the searched frame. This will reduce the candidates by all further frames of this video. Thirdly, in case the current frame is not part of the searched video segment but of the same video the searched segment is part of, the user can fix the requirement to further only consider frames of the same video.

A database query to retrieve the candidate set is compiled from the specified collection of Boolean filtering options of included content descriptors and excluded or fixed videos. The retrieved candidate set includes the amount of occurrences of values inside the frames.

The ranking idea is based on the assumption that frames with more instances of a specified metadata value, are more probable to be the correct ones. Therefore, for each frame in the candidate set, the score is calculated by the sum of the weighted amount of occurrences of specified metadata values. The weight depends on whether the user has expressed weighting critiques for some values.

As an extension, we apply *Term-Frequency Inverse-Document-Frequency (TF-IDF)* [1] on the occurrences to normalize the results before including them in the score calculation. Applying this technique allows reduction of importance of content descriptors that appear in many frames and emphasizes descriptors that are specific for a small amount of frames. The frame with the highest score is returned as suggestion.

3.3 User Interface

The core concept of our *Perfect Match* application UI is inspired by the workflow provided in dating apps like *Tinder*[2]. Those apps suggest a user profile represented by a profile picture to the user, which is able to fastly decide if the suggestion is interesting for him/her by swiping left or right.

Since fast interaction is as well relevant for the VBS competition, we adopted this way of interaction for the video search task. As shown in Fig. 1, the current frame proposal is presented prominently on the left side of the screen. The user has the possibility to use the buttons below to decide what should happen with the frame. If the frame is a match (with the current search task), it can be submitted. If the frame is no perfect match but the user expects the correct video section to be from the same video, they can indicate this. On the other hand, if the frame is not matching the request at all, the user can either decide to skip the current frame, or all further frames of this video. Latter is useful if they expects that the searched video section belongs to another video.

In order to specify the requirements of the current search task, the right side of the UI is used to define criteria for the query. The user can select concepts and objects from all possible options using the drop-down-boxes with auto-completion, as well as specifying OCR content with a free text input.

After choosing the requirements, the user can further refine them by assigning weights. The search query is updated every time the user changes the require-

[2] https://tinder.com.

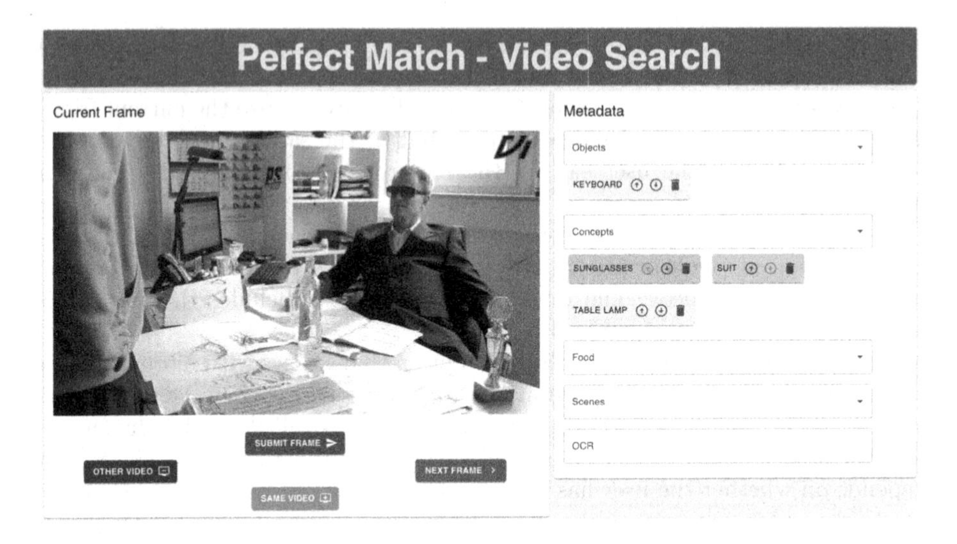

Fig. 1. Screenshot of the user interface.

ments or values, thus every interaction can potentially lead to a new highest ranked suggestion shown on the left side of the UI.

4 Conclusion

In this paper, we introduced the first version of the *Perfect Match* video search application that can be used to search for videos in the V3C dataset and submit solutions for the tasks of the VBS competition. The system indexes the videos and automatically infers visual content descriptors of selected frames. Further, it allows the user to interactively refine the search requirements by expressing critiques in order to find the searched video segment as fast as possible.

Acknowledgements. The presented work has been developed within the research project STREAMDIVER which is funded by the Austrian Research Promotion Agency (FFG) under the project number 886205.

References

1. Aizawa, A.: An information-theoretic perspective of TF-IDF measures. Inf. Process. Manag. **39**(1), 45–65 (2003)
2. Bossard, L., Guillaumin, M., Van Gool, L.: Food-101 – mining discriminative components with random forests. In: Fleet, D., Pajdla, T., Schiele, B., Tuytelaars, T. (eds.) ECCV 2014. LNCS, vol. 8694, pp. 446–461. Springer, Cham (2014). https://doi.org/10.1007/978-3-319-10599-4_29
3. Chen, L., Pu, P.: Interaction design guidelines on critiquing-based recommender systems. User Model. User-Adap. Inter. **19**(3), 167–206 (2009)

4. Chen, L., Pu, P.: Critiquing-based recommenders: survey and emerging trends. User Model. User-Adap. Inter. **22**(1), 125–150 (2012)
5. Jannach, D., Zanker, M., Felfernig, A., Friedrich, G.: Knowledge-Based Recommendation, pp. 81–123. Cambridge University Press, Cambridge (2010). https://doi.org/10.1017/CBO9780511763113.006
6. Jocher, G., et al.: ultralytics/yolov5: v6.2 - YOLOv5 classification models, Apple M1, reproducibility, ClearML and Deci.ai integrations, August 2022. https://doi.org/10.5281/zenodo.7002879
7. Krizhevsky, A., Sutskever, I., Hinton, G.E.: Imagenet classification with deep convolutional neural networks. In: Proceedings of the 25th International Conference on Neural Information Processing Systems. NIPS 2012, vol. 1, pp. 1097–1105. Curran Associates Inc., Red Hook (2012)
8. Lin, T.Y., et al.: Microsoft COCO: common objects in context (2014). https://doi.org/10.48550/ARXIV.1405.0312, https://arxiv.org/abs/1405.0312
9. Lokoč, J., et al.: Is the reign of interactive search eternal? Findings from the video browser showdown 2020. ACM Trans. Multimedia Comput. Commun. Appl. **17**(3) (2021). https://doi.org/10.1145/3445031
10. Radford, A., et al.: Learning transferable visual models from natural language supervision. CoRR abs/2103.00020 (2021). https://arxiv.org/abs/2103.00020
11. Rossetto, L., Schuldt, H., Awad, G., Butt, A.A.: V3C – a research video collection. In: Kompatsiaris, I., Huet, B., Mezaris, V., Gurrin, C., Cheng, W.-H., Vrochidis, S. (eds.) MMM 2019. LNCS, vol. 11295, pp. 349–360. Springer, Cham (2019). https://doi.org/10.1007/978-3-030-05710-7_29
12. Russakovsky, O., et al.: ImageNet large scale visual recognition challenge. Int. J. Comput. Vision **115**(3), 211–252 (2015). https://doi.org/10.1007/s11263-015-0816-y
13. Truong, Q.T., et al.: Marine video kit: a new marine video dataset for content-based analysis and retrieval. In: Dang-Nguyen, D., et al. (eds.) MMM 2023. LNCS, vol. 13833, pp. 539–550. Springer, Cham (2023)
14. Xiao, J., Hays, J., Ehinger, K.A., Oliva, A., Torralba, A.: Sun database: large-scale scene recognition from abbey to zoo. In: 2010 IEEE Computer Society Conference on Computer Vision and Pattern Recognition, pp. 3485–3492 (2010). https://doi.org/10.1109/CVPR.2010.5539970

QIVISE: A Quantum-Inspired Interactive Video Search Engine in VBS2023

Weixi Song[1,2,3], Jiangshan He[1,2,3], Xinghan Li[1,2,3], Shiwei Feng[1,2,3], and Chao Liang[1,2,3(✉)]

[1] National Engineering Research Center for Multimedia Software (NERCMS), Wuhan, China
[2] Hubei Key Laboratory of Multimedia and Network Communication Engineering, Wuhan, China
[3] School of Computer Science, Wuhan University, Wuhan, China
{song.wx,riverhill,xinghan.li,waii2022,cliang}@whu.edu.cn

Abstract. In this paper, we present a quantum-inspired interactive video search engine (QIVISE), which will be tested in VBS2023. QIVISE aims at assisting the user in dealing with Known-Item Search and Ad-hoc Video Search tasks with high efficiency and accuracy. QIVISE is based on a text-image encoder to achieve multi-modal embedding and introduces multiple interaction possibilities, including a novel quantum-inspired interaction on paradigm, label search, and multi-modal search to refine the retrieval results via user's interaction and feedback.

Keywords: Video browser showdown · Interactive video retrieval · Quantum theory

1 Introduction

The Video Browser Showdown [6] is an annual international video search competition held by the International Conference on MultiMedia Modeling (MMM), which aims at evaluating the state-of-the-art video retrieval system and promoting the efficiency of large-scale video search. In VBS2023, the participants are required to solve Known-Item Search (KIS) and Ad-hoc Video Search (AVS) tasks, where the queries are presented as visual clips or textual descriptions. The query can also be extended and formulated by the user via the interaction after the initial search.

In this paper, we present a Quantum-inspired Interactive Video Search Engine (QIVISE). QIVISE is a system that inherits the latest and advanced methods in video retrieval research and previous outstanding systems in VBS as well as introduces an emerging quantum-inspired interaction paradigm. In brief, the contributions of QIVISE are: (1) Motivated by the concepts and keywords search of an existing VBS system [7], QIVISE introduces an interactive label

W. Song, J. He and X. Li—These authors contribute equally to this work.

D.-T. Dang-Nguyen et al. (Eds.): MMM 2023, LNCS 13833, pp. 640–645, 2023.
https://doi.org/10.1007/978-3-031-27077-2_52

search, which allows user to choose the labels extracted from the textual query input and add extra labels to emphasize the key information. (2) QIVISE adopts a quantum-inspired interaction paradigm to model the user's feedback and refine the retrieval results. (3) QIVISE supports video search by muti-modality by which user's can fully use different modal information to emphasize the key information.

2 Interface and Navigation

Fig. 1. Current user interface of QIVISE

An overview of the current interface is shown in Fig. 1.

With regard to human-computer interaction and reference to the existing system like Vibro [3], the demonstrating interface can effectively handle both AVS and KIS tasks. Users can search the targeted video clips by typing in textual descriptions or uploading a video clip. In addition, the system supports two explicit interaction modes. One is that users can add subjective labels for query input, which is introduced in detail in Sect. 3.1. And the other is that users can provide feedback based on query results to improve the final recall rate.

Part 1, refers to labels in Fig 1, is reserved for the customization of the query. A detailed description of all available methods can be found in Sect. 3.

In Part 2, according to the user's query requirements, the system obtains the similarity scores of key frames corresponding to the videos in the database through the vector retrieval engine [11], and filters out 4000 results in descending order. This process takes a few microseconds. To improve searching efficiency, we provide users with a preview in the form of 49 pictures per group in a limited area and by using Masonry Layouts, users can quickly browse all possible results and view detailed information by clicking the picture.

In Part 3, the system introduces a clustering algorithm [2] to select 10 possible most relevant key frames so as to avoid user's long-term selection and comparison. The user evaluates the relevance of demonstrated clips and provides feedback by clicking the favoured or not-favoured button beside the chosen video clip. If user is not satisfied with all the presented results, they can get a new-round results through the refresh button. After the user completes the feedback and click the re-rank button, the system provides an improved secondary sorting result based on the quantum-inspired re-ranking algorithm, which is introduced in detail in Sect. 3.2.

3 Multiple Ways of User Interactions

The system adopts BLIP [5], a Bootstrapping Language-Image Pre-training model, for the text and image embedding [4]. With the initial retrieval results obtained by comparing the cosine similarity between query and image embedding vectors, our system adopts several methods to refine the first-round retrieval results via user interaction, basically including the label search, quantum-inspired re-ranking and multi-modal search.

3.1 Label Search

Compared with a single text query, the labels that are extracted from the query text and chosen by the user can be more accurate in describing the user's searching demand. Our system adopts the latest method [10], an end-to-end framework for extracting labels, or called key phrases, from query sentences. The model extracts the present key phrases from the text and also the ones which is implicit in the text. It makes it possible for the user to catch the missing but useful information.

With present and implicit key phrases extracted from the query, user selects the labels that match their searching demands and ranks them according to their relevance. Besides user can also add the labels that contain the information user wants to emphasize. By embedding the labels into the common space and calculating the cosine similarity between labels and video clips, the system gets an additional relevance score to modify and improve the initial score for accurate rank.

3.2 Quantum-Inspired Interaction for Clips Re-ranking

Following the great leap in quantum-inspired information retrieval [9], our system introduces a novel quantum-inspired interaction paradigm to model the user's interaction.

In quantum theory, the quantum state of the polarization is represented by the linear combination of vertical and horizontal polarization which is denoted as the follow formula:

$$|\varphi\rangle = \alpha|1\rangle + \beta|0\rangle \tag{1}$$

where $|\varphi\rangle$ is the quantum state of polarization, $|1\rangle$ and $|0\rangle$ are the vertical and horizontal polarization, α and β are the coefficients that subject to $|\alpha|^2+|\beta|^2 = 1$.

In quantum-inspired approaches to information retrieval, a document is represented in terms of relevance and non-relevance of a query [8]:

$$|d\rangle = \alpha|\mathrm{r}\rangle + \beta|\neg\mathrm{r}\rangle \tag{2}$$

We have $|\alpha|^2 = p(d|q)$ and $|\beta|^2 = 1 - p(d|q)$, where $p(d|q)$ is the relevance probability in the first-round retrieval, it can be determined by the normalized ranking score.

Then we can get an estimation about $|\mathrm{r}\rangle$ with the equation that $\langle r|d\rangle = \alpha$:

$$|r\rangle = \alpha|\mathrm{d}\rangle + \beta|\neg\mathrm{d}\rangle \tag{3}$$

With the above formulas, we now introduce our strategy for re-ranking. After first-round retrieval, the user can choose the shown video clips that are highly or not consistent with user's demands. With those chosen clips, we can get an estimation about the user's true demands by applying (3). The user's true demands can be regarded as the space spanned by the chosen clip vector and its complement subspace, as shown in Fig. 2. To simplify the calculation and accelerate the interaction procedure, the complement subspace is confirmed by the normalized sum of the basic solution sets of Eq. (4):

$$y_1x_1^i + y_2x_2^i + \cdots + y_{n-1}x_{n-1}^i + y_nx_n^i = 0 \tag{4}$$

The total formulas are as follows:

$$\begin{aligned}
|r\rangle &= \alpha|\mathrm{d}\rangle + \beta|\neg\mathrm{d}\rangle \\
|d\rangle &= (y_1, y_2, ..., y_{n-1}, y_n) \\
|\neg\mathrm{d}\rangle &= \frac{\sum_{i=1}^{n-1}|x^i\rangle}{\|\sum_{i=1}^{n-1}|x^i\rangle\|} \\
|x^i\rangle &= (x_1^i, x_2^i, ..., x_{n-1}^i, x_n^i)
\end{aligned} \tag{5}$$

where $|r\rangle$ is representing the user's demand, n is the dimension of embedding, $|d\rangle$ is representing the chosen video clip, $|x^i\rangle$ is the basis vector of Eq. (4), $|\neg\mathrm{d}\rangle$ represents the complement subspace and α, β are the relevance probabilities in the previous-round retrieval.

With $|\mathrm{r}\rangle$, the final score for re-ranking is defined as follows:

$$Score\,(v_{eva}) = cosine(q, v_{eva}) + \sum cosine\,(r_f^i, v_{eva}) - \sum cosine\,(r_{nf}^i, v_{eva}) \tag{6}$$

where v_{eva} is the video clip to be evaluated in the next-round, $cosine(\cdot)$ is cosine similarity, q is the query embedding vector, r_f^i is the user's demand generated by the ith highly consistent video and r^i_{nf} is the user's demand estimation generated by the ith not-consistent video.

In the case that the relevant probabilities of all chosen video clips equal one, the Eq. (6) is the Rocchio's relevance feedback algorithm [1]. Compared with Rocchio Algorithm, Eq. (6) uses $|r_f\rangle$, which is related to the relevant probability in previous-round retrieval, rather than just using $|d\rangle$ to generate re-ranking score and can improve its robustness against the noise samples to some extent.

For the convenience that the user can evaluate and choose the clips from the entire search results, we apply Hierarchical Clustering to divide the search results into different levels for user's fast interaction.

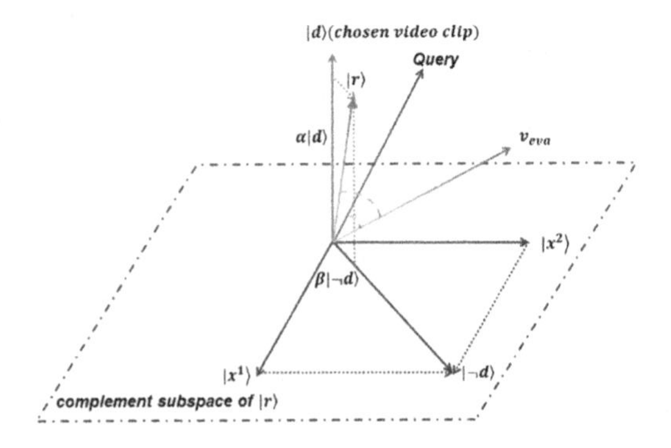

Fig. 2. A quantum-inspired interaction example when n = 3

3.3 Multi-modal Search

The part of multi-modal search can be regarded as an extension of the above search. For example, when dealing with the KIS-V tasks, the user can add a label or a textual description for the query image. Similarly, when doing the KIS-T or AVS tasks, users can input an image as an additional description of the initial query. Then the final score for ranking will be defined as a weighted sum of the different scores calculated independently. In this way, the search parts for different modalities are no longer isolated but combined so tightly that the varied modal information can be fully used.

4 Conclusion

This paper presents a quantum-inspired interactive video search engine for the participation of Video Browser Showdown 2023. The engine adopts a latest text-image embedding method BLIP [5] and introduces several ways for user's interaction including label search, quantum-inspired interaction for re-ranking and multi-modal search.

Acknowledgement. This work is supported by the National Natural Science Foundation of China (No. U1903214, 61876135) and by Ministry of Education Industry-University Cooperation and Collaborative Education Projects (No. 202102246004). The numerical calculations in this paper have been done on the supercomputing system in the Supercomputing Center of Wuhan University.

References

1. Chen, Z., Zhu, B.: Some formal analysis of Rocchio's similarity-based relevance feedback algorithm. Inf. Retrieval **5**, 61–86 (2002). https://doi.org/10.1023/A:1012730924277
2. Gan, J., Tao, Y.: DBSCAN revisited: mis-claim, un-fixability, and approximation. In: Proceedings of the 2015 ACM SIGMOD International Conference on Management of Data (2015). https://doi.org/10.1145/2723372.2737792
3. Hezel, N., Schall, K., Jung, K., Barthel, K.U.: Efficient search and browsing of large-scale video collections with vibro. In: Þór Jónsson, B., et al. (eds.) MMM 2022. LNCS, vol. 13142, pp. 487–492. Springer, Cham (2022). https://doi.org/10.1007/978-3-030-98355-0_43
4. Kenton, J.D., Toutanova, L.K.: BERT: pre-training of deep bidirectional transformers for language understanding. In: Proceedings of NAACL-HLT (2019)
5. Li, J., Li, D., Xiong, C., Hoi, S.: Blip: bootstrapping language-image pre-training for unified vision-language understanding and generation. In: Proceedings of the IEEE Conference on Computer Vision and Pattern Recognition (2022). https://doi.org/10.48550/arXiv.2201.12086
6. Lokoč, J., et al.: Is the reign of interactive search eternal? Findings from the video browser showdown 2020. ACM Trans. Multimedia Comput. Commun. Appl. **17**(3), 1–26 (2021)
7. Rossetto, L., Amiri Parian, M., Gasser, R., Giangreco, I., Heller, S., Schuldt, H.: Deep learning-based concept detection in vitrivr. In: Kompatsiaris, I., Huet, B., Mezaris, V., Gurrin, C., Cheng, W.-H., Vrochidis, S. (eds.) MMM 2019. LNCS, vol. 11296, pp. 616–621. Springer, Cham (2019). https://doi.org/10.1007/978-3-030-05716-9_55
8. Uprety, S., Gkoumas, D., Song, D.: A survey of quantum theory inspired approaches to information retrieval. ACM Comput. Surv. (2021). https://doi.org/10.1145/3402179
9. Wang, P., Hou, Y., Li, Z., Zhang, Y.: QIRM: a quantum interactive retrieval model for session search. Neurocomputing (2021). https://doi.org/10.1016/j.neucom.2021.04.013
10. Wu, H., et al.: UniKeyphrase: a unified extraction and generation framework for keyphrase prediction. In: Findings of the Association for Computational Linguistics: ACL-IJCNLP (2021). https://doi.org/10.48550/arXiv.2106.04847
11. https://github.com/milvus-io/milvus

Exploring Effective Interactive Text-Based Video Search in vitrivr

Loris Sauter[1(✉)], Ralph Gasser[1], Silvan Heller[1], Luca Rossetto[2], Colin Saladin[1], Florian Spiess[1], and Heiko Schuldt[1]

[1] University of Basel, Basel, Switzerland
{loris.sauter,ralph.gasser,silvan.heller,colin.saladin,florian.spiess, heiko.schuldt}@unibas.ch
[2] University of Zurich, Zurich, Switzerland
rossetto@ifi.uzh.ch

Abstract. vitrivr is a general purpose retrieval system that supports a wide range of query modalities. In this paper, we briefly introduce the system and describe the changes and adjustments made for the 2023 iteration of the video browser showdown. These focus primarily on text-based retrieval schemes and corresponding user-feedback mechanisms.

Keywords: Video browser showdown · Interactive video retrieval · Content-based retrieval

1 Introduction

The Video Browser Showdown (VBS) [9,14,19] is a long-running evaluation campaign for interactive multimedia retrieval and user-centric video search. Since 2012 [27], the VBS has provided a highly competitive setup in which systems and their operators are tasked to find video segments within a large collection. The collection currently used is a subset of the Vimeo Creative Commons Collection (V3C) [26], which comprises 2300 h of video material that totals to 1.6 TB in size. In addition to V3C1 [3] and V3C2 [25], the 2023 installment of VBS will feature a homogeneous underwater/scuba diving dataset called the Marine Video Kit [30], of roughly 230 GB and a duration of approximately 11.5 h.

The VBS consists of two types of tasks: Known-Item Search (KIS) and Ad-hoc Video Search (AVS) [12]. The former involves finding a specific video segment based on either a visual preview or a textual description. The latter requires finding items of interest that match a more general description (their correctness is manually judged during the competition).

In this paper, we present vitrivr – an open-source content-based multi-modal multimedia retrieval system – and improvements made to it compared to previous iterations. The 2023 installment marks the 9th time[1] vitrivr participates to VBS in a row [7], with two winning participations in the last four years [8,24].

[1] Including its predecessor, the iMotion system.

D.-T. Dang-Nguyen et al. (Eds.): MMM 2023, LNCS 13833, pp. 646–651, 2023.
https://doi.org/10.1007/978-3-031-27077-2_53

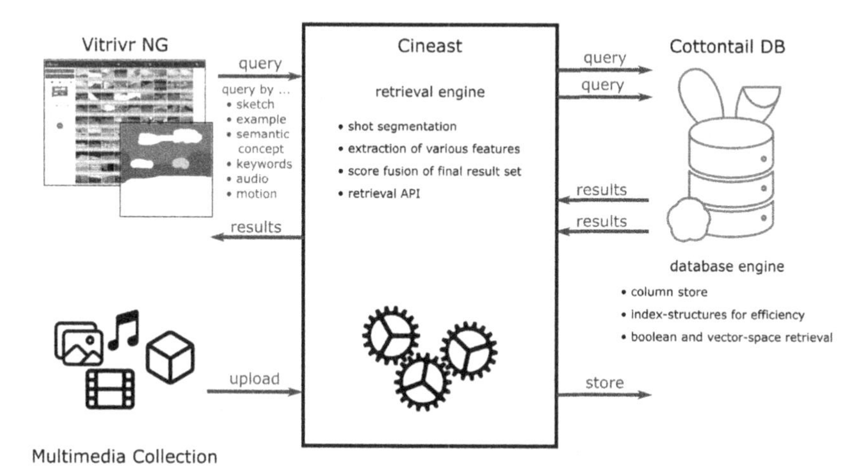

Fig. 1. System overview for vitrivr and its three major components: vitrivr-ng, Cineast and Cottontail DB. Slightly modified version of [8, Fig. 1].

The remainder of this paper is structured as follows: Sect. 2 provides an overview of the vitrivr stack, Sect. 3 highlights the various additions to the stack and we conclude the paper in Sect. 4.

2 vitrivr

vitrivr [23] is an open-source multimedia retrieval stack, capable of supporting a broad range of media and query types including, but not limited to, video (search). An overview of vitrivr's architecture is provided in Fig. 1 [8, Fig 1]. The stack is composed of three main components:

Cottontail DB [5] is the database layer of vitrivr and can be used to store, manage, and query scalar metadata as well as high-dimensional feature vectors. Cottontail DB allows for efficient similarity and Boolean retrieval.

Cineast [17,21,22] is the feature extraction and retrieval engine of the stack. It generates the different feature representation from the input data (videos, user provided queries), orchestrates query execution and implements result aggregation and score fusion.

vitrivr-ng [6] is vitrivr's web-based user interface and facilitates query formulation, result presentation and allows for efficient exploration. In addition, it also enables late-stage filtering and fusion.

All of vitrivr's components are freely available from the project website.[2] Some of the aforenamed components also serve as the basis for other multimedia retrieval systems such as vitrivr-vr [28,29] and Lifegraph [18].

[2] https://vitrivr.org.

3 Novelties for VBS 2023

Since the previous iteration of VBS has shown [9] that most of the top performing systems rely on video-text co-embeddings and support some form of temporal query, we have put some focus on refining and extending means that allow for text-based search.

3.1 Improved Visual-Text Co-embedding

We introduced the first version of our visual-text co-embedding in [29], which consisted of a shallow network that projects the output of two uni-modal pre-trained backbones into a common, semantically aligned space. In order to handle multiple frames of a video rather than only a single image, all frames are passed through their visual backbone and its output is pooled before projection. For this year's iteration of VBS, we have updated several aspects of this approach. The two backbones have been replaced and the aggregation scheme and projection methods have been refined. For textual embedding, we now use a multilingual backbone [31] in order to increase accessibility to non-native English speakers. For the visual frame-level embeddings, we use a more recent convolutional architecture [11] and remove both the final classification as well as the spatial pooling layers. Inspired by [2], we extend the aggregation scheme to not pool the visual embeddings indiscriminately but rather pay attention to the spatial and temporal origin of image-patch embeddings.

3.2 CLIP and vitrivr

The release of the CLIP [16] model by OpenAI in 2021 marked a step-change in the quality achievable when searching for images using text describing their semantic content. In the 2022 edition of VBS, several of the highest scoring teams relied at least in part on feature representations generated by CLIP [1,10,13]. In view of this decisive demonstration of its effectiveness, we have added a CLIP-based feature extractor to vitrivr for the 2023 edition of VBS. The features are extracted based on only one representative frame per shot, as provided by the dataset [26]. During runtime, we provide users the means to chose our co-embeddings or CLIP as query handler.

Since prompt-engineering appears to be a relevant factor in the effective performance of contemporary joint visual language models, as minor changes in the input can lead to rather substantial changes in returned results in some cases, we also employ CLIP-guided image captioning methods, specifically [4] and [15], in order to generate one caption (each) per representative frame of every shot. These captions are not intended to be used for search directly (although such functionality is also supported) but rather to provide feedback to an operator, what a reasonable textual query would be to retrieve any given result. This feedback mechanism can be used by operators to familiarize themselves with the intricacies of the feature in order to help them to construct more effective search prompts.

3.3 SIMD Support for Cottontail DB

Query execution speed is of the essence for interactive video retrieval, especially in competitive settings such as VBS. In the latest iteration of Cottontail DB [5]—the multimedia database layer used by vitrivr—we have therefore started to exploit the use of SIMD instructions to accelerate query execution for brute-force search. The explicit use of SIMD has been enabled by the recent incubation of the new Java Vector API proposed in the JEPs 338, 414 and 417.[3] Even though the current implementation is rather straightforward and despite the feature still being in an early beta stadium, we can report a speed-up of between 20–30% especially for high-dimensional vectors ($d > 1024$). We expect to attain even more acceleration by transitioning the underlying query execution engine from an iterator to a batched processing-model in the (not too distant) future.

3.4 Human-in-the-Loop

A vital component in user-centric video search is the human (retrieval) system operator. The vitrivr team employs regular VBS-style dry-runs since quite some time with a dedicated evaluation setup. In this year's installment, we use our own deployment of the DRES [20] system, with tasks specifically created for that purpose. We analyse each dry-run and, particularly analyse those tasks, where we—the system operator and system—could not find the target. As outlined in Sect. 3.2 we use, among others, the means of textual feature representation of the target in order to learn what search terms would have been purposeful. Specifically for the 2023 installment, we also will have dry-runs in December with peers who did not work on vitrivr, to simulate novice sessions and we might need to adjust some of the user interface functionality based on their feedback.

4 Conclusion

In this paper, we presented the version of vitrivr with which we plan to participate at VBS 2023. As has recent analysis shown, the current trend in user-centric video search goes towards deep learning supported video-text co-embeddings such as CLIP. Thus, we focus on improvements in this domain by expanding our visual-text co-embedding—among others—with a multilingual backbone and introduce the CLIP model in our pipeline as well. Furthermore, various parts of the open-source retrieval system vitrivr have been improved and we will systematically train our system operators with the system as well as simulate novice sessions with our local peers.

Acknowledgements. This work was partly supported by the Swiss National Science Foundation through projects "Participatory Knowledge Practices in Analog and Digital Image Archives" (contract no. 193788) and "MediaGraph" (contract no. 202125).

[3] See https://openjdk.org/jeps/338, accessed September 2022.

References

1. Amato, G., et al.: VISIONE at video browser showdown 2022. In: Þór Jónsson, B., et al. (eds.) MMM 2022. LNCS, vol. 13142, pp. 543–548. Springer, Cham (2022). https://doi.org/10.1007/978-3-030-98355-0_52
2. Bain, M., Nagrani, A., Varol, G., Zisserman, A.: Frozen in time: a joint video and image encoder for end-to-end retrieval. In: International Conference on Computer Vision, ICCV (2021). https://doi.org/10.1109/ICCV48922.2021.00175
3. Berns, F., Rossetto, L., Schoeffmann, K., Beecks, C., Awad, G.: V3C1 dataset: an evaluation of content characteristics. In: International Conference on Multimedia Retrieval. ACM (2019). https://doi.org/10.1145/3323873.3325051
4. Cho, J., Yoon, S., Kale, A., Dernoncourt, F., Bui, T., Bansal, M.: Fine-grained image captioning with CLIP reward. In: Findings of the Association for Computational Linguistics (2022). https://doi.org/10.18653/v1/2022.findings-naacl.39
5. Gasser, R., Rossetto, L., Heller, S., Schuldt, H.: Cottontail DB: an open source database system for multimedia retrieval and analysis. In: International Conference on Multimedia. ACM (2020). https://doi.org/10.1145/3394171.3414538
6. Gasser, R., Rossetto, L., Schuldt, H.: Multimodal multimedia retrieval with vitrivr. In: International Conference on Multimedia Retrieval (2019)
7. Heller, S., et al.: Multi-modal interactive video retrieval with temporal queries. In: Þór Jónsson, B., et al. (eds.) MMM 2022. LNCS, vol. 13142, pp. 493–498. Springer, Cham (2022). https://doi.org/10.1007/978-3-030-98355-0_44
8. Heller, S., et al.: Towards explainable interactive multi-modal video retrieval with vitrivr. In: Lokoč, J., et al. (eds.) MMM 2021. LNCS, vol. 12573, pp. 435–440. Springer, Cham (2021). https://doi.org/10.1007/978-3-030-67835-7_41
9. Heller, S., et al.: Interactive video retrieval evaluation at a distance: comparing sixteen interactive video search systems in a remote setting at the 10th Video Browser Showdown. Int. J. Multimedia Inf. Retrieval, 1–18 (2022). https://doi.org/10.1007/s13735-021-00225-2
10. Hezel, N., Schall, K., Jung, K., Barthel, K.U.: Efficient search and browsing of large-scale video collections with vibro. In: Þór Jónsson, B., et al. (eds.) MMM 2022. LNCS, vol. 13142, pp. 487–492. Springer, Cham (2022). https://doi.org/10.1007/978-3-030-98355-0_43
11. Liu, Z., Mao, H., Wu, C., Feichtenhofer, C., Darrell, T., Xie, S.: A convnet for the 2020s. CoRR abs/2201.03545 (2022)
12. Lokoč, J., et al.: A Task category space for user-centric comparative multimedia search evaluations. In: Þór Jónsson, B., et al. (eds.) MMM 2022. LNCS, vol. 13141, pp. 193–204. Springer, Cham (2022). https://doi.org/10.1007/978-3-030-98358-1_16
13. Lokoč, J., Mejzlík, F., Souček, T., Dokoupil, P., Peška, L.: Video search with context-aware ranker and relevance feedback. In: Þór Jónsson, B., et al. (eds.) MMM 2022. LNCS, vol. 13142, pp. 505–510. Springer, Cham (2022). https://doi.org/10.1007/978-3-030-98355-0_46
14. Lokoč, J., et al.: Is the reign of interactive search eternal? Findings from the video browser showdown 2020. ACM Trans. Multimedia Computi. Commun. Appl. (2021). https://doi.org/10.1145/3445031
15. Mokady, R., Hertz, A., Bermano, A.H.: Clipcap: CLIP prefix for image captioning. CoRR abs/2111.09734 (2021)
16. Radford, A., et al.: Learning transferable visual models from natural language supervision. In: International Conference on Machine Learning (2021)

17. Rossetto, L.: Multi-modal video retrieval. Ph.D. thesis, University of Basel (2018)
18. Rossetto, L., Baumgartner, M., Gasser, R., Heitz, L., Wang, R., Bernstein, A.: Exploring graph-querying approaches in LifeGraph. In: Workshop on Lifelog Search Challenge (2021). https://doi.org/10.1145/3463948.3469068
19. Rossetto, L., et al.: Interactive video retrieval in the age of deep learning – detailed evaluation of VBS 2019. IEEE Trans. Multimedia (2021)
20. Rossetto, L., Gasser, R., Sauter, L., Bernstein, A., Schuldt, H.: A system for interactive multimedia retrieval evaluations. In: Lokoč, J., et al. (eds.) MMM 2021. LNCS, vol. 12573, pp. 385–390. Springer, Cham (2021). https://doi.org/10.1007/978-3-030-67835-7_33
21. Rossetto, L., Giangreco, I., Heller, S., Tănase, C., Schuldt, H.: Searching in video collections using sketches and sample images – the Cineast system. In: Tian, Q., Sebe, N., Qi, G.-J., Huet, B., Hong, R., Liu, X. (eds.) MMM 2016. LNCS, vol. 9517, pp. 336–341. Springer, Cham (2016). https://doi.org/10.1007/978-3-319-27674-8_30
22. Rossetto, L., Giangreco, I., Schuldt, H.: Cineast: a multi-feature sketch-based video retrieval engine. In: International Symposium on Multimedia (2014)
23. Rossetto, L., Giangreco, I., Tanase, C., Schuldt, H.: vitrivr: A flexible retrieval stack supporting multiple query modes for searching in multimedia collections. In: ACM Conference on Multimedia (2016). https://doi.org/10.1145/2964284.2973797
24. Rossetto, L., Amiri Parian, M., Gasser, R., Giangreco, I., Heller, S., Schuldt, H.: Deep learning-based concept detection in vitrivr. In: Kompatsiaris, I., Huet, B., Mezaris, V., Gurrin, C., Cheng, W.-H., Vrochidis, S. (eds.) MMM 2019. LNCS, vol. 11296, pp. 616–621. Springer, Cham (2019). https://doi.org/10.1007/978-3-030-05716-9_55
25. Rossetto, L., Schoeffmann, K., Bernstein, A.: Insights on the V3C2 dataset. CoRR abs/2105.01475 (2021)
26. Rossetto, L., Schuldt, H., Awad, G., Butt, A.A.: V3C – a research video collection. In: Kompatsiaris, I., Huet, B., Mezaris, V., Gurrin, C., Cheng, W.-H., Vrochidis, S. (eds.) MMM 2019. LNCS, vol. 11295, pp. 349–360. Springer, Cham (2019). https://doi.org/10.1007/978-3-030-05710-7_29
27. Schoeffmann, K.: Video browser showdown 2012–2019: a review. In: International Conference on Content-Based Multimedia Indexing (2019)
28. Spiess, F., et al.: Multi-modal video retrieval in virtual reality with vitrivr-VR. In: Þór Jónsson, B., et al. (eds.) MMM 2022. LNCS, vol. 13142, pp. 499–504. Springer, Cham (2022). https://doi.org/10.1007/978-3-030-98355-0_45
29. Spiess, F., Gasser, R., Heller, S., Rossetto, L., Sauter, L., Schuldt, H.: Competitive interactive video retrieval in virtual reality with vitrivr-VR. In: Lokoč, J., et al. (eds.) MMM 2021. LNCS, vol. 12573, pp. 441–447. Springer, Cham (2021). https://doi.org/10.1007/978-3-030-67835-7_42
30. Truong, Q.T., et al.: Marine video kit: a new marine video dataset for content-based analysis and retrieval. In: Dang-Nguyen, D., et al.: MMM 2023. LNCS, vol. 13833, pp. 539–550. Springer. Cham (2023)
31. Yang, Y., et al.: Multilingual universal sentence encoder for semantic retrieval (2019). https://doi.org/10.48550/ARXIV.1907.04307

V-FIRST 2.0: Video Event Retrieval with Flexible Textual-Visual Intermediary for VBS 2023

Nhat Hoang-Xuan[1,3](✉), E-Ro Nguyen[1,3], Thang-Long Nguyen-Ho[1,3], Minh-Khoi Pham[1,3,4], Quang-Thuc Nguyen[1,2,3], Hoang-Phuc Trang-Trung[1,2,3], Van-Tu Ninh[4], Tu-Khiem Le[4], Cathal Gurrin[4], and Minh-Triet Tran[1,2,3](✉)

[1] University of Science, VNU-HCM, Ho Chi Minh City, Vietnam
{hxnhat,nero,nhtlong,pmkhoi,nqthuc,tthphuc,tmtriet}@selab.hcmus.edu.vn
[2] John von Neumann Institute, VNU-HCM, Ho Chi Minh City, Vietnam
[3] Vietnam National University, Ho Chi Minh City, Vietnam
[4] Dublin City University, Dublin, Ireland
tu.ninhvan@adaptcentre.ie, tukhiem.le4@mail.dcu.ie, cathal.gurrin@dcu.ie

Abstract. In this paper, we present a new version of our interactive video retrieval system V-FIRST. Besides the existing features of querying by textual descriptions and visual examples, we propose the usage of an image generator that can generate images from a text prompt as a means to bridge the domain gap. We also include a novel referring expression segmentation module to highlight the objects in an image. This is the first step towards providing adequate explainability to retrieval results, ensuring that the system can be trusted and used in domain-specific and critical scenarios. Searching by a sequence of events is also a new addition, as it proves to be pivotal in finding events from memory. Furthermore, we improved our Optical Character Recognition capability, especially in the case of scene text. Finally, the inclusion of relevant feedback allows the user to explicitly refine the search space. All combined, our system has greatly improved user interaction, leveraging more explicit information and providing more tools for the user to work with.

Keywords: Video retrieval · Interactive system · Joint textual-visual representation · Image generation

1 Introduction

As storage and computational efficiency increase, video data is step-by-step replacing image data in various applications. However, existing video data differs in multiple aspects, from duration, quality, and frame rate, to content type and information density. The information stored in them is vast compared to images. Nonetheless, they are even more difficult to gather due to the extra time dimension. Usually, systems enable intuitive search by letting users search based on

D.-T. Dang-Nguyen et al. (Eds.): MMM 2023, LNCS 13833, pp. 652–657, 2023.
https://doi.org/10.1007/978-3-031-27077-2_54

concepts. A concept can be either coarse (color, shape, etc.) or abstract (action, implications, etc.). Time dimension opens up new concepts that are not present with a single image, for example, a sequence of events like "insert the key then open the door" or implied actions, in the case of waving for greetings or goodbyes. When captured alone, these images could be obscure or ambiguous, hindering retrieval of those events. However, with nearby frames providing the context and the correct direction of time, they can serve to disambiguate and altogether provide a well-defined concept. To simplify the video retrieval problem, it can be beneficial to group the frames into segments and assign them meaning, possibly on multiple scales, to exploit the intrinsic temporal information.

Since the system will be operated by human, it is essential that the different modes of querying are logical and efficient, both in terms of response time and quality of responses. The addition of any features should be carefully considered, as they may overcrowd the interface and bring confusion to the user. To avoid this, it is prudent to thoroughly test the system on users, measure their performance, and collect feedback. The Video Browser Showdown [8] challenge serves to establish an environment where developers can collectively deploy their systems and evaluate on a standard benchmark, with difficulty and complexity increasing each year.

In our second VBS participation, we propose an improved version of our system, V-FIRST 2.0. Our system leverages both text-based information such as scene text, metadata, text descriptions, and visual cues such as colors, sketches, and visual examples. The flexible architecture of our system allows easy integration of new features. The underlying joint-embedding model can also be upgraded or fine-tuned to a specific domain depending on the use case.

Our system is an extended version of FIRST [3,10], a system used for lifelong image retrieval. By making use of the recently popularized diffusion-based models that can generate images from a text prompt, we want to boost further the speed at which we can generate images and tune them to match the target image. It is also very convenient that we can share the embedding between our representation model and the image generation model. Another new feature is the ability to search with a sequence of events. This feature is very practical, as temporal queries are frequent in both the VBS and real-world scenarios. Next, we added a relevance feedback feature to allow the user to manually refine the query vector when the existing one is unsatisfactory. Furthermore, we enhance our OCR model for scene text because such information is helpful in determining the video's content. Lastly, we integrate a referring expression segmentation model to provide a mask of the objects to indicate why they are retrieved. The user can understand what the model is attending to and modify the query in order to guide the system. Other modules can also be built on this, such as pose estimation.

To determine whether normal users (i.e., novice users) can use our system well, we conducted a user study of FIRST [2], where we observed users solving queries in a setting similar to the VBS. In the study, we give a detailed analysis and comparison between novices and experts and give several general "strategies" that novices may employ for better performance. Based on the analysis and feedback, we consider and justify the addition of new features and adjustment of the user interface.

2 Related Work

The vibro [1] team topped the last VBS competition and convincingly achieved the best score in all three categories (textual KIS, video KIS, and AVS). Better representations have helped as teams have been able to answer more queries than before. This puts more emphasis on the features that are built based on the representations and the manner in which they are used.

The dominant interface is still monitor and keyboard, which allows the best input and verification speed. VideoFall [7] proposes a novel multi-user approach, where separate users are responsible for picking promising results and verifying them. Other teams, such as vibro [1] or CVHunter [4], carefully design their interface to pack at much information as possible in a screen. As teams have participated in the VBS and learned from each other, the design and functionalities of systems have influenced each other and slowly converged.

Teams also focus on improving the algorithm for temporal queries and relevance feedback, which are advanced features not commonly found in ordinary search engines. With sufficient proficiency, these tools can help users generate complex queries to quickly reduce the search space, enabling teams to answer queries rapidly.

3 V-FIRST 2.0

3.1 System Overview

Figure 1 shows the overview and the components of our proposed system, V-FIRST 2.0, which is based on our lifelong retrieval systems FIRST [3] and the previous version of V-FIRST [9]. As flexibility was one of our design goals, extending our image retrieval system to support video retrieval is simple.

We have a variety of modules and features that allow searching with textual and visual descriptions, as well as the ability to browse and navigate through the data collection quickly. They are described in detail in the previous version's paper. For this year, we adopt CLIP for both textual and visual representation for its known effectiveness. We focus on other new features and improvements from the last version in the following sections.

3.2 Image Generation via Text Prompt

Previously, we considered using a hand-drawn sketch to decrease the gap between natural language descriptions and images. This approach is logical; however, even experienced illustrators need some time to draw a picture with sufficient detail. With diffusion-based models having the ability to generate realistic images from text prompts, we want to use them as an assistant to augment the image generation process. Enabled by this tool, the illustrator can quickly express the idea and select from a pool of candidates the drawing that best matches their intention and continue from that. This is a step closer to the goal of allowing seamless and accurate expressions of an imagined picture for use as search targets.

3.3 Querying by Sequence of Events

For both textual and video Known-item Search (KIS), an event is usually not depicted as isolated but preceded or succeeded by other events. Furthermore, even when such a description is missing, it is logical to assume that some events must have other preceding/succeeding events; for example, one must open the door before entering the home. Therefore, being able to search using multiple events is critical as it drastically reduces the search space.

We focus on the sequence of two events because it is a good trade-off between the input speed and the searching speed. However, the idea can be easily extended to more. We utilize a simple linear combination weighting scheme where pairs of images are scored based on their relevance to each event in order. The weighting for each event can be modified by the user to account for factors such as the certainty and rareness of each event. This method is independent of the original searching method and hence can work irrespective of the search mode used (e.g., text, visual example, etc.). Practice shows that when the length of the videos is reasonable, this method is fast enough, and with some post-processing to remove similar pairs in a video, it yields good results while keeping query time overhead to a minimum.

3.4 Referring Expression Segmentation for Explainability

Usually, retrieval systems return results with little indication of why they are selected. It takes a quick scan of the result to locate and verify if the results are

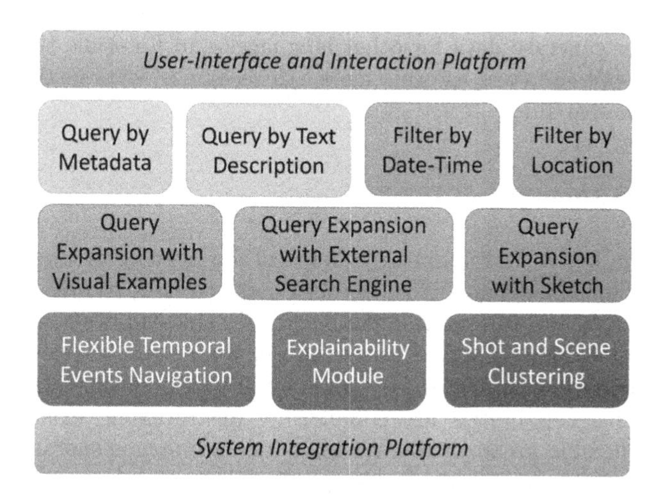

Fig. 1. Overview of the components of our system.

correct. There are cases where this is more challenging (e.g., small objects) or even impossible (e.g., poor lighting, blurry image, or concepts unrecognizable by humans). Therefore, to gain reliability, retrieval systems have to advance beyond just giving raw results.

We propose to use a referring expression segmentation approach to address this problem. It yields a mask for each image, which can assist the user with quickly locating the object or concept. The user can look at the mask to see if the system correctly understands the query and may modify the query accordingly. It is also convenient that the referring expression segmentation problem has the same input as the retrieval system, opening up possibilities for reusing the representations for the segmentation problem.

Our approach [5] is based on recently popular query-based methods. Our learnable queries are fused with language and multi-level visual features to generate mask candidates. The query with the highest confidence score will be selected as the target. This method allows the queries to attend to different features during the training process and simultaneously associates them with visual and linguistic features.

3.5 Street Scene Text Recognition with Auxiliary Synthetic Data

Often, images contain texts that can help identify the location and other information that comes with it. These texts can come from various sources, from signs, brands, and badges, to subtitles and annotations. These naturally occurring texts are usually more challenging to detect than perfectly aligned documents.

The work [6] proposes a dataset that was gathered on the streets. It highlights the challenges of scene text, including dense text, far-away text, rotations and spaces. Based on that dataset, we propose a scene text recognition method that works better with texts from the real world. Our method comprises two stages, text detection and text recognition. Furthermore, we propose a data augmentation pipeline that can generate synthetic data for training. Combined, we can quickly generate data for other languages and fine-tune the text recognition model. We can then use this model to recognize texts in the videos and index them for searching.

4 Conclusion

We introduce a newer and upgraded version of our system for interactive video retrieval. Our focus is to maximize the user's search effectiveness, regardless of system proficiency and competition familiarity. To this end, we conducted a user study and proposed search strategies and subsequently included new user-centric features, namely mask indicator for explainability and relevance feedback.

Our system's capabilities are enhanced while retaining existing features, owing to our flexible architecture. We introduce an image-generation model to generate images that match a prompt quickly. We take advantage of CLIP for better representations. Users can still search with free-form text queries, search by concepts, visual examples, 2D color maps, and sketching. Flexible temporal navigation allows users to browse quickly through a video at varying speeds to explore and verify its contents. Finally, we made changes to the user interface and included new features based on the feedback from our user study.

Acknowledgement. This research was funded by Vingroup and supported by Vingroup Innovation Foundation (VINIF) under project code VINIF.2019.DA19.

References

1. Hezel, N., Schall, K., Jung, K., Barthel, K.U.: Efficient search and browsing of large-scale video collections with vibro. In: Þór Jónsson, B., et al. (eds.) MMM 2022. LNCS, vol. 13142, pp. 487–492. Springer, Cham (2022). https://doi.org/10.1007/978-3-030-98355-0_43
2. Hoang-Xuan, N., et al.: Flexible interactive retrieval SysTem 2.0 for visual lifelog exploration at LSC 2021 Submitted for review
3. Hoang-Xuan, N., et al.: Flexible interactive retrieval SysTem 3.0 for visual lifelog exploration at LSC 2022. In: Proceedings of the 5th Annual on Lifelog Search Challenge. LSC 2022, pp. 20–26. Association for Computing Machinery (2022). https://doi.org/10.1145/3512729.3533013
4. Lokoč, J., Mejzlík, F., Souček, T., Dokoupil, P., Peška, L.: Video search with context-aware ranker and relevance feedback. In: Þór Jónsson, B., et al. (eds.) MMM 2022. LNCS, vol. 13142, pp. 505–510. Springer, Cham (2022). https://doi.org/10.1007/978-3-030-98355-0_46
5. Nguyen, E.R., Hoang-Xuan, N., Tran, M.T.: Visual-language transformer for referring video object segmentation. In: The IEEE/CVF Conference on Computer Vision and Pattern Recognition (CVPR) Workshops, YouTube-VOS
6. Nguyen, N., et al.: Dictionary-guided scene text recognition, pp. 7383–7392. https://openaccess.thecvf.com/content/CVPR2021/html/Nguyen_Dictionary-Guided_Scene_Text_Recognition_CVPR_2021_paper.html
7. Nguyen, T.-N., Puangthamawathanakun, B., Healy, G., Nguyen, B.T., Gurrin, C., Caputo, A.: Videofall - a hierarchical search engine for VBS2022. In: Þór Jónsson, B., et al. (eds.) MMM 2022. LNCS, vol. 13142, pp. 518–523. Springer, Cham (2022). https://doi.org/10.1007/978-3-030-98355-0_48
8. Schoeffmann, K., Lokoč, J., Bailer, W.: 10 years of video browser showdown. In: Proceedings of the 2nd ACM International Conference on Multimedia in Asia. MMAsia 2020, pp. 1–3. Association for Computing Machinery (2021). https://doi.org/10.1145/3444685.3450215
9. Tran, M.-T., et al.: V-FIRST: a flexible interactive retrieval system for video at VBS 2022. In: Þór Jónsson, B., et al. (eds.) MMM 2022. LNCS, vol. 13142, pp. 562–568. Springer, Cham (2022). https://doi.org/10.1007/978-3-030-98355-0_55
10. Trang-Trung, H.P., et al.: Flexible interactive retrieval SysTem 2.0 for visual lifelog exploration at LSC 2021. In: Proceedings of the 4th Annual on Lifelog Search Challenge. LSC 2021, Taipei, Taiwan, pp. 81–87. Association for Computing Machinery (2021). https://doi.org/10.1145/3463948.3469072

VERGE in VBS 2023

Nick Pantelidis, Stelios Andreadis(✉), Maria Pegia, Anastasia Moumtzidou, Damianos Galanopoulos, Konstantinos Apostolidis, Despoina Touska, Konstantinos Gkountakos, Ilias Gialampoukidis, Stefanos Vrochidis, Vasileios Mezaris, and Ioannis Kompatsiaris

Information Technologies Institute/Centre for Research and Technology Hellas, Thessaloniki, Greece
{pantelidisnikos,andreadisst,mpegia,moumtzid,dgalanop,kapost, destousok,gountakos,heliasgj,stefanos,bmezaris,ikom}@iti.gr

Abstract. This paper describes VERGE, an interactive video retrieval system for browsing a collection of images from videos and searching for specific content. The system utilizes many retrieval techniques as well as fusion and reranking capabilities. A Web Application is also part of VERGE, where a user can create queries, view the top results and submit the appropriate data, all in a user-friendly way.

1 Introduction

VERGE is an interactive video retrieval system that provides various search capabilities inside a set of images, along with a Web Application for creating and running queries on the set, viewing the top results and submitting the appropriate ones. Over the past few years VERGE has participated many times in the Video Browser Showdown (VBS) competition [9] trying each time to better adapt to the competition's "Ad-Hoc Video Search" (AVS) and "Known Item Search - Visual/Textual" (KIS-V, KIS-T) tasks. This year the various search modules have been further improved and the Web Application has been updated in order to be even more user-friendly and fast-to-use.

The paper is structured as follows: Sect. 2 describes the overall framework of the system, Sect. 3 continues with a detailed description of the various retrieval modalities, Sect. 4 presents the user interface (UI) and its features, and the paper wraps up with Sect. 5 that briefly describes the future work.

2 The VERGE Framework

As shown in Fig. 1, the VERGE framework is composed of three layers. The first layer contains all the retrieval modalities that are applied on the datasets, i.e. V3C1, V3C2 [18] and the Marine Video Dataset. The outcomes are stored in a database (except for the ones from Text to Video Matching module that only runs on-the-fly). The second layer consists of the various services that accept queries and return the top results. The third layer is the Web Application that allows users to formulate and send queries, connects to the corresponding services and displays the results.

D.-T. Dang-Nguyen et al. (Eds.): MMM 2023, LNCS 13833, pp. 658–664, 2023.
https://doi.org/10.1007/978-3-031-27077-2_55

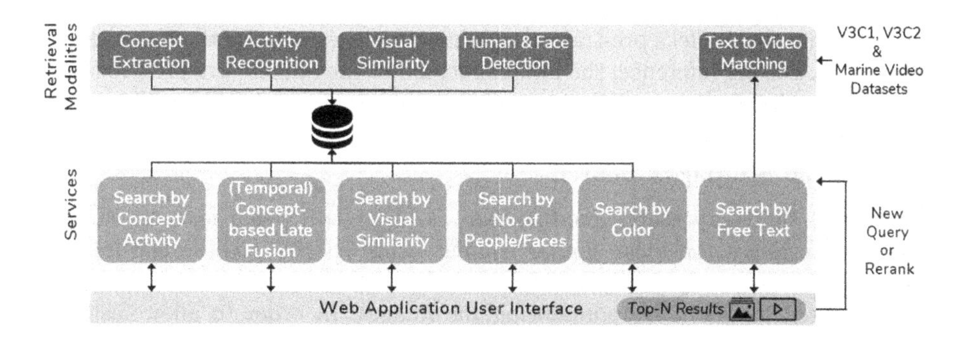

Fig. 1. The VERGE framework

3 Retrieval Modalities

3.1 Concept-Based Retrieval

This module annotates each keyframe of a shot with labels from a pool of concepts, which comprises 1000 ImageNet concepts, a selection of 298 concepts of the TRECVID SIN task [14], 500 event-related concepts, 365 scene classification concepts, 580 object labels as well as 22 sports classification labels. To obtain the annotation scores for the 1000 ImageNet concepts, we used an ensemble method, averaging the concept scores from two pre-trained models that employ different DCNN architectures, i.e. the EfficientNet-B4 [20] and InceptionResNetV2. To obtain scores for the subset from the TRECVID SIN task, we trained and employed a model based on the EfficientNet-B4 architecture on the official SIN task dataset. For the event-related concepts, we used the pre-trained model of EventNet [7]. Regarding the extraction of the scene-related concepts, we utilized the publicly available VGG16 model, fine-tuned on the Places365 dataset. Object detection scores were extracted using models pre-trained on the MS COCO and Open Images V4 datasets, with 80 and 500 detectable objects, respectively. To label sports in video frames, we constructed a custom dataset with Web images from sports and utilized it to train a model of the EfficientNetB3 architecture. Finally, to offer a cleaner representation of the concept-based annotations we employed the sentence-BERT [17] text encoding framework, to measure the text similarity between all concepts' labels. After inspecting the results, we manually formed groups of very similar concepts for which we created a common label and assign the max score of its members.

3.2 Spatio-temporal Activity Recognition

The activity detection and recognition module extracts human-related activities for each shot to enrich the filtering functionalities using the labels of the activities. A list of 400 pre-defined human related-activities and the corresponding scores were extracted for each shot using a 3D-CNN architecture. Especially, the configuration of the 3D-ResNet architecture with 152 layers according to [8] was used and the model's pre-trained weights were learned using the Kinetics-400 dataset [3]. During inference, the input shots fed to the model were pre-processed to be descended to the model's input dimension equal to $16 \times 112 \times 112 \times 3$.

3.3 Visual Similarity Search

The visual similarity search module uses as input the visual features of each shot and retrieves the most similar content using DCNNs. These features are the output of the last pooling layer of the fine-tuned GoogleNet architecture [15] and are used for globally representing the images. In order to allow fast and efficient indexing, an IVFADC index database vector is developed with these vectors [10].

3.4 Human and Face Detection

The human and face detection module aims to detect and count humans and human faces in each keyframe of each shot, so that the user can easily distinguish the results of single-human or multi-human activities. The detection of both human silhouettes (bodies) and human faces (heads) were extracted using a DCNN architecture, YoloV4 [1]. The model's initial weights are learned using the MS COCO [12] dataset and fine-tuned using the CrowdHuman dataset [19] that consists of crowd-center scenes where partial occlusions among humans or between humans and objects are possible to occur. During inference, the total number of humans and human heads is calculated and counted only in case the detected bounding box area is larger than a predefined threshold.

3.5 Text to Video Matching Module

The text-to-video matching module inputs a complex free-text query and retrieves the most relevant video shots from a large set of available video shots. We utilize the T×V model, a cross-modal video retrieval method presented in [6]. The T×V model utilizes multiple textual and visual features along with multiple textual encoders to build multiple cross-modal common latent spaces. The network is trained using video-caption pairs and learns to transform these pairs into multiple common latent spaces. The straightforward comparison between a video and a caption is possible in every individual space. Using a multi-loss-based training approach our network learns the overall similarity by optimizing the individual similarities. Regarding the training, a combination of four large-scale video caption datasets is used (e.g. MSR-VTTT [22], TGIF [11], ActivityNet [2] and Vatex [21]), and the improved marginal ranking loss [4] is used to train the entire network. As initial video shot representation, we utilize three different trained networks, i) the ResNet-152 deep network trained on the ImageNet-11k dataset, ii) the ResNeXt-101 network, pre-trained by weakly supervised learning on web images followed and fine-tuned on ImageNet [13], and iii) the CLIP model (ViT-B/32) [16]. As textual encoders, the networks utilize i) a feedforward encoder utilizing the CLIP-based generated embeddings, and ii) the textual sub-network (ATT) presented in [5].

3.6 Concept-Based Late Fusion

The concept-based late fusion module returns a list of shots, where each shot contains all the queried concepts. Specifically, the method uses two or more visual concepts (Sect. 3.1) and produces a sorted shots' list via a late fusion method. First, for each concept an independent list of shot probabilities $P = \{p_i\}_{i=1}^{i=n}$ is developed. The intersection of the concepts at shot layer is calculated and is sorted using the objective function

$$f(P) = \sum_{i=1}^{|P|} e^{-p_i} + \sum_{i,j=1,i \neq j}^{|P|} e^{-|p_i - p_j|}. \tag{1}$$

This function follows the assumption that the higher the concept probabilities or the more relevant the shots are, the higher their scores are. We deal with it using the difference of the probabilities for all concept pairs' combinations followed by the inverse exponential function to them.

3.7 Temporal Late Fusion

The temporal late fusion module returns a list of tuples of shots, where each element of the list corresponds to the same video and contains all the queried concepts with respect to the given order. In particular, the module incorporates two or more visual concepts (Sect. 3.1) and produces a sorted list without duplicates via a late fusion method. At first, a list of shot probabilities is produced for each concept. Next, the intersection of concepts at video layer is calculated and the first tuple of each video, which respects the ordering of the concepts, is kept. The shots are sorted using an objective function that respects the same assumptions identified in the concept-based late fusion method (Sect. 3.6).

4 User Interface and Interaction Modes

The VERGE UI (Fig. 2) is a Web application that allows a user to easily create and run queries on the dataset and view the top results using the modalities that were described in the previous sections. They can also watch the corresponding video and submit the appropriate data during the VBS competition. The goal of the VERGE UI is to provide a user-friendly, compact, effective and fast tool for searching in image collections.

The UI has two main parts: the menu on the left and the results panel on the right. On the top of the *menu* there is a timer that counts down the remaining time for submission during a VBS task. Below there is a slider where a user can define the size of the images on the results panel, an undo button for restoring previous results, a rerank button for reranking the current results based on the next query, and then follow the various search modules. The first module is the free text search (Sect. 3.5) where the user can type anything in the form of free text and the second one allows the user to search from a list of pre-extracted

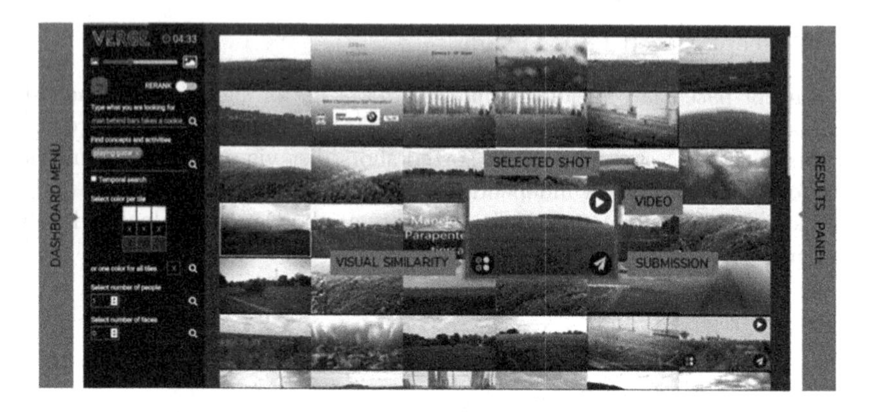

Fig. 2. The VERGE user interface (Color figure online)

concepts and activities (Sects. 3.1, 3.2). Multiple selection is also supported for late fusion (Sect. 3.6) as well as temporal fusion (Sect. 3.7), if the corresponding checkbox is checked. Search by the color of the image is possible by coloring a 3×3 grid. Finally, there are search options based on the number of people or faces (Sect. 3.4) visible in an image.

The *results panel* contains the top results in a grid view. When an image is clicked, a pop-up panel appears that shows all the available shots of the corresponding video. Hovering over an image, three buttons appear. One on the bottom-left corner of the image that, when clicked, returns visually similar images (Sect. 3.3), one on the bottom-right corner for submitting this shot and one on the top-right corner that plays the respective video. Under the video player there is a button to submit directly the time of the video.

To demonstrate the features of VERGE, we shortly describe three use cases. For an AVS task that asks for shots of a single person playing guitar, the user can select the concept "playing guitar" and rerank the results by selecting only one person to appear. For a KIS-V query that searches for a video that shows a grassland on the bottom and the white sky on the top, the user can utilize the search by color, painting the first row white and the third row green (Fig. 2). Lastly, for a KIS-T query that asks for a video that shows "a man behind bars taking a cookie from a tray held" the exact words can be used per se in the free-text search.

5 Future Work

Every year we try to improve the retrieval performance by increasing the response time and the effectiveness of the algorithms, as well as to make the VERGE UI more intuitive, user-friendly and fast. The whole system will be evaluated in the VBS 2023 competition and the participating experience will drive the future steps, regarding both the search methodologies and the UI.

Acknowledgements. This work has been supported by the EU's Horizon 2020 research and innovation programme under grant agreements H2020-101004152 CALLISTO, H2020-833464 CREST, H2020-101070250 XRECO, and H2020 - 101021866 CRiTERIA.

References

1. Bochkovskiy, A., Wang, C.Y., Liao, H.Y.M.: Yolov4: optimal speed and accuracy of object detection. arXiv preprint arXiv:2004.10934 (2020)
2. Caba Heilbron, F., et al.: ActivityNet: a large-scale video benchmark for human activity understanding. In: Proceedings of IEEE CVPR 2015, pp. 961–970 (2015)
3. Carreira, J., Zisserman, A.: Quo Vadis, action recognition? a new model and the kinetics dataset. In: proceedings of the IEEE Conference on Computer Vision and Pattern Recognition, pp. 6299–6308 (2017)
4. Faghri, F., Fleet, D.J., et al.: VSE++: improving visual-semantic embeddings with hard negatives. In: Proceedings of BMVC 2018 (2018)
5. Galanopoulos, D., Mezaris, V.: Attention mechanisms, signal encodings and fusion strategies for improved ad-hoc video search with dual encoding networks. In: Proceedings of ACM ICMR 2020 (2020)
6. Galanopoulos, D., Mezaris, V.: Are all combinations equal? Combining textual and visual features with multiple space learning for text-based video retrieval. In: Karlinsky, L., Michaeli, T., Nishino, K. (eds.) ECCV 2022. LNCS, vol. 13804, pp. 627–643. Springer, Cham (2022). https://doi.org/10.1007/978-3-031-25069-9_40
7. Guangnan, Y., Yitong, L., et al.: Eventnet: a large scale structured concept library for complex event detection in video. In: Proceedings of ACM MM 2015 (2015)
8. Hara, K., et al.: Can spatiotemporal 3D CNNs retrace the history of 2D CNNs and ImageNet? In: Proceedings of IEEE CVPR 2018 (2018)
9. Heller, S., Gsteiger, V., Bailer, W., et al.: Interactive video retrieval evaluation at a distance: comparing sixteen interactive video search systems in a remote setting at the 10th video browser showdown. IJMIR **11**(1), 1–18 (2022)
10. Jegou, H., et al.: Product quantization for nearest neighbor search. IEEE Trans. Pattern Anal. Mach. Intell. **33**(1), 117–128 (2010)
11. Li, Y., Song, Y., Cao, L., Tetreault, J., et al.: TGIF: a new dataset and benchmark on animated gif description. In: Proceedings of IEEE CVPR 2016 (2016)
12. Lin, T.-Y., et al.: Microsoft COCO: common objects in context. In: Fleet, D., Pajdla, T., Schiele, B., Tuytelaars, T. (eds.) ECCV 2014. LNCS, vol. 8693, pp. 740–755. Springer, Cham (2014). https://doi.org/10.1007/978-3-319-10602-1_48
13. Mahajan, D., et al.: Exploring the limits of weakly supervised pretraining. In: Ferrari, V., Hebert, M., Sminchisescu, C., Weiss, Y. (eds.) ECCV 2018. LNCS, vol. 11206, pp. 185–201. Springer, Cham (2018). https://doi.org/10.1007/978-3-030-01216-8_12
14. Markatopoulou, F., Moumtzidou, A., Galanopoulos, D., et al.: ITI-CERTH participation in TRECVID 2017. In: Proceedings of TRECVID 2017 Workshop, USA (2017)
15. Pittaras, N., Markatopoulou, F., Mezaris, V., Patras, I.: Comparison of fine-tuning and extension strategies for deep convolutional neural networks. In: Amsaleg, L., Guðmundsson, G.Þ, Gurrin, C., Jónsson, B.Þ, Satoh, S. (eds.) MMM 2017. LNCS, vol. 10132, pp. 102–114. Springer, Cham (2017). https://doi.org/10.1007/978-3-319-51811-4_9

16. Radford, A., et al.: Learning transferable visual models from natural language supervision. In: Proceedings of the 38th International Conference on Machine Learning (ICML) (2021)
17. Reimers, N., Gurevych, I.: Sentence-BERT: sentence embeddings using Siamese BERT-networks. arXiv preprint arXiv:1908.10084 (2019)
18. Rossetto, L., Schuldt, H., Awad, G., Butt, A.A.: V3C – a research video collection. In: Kompatsiaris, I., Huet, B., Mezaris, V., Gurrin, C., Cheng, W.-H., Vrochidis, S. (eds.) MMM 2019. LNCS, vol. 11295, pp. 349–360. Springer, Cham (2019). https://doi.org/10.1007/978-3-030-05710-7_29
19. Shao, S., Zhao, Z., Li, B., Xiao, T., Yu, G., et al.: CrowdHuman: a benchmark for detecting human in a crowd. arXiv preprint arXiv:1805.00123 (2018)
20. Tan, M., Le, Q.V.: Efficientnet: rethinking model scaling for convolutional neural networks. arXiv preprint arXiv:1905.11946 (2019)
21. Wang, X., et al.: Vatex: a large-scale, high-quality multilingual dataset for video-and-language research. In: Proceedings of IEEE/CVF ICCV 2019, pp. 4581–4591 (2019)
22. Xu, J., Mei, T., et al.: MSR-VTT: a large video description dataset for bridging video and language. In: Proceedings of IEEE CVPR 2016, pp. 5288–5296 (2016)

Vibro: Video Browsing with Semantic and Visual Image Embeddings

Konstantin Schall(✉), Nico Hezel, Klaus Jung, and Kai Uwe Barthel

HTW Berlin, University of Applied Sciences - Visual Computing Group,
Wilhelminenhofstraße 75, 12459 Berlin, Germany
konstantin.schall@htw-berlin.de
http://visual-computing.com/

Abstract. *Vibro* represents a powerful tool for interactive video retrieval and browsing and is the winner of the Video Browser Showdown 2022. Following the saying of "never change a winning system" we did not change any of the underlying concepts nor added any new features. Instead, we focused on improving the three existing cornerstones of the software, which are text-to-image search, image-to-image search and browsing results with 2D sorted maps. The changes to these three parts are summarized in this paper, and in addition, an overview of the *AVS*-mode of *vibro* is given.

Keywords: Content-based video retrieval · Exploration · Visualization · Image browsing · Visual and textual co-embeddings

1 Introduction

The Video Browser Showdown (VBS) [4] is a international competition where participating teams have to solve three task categories: *visual Known-Item Search* (v-KIS), *textual Known-Item Search* (t-KIS) and *Ad-Hoc Video Search* (AVS). In the upcoming 2023 event, the data will consist of the combination of three datasets: V3C1 (7475 video files with a duration of 1000 h hours), V3C2 (9760 videos with 1300 h hours), and a relatively small dataset of diving videos, which are very similar and highly redundant [11].

The large variety of tasks and data lead to very specific requirements for the participating systems. First, the sheer amount of data, with nearly 3 TB of videos, requires a very memory efficient solution and implementation of the software. Second, the systems have to be able to formulate queries in different modalities to solve each of the task categories and browse the results. Third, the underlying analysis of video data must be sufficiently generalizable to work with the diverse videos of the V3C datasets and the low variety of the scuba diving videos.

In the last version of our *vibro* system, we mainly focused on improving the user interface and introduced a CLIP model [8] for joint-embeddings to support full-text search. However, since CLIP embeddings encapsulate information

D.-T. Dang-Nguyen et al. (Eds.): MMM 2023, LNCS 13833, pp. 665–670, 2023.
https://doi.org/10.1007/978-3-031-27077-2_56

of images more on a semantic than a detailed visual level, we used a second embedding for image-based queries. Thanks to these improvements, *vibro* took first place in the overall competition.

This paper presents the next iteration of *vibro* and is structured as follows. For completeness, we want to showcase the main features of *vibro* in the first section. Following the saying of "never change a winning system", we have left the main principles and functionalities of the previous version untouched and only updated the underlying embeddings for textual and image-based queries which is highlighted in the following section. In the last part, we focus on the *AVS* mode of the system, since it uses its own user interface and ranking model and has not been previously described in great detail.

2 GUI and Features

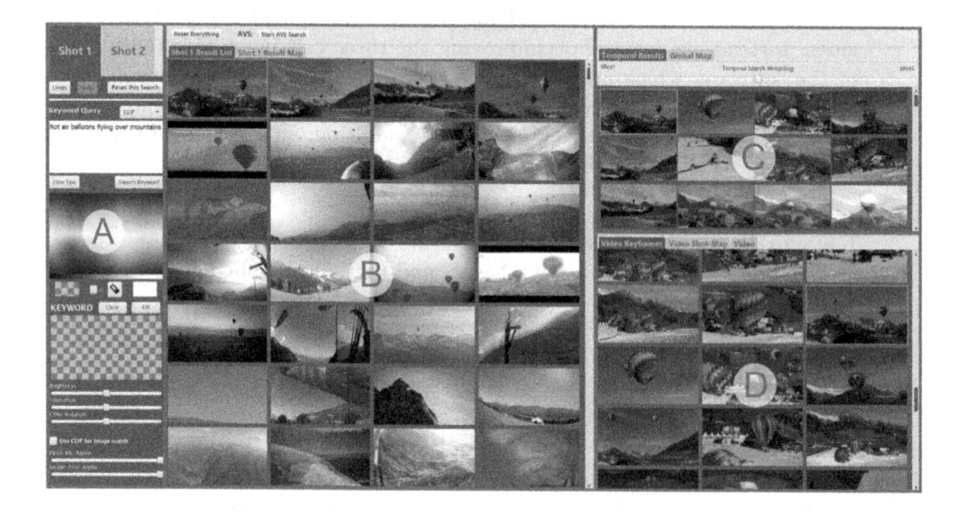

Fig. 1. Graphical user interface for the current version of *Vibro*

Figure 1 gives an overview of *vibro's* user interface. Part A is reserved for query formulation and the results are presented in part B. Here, the top 4000 frames are displayed either in a 2D-sorted map or in a simple descending multi-column list, ordered by relevance to the current query. Searches can be formulated for two consecutive shots by selecting the tab-buttons at the top of part A. If both tabs contain a query, the queries are used in a temporal search. Those results are sorted in part C as a list of five consecutive shots, each representing a segment of a video. Part D is reserved for single video views. One tab displays each shot of the currently selected video, another tab contains a video player. The last view (Global Map) is a further tab in Section C and allows exploring the whole video collection with an image similarity graph [5].

The system supports queries of three modalities: sketch, text and image-by-example. These modalities can be mixed with adjustable weights. Sketch-based queries can be formulated either by drawing on the blank canvas in part A or by selecting an image from any of the other parts and subsequent modification with the drawing tools. Rich-Text queries are supported by employing a CLIP trained joint-embedding model.

Each text input is encoded into a vector by the textual encoder of the model and then a simple similarity search is performed over all shot-vectors, generating a score for each shot. The embeddings for all shots are extracted once and stored locally. Image queries can be initiated by double clicking any shot presented in one of the views. We use a separate set of embeddings produced by a deep neural network, that has been fine-tuned for the specific case of content-based image retrieval in this step and rank all video-shots by its relevance to the query by an exhaustive similarity search. Additionally, out-of-database queries are possible by dragging an image file or URL on the drawing canvas.

For temporal search, we try to arrange sequences of keyframes in such a way that one keyframe matches the query of the first search tab and one of the three following keyframes of the same video matches the query of the second tab. To do this, the similarity values of the search results of tab 1 are combined with the highest of each of the three search results of tab 2 to calculate the harmonic mean of both scores to describe the temporal search result score.

3 Improvements of the Current Version

The three main improvements can be summarized as follows. First, we updated the CLIP model from a ResNet [3] based visual encoder to a visual transformer [2] version (*ViT-L@336*). Second, the content-based retrieval model was significantly improved and last, we updated the sorting algorithm to FLAS [1], which not only slightly improved the arrangement quality, but also led to faster sorting speed.

When we started working on the 2022 version of *vibro*, the largest and best performing version of CLIP, namely *ViT-L@336*, was not publicly available. For this reason we fell back to the second best model, which was the *ResNet50x16* version. Additionally, we conducted heuristic experiments on compression with PCA and concluded that everything below 512 dimensions would harm the text-to-image retrieval qualities. These findings also align with [7]. Since even a modest compression affects the order of results, we have omitted it completely this year. However, the embeddings are still quantized to bytes which is the only post-processing step. With these changes, text-to-image retrieval performs better compared to the last version of *vibro*, since we are not only using a better visual encoder in the first place, but also omit compression, which slightly reduced the retrieval quality.

One of the most challenging aspects of the *VBS* is the high diversity of the video data and because of that, all video-shots have to be encoded by well generalizing deep neural networks. CLIP was trained with over 400 million of image

Table 1. Comparison of different encoders and their embeddings

Encoder	GPR1200 mAP	Data type	Dim	Image size
CLIP Vibro22	71.2	Byte	512	384
CLIP Vibro23	76.1	Byte	768	336
CBIR Vibro22	73.4	Bit	1024	384
CBIR Vibro23	86.1	Bit	1024	336
CBIR Vibro23	87.2	Float	1024	336

and text pairs and as shown by Radfort et al. [8], generalizes to a lot of different tasks. However, since the objective of CLIP is to pair text with images, it is not specialized for the task of content-based image retrieval (CBIR). For example, a nearest neighbor search with a picture of a burger on a plate in a restaurant will yield images of popular fast-food restaurant logos and buildings. Semantically, both the burger and the restaurant incorporate the concept of fast-food and therefore are similar but the particular object instances are different. Networks that specifically have been trained for the task of CBIR, might perform better for image-to-image queries. In the last version of *vibro* a Swin-L network [6] was used for CBIR and we exchanged that model for a ViT-L network, pretrained with CLIP and finetuned with a combination of publicly available image datasets such as ImageNet21k [9] and Google Landmarks v2 [12]. The final training set had over 22 million images from 168 thousand categories. We then evaluated all models with the *mean average precision* score of the GPR1200 dataset [10], which was designed as a benchmark for general-purpose CBIR solutions. Table 1 compares the CLIP and CBIR networks and shows that the current version is significantly superior in image-to-image retrieval settings. The final visual embeddings have 1024 dimensions and are binarized with a threshold of 0, which only marginally harms the retrieval performance but allows us to use the *Hamming distance* for similarity search and reduces the memory requirement by factor 32 compared to the floating point version.

Two of the featured UI parts benefit from the third change we made this year. The first one is the 2D-map arrangement view of the 4000 most relative shots to the query found in section B and the second one is the "Global Map" navigation tab in section C. In both cases video-shots are arranged on a 2D-grid with FLAS [1]. FLAS produces arrangements approximately ten times faster than a traditional SOM and 10% faster than a SSM, while providing better sorting quality.

4 AVS Mode

Whereas the goal of the *v-KIS* and t-KIS tasks is to find one specific video, as many as possible shots fitting a specific textual description have to be submitted for the *ad-hoc video search* (*AVS*) part of the competition. Since videos can only be submitted from part D of the main user interface, we decided to implement a separate interface for the ad-hoc video search. Figures 1 and 2 show how an example task of "find shots with hot air balloons flying over mountains" could be solved with *vibro*. First, the main UI is used to execute a textual query. Next, one of the presented video-shots is selected and the "Start AVS Search" button on the very top of part B of Fig. 1 opens the *AVS* window shown in Fig. 2. Presented are 42 video-shots selected by similarity search with the CBIR embeddings and the selected shot from the main UI as a query. Additionally, all shots are filtered to have an adjustable temporal distance to any of the already presented shots to omit to many results from the same video. The user now has to mark all relevant shots with a click on the image and submits them by pressing the "Send" button (left of Fig. 2). For the next iteration all marked video-shots are used to perform a multi-image similarity search. The scores are merged with a minimum function over each of the query images and the found shots are again filtered temporally, taking into account all previously submitted video-shots. The result of the second iteration is shown in Fig. 2 on the right. This procedure can be repeated until no more relevant images are presented in the result list. The user can then close the *AVS* window, select another initial video-shot from the main UI and open a new *AVS* window. Again, all previously submitted and presented shots are taken into account, unless everything has been reset with the corresponding button.

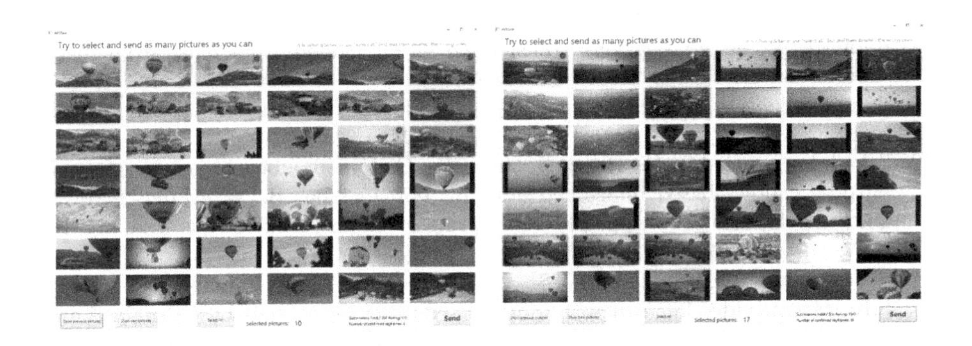

Fig. 2. AVS mode of the current version of *Vibro*

5 Conclusion

This paper presents the 2023 version of *vibro*, a powerful video browsing tool that will participate in the upcoming VBS (Video Browser Showdown). The main improvements can be summarized as enhancements in text-to-image and image-to-image retrieval and also faster and better arrangements of the search results on 2D-grids for easier browsing.

References

1. Barthel, K.U., Hezel, N., Jung, K., Schall, K.: Improved evaluation and generation of grid layouts using distance preservation quality and linear assignment sorting (2022). https://doi.org/10.48550/ARXIV.2205.04255
2. Dosovitskiy, A., et al.: An image is worth 16x16 words: transformers for image recognition at scale. CoRR (2020)
3. He, K., Zhang, X., Ren, S., Sun, J.: Deep residual learning for image recognition. In: 2016 IEEE Conference on Computer Vision and Pattern Recognition (CVPR) (2016)
4. Heller, S., et al.: Interactive video retrieval evaluation at a distance: comparing sixteen interactive video search systems in a remote setting at the 10th video browser showdown. Int. J. Multim. Inf. Retr. **11**(1), 1–18 (2022). https://doi.org/10.1007/s13735-021-00225-2
5. Hezel, N., Barthel, K.U.: Dynamic construction and manipulation of hierarchical quartic image graphs. In: Proceedings of the 2018 ACM on International Conference on Multimedia Retrieval. ICMR 2018, pp. 513–516. Association for Computing Machinery, New York (2018)
6. Liu, Z., et al.: Swin transformer: hierarchical vision transformer using shifted windows. arXiv preprint arXiv:2103.14030 (2021)
7. Lokoč, J., Souček, T.: How many neighbours for known-item search? In: Reyes, N., et al. (eds.) SISAP 2021. LNCS, vol. 13058, pp. 54–65. Springer, Cham (2021). https://doi.org/10.1007/978-3-030-89657-7_5
8. Radford, A., et al.: Learning transferable visual models from natural language supervision. CoRR abs/2103.00020 (2021)
9. Russakovsky, O., et al.: Imagenet large scale visual recognition challenge. Int. J. Comput. Vision (IJCV) (2015)
10. Schall, K., Barthel, K.U., Hezel, N., Jung, K.: GPR1200: a benchmark for general-purpose content-based image retrieval. In: Þór Jónsson, B., et al. (eds.) MMM 2022. LNCS, vol. 13141, pp. 205–216. Springer, Cham (2022). https://doi.org/10.1007/978-3-030-98358-1_17
11. Truong, Q.T., Vu, T.A., Ha, T.S., Lokoč, J., Tim, Y.H.W., Joneja, A., Yeung, S.K.: Marine video kit: A new marine video dataset for content-based analysis and retrieval. In: Dang-Nguyen, D., et al. (eds.) MMM 2023. LNCS, vol. 13833, pp. xx–yy. Springer, Cham (2023)
12. Weyand, T., Araujo, A., Cao, B., Sim, J.: Google landmarks dataset v2 - a large-scale benchmark for instance-level recognition and retrieval. In: CVPR (2020)

VideoCLIP: An Interactive CLIP-based Video Retrieval System at VBS2023

Thao-Nhu Nguyen[1(✉)], Bunyarit Puangthamawathanakun[1], Annalina Caputo[1], Graham Healy[1], Binh T. Nguyen[2], Chonlameth Arpnikanondt[3], and Cathal Gurrin[1]

[1] Dublin City University, Dublin, Ireland
thaonhu.nguyen24@mail.dcu.ie
[2] Vietnam National University, Ho Chi Minh University of Science, Ho Chi Minh City, Vietnam
[3] King Mongkut's University of Technology Thonburi, Bangkok, Thailand

Abstract. In this paper, we present an interactive video retrieval system named VideoCLIP developed for the Video Browser Showdown 2023. To support users in solving retrieval tasks, the system enables search using a variety of modalities, such as rich text, dominant colour, OCR, and query-by-image. Moreover, a new search modality has been added to empower our core engine, which is inherited from the Contrastive Language-Image Pre-training (CLIP) model. Finally, the user interface is enhanced to display results in groups in order to reduce the effort for a user when locating potentially relevant targets.

Keywords: Video browser showdown · Interactive video retrieval · Embedding model

1 Introduction

In the past decade, video has become one of the most widely used forms of media due to its ability to capture the human experiences and provide a user-friendly visual experience. However, browsing through video content to accomplish search targets is a laborious task. Therefore, highly effective systems that are able to manage, store, and search video data efficiently are required.

Annually, the International Conference on Multimedia Modeling (MMM) organises a competition named Video Browser Showdown (VBS) [8], a benchmark for video retrieval systems, in which participants are required to complete three tasks in two categories: Known-Item search (KIS) and Ad-hoc Video Search (AVS). KIS consists of two sub-tasks: visual KIS and textual KIS. Participants have to submit a correct video segment given either a video clip (visual KIS) or a textual description (textual KIS). AVS, in contrast, requires participants to submit as many correct scenes as possible with a given description. In addition

T.-N. Nguyen and B. Puangthamawathanakun—Contributed equally to this research.

D.-T. Dang-Nguyen et al. (Eds.): MMM 2023, LNCS 13833, pp. 671–677, 2023.
https://doi.org/10.1007/978-3-031-27077-2_57

to the two main datasets were used at VBS2022 (V3C1 [2] and V3C2 [12]), the organisers also provide a newly released collection of videos in underwater environment called Marine Video Kit (MVK) [13]. The details of these three datasets can be found in Table 1.

Table 1. Details of the datasets used in VBS2023

Dataset name	# Videos	Duration (hours)
V3C1 [2]	7,475	1,000
V3C2 [12]	9,760	1,300
MVK [13]	1,374	12.38

In this paper, we present VIDEOCLIP, a content-based interactive video retrieval system built on top of the enhanced CLIP-based VIDEOFALL back-end [10]. These enhancements are described in detail in Sect. 3.

2 Related Research

In the recent VBS2022 competition, there were several noteworthy systems. Vibro [5], the VBS2022's winner, was proven to be state-of-the-art by combining key features including temporal search, rich text, relevance feedback, and nearest neighbour search approximation. The system allowed for large-scale search, compromising with slightly lower accuracy. In the case of VISIONE [1], the system provided more options for search modalities, such as objects and colours with spatial relationships, free text, temporal search, and similarity search. In the latest version, the palette of colours used in the system was limited to 11 colours, whereas our system uses 12 colours, with an extra skin colour. CVHunter [9], the winner of the textual KIS task, is a novel system and demonstrated its capability by providing only two key features: context-aware ranking and relevance feedback. For context-aware ranking, simultaneous sub-queries could be defined along with a temporal window to retrieve a result list. Then, users may use relevance feedback by selecting keyframes related to the target. The Contrastive Language-Image Pre-training (CLIP) [11] model has played an important role in all of the above systems, which clearly shows its capabilities in video retrieval challenges.

Our multi-user retrieval system, VIDEOFALL [10], previously proved its potential by obtaining varied insights from different perspectives and was ranked in the top half of the VBS22's participant teams (ranked 7th out of 17 teams). However, due to the network issues between users, the search speed became inconsistent and slower than expected. That is the reason why we have modified our system into a single user interactive retrieval tool with an enhanced back-end model and a novel intuitive user interface.

3 An Overview of VideoCLIP

Our system consists of two parts: a front-end and a back-end. In this iteration, VIDEOCLIP builds upon VIDEOFALL's underlying engine [10] with the following important modifications. Firstly, the hierarchical search has been removed, due to the issues just mentioned. Secondly, we use an updated version of the CLIP [11] model as the backbone. Thirdly, we add a new search modality, meta-search. With meta-search, users may filter results with a list of comparative expressions. Lastly, in the front-end, we re-designed the user interface (UI) to ease the search process by modifying the result presentation and layout, described in Sect. 3.2.

3.1 Search Modalities

The previous system provided search modalities depending on the role of a user. Normal users may use rich text query, dominant colour, and optical character recognition (OCR) search to retrieve the results, while a master user may use query-by-image only. Since we no longer support hierarchical search, in VIDEO-CLIP, we consolidate all earlier options. To address the counting weakness of the CLIP model, we provide a meta-search to filter keyframes with the number of specific objects in each keyframe.

Dominant Colours. Each keyframe is pre-processed in order to retrieve its top dominant colours with the same methodology used in VIDEOFALL [10]. For the KIS task, users may choose the dominant colour(s) that occurred in those keyframes which can be used as a filter for the ranked list of results afterward.

Optical Character Recognition. We exploit Google Vision API[1] to extract any text including written, printed, or typed text visible in the images. This feature is effective for some images, such as labels of goods, banners, or documents.

Rich Text Query. Keywords could be insufficient to describe an image to retrieve a desirable list of results. With the power of the co-embedding space of the CLIP model [11], a natural language query is embedded in the same high-dimensional vector space as an image embedding. Thus, it allows users to search with any textual query.

Query-by-image. Apart from normal textual query-based search, which may be difficult and time-consuming for some images with complicated details, we offer query by image option. Particularly, if a keyframe in the search result meaningfully aligns with the target, users may use it as a reference to find all visually similar keyframes.

[1] https://cloud.google.com/vision/docs/ocr.

Enhanced Embedding Models. The CLIP model [11] has been used in many research fields to date, ranging from image classification, and image similarity, to image captioning due to its ability to connect visual concepts and natural language. It is the state-of-the-art embedding model used by most of the teams, including our team VIDEOFALL [10], in the VBS challenge. To enhance the performance of the embedding model, we create an ensemble model from the two latest large CLIP models, ResNet-50 [4] model (ResNet50 × 64), and a Vision Transformer [3] pretrained at 336-pixel resolution (ViT-L/14@336p).

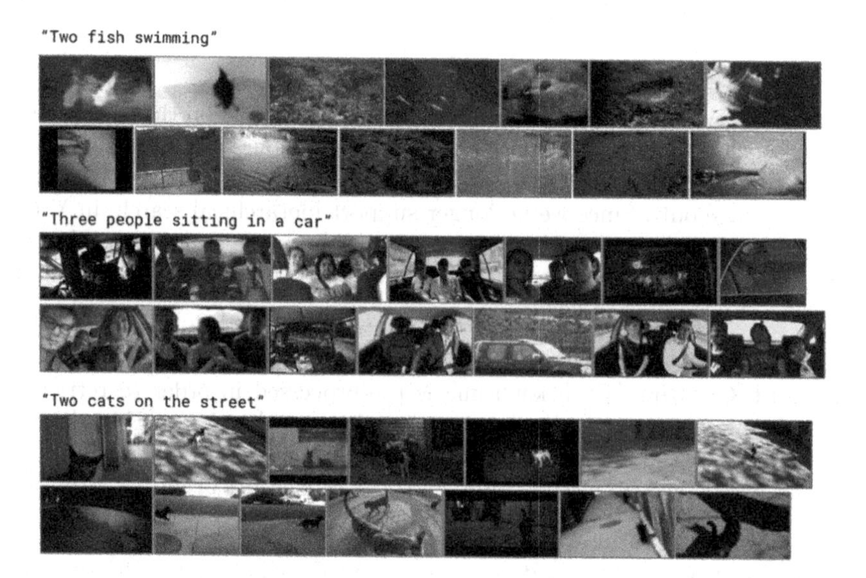

Fig. 1. Examples of failures when using CLIP for tasks that involve counting objects (V3C dataset). With each query, the top 14 keyframes are ranked based on a cosine similarity score. Images with a green border are correct keyframes while those with a red border are incorrect ones. (Color figure online)

Meta-search. Although the utility of the CLIP model has been proven in video retrieval, and in particular in our system last year, it still has some limitations. In particular, it can struggle with tasks such as counting the number of objects or measuring the distance between objects in an image [11]. As can be seen from Fig. 1, more than half of the top 14 images were not the correct scene that matches the input query.

To overcome this problem, we integrate a meta-search function that enables the automatic measurement of the number of objects displayed in keyframes. In particular, users are able to select an available concept with a comparative operator and a specific number to filter the ranked list result. For instance, a user may specify that there is at least one person in the target scene (shown in Fig. 2). The object **human** is selected with a **greater than or equal to** the operator and a number **one**. Utilising a pre-trained You-Only-Look-Once v5 model (YOLOv5) [6], we are able to extract and count objects in the given image, with a total of 80 concepts derived from the categories of Microsoft Common Objects in Context [7].

3.2 User Interface Revision

After participating in VBS2022, we recognise that the time and steps taken for search and submission are critical to the system's overall performance. To enhance this aspect, we have updated the UI as follows.

Fig. 2. The prototype of VIDEOCLIP user interface

User Interface Layout. The layout of the front-end has been rearranged as depicted in Fig. 2. On the left side of the UI, users can formulate search queries (mentioned in Sect. 3.1), while the right side shows the result. Part **A** is the main part where users can input rich text and OCR to search for a result. The new search modal, meta-search, is in part **B**. It allows users to add comparative expressions to filter the result. Next, part **C** contains a collection of 12 colours where users may select the dominant colours appeared in the target. Occupying three-quarters of the entire screen, part **D** is responsible for displaying the final result with details described below.

Result Presentation. In the previous system, all individual keyframes were listed and ranked by their relevance scores, resulting in redundant gazes of the same video from users when looking through the results. However, in this version, the ranked list of the search results is grouped with respect to their video and video segments. In other words, we treat each video as a group of shorter videos and represent the result as a list of the top three keyframes of each shorter video as a group based on its summation relevance score, as shown in Fig. 2.

4 Conclusion

In this paper, we present the latest version of our video interactive retrieval system, VIDEOCLIP, a CLIP-based system. We have updated the user interface, and the search result visualisation, along with upgrading the back-end with the latest CLIP models. In addition, a meta-search is added to alleviate possible weaknesses of the CLIP model.

Acknowledgments. This research was conducted with the financial support of Science Foundation Ireland under Grant Agreement No. 18/CRT/6223, and 13/RC/2106_P2 at the ADAPT SFI Research Centre at DCU. ADAPT, the SFI Research Centre for AI-Driven Digital Content Technology, is funded by Science Foundation Ireland through the SFI Research Centres Programme.

References

1. Amato, G., et al.: VISIONE at video browser showdown 2022. In: Þór Jónsson, B., et al. (eds.) MMM 2022. LNCS, vol. 13142, pp. 543–548. Springer, Cham (2022). https://doi.org/10.1007/978-3-030-98355-0_52
2. Berns, F., Rossetto, L., Schoeffmann, K., Beecks, C., Awad, G.: V3C1 Dataset: an evaluation of content characteristics. In: Proceedings of the 2019 on International Conference on Multimedia Retrieval, ICMR '19 pp. 334–338, New York, NY, USA, 2019. Association for Computing Machinery
3. Dosovitskiy, A., et al.: An image is worth 16x16 words: Transformers for Image Recognition at Scale (2020)
4. He, K., Zhang, X., Ren, S., Sun, J.: Deep residual learning for image recognition (2015)
5. Hezel, N., Schall, K., Jung, K., Barthel, K.U.: Efficient search and browsing of large-scale video collections with vibro. In: Þór Jónsson, B., et al. (eds.) MMM 2022. LNCS, vol. 13142, pp. 487–492. Springer, Cham (2022). https://doi.org/10.1007/978-3-030-98355-0_43
6. Jocher, G., et al.: Ultralytics/YOLOv5: v6.2 - YOLOv5 Classification Models, Apple M1, Reproducibility, ClearML and Deci.ai integrations, Aug. (2022)
7. Lin, T.-Y., et al.: Microsoft COCO: common objects in context. In: Fleet, D., Pajdla, T., Schiele, B., Tuytelaars, T. (eds.) ECCV 2014. LNCS, vol. 8693, pp. 740–755. Springer, Cham (2014). https://doi.org/10.1007/978-3-319-10602-1_48
8. Lokoč, J., et al.: Is the reign of interactive search eternal? Findings from the video browser showdown 2020. ACM Trans. Multimedia Comput. Commun. Appl., 17(3), Jul (2021)

9. Lokoč, J., Mejzlík, F., Souček, T., Dokoupil, P., Peška, L.: Video search with context-aware ranker and relevance feedback. In: Þór Jónsson, B., et al. (eds.) MultiMedia Modeling. Lecture Notes in Computer Science, pp. 505–510. Springer International Publishing, Cham (2022). https://doi.org/10.1007/978-3-030-98355-0_46
10. Nguyen, T.-N., Puangthamawathanakun, B., Healy, G., Nguyen, B.T., Gurrin, C., Caputo, A.: Videofall - a hierarchical search engine for VBS2022. In: Þór Jónsson, B., et al. (eds.) MMM 2022. LNCS, vol. 13142, pp. 518–523. Springer, Cham (2022). https://doi.org/10.1007/978-3-030-98355-0_48
11. Radford, A., et al.: Learning transferable visual models from natural language supervision. In: Meila, M., Zhang, T., eds, Proceedings of the 38th International Conference on Machine Learning, volume 139 of Proceedings of Machine Learning Research, pp. 8748–8763. PMLR, 18–24 Jul 2021
12. Rossetto, L., Schoeffmann, K., Bernstein, A.: Insights on the V3C2 Dataset. CoRR, abs/2105.01475 (2021)
13. Truong, Q.-T., et al.: Marine video kit: A new marine video dataset for content-based analysis and retrieval. In: Marine video kit: a new marine video dataset for content-based analysis and retrieval MMM 2023, Bergen, Norway, January 9–12, 2023. Springer (2023)

Free-Form Multi-Modal Multimedia Retrieval (4MR)

Rahel Arnold, Loris Sauter(✉), and Heiko Schuldt

Databases and Information Systems Research Group, University of Basel, Basel, Switzerland
{rahel.arnold,loris.sauter,heiko.schuldt}@unibas.ch

Abstract. Due to the ever increasing amount of multimedia data, efficient means for multimedia management and retrieval are required. Especially with the rise of deep-learning-based analytics methods, the semantic gap has shrunk considerably, but a human in the loop is still considered mandatory. One of the driving factors of video search is that humans tend to refine their queries after reviewing the results. Hence, the entire process is highly interactive. A natural approach to interactive video search is using textual descriptions of the content of the expected result, enabled by deep learning-based joint visual text co-embedding. In this paper, we present the Multi-Modal Multimedia Retrieval (4MR) system, a novel system inspired by vitrivr, that empowers users with almost entirely free-form query formulation methods. The top-ranked teams of the last few iterations of the Video Browser Showdown have shown that CLIP provides an ideal feature extraction method. Therefore, while 4MR is capable of image and text retrieval as well, for VBS video retrieval is based primarily based on CLIP.

Keywords: Video browser showdown · Interactive video retrieval · Content-based retrieval

1 Introduction

With the ever-growing volume of multimedia data, means for efficient and effective search in such multimedia collections are a necessity. At the annual Video Browser Showdown (VBS) [15] – a competition-style evaluation campaign in the domain of interactive video search – interactive multimedia retrieval systems compete against each other in a pseudo-realistic setting. Particularly, Known-Item Search (KIS) and Ad-hoc Video Search (AVS) are task categories of VBS [8]. The former category consists of two sub-categories, each with a single target video shot of about 20 s with textual and visual hints, respectively. The latter features broader query terms without known ground truth, and human judges decide whether a shot meets the task criteria or not. VBS operates on the Vimeo Creative Commons Collection (V3C) [14], in particular V3C's first and second shards [1,13], culminating in approximately 2300 h video content. In

D.-T. Dang-Nguyen et al. (Eds.): MMM 2023, LNCS 13833, pp. 678–683, 2023.
https://doi.org/10.1007/978-3-031-27077-2_58

addition, a new, more homogeneous "marine video" dataset [17] of more than 11 h complements previous video data for more challenging tasks.

Inspired by vitrivr [3,6], a long-running participant at VBS, we present in this paper 4MR (*M*ulti-*M*odal *M*ulti*M*edia *R*etrieval), a novel open-source multi-modal multimedia retrieval system with a focus on retrieval blocks. In particular, the 4MR system empowers users to express multi-modal queries and to freely combine these with Boolean logic. Like the top-ranked teams from previous iterations [7] that have implemented CLIP [10], we also rely on CLIP as the primary feature extraction method. Furthermore, we support deep learning-based OCR and ASR methods and extracted concept labels with ResNet50 [5].

The remainder of this paper is structured as follows: Sect. 2 introduces the concepts of retrieval blocks and how to formulate queries, Sect. 3 gives insights on the implementation, and Sect. 4 concludes.

2 Retrieval Blocks

In order to efficiently and effectively search in video collections, three search paradigms have proven to be successful [7]: (i) (extended) Boolean search, (ii) vector-based text search, (iii) *k*NN search. We employ all three paradigms and build multi-modal queries on them: Boolean search in order to efficiently use provided metadata such as collection affiliation. We plan to use this feature to specify V3C shards one and two or the Marine Video Kit. However, vector search is used to search in deep learning-based extracted feature data such as CLIP representations. In addition, textual vector search is applied to search in OCR data, for example.

At its core, our multi-modal query model consists of so-called *Query Statements*. A Query Statement is the smallest unit in our model and might be one of the following: (i) textual search, either Boolean or *k*NN search, (ii) metadata search, using Boolean search (iii) visual search, using *k*NN search.

Two Query Statements can be linked with a so-called *Statement Linking Operator*, either the logical AND or OR. One or two Query Statements form a *Query Block* with or without a Statement Linking Operator. Query Blocks can subsequently be arbitrarily nested, and their relationship to each other is also defined by Statement Linking Operators. Ultimately, one or more Query Blocks are called *Retrieval Statement*, put into relation with the so-called *Retrieval Linking Operator*. Retrieval Linking Operators define an order so that the prerequisites of one Retrieval Statement must be met first before the next one can be applied, effectively creating stages. Another such Retrieval Linking Operator could represent some temporal relation to the Retrieval Statements.

This nesting of Query Blocks and their defined relation to one another is sufficient to formulate arbitrarily complex queries as needed for interactive search. While the model would allow for interactive search within text-based and inherently multi-media collections, this functionality cannot be used in VBS as the competition is video search only. However, we intend to use the Query Blocks in order to express complex queries.

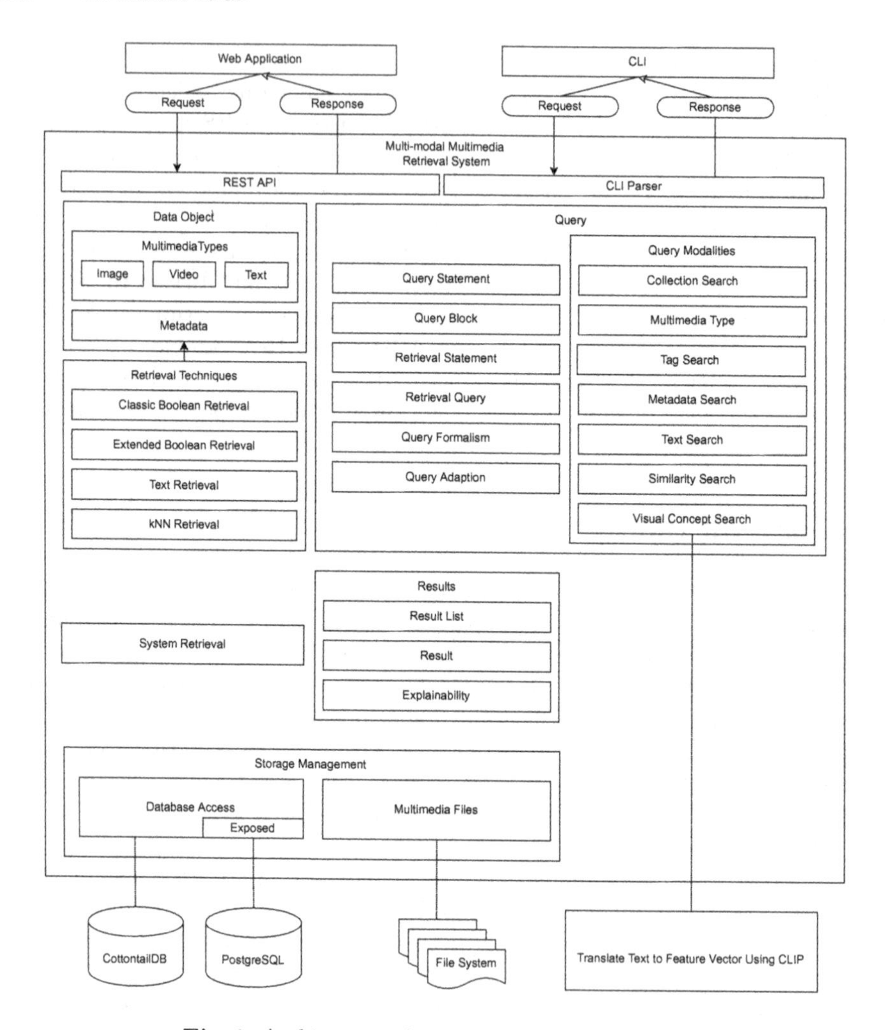

Fig. 1. Architecture diagram of the 4MR system.

3 Implementation

The 4MR system follows a three-tier architecture [16], as shown in Fig. 1. The storage layer consists of a Postgres database[1] (textual and Boolean search) for textual information, a CottontailDB database [2] for vectors (and kNN search) and the file system to store the actual media files. Multiple storage systems are used to exploit their strengths: Postgres for efficient Boolean search and CottontailDB for kNN search. In what follows, we describe the retrieval engine (Sect. 3.1), explain which features we search for (Sect. 3.2) and finally introduce the user interface (Sect. 3.3).

[1] https://www.postgresql.org.

3.1 Retrieval Engine

Written in Kotlin, the retrieval engine communicates with the storage tier using a JDBC adapter and a gRPC client for Postgres and CottontailDB, respectively. The entire parsing of queries, as well as relaying the appropriate parts to the underlying storage systems in correspondence to the query type, is handled in the retrieval engine. Ultimately, the results of the storage components are fused into a single ranked result list and sent to the user interface. All functionality of the retrieval engine is made accessible to the front end by a REST API.

3.2 Video Analysis

Due to the success of deep learning-based video analysis methods, or feature extraction, our system employs four such models. In particular, we use Contrastive Language-Image Pre-training, better known as CLIP, based on recognising image elements using our natural language to describe them [9–11]. It builds on a large body of work on zero-shot transfer, natural language supervision, and multi-modal learning. We use CLIP, or more precisely the "ViT-B/32"-model from Open-AI, to provide visual concept and similarity search. The visual concept search encodes the text input with the model. The resulting vector is then used for a kNN search, where the best fitting images with small distances are returned. In the concept of similarity search, kNN search is directly performed on the 512-feature-long vectors. Besides CLIP, a video's textual and audio content is often interesting for queries. We support optical character recognition (OCR) queries in these domains and automatic speech recognition (ASR), respectively. The features used for these features are the same as presented by our inspiration, vitrivr [12] and rely on text search entirely.

Last but not least, we support the notion of concepts recognised within videos, so-called tags. Using a residual network with 50 layers, ResNet50 [4], we retain tags and the confidence that the neural net classified them, which we can use in the query formulation.

3.3 User Interface

Heavily inspired by vitrivr's frontend vitrivr-ng[2], we also have an Angular[3] based frontend divided into a query sidebar and central result presentation area (see Fig. 2).

[2] https://github.com/vitrivr/vitrivr-ng.

[3] https://angular.io.

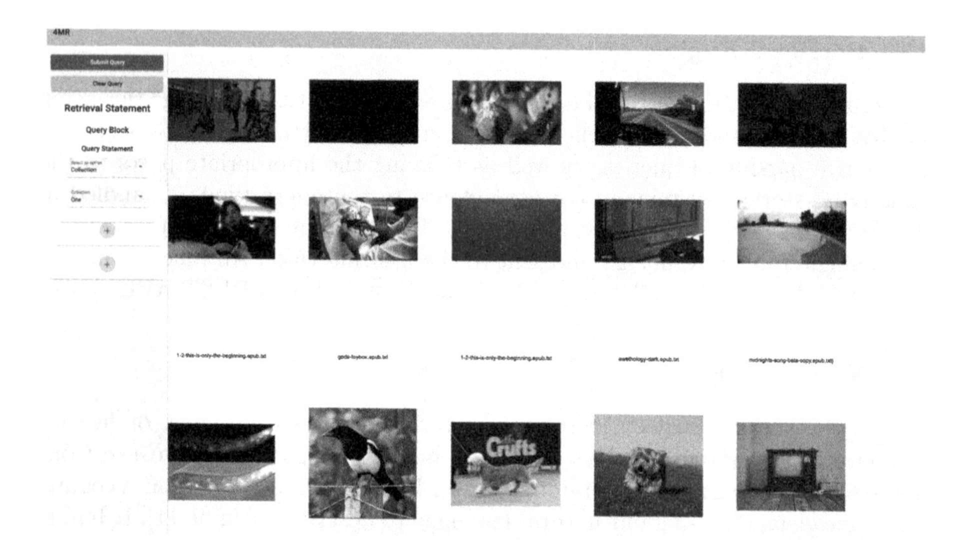

Fig. 2. A screenshot of the 4MR system in action. The left features the query formulation area and the center is used for result presentation.

4 Conclusion

We introduce 4MR, a new system focusing on query formulation based on vitrivr, to participate in the Video Browser Showdown 2023. The contribution is twofold: On one hand, we describe a query formulation concept which empowers users to combine query blocks and freely define their relationship. On the other hand, we provide an implementation of the concept in order to be able to evaluate our system in the competitive setting of VBS. Our query formulation methodology uses deep learning-based features, particularly CLIP, like the top-ranked teams in the last instances of VBS.

References

1. Berns, F., Rossetto, L., Schoeffmann, K., Beecks, C., Awad, G.: V3C1 dataset: an evaluation of content characteristics. In: International Conference on Multimedia Retrieval. ACM (2019). https://doi.org/10.1145/3323873.3325051
2. Gasser, R., Rossetto, L., Heller, S., Schuldt, H.: Cottontail DB: an open source database system for multimedia retrieval and analysis, pp. 4465–4468. Association for Computing Machinery, New York, USA (2020). https://doi.org/10.1145/3394171.3414538
3. Gasser, R., Rossetto, L., Schuldt, H.: Multimodal multimedia retrieval with vitrivr. In: International Conference on Multimedia Retrieval (2019)
4. He, K., Zhang, X., Ren, S., Sun, J.: Deep residual learning for image recognition (2015). https://doi.org/10.48550/ARXIV.1512.03385, https://arxiv.org/abs/1512.03385

5. He, K., Zhang, X., Ren, S., Sun, J.: Deep residual learning for image recognition. In: 2016 IEEE Conference on Computer Vision and Pattern Recognition (CVPR) 2016, Las Vegas, NV, USA, 27–30 Jun 2016, pp. 770–778. IEEE Computer Society (2016). https://doi.org/10.1109/CVPR.2016.90
6. Heller, S., et al.: Multi-modal interactive video retrieval with temporal queries. In: Þór Jónsson, B. (ed.) MMM 2022. LNCS, vol. 13142, pp. 493–498. Springer, Cham (2022). https://doi.org/10.1007/978-3-030-98355-0_44
7. Heller, S., et al.: Towards explainable interactive multi-modal video retrieval with vitrivr. In: Lokoč, J. (ed.) MMM 2021. LNCS, vol. 12573, pp. 435–440. Springer, Cham (2021). https://doi.org/10.1007/978-3-030-67835-7_41
8. Lokoč, J., et al.: A task category space for user-centric comparative multimedia search evaluations. In: MultiMedia Modeling (2022). https://doi.org/10.1007/978-3-030-98358-1_16
9. OpenAI: Github repository clip. https://github.com/openai/CLIP. Accessed 10 Oct 2022
10. Radford, A., et al.: Learning transferable visual models from natural language supervision (2021). https://doi.org/10.48550/ARXIV.2103.00020,https://arxiv.org/abs/2103.00020
11. Radford, A., Sutskever, I., Kim, J.W., Krueger, G., Agarwal, S.: CLIP: connecting text and images. https://openai.com/blog/clip/. Accessed 10 Oct 2022
12. Rossetto, L., Amiri Parian, M., Gasser, R., Giangreco, I., Heller, S., Schuldt, H.: Deep learning-based concept detection in vitrivr. In: Kompatsiaris, I., Huet, B., Mezaris, V., Gurrin, C., Cheng, W.-H., Vrochidis, S. (eds.) MMM 2019. LNCS, vol. 11296, pp. 616–621. Springer, Cham (2019). https://doi.org/10.1007/978-3-030-05716-9_55
13. Rossetto, L., Schoeffmann, K., Bernstein, A.: Insights on the V3C2 dataset. CoRR abs/2105.01475 (2021). https://arxiv.org/abs/2105.01475
14. Rossetto, L., Schuldt, H., Awad, G., Butt, A.A.: V3C – a research video collection. In: Kompatsiaris, I., Huet, B., Mezaris, V., Gurrin, C., Cheng, W.-H., Vrochidis, S. (eds.) MMM 2019. LNCS, vol. 11295, pp. 349–360. Springer, Cham (2019). https://doi.org/10.1007/978-3-030-05710-7_29
15. Schoeffmann, K.: Video browser showdown 2012–2019: a review. In: International Conference on Content-Based Multimedia Indexing (2019). https://doi.org/10.1109/CBMI.2019.8877397
16. Schuldt, H.: Multi-tier architecture. In: LIU, L., ÖZSU, M.T. (eds.) Encyclopedia of Database Systems. Springer, Boston, MA (2009). https://doi.org/10.1007/978-0-387-39940-9_652
17. Truong, Q.T., Vu, T.A., Ha, T.S., Lokoc, J., Tim, Y.H.W., Joneja, A., Yeung, S.K.: Marine video kit: a new marine video dataset for content-based analysis and retrieval. In: MultiMedia Modeling - 29th International Conference, MMM 2023, Bergen, Norway, 9–12 Jan 2023. Lecture Notes in Computer Science, Springer (2023)

diveXplore at the Video Browser Showdown 2023

Klaus Schoeffmann(✉), Daniela Stefanics, and Andreas Leibetseder

Institute of Information Technology (ITEC), Klagenfurt University, Klagenfurt, Austria
{klaus.schoeffmann,daniela.stefanics,andreas.leibetseder}@aau.at

Abstract. The diveXplore system has been participating in the VBS since 2017 and uses a sophisticated content analysis stack as well as an advanced interface for concept, object, event, and texts search. This year, we perform several changes in order to make the system both much easier to use as well as more efficient. These changes include using shot-based analysis over uniform sampled segments, integration of textual and image transformers (CLIP), as well as optimizing the user interface for faster browsing of video summaries.

Keywords: Video retrieval · Interactive video search · Video analysis

1 Introduction

The Video Browser Showdown (VBS) [6,9] is an international retrieval competition that aims at evaluating large-scale video search and retrieval in a fair and reproducible setting. It is held as an annual large event at the International Conference on Multimedia (MMM), where it is performed in live manner over many hours with numerous tasks and different task types (Known-Item Search/KIS and Ad-hoc Video Search/AVS). Several teams from many different countries compete for solving video search tasks on a pre-known large video dataset with on-demand and time-limited queries. KIS tasks evaluate the systems' search performance for specific unique target segmentes, whereas AVS tasks evaluate their performance for finding multiple clip examples of a given topic. The developed systems are typically used by people familiar with them (experts) as well as inexperienced users (novices). In its 12th iteration, VBS2023 is a very challenging event, since tasks will be more difficult (smaller target segments) and the current dataset (V3C1 and V3C2 [3,10], which entails about 2300 h hours of diverse video content) will be extended further with another small dataset from scuba-diving (the *marine* dataset).

The diveXplore system has been developed at Klagenfurt University over the last few years and consists of three major components (backend, middleware, and frontend), as described in more detail in Sect. 2. Version 2023 of this system builds on previous versions [5] but has also been significantly re-engineered,

D.-T. Dang-Nguyen et al. (Eds.): MMM 2023, LNCS 13833, pp. 684–689, 2023.
https://doi.org/10.1007/978-3-031-27077-2_59

in order to make it both more competitive and easier to use for novices. Most importantly, a transformer-based video-text embedding [8,12] has been integrated. In this extended demo paper we detail the architecture and integrated changes of diveXplore 2023.

2 diveXplore 2023

2.1 Architecture

The architecture of diveXplore 2023 is visible in Fig. 1. It consists of three major components: the analysis backend, the data management middleware, and the web-based user interface as a frontend. All the videos in the entire collection (V3C and Marine) are analyzed on the backend, which stores the analysis results in the metadata database (with MongoDB[1]). Database queries are processed via a dedicated Node.js server that is connected to the frontend via the websocket protocol. All the thumbnails, images (keyframes), and videos are served by another Node.js/Webserver that retrieves requested items directly from the filesystem. These middleware servers may run either at the same machine as the user interface does, or they may be distributed across the network for scalability.

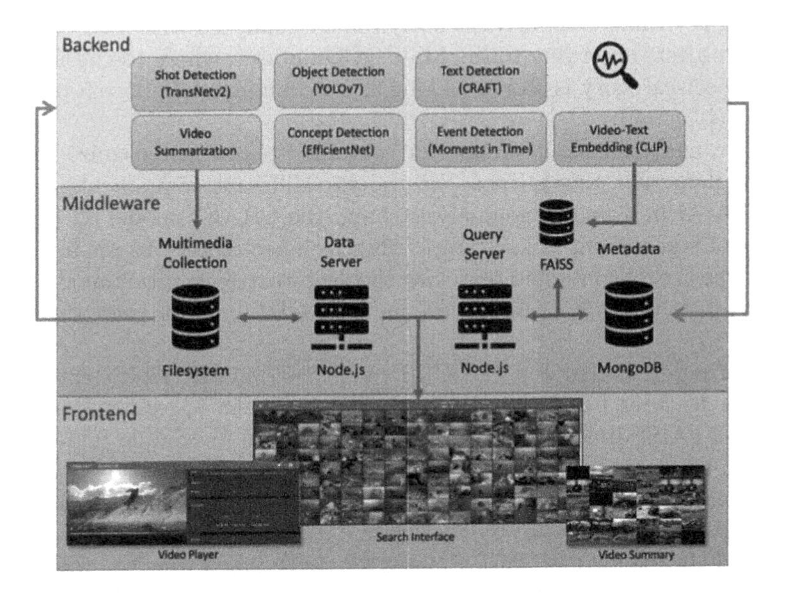

Fig. 1. diveXplore 2023 architecture

[1] https://www.mongodb.com.

2.2 Video Analysis/Backend

The analysis starts with segmenting the video files into smaller units. Version 2023 of diveXplore does no longer use the user-controlled uniform sampling for this purpose, since it has turned out as a major problem with the current dataset size (2300 h of video content for V3C1 and V3C2), which either results in too many similar segments (e.g., when using one second for every segment), or overlooks too much content when using too large segment sizes (e.g., 20 s). Instead, the V3C1 and V3C2 content is segmented with the TransNetv2 model [13], which has been developed at Charles University and achieves high accuracy for shot boundaries segmentation in general videos. As keyframes we simply use the middle frame of every shot and store them with their position in the database (the actual image is on the filesystem). However, since VBS 2023 also uses the Marine dataset, which consists of unedited content and lacks shot transitions, we still use uniform-subsampling for segmentation of these videos. In particular, we create equidistant 10-seconds segments for the videos in the Marine dataset and use their middle frames as keyframes.

All the extracted keyframes are further analyzed for full-frame concepts with EfficientNet [14], trained on the GPR1200 dataset [11] as well as on the Places-365 dataset [17]. Detected concepts with a specific confidence value above a pre-defined threshold ($t_c = 0.6$) are stored in the metadata database for the corresponding keyframe. Similarly, each keyframe is analyzed with YOLOv7 [16] for contained objects according to the MS-COCO dataset, where the best detection of each object-category is stored as long as its confidence is above a pre-defined threshold ($t_o = 0.7$).

We also analyze each keyframe for contained text and recognize the words in text regions. For this purpose we use two different components developed by CLOVA AI in a two-stage analysis chain: the CRAFT model [2] is used for detection of text-regions in keyframes, whereas detected regions are further analyzed for the actual contained text with the deep-text-recognition model [1]. The recognized text is finally stored in the metadata database for the corresponding keyframe.

Furthermore, we employ the CLIP ViT-B/32 model [8] to extract text-embedding features from each keyframe and store the 512-dimensional vector in a flat L2 FAISS index [4]. This index is later used for simple text-queries as well as for keyframe-based similarity search. For detecting video-events we use the Moments in Time model [7] and analyze every video shot for 304 events. Detected events are stored in the metadata database if the confidence is above a pre-defined threshold ($t_e = 0.5$). Finally, for every video we also create a video summary, which is stored on the filesystem of the multimedia collection. Table 1 contains an overview of all analysis components.

2.3 Data Management/Middleware

All the analysis results are stored in the metadata database (MongoDB), as well as their references to thumbnails, keyframes, and videos, which are stored on

Table 1. Overview of analysis components used by diveXplore 2023

Type	Model	Ref.	Description
Shots	TransNetv2	[13]	Deep model to detect shot boundaries in V3C
Segments	uniform subsampling		10s segments as units in Marine videos [15]
Concepts	EfficientNet B2	[14]	GPR1200 [11] and Places365 [17] categories
Objects	YOLOv7	[16]	80 MS COCO categories
Events	Moments in Time	[7]	304 Moments in Time action events
Texts	CRAFT	[1,2]	Text region localization and OCR
Text embedding	CLIP ViT-B/32	[8]	512 CLIP features stored in a FAISS index
Similarity	CLIP ViT-B/32	[8]	512 CLIP features stored in a FAISS index

the filesystem. To make this data available to the frontend, we use two different Node.js servers as a middleware: the *query server* listens on a websocket for queries from the frontent, and the *data server* basically provides HTTP access to all multimedia files on the filesystem. The query server forwards (and translates) queries to the metadata database and returns its response back to the frontend. This additional middlelayer abstracts the database access, keeps the user interface slim, and easily allows to exchange the actual database without any changes on the interfaces (e.g., changing from MongoDB to Neo4j). Also, it distinguishes between queries to the metadata database and queries that need to be forwarded to the FAISS index, such as CLIP freetext and similarity queries. Finally, it easily allows for scaling the data server, e.g., by using several data servers that are integrated by the query server for URLs in the response to the frontend in a round-robin fashion.

2.4 User Interface (UI)/Frontend

The user interface of diveXplore 2023 is shown in Fig. 2. It consists of a search bar, different filter settings, and a result grid below. The search bar allows for simple freetext search with CLIP, which is great for novice users, but also provides an expert mode that can combine the many different search modalities (e.g., concepts, objects, events, texts, etc.) with simple prefixes (e.g., `-c`, `-o`, `-e`, `-t`, etc.). The search bar defaults to CLIP-based freetext search but supports the expert user in selecting prefixes by always showing matching components after something has been entered. The example in Fig. 2 shows these suggestions after `car` has been entered – we can see that there are several concepts matching with car, but also a lot of events that contain the word `car`. The expert search bar also allows to enter boolean and temporal queries (e.g., `-c (car || street),tree`, which would search for *car* or *street* together with *tree*, or `-c (car || street) < airport`, which would search for the concept *street* or *car* happening in any video before the concept *airport*.

Finally, it should be mentioned that the result grid can be switched to video summaries instead of shots, where each result would be shown with its context in the corresponding video (i.e., within the video summary). This should facilitate AVS queries, where it is important to find as many diverse videos as possible.

Fig. 2. diveXplore 2023 user interface

3 Conclusion

We introduce diveXplore 2023 for the VBS2023 competition. In contrast to previous years we strongly build on shots instead of uniformly sampled segments, integrate state-of-the-art analysis of concepts, objects, and texts, and differentiate between novice and expert users. The first type of users can rely on the easier CLIP-based search, while experts have a rich selection and combination features, including boolean and temporal combination of different modalities.

References

1. Baek, J., et al.: What is wrong with scene text recognition model comparisons? dataset and model analysis. In: Proceedings of the IEEE/CVF International Conference on Computer Vision, pp. 4715–4723 (2019)
2. Baek, Y., Lee, B., Han, D., Yun, S., Lee, H.: Character region awareness for text detection. In: Proceedings of the IEEE/CVF Conference on Computer Vision and Pattern Recognition, pp. 9365–9374 (2019)

3. Berns, F., Rossetto, L., Schoeffmann, K., Beecks, C., Awad, G.: V3c1 dataset: An evaluation of content characteristics. In: Proceeding of the 2019 on International Confernce on Multimedia Retrieval, pp. 334–338. ACM (2019)
4. Johnson, J., Douze, M., Jégou, H.: Billion-scale similarity search with GPUs. IEEE Trans. Big Data **7**(3), 535–547 (2019)
5. Leibetseder, A., Schoeffmann, K.: diveXplore 6.0: ITEC's interactive video exploration system at VBS 2022. In: Þór Jónsson, B., Gurrin, C., Tran, M.-T., Dang-Nguyen, D.-T., Hu, A.M.-C., Huynh Thi Thanh, B., Huet, B. (eds.) MMM 2022. LNCS, vol. 13142, pp. 569–574. Springer, Cham (2022). https://doi.org/10.1007/978-3-030-98355-0_56
6. Lokoč, J., et al.: Interactive search or sequential browsing? a detailed analysis of the video browser showdown 2018. ACM Trans. Multimedia Comput. Commun. Appl. **15**(1), 29:1–29:18 (Feb 2019). https://doi.org/10.1145/3295663,http://doi.acm.org/10.1145/3295663
7. Monfort, M., et al.: Moments in time dataset: One million videos for event understanding. IEEE Trans. Pattern Anal. Mach. Intell. **42**(2), 502–508 (2020). https://doi.org/10.1109/TPAMI.2019.2901464
8. Radford, A.,et al.: Learning transferable visual models from natural language supervision. In: International Conference on Machine Learning, pp. 8748–8763. PMLR (2021)
9. Rossetto, L., et al.: Interactive video retrieval in the age of deep learning - detailed evaluation of vbs 2019. IEEE Trans. Multimedia **23**, 243–256 (2021). https://doi.org/10.1109/TMM.2020.2980944
10. Rossetto, L., Schoeffmann, K., Bernstein, A.: Insights on the v3c2 dataset. arXiv preprint arXiv:2105.01475 (2021)
11. Schall, K., Barthel, K.U., Hezel, N., Jung, K.: GPR1200: a benchmark for general-purpose content-based image retrieval. In: Þór Jónsson, B., Gurrin, C., Tran, M.-T., Dang-Nguyen, D.-T., Hu, A.M.-C., Huynh Thi Thanh, B., Huet, B. (eds.) MMM 2022. LNCS, vol. 13141, pp. 205–216. Springer, Cham (2022). https://doi.org/10.1007/978-3-030-98358-1_17
12. Singh, A., et al.: Flava: A foundational language and vision alignment model. In: Proceedings of the IEEE/CVF Conference on Computer Vision and Pattern Recognition, pp. 15638–15650 (2022)
13. Souček, T., Lokoč, J.: Transnet v2: an effective deep network architecture for fast shot transition detection. arXiv preprint arXiv:2008.04838 (2020)
14. Tan, M., Le, Q.: Efficientnet: Rethinking model scaling for convolutional neural networks. In: International Conference on Machine Learning, pp. 6105–6114. PMLR (2019)
15. Truong, Q.T., et al.: Marine video kit: A new marine video dataset for content-based analysis and retrieval. In: MultiMedia Modeling - 29th International Conference, MMM 2023, Bergen, Norway, January 9–12, 2023. Springer (2023)
16. Wang, C.Y., Bochkovskiy, A., Liao, H.Y.M.: Yolov7: Trainable bag-of-freebies sets new state-of-the-art for real-time object detectors. arXiv preprint arXiv:2207.02696 (2022)
17. Zhou, B., Lapedriza, A., Khosla, A., Oliva, A., Torralba, A.: Places: A 10 million image database for scene recognition. IEEE Trans. Pattern Anal. Mach. Intell. **40**(6), 1452–1464 (2018). https://doi.org/10.1109/TPAMI.2017.2723009

Reinforcement Learning Enhanced PicHunter for Interactive Search

Zhixin Ma[1(✉)], Jiaxin Wu[2], Weixiong Loo[1], and Chong-Wah Ngo[1]

[1] School of Computing and Information Systems, Singapore Management University, Singapore, Singapore
zxma.2020@phdcs.smu.edu.sg, {wxloo,cwngo}@smu.edu.sg
[2] Department of Computer Science, City University of Hong Kong, Hong Kong, China
jiaxin.wu@my.cityu.edu.hk

Abstract. With the tremendous increase in video data size, search performance could be impacted significantly. Specifically, in an interactive system, a real-time system allows a user to browse, search and refine a query. Without a speedy system quickly, the main ingredient to engage a user to stay focused, an interactive system becomes less effective even with a sophisticated deep learning system. This paper addresses this challenge by leveraging approximate search, Bayesian inference, and reinforcement learning. For approximate search, we apply a hierarchical navigable small world, which is an efficient approximate nearest neighbor search algorithm. To quickly prune the search scope, we integrate PicHunter, one of the most popular engines in Video Browser Showdown, with reinforcement learning. The integration enhances PicHunter with the ability of systematic planning. Specifically, PicHunter performs a Bayesian update with a greedy strategy to select a small number of candidates for display. With reinforcement learning, the greedy strategy is replaced with a policy network that learns to select candidates that will result in the minimum number of user iterations, which is analytically defined by a reward function. With these improvements, the interactive system only searches a subset of video datasets relevant to a query while being able to quickly perform Bayesian updates with systematic planning to recommend the most probable candidates that can potentially lead to minimum iteration rounds.

Keywords: Reinforcement learning · Bayesian method · Relevance feedback · Interactive video retrieval

1 Introduction

As response time is critically important for a user-in-the-loop system, we focus on improving speed efficiency in terms of search space pruning and candidate selection for user feedback. The former is based on the state-of-the-art approximate near-neighbor search [10], which only traverses a small set of candidates for search result ranking. The latter is based on reinforcement learning to train a policy network that can recommend candidates to minimize user iterations.

D.-T. Dang-Nguyen et al. (Eds.): MMM 2023, LNCS 13833, pp. 690–696, 2023.
https://doi.org/10.1007/978-3-031-27077-2_60

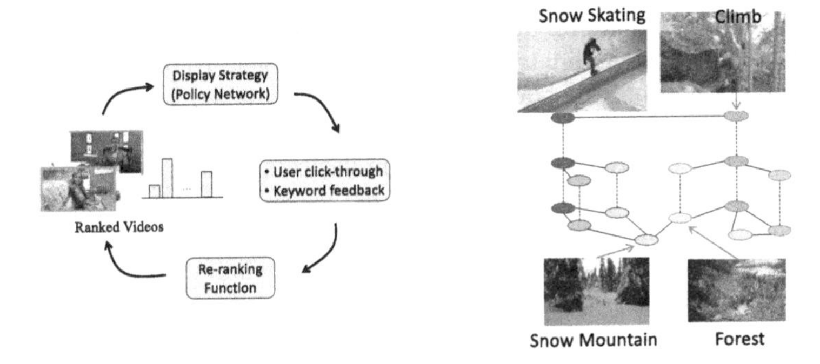

Fig. 1. Improvements over VIREO search system [9]: PicHunter++ (left), hierarchical navigable small world where more similar color shows higher similarity (right).

Intuitively, search space pruning addresses the issue of real-time system response. In contrast, the candidate selection strategy addresses the issue of mental tiredness by assisting a user to reach the search targets with minimum possible browsing efforts.

As most of the improvements are based on the integration of off-the-shelf techniques, this paper presents mainly our findings on PicHunter [2], which was proposed more than twenty years ago but remained one of the most competitive baselines [3,5,6] in video interactive search. For example, SOM-Hunter [4], a variant of PicHunter with a self-organizing map (SOM), has demonstrated strong performance in Video Browser Showdown (VBS). Empirically, we show that, even for the current VBS benchmark dataset with 1 million clips [1], PicHunter can achieve recall@1 >50% with an average of 5 user iterations for 1 million visual known item queries under the ideal scenario that will be discussed later. Nevertheless, in practice, such high performance is not attainable due to the fact that human perception of visual similarity is different from what is defined in an ideal scenario. Instead of modeling human perception, we apply reinforcement learning to replace the greedy update in PicHunter with the goal of reaching the search target as soon as possible. This modification results in PicHunter recommending the candidates likely to speed up the search and browsing process rather than the candidates with the highest Bayesian scores. Figure 1 illustrates the two improvements made over the search system presented in VBS 2022 [9].

2 PicHunter

PicHunter [2] is an interactive video search system using relevance feedback. In each round of user iterations, the display strategy recommends user a set of video clips for selection. Based on the feedback, PicHunter uses Bayesian rules to re-rank the candidates and predict the search target. Specifically, let $D = \{d\}_{i=1}^{|D|}$ denote the set of video clips displayed to a user. The user will select video clips D^{+}, which are visually closer to the search target, and the remaining clips D^{-} are assumed irrelevant. Given the user's action, the system applies Bayesian rules to update the underlying probability distribution $P = \{p_i\}$, where p_i denotes the probability of the i-th candidate being the search target. To test the performance of PicHunter on visual known item search, we designed a user simulator to provide feedback to the recommended candidate clips. It is worth mentioning that the search is purely based on visual similarity and does not involve textual query. The probability $P = \{p_i\}$ is initialized as 1, indicating that all clips are equally likely to be the search target. The system displays $|D|$ video clips in each turn for a user to select $|D^+|$ clips, which are more similar to the search target than the remaining $|D| - |D^+|$ clips. Both the selected and unselected clip sets will be used to update the probability distribution P as follows

$$p'_i = p_i \cdot \prod_{l \in D^+} \frac{exp^{\frac{-\delta_{cos}(v_f, v_l)}{\sigma}}}{\sum_{x \in D^- \cup \{l\}} exp^{\frac{-\delta_{cos}(v_u, v_l)}{\sigma}}} \tag{1}$$

where σ is a hyper-parameter to control the temperature scaling.

We conduct the experiments on two datasets, MSR-VTT [14] and V3C1 [1], to examine the effectiveness of PicHunter for visual known-item search (VKIS). There are 7,000 video clips in MSR-VTT and 100,000 clips in V3C1. In each iteration, the system will display a set of video clips D with the highest probabilities in P to a user. We employ a user simulator in the simulated experiments to provide relevance feedback. The simulator imitates an ideal situation where a user can always select the more similar clips from D by following the visual perception of the machine. Specifically, the machine vision of similarity is defined based on the cosine similarity of CLIP4Clip [7] features. The $|D^+|$ selected clips are more similar to the search target than the $|D|-|D^+|$ clips based on this machine vision. In the experiment, we set the display size, i.e., $|D| = 8$.

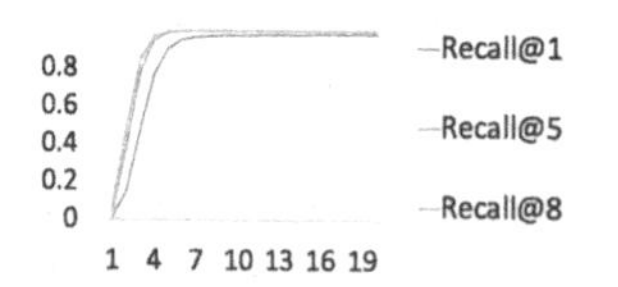

Fig. 2. The search performance (y-axis) over different rounds of iteration (x-axis) on MSR-VTT.

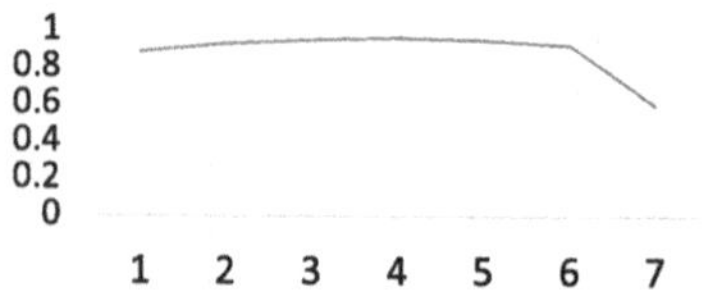

1
0.8
0.6
0.4
0.2
0
1 2 3 4 5 6 7

Fig. 3. Recall@1 (y-axis) versus like number (x-axis) within a maximum of 7 iteration rounds on MSR-VTT.

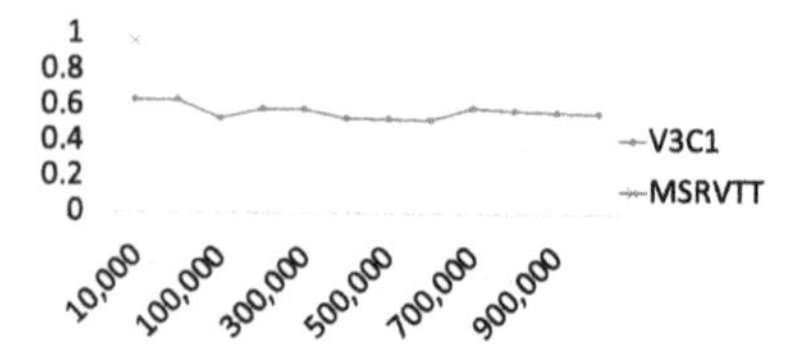

Fig. 4. Scalability test of PicHunter: recall@1 (y-axis) versus the number of clips (x-axis) within 7 rounds of user iterations. Note that the number of queries is set to 10,000.

We investigate the search performance in terms of recall@$\{1, 5\}$ by allowing a simulator to select $|D^+| < |D|$ "like number" of similar clips.

Figure 2 shows the search performance averaged over 7,000 queries on the MSR-VTT dataset. The like number is set to $|D^+| = 4$. As noted, the recall performance increased sharply to more than 80% and saturated after 5 and 7 iterations for recall@5 and recall@1, respectively. To investigate the optimal number of clips to be feedbacked to PicHunter, Fig. 3 shows the performance trend of recall@1 by varying the like number. The result shows that the impact of likes is not significant. Even by selecting only one clip, the recall@1 can reach 88% within seven rounds of iterations. The performance gradually increases to 96% by providing four likes.

To access the scalability of PicHunter, we test the recall@1 performance by increasing the data size, as shown in Fig. 4. Surprisingly, the performance trend does not vary proportional to the data size and stays around 53%-64% across different scales. On the V3C1 dataset with 1 million clips, recall@1 reaches around 56% within 7 iteration rounds with $|D^+| = 4$. Compared to MSR-VTT of 7,000 clips with recall@1=96%, the performance drop is considered significant. Nevertheless, the result shows that the performance depends more on the query difficulty than the data size. As observed, MSR-VTT queries are relatively easier to search compared to V3C1 with diverse visual content.

While PicHunter shows superior retrieval performance and resilience to data scale, the performance is highly dependent on whether the selected clips follow the perception of machine vision. We conduct another experiment to disrupt the ideal experimental setting by introducing noise to clip selection. Specifically, the simulator randomly selects one clip out of the four most similar clips to a search target in each round of user iteration. By doing this, recall@1 remains almost 0 on average after 7 iteration rounds. Basically, when the most similar clip, as perceived by the machine, is not selected, the Bayesian update cannot reflect the actual similarity distribution of the clips to a search target. In this situation, the recommended clips do not lead to the convergence of the search loop. In the experiment, even by randomly selecting three out of the four most similar clips as feedback, recall@1 barely reaches 20%.

3 PicHunter++

To remedy the limitations of PicHunter, we integrate the Bayesian update with reinforcement learning (RL). Specifically, the original display strategy in PicHunter, which greedily recommends the top $|D|$ clips most similar to a search target based on the probability distribution P, is replaced by RL. We train a policy network, a convolutional neural network, to take in the sorted probability P as input and output the probability $P^{'}$ as the distribution to sample $|D|$ clips for displaying. The network is optimized with the advantage actor-critic algorithm (A2C) with a reward function which is set to minimize the number of user iterations and maximize the recall performance. Unlike the greedy selection, which depends only on the historical update of P, RL provides a systematic mechanism for planning the navigation path to maximize future reward based on the current P. Furthermore, with RL, PicHunter can be more easily modified for textual known-item search (TKIS). For example, the policy network considers not only the liked clips selected by a user but also the textual query and user feedback interactively provided in different rounds for RL [8]. The probability distribution of clips, P, is still updated with the original Bayesian formula as in Eq. 1.

4 Interactive Search System

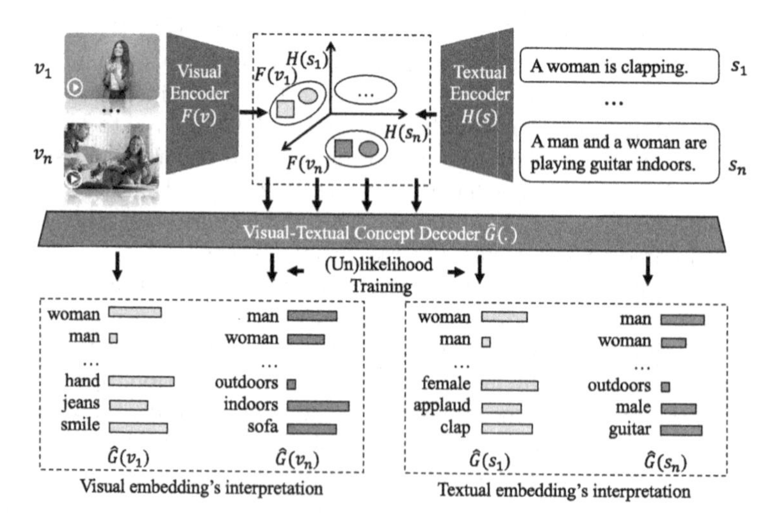

Fig. 5. The backend of VIREO engine which learns interpretable embeddings for both concept-free and concept-based searches.

Figure 5 shows the backend of VIREO search engine [13] where PicHunter++ operates on. Compared to our system in [9], the engine considers both likelihood and unlikelihood loss functions for learning interpretable embeddings. The engine produces an embedding of 2,048 dimensions for a video clip. For V3C1 [1], V3C2 [11] and the new-released marine dataset [12], such high-dimensional embeddings inhibit real-time interaction if linear search is performed on a typical laptop with 4-core CPU and memory size of 16G bytes. To improve system response time, we employ the state-of-the-art approximate k-nearest neighbor search, hierarchical navigable small world (HNSW) algorithm [10], to organize the embeddings as a hierarchical graph. As shown in Fig. 1, the graph provides multi-layer indexing of embeddings for coarse-to-fine traversing of similar clips. The bottom layer indexes all the clips while the upper layers index only the representative clips. The connections from a representative clip to its children in the lower layer are linked (shown as dotted lines in Fig. 5) for progressive search.

Given a textual query, the search starts by comparing the query embedding with the embeddings of representative clips indexed at the top layer. The search progresses to the next layer by comparing only to the children of representative clips retained at the upper layer of the graph. The progressive traversal and search of candidates across layers avoid exhaustive comparison and hence significantly cut short the processing time.

5 Conclusion

We have presented the major improvements over our interactive system [9]. Specifically, the Bayesian inference from PicHunter [2] helps to narrow down the search space. The policy neural network improves the display strategy in PicHunter, boosts the recall performance and further reduces the interaction rounds. For efficient search, we employ HNSW algorithm to index the video embeddings. We also improve the backend of the VIREO search engine using unlikelihood training.

Acknowledgment. This research was supported by the Singapore Ministry of Education (MOE) Academic Research Fund (AcRF) Tier 1 grant.

References

1. Berns, F., Rossetto, L., Schoeffmann, K., Beecks, C., Awad, G.: V3c1 dataset: An evaluation of content characteristics. In: Proceedings of the 2019 on International Conference on Multimedia Retrieval, pp. 334–338. ICMR '19 (2019)
2. Cox, I.J., Miller, M.L., Minka, T.P., Papathomas, T.V., Yianilos, P.N.: The bayesian image retrieval system, pichunter: theory, implementation, and psychophysical experiments. IEEE Trans. Image Process. **9**(1), 20–37 (2000)
3. Heller, S., et al.: Interactive video retrieval evaluation at a distance: comparing sixteen interactive video search systems in a remote setting at the 10th video browser showdown. Int. J. Multimed. Inform. Retrieval **11**, 1–18 (2022)
4. Kratochvíl, M., Mejzlík, F., Veselý, P., Soucek, T., Loko, J.: Somhunter: Lightweight video search system with som-guided relevance feedback. In: Proceedings of the 28th ACM International Conference on Multimedia (2020)
5. Kratochvíl, M., Veselý, P., Mejzlík, F., Lokoč, J.: SOM-Hunter: video browsing with relevance-to-SOM feedback loop. In: Ro, Y.M., et al. (eds.) MMM 2020. LNCS, vol. 11962, pp. 790–795. Springer, Cham (2020). https://doi.org/10.1007/978-3-030-37734-2_71
6. Loko, J., et al.: DIs the reign of interactive search eternal? findings from the video browser showdown 2020. ACM Trans. Multimed. Comput., Commun. Appl. (TOMM) **17**, 1–26 (2021)
7. Luo, H., et al.: CLIP4Clip: An empirical study of clip for end to end video clip retrieval. arXiv:2104.08860 (2021)
8. Ma, Z., Ngo, C.W.: Interactive video corpus moment retrieval using reinforcement learning, pp. 296–306. MM '22, Association for Computing Machinery, New York, NY, USA (2022). https://doi.org/10.1145/3503161.3548277

9. Ma, Z., Wu, J., Hou, Z., Ngo, C.-W.: Reinforcement learning-based interactive video search. In: Þór Jónsson, B., Gurrin, C., Tran, M.-T., Dang-Nguyen, D.-T., Hu, A.M.-C., Huynh Thi Thanh, B., Huet, B. (eds.) MMM 2022. LNCS, vol. 13142, pp. 549–555. Springer, Cham (2022). https://doi.org/10.1007/978-3-030-98355-0_53
10. Malkov, Y.A., Yashunin, D.A.: Efficient and robust approximate nearest neighbor search using hierarchical navigable small world graphs. IEEE Trans. Pattern Anal. Mach. Intell. **42**, 824–836 (2020)
11. Rossetto, L., Schoeffmann, K., Bernstein, A.: Insights on the v3c2 dataset. arXiv preprint arXiv:2105.01475 (2021)
12. Truong, Q.T., et al.: Marine video kit: A new marine video dataset for content-based analysis and retrieval. In: MultiMedia Modeling, MMM 2023 (2023)
13. Wu, J., Ngo, C.W., Chan, W.K., Hou, Z.: (un)likelihood training for interpretable embedding (2022). https://doi.org/10.48550/ARXIV.2207.00282
14. Xu, J., Mei, T., Yao, T., Rui, Y.: Msr-vtt: A large video description dataset for bridging video and language. In: 2016 IEEE Conference on Computer Vision and Pattern Recognition (CVPR), pp. 5288–5296 (2016)

Author Index

D.-T. Dang-Nguyen et al. (Eds.): MMM 2023, LNCS 13833, pp. 697–702, 2023.
https://doi.org/10.1007/978-3-031-27077-2

Printed in the United States
by Baker & Taylor Publisher Services